Contributions to Finance and Accounting

The book series 'Contributions to Finance and Accounting' features the latest research from research areas like financial management, investment, capital markets, financial institutions, FinTech and financial innovation, accounting methods and standards, reporting, and corporate governance, among others. Books published in this series are primarily monographs and edited volumes that present new research results, both theoretical and empirical, on a clearly defined topic. All books are published in print and digital formats and disseminated globally. This book series is indexed in Scopus.

Peter Brusov • Tatiana Filatova

Business Valuation and Company Value

New Qualitative Approaches Considering Business and Financial Risks

Springer

Peter Brusov
Financial University under the Government of Russian Federation
Moscow, Russia

Tatiana Filatova
Financial University under the Government of Russian Federation
Moscow, Russia

ISSN 2730-6038 ISSN 2730-6046 (electronic)
Contributions to Finance and Accounting
ISBN 978-3-031-90846-0 ISBN 978-3-031-90847-7 (eBook)
https://doi.org/10.1007/978-3-031-90847-7

This Springer imprint is published by the registered company Springer Nature Switzerland AG
The registered company address is: Gewerbestrasse 11, 6330 Cham, Switzerland

Preface

Dedicated to the eternal love
that touched us with its wing
and made us happy,
and without which life is meaningless.

Business and company valuation plays a very important role in making an informed investment decision; when buying and selling shares and bonds of an enterprise on the stock market; when buying and selling an enterprise; when restructuring an enterprise; when merging, liquidating, taking over, spinning off an independent company; when determining the creditworthiness of an enterprise and the cost of collateral for lending; when insuring; when taxing, etc. The position of an enterprise on the market is determined by the market capitalization of the enterprise and determines how much investors are willing to pay for its shares.

The value of any asset (shares, bonds, companies, etc.) is equal to the current value of all income generated by this asset. Thus, to find the value of a company, it is necessary to find the present value of all income received by the company from the moment of its creation to the current moment, that is, in the retrospective period. And to evaluate a business, it is necessary to find the current value of all future income received by the enterprise in the representative and terminate periods.

In monograph a new approach to business valuation and company value has been developed. As is well known, there are three main methods in business valuation: cost, comparative, and income, with the latter being the most adequate and accurate. In monograph two of the three approaches to the income method have been modified: CAPM and WACC.

The existing methods of business and company valuation have many shortcomings, two of which are global, related to the impossibility of estimating one of the fundamental parameters of valuation—the discount rate, and the impossibility of correctly accounting for the retrospective period of valuation. Both of these shortcomings can currently be overcome only within the framework of the modern theory of Brusov–Filatova–Orekhova (BFO), which is done in this monograph, devoted to the development of a qualitatively new approach to the valuation of businesses and companies. Correct determination of the discount rate will prevent abuses in business evaluation, in investments, and in determining the fair value of shareholders' dividend income. The second most important method of business valuation—CAPM—was generalized by the authors to include financial risk along with business risk, and for the first time this was done correctly. A new model—CAPM 2.0—was created and applied to many companies. And, finally, CAPM 2.0 was included in the modern theory of Brusov–Filatova–Orekhova (BFO), which allows for the correct valuation of a business and a company, taking into account both business and financial risks.

Incorporation of CAPM 2.0 and Fama–French models into the main theory of capital structure—the Brusov–Filatova–Orekhova (BFO) theory—is important for the correct assessment of the financial performance of companies. The developed approach allows using the powerful tools of the highly developed BFO theory (and its eternal limit—the Modigliani–Miller (MM) theory) for the correct assessment of the main financial indicators of the company and their forecasting, as well as for the assessment of the business and the company taking into account both types of risks: business and financial.

Recently, the WACC formulas within the framework of the Brusov–Filatova–Orekhova (BFO) theory and the Modigliani–Miller (MM) theory have been modified, taking into account the conditions of the actual operation of the company, such as frequent payment of income tax, the method of paying income tax (at the end of periods or in advance), with variable income, etc. To improve the accuracy of business valuation, it is necessary to take these effects into account.

Krondurg, Moscow, Russia
25 January, 2025

Peter Brusov
Tatiana Filatova

Competing Interests The authors have no competing interests to declare that are relevant to the content of this manuscript.

About the Book

Business valuation and company value play a very important role in making an informed investment decision; buying and selling shares and bonds of a company on the stock market; insurance; taxation, etc. The position of a company in the market is determined by the market capitalization of the company and determines how much investors are willing to pay for its shares.

The existing methods of business and company valuation have many shortcomings, two of which are global, related to the impossibility of estimating one of the fundamental parameters of valuation—the discount rate and the impossibility of correctly accounting for the retrospective period of valuation. Both of these shortcomings can currently be overcome only within the framework of the modern theory of Brusov–Filatova–Orekhova (BFO), which is done in this monograph, devoted to the development of a qualitatively new approach to the valuation of businesses and companies. The second most important method of business valuation—CAPM—was generalized by the authors to include financial risk along with business risk, and for the first time this was done correctly. A new model—CAPM 2.0—was created and applied to many companies. And, finally, CAPM 2.0 was included in the modern theory of Brusov–Filatova–Orekhova (BFO), which allows for the correct valuation of a business and a company, taking into account both business and financial risks.

Contents

About the Authors

Peter Brusov is professor at the Financial University under the Government of the Russian Federation in Moscow (Russia). He is an expert in financial management, corporate finance, investments, ratings, taxation, and business valuation. He is a co-author of the famous Brusov–Filatova–Orekhova theory, which replaced the well-known Modigliani–Miller theory. As well he is a co-author of the CAPM 2.0 model, which generalized the famous CAPM model for accounting the financial risk along with business risk.

He has been a visiting Professor of Northwestern University (USA), Cornell University (USA), Osaka City University (Japan), and National Chung Cheng University (Taiwan) among other places. He is the author of over 550 research publications including nine monographs, numerous textbooks, and articles.

Tatiana Filatova is professor at the Financial University under the Government of the Russian Federation in Moscow (Russia). She is an expert in financial management, corporate finance, investments, ratings, taxation, and business valuation. Tatiana Filatova is a co-author of the famous Brusov–Filatova–Orekhova theory, which replaced the well-known Modigliani–Miller theory. As well she is a co-author of the CAPM 2.0 model, which generalized the famous CAPM model for accounting the financial risk along with business risk.

In the past 20 years, she has been Dean of the faculties of financial management, management, state and municipal government at the Financial University. Tatiana Filatova is the author of over 300 research publications including eight monographs, numerous textbooks, and articles.

Part I
Introduction

Chapter 1
Introduction

This monograph is devoted to the development of a qualitatively new approach to assessing the value of a business and a company.

Business and company valuation plays a very important role in making an informed investment decision; when buying and selling shares and bonds of an enterprise on the stock market; when buying and selling an enterprise; when restructuring an enterprise; when merging, liquidating, taking over, spinning off an independent company; when determining the creditworthiness of an enterprise and the cost of collateral for lending; when insuring; when taxing, etc. The position of an enterprise on the market is determined by the market capitalization of the enterprise and determines how much investors are willing to pay for its shares.

As is well known, there are three main methods in business valuation: cost, comparative and income, with the latter being the most adequate and accurate. There are three approaches to the income method: CAPM, APT, and WACC.

In the income method the value of any asset (shares, bonds, companies, etc.) is equal to the current value of all income generated by this asset. Thus, to find the value of a company, it is necessary to find the current value of all income received by the company from the moment of its creation to the current moment, that is, in the retrospective period. And to evaluate a business, it is necessary to find the current value of all future income received by the enterprise in the representative and terminate periods.

The existing methods of business and company valuation have many shortcomings, two of which are global, related to the impossibility of estimating one of the fundamental parameters of valuation—the discount rate, and the impossibility of correctly accounting for the retrospective period of valuation. Both of these shortcomings can currently be overcome only within the framework of the modern theory of Brusov–Filatova–Orekhova (BFO), which is done in this monograph, devoted to the development of a qualitatively new approach to the valuation of businesses and companies. Correct determination of the discount rate will prevent abuses in business evaluation, in investments, and in determining the fair value of shareholders' dividend income The second most important method of business valuation—

P. Brusov, T. Filatova, *Business Valuation and Company Value*, Contributions to Finance and Accounting, https://doi.org/10.1007/978-3-031-90847-7_1

CAPM—was generalized by the authors to include financial risk along with business risk, and for the first time this was done correctly. A new model—CAPM 2.0—was created and applied to many companies. And, finally, CAPM 2.0 was included in the modern theory of Brusov–Filatova–Orekhova (BFO), which allows for the correct valuation of a business and a company, taking into account both business and financial risks.

The structure of the book follows the main methods of the income approach: CAPM, APT, and WACC. The first part is Introduction. The second part is devoted to CAPM and its modification CAPM 2.0, created by the authors (Chap. 2). The Chap. 3 in the second part is devoted to the Arbitrage Pricing Theory (APT), which is a multi–factor model in contrast to the single–factor model CAPM.

The third part is related to the use of one of the two main theories of capital structure (the Brusov–Filatova–Orekhova (BFO) Theory and the Modigliani–Miller (MM) Theory), which take into account only the financial risk associated with the use of debt financing and allow calculating all financial indicators of a company of any age (BFO Theory) or perpetual (MM Theory). The new approach developed in this book allows taking into account the following conditions of the real functioning of the company: variable income; inflation; frequency and method of paying income tax (advance payments or payments at the end of periods) and the principal debt. These conditions have been taken into account by the authors and discussed in detail.

The fourth part discusses some important effects that should be taken into account in the WACC method (golden age, dividend policy, etc.).

The fifth part considers the inclusion of CAPM and CAPM 2.0 models in theories of capital structure, which allows taking into account business and financial risks simultaneously. The application of this approach to some companies is discussed. In order to create a methodology for assessing the main financial indicators of a company, taking into account both business and financial risks it has been made the incorporation of CAPM and Fama–French models into two major theories of capital structure—Brusov–Filatova–Orekhova (BFO) Theory and Modigliani–Miller (MM) Theory. This turns out to be important for a correct assessment of the financial performance of companies. The developed approach makes it possible to use the powerful tools of these highly developed theories (BFO and MM) for the correct assessment of the main financial indicators of the company and their forecasting, taking into account both types of risks.

The sixth part describes two modifications of a qualitatively new approach to business valuation and company value. The developed approach corrects some of the main shortcomings of the income approach. The new methodology significantly improves the accuracy of the valuation.

The conclusions are described in seventh part.

Part II
CAPM; CAPM 2.0 and APT

Chapter 2
Capital Asset Pricing Model 2.0: Account of Business and Financial Risk

Abbreviations

CAPM	Capital Asset Pricing Model
MM	The Modigliani–Miller theory
k_i; k_{I}; k_{m}	Stand for company, industry and market returns
σ_i; σ_{I}; σ_{m}	Are standard deviation for company, industry and market returns
$\beta_{i\mathrm{I}}$; β_{im}	Are Beta coefficients company to industry and to market
L	Is leverage level
k_{d}	Is the cost of debt

2.1 Introduction

In real economy financial and business risks exist. Financial risks are related to use of debt financing and are described by capital structure theories. Business risks associated with investments into specific company (and not to the entire market (industry)) and are described by CAPM (market or industry version).

Based on the portfolio theory by Harry Markowitz, the Capital Asset Pricing Model (CAPM) was developed independently [1–5] by Jack Traynor (1961), William F. Sharp (1964), John Lintner (1965), and Jan Mossin (1966).

2.1.1 CAPM (Capital Asset Pricing Model)

2.1.1.1 Market Approach

CAPM is a simple, but widely used, one-factor model that describes the relationship between the expected return on assets (stocks, investments, etc.) and the risk-free

P. Brusov, T. Filatova, *Business Valuation and Company Value*, Contributions to Finance and Accounting, https://doi.org/10.1007/978-3-031-90847-7_2

rate, taking into account systematic (business) risk. This relationship is described by the equity risk premium, which depends on the asset's beta (which describes the asset's correlation or sensitivity to the market), the risk-free rate (say, the Treasury bill rate or the central bank's key rate), and the expected return in the market. CAPM assumes an idealized open market structure where all risky assets refer to all tradable shares available to everyone. In addition, we have a risk-free asset (for borrowing and/or lending in unlimited quantities) with an interest rate of k_f. One assumes that all information is available to everyone, such as covariances, variances, average stock returns, and so on. One also assume that investor is a rational, risk-averse, who uses the same Markowitz portfolio theory.

The following assumptions are made within the CAPM model:

1. All investors are risk averse and have the same time frame to evaluate information.
2. Unlimited capital exists to borrow at the risk-free rate.
3. Investments can be divided into unlimited parts and sizes.
4. Taxes, inflation and transaction costs are absent.
5. Return and risk are linearly related.

CAPM (Capital Asset Pricing Model) describes the profitability of asset and is described by the following formula

$$k_i = k_{\mathrm{f}} + \beta_i(k_{\mathrm{m}} - k_{\mathrm{f}}). \tag{2.1}$$

Here, k_{f} is risk free profitability, β is the β-coefficient of the company. It shows the dependence of the return on the asset and the return on the market as a whole. The β-coefficient is described by the following formula

$$\beta_i = \frac{\mathrm{cov}_{im}}{\sigma_{\mathrm{m}}^2} = \rho_{\mathrm{im}} \frac{\sigma_i}{\sigma_{\mathrm{m}}}. \tag{2.2}$$

Here σ_i is the risk (standard deviation) of i-th asset, σ_{m} is market risk (standard deviation of market index), cov_{im} is covariance between i-th asset and market portfolio.

An investor invests in risky securities only if their return is higher than the return on risk-free securities, so always $k_i > k_{\mathrm{f}}$ and $k_{\mathrm{m}} > k_{\mathrm{f}}$.

The beta-coefficient of a security, β, has the meaning of the amount of riskiness of this security. It follows from formula (2.1) that:

1. If $\beta = 1$ the yield of the security is equal to the yield of the average market portfolio ($k_i = k_{\mathrm{f}}$);
2. If $\beta > 1$, the security is more risky than the average on the security market ($k_i > k_{\mathrm{f}}$);
3. If $\beta < 1$, the security is less risky than the average on the security market ($k_i < k_{\mathrm{f}}$).

Securities betas are calculated using statistical data on returns on specific securities and the average market returns on securities traded on the market.

2.1.1.2 Disadvantages of the CAPM Model

CAPM has some well-known disadvantages.

1. The CAPM formula only works under assumption that the market is dominated by purely rational players who make decisions that favor only investment returns. This, of course, is not always true.
2. CAPM assumes that each market participant acts on the basis of the same information. In reality, relevant information is distributed unevenly among the public, so some participants may make decisions based on information that others do not.
3. Using beta as the main part of the formula. But beta takes into account only changes in the stock price in the market. However the share price can change for reasons other than the market. Stocks can rise or fall in value for deliberate reasons, not just volatility.
4. CAPM only uses historical data. But historical stock price changes are not enough to determine the overall risk of an investment. Other factors should be considered, such as economic conditions, industry peculiarities and competitor characteristics, and internal and external activities of the company itself.

So, the model has a number of limitations: the model does not take into account taxes, transaction costs, non-transparency of the financial market, etc.

Finally, to predict future returns, a retrospective level of market risk is used, which leads to a forecast error.

2.1.1.3 Derivation of the CAPM Formula

To derive the CAPM formula, we create a portfolio of asset i and the market portfolio M; $(\alpha, 1-\alpha)$, with the price share of the security in the portfolio $\alpha \in [0, 1]$. The rate of return is thus $\mu(\alpha) = \alpha\mu_i + (1-\alpha)\mu_{\mathrm{M}}$. Asset i is assumed not efficient; it's point lies in the feasible region but not on the efficient frontier. Thus as αvaries this portfolio's point traces out a smooth curve in the feasible (σ, μ) region $\alpha \to (\sigma(\alpha), \mu(\alpha))$ parametrized by α, where

$$\mu(\alpha) = \alpha\mu_i + (1-\alpha)\mu_{\mathrm{M}} = \alpha(\mu_i - \mu_{\mathrm{M}}) + \mu_{\mathrm{M}} \tag{2.3}$$

Using well known formula for portfolio dispersion

$$\sigma^2(\alpha) = \alpha^2\sigma_i^2 + (1-\alpha)^2\sigma_{\mathrm{M}}^2 + 2\alpha(1-\alpha)\sigma_{\mathrm{M},i},$$

one gets

$$\begin{aligned}\sigma(\alpha) &= \sqrt{\alpha^2\sigma_i^2 + (1-\alpha)^2\sigma_{\mathrm{M}}^2 + 2\alpha(1-\alpha)\sigma_{\mathrm{M},i}} = \\ &= \sqrt{\alpha^2(\sigma_i^2 + \sigma_{\mathrm{M}}^2 - 2\sigma_{\mathrm{M},i}) + \sigma_{\mathrm{M}}^2 + 2\alpha(\sigma_{\mathrm{M},i} - \sigma_{\mathrm{M}}^2)}\end{aligned} \tag{2.4}$$

When $\alpha = 0$ $(\sigma(0), \mu(0)) = (\sigma_{\mathrm{M}}, \mu_{\mathrm{M}})$ and when $\alpha = 1$ $(\sigma(0), \mu(0)) = (\sigma_i, \mu_i)$. Thus the curve touches the capital market line at the market point $(\sigma_{\mathrm{M}}, \mu_{\mathrm{M}})$, but otherwise remains off the capital market line but (of course) within the feasible region where it also hits the point (σ_i, μ_i). We conclude that the curve is tangent to the capital asset line at point $(\sigma_{\mathrm{M}}, \mu_{\mathrm{M}})$ and therefore when $\alpha = 0$ the curve's derivative

$$\frac{d\mu(\alpha)}{d\sigma(\alpha)}_{\alpha=0}$$

is identical to the slope of the capital asset line at point M. The slope (the price of risk) of the line is given by (2.2). We thus conclude that when $\alpha = 0$

$$\frac{\mathrm{d}\mu(\alpha)}{\mathrm{d}\sigma(\alpha)} = \frac{\mu_{\mathrm{M}} - \mu_{\mathrm{f}}}{\sigma_{\mathrm{M}}}. \tag{2.5}$$

The composition rule from calculus yields

$$\frac{\mathrm{d}\mu(\alpha)}{\mathrm{d}\sigma(\alpha)} = \frac{\mathrm{d}\mu(\alpha)/\mathrm{d}\alpha}{\mathrm{d}\sigma(\alpha)/\mathrm{d}\alpha}$$

Differentiating (2.3) and (2.4) and evaluating at $\alpha = 0$ yields

$$\frac{\mathrm{d}\mu(\alpha)}{\mathrm{d}\sigma(\alpha)} = \frac{\mu_i - \mu_{\mathrm{M}}}{(\sigma_{\mathrm{M},i} - \sigma_{\mathrm{M}}^2)/\sigma_{\mathrm{M}}}$$

From (2.5) we deduce that

$$\frac{\mu_i - \mu_{\mathrm{M}}}{(\sigma_{\mathrm{M},i} - \sigma_{\mathrm{M}}^2)/\sigma_{\mathrm{M}}} = \frac{\mu_{\mathrm{M}} - \mu_{\mathrm{f}}}{\sigma_{\mathrm{M}}} \tag{2.6}$$

Solving for μ_i then yields the CAPM formula for asset i, $\mu_i - \mu_{\mathrm{f}} = \beta_i \cdot (\mu_{\mathrm{m}} - \mu_{\mathrm{f}})$, as was to be shown. For the more general portfolio result, observe that

$$\mu_{\mathrm{p}} - \mu_{\mathrm{f}} = -\mu_{\mathrm{f}} + \sum_{i=1}^{n} \alpha_i \mu_i = \sum_{i=1}^{n} \alpha_i (\mu_i - \mu_{\mathrm{f}}) = \sum_{i=1}^{n} \alpha_i \beta_i (\mu_{\mathrm{M}} - \mu_{\mathrm{f}}) = (\mu_{\mathrm{M}} - \mu_{\mathrm{f}}) \times \sum_{i=1}^{n} \alpha_i \beta_i$$

(CAPM formula for single asset *i*)

Note that when $\beta_{\mathrm{p}} = 1$ then $\mu_{\mathrm{p}} = \mu_{\mathrm{M}}$; the expected rate of return is the same as for the market portfolio. When $\beta_{\mathrm{p}} > 1$, then $\mu_{\mathrm{p}} > \mu_{\mathrm{M}}$; when $\beta_{\mathrm{p}} < 1$, then $\mu_{\mathrm{p}} < \mu_{\mathrm{M}}$.

In case of negatively correlated asset *i* with *M*, $\sigma_{\mathrm{M},\, i} < 0$, then $\beta_i < 0$ and $\mu_i < \mu_{\mathrm{f}}$; the expected rate of return is less than the risk-free rate. Effectively, such a negatively correlated asset serves as insurance against a drop in the market returns, and might be viewed by some investors as having enough such advantages so as to make it worth the low return. Such an example could be investing in gold.

2.1.1.4 Modifications of CAPM: The Multiple Factors Models

The CAPM operates on only one factor that affects the future performance of a stock.

There are several models with multiple factors that modify the CAPM in this regard. Among them are Fama-French (three- and five-factor models) and APT (Arbitrage Pricing Theory) models [6].

2.1.1.4.1 Fama-French Model

In 1992, Y. Fama K. and French [7–10] proved that future returns are also affected by factors such as company size and industry affiliation. They have developed three- and five-factor models.

Fama–French Three–Factor Model

Fama–French three-factor model takes into account two additional risk factors, namely, size and book to market equity along with market beta

$$k_e = k_f + \beta_U \left(k_m - k_f\right) + s \cdot \mathrm{SMB} + h \cdot \mathrm{HML} \tag{2.7}$$

where SMB—the difference between the returns of companies with large and small capitalization;

HML—the difference between the returns of companies with low and high intrinsic value (*indicator B/P*)

Fama–French Five-factor Model

$$k_e = k_f + \beta_U \left(k_m - k_f\right) + s \cdot \mathrm{SMB} + h \cdot \mathrm{HML} + r \cdot \mathrm{RMW} + c \cdot \mathrm{CMA} \tag{2.8}$$

where RMW—return on equity; CMA—company capital expenditure.

2.1.1.4.2 Arbitrage Pricing Theory (APT)

In the APT model, the return on an asset can be expressed by the following formula:

$$r_i = a_i + \beta_{i1}F_1 + \beta_{i2}F_2 + \ldots + \beta_{in}F_n + \varepsilon_i, \tag{2.9}$$

where a_i is a constant per asset; F_i is a systematic factor, such as a macroeconomic or company-specific factor; β_i is the sensitivity of the asset in relation to the factor F_i; and ε_i is a random variable with an expected mean of zero.

APT formula has the form:

$$E(r_i) = r_f + \beta_{i1}\widehat{F}_1 + \beta_{i2}\widehat{F}_2 + \ldots + \beta_{in}\widehat{F}_n, \tag{2.10}$$

where r_f is the risk-free rate of return, β_{ik} is the sensitivity of the asset i with respect to factor k, $\widehat{F}_k$ is the risk premium for factor k.

In opposite to the CAPM, which has only one factor and one beta, The APT formula has multiple factors that include non-company factors, which requires the asset's beta with respect to each separate factor. The APT does not explain what these factors are, and APT model users should analytically determine factors that might affect the asset's returns. The factor used in the CAPM is the difference between the market rate of return and the risk-free rate of return.

The CAPM is a one-factor model and is simpler to use, thus investors prefer to use it to valuate the expected rate of return rather than using APT, which requires users valuate the multiple factors.

2.1.1.5 Industry Approach

CAPM has an alternative approach that refers to the industrial index rather than the market.

$$k_i = k_\mathrm{f} + \beta_i(k_\mathrm{I} - k_\mathrm{f}). \tag{2.11}$$

Here, k_f is risk free profitability, β is the β-coefficient of the company. In this case it shows the dependence of the return on the asset and the return on the industry as a whole. The β-coefficient now is described by the following formula

$$\beta_i = \frac{\mathrm{cov}_{i\mathrm{I}}}{\sigma_\mathrm{I}^2} = \rho_{i\mathrm{I}}\frac{\sigma_i}{\sigma_\mathrm{I}}. \tag{2.12}$$

Here σ_i is the risk of i-th asset, σ_I is industry risk (standard deviation of industry index), $\mathrm{cov}_{i\mathrm{I}}$ is covariance between i-th asset and industry index. Note, that the industry approach better describes the return on an asset than the market approach.

The CAPM approach is still evolving and we will describe one of the directions of this development below.

2.1.1.6 The Symmetric CAPM

One of the remaining internal problems of CAPM is the distribution function. The Capital Asset Pricing Model (CAPM) assumes a Gaussian or Normal distribution. In practice, this assumption may be violated. In [11], a symmetric CAPM is proposed, assuming distributions with lighter or heavier tails than the normal distribution. Elliptic distributions (normal, exponential and Student-t) are considered. This consideration is of a general nature. Authors conducted a detailed case study to apply the obtained results estimating the systematic risk of the financial assets of a Chilean company with real data. A Chilean company is just illustration of obtained results.

In addition, the authors of [11] study the methods of leverage and local impact for diagnostics in a symmetric CAPM. It is concluded that the considered models give better results than the CAPM with Gaussian distribution.

In [12–14], empirical studies were carried out under the assumption that stock returns have distributions with heavier tails than the normal distribution.

The Student-t distribution instead of the normal distribution was considered in [13] and in [15], taking into account the maximum likelihood method for estimating its parameters. Paper [14] concluded that asset valuation should be carried out within the framework of the CAPM and the discounted dividend model.

2.2 Hamada Model

The Modigliani–Miller theory [16, 17], with the accounting of taxes has been united with CAPM (Capital asset pricing model) in 1961 by Hamada [18, 19]. For the cost of equity of a leveraged company, the below formula has been derived.

$$k_\mathrm{e} = k_\mathrm{f} + \beta_\mathrm{U}(k_\mathrm{m} - k_\mathrm{f}) + \beta_\mathrm{U}(k_\mathrm{m} - k_\mathrm{f})\frac{D}{S}(1-t) \tag{2.13}$$

The first term represents risk-free profitability k_f, the second term is business risk premium, $\beta_\mathrm{U}(k_\mathrm{m} - k_\mathrm{f})$, and the third term is financial risk premium $\beta_\mathrm{U}(k_\mathrm{m} - k_\mathrm{f})\frac{D}{S}(1-t)$.

In the case of an unlevered company ($D = 0$), the financial risk (the third term) is zero, and its shareholders receive only a business risk premium.

Hamada used an empirical approach in incorporating the level of leverage into the CAPM. One of the main objectives in his research was to distinguish companies without leverage from companies with leverage. The latter make up almost the majority of real companies. In 1972, he surveyed 304 companies, among which he found 102 non-leveraged and 202 leveraged [19]. Comparing equity returns of two types of companies, he got his formula for the β-factor, which takes into account the level of leverage.

The incorrectness of the Hamada approximation will be shown below in the framework of the new approach we have developed, which describes both business and financial risk.

2.3 CAPM 2.0

In this section, we develop a new approach that describes both business and financial risk. We call this approach CAPM 2.0. as opposed to conventional CAPM, which accounts only business risk.

2.3.1 *Derivation of the Main Formula CAPM 2.0*

Let's combine CAPM (Capital asset pricing model) and the Modigliani–Miller theory [16, 17, 20–22] not phenomenologically, like Hamada [18, 19], but analytically.

Substituting the CAPM formula

$$k_0 = k_f + \beta_U (k_m - k_f) \tag{2.14}$$

into Modigliani–Miller formula for equity cost

$$k_e = k_0 + L(k_0 - k_d)(1 - t), \tag{2.15}$$

one gets the following result

$$\begin{aligned} &k_e = k_0 + L(k_0 - k_d)(1 - t) = \\ &k_f + \beta_U (k_m - k_f) + L(k_f + \beta_U (k_m - k_f) - k_d)(1 - t) = \\ &k_f(1 + L(1 - t)) + \beta_U (k_m - k_f)(1 + L(1 - t)) - Lk_d(1 - t) \end{aligned} \tag{2.16}$$

The second term is the same as in Hamada's formula (3), but the first term is renormalized value of risk free profitability and the last term, which depends on the cost of debt k_d, is missing from Hamada's formula (2.13).

So the difference with Hamada's formula is:

while in Hamada's formula only beta coefficient is renormalized, in formula (8) the first term (risk-free return) is also renormalized by the same factor $(1 + L(1 - t))$ and the last term, depending on the cost of debt k_d, appears, which is absent in Hamada's formula. Factor $(1 - t)$ (tax corrector) exists due to the tax shield.

The incorrectness of Hamada's approximation becomes obvious.

We could rewrite expression (2.16) as a sum of two parts, one of which is the Hamada expression, and the second is an additional term that we received

$$\begin{aligned} k_e &= k_f + \beta_U(k_m - k_f)(1 + L(1-t)) + k_f L(1-t) - L k_d(1-t) = \\ &= k_f + \beta\cdot(k_m - k_f) + L(1-t)(k_f - k_d) = k_{e;\text{CAPM}} + \Delta \end{aligned} \tag{2.17}$$

Here

$$\Delta = L(1-t)(k_f - k_d). \tag{2.18}$$

This formula takes into account both business and financial risk and is the main result of the work. Below in Sect. 2.4 we will apply developed by us approach to several companies, calculate their profitability using formula (2.17) and compare the results with conventional CAPM that take into account only business risks (with some notes).

From the formula (2.18) it follows, that the value added to the company's return (Δ) with respect to results of conventional CAPM does not depend on the industry or market version of CAPM and turns out to be the same for both cases.

2.3.2 *Methodology and Features of the Application of the New Approach*

It is clear that the vast majority of companies are leveraged ones because they use debt financing. The use of debt financing is determined by several factors.

1. All listing companies are quite large and participate in expanded reproduction, which requires the attraction of borrowed capital.
2. The use of debt financing allows you to take advantage from the tax shield: reduce the cost of capital raised and increase the company value.

This means that the standard CAPM formula takes into account business risk and part of the financial risk accounted for by the leveraged beta coefficient in the form of Hamada. This reduces the importance of Hamada's formula, since the covariance found from the statistical reporting and Beta-coefficient already contains the level of leverage and does not need to be renormalized. While the additional term(s) found by us must be taken into account in order to correctly determine the premium for financial risk.

If we consider the almost never occurring case of a non-leveraged company, then we need to apply the standard CAPM formula with a non-leveraged beta. However, if we want to make a forecast for assessing the profitability of the company, taking into account the future level of leverage, it is necessary to take into account all three of the above additives related to taking into account debt financing.

2.4 Application of Developed by Us Approach to Several Companies

The application of the new approach is carried out through the following steps:

At the first stage, it is necessary to collect and process statistical data at three levels: company, industry and market.

We need the following parameters:

1. For company k_i; L, k_d; σ_i; β_{iI}; β_{im};
2. For industry k_I; L, k_d; σ_I; β_{iI};
3. Market k_m; σ_m; β_{im}.

Here k_i; k_I; k_m stand for company, industry, and market returns;
σ_i; σ_I; σ_m are standard deviation for company, industry, and market returns;
β_{iI}; β_{im} are beta coefficient company to industry and to market;
L is leverage level; k_d is the cost of debt.

It is also necessary to use methods for their processing, since we will operate with average annual values, and the data on the sites usually give daily quotes.

- *At the second stage* we evaluate the company's profitability within the traditional CAPM.
- *At the third stage* one needs to use formulas (2.17) and (2.18) to estimate the company return under taking into account both business and financial risk.
- *At the fourth stage*, the company's profitability is compared on the traditional CAPM approach and on the new CAPM 2.0 approach.

We have a database of dozens of companies from different countries, which can be accessed upon request. Five companies (PJSC Severstal, PJSC Polymetal, PJSC Rosneft, Pfizer INC., Walt Disney Company) were selected to illustrate the results obtained. As will be seen below, the results of the five selected companies, as well as other companies, are highly dependent on the level of leverage and the difference between k_d and k_F. Below we present and compare the results of a sample of five companies within the traditional CAPM approach and within the new CAPM 2.0 approach.

2.4.1 Estimation of the Return of PJSC Severstal for the Period 2018–2021 by CAPM (Ticker PJSC Severstal on the Moscow Exchange Is CHMF)

In Table 2.2 comparison of PJSC Severstal profitability estimates for the period 2018–2021 on CAPM and on the new CAPM 2.0 approach is shown.

From the Tables 2.1 and 2.2 it follows that that accounting for financial risk properly significantly affects the assessment of the return on assets in both versions of CAPM: industry and market.

Table 2.1 Estimation of indicators and of the return of PJSC Severstal for the period 2018–2021 by CAPM

	Indicators	2018	2019	2020	2021
k_F		8.02%	7.59%	6.27%	7.34%
Company LKOH	k_i	6.23%	−0.53%	41.01%	21.27%
	L	1.21	1.75	1.51	1.37
	k_d	3.77%	3.98%	3.38%	3.36%
	σ_i	0.24	0.19	0.26	0.28
Industry MOEXOG	k_I	8.71%	10.68%	47.75%	7.37%
	L	0.41	0.81	0.66	0.68
	σ_I	0.19	0.11	0.24	0.17
	$\beta_{i.\ I}$	0.40	−0.19	0.91	1.42
Market IMOEX	μ_m	12.20%	28.58%	8.06%	15.08%
	σ_m	0.17	0.11	0.26	0.16
	$\beta_{i.\ m}$	0.77	−0.73	0.69	0.85
k_i		**6.23%**	**−0.53%**	**41.01%**	**21.27%**
k_i CAPM (industry)		**8.29%**	**6.99%**	**44.22%**	**7.38%**
k_i CAPM (market)		**11.22%**	**−7.64%**	**7.51%**	**13.93%**

Table 2.2 Comparison of PJSC Severstal profitability estimates for the period 2018–2021 on CAPM and on the new CAPM 2.0 approach

	2018 (%)	2019 (%)	2020 (%)	2021 (%)
k_i	**6.23**	**−0.53**	**41.01**	**21.27**
k_i CAPM (industry)	**8.29**	**6.99**	**44.22**	**7.38**
k_i CAPM (market)	**11.22**	**−7.64**	**7.51**	**13.93**
Δ	4.14	5.05	3.49	4.36
k_i CAPM (industry) New approach	12.43	12.04	47.71	11.76
k_i CAPM (market) New approach	15.36	−2.61	11.0	18.29

In industry CAPM/CAPM 2.0 return in 2018 was **8.29%**/12.43%;

in 2019 **6.99%**/12.04%;
in 2020 **44.22%**/47.71%;
in 2021 **7.38%**/11.76%.

In market CAPM/CAPM 2.0 return in 2018 was **11.22%**/15.36%;

in 2019–**7.64%**/–2.61%;
in 2020 **7.51%**/11.0%;
in 2021 **13.93%**/18.29%.

It can be seen that the financial risk premium increases the company's income each year.

2.4.2 *Estimation of the Return of PJSC Polymetal for the Period 2018–2022 by CAPM Capital Asset Pricing Model (Ticker PJSC Polymetal on the Moscow Exchange Is POLY)*

Summary of indicators for Polymetal shares, the RTS mining and metal index and the MICEX index in the period 2018–2022 could be found in Table 2.3.

Table 2.3 gives: (1) the company's average annual return; (2) the company's profitability with an industry business risk premium; (3) profitability of a company with a market business risk premium.

Comparison of PJSC Polymetal profitability estimates for the period 2018–2022 on CAPM and on the new CAPM 2.0 approach are shown in Table 2.4.

From the Tables 2.3 and 2.4 it follows that that accounting for financial risk properly significantly affects the assessment of the return on assets in both versions of CAPM: industry and market.

In industry CAPM/CAPM 2.0 return in 2018 was 2.11%/19.32%;

in 2019 40.20%/47.57%;
in 2020 56.08%/58.93%;
in 2021 12.37%/14.30%;
in 2022–9.80%/−1.36%.

Table 2.3 Summary table of indicators for Polymetal shares, the RTS mining and metal index and the MICEX index in the period 2018–2022

Year	2018	2019	2020	2021	2022
Company level (Polymetal)					
Profitability actual	3.48%	32.80%	78.71%	−24.39%	−71.71%
Standard deviation	0.304	0.242	0.454	0.264	0.705
Average debt cost	3.52%	4.89%	4.00%	2.88%	3.28%
Leverage level	4.78	3.41	1.57	0.54	1.6
Industry level (RTS mining and metal index)					
Profitability actual	2.11%	40.20%	56.08%	12.37%	−9.80%
Standard deviation	0.233	0.144	0.390	0.230	0.580
Average leverage level	0.408	0.370	0.351	1.128	0.818
Beta with Polymetal	0.208	0.107	0.436	0.349	0.250
Profitability (industry CAPM)	6.79%	11.07%	27.97%	9.10%	4.96%
Correlation with Polymetal	0.27	0.18	0.51	0.40	0.30
Market level (Moscow exchange index MICEX)					
Profitability actual	18.15%	36.24%	22.57%	3.25%	−16.78%
Standard deviation	0.167	0.120	0.271	0.163	0.497
Beta with poly	0.443	0.189	0.516	0.307	0.508
Profitability (market CAPM)	12.51%	12.99%	14.67%	6.09%	−3.68%
Correlation with Polymetal	0.24	0.09	0.31	0.19	0.36
k_F	8.02%	7.59%	6.27%	7.34%	9.87%

Table 2.4 Comparison of PJSC Polymetal profitability estimates for the period 2018–2022 on CAPM and on the new CAPM 2.0 approach

Year	2018 (%)	2019 (%)	2020 (%)	2021 (%)	2022 (%)
Profitability actual	3.48	32.80	78.71	−24.39	−71.71
Profitability (industry CAPM)	2.11	40.20	56.08	12.37	−9.80
Profitability (market CAPM)	12.51	12.99	14.67	6.09	−3.68
Δ	17.21	7.37	2.85	1.93	8.44
Profitability (industry CAPM) New approach	19.32	47.57	58.93	14.30	−1.36
Profitability (market CAPM) New approach	29.72	20.36	17.52	8.02	4.76

In market CAPM/ CAPM 2.0 return in 2018 was 12.51%/29.72%;

in 2019 12.99%/20.36%;
in 2020 14.67%/17.52%;
in 2021 6.09%/8.02%;
in 2022–3.68%/4.76%.

It can be seen that the financial risk premium increases the company's income each year.

2.4.3 *Estimation of the Return of PJSC Rosneft, ROSN for the Period 2018–2021 by CAPM*

Summary of indicators for PJSC Rosneft, ROSN shares, industry an market indexes in the period 2018–2021 could be found in Table 2.5.

2.4.4 *Estimation of the Return of Pfizer INC., PFE for the Period 2018–2022 by CAPM*

Summary of indicators for Pfizer INC., PFE shares, industry an market indexes in the period 2018–2022 could be found in Table 2.7.

2.4.5 *Estimation of the Return of Walt Disney Company: DIS for the Period 2018–2022 by CAPM*

Summary of indicators for Walt Disney Company: DIS shares, industry an market indexes in the period 2018–2022 could be found in Table 2.9.

Table 2.5 Summary table of indicators for PJSC Rosneft, ROSN shares, industry an market indexes in the period 2018–2021

Level	Index	2018	2019	2020	2021
Company	k_i	53.3%	7.2%	4.4%	42.0%
	σ_i	0.255561	0.173661	0.424433	0.256377
	L_i	5.1	4.4	5.1	4.8
	k_d	1.55%	0.86%	2.73%	2.34%
Industry	k_I	36%	24%	−16%	25%
	σ_I	0.164794	0.130976	0.318362	0.194733
	L_I	0.415428	0.361656	0.359434	1.135839
	β_{iI}	0.690794	0.817217	1.19458	0.964128
Market	μ_m	12%	29%	8%	15%
	σ_m	0.171552	0.111067	0.259559	0.163953
	β_{im}	0.450119	0.829661	1.382028	0.987169
	k_i, %	53.3%	7.2%	4.4%	42.0%
	k_i, industry %	27.4%	21.4%	−20.3%	24.1%
	k_i, market %	9.93%	24.96%	8.66%	15.05%

Table 2.6 Comparison of PJSC Rosneft, ROSN profitability estimates for the period 2018–2021 on CAPM and on the new CAPM 2.0 approach

Year	2018 (%)	2019 (%)	2020 (%)	2021 (%)
k_F	8.02	7.59	6.27	7.34
Δ	26.40	23.69	14.44	19.20
k_i, %	53.3	7.2	4.4	42.0
k_i, industry %	27.4	21.4	−20.3	24.1
k_i, market %	9.93	24.96	8.66	15.05
k_i CAPM (industry) New approach	53.8	45.09	−5.86	43.3
k_i CAPM (market) New approach	36.33	48.65	23.10	34.25

It can be seen from Tables 2.5, 2.6, 2.7, 2.8, 2.9 and 2.10 that when kd exceeds kF, the financial risk premium becomes negative. It can also be seen Tables 2.1, 2.2, 2.3, 2.4, 2.5 and 2.6 that for companies from the extractive industries, such as Rosneft, Polymetal, and to a lesser extent Severstal, whose leverage level, due to the specifics of the industry is quite high, the premium for financial risk is high as well.

At the same time, companies with a typical level of leverage from 0.5 to 1 (1.5) have a low financial risk premium compared to a business risk (see Tables 2.7, 2.8, 2.9 and 2.10). This is well seen for Walt Disney Company (L is of order 0.5) and for Pfizer INC. (L is of order 1–1.5)

Table 2.7 Summary table of indicators for Pfizer INC., PFE shares, industry an market indexes in the period 2018–2022

Level	Index	2018	2019	2020	2021	2022
Company	k_i	17.86%	−7. 70%	4. 28%	64.48%	2.83%
	σ_i	0.047	0.050	0.10	0.08	0.072
	L_i	1.5	1.64	1.43	1.34	1.06
	k_d	3.26%	3.54%	3.36%	3.57%	3.67%
	β_{im}	0.92	0.298	0.87	0.38	0.51
	ρ_{im}	0.78	0.20	0.63	0.15	0.47
Industry	k_I	−1.76%	13.39%	14.74%	22.60%	– 3.47%
	σ_I	0.05	0.03	0.056	0.04	0.045
	L_I	1.55	1.63	1.72	1.77	1.59
	k_d	4.56%	3.67%	3.00%	3.58%	5.88%
	β_{iI}	0.87	1.03	1.53	0.53	1.16
	ρ_{iI}	0.92	0.64	0.83	0.26	0.71
Market	k_m	– 11.22%	19.48%	16.45%	28.32%	17.61%
	σ_m	0.04	0.033	0.075	0.030	0.066
	β_{im}	0.92	0.298	0.87	0.38	0.51
k_F		3. 02%	2.39%	1.00%	1. 91%	3.98%
k_i (Industry)		−1.13%	13.73%	21.98%	12.91%	– 4.64%
k_i (Market)		– 10. 04%	7.49%	14.44%	12.05%	10.98%

Table 2.8 Comparison of Pfizer INC., PFE shares profitability estimates for the period 2018–2022 on CAPM and on the new CAPM 2.0 approach

Year	2018 (%)	2019 (%)	2020 (%)	2021 (%)	2022 (%)
k_F	3.02	2.39	1.00	1.91	3.98
Δ	−0.29	−1.49	−2.67	−1.76	0.26
k_i	17.86	−7.70	4.28	64.48	2.83
k_i (industry)	−1.13	13.73	21.98	12.91	-4.64
k_i (market)	−10.04	7.49	14.44	12.05	10.98
k_i CAPM (industry) New approach	−1.42	12.24	19.31	11.15	-4.38
k_i CAPM (market) New approach	−10.33	6.05	11.77	10.29	11.24

2.5 Conclusions

New approach has been developed to a return on assets assessment that generalizes CAPM to account for both business and financial risks. We combine the theory of CAPM and the Modigliani–Miller (MM) theory. The first is based on portfolio analysis and accounting for business risks in relation to the market (or industry). The second one (the Modigliani–Miller (MM) theory) describes a specific company and takes into account the financial risks associated with the use of debt financing. The combination of these two different approaches makes it possible to take into

Table 2.9 Summary table of indicators for Walt Disney Company: DIS shares, industry (SPLRCD) and market (S&P500) indexes in the period 2018–2022

Level	Index	2018	2019	2020	2021	2022
Company	k_i	1.99%	31.90%	25.27%	−14.51%	−43.91%
	σ_i (по μ)	0.041	0.088	0.132	0.069	0.108
	L_i	0.43	0.53	0.70	0.61	0.51
	k_d	3.27%	2.65%	2.81%	2.84%	3.20%
	β_{im}	0.57	0.93	1.54	1.23	1.18
Industry (SPLRCD)	k_I	−0.49%	26.20%	32.07%	23.66%	−37.58%
	σ_I(по μ)	0.057	0.046	0.090	0.041	0.089
	L_I	0.20	0.20	0.15	0.15	0.33
	β_{iI}	0.355	0.695	1.081	0.049	0.816
	k_i	1.78%	18.94%	34.55%	2.96%	−29.94%
Market (S&P500)	k_m	−6.24%	28.88%	16.26%	26.89%	−19.44%
	σ_m	0.044	0.037	0.075	0.032	0.066
	β_{im}	0.57	0.93	1.54	1.23	1.18
	k_F	3.020%	2.389%	1.646%	1.905%	3.975%
	k_i	−2.28%	27.13%	24.14%	32.61%	−23.64%
	k_i, company	1.99%	31.90%	25.27%	−14.51%	−43.91%
	k_i, industry	1.78%	18.94%	34.55%	2.96%	−29.94%
	k_i, market	−2.28%	27.13%	24.14%	32.61%	−23.64%

Table 2.10 Comparison of Walt Disney Company: DIS shares profitability estimates for the period 2018–2022 on CAPM and on the new CAPM 2.0 approach

Year	2018 (%)	2019 (%)	2020 (%)	2021 (%)	2022 (%)
k_F	3. 02	2.39	1.646	1. 91	3.98
Δ	−0.08	−0.11	−0.64	−0.45	0.31
k_i, company	1.99	31.90	25.27	−14.51	−43.91
k_i, industry	1.78	18.94	34.55	2.96	−29.94
k_i, market	−2.28	27.13	24.14	32.61	−23.64
k_i CAPM (industry) New approach	1.70	18.83	33.91	2.51	−29.63
k_i CAPM (market) New approach	−2.36	27.02	23.50	32.16	−23.33

account both types of risks: business and financial ones. We combine these two approaches analytically, while Hamada did it phenomenologically. Both approaches are quite well developed and tested, so that the validity of the model built on these two approaches is beyond doubt. It is shown that in addition to the renormalization of the beta-coefficient obtained in the Hamada model, two additional terms are found: the renormalized risk-free income and the term depending on the cost of debt kd. A critical analysis of the Hamada model was carried out, which showed that the Hamada model is not applicable in practice. Two versions of CAPM (market or industry) are considered. It has been shown, that the value added to the company's return (Δ) with respect to results of conventional CAPM does not depend on the industry or market version of CAPM and turns out to be the same for both cases. The results obtained show that accounting for financial risk properly significantly affects the assessment of the return on assets. It can be seen from Tables 2.5, 2.6, 2.7, 2.8, 2.9 and 2.10 that when kd exceeds kF, the financial risk premium becomes negative. It can also be seen Tables 2.1, 2.2, 2.3, 2.4, 2.5 and 2.6 that for companies from the extractive industries, such as Rosneft, Polymetal, and to a lesser extent Severstal, whose leverage level, due to the specifics of the industry is quite high, the premium for financial risk is high as well.

At the same time, companies with a typical level of leverage from 0.5 to 1 (1.5) have a low financial risk premium compared to a business risk (see Tables 2.7, 2.8, 2.9 and 2.10). This is well seen for Walt Disney Company (L is of order 0.5) and for Pfizer INC. (L is of order 1–1.5).

The proposed approach allows making forecasts on the company's profitability without using the theory of capital structure. Estimated financial risk premiums depend on the level of leverage (capital structure) and the cost of borrowings. By planning the values of these parameters, the manager can predict the profitability of the company in the future. The novelty of the article lies in the development of a new approach that generalizes CAPM to account for both business and financial risks.

References

1. Treynor, J. L. (1965). How to evaluate the management of investment funds. *Harvard Business Review, 43*, 63–75.
2. French, C. W. (2003). The Treynor capital asset pricing model. *Journal of Investment Management, 1*(2), 60–72. Available at SSRN: https://ssrn.com/abstract=447580
3. Sharpe, W. F. (1964). Capital asset prices: A theory of market equilibrium under conditions of risk. *The Journal of Finance, 19*(3), 425–442.
4. Mossin, J. (1966). Equilibrium in a capital asset market. *Econometrica, 34*, 768–783.
5. Lintner, J. (1965). The valuation of risk assets and the selection of risky investments in stock portfolios and capital budgets. *Review of Economics and Statistics, 47*, 13–37.
6. Ross, S. (1976). The arbitrage theory of capital asset pricing. *Journal of Economic Theory, 13*(3), 341–360. https://doi.org/10.1016/0022-0531(76)90046-6
7. Fama, E. F., & French, K. R. (1992). The crosssection of expected stock returns. *The Journal of Finance, 47*(2), 427–465.
8. Fama, E. F., & French, K. R. (1993). Common risk factors in the returns on stocks and bonds. *Journal of Financial Economics, 33*(1), 3–56.
9. Fama, E. F., & French, K. R. (1995). Size and book-to-market factors in earnings and returns. *The Journal of Finance, 50*(1), 131–155.
10. Fama, E. F. (1965). The behavior of stock–market prices. *Journal of Business, 38*, 34–105.
11. Leal, D., Jiménez, R., Riquelme, M., & Leiva, V. (2023). Elliptical capital asset pricing models: Formulation, diagnostics. *Case Study with Chilean Data, and Economic Rationale Mathematics, 11*(6), 1394. https://doi.org/10.3390/math11061394
12. Zhou, G. (1993). Asset–pricing tests under alternative distributions. *The Journal of Finance, 48*, 1927–1942.
13. Lange, K. L., Little, R. J., & Taylor, J. M. (1989). Robust statistical modeling using the t distribution. *Journal of the American Statistical Association, 84*, 881–896.

14. Chen, J., Wu, Y., & Xu, Y. (2021). Research and analysis of asset pricing model based on the empirical test of stock price. In *Proceedings of the Third International Conference on Economic Management and Cultural Industry, Guangzhou, China* (pp. 1288–1293).
15. Blattberg, R. C., & Gonedes, N. J. (1974). A comparison of the stable and student distributions as statistical models for stock prices. *Journal of Business, 47*, 244–280.
16. Modigliani, F., & Miller, M. H. (1958). The cost of capital, corporation finance and the theory of investment. *The American Economic Review, 48*, 261–296.
17. Modigliani, F., & Miller, M. H. (1963). Corporate income taxes and the cost of capital: A correction. *The American Economic Review, 53*, 433–443.
18. Hamada, R. (1969). Portfolio analysis, market equilibrium, and corporate finance. *J. Finance, 24*, 13–31.
19. Hamada, R. (1972). The effect of the firm's capital structure on the systematic risk of common stocks. *The Journal of Finance, 27*, 435–452.
20. Brusov, P., Filatova, T., & Orekhova N. (2022). *Generalized Modigliani–Miller theory: Applications in corporate finance, investments, taxation and ratings*. Springer Nature, 362 pp.
21. Brusov, P., & Filatova, T. (2023). Capital Structure Theory: Past, Present, Future. *Mathematics, 11*(3), 616. https://doi.org/10.3390/math11030616
22. Brusov, P., Filatova, T., & Orekhova N. (2023). *The Brusov–Filatova–Orekhova theory of capital structure: Applications in corporate finance, investments, taxation and ratings*. Springer Nature, 769 pp.

Chapter 3
The Arbitrage Pricing Theory (APT)

Arbitrage Pricing Theory (APT) is one of three methods of the income approach to business valuation, along with CAPM and WACC. APT is a multi-factor model for asset pricing which relates various macro-economic (systematic) risk variables to the pricing of financial assets. Proposed by Stephen Ross in 1976 [1] it is widely believed to be an improved alternative to its predecessor, the capital asset pricing model (CAPM) [2]. APT is founded upon the law of one price, which suggests that within an equilibrium market, rational investors will implement arbitrage such that the equilibrium price is eventually realized [2]. As such, APT argues that when opportunities for arbitrage are exhausted in a given period, then the expected return of an asset is a linear function of various factors or theoretical market indices, where sensitivities of each factor is represented by a factor-specific beta coefficient or factor loading. Consequently, it provides traders with an indication of "true" asset value and enables exploitation of market discrepancies via arbitrage. The linear factor model structure of the APT is used as the basis for evaluating asset allocation, the performance of managed funds as well as the calculation of cost of capital [3]. Furthermore, the newer APT model is more dynamic being utilized in more theoretical application than the preceding CAPM model. A 1986 article written by Gregory Connor and Robert Korajczyk, utilized the APT framework and applied it to portfolio performance measurement suggesting that the Jensen coefficient is an acceptable measurement of portfolio performance [4].

3.1 Arbitrage Pricing Theory (APT) Model

APT is a single-period static model, which helps investors understand the trade-off between risk and return. The average investor aims to optimize the returns for any given level or risk and as such, expects a positive return for bearing greater risk. As per the APT model, risky asset returns are said to follow a factor intensity structure if they can be expressed as:

P. Brusov, T. Filatova, *Business Valuation and Company Value*, Contributions to Finance and Accounting, https://doi.org/10.1007/978-3-031-90847-7_3

$$r_i = a_i + \beta_{i1} \cdot f_1 + \beta_{i2} \cdot f_2 + \ldots + \beta_{in} \cdot f_n + \varepsilon_i \tag{3.1}$$

where

a_i is a constant for asset; f_i is a systematic factor for $1 \leq i \leq n$; β_{ik} is the sensitivity of the ith asset to factor f_k, $1 \leq k \leq n$, also called factor loading; ε_i is the risky asset's idiosyncratic random shock with mean zero.

Idiosyncratic shocks are assumed to be uncorrelated across assets and uncorrelated with the factors.

The APT model states that if asset returns follow a factor structure then the following relation exists between expected returns and the factor sensitivities:

$$E(r_i) = r_f + \beta_{i1} \cdot \mathrm{RP}_1 + \beta_{i2} \cdot \mathrm{RP}_2 + \ldots + \beta_{in} \cdot \mathrm{RP}_n \tag{3.2}$$

Here RP_k is the risk premium of factor f_k; r_f is the risk-free rate.

That is, the expected return of an asset j is a linear function of the asset's sensitivities to the n factors.

For the latter to be true, certain assumptions and requirements must be met: there must be perfect competition in the market, and the total number of factors must never exceed the total number of assets (to avoid the matrix singularity problem.

3.2 Assumptions of APT Model

The APT model for asset valuation is founded on the following assumptions [2]:

1. Investors are risk-averse in nature and possess the same expectations.
2. Efficient markets with limited opportunity for arbitrage.
3. Perfect capital markets.
4. Infinite number of assets.

Risk factors are indicative of systematic risks that cannot be diversified away and thus impact all financial assets, to some degree. Thus, these factors must be:

1. Non-specific to any individual company or industry.
2. Compensated by the market via a risk premium.
3. Must be random variables.
4. Arbitrage (with limited opportunity) should exist.

Arbitrage is a practice in which investors take advantage of small deviations in the valuation of an asset from its fair value to make a profit. It is the realization of positive expected returns on overvalued or undervalued securities in an inefficient market without any additional risk and zero additional investment.

3.3 Arbitrage's Mechanics

In APT, arbitrage involves trading two assets, at least one of which is mispriced. The arbitrageur sells an asset that is relatively too expensive and uses the proceeds to buy an asset that is relatively too cheap. In APT, an asset is mispriced if its current price differs from the price predicted by the model. The price of an asset today should equal the sum of all future cash flows discounted at the APT rate, where the expected return of the asset is a linear function of various factors and the sensitivity to changes in each factor is represented by a factor-specific beta.

A correctly priced asset here may be in fact a synthetic asset—a portfolio consisting of other correctly priced assets. This portfolio has the same exposure to each of the macroeconomic factors as the mispriced asset. The arbitrageur creates the portfolio by identifying n correctly priced assets (one per risk-factor, plus one) and then weighting the assets such that portfolio beta per factor is the same as for the mispriced asset.

When the investor is long the asset and short the portfolio (or vice versa) he has created a position which has a positive expected return (the difference between asset return and portfolio return) and which has a net zero exposure to any macroeconomic factor and is therefore risk free (other than for firm specific risk). The arbitrageur is thus in a position to make a risk-free profit:

Where Today's Price is Too Low

The implication is that at the end of the period the portfolio would have appreciated at the rate implied by the APT, whereas the mispriced asset would have appreciated at more than this rate. The arbitrageur could therefore:

Today:

1. Short sell the portfolio.
2. Buy the mispriced asset with the proceeds.

At the End of the Period:

1. Sell the mispriced asset.
2. Use the proceeds to buy back the portfolio.
3. Pocket the difference.

Where Today's Price Is Too High

The implication is that at the end of the period the portfolio would have appreciated at the rate implied by the APT, whereas the mispriced asset would have appreciated at less than this rate. The arbitrageur could therefore:

Today:

1. Short sell the mispriced asset.
2. Buy the portfolio with the proceeds.

At the End of the Period:

1. Sell the portfolio.
2. Use the proceeds to buy back the mispriced asset.
3. Pocket the difference.

3.4 Difference Between the Capital Asset Pricing Model (CAPT) and APT

The APT along with the capital asset pricing model (CAPM) is one of two influential theories on asset pricing.

At first glance, the CAPM and APT formulas appear identical, but in CAPM there is only one factor and one beta. In contrast, in the APT formula there are multiple factors that include non-corporate factors, which requires the beta of the asset with respect to each individual factor. However, APT does not provide any insight into what these factors might be, so users of the APT model must analytically identify relevant factors that might affect the asset's return. On the other hand, the factor used in CAPM is the difference between the expected market rate of return and the risk-free rate of return.

The APT differs from the CAPM in that it is less restrictive in its assumptions, making it more flexible for use in a wider range of application. Thus, it has greater explanatory power (as opposed to statistical power) for expected asset returns. It assumes that each investor will hold a unique portfolio with its own particular array of betas, as opposed to the identical "market portfolio." In some ways, the CAPM can be considered a "special case" of the APT in that the securities market line represents a single-factor model of the asset price, where beta is exposed to changes in value of the market.

Fundamentally, the CAPM is derived on the premise that all factors in the economy can be reconciled into one factor represented by a market portfolio, thus implying they all have equivalent weight on the asset's return. In contrast, the APT model suggests that each stock reacts uniquely to various macroeconomic factors and thus the impact of each must be accounted for separately [2].

A disadvantage of APT is that the selection and the number of factors to use in the model is ambiguous. Most academics use three to five factors to model returns, but the factors selected have not been empirically robust. In many instances the CAPM, as a model to estimate expected returns, has empirically outperformed the more advanced APT [5].

Additionally, the APT can be seen as a "supply-side" model, since its beta coefficients reflect the sensitivity of the underlying asset to economic factors. Thus, factor shocks would cause structural changes in assets' expected returns, or in the case of stocks, in company profitabilities. On the other side, the capital asset pricing model is considered a "demand side" model. Its results, although similar to those of the APT, arise from a maximization problem of each investor's utility function, and from the resulting market equilibrium (investors are considered to be the "consumers" of the assets).

3.5 Implementation of APT

As with CAPM, factor-specific betas are estimated by linearly regressing historical security returns on the factor in question. Unlike CAPM, however, APT itself does not reveal the identity of its estimated factors—the number and nature of these factors are likely to vary over time and across economies. As a result, the question is essentially empirical in nature. However, several a priori guidelines are offered regarding the characteristics required of potential factors:

1. Their impact on asset prices is manifested by their unexpected movements, and they are completely unpredictable to the market at the beginning of each period [2].
2. They should represent non-diversifiable effects (they are obviously likely to be macroeconomic rather than firm-specific in nature) on expected returns and should therefore be quantified with non-zero prices [2].
3. Timely and accurate information on these variables is required.
4. The relationship should be theoretically justifiable on economic grounds.

Chen, Roll, and Ross identified the following macro-economic factors as significant in explaining security returns [6]:

1. Surprises in inflation.
2. Surprises in GNP as indicated by an industrial production index.
3. Surprises in investor confidence due to changes in default premium in corporate bonds.
4. Surprise shifts in the yield curve.

As a practical matter, indices or spot or futures market prices may be used in place of macro-economic factors, which are reported at low frequency (e.g. monthly) and often with significant estimation errors. Market indices are sometimes derived by means of factor analysis. More direct "indices" that might be used are:

1. Short-term interest rates.
2. The difference in long-term and short-term interest rates.
3. A diversified stock index such as the S&P 500 or NYSE Composite.
4. Oil prices;
5. Gold or other precious metal prices.
6. Currency exchange rates.

Note, that because CAPM is a single-factor model and is easier to use, investors may want to use it to determine the expected theoretical appropriate rate of return rather than using APT, which requires users to quantify multiple factors.

3.6 International Arbitrage Pricing Theory

International arbitrage pricing theory (IAPT) is an important extension of the base idea of arbitrage pricing theory which further considers factors such as exchange rate risk. In 1983 Bruno Solnik [7] created an extension of the original arbitrage pricing theory to include risk related to international exchange rates hence making the model applicable international markets with multi-currency transactions. Solnik suggested that there may be several factors common to all international assets, and conversely, there may be other common factors applicable to certain markets based on nationality [7].

Fama and French [8] originally proposed a three-factor model in 1995 which, consistent with the suggestion from Solnik above suggests that integrated international markets may experience a common set of factors, hence making it possible to price assets in all integrated markets using their model. The Fama and French three factor model attempts to explain stock returns based on market risk, size, and value [8].

A 2012 paper [9] aimed to empirically investigate Solnik's IAPT model and the suggestion that base currency fluctuations have a direct and comprehendible effect on the risk premiums of assets. This was tested by generating a returns relation which broke down individual investor returns into currency and non-currency (universal) returns. The paper utilized Fama and French's three factor model (explained above) to estimate international currency impacts on common factors. It was concluded that the total foreign exchange risk in international markets consisted of the immediate exchange rate risk and the residual market factors. This, along with empirical data tests validates the idea that foreign currency fluctuations have a direct effect on risk premiums and the factor loadings included in the APT model, hence, confirming the validity of the IAPT model [9].

References

1. Ross, S. A. (1976). The arbitrage theory of capital asset pricing. *Journal of Economic Theory, 13*(3), 341–360. https://doi.org/10.1016/0022-0531(76)90046-6. ISSN 0022-0531.
2. Basu, D., & Chawla, D. (2012). An empirical test of the arbitrage pricing theory—The case of Indian stock market. *Global Business Review, 13*(3), 421–432. https://doi.org/10.1177/097215091201300305. ISSN 0972-1509.
3. Huberman, G., & Wang, Z. (2005). *Arbitrage pricing theory*. Federal Reserve Bank of New York Staff Reports.
4. Connor, G., & Korajczyk, R. (1986). Performance measurement with the arbitrage pricing theory: A new framework for analysis. *Journal of Financial Economics, 15*(3), 373–394. https://doi.org/10.1016/0304-405X(86)90027-9
5. French, J. (2017). Macroeconomic forces and arbitrage pricing theory. *Journal of Comparative Asian Development, 16*(1), 1–20. https://doi.org/10.1080/15339114.2017.1297245
6. Chen, N.-F., Roll, R., & Ross, S. A. (1986). Economic forces and the stock market. *The Journal of Business, 59*(3), 383–403. https://doi.org/10.1086/296344. ISSN 0021-9398. JSTOR 2352710.

7. Solnik, B. (1983). International arbitrage pricing theory. *The Journal of Finance, 38*(2), 449–457. https://doi.org/10.2307/2327978.JSTOR2327978
8. Fama, E., & French, K. (1996). Multifactor explanations of asset pricing anomalies. *The Journal of Finance, 51*(1), 55–84. https://doi.org/10.1111/j.1540-6261.1996.tb05202.x
9. Armstrong, W., Knif, J., Kolari, J., & Pynnonen, S. (2012). Exchange risk and universal returns: A test of international arbitrage pricing theory. *Pacific Basin Finance Journal, 20*(1), 24–40. https://doi.org/10.1016/j.pacfin.2011.08.003

Part III
Capital Structure Theories

Chapter 4
Capital Structure

4.1 Introduction

Under the capital structure, one understands the relationship between equity and debt capital of the company. Does capital structure affect the company's main settings, such as the cost of capital, profit, value of the company, and the others, and, if affects, how? Choice of an optimal capital structure, i.e., a capital structure, which minimizes the weighted average cost of capital, WACC, and maximizes the value of the company, *V*, is one of the most important tasks solved by financial manager and by the management of a company. The first serious study (and first quantitative study) of influence of capital structure of the company on its indicators of activities was the work by Modigliani and Miller [1]. Until this study, the approach existed (let us call it traditional), which was based on empirical data analysis.

4.2 Basic Theories of Capital Structure

4.2.1 A Historical Point of View

From a historical point of view, *five stages* in the development of the capital structure theory can be distinguished: *First* (before 1958)—the traditional approach, based on practical experience and existed before the appearance of the first quantitative theory by Modigliani and Miller (*the second stage)* (1958–1963) (3–5). *The third stage* (1964–2008) is the numerous attempts of scientists to modify the Modigliani–Miller theory. *The fourth stage* (2008–2019) is the appearance of Brusov–Filatova–Orekhova (BFO) theory, which removed the main restriction of the Modigliani–Miller (MM) theory associated with the infinite life-time of the company (6). And, finally, *the fifth stage* (2019 up to now), which began a couple years ago and is associated with the adaptation of the two main theories of the capital structure

P. Brusov, T. Filatova, *Business Valuation and Company Value*, Contributions to Finance and Accounting, https://doi.org/10.1007/978-3-031-90847-7_4

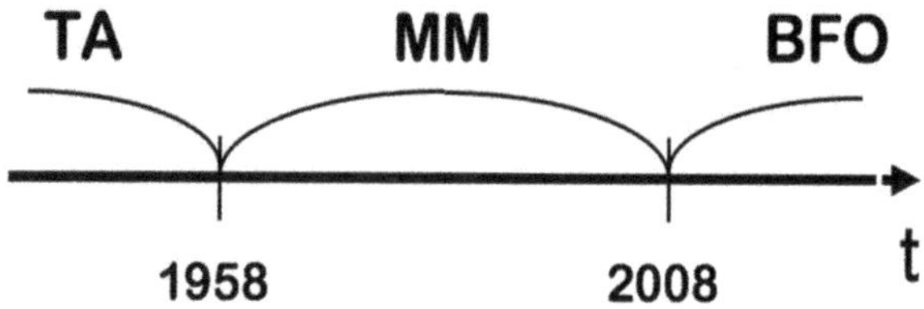

Fig. 4.1 The historical development of the theory of capital structure from the empirical traditional approach (TA) to the general theory of capital structure, BFO (Brusov–Filatova–Orekhova), through the perpetuity Modigliani–Miller theory

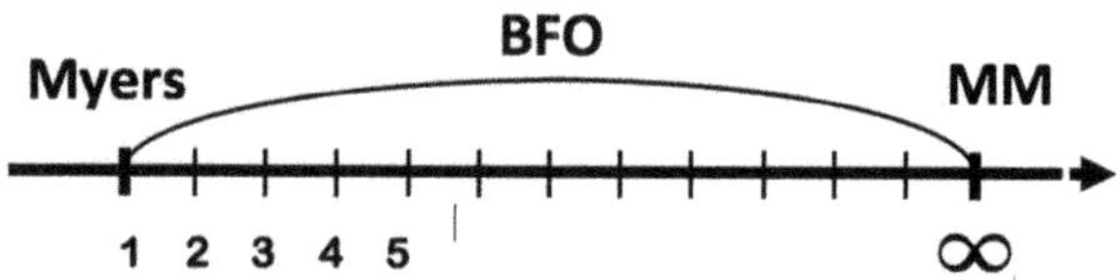

Fig. 4.2 Only two results for capital structure of the company were known before BFO theory: Modigliani–Miller (MM) for perpetuity companies and Myers for 1-year company, while the created BFO theory filled the entire interval between $n = 1$ and $n = \infty$

(Brusov–Filatova–Orekhova and Modigliani–Miller) to the established financial practice of the company's functioning by taking into account the real conditions of their work [2–8].

One of the most important assumptions of the Modigliani–Miller theory is that all financial flows and all companies are perpetuity. This limitation was lift out by Brusov–Filatova–Orekhova in 2008 [9], who have created BFO (Brusov–Filatova–Orekhova) theory—modern theory of capital cost and capital structure for companies of arbitrary age (BFO-1 theory) and for companies of arbitrary life time (BFO-2 theory) [10]. Figure 4.1 shows the historical development of the theory of capital structure from the empirical traditional approach to the general theory of capital structure, BFO, through the perpetuity Modigliani–Miller theory.

One year companies were studied by Steve Myers in 2001, who showed that the weighted average cost of capital WACC is greater in this case than in the perpetual Modigliani–Miller case, and the value of company V is therefore less. Only two results for capital structure of the company were known up to 2008, when BFO theory has appeared: Modigliani–Miller for perpetuity companies and Myers for one-year company (see Fig. 4.2). The created BFO theory filled the entire interval between $t = 1$ and $t = \infty$. This expands capital structure theory for the companies of arbitrary ages and/or arbitrary life-time. Many new meaningful effects have been discovered in the BFO theory.

4.2.2 The Empirical (Traditional) Approach

In the traditional approach, based on practical experience and existing before the advent of the first quantitative theory of Modigliani and Miller, the weighted average cost of capital WACC and the associated capitalization of the company, $V = \mathrm{CF}/\mathrm{WACC}$, depend on the capital structure, the level of debt load, L. The cost of debt is always lower than the cost of equity, because the former has less risk due to the fact that the claims of creditors are satisfied before the claims of shareholders in the event of bankruptcy. As a result, an increase in the share of cheaper borrowed capital in the total capital structure to the limit that does not cause a violation of financial stability and an increase in the risk of bankruptcy leads to a decrease in the weighted average cost of capital WACC.

The return required by investors (equal to the cost of equity) is growing; however, its growth did not offset the benefits of using cheaper borrowed capital. Therefore, the traditional approach welcomes the increase in leverage $L = D/S$ and the associated increase in the value of the company $V = \mathrm{CF}/\mathrm{WACC}$. The empirical traditional approach existed until the appearance of the first quantitative theory by Modigliani and Miller [1].

Within the based on existing practical experience traditional approach the competition between the advantages of debt financing at a low leverage level and its disadvantages at a high leverage level forms the optimal capital structure, defined as the leverage level, at which WACC is minimal and company value, V, is maximum.

4.2.3 The Modigliani–Miller Theory

4.2.3.1 The Modigliani–Miller Theory with Taxes

There are two versions of the Modigliani–Miller theory: without taxes and with taxes. Without taxes the following expressions for V, $WACC$ and k_e

$$V = V_0 = \frac{\mathrm{EBIT}}{k_0}, \tag{4.1}$$

were V_0 stands for the unlevered company value, EBIT stands for earnings before interest and taxes and k_0 stands for the equity cost at zero leverage level L.

From (4.1) one gets for the weighted average cost of capital WACC:

$$\mathrm{WACC} = k_0. \tag{4.2}$$

From the expression for WACC

$$\mathrm{WACC}=k_0=k_e w_e+k_d w_d. \tag{4.3}$$

and accounting (4.1), one get for the equity cost, k_e,

$$k_e=\frac{k_0}{w_e}-k_d\frac{w_d}{w_e}=\frac{k_0(S+D)}{S}-k_d\frac{D}{S}=k_0+(k_0-k_d)\frac{D}{S}=k_0+(k_0-k_d)L \tag{4.4}$$

Here, D stands for debt capital value; S stands for equity capital value; k_d and w_d stand for the cost and share of the company's debt capital; k_e and w_e stand for the equity capital cost and share. It is seen from (4.4), that the equity increases linearly with the leverage level.

4.2.3.2 The Modigliani–Miller Theory with Taxes

Within the framework of the Modigliani–Miller theory with taxes (4), the following expression was postulated for the value of a company using borrowed funds, V,

$$V=V_0+D{\cdot}t. \tag{4.5}$$

The expression for WACC immediately follows from (4.5):

$$\mathrm{WACC}=k_0{\cdot}(1-w_d t) \tag{4.6}$$

The following formula for the cost of equity k_e can be obtained from (4.6) within the framework of the Modigliani–Miller theory with taxes:

$$k_e=k_0+L{\cdot}(k_0-k_d)(1-t) \tag{4.7}$$

Two formulas (4.7) (MM with taxes) and (4.4) (MM without taxes) differs by the multiplier (1-t), called the tax corrector. It is less than unit, thus the $k_e(L)$ curve slope decreases with accounting the taxes.

4.2.4 *Modifications of Modigliani–Miller Theory*

Since the creation of Modigliani–Miller theory, numerous attempts have been made to improve and develop this theory (see [11–95]). We will discuss some of these modifications below. One of them is Hamada model.

4.2.4.1 Hamada Model

The Modigliani–Miller theory, with accounting taxes has been united with CAPM (Capital Asset Pricing Model) in 1961 by Hamada. For the cost of equity of leveraged company the below formula has been derived:

$$k_e = k_F + (k_M - k_F)b_U + (k_M - k_F)b_U \frac{D}{S}(1 - T), \tag{4.8}$$

Here b_U is the β-coefficient of the unlevered company. First term represents risk-free profitability k_F, second term—business risk premium, $(k_M - k_F)b_U$, and third term—financial risk premium $(k_M - k_F)b_U \frac{D}{S}(1 - T)$.

In the case of a unlevered company ($D = 0$), the financial risk (the third term) is zero, and its shareholders receive only a business risk premium.

Equating CAPM formula to right side of (4.8), one gets:

$$k_F + (k_M - k_F)b_U = k_F + (k_M - k_F)b_U + (k_M - k_F)b_U \frac{D}{S}(1 - T) \tag{4.9}$$

or

$$b = b_U\left(1 + \frac{D}{S}(1 - T)\right). \tag{4.10}$$

Below are the formulas for the equity cost, k_e, debt cost, k_d and WACC in the CAPM model and (in parenthesis) in the Modigliani–Miller theory [1, 96, 97].

The equity cost for unlevered company:

$$k_e = k_F + \left(\bar{k}_M - k_F\right)\beta_U, \quad (k_e = k_0). \tag{4.11}$$

The equity cost for levered company:

$$k_e = k_F + \left(\bar{k}_M - k_F\right)\beta_e, \quad (k_e = k_0 + (1 - T)(k_0 - k_d)L). \tag{4.12}$$

The debt cost:

$$k_d = k_F + \left(\bar{k}_M - k_F\right)\beta_d, (k_d = k_F; \beta_d = 0).$$

The weighted average cost of capital WACC

$$\text{WACC} = k_e w_e + k_d w_d(1 - T), \quad (\text{WACC} = k_0(1 - T w_d)). \tag{4.13}$$

4.2.4.2 The Cost of Capital under Risky Debt

In the Modigliani–Miller theory there are two assets types: risky equity and risk-free debt. But assuming about the risk of bankruptcy of company and the ability to nonpayment of debt may change the situation with debt. It has been shown by Stiglitz [98] and Rubinstein [99] that the assumption concerning risky debt does not change the company value with respect to the Modigliani and Miller results under free of risk debt [1, 96, 97]. However, the debt cost is changed from a constant value $k_d = k_F$ to variable one. As Hsia [100], based on the models of pricing options, Modigliani–Miller and CAPM, has shown in the formula for the net discount profit, a term, reflecting tax shield should be discounted at the rate:

$$k_d' = k_F + (k_0 - k_F)N(-d_1)\frac{1}{w_d}, \tag{4.14}$$

where

$$d_1 = \frac{-\ln w_d + k_F t}{\sigma\sqrt{t}} k_F + \frac{1}{2}\sigma\sqrt{t}, \tag{4.15}$$

$N(-d_1)$-cumulative normal distribution of probability of random value d_1 and t-a moment of payment a credit.

4.2.4.3 The Account of Corporate and Individual Taxes (Miller Model)

Modigliani and Miller [97] have considered only corporate taxes, but they did not take into account the individual taxes of investors.

Miller [101] has developed the model with account of the corporate and individual taxes and has studied the impact of leverage on the company value.

The model is described by the following formula for levered company value:

$$V_L = V_U + \frac{I(1 - T_d)}{k_d}\left[1 - \frac{(1 - T_C)(1 - T_S)}{(1 - T_d)}\right]. \tag{4.16}$$

Here first term means the unlevered company value:

$$V_U = \frac{\text{EBIT}(1 - T_C)(1 - T_S)}{k_0}. \tag{4.17}$$

where, T_S stands for the tax on income of an individual investor from his ownership by corporation stock rate, T_C stands for the tax rate on corporate income, T_D stands for the tax rate on interest income from the provision of investor-individuals of credits to other investors and companies.

The Miller model estimates the value of levered company, accounting the corporate tax, as well as tax on individuals. Second term of sum in the Miller formula:

$$\left[1-\frac{(1-T_{\mathrm{C}})(1-T_{\mathrm{S}})}{(1-T_{\mathrm{d}})}\right]D, \tag{4.18}$$

represents the gains from use of debt capital. This term replaces the tax shield in the Modigliani–Miller model with corporate taxes:

$$V_{\mathrm{L}}=V_{\mathrm{U}}+TD. \tag{4.19}$$

4.2.4.4 Alternative Expression for WACC

Alternative formula for the WACC, different from Modigliani–Miller one has been derived in [102–105] from the WACC definition and the balance identity (see as well [102]):

$$\mathrm{WACC}=k_0(1-w_{\mathrm{d}}T)-k_{\mathrm{d}}tw_{\mathrm{d}}+k_{\mathrm{TS}}tw_{\mathrm{d}}, \tag{4.20}$$

where k_0, k_{d} and k_{TS} are the expected returns respectively on the unlevered company, the debt and the tax shield.

Some additional conditions are required Eq. (4.20) practical applicability. If the WACC is constant over time, as it stated in [103] the levered company capitalization is found by discounting with the WACC of the unlevered company.

In textbooks [102, 105] formulas for the special cases, where the WACC is constant, could be found.

In 1963 Modigliani and Miller assume that the debt value D is constant. Then, as the expected after-tax cash-flow of the unlevered firm is fixed, V_0 is constant as well. By assumption, $k_{\mathrm{TS}} = k_{\mathrm{D}}$ and the value of the tax shield is TS = tD. Thus, the capitalization of the company V is a constant and the alternative formula (4.20) becomes a formula for a constant WACC:

$$\mathrm{WACC}=k_0(1-w_{\mathrm{d}}T) \tag{4.21}$$

Because the debt k_{d} and the tax shield k_{TS} have debt nature it seems reasonable that the expected returns on they are equals as suggested by "classical" Modigliani–Miller (MM) theory, which has been modified by Brusov et al. for cases of practical meanings.

4.2.4.5 The Miles–Ezzell Model Versus the Modigliani–Miller Theory

Denis M. Becker [106] discussed the differences between the Modigliani–Miller theory [1, 97] and the Miles–Ezzell model ([107]), which deal with the stochasticity of free cash flows. The Modigliani–Miller theory considers a stationary process, while in the Miles–Ezzell model the process is stochastic. The author conducts a numerical experiment that allows you to determine the values and discount rates using a risk-neutral approach. He analyzes three formulas

Modigliani–Miller theory [1, 97]:

$$\mathrm{WACC} = k_0 \cdot (1 - w_{\mathrm{d}} t) \tag{4.22}$$

Miles–Ezzell model ([107]):

$$\mathrm{WACC} = k_0 - t \cdot w_{\mathrm{d}} \cdot k_{\mathrm{f}} \cdot \frac{1 + k_0}{1 + k_{\mathrm{f}}} \tag{4.23}$$

Cooper, I. A., & Nyborg, K. G. (2006):

$$\mathrm{WACC} = k_0 - k_{\mathrm{f}} \cdot w_{\mathrm{d}} t \tag{4.24}$$

where k_{f} stands for the risk-free rate, which equals the required return of the debt holders.

The author shows that in the Miles–Ezzell model, all cash flows and the values depend on the path, in contrast to the Modigliani–Miller theory. Also, in the Miles–Ezzell model, all discount rates are time independent, with the exception of the discount rate used to discount tax shields, which depends on the duration of the cash flows. Conversely, in the Modigliani–Miller theory, all discount rates change over time, except for the constant tax shield discount rate. This affects the applicability of the well-known formula for annuities and the development of models for estimating both finite and perpetual cash flows.

In this paper Becker [106] raises the issue of paying the debt body together with payment of interest on the debt. Regarding this issue we would like to note that in both classical MM and BFO theories, the body of the debt is not paid. In the framework of the Modigliani–Miller theory, such an account is fundamentally impossible, while in the BFO theory it can be done and was done in the framework of the BFO-2 theory, where the amount of debt D decreases with time. This decrease in the value of debt D results in a decrease in the tax shield (see BFO-2 theory).

4.3 Trade-Off Theory

During decades one of the main theory of capital structure was the trade-off theory. There are two modifications of trade-off theory: the static and dynamic [108–110].

4.3.1 Static Theory

The static trade-off theory accounts the income tax and cost of bankruptcy. Within this theory the optimal capital structure is formed by the balancing act between the benefits of debt financing at low leverage level (from the tax shield from interest deduction) and the disadvantage of debt financing at high leverage level (from the increased financial distress and expected bankruptcy costs). The tax shield benefit is equal to the product of corporate income tax rate and the market value of debt and the expected bankruptcy costs are equal to the product of probability of bankruptcy and the estimated bankruptcy costs.

The static version of the trade-off theory does not take into account the costs of the adapting of the capital structure to the optimal one, economic behavior of managers, owners, and other participants of economical process, as well as a number of other factors.

4.3.2 Dynamic Theory

The dynamic version of trade-off theory suggests that the costs of adjusting the capital structure are high, and therefore companies will only change their capital structure if the benefits outweigh the costs. Therefore, there is an optimal range that varies on the outside of each lever but remains the same on the inside. Companies try to adjust their leverage when it reaches the edge of the optimal range. Depending on the type of adaptation costs, companies reach the target ratio faster or slower. Proportional changes involve a small adjustment, while fixed changes imply significant costs.

In the dynamic version of the trade-off theory, company capital structure decision in current period depends on the waiting company income in the next period.

As it has been shown within BFO theory [111, 112], under increased financial distress costs and bankruptcy risk, the optimal capital structure is absent. This means that the trade-off theory does NOT work in either the static or dynamic version.

4.3.3 Proof of the Bankruptcy of the Trade-off Theory

Brusov et al. [111] tested whether assumptions about risky debt financing (and about rising lending rates in the run-up to bankruptcy) do not lead to an increase in the weighted average cost of capital, WACC. They used the following model:

$$k_0 = 24\%; \quad \overline{k}_{\mathrm{d}} = \left\{ \begin{array}{c} 0.07; \quad \text{at } L \le 2 \\ 0.07 + 0.1(L-2); \quad \text{at } L > 2 \end{array} \right\}. \tag{4.25}$$

and showed that under the above conditions, WACC still decreases with leverage (Fig. 4.3).

This means there is no minimum for WACC with leverage change, and no maximum for company capitalization with leverage change. Thus, this means that there is no optimal capital structure in the famous theory of trade-offs. This fact has been explained by Brusov et al. [10, 111]. The explanation to this fact has been done within the Brusov–Filatova–Orekhova theory by studying the dependence of the equity cost k_e on leverage. The explanation for this fact is as follows. The dependence of the cost of equity k_e on the level of leverage undergoes significant changes when the growth in the cost of debt capital k_d with the level of leverage is included. The linear growth of the cost of equity k_e at a low level of leverage is replaced by its fall, starting from a certain value L_0. The value of L_0 sometimes correlates exactly with the initial growth point k_d with leverage level, but sometimes it takes values much higher.

The rate of decline in the cost of equity k_e with leverage increases with an increase in the growth rate of the cost of debt k_d, as well as when moving from linear to quadratic growth and exponential growth. So, we come to the conclusion that the increase in the cost of debt capital k_d with leverage leads to the decrease of equity

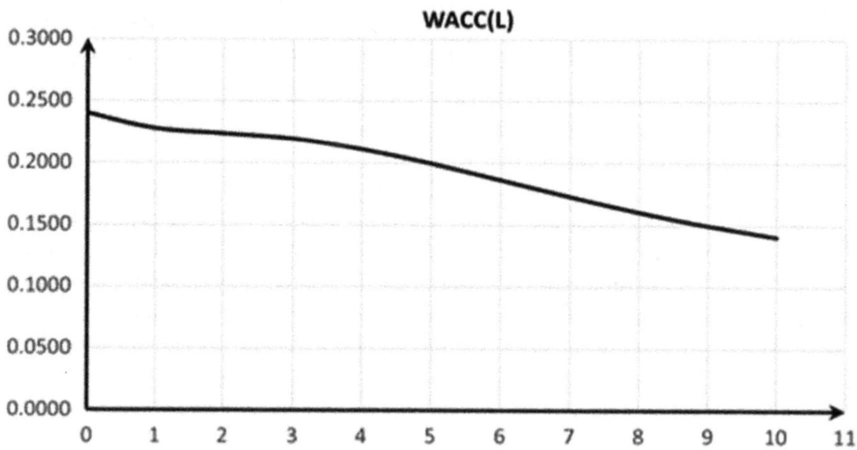

Fig. 4.3 Dependence of WACC on L

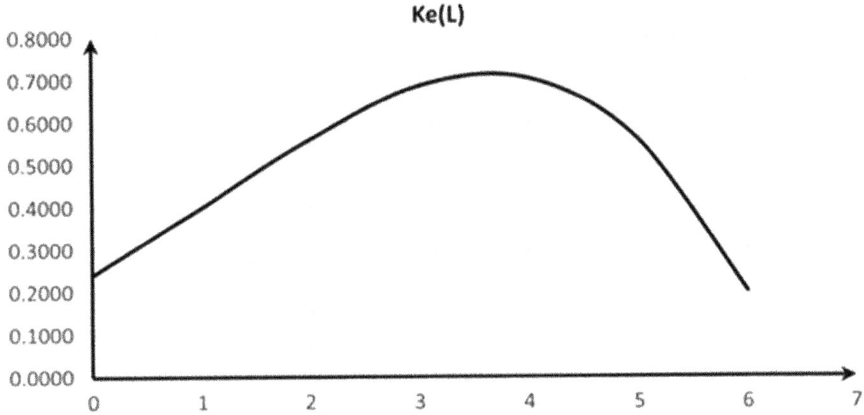

Fig. 4.4 Dependence of equity cost k_e on L

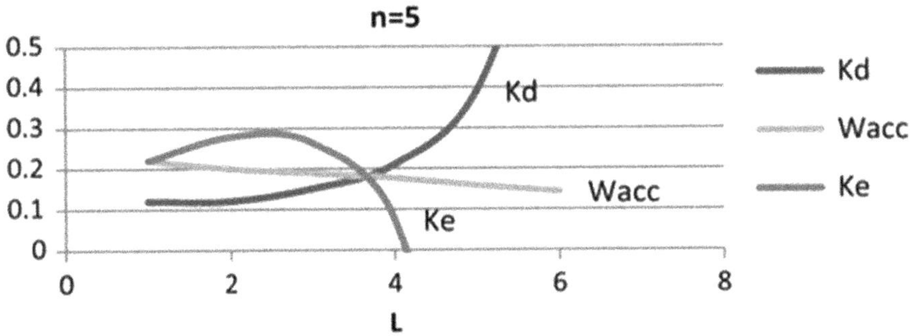

Fig. 4.5 Dependence of equity cost k_e, debt cost k_d, and WACC on leverage L

cost k_e with leverage, starting with some value L_0. *This is the cause of the absence of weighted average capital cost growth with leverage at all its values.* The conclusions made are independent of rate of growth of k_d with leverage (Figs. 4.4 and 4.5):

Thus, this means that there is no optimal capital structure in the famous theory of trade-offs. This paradoxical conclusion explains the absence of an optimal capital structure in the famous trade-off theory. This means that the competition between the benefits of leverage and the cost of financial hardship (or the cost of bankruptcy) is NOT balanced, and the hope that trade-off theory gives us an optimal capital structure is unfortunately not realized. Mechanism of formation of the company optimal capital structure, different from suggested by trade off theory has been developed by Brusov et al. [113].

4.4 Accounting of Transaction Cost

If the cost of changing the capital structure to its optimal value is high, the company may decide not to change the capital structure and maintain the current capital structure.

The company may decide that it is more cost effective not to change the capital structure, even if it is not optimal, for a certain period of time. Because of this, the actual and target capital structure may differ.

4.5 Accounting of Asymmetry of Information

In real financial markets, information is asymmetric (company managers have more reliable information than investors and creditors), and the rationality of economic entities is limited.

4.6 Signaling Theory

Information asymmetry can be reduced based on certain signals for lenders and investors related to the behavior of managers in the capital market. It should take into account the previous company development and the current and projected profitability of the activity [114].

4.7 Pecking Order Theory

The pecking order theory [115–117], describes a preferred sequence of funding types for raising capital. That is, companies first use financing from retained earnings (internal equity), the second source is debt, and the last source is the issuance of new shares of common stock (external equity). Empirical data on the leverage level of non-financial companies, combined with the decision-making process of top management and the board of directors, indicate a greater commitment to pecking order theory. Note that this theory contradicts both MM and BFO theory, which underline the quite importance of debt financing.

4.8 Behavioral Theories

Behavioral theories have been developed by Filbeck et al., Jenter, Baker et al., Graham et al.

4.8.1 Manager Investment Autonomy

Company managers implement those decisions that, from their point of view, will be positively perceived by investors and, accordingly, will positively affect the market value of companies: when the market value of the company's shares and the degree of consensus between the expectations of managers and investors is high, the company conducts an additional share issue, and in the opposite situation uses debt instruments. Thus, the company capital structure is more influenced by investors, whose expectations are taken into account by company managers.

4.8.2 The Equity Market Timing Theory

The level of leverage is determined by market dynamics. Equity market timing theory means that a company should issue shares at a high price and repurchase them at a low price. The idea is to take advantage of the temporary fluctuations in the value of equity relative to the value of other forms of capital.

4.8.3 Information Cascades

To save costs and avoid mistakes, the company's capital structure can be formed not on the basis of calculations of the optimal capital structure (this is a non-trivial task and its correct solution could be found within the framework of the BFO theory) or depending on the available sources of financing of the company, but borrowed from other companies, having successful, proven managers (heads of companies), as well as using (following the majority) the most popular methods of managing the capital structure, or even simply by copying of the capital structure of successful companies in a similar industry.

4.9 Theories of Conflict of Interests

4.9.1 Theory of Agency Costs

The company management may make decisions that are contrary to the interests of shareholders or creditors, respectively; expenses are necessary to control its actions. For solving the agency problem the correct choice of the compensation package (the agent's share in the property, bonus, stock options) is needed, which allows you to link the managers profit with the dynamics of equity capital and ensure the motivation of managers for preservation and growth of equity capital (Jensen et al., 1973).

4.9.2 Theory of Corporate Control and Costs Monitoring

In the presence of information asymmetry, creditors providing capital are interested in the possibility of self-control of the effectiveness of its use and return. Monitoring costs are usually passed on to company owners by being included in the loan rate. The level of monitoring costs depends on the scale of the business; therefore, with the increase in the scale of the business, the company weighted average capital cost increases, and the company market value decreases.

4.9.3 Theory of Stakeholders

Stakeholder theory is a theory that defines and models the groups that are the stakeholders of a company. The diversity and intersection of interests of stakeholders, their different assessment of acceptable risk give rise to conditions for a conflict of their interests, that is, they make adjustments to the process of the capital structure optimizing.

4.10 BFO Theory

The restriction associated with the infinite life of companies and the eternity of cash flows within the framework of the MM theory was removed in 2008 by Brusov, Filatova and Orekhova [9], who created the modern theory of the cost of capital and capital structure—the BFO theory, which is valid both for companies of arbitrary age, as well as for companies with an arbitrary lifetime. A generalization of the assessment of the tax shield TS and the value of the company: without leverage V_0 and with leverage V was required to modify the theory of MM (see the formulas below):

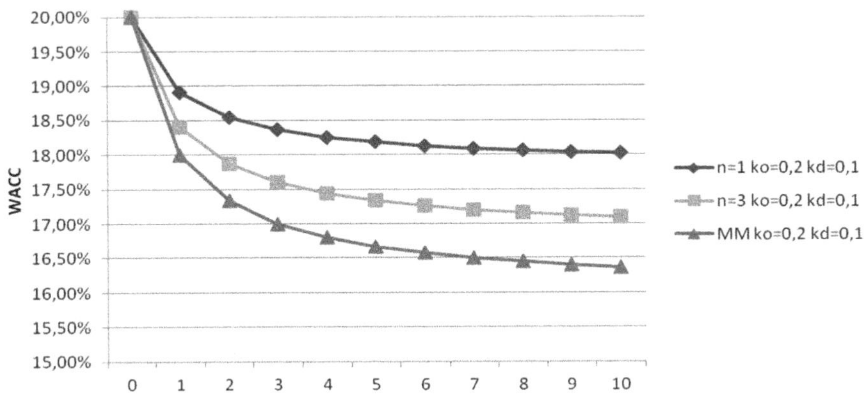

Fig. 4.6 Dependence of WACC on leverage level for $n = 1$, $n = 3$ and $n = \infty$

$$\mathrm{TS} = k_{\mathrm{d}}\mathrm{DT}\sum_{t=1}^{n}(1+k_{\mathrm{d}})^{-t} = \mathrm{DT}[1-(1+k_{\mathrm{d}})^{-n}]. \tag{4.26}$$

$$V_0 = \mathrm{CF}[1-(1+k_0)^{-n}]/k_0; \quad V = \mathrm{CF}[1-(1+\mathrm{WACC})^{-n}]/\mathrm{WACC}. \tag{4.27}$$

$$\frac{1-(1+\mathrm{WACC})^{-n}}{\mathrm{WACC}} = \frac{1-(1+k_0)^{-n}}{k_0[1-\omega_{\mathrm{d}}T(1-(1+k_{\mathrm{d}})^{-n})]}. \tag{4.28}$$

Here, S stands for the equity capital value, $w_{\mathrm{d}} = \frac{D}{D+S}$ stands for the debt capital share, $k_{\mathrm{e}}, w_{\mathrm{e}} = \frac{S}{D+S}$ stands for the equity cost and the equity capital share, and $L = D/S$ stands for the value of financial leverage, D stands for the debt capital value.

For a 1-year company Steve Myers in 2001 has derived formula, which could be obtained easily from formula (4.28) by substituting $n = 1$:

$$\mathrm{WACC} = k_0 - \frac{(1+k_0)k_{\mathrm{d}}}{1+k_{\mathrm{d}}}w_{\mathrm{d}}T$$

and by substituting $n = \infty$ we arrive to the Modigliani–Miller formula for WACC [97]:

$$\mathrm{WACC} = k_0 \cdot (1 - w_{\mathrm{d}}t)$$

The methods and the conclusions of the Brusov–Filatova–Orekhova (BFO) theory are well-known in the scientific literature (for example, see references). In a number of works the BFO theory is used in practical calculations (Fig. 4.6).

4.10.1 Brusov–Filatova–Orekhova Theorem

4.10.1.1 Case of Absence of Corporate Taxes

Modigliani–Miller theory in case of absence of corporate taxes gives the following results for dependence of WACC and equity cost k_e on leverage:

$$V_0 = V_L; \quad CF/k_0 = CF/\text{WACC}, \quad \text{and thus WACC} = k_0. \tag{4.29}$$

$$\text{WACC} = w_e \cdot k_e + w_d \cdot k_d; \tag{4.30}$$

and thus:

$$\begin{aligned} k_e &= \frac{\text{WACC} - w_d \cdot k_d}{w_e} = \frac{k_0 - \frac{L}{1+L} k_d}{\frac{1}{1+L}} \\ &= k_0 + L(k_0 - k_d). \end{aligned} \tag{4.31}$$

For the finite life-time (finite age) companies, Modigliani–Miller theorem about equality of value of financially independent and financially dependent companies ($V_0 = V_L$) has the following view [111–113, 118, 119]:

$$\begin{aligned} &V_0 = V_L; \\ &CF \cdot \frac{[1 - (1 + k_0)^{-n}]}{k_0} = CF \cdot \frac{[1 - (1 + \text{WACC})^{-n}]}{\text{WACC}}. \end{aligned} \tag{4.32}$$

Using this relation, we prove an important Brusov–Filatova–Orekhova theorem:

Under the absence of corporate taxes, the equity cost of the company, k_e, as well as its weighted average cost of capital, WACC, does not depend on the life-time (age) of the company and is equal, respectively, to

$$k_e = k_0 + L(k_0 - k_d); \quad \text{WACC} = k_0. \tag{4.33}$$

4.11 BFO Theory and Modigliani–Miller Theory Under Inflation

The influence of inflation on cost of raising capital and the company value has been studied within modern theory of capital cost and capital structure–Brusov–Filatova–Orekhova theory (BFO theory) [9, 111–113, 118–120] and within its perpetuity limit–Modigliani–Miller theory [1, 96, 97]. For the first time it was shown [119] by

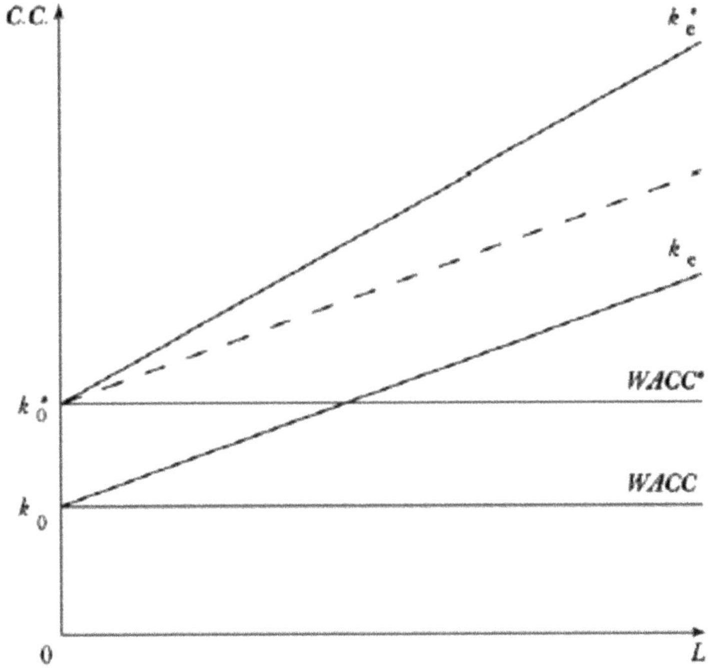

Fig. 4.7 Dependence of the equity cost and the weighted average cost of capital on leverage in the Modigliani–Miller theory without taxing under inflation. It is seen that growing rate of equity cost increases with leverage. Axis y means capital costs-CC

the direct incorporation of inflation into both theories that inflation not only increases the cost of raising capital (the equity cost and the weighted average cost of capital), but as well it changes their dependence on leverage. In particular, it increases the growth rate of equity cost with leverage. The company value is decreased under inflation (Fig. 4.7).

$$\frac{1-(1+\text{WACC}^*)^{-n}}{\text{WACC}^*} = \frac{1-[(1+k_0)(1+\alpha)]^{-n}}{(k_0(1+\alpha)+\alpha)\cdot[1-\omega_d T(1-((1+k_d)(1+\alpha))^{-n})]}. \tag{4.34}$$

Formula (4.34) is the generalized Brusov–Filatova–Orekhova formula under inflation.

Here, WACC* = WACC · (1 + α) + α.

In the Modigliani–Miller theory (Fig. 4.8):

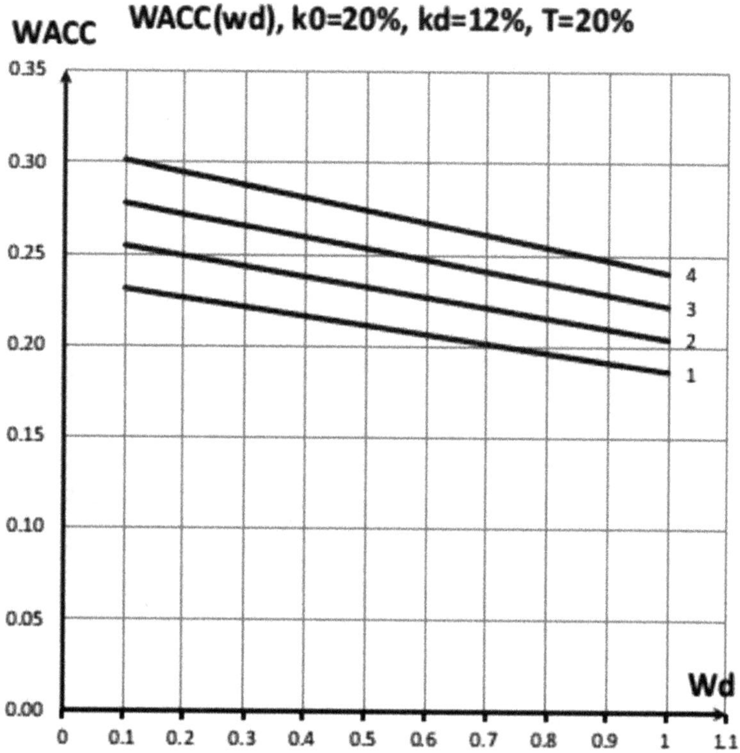

Fig. 4.8 Dependence of the weighted average cost of capital WACC on debt fraction w_d at different inflation rate α (1, $\alpha = 3\%$; 2, $\alpha = 5\%$; 3, $\alpha = 7\%$; and 4, $\alpha = 9\%$) for 5-year company

$$\mathrm{WACC}^* = (k_0(1+\alpha)+\alpha)\cdot(1-w_d t)$$

$$\begin{aligned} k_e^* &= k_0^* + L(1-T)\left(k_0^* - k_d^*\right) \\ &= [k_0(1+\alpha)+\alpha] + L(1-T)(k_0 - k_d)(1+\alpha). \end{aligned} \tag{4.35}$$

The modified equation for the weighted average cost of capital WACC, applicable to companies of arbitrary age under inflation, was derived in the framework of the modern theory of cost and capital structure—the theory of Brusov–Filatova–Orekhova (BFO theory). The modified BFO equation allows investigate the dependence of the weighted average cost of capital, WACC, and the cost of equity, k_e, on the level of financial leverage, L, on income tax rate, t, on the age of the company, n, on the cost of equity, k_0, and the cost of debt, k_d, for various inflation rates α.

Using the modified BFO equation, authors analyzed the dependence of the weighted average cost of capital WACC on the share of debt w_d at various income taxes t, as well as the inflation rate α.

It was shown that WACC decreases with an increase leverage level L (or the share of debt, w_d), and the faster, the higher the income tax rate t. The distance between the lines, corresponding to different values of the income tax rate at the same step (10%), increases with the growth of the inflation rate α. The area of change (with a change in the income tax rate t) of the weighted average cost of capital WACC increases with the inflation rate α, as well as with the age of the company n.

4.12 BFO Theory for the Companies Ceased to Exist at the Time Moment *n* (BFO-2 Theory)

The MM theory is only applicable to perpetual companies. As follows from the derivation of the BFO formula, the developed methodology is applied to companies of arbitrary age n (which have reached the age of n-years and continue to operate in the market). In other words, the BFO is applicable to the most interesting practical cases and allows you to analyze the financial condition of operating companies. However, the BFO theory also allows us to consider the financial condition of companies that have ceased to exist, i.e. those for which n means not age, but lifetime, i.e. the time of existence. There can be many schemes for terminating a company: bankruptcy, merger, takeover, etc. Below we will consider one of these schemes, when the value of debt capital D becomes equal to zero at the time of the termination of the company n: in this case, the BFO theory requires minimal updates, as shown below [10].

From the formula for the company's capitalization, it is easy to obtain an estimate of the "residual capitalization" of the company, discounted to the time moment k:

$$V_k = \sum_{t=k+1}^{n} \frac{\mathrm{CF}}{(1+\mathrm{WACC})^t} = \frac{\mathrm{CF}}{\mathrm{WACC}}\left[1-(1+\mathrm{WACC})^{-(n-k)}\right]. \tag{4.36}$$

Substituting

$$V_k = w_\mathrm{d} D, \tag{4.37}$$

one gets the expression for the tax shield for n years, subject to the termination of the company's activities at time moment n:

$$\begin{aligned} \mathrm{TS}_n &= t k_\mathrm{d} w_\mathrm{d} \sum_{k=1}^{n} \frac{V_{k-1}}{(1+k_\mathrm{d})^k} = \frac{t k_\mathrm{d} w_\mathrm{d} \mathrm{CF}}{\mathrm{WACC}} \sum_{k=1}^{n} \frac{1-(1+\mathrm{WACC})^{-(n-k+1)}}{(1+k_\mathrm{d})^k} \\ &= \frac{t k_\mathrm{d} w_\mathrm{d}}{\mathrm{WACC}}\left[\frac{1-(1+k_\mathrm{d})^{-n}}{k_\mathrm{d}} - \frac{(1+k_\mathrm{d})^{-n}-(1+\mathrm{WACC})^{-n}}{\mathrm{WACC}-k_\mathrm{d}}\right]. \end{aligned} \tag{4.38}$$

Substituting this formula into equation

$$V_L = V_0 + (\mathrm{TS})_n \tag{4.39}$$

we arrive to the following equation (which is determines *the BFO-2 theory*):

$$\frac{1-(1+\mathrm{WACC})^{-n}}{\mathrm{WACC}} = \frac{1-(1+k_0)^{-n}}{k_0} + \frac{tk_\mathrm{d}w_\mathrm{d}}{\mathrm{WACC}} \times \left[\frac{1-(1+k_\mathrm{d})^{-n}}{k_\mathrm{d}} - \frac{(1+k_\mathrm{d})^{-n}-(1+\mathrm{WACC})^{-n}}{\mathrm{WACC}-k_\mathrm{d}}\right], \tag{4.40}$$

whence it is possible to find WACC for companies with an arbitrary life-time n, under condition, that the company ceases to operate at time moment n.

4.13 The Modigliani–Miller Theory with Advance Payments of Tax on Profit

The first serious study (and first quantitative study) of influence of capital structure of the company on its indicators of activities was the work by Nobel Prize Winners Modigliani and Miller. Their theory has a lot of limitations. One of the most important and serious assumptions of the Modigliani–Miller theory is that all financial flows as well as all companies are perpetuity. This limitation was lift out by Brusov–Filatova–Orekhova in 2008 [9], who have created BFO theory–modern theory of capital cost and capital structure for companies of arbitrary age. Despite the fact that the Modigliani–Miller theory is currently a particular case of the general theory of capital cost and capital structure—Brusov–Filatova–Orekhova (BFO) theory—it is still widely used at the West.

Brusov et al. [121] discuss another limitation of the Modigliani–Miller theory: the method of paying income tax. The Modigliani–Miller theory accounts for these payments as immediate annuities, while in practice these payments are often made in advance and therefore must be accounted for as annuity–due. The authors generalized the Modigliani–Miller theory to the case of advance payments on income tax, which is widely used in practice. This generalization leads to a serious change in all the main provisions of Modigliani and Miller. These changes are as follows: WACC begins to depend on the cost of debt k_d, WACC turns out to be lower than in the case of the classical Modigliani–Miller theory, and the value of the company becomes higher. The formula for WACC has the following form:

$$\mathrm{WACC} = k_0\left(1 - w_\mathrm{d}t(1+k_\mathrm{d})\right) \tag{4.41}$$

Although the dependence of the cost of equity k_e on the level of leverage L remains linear, the angle of inclination relative to the L axis turns out to be smaller: this may lead to a modification of the company's dividend policy. Correct accounting for the method of paying income tax shows that the shortcomings of the

Modigliani–Miller theory are more serious than everyone expected: the underestimation of WACC is indeed greater, as is the overestimation of the company's capitalization. This means that the systematic risks arising from the use of the modified Modigliani–Miller theory (the MM theory) (more correct than the "classical" theory) are in practice higher than it was supposed by the "classical" version of this theory.

4.14 The Modigliani–Miller Theory with Arbitrary Frequency of Payment of Tax on Profit

One of the conditions for the real functioning of companies is the payment of income tax with an arbitrary frequency (monthly, quarterly, every 6 months or annually). The return is not required more than once a year, but companies may be required to file estimated taxes based on profits earned. The authors of (1) generalized the Modigliani–Miller theory to the case of paying income tax with arbitrary periodicity. The formula for WACC has the following form Brusov et al. (2021c):

$$\mathrm{WACC} = k_0 \cdot \left(1 - \frac{k_d w_d t}{p} \frac{1}{(1 + k_d)^{\frac{1}{p}} - 1}\right) \tag{4.42}$$

and for company value:

$$V = \frac{\mathrm{WACC}}{k_0 \cdot \left(1 - \frac{k_d w_d t}{p} \frac{1}{(1+k_d)^{\frac{1}{p}} - 1}\right)} \tag{4.43}$$

They showed that the frequency of income tax payment affects all the main financial indicators of the company and leads to a number of important consequences. Theoretically, the authors derived all the main formulas of the modified Modigliani–Miller theory and used them to obtain all the main financial indicators of the company at arbitrary frequencies of income tax payment. They showed that: (1) all Modigliani–Miller theorems, assertions, and formulas have been changed; (2) all major financial indicators, such as the weighted average cost of capital (WACC), the value of the company, V and the cost of equity, k_e, depend on the frequency of income tax payments; (3) in the case of paying tax on profit more than once a year (with $p \neq 1$), as is the case in practice, WACC, company value, V and equity value, as well as equity cost k_e start depend on the cost of debt, k_d, and in the usual (classical) Modigliani–Miller theory, all these quantities do not depend on k_d; (4) the results obtained allow the company to choose the number of tax on profit payments per year (of course, within the framework of the current tax legislation): more frequent tax on profit payments are beneficial to both parties, the company and the tax regulator.

4.15 Generalization of the Modigliani–Miller Theory for the Case of Variable Profit

In [3], for the first time, the world-famous theory of Nobel laureates Modigliani and Miller was generalized to the case of variable income, which significantly expands the applicability of the theory in practice, in particular, in corporate finance, business valuation, investments, ratings etc. The following formulas for WACC and company value have been obtained:

$$\text{WACC} - g = (k_0 - g) \cdot (1 - w_d t) \tag{4.44}$$

$$V = \frac{\text{CF}}{\text{WACC} - g} = \frac{\text{CF}}{(k_0 - g) \cdot (1 - w_d t)} \tag{4.45}$$

It was shown that all the theorems, assertions, and formulas of Modigliani and Miller have been essentially changed. The authors obtained the following results: (1) WACC and k_0 are no longer the discount rates as it takes place in case of classical Modigliani–Miller theory with constant profit. The role of discount rate for leverage company pass from the weighted average cost of capital, WACC, to WACC–g (where g is growing rate), for a financially independent company from k_0 to k_0–g. The real discount rates WACC–g and k_0–g decrease with g, while WACC grows with g. This decrease leads to an increase of company value with g. (2) The slope of the equity cost $k_e(L)$ grows with g. Via the fact, that the economically justified value of dividends is equal to equity cost this will modify the company's dividend policy. (3) It has been discovered the qualitatively new effect in corporate finance: the slope of the curve $k_e(L)$ at rate $g < g^*$ turns out to be negative. This effect significantly alter the company's dividend policy principles.

4.16 The Generalization of the Brusov–Filatova–Orekhova Theory for the Case of Payments of Tax on Profit with Arbitrary Frequency

The Brusov–Filatova–Orekhova (BFO) theory and its limit of eternity, the Modigliani–Miller theory—both the main theories of the cost of capital and capital structure—consider income tax payments once a year. These payments in the real economy are made more frequently (every 6 months, quarter, month, etc.). In [3] the Modigliani–Miller theory was generalized by Brusov et al. for the case of paying income tax with arbitrary periodicity. In [8], for the first time, authors generalized the theory of Brusov–Filatova–Orekhova (BFO) to this case. Taking into account one of the features of the real functioning of companies—frequent payments of income tax—the authors brought the BFO theory closer to economic practice. Authors derived modified BFO formula

$$\frac{1-(1+\mathrm{WACC})^{-n}}{\mathrm{WACC}}=\frac{1-(1+k_0)^{-n}}{k_0\cdot\left(1-\frac{k_\mathrm{d}w_\mathrm{d}t}{p}\frac{[1-(1+k_\mathrm{d})^{-n}]}{(1+k_\mathrm{d})^{1/p}-1}\right)} \tag{4.46}$$

and showed that:

(1) All BFO formulas change; (2) all major financial parameters of a company, such as company capitalization, V, weighted average cost of capital, WACC, cost of equity, k_e, etc. depend on the frequency of tax on profit payments. The cost of raising capital WACC decreases with the frequency of tax on profit payments and the company value, V increases with the frequency of tax on profit payments. A qualitatively new anomalous effect takes place at some age of the company, n, and at some frequency of income tax payments p: the cost of equity $k_\mathrm{e}(L)$ decreases with leverage level L.

Since the economically justified amount of dividends is equal to the cost of equity, this fundamentally changes the company's dividend policy. More frequent income tax payments are beneficial to both parties—the company and the tax regulator: the company, because it increases the value of the company, and the tax regulator, because it benefits from earlier payments due to the time value of money.

4.17 Benefits of Advance Payments of Tax on Profit: Consideration within the Brusov–Filatova–Orekhova (BFO) Theory

Both theories of capital structure—the Brusov–Filatova–Orekhova (BFO) and its limit of eternity, the theory of Modigliani–Miller—consider the case of paying tax on profit at the end of the year. However, in practice, companies could make these payments in advance. In [121] the Modigliani–Miller theory was modified for the case of advance payments of tax on profit, and the results turned out to be completely different from the results of the usual Modigliani–Miller theory. In [5], the Brusov–Filatova–Orekhova (BFO) theory was modified for the first time for advance payments on income tax. The impact of such a transition to advance payments is much more significant than in the theory of MM, and even leads to a qualitatively new effect in the cost of equity dependence on leverage. Authors derived modified BFO formula for WACC:

$$\frac{1-(1+\mathrm{WACC})^{-n}}{\mathrm{WACC}}=\frac{1-(1+k_0)^{-n}}{k_0\cdot\left(1-w_\mathrm{d}t[1-(1+k_\mathrm{d})^{-n}](1+k_\mathrm{d})^{1/p}\right)} \tag{4.47}$$

It is concluded that the tax shield and the way it is formed (payments at the end of the year or in advance) leads to very important consequences, changing all the company's financial indicators, such as the company's value and capital raising costs, and also significantly changing the company's dividend policy.

4.18 Influence of Method and Frequency of Profit Tax Payments on Company Financial Indicators

In practice, income tax payments may (1) be made in advance and (2) more frequently than once a year. In [5], the simultaneous influence of these two factors on the company's financial performance was studied. To this end, the Brusov–Filatova–Orekhova (BFO) theory was first generalized to the case of advance payments of income tax with arbitrary periodicity. Authors derived the following modified BFO formula for WACC:

$$\frac{1-(1+\mathrm{WACC})^{-n}}{\mathrm{WACC}}=\frac{1-(1+k_0)^{-n}}{k_0\cdot\left(1-\frac{k_d w_d t}{p}\frac{\left[1-(1+k_d)^{-n}\right]}{\left((1+k_d)^{1/p}-1\right)}(1+k_d)^{1/p}\right)} \tag{4.48}$$

It is concluded that all financial indicators of the company, such as the weighted average cost of capital (WACC) and the cost of equity (k_e), the value of the company (V) depend on the frequency of income tax payment. WACC increased with the frequency of payments, and the value of the company decreased. This meant that infrequent payments could be beneficial to the company in the case of advance income tax payments. The slope of the cost of equity, $k_e(L)$, increased with the frequency of payments. Depending on the age of the company, $k_e(L)$ either increased for some payment frequencies or decreased with L for all payment frequencies. The authors compared these results with their results for income tax payments at the end of the periods (see formula (4.47)) and found a huge difference: while in the latter case, WACC decreased with payment frequency, and the company capitalization increased with payment frequency, the value of WACC in this case remains bigger, and the company capitalization remains less than in the case of advance income tax payments of any frequency. The importance of advance income tax payments has been underlined by this fact. Authors developed the recommendations to regulator to increase the practice of advance payments of income tax due to the benefits of this for both parties: the companies and the state. A new effect—the decrease in equity cost with the level of leverage (L)—has been discovered.

4.19 The Brusov–Filatova–Orekhova (BFO) Theory with Variable Income

Two main theories of capital structure—Brusov–Filatova–Orekhova (BFO) and Modigliani–Miller consider the case of constant profit. But in practice, the company's income, of course, is variable. Both theories have recently been generalized to the case of variable income [3, 8], which significantly expands the applicability of both theories of capital structure in practice. The following generalized BFO formula for WACC has been discovered:

$$\frac{1-\left(\frac{1+g}{1+\mathrm{WACC}}\right)^n}{\mathrm{WACC}-g}=\frac{1-\left(\frac{1+g}{1+k_0}\right)^n}{(k_0-g)\cdot(1-w_{\mathrm{d}}t[1-(1+k_{\mathrm{d}})^{-n}])} \tag{4.49}$$

In the theory of Brusov–Filatova–Orekhova (BFO) [8], generalized to the case of variable income, income tax payments are made at the end of periods, while in practice these payments can also be made in advance. In [6] the authors considered two modifications of the Brusov–Filatova–Orekhova (BFO) theory with variable income: (1) with the payment of income tax at the end of periods and (2) with advance payments of income tax.

For these two cases, BFO formulas were derived for the weighted average cost of capital, WACC, for the value of the company, V, and within these formulas, a comprehensive analysis of the dependence of WACC, of the discount rate, WACC–g (here g is the growth rate), company capitalization, V, cost of equity, k_e, on debt financing at different values of the growth rate, g, at different values of the cost of debt capital, k_d, and at different values of the age of the company, n. The results for cases (1) and (2) are compared, which allows us to conclude that case (2) is always preferable for both the company and the regulator. This makes it possible to develop recommendations for both parties to expand the practice of advance income tax payments in the framework of real economic practice.

Brusov et al. emphasized an important observation. If in the classical versions of the Brusov–Filatova–Orekhova (BFO) theory and its perpetual limit—the theory of Nobel laureates Modigliani and Miller, where the case of constant profit was considered, and where the gap between these two theories is huge (many qualitative effects that take place in the first theory, missing in the second), when taking into account the variable profit, some effects of the BFO theory also take place in the Modigliani–Miller theory. This means that taking into account some effects that are present in economic practice (for example, variable income) brings both theories closer, and even the Modigliani–Miller theory, with all its many limitations, becomes more applicable in economic practice. However, it should be remembered that the Modigliani–Miller theory is only true for perpetual companies, while the BFO theory is valid for companies of any age and from this point of view they never coincide.

4.20 Qualitatively New Effects in the Theory of Capital Structure

Among the many new effects discovered in the framework of the BFO theory, we will describe only two qualitatively new ones:

1. Golden and silver ages of the company.
2. Anomalous dependence of the cost of equity on the leverage level.

4.20.1 Golden and Silver Ages of the Company

To determine the minimum cost of raising capital depending on the age of the company, Brusov et al. In 2015, studied the dependence of the cost of raising capital on the age of the company, *n*, at different levels of leverage, *L*, at different values of capital costs. The authors used the Brusov–Filatov–Orekhova theory [9, 111–113, 118–120].

It was shown for the first time that the value of WACC in the Modigliani–Miller theory [1, 96, 97] is not minimal, and the value of the company is not maximal, contrary to the opinion of all financiers: it turned out that at a certain age of the company, the value of WACC is lower than in the eternal Modigliani–Miller theory, and the value of company *V* is greater than the value of company *V* in the Modigliani–Miller theory.

Authors discovered two types of WACC dependences on the age of company n: monotonic decreasing with n and WACC decreasing with passage through minimum, followed by a limited growth. The latter type of dependence of WACC on the age of company *n* allows the companies take advantage of the benefits given at a certain stage of their development.

Moreover, since capital costs, k_e and k_d, affect the company's golden age by changing it (for example, changing the value of dividend payments reflecting the cost of equity, etc.), the company can postpone or extend its golden age. at which the cost of raising capital is minimal (and less than its perpetual limit), and the value of companies is maximum (above their perpetual value) (Fig. 4.9).

It has been shown that existence of the company "golden age" depends not on the capital costs values, but on the difference between equity k_0 and debt k_d costs. The effect of the company "golden age" takes place at small enough difference between k_0 and k_d costs, while at this difference high value this effect is absent: curve WACC (*n*) monotonic decreases with company age *n*. Curve WACC(*L*) for perpetuity limit ($n = \infty$) lies between curves WACC(*L*) for 1 year ($n = 1$) and 3 years ($n = 3$) companies (Fig. 4.10) for the companies with the "golden age", while curve WACC (*L*) for perpetuity limit is the lowest one for the companies without the "golden age" (Fig. 4.11).

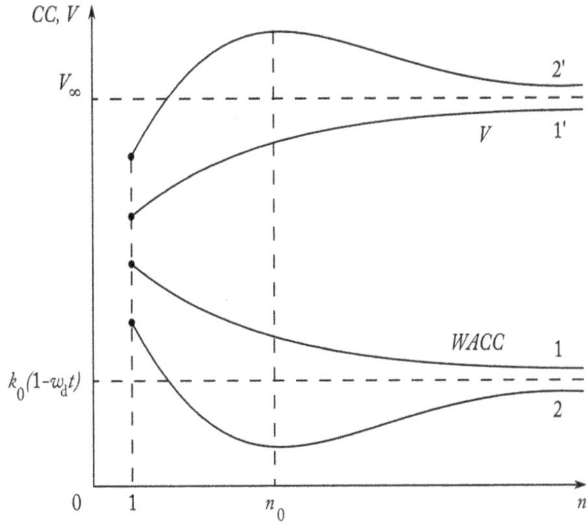

Fig. 4.9 Two types of dependences of WACC, and company value, V, on the company age n: 1–1′—monotonic decreasing of WACC, and company value, V, increasing with the company age n; 2–2′—WACC decreasing with n, passage through minimum (at n_0), followed by a limited growth and increase of V with the passage through a maximum (at n_0) and then a limited decreasing

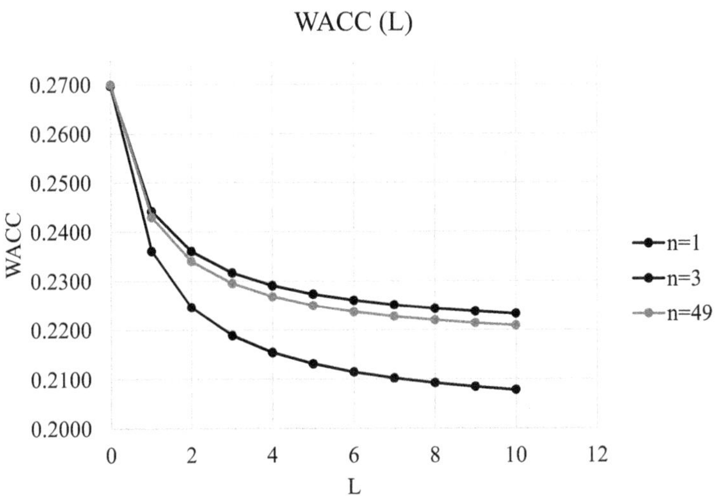

Fig. 4.10 The dependence of WACC on leverage level L at different the company age n ($n = 1$, 3, 49)

4.20.2 Silver Age of the Company (Fig. 4.12)

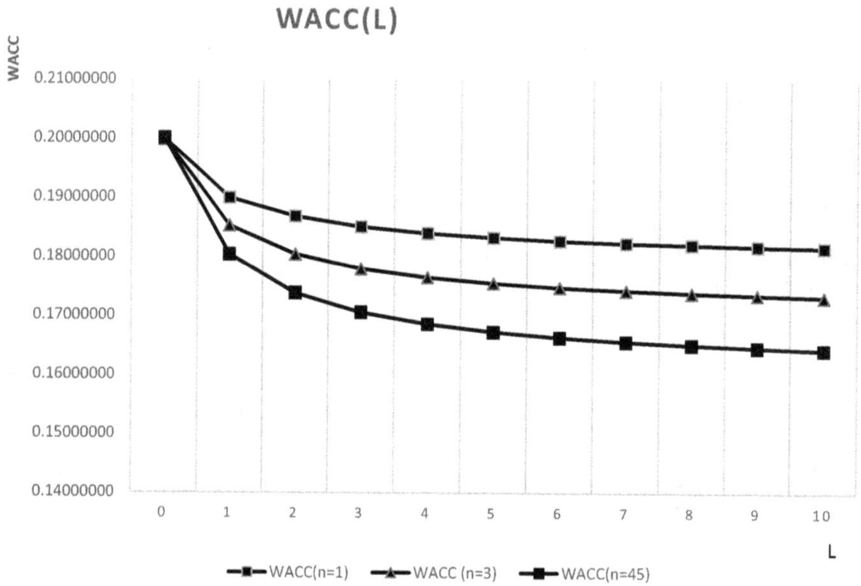

Fig. 4.11 The dependence of WACC on the leverage level, L, at different company age n ($n = 1$, 3, 45)

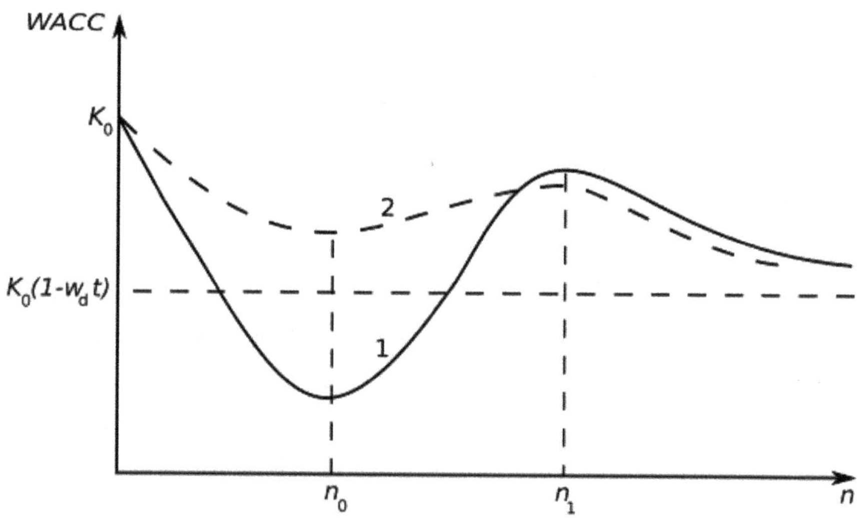

Fig. 4.12 Dependence of WACC on company age n: company "golden age" (curve 1) and company "silver age" (curve 2) under existence of "Kulik" effect, where n_0 is company "golden (silver) age" and n_1 is the company age of local maximum in the dependence of WACC(n)

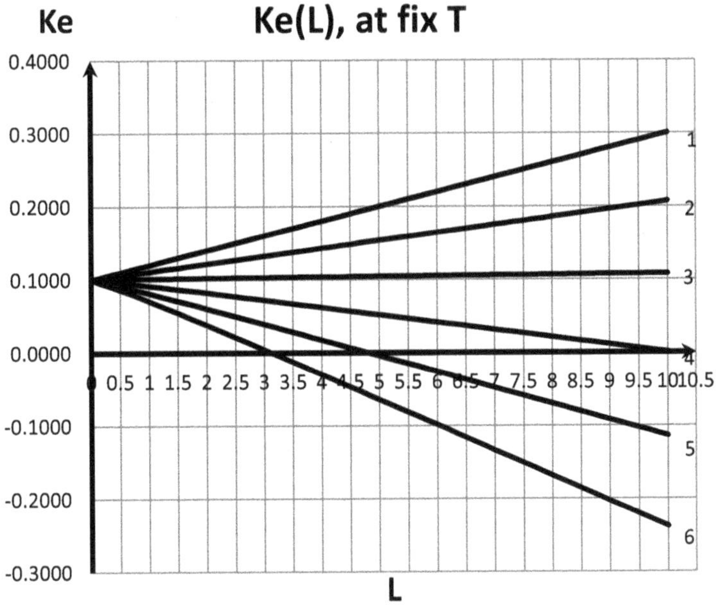

Fig. 4.13 Dependence of equity cost, k_e, on leverage level L at different tax on income rate T ($n = 5$, $k_0 = 10$ % , $k_d = 8$%)(1—$T = 0$; 2—$T = 0.2$; 3—$T = 0.4$; 4—$T = 0.6$; 5—$T = 0.8$; 6—$T = 1$)

4.20.3 Anomalous Dependence of the Company's Equity Value on Leverage

Brusov et al. [112] found a qualitatively new effect in the capital structure: a decrease in the cost of equity at a leverage level L. This effect is absent in the perpetual Modigliani–Miller limit, but appears in companies of finite age at an income tax rate exceeding a certain value T^*.

For some values of the cost of equity and the cost of debt, this effect exists at the tax on income values existing in financial practice. This ensures the practical significance of the discovered effect. Accounting for this effect is important when changing tax legislation and can significantly change the company's dividend policy (Fig. 4.13).

4.21 A Stochastic Extension the Modigliani–Miller Theory

A Stochastic Extension of the Modigliani–Miller theory has been done in a few papers.

De Marzo [122] showed that the Modigliani-Miller theory is correct in a multi-period stochastic economy with incomplete markets. In the model author developed, companies can trade securities, and stock prices and dividends become completely interdependent. Several corporate control mechanisms were considered, including maximizing the value of the company and shareholder voting. It is shown that in equilibrium companies have no incentive to trade securities, and the set of equilibrium distributions does not change as a result of such trading.

Sethi et al. [123] study the company value valuation problem under a partial equilibrium, a time-dependent discount rate and a general stochastic environment in a discrete-time setting. The time path of share price and the company value are obtained upon the assumption of risk neutrality.

4.22 Conclusions

This Chapter analyzes all the main existing theories of capital structure, including Modigliani–Miller and BFO. Various modifications of these two and other theories, including the latest ones, related to the conditions of the real functioning of companies and developed in recent years are discussed in detail.

The generalization of Modigliani–Miller theory for the cases of practical interest lead to a much more significant change in Modigliani–Miller theory than all previous modifications, studied over the previous decades. Obtained results showed that under such a generalization there is some convergence of Modigliani–Miller theory and BFO theory: the former theory begins to have some new properties similar to those of BFO theory, which was not there in the classical Modigliani–Miller theory, between the properties of which and of the BFO theory was a huge gap. Most of the innovative effects existing in the BFO theory are absent in the classical Modigliani–Miller theory.

All of the above means that the considered effects, such as the variable income of the company, frequent paying income tax, advance payment of income tax, the combinations of this conditions and others, are more important than the effects studied earlier and having much less influence on the Modigliani–Miller theory. Taking into account these effects leads to some convergence between Modigliani–Miller theory and the BFO theory. At the same time, one must understand that Modigliani–Miller theory remains perpetual and will never describe a company of arbitrary age, which is extremely important for the practical application of the theory.

References

1. Modigliani, F., & Miller, M. H. (1958). The cost of capital, corporation finance and the theory of investment. *The American Economic Review, 48*(3), 261–296.
2. Brusov, P., Filatova, T., Chang, S.-I., & Lin, G. (2021). Innovative investment models with frequent payments of tax on income and of interest on debt. *Mathematics, 9*(13), 1491.

3. Brusov, P., Filatova, T., Orekhova, N., Kulik, V., Chang, S.-I., & Lin, G. (2021). Generalization of the Modigliani–Miller theory for the case of variable profit. *Mathematics, 9*(11), 1286.
4. Brusov, P., & Filatova, T. (2021). The Modigliani–Miller theory with arbitrary frequency of payment of tax on profit. *Mathematics, 9*(11), 1198.
5. Brusov, P., Filatova, T., Orekhova, N., Kulik, V., Chang, S.-I., & Lin, G. (2022). The generalization of the Brusov–Filatova–Orekhova theory for the case of payments of tax on profit with arbitrary frequency. *Mathematics, 10*(8), 1343. https://doi.org/10.3390/math10081343
6. Brusov, P., Filatova, T., & Kulik, V. (2022). Benefits of advance payments of tax on profit: Consideration within the Brusov–Filatova–Orekhova (BFO) theory. *Mathematics, 10*(12), 2013. https://doi.org/10.3390/math10122013
7. Brusov, P., & Filatova, T. (2022). Influence of method and frequency of profit tax payments on company financial indicators. *Mathematics, 10*(14), 2479. https://doi.org/10.3390/math10142479
8. Brusov, P., & Filatova, T. (2022). Generalization of the Brusov–Filatova–Orekhova theory for the case of variable income. *Mathematics, 10*(19), 3661. https://doi.org/10.3390/math10193661
9. Filatova, T., Orehova, N., & Brusova, A. (2008). Weighted average cost of capital in the theory of Modigliani–Miller, modified for a finite life–time company. *Bull FU, 48*, 68–77.
10. Brusov, P., Filatova, T., Orehova, N., & Eskindarov, M. (2018). *Modern corporate finance, investments, taxation and ratings* (2nd ed., pp. 1–571). Springer Nature Publishing.
11. Acharya, V., Myers, S. C., & Rajan, R. (2011). The internal governance of firms. *Journal of Finance, 66*(3), 689–720.
12. Arrow, K. J. (1964). The role of securities in the optimal allocation of risk–bearing. *The Review of Economic Studies, 31*(2), 91–96.
13. Asquith, P., & Mullins, D. W. (1986). Equity issues and offering dilution. *Journal of Financial Economics, 15*(1–2), 61–89.
14. Baumol W, Panzar JG, Willig RD. 1982. Contestable markets and the theory of industry structure.
15. Black, F. (1986). Noise. *Journal of Finance, 41*(3), 529–543.
16. Brealey, R. A., & Myers, S. C. (1981). *Principles of corporate finance* (1st ed.). McGraw-Hill.
17. Brealey, R. A., Myers, S. C., & Allen, F. (2014). *Principles of corporate finance* (11th ed.). McGraw-Hill.
18. Bhattacharya, S. (1978). Project valuation with mean–reacting cash flows. *Journal of Finance, 33*, 1317–1331.
19. Brusov, P. N., Filatova, T. V., Orekhova, N. P., Kulik, V. L., Chang, S.-I., & Lin, Y. C. G. (2020). Application of the Modigliani–Miller theory, modified for the case of advance payments of tax on profit, in rating methodologies. *Journal of Reviews on Global Economics, 9*, 282–292.
20. Derrig, R. (1994). Theoretical considerations of the effect of federal income taxes on investment income in property–liability ratemaking. *The Journal of Risk and Insurance, 61*, 691–709.
21. Diamond, D. A., & He, Z. (2014). A theory of debt maturity: The long and short of debt overhang. *Journal of Finance, 69*(2), 719–762.
22. Donaldson, G. (1961). *Debt capacity: A study of corporate debt policy and the determination of debt capacity*. Div. Res. Harvard Grad. Stud. Bus. Admin.
23. Erel, I., Myers, S. C., & Read, J. A. Jr. (2015). A theory of risk capital. *Journal of Financial Economics*. In press. https://doi.org/10.1016/j.jfineco.2014.10.006
24. Fairley, W. B. (1979). Investment income and profit margins in property–liability insurance. *Bell Journal of Economics, 10*, 191–210.
25. Fama, E. F. (1977). Risk adjusted discount rates and capital budgeting under uncertainty. *Journal of Financial Economics, 5*(1), 3–24.

26. Fama, E. F. (1996). Discounting under uncertainty. *Journal of Business, 69*(4), 415–429.
27. Fisher, F. M., & McGowan, J. J. (1983). On the misuse of accounting rates of return to infer monopoly profits. *The American Economic Review, 73*, 82–97.
28. Filatova, T., Brusov, P., & Orekhova, N. (2022). Impact of advance payments of tax on profit on effectiveness of investments. *Mathematics, 10*(4), 666. https://doi.org/10.3390/math10040666
29. Frank, M. Z., & Goyal, V. K. (2003). Testing the pecking order theory of capital structure. *Journal of Financial Economics, 67*, 217–248.
30. Fulghieri, P., Garcia, D., & Hackbarth, D. (2014). *Asymmetric information and the pecking (dis-)order*. Working Paper, Boston University, Univ. N.C.
31. Gordon, M. J., & Shapiro, E. (1956). Capital equipment analysis: The required rate of profit. *Management Science, 3*, 102–110.
32. Graham, J. R., & Leary, M. T. (2011). A review of empirical capital structure research and directions for the future. *Annual Review of Financial Economics, 3*, 309–345.
33. Hamada, R. (1969). Portfolio analysis, market equilibrium, and corporate finance. *The Journal of Finance, 24*, 13–31.
34. Hausman, J., & Myers, S. C. (2002). Regulating U.S. railroads: The effects of sunk costs and asymmetric risk. *Journal of Regulatory Economics, 22*(3), 287–310.
35. Healy, P., Howe, C., & Myers, S. C. (2002). R&D accounting and the tradeoff between relevance and objectivity: A pharmaceutical industry simulation. *Journal of Accounting Research, 40*(3), 677–710.
36. Hirshleifer, J. (1965). Investment decision under uncertainty: Choice–theoretic approaches. *Quarterly Journal of Economics, 79*(4), 509–536.
37. Hirshleifer, J. (1966). Investment decision under certainty: Applications of the state–preference approach. *Quarterly Journal of Economics, 80*(2), 252–277.
38. Jensen, M. C. (1986). Agency costs of free cash flow, corporate finance and takeovers. *The American Economic Review, 76*(2), 323–329.
39. Jin, L., & Myers, S. C. (2006). R2 around the world: New theory and new tests. *Journal of Financial Economics, 79*(2), 257–292.
40. Kaplan, S., & Ruback, R. (1995). The valuation of cash flow forecasts: An empirical analysis. *Journal of Finance, 50*(4), 1059–93 32.
41. Myers Annual Review of Financial Economics. 2015.7:1–34. Downloaded from www.annualreviews.org. Access provided by University of California – Los Angeles UCLA on 12/11/18. For personal use only.
42. Keynes, J. M. (1936). *The general theory of employment, interest and money*. Macmillan.
43. Kolbe, A. L., Tye, W. B., & Myers, S. C. (1993). *Regulatory risk: Economic principles and applications to natural gas pipelines and other industries*. Springer.
44. Lambrecht, B., & Myers, S. C. (2007). A theory of takeovers and disinvestment. *Journal of Finance, 62*(2), 809–845.
45. Lambrecht, B., & Myers, S. C. (2008). Debt and managerial rents in a real–options model of the firm. *Journal of Financial Economics, 89*(2), 209–231.
46. Lambrecht, B., & Myers, S. C. (2012). A Lintner model of dividends and managerial rents. *Journal of Finance, 67*(5), 1761–1810.
47. Lambrecht, B., & Myers, S. C. (2015). *The dynamics of investment, payout and debt*. Working Paper. Massachusetts Institute of Technology.
48. Lemmon, M. L., Roberts, M. R., & Zender, J. (2008). Back to the beginning: Persistence and the cross–section of corporate capital structure. *Journal of Finance, 63*(4), 1575–1608.
49. Lemmon, M. L., & Zender, J. (2010). Debt capacity and tests of capital structure theories. *Journal of Financial and Quantitative Analysis, 45*, 1161–1187.
50. Lintner, J. (1956). Distribution of incomes of corporations between dividends, retained earnings and taxes. *The American Economic Review, 46*, 97–113.
51. Lintner, J. (1965). Optimal dividends and corporate growth under uncertainty. *Quarterly Journal of Economics, 77*, 59–95.

52. Majd, S., & Myers, S. C. (1987). Tax asymmetries and corporate income tax reform. In M. Feldstein (Ed.), *The effects of taxation on capital accumulation* (pp. 343–373). University of Chicago Press.
53. Majluf, N. S. (1978). *Study on mergers: A rationale for conglomerate mergers*. PhD Thesis, Massachusetts Institute of Technology.
54. Merton, R. C., & Perold, A. F. (1993). Theory of risk capital in financial firms. *Journal of Applied Corporate Finance, 6*, 16–32.
55. Miller, M. H. (1977). Debt and taxes. *Journal of Finance, 32*, 261–276.
56. Miller, M. H., & Modigliani, F. (1961). Dividend policy, growth and the valuation of shares. *Journal of Business, 34*, 411–433.
57. Morck, R., Yeung, B. Y., & Yu, W. (2000). The information content on stock markets: Why do emerging markets have synchronous stock price movements? *Journal of Financial Economics, 58*, 215–260.
58. Mullins, D. W. Jr. (1976). *Communications Satellite Corp. Case Study 276195*. Harvard Business School.
59. Myers, S. C. (1967). *Effects of uncertainty on the valuation of securities and the financial decisions of the firm*. Ph.D. Thesis, Stanford University, Stanford, CA.
60. Myers, S. C. (1968). A time–state–preference model of security valuation. *Journal of Financial and Quantitative Analysis, 3*, 1–33.
61. Myers, S. C. (1968). Procedures for capital budgeting under uncertainty. *Industrial Management Review, 9*(3), 1–20.
62. Myers, S. C. (1972). Application of finance theory to public utility rates cases. *Bell Journal of Economics, 3*, 58–97.
63. Myers, S. C. (1973). A simple model of firm behavior under regulation and uncertainty. *Bell Journal of Economics, 4*, 304–315.
64. Myers, S. C. (1973). On public utility regulation under uncertainty. In R. H. Howard (Ed.), *Risk and regulated firms* (pp. 32–46). Div. Res. Grad. Sch. Bus., Mich. State Univ.
65. Myers, S. C. (1974). Interactions of corporate financing and investment decisions: Implications for capital budgeting. *Journal of Finance, 29*, 1–25.
66. Myers, S. C. (1977). Determinants of corporate borrowing. *Journal of Financial Economics, 5* (2), 147–175.
67. Myers, S. C. (1988). Fuzzy efficiency. *Institutional Investor*, 8–9.
68. Myers, S. C. (1989). Still searching for optimal capital structure. In R. W. Kopke & E. S. Rosengren (Eds.), *Are the distinctions between debt and equity disappearing?* (pp. 80–95). Federal Reserve Bank.
69. Myers, S. C. (1999). Financial architecture. *European Financial Management Association, 5* (2), 133–141.
70. Myers, S. C. (2000). Outside equity. *Journal of Finance, 55*(3), 1005–1037.
71. Myers, S. C., & Cohn, R. (1987). A discounted cash flow approach to property–liability insurance rate regulation. In J. D. Cummings & S. Harrington (Eds.), *Fair rate of return in property liability insurance* (pp. 55–78). Kluwer-Nijhoff.
72. Myers, S. C., Dill, D. A., & Bautista, A. J. (1976). Valuation of financial lease contracts. *Journal of Finance, 31*(3), 799–819.
73. Myers, S. C., & Howe, C. (1997). *A life cycle model of pharmaceutical R&D*. Working Paper, Program on the Pharmaceutical Industry. Massachusetts Institute of Technology.
74. Myers, S. C., Kolbe, A. L., & Tye, W. B. (1985). Inflation and rate of return regulation. *Research in Transportation Economics, 2*, 83–119.
75. Myers, S. C., & Majd, S. (1990). Abandonment value and project life. In F. Fabozzi (Ed.), *Advances in futures and options research* (Vol. 4, pp. 1–21). JAI.
76. Myers, S. C., & Majluf, N. S. (1984). Corporate financing and investment decisions when firms have information that investors do not have. *Journal of Financial Economics, 13*(2), 187–221.

77. Myers, S. C., & Pogue, G. A. (1974). A programming approach to corporate financial management. *Journal of Finance, 29*(2), 579–599.
78. Myers, S. C., & Rajan, R. (1998). The paradox of liquidity. *Quarterly Journal of Economics, 113*(3), 733–771.
79. Myers, S. C., & Read, J. A., Jr. (2001). Capital allocation for insurance companies. *The Journal of Risk and Insurance, 68*(4), 545–580.
80. Myers, S. C., & Read, J. A. Jr. (2014). *Real options, taxes and leverage*. Working Paper. Massachusetts Institute of Technology.
81. Myers, S. C., & Shyam-Sunder, L. (1996). Measuring pharmaceutical industry risk and the cost of capital. In R. B. Helms (Ed.), *Competitive strategies in the pharmaceutical industry* (pp. 208–237). American Enterprise Institute.
82. Myers, S. C., & Turnbull, S. M. (1977). Capital budgeting and the capital asset pricing model: Good news and bad news. *Journal of Finance, 32*(2), 321–333.
83. Robichek, A. A., & Myers, S. C. (1965). *Optimal financing decisions*. Prentice-Hall.
84. Robichek, A. A., & Myers, S. C. (1966). Conceptual problems in the use of risk–adjusted discount rates. *Journal of Finance, 21*(4), 727–730.
85. Robichek, A. A., & Myers, S. C. (1966). Problems in the theory of optimal capital structure. *Journal of Financial & Quantitative Analysis, 1*(2), 1–35.
86. Robichek, A. A., & Myers, S. C. (1966). Valuation of the firm: Effects of uncertainty in a market context. *Journal of Finance, 21*(2), 215–227.
87. Rosenbaum, J., & Pearl, J. (2013). *Investment banking* (p. 2nded). Wiley.
88. Schwartz, E. (1959). The theory of the capital structure of the firm. *Journal of Finance, 14*, 18–39.
89. Sharpe, W. F. (1965). Capital asset prices: A theory of market equilibrium under conditions of risk. *Journal of Finance, 19*(3), 425–442.
90. Shleifer, A., & Summers, L. H. (1988). Breach of trust in hostile takeovers. In A. J. Auerbach (Ed.), *Corporate takeovers: Causes and consequences* (pp. 33–56). University of Chicago Press.
91. Shyam-Sunder, L., & Myers, S. C. (1999). Testing static tradeoff against pecking order models of capital structure. *Journal of Financial Economics, 51*(2), 219–244.
92. Skinner, D. J. (2008). The evolving relation between earnings, dividends and stock repurchases. *Journal of Financial Economics, 87*, 582–609.
93. Solomon, E., & Laya, J. (1967). Measurement of company profitability: Some systematic errors in the accounting rate of return. In A. A. Robichek (Ed.), *Financial research and management decisions* (pp. 152–183). Wiley.
94. Van Horne J. (1966). Capital budgeting decisions involving combinations of risky assets. *Management of Science 19*, B84–92.
95. Williams, J. B. (1938). *The theory of investment value*. Harvard University Press.
96. Miller, M. H., & Modigliani, F. (1966). Some estimates of the cost of capital to the electric utility industry, 1954–57. *The American Economic Review, 56*, 333–391.
97. Modigliani, F., & Miller, M. H. (1963). Corporate income taxes and the cost of capital: A correction. *The American Economic Review, 53*, 433–443.
98. Stiglitz, J. (1969). A re-examination of the Modigliani-Miller theorem. *American Economic Review, 59*, 784–793.
99. Rubinstein, M. E. (1973). A mean-variance synthesis of corporate financial theory. *The Journal of Finance, 28*, 167–181.
100. Hsia, C.-C. (1981). Optimal debt of a firm: An option pricing approach. *Journal of Financial Research, Southern Finance Association; Southwestern Finance Association, 4*(3), 221–231.
101. Miller, M. (1997). Debt and taxes. *The Journal of Finance, 32*, 261–276.
102. Berk, J., & DeMarzo, P. (2007). *Corporate finance*. Pearson-Addison Wesley.
103. Farber, A., Gillet, R., & Szafarz, A. (2006). A general formula for the WACC. *International Journal of Business, 11*(2), 211–218.

104. Fernandez, P. (2006). A general formula for the WACC: A comment. *International Journal of Business, 11*(2), 219.
105. Harris, R., & Pringle, J. (1985). Risk–adjusted discount rates—Extension form the average–risk case. *The Journal of Financial Research, 1985*(8/3), 237–244.
106. Becker, D. M. (2021). The difference between Modigliani–Miller and Miles–Ezzell and its consequences for the valuation of annuities. *Cogent Economics & Finance, 9*, 1. https://doi.org/10.1080/23322039.2020.1862446
107. Miles, J., & Ezzell, R. (1980). The weighted average cost of capital, perfect capital markets and project life: A clarification. *Journal of Financial and Quantitative Analysis, 15*(3), 719–730.
108. Brennan, M., & Schwartz, E. (1984). Optimal financial policy and firm valuation. *The Journal of Finance, 39*, 593–607.
109. Leland, H. (1994). Corporate debt value, bond covenants and optimal capital structure. *The Journal of Finance, 49*, 1213–1252.
110. Kane, A., Marcus, A., & McDonald, R. (1984). How big is the Tax advantage to Debt? *The Journal of Finance, 39*, 841–853.
111. Brusov, P., Filatova, P., & Orekhova, N. (2013). Absence of an optimal capital structure in the famous tradeoff theory! *Journal of Reviews on Global Economics, 2*, 94–116.
112. Brusov, P., Filatova, T., & Orehova, N. (2013). A qualitatively new effect in corporative finance: Abnormal dependence of cost of equity of company on leverage. *Journal of Reviews on Global Economics, 2*, 183–193.
113. Brusov, P., Filatova, P., & Orekhova, N. (2014). Mechanism of formation of the company optimal capital structure, different from suggested by trade off theory. *Cogent Economics & Finance, 2*, 1–13. https://doi.org/10.1080/23322039.2014.946150
114. Myers, S. C., Kolbe, A. L., & Tye, W. B. (1984). Regulation and capital formation in the oil pipeline industry. *Transportation Journal, 23*(4), 25–49.
115. Myers, S. C. (1984). Finance theory and financial strategy. *Interfaces, 14*, 126–137.
116. Myers, S. C. (1984). The capital structure puzzle. *Journal of Finance, 39*(3), 575–592.
117. Fama, E. F., & French, K. (2005). Financing decisions: Who issues stock? *Journal of Financial Economics, 76*(3):549–82 (Federal Power Commission et al. v. Hope Natural Gas Co., 320 U.S. 591 (1944)).
118. Brusov, P., Filatova, T., Eskindarov, M., & Orehova, N. (2012). Hidden global causes of the global financial crisis. *Journal of Reviews on Global Economics, 1*, 106–111.
119. Brusov, P., Filatova, T., & Orehova, N. (2014). Inflation in Brusov–Filatova–Orekhova theory and in its perpetuity limit–Modigliani–Miller theory. *Journal of Reviews on Global Economics, 3*, 175–185.
120. Brusova, A. (2011). A comparison of the three methods of estimation of weighted average cost of capital and equity cost of company. *Financial Analysis Problems and Their Solutions, 34* (76), 36–42.
121. Brusov, P. N., Filatova, T. V., Orekhova, N. P., Kulik, V. L., Chang, S. I., & Lin, Y. C. G. (2020). Modification of the Modigliani–Miller theory for the case of advance payments of tax on profit. *Journal of Reviews on Global Economics, 9*, 257–267.
122. De Marzo, P. M. (1988). An extension of the Modigliani-Miller theorem to stochastic economies with incomplete markets and fully interdependent securities. *Journal of Economic Theory, 45*, 353–369.
123. Sethi, S., Derzko, N., & Lehoczky, J. (1991). A stochastic extension of the Miller-Modigliani framework. *Mathematical Finance, 1*(4), 57–76. SSRN: https://ssrn.com/abstract=1082940

Chapter 5
Modigliani–Miller Theory

5.1 Introduction

Under the capital structure, one understands the relationship between equity and debt capital of the company. Does capital structure affect the company's main settings, such as the cost of capital, profit, value of the company, and the others, and, if affects, how? Choice of an optimal capital structure, i.e., a capital structure, which minimizes the weighted average cost of capital, WACC, and maximizes the value of the company, V, is one of the most important tasks solved by financial manager and by the management of a company. The first serious study (and first quantitative study) of influence of capital structure of the company on its indicators of activities was the work by Modigliani and Miller [1]. Until this study, the approach existed (let us call it traditional), which was based on empirical data analysis.

One of the most important assumptions of the Modigliani–Miller theory is that all financial flows are perpetuity. This limitation was lift out by Brusov–Filatova–Orekhova in 2008 [2], who have created BFO theory—modern theory of capital cost and capital structure for companies of arbitrary age (BFO-1 theory) and for companies of arbitrary life time (BFO-2 theory) [3]. In Fig. 5.1 the historical development of capital structure theory from the traditional (empirical) approach, through perpetuity Modigliani–Miller approach to general capital structure theory—Brusov–Filatova–Orekhova (BFO) theory is shown.

In 2001 Steve Myers [4] has considered the case of one-year company and show that in this case the weighted average cost of capital, WACC, is higher than in Modigliani–Miller case, and the capitalization of the company, V, is less than in Modigliani–Miller case.

So, before 2008 only two results for capital structure of the company were available: Modigliani–Miller for perpetuity company and Myers for one-year company (see Fig. 5.2). BFO theory has filled out whole interval between $t = 1$ and $t = 8$. It gives the possibility to calculate the capitalization V, the weighted average cost of capital, $WACC$, equity cost k_e and other financial parameters for companies of

P. Brusov, T. Filatova, *Business Valuation and Company Value*, Contributions to Finance and Accounting, https://doi.org/10.1007/978-3-031-90847-7_5

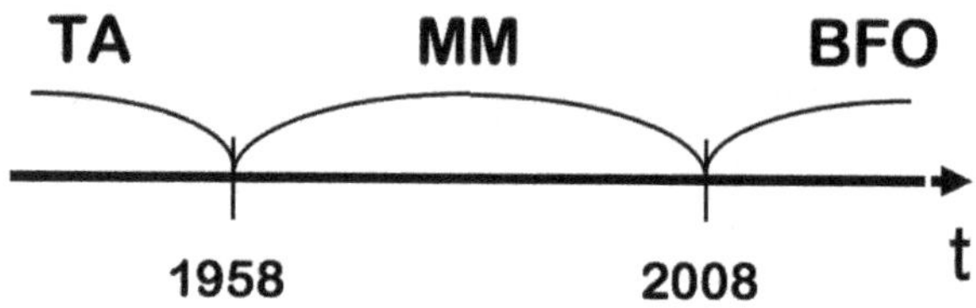

Fig. 5.1 Historical development of capital structure theory (here TA—traditional (empirical) approach, MM—Modigliani–Miller approach, BFO—Brusov–Filatova–Orekhova theory)

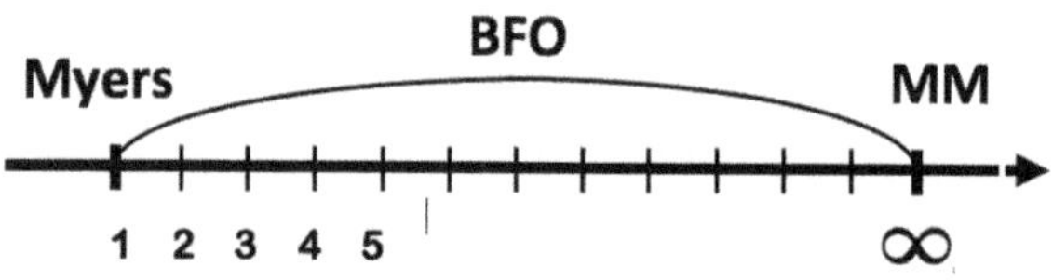

Fig. 5.2 MM theory describes perpetuity limit, Myers paper describes one-year company while BFO theory fills the whole numeric axis (from $n = 1$ up to perpetuity limit $n = 8$)

arbitrary age and for companies of arbitrary life-time. BFO theory has led to a lot of new meaningful effects in modern capital structure theory, discussed in this monograph.

5.2 The Traditional Approach

The traditional (empirical) approach told that weighted average cost of capital, WACC, and the associated company capitalization, $V = \mathrm{CF/WACC}$, depend on the capital structure, the level of leverage, L. Debt cost always turns out to be lower than equity cost because first one has lower risk, via the fact, that in the event of bankruptcy creditor claims are met prior to shareholders claims.

As a result, an increase in the proportion of lower-cost debt capital in the overall capital structure up to the limit which does not cause violation of financial sustainability and growth of risk of bankruptcy leads to lower weighted average cost of capital, WACC.

The profitability required by investors (the equity cost) is growing; however, its growth has not led to compensation of benefits from use of lower-cost debt capital. Therefore, the traditional approach welcomes the increased leverage $L = D/S$ and the associated increase of company capitalization. The traditional (empirical) approach has existed up to appearance of the first quantitative theory by Modigliani and Miller ([1]).

5.3 Modigliani–Miller Theory

5.3.1 Modigliani–Miller Theory Without Taxes

Modigliani and Miller (??) in their first paper [1] have come to the conclusions which were fundamentally different from the conclusions of traditional approach. Under assumptions (see Sect. 5.3.3 for details) that there are no taxes, no transaction costs, no bankruptcy costs, perfect financial markets exist with symmetry information, equivalence in borrowing costs for both companies and investors, etc., they have showed that choosing of the ratio between the debt and equity capital does not affect company value as well as capital costs (Fig. 5.3).

Under above assumptions, Modigliani and Miller have analyzed the impact of financial leverage, supposing the absence of any taxes (on corporate profit as well as individual one). They have formulated and proven two following statements.

Without taxes, the total cost of any company is determined by the value of its EBIT–Earnings Before Interest and Taxes, discounted with fixed rate k_0, ***corresponding to group of business risk of this company:***

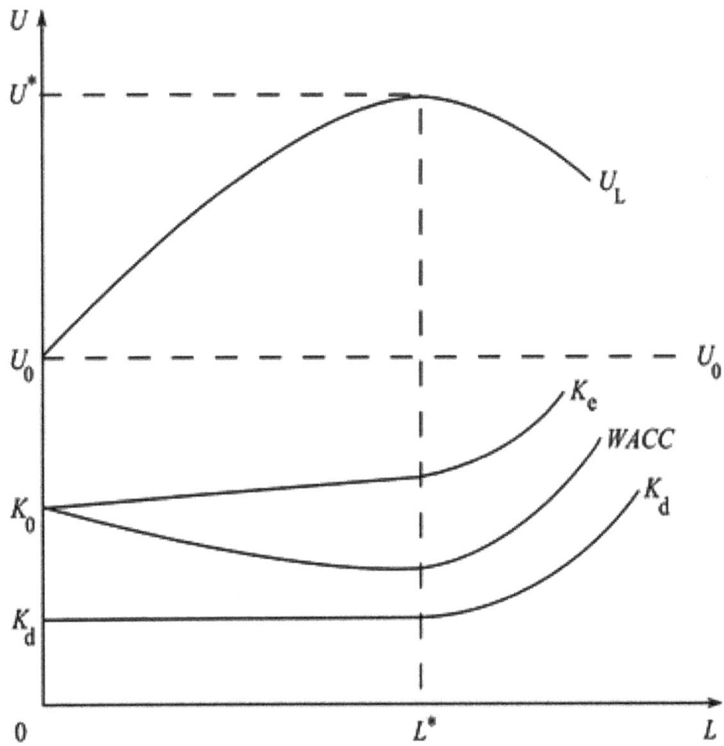

Fig. 5.3 Dependence of company capitalization, U_L, equity cost, k_e, debt cost, k_d, and weighted average cost of capital, WACC, in traditional (empirical) approach

$$V_L = V_U = \frac{\text{EBIT}}{k_0}. \tag{5.1}$$

Index L means financially dependent company (using debt financing), while index U means a financially independent company.

Authors supposed that both companies belong to the same group of business risk, and k_0 corresponds to required profitability of financially independent company, having the same business risk.

Because, as it follows from the formula (Eq. 5.1), value of the company does not depend on the value of debt, than according to Modigliani–Miller theorem [1], in the absence of taxes, value of the company is independent of the method of its funding. This means as well that weighted average cost of capital, W???, of this company does not depend on its capital structure and is equal to the capital cost, which this company will have under the funding by equity capital only.

$$V_0 = V_L; \quad \text{CF}/k_0 = \text{CF}/\text{WACC}, \quad \text{and thus WACC} = k_0.$$

Note that first Modigliani–Miller theorem is based on suggestion about independence of weighted average cost of capital and debt cost on leverage level.

From the first Modigliani–Miller theorem [1], it is easy to derive an expression for the equity capital cost:

$$\text{WACC} = k_0 = k_e w_e + k_d w_d. \tag{5.2}$$

Finding from here k_e, one gets:

$$k_e = \frac{k_0}{w_e} - k_d \frac{w_d}{w_e} = \frac{k_0(S+D)}{S} - k_d \frac{D}{S} = k_0 + (k_0 - k_d)\frac{D}{S} = k_0 + (k_0 - k_d)L \tag{5.3}$$

Here,

D	Value of debt capital of the company
S	Value of equity capital of the company
$k_d, w_d = \frac{D}{D+S}$	Cost and fraction of debt capital of the company:
$k_e, w_e = \frac{S}{D+S}$	Cost and fraction of equity capital of the company:
$L = D/S$	Financial leverage:

Thus, we come to second statement (theorem) of Modigliani–Miller theory about the equity cost of financially dependent (leverage) company [1].

Equity cost of leverage company k_e could be found as equity cost of financially independent company k_0 of the same group of risk, plus premium for risk, the value which is equal to production of difference (k_0 - k_d) on leverage level L:

$$k_e = k_0 + (k_0 - k_d)L. \tag{5.4}$$

Formula (Eq. 5.4) shows that equity cost of the company increases linearly with leverage level (Fig. 5.3).

The combination of these two Modigliani–Miller statements implies that the increasing of level of debt in the capital structure of the company does not lead to increased value of firms, because the benefits gained from the use of more low-cost debt capital markets will be exactly offset by an increase in risk (we are speaking about the financial risk, the risk of bankruptcy) and, therefore, by an increase in cost of equity capital of firms: investors will increase the required level of profitability under increased risk, by which a higher level of debt in the capital structure is accompanied.

In this way, the Modigliani–Miller theorem argues that in the absence of the taxes, the capital structure of the company does not affect the value of the company and its weighted average cost of capital, WACC, and equity cost increases linearly with the increase of financial leverage.

Explanations, given by Modigliani and Miller under receiving of their conclusions, are the following [1]. Value of the company depends on profitability and risk only and does not depend on the capital structure. Based on the principle of preservation of the value, they postulated that the value of the company, which is equal to the sum of the equity and debt funds, is not changed when the ratio between its parts is changed. An important role in justification of Modigliani–Miller statements an existence of an arbitral awards opportunities for the committed markets plays. Two identical companies, differing only by the leverage level, must have the same value. If this is not the case, the arbitration aligns business cost: investors of less cost company can invest capital in a company of more value. Selling of shares of the first company and buying of stock of the second company will continue until the values of both companies are not equalized.

Most of Modigliani and Miller assumptions [1], of course, are unrealistic. Some assumptions can be removed without changing the conclusions of the model. However, assuming no costs of bankruptcy and the absence of taxes (or the presence of corporate taxes only) are crucial—the change of these assumptions alters conclusions. The last two assumptions rule out the possibility of signaling theory and agency costs theory and, thus, also constitute a critical prerequisite (Fig. 5.4).

5.3.2 *Modigliani–Miller Theory with Taxes*

In the real situation, taxes on profit of companies always exist. Since the interest paid on debt are excluded from the tax base-it leads to the so-called effect of "tax shield": value of the company that used the borrowed capital (leverage company) is higher than the value of the company that financed entirely by the equity (non-leverage company). The value of the "tax shield" for 1 year is equal to $k_d\,D\,T$, where D—the

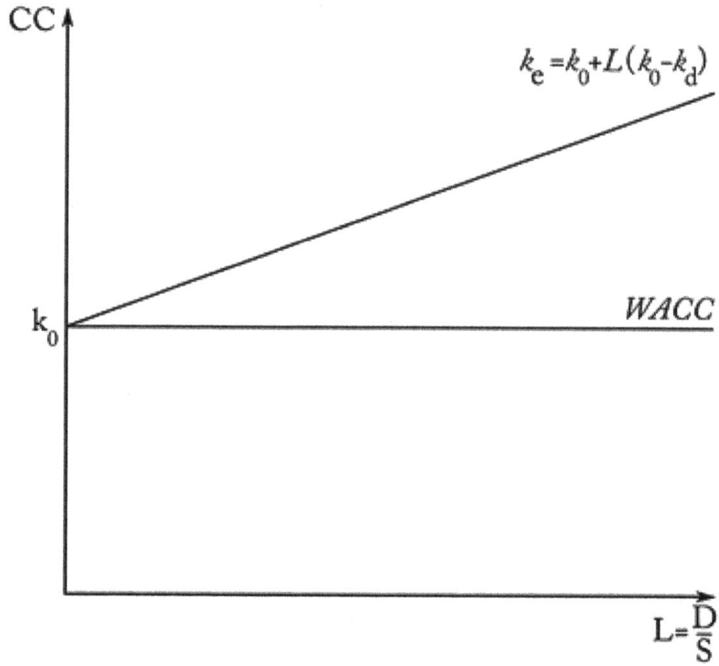

Fig. 5.4 Dependence of equity cost k_e and WACC on leverage level L within Modigliani–Miller theory without taxes

value of debt, T—the income tax rate, and k_d—the interest on the debt (or debt capital cost) [5]. The value of the "tax shield" for perpetuity company for all time of its existence is equal to (we used the formula for the sum of terms of an infinitely decreasing geometric progression).

$$(\mathrm{PV})_{\mathrm{TS}} = k_d DT \overset{8}{\underset{t=1}{?}} (1 + k_d)^{-t} = DT \tag{5.5}$$

and the cost of leverage company is equal to:

$$V = V_0 + DT, \tag{5.6}$$

where V_0 is the value of financially independent company.

Thus, we obtain the following result obtained by ?odigliani and ?iller [5]:

The value of financially dependent company is equal to the value of the company of the same risk group used no leverage, increased by the value of tax shield arising from financial leverage, and equal to the product of rate of corporate income tax T and the value of debt D.

Let us now get the expression for the equity capital cost of the company under the existence of corporate taxes.

Accounting that $V_0 = \mathrm{CF}/k_0$ and that the ratio of debt capital $w_\mathrm{d} = D/V$, one gets

$$V = \mathrm{CF}/k_0 + w_\mathrm{d} VT. \tag{5.7}$$

Because the value of leverage company is $V = \mathrm{CF}/\mathrm{WACC}$, for weighted average cost of capital, WACC, we get:

$$\mathrm{WACC} = k_0(1 - w_\mathrm{d}T). \tag{5.8}$$

From here the dependence of WACC on leverage $L = D/S$ becomes the following:

$$\mathrm{WACC} = k_0(1 - LT/(1 + L)). \tag{5.9}$$

On the other hand, on definition of the weighted average cost of capital with "tax shield" accounting, we have:

$$\mathrm{WACC} = k_0 w_\mathrm{e} + k_\mathrm{d} w_\mathrm{d}(1 - T). \tag{5.10}$$

Equating Eqs. (5.9) and (5.11), one gets:

$$k_0(1 - w_\mathrm{d}T) = k_0 w_\mathrm{e} + k_\mathrm{d} w_\mathrm{d}(1 - T) \tag{5.11}$$

and from here, for equity cost, we get the following expression:

$$\begin{aligned} k_\mathrm{e} &= k_0 \frac{(1 - w_\mathrm{d}T)}{w_\mathrm{e}} - k_\mathrm{d}\frac{w_\mathrm{d}}{w_\mathrm{e}}(1 - T) = k_0\frac{1}{w_\mathrm{e}} - k_0\frac{w_\mathrm{d}}{w_\mathrm{e}}T - k_\mathrm{d}\frac{D}{S}(1 - T) \\ &= k_0\frac{D + S}{S} - k_0\frac{D}{S}T - k_\mathrm{d}\frac{D}{S}(1 - T) = k_0 + L(1 - T)(k_0 - k_\mathrm{d}). \end{aligned} \tag{5.12}$$

So, we get the following statement obtained by ?odigliani and ?iller [5]:

Equity cost of leverage company k_e paying tax on profit could be found as equity cost of financially independent company k_0 of the same group of risk, plus premium for risk, the value which is equal to production of difference (k_0 - k_d) on leverage level L and on tax shield (1—T).

It should be noted that the formula (Eq. 5.12) is different from the formula (Eq. 5.4) without tax only by the multiplier (1–T) in term, indicating a premium for risk. As the multiplier is less than unit, the corporate tax on profits leads to the fact that capital is growing with the increasing of financial leverage, slower than it would have been without them.

Analysis of formulas (Eqs. 5.4, 5.9, and 5.12) leads to following conclusions. When leverage grows:

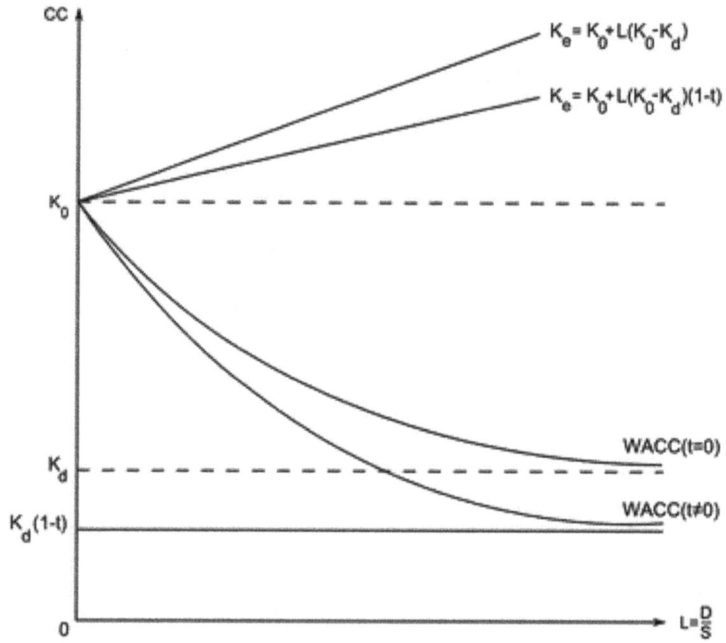

Fig. 5.5 Dependence of equity capital cost, debt cost, and WACC on leverage in Modigliani–Miller theory without taxes ($t = 0$) and with taxes (t ? 0)

1. Value of company increases.
2. Weighted average cost of capital WACC decreases from k_0 (at $L = 0$) up to $k_0(1 - T)$ (at $L = 8$) (when the company is funded solely by borrowed funds).
3. Equity cost increases linearly from k_0 (at $L = 0$) up to 8 (at $L = 8$).

Within their theory, ?odigliani and ?iller [5] had come to the following conclusions. With the growth of financial leverage (Fig. 5.5):

1. The company value increases.
2. The weighted average cost of capital decreases from k_0 (for $L = 0$) up to $k_0(1 - T)$ (for $L = 8$, when the company is financed entirely with borrowed funds).
3. The cost of equity capital increases linearly from k_0 (for $L = 0$) up to 8 (for $L = 8$).

5.3.3 Main Assumptions of Modigliani–Miller Theory

The most important assumptions of the Modigliani–Miller theory are as following:

1. Investors are behaving rationally and instantaneously, see profit opportunity, inadequate investment risk. Therefore, the possibility of a stable situation of the arbitration, i.e., obtain the risk-free profit on the difference in prices for the same asset cannot be kept any length of time-reasonable investors quickly take

advantage of it for their own purposes and equalize conditions in the market. This means that in a developed financial market capital, the same risk should be rewarded by the same rate of return.

2. Investment and financial market opportunities should be equally accessible to all categories of investors—whether institutional or individual investors, large or small, rapidly growing or stable, or experienced or relatively inexperienced.
3. Transaction costs associated with funding are very small. In practice, the magnitude of transaction costs is inversely proportional to the amount of finance involved, so this assumption is more consistent with reality than the large sums involved: i.e., in attracting small amounts, the transaction costs can be high, while, as in attracting large loans, as well as during placement of shares at a significant amount, the transaction costs can be ignored.
4. Investors get money and provide funds to borrowers at risk-free rate. In all probability, this assumption is due to the fact that the lender seeks to protect himself by using one or other guarantees, pledge of assets, the right to pay claims on third parties, and the treaty provisions restricting the freedom of the borrower to act to the detriment of the creditor. Lender's risk is really small, but its position can be considered risk free with respect to the position of the borrower and, accordingly, should be rewarded by a risk-free rate of return.
5. Companies have only two types of assets: risk-free debt capital and risky equity capital.
6. There is no possibility of bankruptcy, i.e., irrespective of what the level of financial leverage of the company—borrowers are reached—bankruptcy is not threatening them. Thus, bankruptcy costs are absent.
7. There are no corporate taxes and taxes on personal income of investors. If the personal income tax can indeed be neglected, because the assets of the company separated from the assets of shareholders, the corporate income taxes should be considered in the development of more realistic theories (which was done by Modigliani and Miller in their second paper devoted to the capital structure [5].
8. Companies are in the same class of risky companies.
9. All financial flows are perpetuity.
10. Companies have the same information.
11. Management of the company maximizes the capitalization of the company.

5.3.4 Modifications of Modigliani–Miller Theory

Taking into Account Market Risk: Hamada Model Robert H?m?d? [6] unites Capital Asset Pricing Model (CAPM) with Modigliani–Miller model taxation. As a result, he derived the following formula for calculation of the equity cost of financially dependent company, including both financial and business risk of company:

$$k_e = k_F + (k_M - k_F)b_U + (k_M - k_F)b_U\frac{D}{S}(1 - T), \tag{5.13}$$

where b_U is the β-coefficient of the company of the same group of business risk, that the company under consideration, but with zero financial leverage. The formula (Eq. 5.13) represents the desired profitability of equity capital k_e as a sum of three components: risk-free profitability k_F, compensating to shareholders a temporary value of their money, premium for business risk $(k_M$ - $k_F)b_U$, and premium for financial risk $(k_M - k_F)b_U\frac{D}{S}(1 - T)$.

If the company does not have borrowing ($D = 0$), the financial risk factor will be equal to zero (the third term is drawn to zero), and its owners will only receive the premium for business risk.

To apply the Hamada equation, specialists in practice, in most cases, use book value of equity capital as its approach of market value. Nevertheless, the Hamada formula implies the use of market value of the assets.

It should be noted also that the formula (Eq. 5.13) can be used to derive other equation, using which you can analyze the impact of financial leverage on β-factor of company shares.

Equating CAPM formula to equity cost, we get:

$$k_F + (k_M - k_F)b_U = k_F + (k_M - k_F)b_U + (k_M - k_F)b_U\frac{D}{S}(1 - T) \tag{5.14}$$

or

$$b = b_U\left(1 + \frac{D}{S}(1 - T)\right). \tag{5.15}$$

In this way, the assumptions on which Modigliani–Miller theory and CAPM are based, β-factor of equity capital of financially dependent company is equal to β-factor of financially independent company, corrected on tax on profit rate and applied leverage level. Consequently, market risk of the company, measured by a factor b, depends on both the business risk of the company, a measure of which is b_U,, and on the financial risk b, which is calculated by the formula (Eq. 5.15).

In conclusion, here are the formulas for calculating the capital costs within the CAPM model [in parenthesis, there are formulas within the Modigliani–Miller theory [1, 5, 7].

The equity cost for company without debt capital:

$$k_e = k_F + (\overline{k}_M - k_F)\beta_U, \quad (k_e = k_0). \tag{5.16}$$

The equity cost for company with debt capital:

$$k_e = k_F + (\bar{k}_M - k_F)\beta_e, \quad (k_e = k_0 + (1 - T)(k_0 - k_d)L). \tag{5.17}$$

The debt cost:

$$k_d = k_F + (\bar{k}_M - k_F)\beta_d, (k_d = k_F; \beta_d = 0). \tag{5.18}$$

The weighted average cost of capital WACC:

$$\text{WACC} = k_e w_e + k_d w_d(1 - T), \quad (\text{WACC} = k_0(1 - T w_d)). \tag{5.19}$$

The Cost of Capital Under Risky Debt Another hypothesis of Modigliani and Miller was the suggestion about free of risk debt (in their theory, there are two types of assets: risky equity and free of risk debt). However, if we assume the risk of bankruptcy of company (and, accordingly, the ability to nonpayment of loans), the situation may change. Stiglitz [8] and Rubinstein [9] have shown that the conclusions concerning the total value of company do not change as compared to the findings derived by Modigliani and Miller under assumptions about free of risk debt [1, 5, 7]. However, the debt cost is changed. If previously, under assumption about the free of risk debt, it (debt cost) was regarded as a constant $k_d = k_F$, now it is not a constant. This claim is based on the work by Hsia [10], where based on the models of pricing options, Modigliani–Miller and CAPM, it was shown that if one uses the formula for the net discount income, a term, reflecting tax protection on debt, should be discounted at the rate:

$$k'_d = k_F + (k_0 - k_F)N(-d_1)\frac{1}{w_d}, \tag{5.20}$$

where

$$d_1 = \frac{-\ln w_d + k_F t}{s\sqrt{t}} k_F + \frac{1}{2} s\sqrt{t}, \tag{5.21}$$

here t–a moment of payment a credit, $N(-d_1)$—cumulative normal distribution of probability of random value d_1.

The Account of Corporate and Individual Taxes (Miller Model) In the second article, Modigliani and Miller [5] considered taxation of corporate profits, but did not take into account the presence in the economy of individual taxes of investors.

Merton Miller [11] has introduced the model, demonstrating impact of leverage on the company value with account of the corporate and individual taxes [12].

To describe his model, we will enter the following legends: T_C—tax on corporate profits rate, T_S—the tax rate on income of an individual investor from his ownership by stock of corporation, T_D—tax rate on interest income from the provision of investor-individuals of credits to other investors and companies. Income from shares partly comes in the form of a dividend and, in part, as capital profits, so that T_S is a weighted average value of effective rates of tax on dividends and capital profits on shares, while the income from the provision of loans usually comes in the form of the interests. The last are usually taxed at a higher rate. In the light of the individual taxes, and with the same assumptions that have been made for Modigliani–Miller models previously, the financially independent company value can be determined as follows:

$$V_{\mathrm{U}}=\frac{\mathrm{EBIT}(1-T_{\mathrm{C}})(1-T_{\mathrm{S}})}{k_{0}}. \tag{5.22}$$

A term (1 - T_{S}) allows to take into account the individual taxes in the formula. In this way, numerator indicates which part of the operating company's profit remains in the possession of the investors, after the company will pay taxes on their profits, and its shareholders then will pay individual taxes on income from stock ownership. Since individual taxes reduce profits, remaining in the disposal of investors, the last, at other things being equal circumstances, also reduce and an overall assessment of the financially independent company value.

We will assess the financially dependent company under condition of a double taxation of income investors. To start, let us divide the annual cash flows of financially dependent company $\mathrm{CF_L}$ into flows sent to its shareholders $\mathrm{CF_e}$ and the flows belonging to debt owners $\mathrm{CF_d}$, with account of both corporation tax on profits and on the income of individuals:

$$\mathrm{CF_L}=\mathrm{CF_e}+\mathrm{CF_d}=(\mathrm{EBIT}-I)(1-T_{\mathrm{C}})(1-T_{\mathrm{S}})+I(1-T_{\mathrm{D}}), \tag{5.23}$$

where I is the annual interest payments on debt.

The formula (Eq. 5.23) can also be rewritten as follows:

$$\begin{aligned}\mathrm{CF_L}=\mathrm{CF_e}+\mathrm{CF_d}&=\mathrm{EBIT}(1-T_{\mathrm{C}})(1-T_{\mathrm{S}})-I(1-T_{\mathrm{C}})(1-T_{\mathrm{S}})\\&+I(1-T_{\mathrm{D}}).\end{aligned} \tag{5.24}$$

The first term of the equation (Eq. 5.24) corresponds to cash flow after taxes for financially independent company, shown in equation (Eq. 5.22), which shows its present value. The second and the third terms of the equation, reflecting the financial dependence, corresponds to cash flows related to the debt financing, which, as previously, is considered as free of risk. Their present values are obtained by discounting by risk-free nominal rate on debt k_{d}.

By combining the present values of all three terms, we get the company value under using the debt financing and in the presence of all types of taxation:

$$V_{\mathrm{U}}=\frac{\mathrm{EBIT}(1-T_{\mathrm{C}})(1-T_{\mathrm{S}})}{k_{0}}-\frac{I(1-T_{\mathrm{C}})(1-T_{\mathrm{S}})}{k_{\mathrm{d}}}+\frac{I(1-T_{\mathrm{d}})}{k_{\mathrm{d}}}. \tag{5.25}$$

First term in (Eq. 5.25) is identical to V_U in formula (Eq. 5.22). Accounting this and combining two last terms, we get the following formula:

$$V_L = V_U + \frac{I(1 - T_d)}{k_d}\left[1 - \frac{(1 - T_C)(1 - T_S)}{(1 - T_d)}\right]. \tag{5.26}$$

The amount of paid interests with taking into account the taxation, divided by the desired profitability of debt capital,

$$\frac{I(1 - T_d)}{k_d} \tag{5.27}$$

is equal to market value of the debt D. Substituting D in the previous formula, we get the final expression, which is known as a formula of a Miller model:

$$V_L = V_U + \left[1 - \frac{(1 - T_C)(1 - T_S)}{(1 - T_d)}\right]D. \tag{5.28}$$

The Miller model allows you to obtain an estimate of the value of financially dependent company, taking into account the corporate tax, as well as tax on individuals. The Miller formula (Eq. 5.28) has several important consequences:

1. Second term of sum,

$$\left[1 - \frac{(1 - T_C)(1 - T_S)}{(1 - T_d)}\right]D, \tag{5.29}$$

represents the gains from use of debt capital. This term replaces the tax on profit of corporation rate in the Modigliani–Miller model with corporate taxes:

$$V_L = V_U + TD. \tag{5.30}$$

2. If we ignore taxes, a term (Eq. 5.29) will be equal to zero. Thus, in this case, the formula (Eq. 5.28) is transformed into the original version of the Modigliani–Miller model without taxes.
3. If we neglect taxes on individuals, the considering term becomes 1 - (1 - T_C) = T_C, so, in this case, (Eq. 5.28) becomes a Modigliani–Miller model with corporate taxes (Eq. 5.30).
4. If the shareholder receives profit only in the form of dividend, and if effective tax rates on income from shares and bonds are equal ($T_S = T_D$), the terms 1 - T_S and 1 - T_D are shrinking, and the factor for D in (Eq. 5.29) again is equal to T_C.
5. If the shareholder receives dividends, and income from capital, the situation is changed.

In this case, effective tax rates on income from shares and bonds are not equal. Let's take a look at common case, when individual taxes on income for the company shares are less than individual taxes of creditors. This encourages investors to purchase the shares of the company compared to purchasing the bonds of the company. In this case, $T_S < T_D$. Then factor in D in (Eq. 5.28) β has a look

$$\begin{aligned}\beta &= 1 - \frac{(1-T_C)(1-T_S)}{(1-T_d)} = 1-(1-T_C)a = 1-(1-T_C)(1+?)\\ &= T_C + ?(T_C - 1) < T_C.\end{aligned} \tag{5.31}$$

It is less than T_C, because $? > 0$, $T_C < 1$, therefore, in this case, the effect of using of debt financing, although there is, but it is less, than in the absence of individual taxes. In other words, the effect of tax shields for the company in this case decreases, and it becomes less than the above individual taxes of creditors (individual taxes for the obligations of the company) in comparison with the individual income tax on shares.

6. Let's take a look at case $T_S > T_D$, when individual income taxes on shares are bigger than individual taxes creditors. The factor β takes view:

$$\begin{aligned}\beta &= 1 - \frac{(1-T_C)(1-T_S)}{(1-T_d)} = 1-(1-T_C)a = 1-(1-T_C)(1-?)\\ &= T_C + ?(1-T_C) > T_C.\end{aligned} \tag{5.32}$$

It is bigger than T_C, because $? > 0$, $T_C < 1$; therefore, in this case, the effect of use of debt financing is increased compared with the case of the absence of individual taxes.

7. If $(1 - T_C)(1 - T_S) = 1 - T_D$, then this term is zero, and the effect of using debt financing will also be zero. This means that the benefits of the use of tax shields as a result of the application of debt financing will be fully offset by additional losses of investors, associated with a higher tax rate on interest on income of individuals. In this case, the capital structure will not affect the company value and its capital cost-in other words, you can apply Modigliani–Miller theory without tax [1].

In his report, Miller [12] claimed that companies on average will use issuance of shares and debt securities in such a way as to result in taxation of investors' income to be optimal. In such an equilibrium state will occur equality:

$$(1-T_C)(1-T_S) = 1 - T_D, \tag{5.33}$$

and thus, as we have pointed out above, capital structure will not affect the market company value and its capital cost. Thus, by Miller, the conclusions on the irrelevance of the capital structure, made on the base of the original Modigliani–Miller model with zero taxes, remain in force.

Subsequently, researchers adapted and checked the Miller results. Their works, as a rule, have been devoted to the Miller's conclusion concerning the absence of the gains from the use of the debt capital by the company. In the United States, an effective tax rate on the income of shareholders is lower than the one on the income of creditors, but, nevertheless, the product $(1 - T_C)(1 - T_S)$ is less than $1 - T_D$.

Consequently, the companies may receive the benefit from use of debt financing. However, in the Miller's work, in fact, has been shown that the distinction of rates of individual taxes on income of shareholders and creditors to some extent compensates the advantages of use of debt financing, and, in this way, the tax benefits of debt are less than anticipated at a more earlier Modigliani–Miller model, where only corporate taxes have been taken into account.

In conclusion, we present in Table 5.1 classification and summary of main theories of capital structures of company.

Table 5.1 Classification and summary of main theories of capital structures of company

Theory		Main thesis
Traditional theory		Empirical theory, existing before appearance of the first quantitative theory of capital structures (Modigliani–Miller theory) in 1958 [1, 5, 7]. Weighted average cost of capital depends on capital structures of company. There is an optimal dependence on capital structures of company
Modigliani–Miller theory (??)	Without taxes	Capital cost and capitalization of the company are irrelevant on the capital structures of company
	With taxes	Weighted average cost of capital is decreased with leverage level, equity cost is increased linearly with leverage level, and capitalization of the company is increased with leverage level continuously
Brusov–Filatova–Orekhova theory (BFO-1)	For arbitrary age Without inflation	BFO theory [2, 13–27] has replaced the famous theory of capital cost and capital structure by Nobel laureates, Modigliani and Miller [1, 5, 7]. The authors have moved from the assumption of Modigliani–Miller concerning the perpetuity (infinite time of life) of companies and further elaborated quantitative theory of valuation of core parameters of financial activities of companies with arbitrary age. Results of modern BFO theory turn out to be quite different from that of Modigliani–Miller theory. It shows that later, via its perpetuity, underestimates the assessment of weighted average cost of capital and the equity cost of the company and substantially overestimates the assessment of the capitalization of the company. Such an incorrect assessment of key performance indicators of financial activities of companies has led to an underestimation of risks involved, and impossibility, or serious difficulties in adequate managerial decision-making, which was one of the implicit reasons of global financial crisis of 2008 year. In the BFO theory, in investments at certain values of return-on-investment, there is an optimum investment structure. As well authors have developed a new mechanism of formation of the company optimal capital structure, different from suggested by trade-off theory
	For arbitrary age With inflation	Inflation not only increases the equity cost and the weighted average cost of capital, but as well it changes their dependence on

(continued)

Table 5.1 (continued)

Theory		Main thesis
		leverage. In particular, it increases growing rate of equity cost with leverage. Capitalization of the company is decreased under accounting of inflation
	For arbitrary age With increased financial distress costs and risk of bankruptcy	In BFO theory, with increased financial distress costs and risk of bankruptcy, the optimal capital structure is absent, which means that trade-off theory does NOT work
Brusov–Filatova–Orekhova theory (BFO-2)	For arbitrary lifetime	In perpetuity limit (Modigliani–Miller theory) time of life of company and company age turns out to be the same: both of them are infinite. When we move from the assumption of Modigliani–Miller concerning the perpetuity of companies these concepts (time of life of company and company age) become different and one should distinct them under generalization of Modigliani–Miller theory with respect to finite n. Thus we have developed two kinds of finite n-theories: BFO-1 and BFO-5. BFO-1 theory is related to companies with arbitrary age and BFO-2 theory is related to companies with arbitrary life-time companies. By other words, BFO-1 is applicable for most interesting case of companies that reached the age of n-years and continue to exist on the market, and allows to analyze the financial condition of the operating companies. BFO-2 theory allows examine the financial status of the companies which ceased to exist, i.e., of those for which n means not age, but a life-time, i.e., the time of existence. A lot of schemes of termination of activities of the company can exist: bankruptcy, merger, acquisition, etc. One of those schemes, when the value of the debt capital D becomes zero at the time of termination of activity of company n, is considered in Chap. 3 and comparison of results of BFO-1 and BFO-2 has been done.
Brusov–Filatova–Orekhova theory (BFO-3)	For rating needs	A new approach to rating methodology has been developed in Chaps. 21–23. Chapters 21 and 22 are devoted to rating of non-financial issuers, while Chap. 23 is devoted to long-term project rating. The key factors of a new approach are: (1) The adequate use of discounting of financial flows virtually not used in existing rating

(continued)

Table 5.1 (continued)

Theory		Main thesis
		methodologies, (2) The incorporation of rating parameters (financial "ratios") into the modern theory of capital structure (Brusov–Filatova–Orekhova (BFO-3 theory) (in Chaps. 21 into its perpetuity limit). This on the one hand allows use the powerful tools of this theory in the rating, and on the other hand it ensures the correct discount rates when discounting of financial flows. The interplay between rating ratios and leverage level which can be quite important in rating is discussed as well. All these create a new base for rating methodologies. New approach to ratings and rating methodologies allows to issue more correct ratings of issuers, makes the rating methodologies more understandable and transparent
Trade-off theory	Static	The static trade-off theory is developed with accounting of tax on profit and bankruptcy cost. It attempts to explain the optimal capital structure in terms of the balancing act between the benefits of debt (tax shield from interest deduction) and the disadvantage of debt (from the increased financial distress and expected bankruptcy costs). The tax shield benefit is the corporate income tax rate multiplied by the market value of debt and the expected bankruptcy costs are the probability of bankruptcy multiplied by the estimated bankruptcy costs
		Does not take into account the costs of the adaptation of financial capital structure to the optimal one, economic behavior of managers, owners, and other participants of economical process, as well as a number of other factors
		As it has been shown in BFO theory, the optimal capital structure is absent in trade-off theory
	Dynamic	The dynamic trade-off models assume that costs of constant capital adjustment are high and thus firms will change capital structure only if benefits exceed costs. Therefore, there is an optimal range, outside of each leverage changes but remains unchanged inside. Companies try to adjust their leverage when it reaches the boundary of the

(continued)

Table 5.1 (continued)

Theory		Main thesis
		optimal range. Subject to types of adjustment costs firms reach target ratio faster or slower. Proportional changes imply slight correction, whereas fixed changes imply considerable costs. In the dynamic model, correct decision on financial structure capital of the company in this period depends on the profit, which the company hopes to receive in the next period
		In BFO theory, with increased financial distress costs and risk of bankruptcy, the optimal capital structure is absent, which means that trade-off theory does NOT work: in static version as well as in dynamic one
Accounting of transaction cost		Accounting of the recapitalization transaction costs for the company, in which these costs are high, leads to the conclusion that a more cost-effective is not to modify financial capital structure, even if it is not optimal, during a certain period of time. The actual and target capital structure may vary because of the tool costs
Accounting of asymmetry of information		At the real financial markets, information is asymmetric (managers of the companies have owned more reliable information than investors and creditors), and rationality of economic subjects is limited
Signaling theory		Information asymmetry may be reduced on the basis of certain signals for creditors and investors, related to the behavior of managers on the capital market. It should take into account the previous development of the company and the current and projected cost-effectiveness of activities
Pecking order theory		The pecking order theory is the preferred, and empirically observed, sequence of financing type to raise capital. That is, firms first tap retained earnings (internal equity) finance, second source is debt, and the last source is issuing new common stock shares (external equity). The empirical evidence of nonfinancial firm debt ratios coupled with the decision-making process of top management and the board of directors point to greater adherence to the pecking order theory
Theories of conflict of interests	Theory of agency costs	Management of the company may take decisions that are contrary to the interests of

(continued)

Table 5.1 (continued)

Theory		Main thesis
		the shareholders or creditors, respectively; the costs are necessary to monitor its actions. An effective tool for resolving agent problem is the correct selection of compensation package (the share of participation of agent in property, bonus, stock options), allowing to link revenue of managers with the dynamics of equity capital and to provide motivation for managers to its (equity capital) conservation and growth
	Theory of corporate control and costs monitoring	If asymmetries of information exist, creditors, providing the capital, are interested in the possibility of the implementation of the self-monitoring of the effectiveness of its use and return. Costs for monitoring, as a rule, put on the company owners by their inclusion into credit rate. The level of monitoring costs depends on the scale of the business; therefore, with the increase of the business scale, the weighted average cost of capital of the company grows and company market value is reduced
	Theory of stakeholders	Stakeholder theory is a theory that identifies and models the groups that are stakeholders of a corporation or project. The diversity and the intersection of stakeholders' interests and different assessment by them of acceptable risk generate conditions for conflict of their interest, that is, making some corrections into the process of optimizing financial capital structure
Behavioral theories	Manager investment autonomy	Managers implement those decisions, which, from their point of view, will be positively perceived by investors and, respectively, positively affect the market value of companies: when the market value of shares of a company and the degree of consensus of expectations of managers and investors are high, the company has an additional issue of shares, and in the opposite situation, it uses debt instruments. In this way, the financial capital structure is more influenced by investors, the expectations of which are taken into account by managers
	The equity market timing theory	Leverage level is determined by market dynamics. Equity market timing theory means that company should issue shares at high price and repurchase them at low price.

(continued)

Table 5.1 (continued)

Theory		Main thesis
		The idea is to exploit temporary fluctuations in the equity cost relative to the cost of other forms of capital
	Information cascades	In order to save costs and to avoid errors, financial capital structure can be formed not on the basis of the calculations of optimal capital structure or depending on available in different periods of company life funding sources, but borrow from other companies that have successful, reputable managers (companies' leaders), as well as using (in the wake of the majority) the most popular methods of management of capital structure

References

1. ?odigliani, F., & ?iller, M. (1958). The cost of capital, corporate finance, and the theory of investment. *The American Economic Review, 48*, 261–297.
2. Filatova, ?., Orehova, N., & Brusova, ?. (2008). Weighted average cost of capital in the theory of Modigliani–Miller, modified for a finite life–time company. *Bull FU, 48*, 68–77.
3. Brusov, P., Filatova, T., Orehova, N., & Eskindarov, M. (2015). *Modern corporate finance, investment and taxation* (1st ed., pp. 1–368). Springer International Publishing.
4. Myers, S. (2001). Capital structure. *The Journal of Economic Perspectives, 15*(2), 81–102.
5. ?odigliani, F., & ?iller, M. (1963). Corporate income taxes and the cost of capital: A correction. *The American Economic Review, 53*, 147–175.
6. Hamada, R. (1969). Portfolio analysis, market equilibrium, and corporate finance. *The Journal of Finance, 24*(1), 13–31.
7. Modigliani, F., & Miller, M. (1966). Some estimates of the cost of capital to the electric utility industry 1954–1957. *The American Economic Review, 56*, 333–391.
8. Stiglitz, J. (1969). A re–examination of the Modigliani–Miller theorem. *The American Economic Review, 59*(5), 784–793.
9. Rubinstein, M. (1973). A mean–variance synthesis of corporate financial theory. *The Journal of Finance, 28*, 167–181.
10. Hsia, ?. (1981). Coherence of the modern theories of finance. *Financial Review, 16*(1), 27–42.
11. Miller, M. (1997). *Merton Miller on derivatives*. Wiley.
12. Miller, M. (1976). Debt and taxes. *Journal of Financ, 32*(2), 261–275.
13. Brusov, P. N., & Filatova, ?. V. (2011). From Modigliani–Miller to general theory of capital cost and capital structure of the company. *Finance and Credit, 435*, 2–8.
14. Brusov, P., Filatova, T., Orehova, N., & Brusova, A. (2011a). Weighted average cost of capital in the theory of Modigliani–Miller, modified for a finite life–time company. *Applied Financial Economics, 21*(11), 815–824.
15. Brusov, P., Filatova, T., Orehova, N., Brusov, P. P., & Brusova, N. (2011b). From Modigliani–Miller to general theory of capital cost and capital structure of the company. *Research Journal of Economics, Business and ICT, 2*, 16–21.
16. Brusov, P., Filatova, T., Orehova, N., et al. (2011c). Influence of debt financing on the effectiveness of the investment project within the Modigliani–Miller theory. *Research Journal of Economics, Business and ICT, 2*, 11–15.
17. Brusov, P., Filatova, T., Eskindarov, M., & Orehova, N. (2012a). Influence of debt financing on the effectiveness of the finite duration investment project. *Applied Financial Economics, 22*(13), 1043–1052.
18. Brusov, P., Filatova, T., Eskindarov, M., & Orehova, N. (2012b). Hidden global causes of the global financial crisis. *Journal of Reviews on Global Economics, 1*, 106–111.
19. Brusov, P., Filatova, P., & Orekhova, N. (2013a). Absence of an optimal capital structure in the famous tradeoff theory! *Journal of Reviews on Global Economics, 2*, 94–116.
20. Brusov, P., Filatova, T., & Orehova, N. (2013b). A qualitatively new effect in corporative finance: Abnormal dependence of cost of equity of company on leverage. *Journal of Reviews on Global Economics, 2*, 183–193.
21. Brusov, P., Filatova, P., & Orekhova, N. (2014a). Mechanism of formation of the company optimal capital structure, different from suggested by trade off theory. *Cogent Economics & Finance, 2*, 1–13. https://doi.org/10.1080/23322039.2014.946150
22. Brusov, P., Filatova, T., & Orehova, N. (2014b). Inflation in Brusov–Filatova–Orekhova theory and in its perpetuity limit–Modigliani–Miller theory. *Journal of Reviews on Global Economics, 3*, 175–185.
23. Brusova, A. (2011). ? comparison of the three methods of estimation of weighted average cost of capital and equity cost of company. *Financial Analysis Problems and Solutions, 34*(76), 36–42.

24. Brusov, P., Filatova, T., Orehova, N., & Kulk, W. I. (2018a). New meaningful effects in modern capital structure theory. *Journal of Reviews on Global Economics, 7*, 104–122.
25. Brusov, P., Filatova, T., Orehova, N., & Kulk, V. (2018b). A "Golden age" of the companies: Conditions of its existence. *Journal of Reviews on Global Economics, 7*, 88–103.
26. Brusov, P., Filatova, T., Orehova, N., & Kulk, V. (2018c). Rating methodology: New look and new horizons. *Journal of Reviews on Global Economics, 7*, 63–87.
27. Brusov, P., Filatova, T., Orehova, N., & Kulk, V. (2018d). Rating: New approach. *Journal of Reviews on Global Economics, 7*, 37–62.

Chapter 6
The Modigliani–Miller Theory with Arbitrary Frequency of Payment of Tax on Profit

6.1 Introduction

6.1.1 Literature Review

In their first paper Modigliani and Miller (1958) [1–4] consider the problem of capital cost and capital structure for perpetuity companies without any taxes. In the second article, Modigliani and Miller (1963) [1–3, 5] considered taxation of corporate profits, but did not take into account the presence in the economy of individual taxes of investors. Merton Miller (1977) [1–3, 6] has introduced the model, demonstrating impact of leverage on the company value with account of the corporate and individual taxes. To describe his model, we will enter the following legends: T_C—tax on corporate profits rate, T_S—the tax rate on income of an individual investor from his ownership by stock of corporation, T_D—tax rate on interest income from the provision of investor–individuals of credits to other investors and companies. Income from shares partly comes in the form of a dividend and, in part, as capital profits, so that T_S is a weighted average value of effective rates of tax on dividends and capital profits on shares, while the income from the provision of loans usually comes in the form of the interests. The last are usually taxed at a higher rate. In the light of the individual taxes, and with the same assumptions that have been made for Modigliani–Miller models previously, the financially independent company value can be determined as follows:

$$V_U = \frac{\text{EBIT}(1 - T_C)(1 - T_S)}{k_0}. \tag{6.1}$$

A term $(1 - T_S)$ allows to take into account the individual taxes in the formula. In this way, numerator indicates which part of the operating company's profit remains in the possession of the investors, after the company will pay taxes on their profits, and its shareholders then will pay individual taxes on income from stock ownership.

P. Brusov, T. Filatova, *Business Valuation and Company Value*, Contributions to Finance and Accounting, https://doi.org/10.1007/978-3-031-90847-7_6

Since individual taxes reduce profits, remaining in the disposal of investors, the last, at other things being equal circumstances, also reduce and an overall assessment of the financially independent company value.

Robert Hamada (1969) [1–3, 7] unites Capital Asset Pricing Model (CAPM) with Modigliani–Miller model taxation. As a result, he derived the following formula for calculation of the equity cost of financially dependent company, including both financial and business risk of company:

$$k_e = k_F + (k_M - k_F)b_U + (k_M - k_F)b_U \frac{D}{S}(1 - T), \tag{6.2}$$

where b_U is the β-coefficient of the company of the same group of business risk, that the company under consideration, but with zero financial leverage. The formula (6.2) represents the desired profitability of equity capital k_e as a sum of three components: risk-free profitability k_F, compensating to shareholders a temporary value of their money, premium for business risk $(k_M - k_F)b_U$, and premium for financial risk $(k_M - k_F)b_U \frac{D}{S}(1 - T)$.

If the company does not have borrowing ($D = 0$), the financial risk factor will be equal to zero (the third term is drawn to zero), and its owners will only receive the premium for business risk.

More general formula for the WACC, then famous Modigliani—Miller (MM) one has been derived and discussed by a few authors (Farber, A., R. Gillet, and A. Szafarz, 2006, Fernandez, P., 2007, Harris, R., and J. Pringle, Miles, J.A., and J.R. Ezzel, Peter MDeMarzo et al) [8–13]. It takes the following form (Eq. (18) in FGS, 2006 [8])

$$\mathrm{WACC} = k_0(1 - w_d T) - k_d \mathrm{t} w_d + k_{TS} \mathrm{t} w_d \tag{6.3}$$

where k_0, k_d and k_{TS} are the expected returns respectively on the unlevered firm, the debt and the tax shield, V is the value of the levered firm, VTS is the value of the tax shield, D is the value of the debt and TC is the corporate tax rate.

This formula is derived from the balance sheet identity and the definition of the weighted average cost of capital (for a similar presentation, see [14]). It should therefore be verified at any point in time, whatever returns are annually or continuously compounded.

Although Eq. (6.3) is fully general, its practical applicability requires additional conditions. Indeed, when the WACC is constant over time, the value of a levered firm can be computed by discounting with the WACC the unlevered free cash-flows. Therefore, we pay a special attention to the special cases that make the WACC constant. The resulting particular formulas can also be found in textbooks (Brealey et al., 2006; Ross et al., 2005 [12]).

First, Modigliani and Miller (1963) [5] assume that the level of debt D is constant. Then, as the expected after-tax cash-flow of the unlevered firm is fixed, V_0 is also constant. By assumption, $k_{TS} = k_D$ and the value of the tax shield is TS = tD.

Therefore, the value of the firm V is a constant and the general WACC formula (6.15) simplifies to a constant WACC:

$$\text{WACC} = k_0(1 - w_d T) \tag{6.4}$$

But our opinion is that "classical" Modigliani–Miller (MM) theory, which coming from the suggestion that the expected returns on the debt k_d and the tax shield k_{TS} are equals (because both of them have debt nature), is much more reasonable and in our paper we modify namely "classical" Modigliani–Miller (MM) theory, which is still widely used in practice.

One of the most serious limitations of the Modigliani–Miller theory is the suggestion about perpetuity of the companies. In 2008, Brusov–Filatova–Orekhova [1–3, 15–25] have lifted up this limitation and have shown that the accounting of the finite life-time (or finite age) of the company leads to significant changes of all Modigliani–Miller results [4–6, 26]: capitalization of the company is changed, as well as the equity cost, k_e, and the weighted average cost of capital, WACC, in the presence of corporative taxes. Besides, a number of qualitatively new effects in corporate finance, obtained in Brusov–Filatova–Orekhova theory [1–3, 15–25], are absent in Modigliani–Miller theory [4–6, 26] and in its modifications [27–30].

The famous Brusov–Filatova–Orekhova formula for weighted average cost of capital, WACC, has the following form:

$$\frac{1 - (1 + \text{WACC})^{-n}}{\text{WACC}} = \frac{1 - (1 + k_0)^{-n}}{k_0[1 - \omega_d T(1 - (1 + k_d)^{-n})]}. \tag{6.5}$$

Here, S—the value of own (equity) capital of the company, $w_d = \frac{D}{D+S}$—the share of debt capital; $k_e, w_e = \frac{S}{D+S}$—the cost and the share of the equity of the company, $L = D/S$—financial leverage.

A lot of meaningful effects has been discovered within BFO theory: these effects are absent within MM theory. BFO theory has destroyed some main existing principles of financial management: among them trade off theory, which was considered as keystone of formation of optimal capital structure of the company during many decades and *bankruptcy of which has been proven* by Brusov–Filatova–Orekhova [1–3, 15–25].

6.1.2 Focus and Objective of the Chapter

The tax shield plays a crucial role in both main capital structure theories: Brusov–Filatova–Orekhova (BFO theory) and its perpetuity limit Modigliani–Miller theory. How it is formed influences the results of both theories. Both of them describes the case of annual payments of tax of profit, but in practice these payments are made more frequent: semiannually, quarterly, monthly. While a return is not required more than once a year, businesses may be responsible for filing estimated taxes based on profits earned. This requirement is dependent on showing a profit. For example, sole proprietors must file estimated taxes on profits quarterly, on the 15th day of April, June, September and January. In Russia tax on profit payments could be made annually, quarterly, monthly. In current paper we show that the account of frequency of payment of tax on profit within the Modigliani–Miller theory turns out to be very important and leads to significant consequences.

To study this problem we modify the Modigliani–Miller theory for the case of arbitrary frequency of payment of tax on profit and show that:

1. All Modigliani–Miller Theorems, statements and all formulas change.
2. All main financial indicators, such as the weighted average cost of capital, WACC, company value, V and equity cost, k_e depend on the frequency of tax on profit payments. This allows to company manage WACC, V, k_e etc. by choosing the number of payments of tax of profit p per year.
3. Increase the number of payments of tax of profit per year leads to decrease of the cost of attracting capital and increase of the company value.
4. Obtained results allow to company choose the number of payments of tax of profit per year, as many, as it is profitable to it (of course, within actual tax legislation).
5. In case of income tax payments more than once per year (at $p \neq 1$), as it takes place in practice, the weighted average cost of capital, WACC, company value, V and equity cost, k_e start depend on k_d, while in ordinary (classical) Modigliani–Miller theory all these values DO NOT depend on k_d.
6. The tilt angle of the curve of equity cost, k_e (L), decreases with the number of payments of tax of profit p, this modifies the dividend policy of the company, because the economically justified value of dividends is equal to equity cost.

6.2 Modification of the Modigliani–Miller Theory for the Case of Arbitrary Frequency of Payments of Tax on Profit

6.2.1 Tax Shield

Let us calculate first the tax shield in the Modigliani–Miller theory for the case of p payments of tax on profit per year:

$$\mathrm{TS} = \frac{k_d D t}{p(1+k_d)^{1/p}} + \frac{k_d D t}{p(1+k_d)^{2/p}} + \frac{k_d D t}{p(1+k_d)^{3/p}} + \ldots \tag{6.6}$$

TS represents a geometric progression with denominator: $q = \frac{1}{(1+k_d)^{\frac{1}{p}}}$.

Summing the progression, one gets:

$$\mathrm{TS} = \frac{k_\mathrm{d}Dt}{p(1+k_\mathrm{d})^{1/p}\left(1-(1+k_\mathrm{d})^{-1/p}\right)} = \frac{k_\mathrm{d}Dt}{p\left((1+k_\mathrm{d})^{1/p}-1\right)}. \tag{6.7}$$

Note, that in case of classical Modigliani–Miller theory:

$$\mathrm{TS} = Dt. \tag{6.8}$$

It is easy to get this result from (6.7), putting the frequency of payments of tax on profit $p = 1$.

6.2.2 The Weighted Average Cost of Capital, WACC

Let us now calculate the weighted average cost of capital, WACC, in the Modigliani–Miller theory for the case of p payments of tax on profit per year.

From the Modigliani–Miller theorem we have for the company value V

$$V = V_0 + \mathrm{TS}. \tag{6.9}$$

Putting expression (6.7) for TS, we get:

$$V = V_0 + \frac{k_\mathrm{d}Dt}{p\left((1+k_\mathrm{d})^{1/p}-1\right)}. \tag{6.10}$$

Substituting $D = w_\mathrm{d}V$, one gets:

$$V\left(1 - \frac{k_\mathrm{d}wtd}{p\left((1+k_\mathrm{d})^{1/p}-1\right)}\right) = V_0. \tag{6.11}$$

Accounting, that:

$$V = \frac{\mathrm{CF}}{\mathrm{WACC}};\, V_0 = \frac{\mathrm{CF}}{k_0}, \tag{6.12}$$

we have:

$$\frac{\mathrm{CF}}{\mathrm{WACC}} \cdot \left(1 - \frac{k_{\mathrm{d}} w t \mathrm{d}}{p\left((1+k_{\mathrm{d}})^{1/p} - 1\right)}\right) = \frac{\mathrm{CF}}{k_0}. \tag{6.13}$$

From here we arrive to the expression for the weighted average cost of capital, WACC, for the case of p payments of tax on profit per year:

$$\mathrm{WACC} = k_0 \cdot \left(1 - \frac{k_{\mathrm{d}} w_{\mathrm{d}} t}{p\left((1+k_{\mathrm{d}})^{1/p} - 1\right)}\right). \tag{6.14}$$

Note, that in case of classical Modigliani–Miller theory:

$$\mathrm{WACC} = k_0 \cdot (1 - w_{\mathrm{d}} t). \tag{6.15}$$

It is easy to get this result from (6.14), putting the frequency of payments of tax on profit per year $p = 1$.

6.2.3 Company Value, V

Let us now calculate the company value, V, in the Modigliani–Miller theory for the case of p payments of tax on profit per year.

Putting WACC value into first formula of (6.12) one gets for company value, V

$$V = \frac{\mathrm{CF}}{\mathrm{WACC}} = \frac{\mathrm{CF}}{k_0 \cdot \left(1 - \frac{k_{\mathrm{d}} w_{\mathrm{d}} t}{p\left((1+k_{\mathrm{d}})^{1/p} - 1\right)}\right)}. \tag{6.16}$$

In case of "classical" Modigliani–Miller theory one has

$$V = \frac{\mathrm{CF}}{\mathrm{WACC}} = \frac{\mathrm{CF}}{k_0 \cdot (1 - w_{\mathrm{d}} t)}. \tag{6.17}$$

It is easy to get this result from (6.16), putting the frequency of payments of tax on profit per year $p = 1$.

6.2.4 Equity Cost, k_e

Let us now calculate the equity cost, k_{e}, in the Modigliani–Miller theory for the case of p payments of tax on profit per year.

General expression for the weighted average cost of capital, WACC, upon definition has the following form

$$\mathrm{WACC} = k_e w_e + k_d w_d (1 - t). \tag{6.18}$$

We get from (6.18) the following formula for the equity cost, k_e,

$$k_e = \frac{\mathrm{WACC}}{w_e} - \frac{k_d w_d (1 - t)}{w_e}. \tag{6.19}$$

Accounting from (6.14) the expression for the weighted average cost of capital, WACC,

$$\mathrm{WACC} = k_0 \cdot \left(1 - \frac{k_d w_d t}{p\left((1 + k_d)^{1/p} - 1\right)}\right), \tag{6.20}$$

one gets:

$$\begin{aligned} k_e &= \frac{k_0}{w_e} \cdot \left(1 - \frac{k_d w_d t}{p\left((1 + k_d)^{1/p} - 1\right)}\right) - \frac{k_d w_d (1 - t)}{w_e} \\ &= k_0 \cdot \left(1 + L - \frac{k_d L t}{p\left((1 + k_d)^{1/p} - 1\right)}\right) - k_d L (1 - t) \\ &= k_0 + L\left(k_0 - k_d (1 - t) - \frac{k_0 k_d t}{p\left((1 + k_d)^{1/p} - 1\right)}\right). \end{aligned} \tag{6.21}$$

Thus, we have the following formula for the equity cost, k_e, in case of arbitrary frequency of payment of tax on profit:

$$k_e = k_0 + L\left(k_0 - k_d (1 - t) - \frac{k_0 k_d t}{p\left((1 + k_d)^{1/p} - 1\right)}\right) \tag{6.22}$$

From this formula it is easy to get "classical" Modigliani–Miller expression for the equity cost, k_e, in case of annual payment of tax on profit, putting $p = 1$

$$k_e = k_0 + L(k_0 - k_d)(1 - t) \tag{6.23}$$

Thus, from Sects. 6.2.1–6.2.4 we come to conclusion, that in case of income tax payments more than once per year (at $p \neq 1$), as it takes place in practice, the weighted average cost of capital, WACC, company value, V and equity cost, k_e start depend on k_d, while in ordinary (classical) Modigliani–Miller theory all these values DO NOT depend on k_d.

In subsequent paragraphs (6.3–6.5) we will study numerically with use of Microsoft Excel the dependence of the main financial indicators of the company (the weighted average cost of capital, WACC, the company value, V, the cost of equity, k_d, on leverage level L for cases of practical interest with respect to equity and debt costs and frequency of payment of tax on profit p.

6.3 Dependence of the Weighted Average Cost of Capital, WACC, on Leverage Level L

6.3.1 Dependence of the Weighted Average Cost of Capital, WACC, on Leverage Level L *at Different* p *and Fixed* k_d

Let us start from the study numerically with use of Microsoft Excel of the dependence of the weighted average cost of capital, WACC, on leverage level L at different p and fixed k_d, using obtained by us formula (6.14)

$$\mathrm{WACC} = k_0 \cdot \left(1 - \frac{k_d w_d t}{p\left((1 + k_d)^{1/p} - 1\right)}\right).$$

Table 6.1 Dependence of WACC on leverage level L at $k_0 = 0.2$, $k_d = 0.16$ and $p = 1$

L	T	k_0	k_d	p	WACC
0	0.2	0.2	0.16	1	0.200000
1	0.2	0.2	0.16	1	0.180000
2	0.2	0.2	0.16	1	0.173333
3	0.2	0.2	0.16	1	0.170000
4	0.2	0.2	0.16	1	0.168000
5	0.2	0.2	0.16	1	0.166667
6	0.2	0.2	0.16	1	0.165714
7	0.2	0.2	0.16	1	0.165000
8	0.2	0.2	0.16	1	0.164444
9	0.2	0.2	0.16	1	0.164000
10	0.2	0.2	0.16	1	0.163636

Table 6.2 Dependence of WACC on leverage level L at $k_0 = 0.2$, $k_d = 0.16$ and $p = 2$

L	T	k_0	k_d	p	WACC
0	0.2	0.2	0.16	2	0.200000
1	0.2	0.2	0.16	2	0.179230
2	0.2	0.2	0.16	2	0.172306
3	0.2	0.2	0.16	2	0.168845
4	0.2	0.2	0.16	2	0.166767
5	0.2	0.2	0.16	2	0.165383
6	0.2	0.2	0.16	2	0.164394
7	0.2	0.2	0.16	2	0.163652
8	0.2	0.2	0.16	2	0.163075
9	0.2	0.2	0.16	2	0.162613
10	0.2	0.2	0.16	2	0.162236

Table 6.3 Dependence of WACC on leverage level L at $k_0 = 0.2$, $k_d = 0.16$ and $p = 4$

L	T	k_0	k_d	p	WACC
0	0.2	0.2	0.16	4	0.200000
1	0.2	0.2	0.16	4	0.178837
2	0.2	0.2	0.16	4	0.171783
3	0.2	0.2	0.16	4	0.168256
4	0.2	0.2	0.16	4	0.166139
5	0.2	0.2	0.16	4	0.164728
6	0.2	0.2	0.16	4	0.163721
7	0.2	0.2	0.16	4	0.162965
8	0.2	0.2	0.16	4	0.162377
9	0.2	0.2	0.16	4	0.161907
10	0.2	0.2	0.16	4	0.161522

Table 6.4 Dependence of WACC on leverage level L at $k_0 = 0.2$, $k_d = 0.16$ and $p = 12$

L	T	k_0	k_d	p	WACC
0	0.2	0.2	0.16	12	0.200000
1	0.2	0.2	0.16	12	0.178573
2	0.2	0.2	0.16	12	0.171430
3	0.2	0.2	0.16	12	0.167859
4	0.2	0.2	0.16	12	0.165716
5	0.2	0.2	0.16	12	0.164288
6	0.2	0.2	0.16	12	0.163267
7	0.2	0.2	0.16	12	0.162502
8	0.2	0.2	0.16	12	0.161907
9	0.2	0.2	0.16	12	0.161431
10	0.2	0.2	0.16	12	0.161041

From Tables 6.1, 6.2, 6.3, 6.4 and Figs. 6.1, 6.2 it is seen, that the weighted average cost of capital, WACC, decreases with leverage level L at all values of frequency of payments of tax on profit p. With increase of p WACC decreases. As we'll see below, this will lead to increase of company value, V, with p.

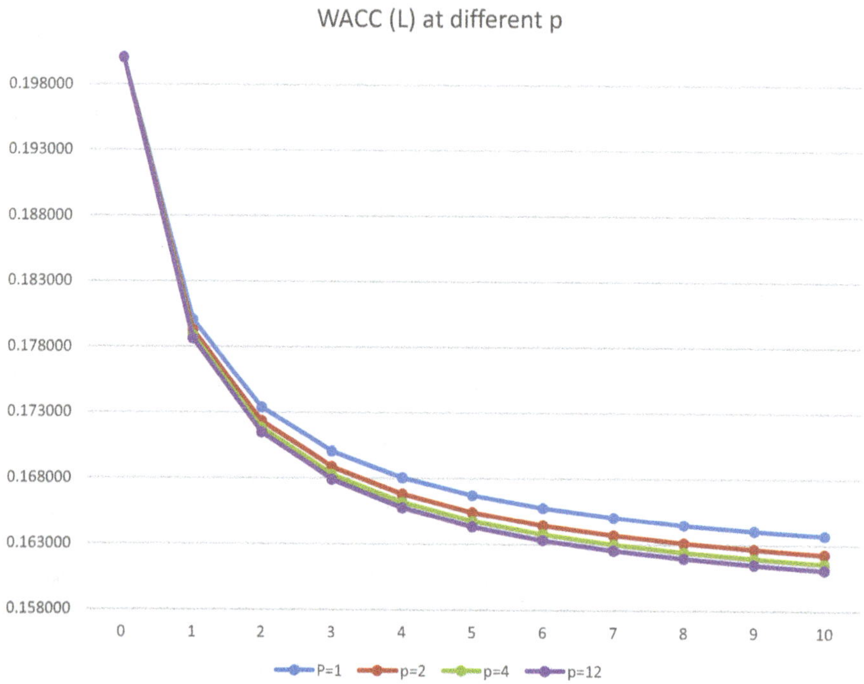

Fig. 6.1 Dependence of the weighted average cost of capital, WACC, on leverage level L at different p and fixed $k_d = 0.16$

To demonstrate, that above conclusion takes place for all values of equity cost (at zero leverage level L) k_0, we consider one more example: $k_0 = 0.22$ and different $p = 2; 4; 12$ (Fig. 6.3).

6.3.2 *Dependence of the Weighted Average Cost of Capital, WACC, on Leverage Level* L *at Different* k_d *and Fixed* p

In this paragraph we investigate the dependence of the weighted average cost of capital, WACC, on leverage level L at different values of debt cost k_d and fixed p.

From Tables 6.5, 6.6, 6.7, 6.8, 6.9, 6.10, 6.11, 6.12, 6.13 and Figs. 6.4, 6.5, 6.6, 6.7 it is seen, that in case of income tax payments more than once per year (at $p \neq 1$), as it takes place in practice, the weighted average cost of capital, WACC, (and as we'll see below, company value, V and equity cost, k_e start depend on debt cost, k_d, while in ordinary (classical) Modigliani–Miller theory all these values DO NOT depend on k_d). The weighted average cost of capital, WACC, decreases with leverage level L at all values of debt cost k_d and fixed frequency of payments of tax on profit p. With increase of k_d WACC decreases. This is connected to the fact

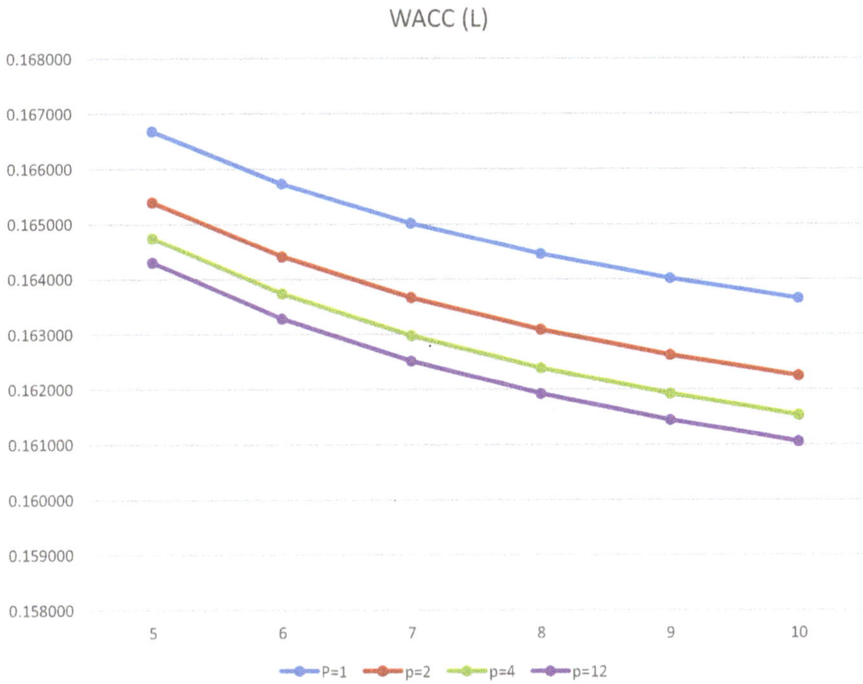

Fig. 6.2 Dependence of the weighted average cost of capital, WACC, on leverage level L at different p and fixed $k_d = 0.16$ (larger scale)

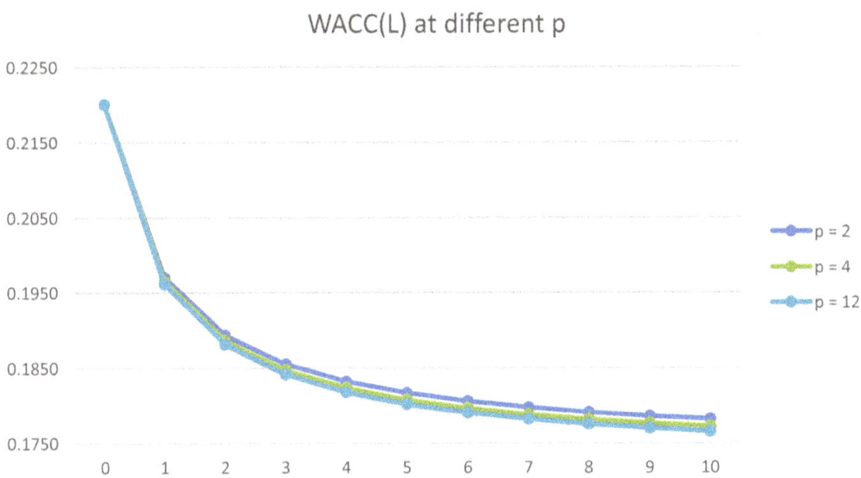

Fig. 6.3 Dependence of the weighted average cost of capital, WACC, on leverage level L at different p and fixed $k_0 = 0.22$

Table 6.5 Dependence of WACC on leverage level L at $k_0 = 0.22$ and different $p = 2$; 4; 12

L	WACC		
	$p = 2$	$p = 4$	$p = 12$
0	0.2200	0.2200	0.2200
1	0.1970	0.1964	0.1960
2	0.1893	0.1886	0.1881
3	0.1854	0.1846	0.1841
4	0.1831	0.1823	0.1817
5	0.1816	0.1807	0.1801
6	0.1805	0.1796	0.1789
7	0.1797	0.1787	0.1781
8	0.1790	0.1781	0.1774
9	0.1785	0.1775	0.1769
10	0.1781	0.1771	0.1765

Table 6.6 Dependence of WACC on leverage level L at $k_0 = 0.2$, $k_d = 0.18$ and $p = 1$

L	T	k_0	k_d	p	WACC
0	0.2	0.2	0.18	12	0.20000
1	0.2	0.2	0.18	12	0.17840
2	0.2	0.2	0.18	12	0.17120
3	0.2	0.2	0.18	12	0.16760
4	0.2	0.2	0.18	12	0.16544
5	0.2	0.2	0.18	12	0.16400
6	0.2	0.2	0.18	12	0.16297
7	0.2	0.2	0.18	12	0.16220
8	0.2	0.2	0.18	12	0.16160
9	0.2	0.2	0.18	12	0.16112
10	0.2	0.2	0.18	12	0.16073

Table 6.7 Dependence of WACC on leverage level L at $k_0 = 0.2$, $k_d = 0.16$ and $p = 12$

L	T	k_0	k_d	p	WACC
0	0.2	0.2	0.16	12	0.20000
1	0.2	0.2	0.16	12	0.17857
2	0.2	0.2	0.16	12	0.17143
3	0.2	0.2	0.16	12	0.16786
4	0.2	0.2	0.16	12	0.16572
5	0.2	0.2	0.16	12	0.16429
6	0.2	0.2	0.16	12	0.16327
7	0.2	0.2	0.16	12	0.16250
8	0.2	0.2	0.16	12	0.16191
9	0.2	0.2	0.16	12	0.16143
10	0.2	0.2	0.16	12	0.16104

Table 6.8 Dependence of WACC on leverage level L at $k_0 = 0.2$, $k_d = 0.14$ and $p = 12$

L	T	k_0	k_d	p	WACC
0	0.2	0.2	0.14	12	0.20000
1	0.2	0.2	0.14	12	0.17875
2	0.2	0.2	0.14	12	0.17166
3	0.2	0.2	0.14	12	0.16812
4	0.2	0.2	0.14	12	0.16600
5	0.2	0.2	0.14	12	0.16458
6	0.2	0.2	0.14	12	0.16357
7	0.2	0.2	0.14	12	0.16281
8	0.2	0.2	0.14	12	0.16222
9	0.2	0.2	0.14	12	0.16174
10	0.2	0.2	0.14	12	0.16136

Table 6.9 Dependence of WACC on leverage level L at $k_0 = 0.2$, $k_d = 0.12$ and $p = 12$

L	T	k_0	k_d	p	WACC
0	0.2	0.2	0.12	12	0.20000
1	0.2	0.2	0.12	12	0.17892
2	0.2	0.2	0.12	12	0.17190
3	0.2	0.2	0.12	12	0.16838
4	0.2	0.2	0.12	12	0.16628
5	0.2	0.2	0.12	12	0.16487
6	0.2	0.2	0.12	12	0.16387
7	0.2	0.2	0.12	12	0.16311
8	0.2	0.2	0.12	12	0.16253
9	0.2	0.2	0.12	12	0.16206
10	0.2	0.2	0.12	12	0.16168

Table 6.10 Dependence of WACC on leverage level L at $k_0 = 0.2$, $k_d = 0.18$ and $p = 4$

L	T	k_0	k_d	p	WACC
0	0.2	0.2	0.18	4	0.20000
1	0.2	0.2	0.18	4	0.17870
2	0.2	0.2	0.18	4	0.17160
3	0.2	0.2	0.18	4	0.16804
4	0.2	0.2	0.18	4	0.16591
5	0.2	0.2	0.18	4	0.16449
6	0.2	0.2	0.18	4	0.16348
7	0.2	0.2	0.18	4	0.16272
8	0.2	0.2	0.18	4	0.16213
9	0.2	0.2	0.18	4	0.16165
10	0.2	0.2	0.18	4	0.16127

that the increase of k_d increases the value of tax shield in case of income tax payments more than once per year (at $p \neq 1$). Note, that in "classical" Modigliani–Miller theory with one income tax payment per year all financial indicators don't depend on debt cost, k_d.

Table 6.11 Dependence of WACC on leverage level L at $k_0 = 0.2$, $k_d = 0.16$ and $p = 4$

L	T	k_0	k_d	p	WACC
0	0.2	0.2	0.16	4	0.20000
1	0.2	0.2	0.16	4	0.17884
2	0.2	0.2	0.16	4	0.17178
3	0.2	0.2	0.16	4	0.16826
4	0.2	0.2	0.16	4	0.16614
5	0.2	0.2	0.16	4	0.16473
6	0.2	0.2	0.16	4	0.16372
7	0.2	0.2	0.16	4	0.16296
8	0.2	0.2	0.16	4	0.16238
9	0.2	0.2	0.16	4	0.16191
10	0.2	0.2	0.16	4	0.16152

Table 6.12 Dependence of WACC on leverage level L at $k_0 = 0.2$, $k_d = 0.14$ and $p = 4$

L	T	k_0	k_d	p	WACC
0	0.2	0.2	0.14	4	0.20000
1	0.2	0.2	0.14	4	0.17898
2	0.2	0.2	0.14	4	0.17197
3	0.2	0.2	0.14	4	0.16847
4	0.2	0.2	0.14	4	0.16637
5	0.2	0.2	0.14	4	0.16496
6	0.2	0.2	0.14	4	0.16396
7	0.2	0.2	0.14	4	0.16321
8	0.2	0.2	0.14	4	0.16263
9	0.2	0.2	0.14	4	0.16216
10	0.2	0.2	0.14	4	0.16178

Table 6.13 Dependence of WACC on leverage level L at $k_0 = 0.2$, $k_d = 0.12$ and $p = 4$

L	T	k_0	k_d	p	WACC
0	0.2	0.2	0.12	4	0.20000
1	0.2	0.2	0.12	4	0.17912
2	0.2	0.2	0.12	4	0.17216
3	0.2	0.2	0.12	4	0.16868
4	0.2	0.2	0.12	4	0.16659
5	0.2	0.2	0.12	4	0.16520
6	0.2	0.2	0.12	4	0.16421
7	0.2	0.2	0.12	4	0.16346
8	0.2	0.2	0.12	4	0.16288
9	0.2	0.2	0.12	4	0.16242
10	0.2	0.2	0.12	4	0.16204

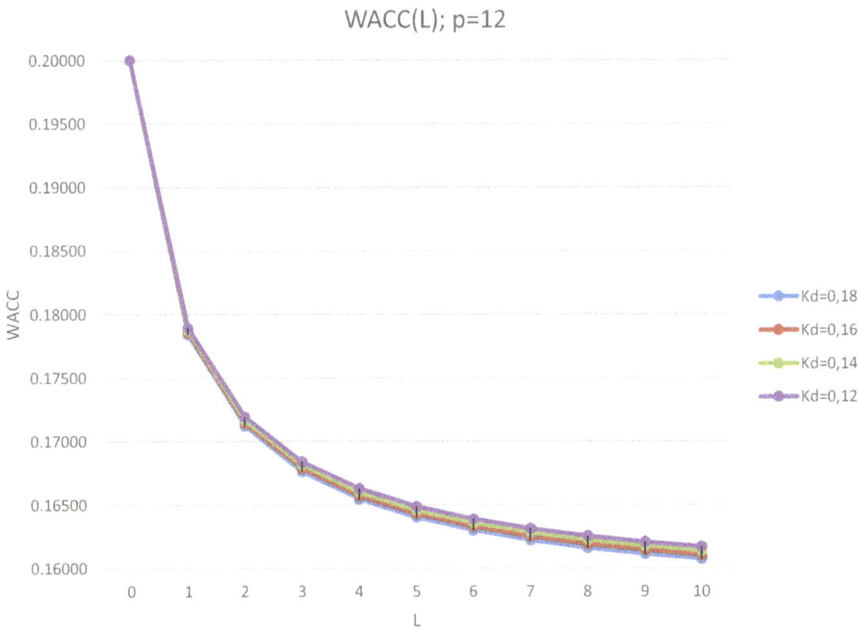

Fig. 6.4 Dependence of the weighted average cost of capital, WACC, on leverage level L at different k_d (12%; 14%; 16%; 18%) and fixed $p = 12$

6.4 Dependence of the Company Capitalization, V, on Leverage Level L

In this section we study the dependence of the company capitalization, V, on leverage level L for the following situations: (1) at different $p = 1;2;4;12$ and fixed $k_d = 0.16$; (2) for fixed p and at different values of debt cost $k_d = 0.18;0.16;0.14;0.12$.

6.4.1 Dependence of the Company Capitalization, V, on Leverage Level L at Different p and Fixed k_d

Obtained results on the dependence of the company capitalization, V, on leverage level L at different $p = 1;2;4;12$ and fixed $k_d = 0.16$ (Tables 6.14, 6.15, 6.16, 6.17 and Figs. 6.8, 6.9) show that the company capitalization, V, increases with p and reach maximum at $p = 12$. This means, that for company is more profitable to pay tax on profit as much more frequently as tax legislation allows.

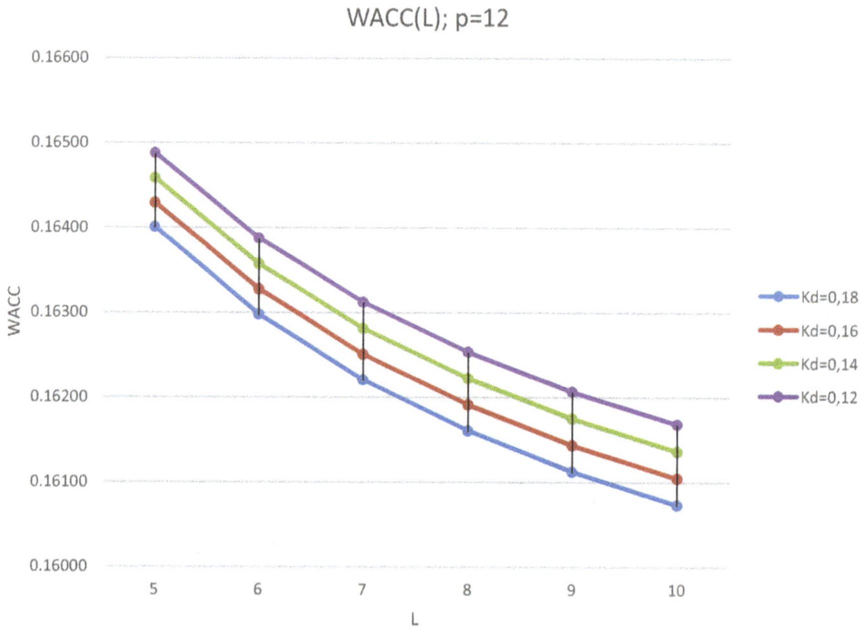

Fig. 6.5 Dependence of the weighted average cost of capital, WACC, on leverage level L at different k_d (12%; 14%; 16%; 18%) and fixed $p = 12$ (larger scale)

6.4.2 Dependence of the Company Capitalization, V*, on Leverage Level* L *at Different* k_d *and Fixed* p

Let us now study the dependence of the company capitalization, V, on leverage level L for fixed p and at different values of debt cost $k_d = 0.18;0.16;0.14;0.12$.

Obtained results on the dependence of the company capitalization, V, on leverage level L at fixed p (but for $p = 1;2;4;12$) and different $k_d = 12\%;\ 14\%;\ 16\%;\ 18\%$ (Tables 6.18, 6.19, 6.20, 6.21, 6.22, 6.23, 6.24, 6.25, 6.26, 6.27, 6.28, 6.29 and Figs. 6.10, 6.11, 6.12, 6.13, 6.14, 6.15) show that the company capitalization, V, increases with k_d at each value of fixed p and reach maximum at $k_d = 18\%$. This corresponds to decrease of the weighted average cost of capital, WACC, with k_d at fixed p, obtained by us in Sect. 6.3. This is connected to the fact that the increase of k_d increases the value of tax shield in case of income tax payments more than once per year (at $p \neq 1$). Note, that in "classical" Modigliani–Miller theory with one income tax payment per year all financial indicators, including company capitalization, V, don't depend on debt cost, k_d.

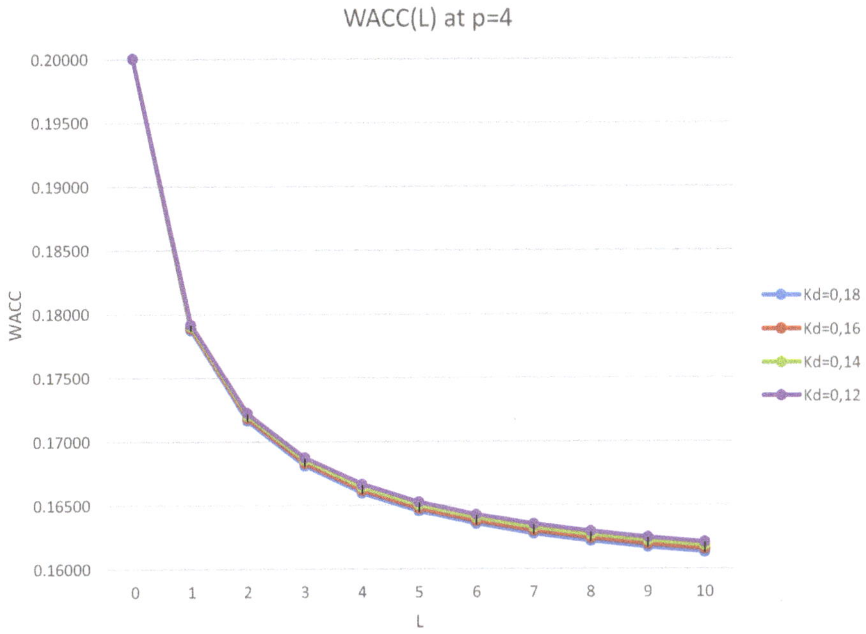

Fig. 6.6 Dependence of the weighted average cost of capital, WACC, on leverage level L at different k_d (12%; 14%; 16%; 18%) and fixed $p = 4$

6.5 Dependence of the Equity Cost, k_e, on Leverage Level L

In this section we study the dependence of the equity cost, k_e, on leverage level L for the following situations: (1) at different $p = 1;2;4;12$ and fixed $k_d = 0.16$; (2) for fixed p and at different values of debt cost $k_d = 0.18;0.16;0.14;0.12$.

6.5.1 Dependence of the Equity Cost, k$_e$*, on Leverage Level* L *at Different* p *and Fixed* K$_d$ *= 0.16*

Let us start from the study of the dependence of the equity cost, k_e, on leverage level L at different p and fixed $k_d = 0.16$.

Obtained results on the dependence of the equity cost, k_e, on leverage level L at fixed $k_d = 16\%$ and different $p = 1;2;4;12$ (Tables 6.30, 6.31, 6.32, 6.33 and Figs. 6.16, 6.17) show that the tilt angle of the curve of equity cost, k_e (L), decreases with the number of payments of tax of profit p, this modifies the dividend policy of the company, because the economically justified value of dividends is equal to equity cost. Note, that in "classical" Modigliani–Miller theory with one income

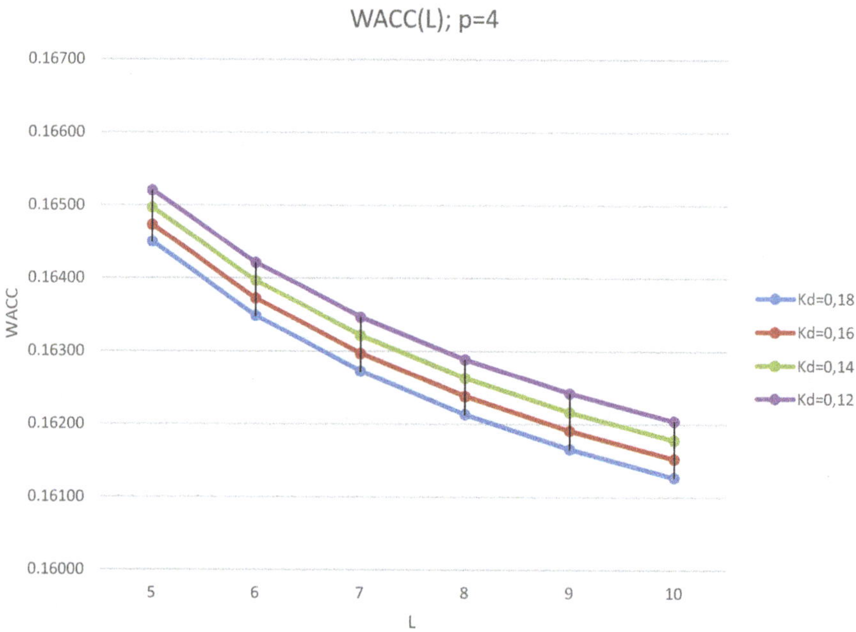

Fig. 6.7 Dependence of the weighted average cost of capital, WACC, on leverage level L at different k_d (12%; 14%; 16%; 18%) and fixed $p = 4$ (larger scale)

Table 6.14 Dependence of the company capitalization, V, on leverage level L at $p = 1$, $k_d = 0.16$

L	T	k_0	k_d	CF	p	V
0	0.2	0.2	0.16	1000	1	5000.00
1	0.2	0.2	0.16	1000	1	5555.56
2	0.2	0.2	0.16	1000	1	5769.23
3	0.2	0.2	0.16	1000	1	5882.35
4	0.2	0.2	0.16	1000	1	5952.38
5	0.2	0.2	0.16	1000	1	6000.00
6	0.2	0.2	0.16	1000	1	6034.48
7	0.2	0.2	0.16	1000	1	6060.61
8	0.2	0.2	0.16	1000	1	6081.08
9	0.2	0.2	0.16	1000	1	6097.56
10	0.2	0.2	0.16	1000	1	6111.11

tax payment per year all financial indicators, including the equity cost, k_e, don't depend on debt cost, k_d.

Table 6.15 Dependence of the company capitalization, V, on leverage level L at $p = 2$, $k_d = 0.16$

L	T	k_0	k_d	CF	p	V
0	0.2	0.2	0.16	1000	2	5000.00
1	0.2	0.2	0.16	1000	2	5579.43
2	0.2	0.2	0.16	1000	2	5803.62
3	0.2	0.2	0.16	1000	2	5922.61
4	0.2	0.2	0.16	1000	2	5996.37
5	0.2	0.2	0.16	1000	2	6046.58
6	0.2	0.2	0.16	1000	2	6082.96
7	0.2	0.2	0.16	1000	2	6110.53
8	0.2	0.2	0.16	1000	2	6132.15
9	0.2	0.2	0.16	1000	2	6149.55
10	0.2	0.2	0.16	1000	2	6163.87

Table 6.16 Dependence of the company capitalization, V, on leverage level L at $p = 4$, $k_d = 0.16$

L	T	k_0	k_d	CF	p	V
0	0.2	0.2	0.16	1000	4	5000.00
1	0.2	0.2	0.16	1000	4	5591.68
2	0.2	0.2	0.16	1000	4	5821.31
3	0.2	0.2	0.16	1000	4	5943.34
4	0.2	0.2	0.16	1000	4	6019.04
5	0.2	0.2	0.16	1000	4	6070.60
6	0.2	0.2	0.16	1000	4	6107.96
7	0.2	0.2	0.16	1000	4	6136.29
8	0.2	0.2	0.16	1000	4	6158.51
9	0.2	0.2	0.16	1000	4	6176.39
10	0.2	0.2	0.16	1000	4	6191.11

Table 6.17 Dependence of the company capitalization, V, on leverage level L at $p = 12$, $k_d = 0.16$

L	T	k_0	k_d	CF	p	V
0	0.2	0.2	0.16	1000	12	5000.00
1	0.2	0.2	0.16	1000	12	5599.96
2	0.2	0.2	0.16	1000	12	5833.28
3	0.2	0.2	0.16	1000	12	5957.38
4	0.2	0.2	0.16	1000	12	6034.41
5	0.2	0.2	0.16	1000	12	6086.88
6	0.2	0.2	0.16	1000	12	6124.92
7	0.2	0.2	0.16	1000	12	6153.77
8	0.2	0.2	0.16	1000	12	6176.39
9	0.2	0.2	0.16	1000	12	6194.61
10	0.2	0.2	0.16	1000	12	6209.59

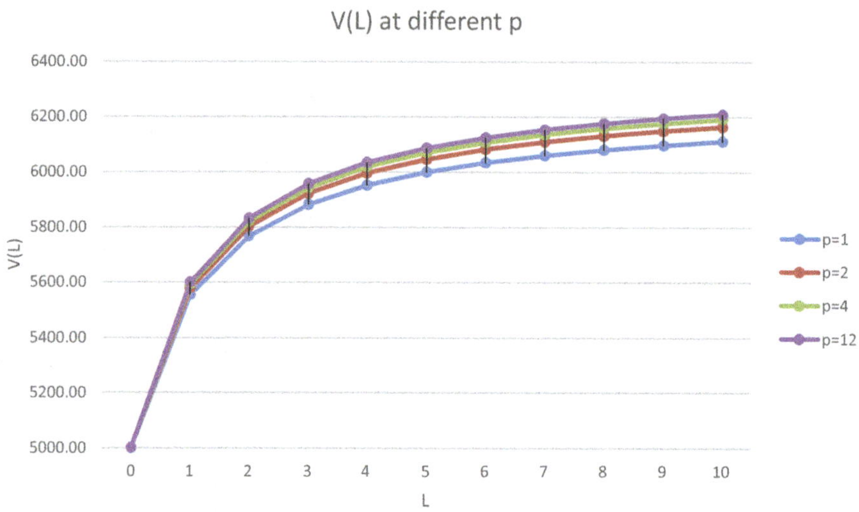

Fig. 6.8 Dependence of the company capitalization, V, on leverage level L at difference p (1; 2; 4; 12) and fixed $k_d = 0.16$

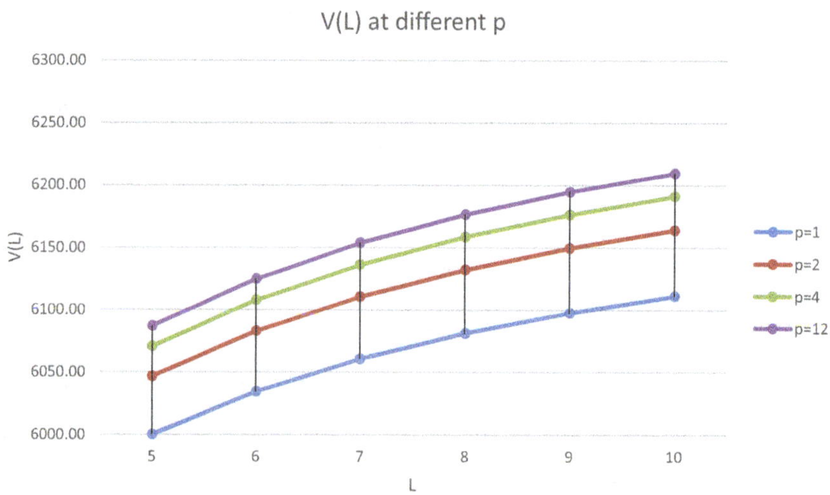

Fig. 6.9 Dependence of the company capitalization, V, on leverage level L at difference p (1; 2; 4; 12) and fixed $k_d = 0.16$ (larger scale)

Table 6.18 Dependence of the company capitalization, V, on leverage level L at $p = 2$ and $k_d = 0.18$

L	T	k_0	k_d	CF	p	V
0	0.2	0.2	0.18	1000	2	5000.00
1	0.2	0.2	0.18	1000	2	5582.31
2	0.2	0.2	0.18	1000	2	5807.78
3	0.2	0.2	0.18	1000	2	5927.48
4	0.2	0.2	0.18	1000	2	6001.70
5	0.2	0.2	0.18	1000	2	6052.22
6	0.2	0.2	0.18	1000	2	6088.83
7	0.2	0.2	0.18	1000	2	6116.58
8	0.2	0.2	0.18	1000	2	6138.34
9	0.2	0.2	0.18	1000	2	6155.85
10	0.2	0.2	0.18	1000	2	6170.26

Table 6.19 Dependence of the company capitalization, V, on leverage level L at $p = 2$ and $k_d = 0.16$

L	T	k_0	k_d	CF	p	V
0	0.2	0.2	0.16	1000	2	5000.00
1	0.2	0.2	0.16	1000	2	5579.43
2	0.2	0.2	0.16	1000	2	5803.62
3	0.2	0.2	0.16	1000	2	5922.61
4	0.2	0.2	0.16	1000	2	5996.37
5	0.2	0.2	0.16	1000	2	6046.58
6	0.2	0.2	0.16	1000	2	6082.96
7	0.2	0.2	0.16	1000	2	6110.53
8	0.2	0.2	0.16	1000	2	6132.15
9	0.2	0.2	0.16	1000	2	6149.55
10	0.2	0.2	0.16	1000	2	6163.87

Table 6.20 Dependence of the company capitalization, V, on leverage level L at $p = 2$ and $k_d = 0.14$

L	T	k_0	k_d	CF	p	V
0	0.2	0.2	0.14	1000	2	5000.00
1	0.2	0.2	0.14	1000	2	5576.53
2	0.2	0.2	0.14	1000	2	5799.44
3	0.2	0.2	0.14	1000	2	5917.71
4	0.2	0.2	0.14	1000	2	5991.01
5	0.2	0.2	0.14	1000	2	6040.90
6	0.2	0.2	0.14	1000	2	6077.05
7	0.2	0.2	0.14	1000	2	6104.44
8	0.2	0.2	0.14	1000	2	6125.92
9	0.2	0.2	0.14	1000	2	6143.21
10	0.2	0.2	0.14	1000	2	6157.43

Table 6.21 Dependence of the company capitalization, V, on leverage level L at $p = 2$ and $k_d = 0.12$

L	T	k_0	k_d	CF	p	V
0	0.2	0.2	0.12	1000	2	5000.00
1	0.2	0.2	0.12	1000	2	5573.61
2	0.2	0.2	0.12	1000	2	5795.22
3	0.2	0.2	0.12	1000	2	5912.77
4	0.2	0.2	0.12	1000	2	5985.62
5	0.2	0.2	0.12	1000	2	6035.19
6	0.2	0.2	0.12	1000	2	6071.10
7	0.2	0.2	0.12	1000	2	6098.31
8	0.2	0.2	0.12	1000	2	6119.65
9	0.2	0.2	0.12	1000	2	6136.83
10	0.2	0.2	0.12	1000	2	6150.96

Table 6.22 Dependence of the company capitalization, V, on leverage level L at $k_d = 0.18$ and fixed $p = 4$

L	T	k_0	k_d	CF	p	V
0	0.2	0.2	0.18	1000	4	5000.00
1	0.2	0.2	0.18	1000	4	5596.08
2	0.2	0.2	0.18	1000	4	5827.66
3	0.2	0.2	0.18	1000	4	5950.79
4	0.2	0.2	0.18	1000	4	6027.20
5	0.2	0.2	0.18	1000	4	6079.24
6	0.2	0.2	0.18	1000	4	6116.97
7	0.2	0.2	0.18	1000	4	6145.57
8	0.2	0.2	0.18	1000	4	6168.00
9	0.2	0.2	0.18	1000	4	6186.06
10	0.2	0.2	0.18	1000	4	6200.92

Table 6.23 Dependence of the company capitalization, V, on leverage level L at $k_d = 0.16$ and fixed $p = 4$

L	T	k_0	k_d	CF	p	V
0	0.2	0.2	0.16	1000	4	5000.00
1	0.2	0.2	0.16	1000	4	5591.68
2	0.2	0.2	0.16	1000	4	5821.31
3	0.2	0.2	0.16	1000	4	5943.34
4	0.2	0.2	0.16	1000	4	6019.04
5	0.2	0.2	0.16	1000	4	6070.60
6	0.2	0.2	0.16	1000	4	6107.96
7	0.2	0.2	0.16	1000	4	6136.29
8	0.2	0.2	0.16	1000	4	6158.51
9	0.2	0.2	0.16	1000	4	6176.39
10	0.2	0.2	0.16	1000	4	6191.11

Table 6.24 Dependence of the company capitalization, V, on leverage level L at $k_d = 0.14$ and fixed $p = 4$

L	T	k_0	k_d	CF	p	V
0	0.2	0.2	0.14	1000	4	5000.00
1	0.2	0.2	0.14	1000	4	5587.26
2	0.2	0.2	0.14	1000	4	5814.92
3	0.2	0.2	0.14	1000	4	5935.85
4	0.2	0.2	0.14	1000	4	6010.85
5	0.2	0.2	0.14	1000	4	6061.91
6	0.2	0.2	0.14	1000	4	6098.92
7	0.2	0.2	0.14	1000	4	6126.98
8	0.2	0.2	0.14	1000	4	6148.98
9	0.2	0.2	0.14	1000	4	6166.69
10	0.2	0.2	0.14	1000	4	6181.26

Table 6.25 Dependence of the company capitalization, V, on leverage level L at different $k_d = 0.12$ and fixed $p = 4$

L	T	k_0	k_d	CF	p	V
0	0.2	0.2	0.12	1000	4	5000.00
1	0.2	0.2	0.12	1000	4	5582.81
2	0.2	0.2	0.12	1000	4	5808.49
3	0.2	0.2	0.12	1000	4	5928.32
4	0.2	0.2	0.12	1000	4	6002.62
5	0.2	0.2	0.12	1000	4	6053.19
6	0.2	0.2	0.12	1000	4	6089.84
7	0.2	0.2	0.12	1000	4	6117.62
8	0.2	0.2	0.12	1000	4	6139.41
9	0.2	0.2	0.12	1000	4	6156.94
10	0.2	0.2	0.12	1000	4	6171.37

Table 6.26 Dependence of the company capitalization, V, on leverage level L at $k_d = 0.18$ and fixed $p = 12$

L	T	k_0	k_d	CF	p	V
0	0.2	0.2	0.18	1000	12	5000.00
1	0.2	0.2	0.18	1000	12	5605.40
2	0.2	0.2	0.18	1000	12	5841.15
3	0.2	0.2	0.18	1000	12	5966.63
4	0.2	0.2	0.18	1000	12	6044.53
5	0.2	0.2	0.18	1000	12	6097.61
6	0.2	0.2	0.18	1000	12	6136.09
7	0.2	0.2	0.18	1000	12	6165.28
8	0.2	0.2	0.18	1000	12	6188.17
9	0.2	0.2	0.18	1000	12	6206.60
10	0.2	0.2	0.18	1000	12	6221.77

Table 6.27 Dependence of the company capitalization, V, on leverage level L at $k_d = 0.16$ and fixed $p = 12$

L	T	k_0	k_d	CF	p	V
0	0.2	0.2	0.16	1000	12	5000.00
1	0.2	0.2	0.16	1000	12	5599.96
2	0.2	0.2	0.16	1000	12	5833.28
3	0.2	0.2	0.16	1000	12	5957.38
4	0.2	0.2	0.16	1000	12	6034.41
5	0.2	0.2	0.16	1000	12	6086.88
6	0.2	0.2	0.16	1000	12	6124.92
7	0.2	0.2	0.16	1000	12	6153.77
8	0.2	0.2	0.16	1000	12	6176.39
9	0.2	0.2	0.16	1000	12	6194.61
10	0.2	0.2	0.16	1000	12	6209.59

Table 6.28 Dependence of the company capitalization, V, on leverage level L at $k_d = 0.14$ and fixed $p = 12$

L	T	k_0	k_d	CF	p	V
0	0.2	0.2	0.14	1000	12	5000.00
1	0.2	0.2	0.14	1000	12	5594.50
2	0.2	0.2	0.14	1000	12	5825.38
3	0.2	0.2	0.14	1000	12	5948.11
4	0.2	0.2	0.14	1000	12	6024.27
5	0.2	0.2	0.14	1000	12	6076.13
6	0.2	0.2	0.14	1000	12	6113.73
7	0.2	0.2	0.14	1000	12	6142.23
8	0.2	0.2	0.14	1000	12	6164.58
9	0.2	0.2	0.14	1000	12	6182.59
10	0.2	0.2	0.14	1000	12	6197.39

Table 6.29 Dependence of the company capitalization, V, on leverage level L at $k_d = 0.12$ and fixed $p = 12$

L	T	k_0	k_d	CF	p	V
0	0.2	0.2	0.12	1000	12	5000.00
1	0.2	0.2	0.12	1000	12	5589.01
2	0.2	0.2	0.12	1000	12	5817.45
3	0.2	0.2	0.12	1000	12	5938.82
4	0.2	0.2	0.12	1000	12	6014.10
5	0.2	0.2	0.12	1000	12	6065.35
6	0.2	0.2	0.12	1000	12	6102.50
7	0.2	0.2	0.12	1000	12	6130.67
8	0.2	0.2	0.12	1000	12	6152.75
9	0.2	0.2	0.12	1000	12	6170.53
10	0.2	0.2	0.12	1000	12	6185.16

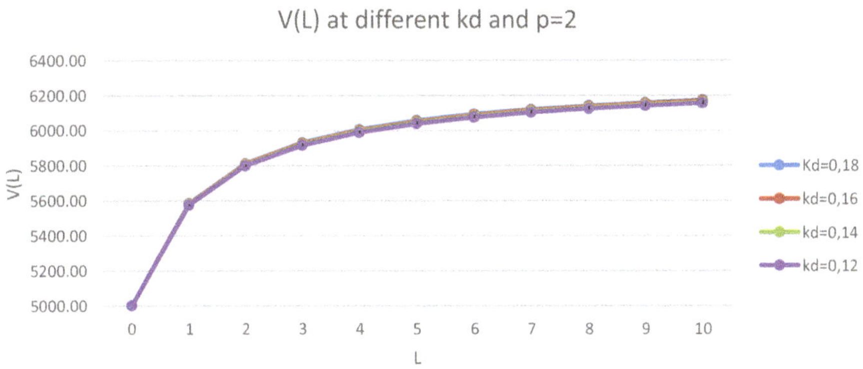

Fig. 6.10 Dependence of the company capitalization, V, on leverage level L at different k_d (12%; 14%; 16%; 18%) and fixed $p = 2$

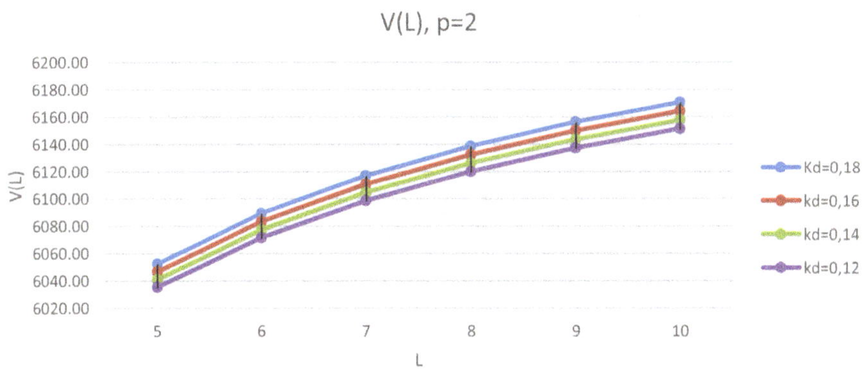

Fig. 6.11 Dependence of the company capitalization, V, on leverage level L at different k_d (12%; 14%; 16%; 18%) and fixed $p = 2$ (larger scale)

6.5.2 *Dependence of the Equity Cost,* k_e*, on Leverage Level* L *at Different* k_d *and Fixed* $p = 2$

In this section we study the dependence of the equity cost, k_e, on leverage level L at different $k_d = 18\%$; 16%; 14%; 12% and fixed number of income tax payments per year $p = 2$.

Conclusions Obtained results on the dependence of the equity cost, k_e, on leverage level L at fixed but different $p = 1;2;4;12$ and different values of debt cost k_d (Tables 6.34, 6.35, 6.36, 6.37, 6.38, 6.39, 6.40, 6.41, 6.42, 6.43, 6.44, 6.45 and Figs. 6.18, 6.19, 6.20) show that the tilt angle of the curve of equity cost, k_e (L), decreases with the increase of values of debt cost k_d: this modifies the dividend policy of the company, because the economically justified value of dividends is equal to equity cost. Note, that in "classical" Modigliani–Miller theory with one

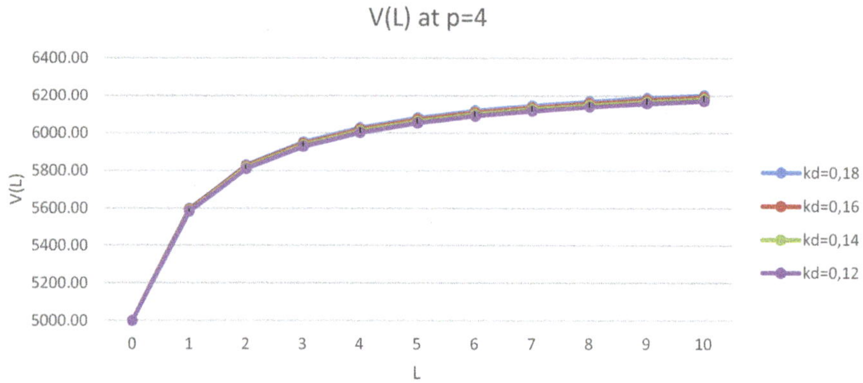

Fig. 6.12 Dependence of the company capitalization, V, on leverage level L at different k_d (12%; 14%; 16%; 18%) and fixed $p = 4$

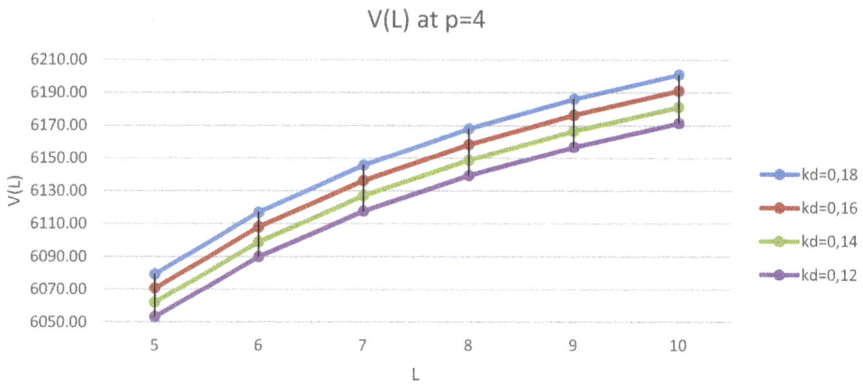

Fig. 6.13 Dependence of the company capitalization, V, on leverage level L at different k_d (12%; 14%; 16%; 18%) and fixed $p = 4$ (larger scale)

income tax payment per year all financial indicators, including the equity cost, k_e, don't depend on debt cost, k_d.

6.6 Conclusions

The tax shield plays a crucial role in both main capital structure theories: Brusov–Filatova–Orekhova (BFO theory) and its perpetuity limit Modigliani–Miller theory. The way it is formed affects the results of the theories. Both of them describes the

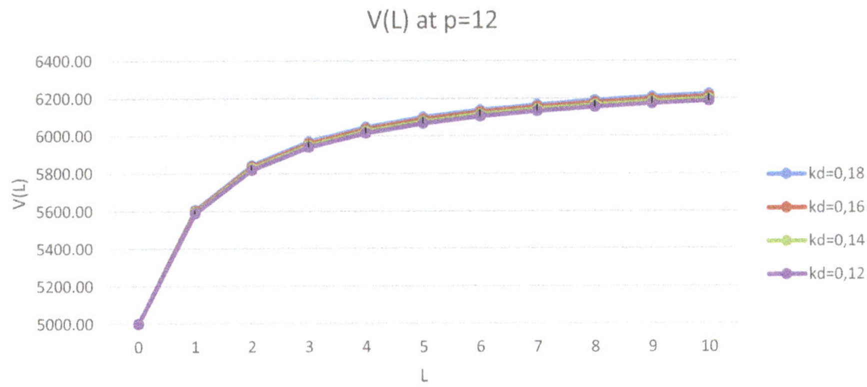

Fig. 6.14 Dependence of the company capitalization, V, on leverage level L at different k_d (12%; 14%; 16%; 18%) and fixed $p = 12$

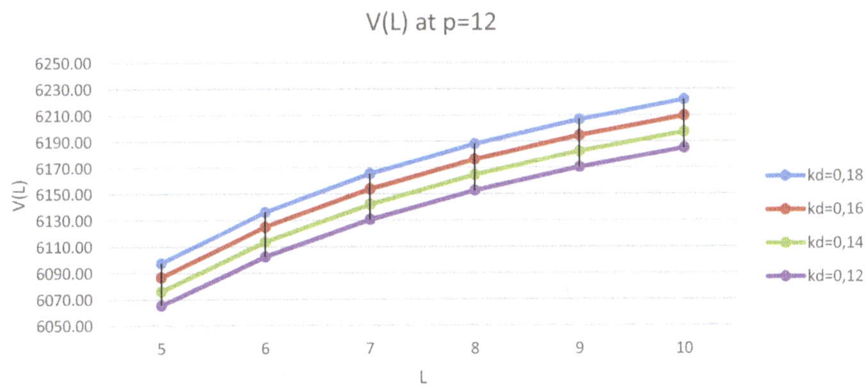

Fig. 6.15 Dependence of the company capitalization, V, on leverage level L at different k_d (12%; 14%; 16%; 18%) and fixed $p = 12$ (larger scale)

Table 6.30 Dependence of the equity cost, k_e, on leverage level L at $p = 1$ and fixed $k_d = 0.16$

L	T	k_0	k_d	p	k_e
0	0.2	0.2	0.16	1	0.2000
1	0.2	0.2	0.16	1	0.2320
2	0.2	0.2	0.16	1	0.2640
3	0.2	0.2	0.16	1	0.2960
4	0.2	0.2	0.16	1	0.3280
5	0.2	0.2	0.16	1	0.3600
6	0.2	0.2	0.16	1	0.3920
7	0.2	0.2	0.16	1	0.4240
8	0.2	0.2	0.16	1	0.4560
9	0.2	0.2	0.16	1	0.4880
10	0.2	0.2	0.16	1	0.5200

Table 6.31 Dependence of the equity cost, k_e, on leverage level L at $p = 2$ and fixed $k_d = 0.16$

L	T	k_0	k_d	p	k_e
0	0.2	0.2	0.16	2	0.2000
1	0.2	0.2	0.16	2	0.2305
2	0.2	0.2	0.16	2	0.2609
3	0.2	0.2	0.16	2	0.2914
4	0.2	0.2	0.16	2	0.3218
5	0.2	0.2	0.16	2	0.3523
6	0.2	0.2	0.16	2	0.3828
7	0.2	0.2	0.16	2	0.4132
8	0.2	0.2	0.16	2	0.4437
9	0.2	0.2	0.16	2	0.4741
10	0.2	0.2	0.16	2	0.5046

Table 6.32 Dependence of the equity cost, k_e, on leverage level L at $p = 4$ and fixed $k_d = 0.16$

L	T	k_0	k_d	p	k_e
0	0.2	0.2	0.16	4	0.2000
1	0.2	0.2	0.16	4	0.2297
2	0.2	0.2	0.16	4	0.2593
3	0.2	0.2	0.16	4	0.2890
4	0.2	0.2	0.16	4	0.3187
5	0.2	0.2	0.16	4	0.3484
6	0.2	0.2	0.16	4	0.3780
7	0.2	0.2	0.16	4	0.4077
8	0.2	0.2	0.16	4	0.4374
9	0.2	0.2	0.16	4	0.4671
10	0.2	0.2	0.16	4	0.4967

Table 6.33 Dependence of the equity cost, k_e, on leverage level L at $p = 12$ and fixed $k_d = 0.16$

L	T	k_0	k_d	p	k_e
0	0.2	0.2	0.16	12	0.2000
1	0.2	0.2	0.16	12	0.2291
2	0.2	0.2	0.16	12	0.2583
3	0.2	0.2	0.16	12	0.2874
4	0.2	0.2	0.16	12	0.3166
5	0.2	0.2	0.16	12	0.3457
6	0.2	0.2	0.16	12	0.3749
7	0.2	0.2	0.16	12	0.4040
8	0.2	0.2	0.16	12	0.4332
9	0.2	0.2	0.16	12	0.4623
10	0.2	0.2	0.16	12	0.4915

case of annual payments of tax of profit, but in practice these payments are made more frequent: semiannually, quarterly, monthly. While a return is not required more than once a year, businesses may be responsible for filing estimated taxes based on profits earned. This requirement is dependent on showing a profit. For example, sole

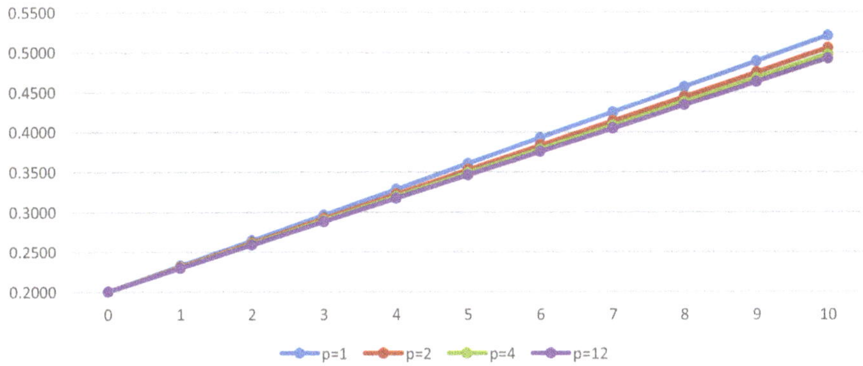

Fig. 6.16 Dependence of the equity cost, k_e, on leverage level L at different p (1; 2; 4; 12) and fixed $k_d = 0.16$

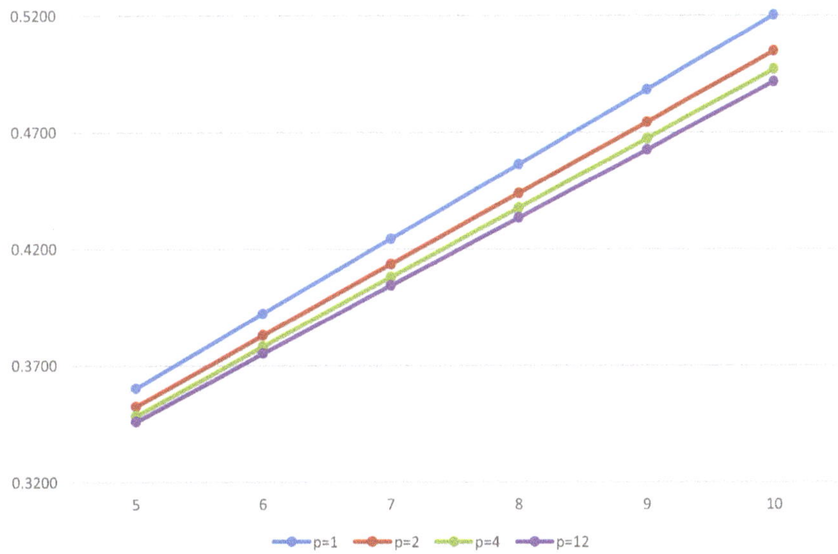

Fig. 6.17 Dependence of the equity cost, k_e, on leverage level L at different p (1; 2; 4; 12) and fixed $k_d = 0.16$ (lager scale)

proprietors must file estimated taxes on profits quarterly, on the 15th day of April, June, September and January. In Russia tax on profit payments could be made annually, quarterly, monthly. We show in current paper that the account of frequency of payment of tax on profit turns out to be very important and leads to significant consequences.

Table 6.34 Dependence of the equity cost, k_e, on leverage level L at $p = 2$ and fixed $k_d = 0.18$

L	T	k_0	k_d	p	k_e
0	0.2	0.2	0.18	2	0.2000
1	0.2	0.2	0.18	2	0.2143
2	0.2	0.2	0.18	2	0.2285
3	0.2	0.2	0.18	2	0.2428
4	0.2	0.2	0.18	2	0.2571
5	0.2	0.2	0.18	2	0.2714
6	0.2	0.2	0.18	2	0.2856
7	0.2	0.2	0.18	2	0.2999
8	0.2	0.2	0.18	2	0.3142
9	0.2	0.2	0.18	2	0.3285
10	0.2	0.2	0.18	2	0.3427

Table 6.35 Dependence of the equity cost, k_e, on leverage level L at $p = 2$ and fixed $k_d = 0.16$

L	T	k_0	k_d	p	k_e
0	0.2	0.2	0.16	2	0.2000
1	0.2	0.2	0.16	2	0.2305
2	0.2	0.2	0.16	2	0.2609
3	0.2	0.2	0.16	2	0.2914
4	0.2	0.2	0.16	2	0.3218
5	0.2	0.2	0.16	2	0.3523
6	0.2	0.2	0.16	2	0.3828
7	0.2	0.2	0.16	2	0.4132
8	0.2	0.2	0.16	2	0.4437
9	0.2	0.2	0.16	2	0.4741
10	0.2	0.2	0.16	2	0.5046

Table 6.36 Dependence of the equity cost, k_e, on leverage level L at $p = 2$ and fixed $k_d = 0.14$

L	T	k_0	k_d	p	k_e
0	0.2	0.2	0.14	2	0.2000
1	0.2	0.2	0.14	2	0.2466
2	0.2	0.2	0.14	2	0.2933
3	0.2	0.2	0.14	2	0.3399
4	0.2	0.2	0.14	2	0.3866
5	0.2	0.2	0.14	2	0.4332
6	0.2	0.2	0.14	2	0.4799
7	0.2	0.2	0.14	2	0.5265
8	0.2	0.2	0.14	2	0.5732
9	0.2	0.2	0.14	2	0.6198
10	0.2	0.2	0.14	2	0.6665

Table 6.37 Dependence of the equity cost, k_e, on leverage level L at $p = 2$ and fixed $k_d = 0.12$

L	T	k_0	k_d	p	k_e
0	0.2	0.2	0.12	2	0.2000
1	0.2	0.2	0.12	2	0.2628
2	0.2	0.2	0.12	2	0.3257
3	0.2	0.2	0.12	2	0.3885
4	0.2	0.2	0.12	2	0.4513
5	0.2	0.2	0.12	2	0.5142
6	0.2	0.2	0.12	2	0.5770
7	0.2	0.2	0.12	2	0.6398
8	0.2	0.2	0.12	2	0.7027
9	0.2	0.2	0.12	2	0.7655
10	0.2	0.2	0.12	2	0.8283

Table 6.38 Dependence of the equity cost, k_e, on leverage level L at $k_d = 0.18$ and fixed $p = 4$

L	T	k_0	k_d	p	k_e
0	0.2	0.2	0.18	4	0.2000
1	0.2	0.2	0.18	4	0.2134
2	0.2	0.2	0.18	4	0.2268
3	0.2	0.2	0.18	4	0.2402
4	0.2	0.2	0.18	4	0.2536
5	0.2	0.2	0.18	4	0.2670
6	0.2	0.2	0.18	4	0.2804
7	0.2	0.2	0.18	4	0.2938
8	0.2	0.2	0.18	4	0.3071
9	0.2	0.2	0.18	4	0.3205
10	0.2	0.2	0.18	4	0.3339

Table 6.39 Dependence of the equity cost, k_e, on leverage level L at $k_d = 0.16$ and fixed $p = 4$

L	T	k_0	k_d	p	k_e
0	0.2	0.2	0.16	4	0.2000
1	0.2	0.2	0.16	4	0.2297
2	0.2	0.2	0.16	4	0.2593
3	0.2	0.2	0.16	4	0.2890
4	0.2	0.2	0.16	4	0.3187
5	0.2	0.2	0.16	4	0.3484
6	0.2	0.2	0.16	4	0.3780
7	0.2	0.2	0.16	4	0.4077
8	0.2	0.2	0.16	4	0.4374
9	0.2	0.2	0.16	4	0.4671
10	0.2	0.2	0.16	4	0.4967

Table 6.40 Dependence of the equity cost, k_e, on leverage level L at $k_d = 0.14$ and fixed $p = 4$

L	T	k_0	k_d	p	k_e
0	0.2	0.2	0.14	4	0.2000
1	0.2	0.2	0.14	4	0.2460
2	0.2	0.2	0.14	4	0.2919
3	0.2	0.2	0.14	4	0.3379
4	0.2	0.2	0.14	4	0.3838
5	0.2	0.2	0.14	4	0.4298
6	0.2	0.2	0.14	4	0.4757
7	0.2	0.2	0.14	4	0.5217
8	0.2	0.2	0.14	4	0.5677
9	0.2	0.2	0.14	4	0.6136
10	0.2	0.2	0.14	4	0.6596

Table 6.41 Dependence of the equity cost, k_e, on leverage level L at $k_d = 0.12$ and fixed $p = 4$

L	T	k_0	k_d	p	k_e
0	0.2	0.2	0.12	4	0.2000
1	0.2	0.2	0.12	4	0.2622
2	0.2	0.2	0.12	4	0.3245
3	0.2	0.2	0.12	4	0.3867
4	0.2	0.2	0.12	4	0.4490
5	0.2	0.2	0.12	4	0.5112
6	0.2	0.2	0.12	4	0.5735
7	0.2	0.2	0.12	4	0.6357
8	0.2	0.2	0.12	4	0.6979
9	0.2	0.2	0.12	4	0.7602
10	0.2	0.2	0.12	4	0.8224

Table 6.42 Dependence of the equity cost, k_e, on leverage level L at $k_d = 0.18$ and fixed $p = 12$

L	T	k_0	k_d	p	k_e
0	0.2	0.2	0.18	12	0.2000
1	0.2	0.2	0.18	12	0.2128
2	0.2	0.2	0.18	12	0.2256
3	0.2	0.2	0.18	12	0.2384
4	0.2	0.2	0.18	12	0.2512
5	0.2	0.2	0.18	12	0.2640
6	0.2	0.2	0.18	12	0.2768
7	0.2	0.2	0.18	12	0.2896
8	0.2	0.2	0.18	12	0.3024
9	0.2	0.2	0.18	12	0.3152
10	0.2	0.2	0.18	12	0.3280

Table 6.43 Dependence of the equity cost, k_e, on leverage level L at $k_d = 0.16$ and fixed $p = 12$

L	T	k_0	k_d	p	k_e
0	0.2	0.2	0.16	12	0.2000
1	0.2	0.2	0.16	12	0.2291
2	0.2	0.2	0.16	12	0.2583
3	0.2	0.2	0.16	12	0.2874
4	0.2	0.2	0.16	12	0.3166
5	0.2	0.2	0.16	12	0.3457
6	0.2	0.2	0.16	12	0.3749
7	0.2	0.2	0.16	12	0.4040
8	0.2	0.2	0.16	12	0.4332
9	0.2	0.2	0.16	12	0.4623
10	0.2	0.2	0.16	12	0.4915

Table 6.44 Dependence of the equity cost, k_e, on leverage level L at $k_d = 0.12$ and fixed $p = 12$

L	T	k_0	k_d	p	k_e
0	0.2	0.2	0.14	12	0.2000
1	0.2	0.2	0.14	12	0.2455
2	0.2	0.2	0.14	12	0.2910
3	0.2	0.2	0.14	12	0.3365
4	0.2	0.2	0.14	12	0.3820
5	0.2	0.2	0.14	12	0.4275
6	0.2	0.2	0.14	12	0.4730
7	0.2	0.2	0.14	12	0.5185
8	0.2	0.2	0.14	12	0.5640
9	0.2	0.2	0.14	12	0.6094
10	0.2	0.2	0.14	12	0.6549

Table 6.45 Dependence of the equity cost, k_e, on leverage level L at $k_d = 0.12$ and fixed $p = 12$

L	T	k_0	k_d	p	k_e
0	0.2	0.2	0.12	12	0.2000
1	0.2	0.2	0.12	12	0.2618
2	0.2	0.2	0.12	12	0.3237
3	0.2	0.2	0.12	12	0.3855
4	0.2	0.2	0.12	12	0.4474
5	0.2	0.2	0.12	12	0.5092
6	0.2	0.2	0.12	12	0.5711
7	0.2	0.2	0.12	12	0.6329
8	0.2	0.2	0.12	12	0.6948
9	0.2	0.2	0.12	12	0.7566
10	0.2	0.2	0.12	12	0.8185

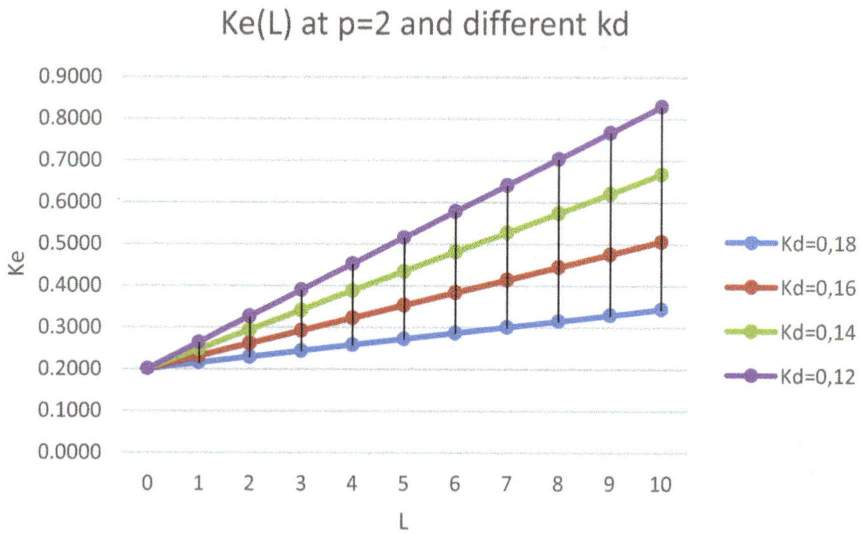

Fig. 6.18 Dependence of the equity cost, k_e, on leverage level L at different k_d (12%; 14%; 16%; 18%) and fixed $p = 2$

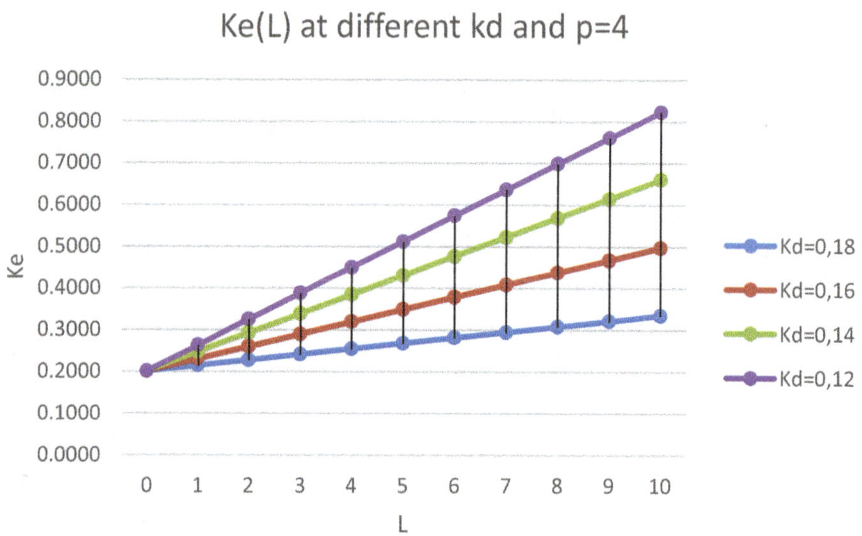

Fig. 6.19 Dependence of the equity cost, k_e, on leverage level L at different k_d (12%; 14%; 16%; 18%) and fixed $p = 4$

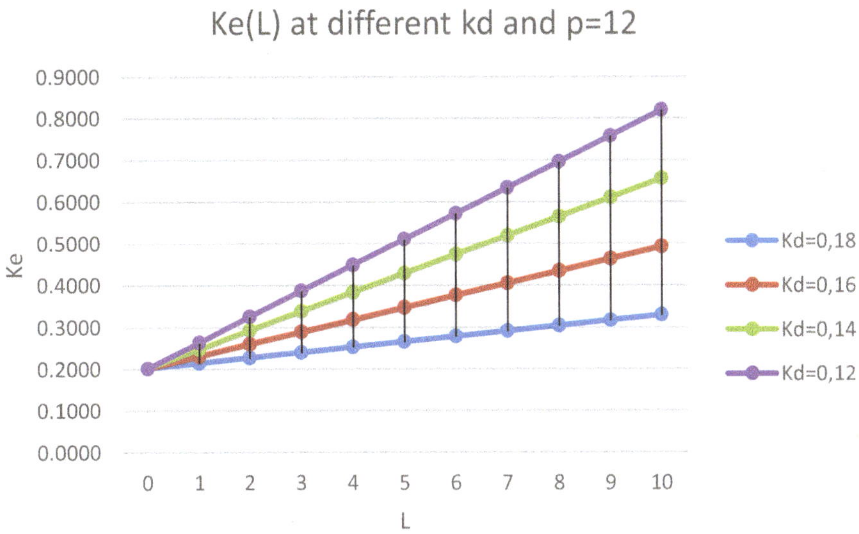

Fig. 6.20 Dependence of the equity cost, k_e, on leverage level L at different k_d (12%; 14%; 16%; 18%) and fixed $p = 12$

To study this problem we modify the Modigliani–Miller theory for the case of arbitrary frequency of payment of tax on profit and show that:

- all Modigliani–Miller statements and all formulas change,
- in case of income tax payments more than once per year (at $p \neq 1$), as it takes place in practice, the weighted average cost of capital, WACC, company value, V and equity cost, k_e start depend on k_d, while in ordinary (classical) Modigliani–Miller theory all these values DO NOT depend on k_d.
- obtained results allow to company choose the number of payments of tax of profit per year, as many, as it is profitable to it (of course, within actual tax legislation);
- the tilt angle of the curve of equity cost, $k_e(L)$, decreases with p, this modifies the dividend policy of the company, because the economically justified value of dividends is equal to equity cost.
- one more influence of taxes within Modigliani–Miller theory is found: under accounting the number of payments of the tax of profit influent on all main indicators of the company: the weighted average cost of capital, WACC, company value, V, equity cost, k_e etc. This allows to company manage WACC, V, k_e etc. by choosing the number of payments of tax of profit p.

It turns out that increase the number of payments of tax of profit per year leads to decrease of the cost of attracting capital and increase of the company value. More frequent payments of income tax are beneficial for both parties: for the company and for the tax regulator: for the company, this leads to an increase in the value of the company, and for the tax regulator, earlier payments are beneficial due to the time value of money.

References

1. Brusov, P., Filatova, T., & Orehova, N. (2020). *Ratings: Critical analysis and new approaches of quantitative and qualitative methodology*. Springer Nature Publishing, monograph. https://www.springer.com/de/book/9783030562427
2. Brusov, P., Filatova, T., Orehova, N., & Eskindarov, M. (2015). *Modern corporate finance, investments and taxation*. Springer International Publishing, 373 p. monograph. https://www.springer.com/gp/book/9783319147314
3. Brusov, P., Filatova, T., Orehova, N., & Eskindarov, M. (2018). *Modern corporate finance, investments, taxation and ratings*. Springer Nature Publishing, 571 p. monograph.
4. Modigliani, F., & Miller, M. (1958). The cost of capital, corporate finance, and the theory of investment. *The American Economic Review, 48*, 261–297.
5. Modigliani, F., & Miller, M. (1963). Corporate income taxes and the cost of capital: A correction. *The American Economic Review, 53*, 147–175.
6. Miller, M. (1977). Debt and taxes. *The Journal of Finance, 32*(2), 261–275. https://doi.org/10.1111/j.1540-6261.1977.tb03267.x
7. Hamada, R. (1969). Portfolio analysis, market equilibrium, and corporate finance. *Journal of Finance, 24*, 13–31.
8. Farber, A., Gillet, R., & Szafarz, A. (2006). A general formula for the WACC. *International Journal of Business, 11*(2), 211–218.
9. Fernandez, P. (2007). A general formula for the WACC: A comment. *International Journal of Business* (this issue).
10. Harris, R., & Pringle, J. (1985). Risk–Adjusted Discount Rates—Extension form the Average–Risk Case. *The Journal of Financial Research, 8*(3), 237–244.
11. Miles, J. A., & Ezzel, J. R. (1980). The weighted average cost of capital, perfect capital markets and project life: A clarification. *Journal of Financial and Quantitative Analysis, 15*(3), 719–730.
12. Ross, S., Westerfiel, R., & Jaffee, D. (2005). *Corporate finance* (7th ed.). McGraw Hill.
13. Taggart, R. A. (1991). Consistent valuation and cost of capital expressions with corporate and personal taxes. *Financial Management, 20*, 8–20.
14. Berk, J., & DeMarzo, P. (2007). *Corporate finance*. Pearson-Addison Wesley.
15. Brusov, P. N., Filatova, T. V., Orekhova, N. P., Kulik, V. L., & Weil, I. (2019). Ratings of the investment projects of arbitrary durations: New methodology. *Journal of Reviews on Global Economics, 8*, 437–448. https://doi.org/10.6000/1929-7092.2019.08.37
16. Brusov, P. (2018). Editorial: Introduction on special issue on the banking system and financial markets of Russia and other countries: Problems and prospects. *Journal of Reviews on Global Economics, 7*, i–vi.
17. Filatova, T. V., Brusov, P. N., Orekhova, N. P., & Kulik, V. L. (2018). Ratings of the long–term projects: New approach. *Journal of Reviews on Global Economics, 7*, 645–661. https://doi.org/10.6000/1929-7092.2018.07.59
18. Brusov, P. N., Filatova, T. V., Orekhova, N. P., & Kulik, V. L. (2018). Rating: New approach. *Journal of Reviews on Global Economics, 7*, 37–62. https://doi.org/10.6000/1929-7092.2018.07.05
19. Brusov, P. N., Filatova, T. V., Orekhova, N. P., & Kulik, V. L. (2018). A "golden age" of the companies: Conditions of its existence. *Journal of Reviews on Global Economics, 7*, 88–103. https://doi.org/10.6000/1929-7092.2018.07.07
20. Brusov, P. N., Filatova, T. V., Orekhova, N. P., & Kulik, V. L. (2018). Rating methodology: New look and new horizons. *Journal of Reviews on Global Economics, 7*, 63–87. https://doi.org/10.6000/1929-7092.2018.07.06
21. Brusov, P. N., Filatova, T. V., Orekhova, N. P., Kulik, V. L., & Weil, I. (2018). New meaningful effects in modern capital structure theory. *Journal of Reviews on Global Economics, 7*, 104–122. https://doi.org/10.6000/1929-7092.2018.07.08
22. Brusov, P. (2018). Editorial. *Journal of Reviews on Global Economics, 7*, i–v, SCOPUS.

23. Brusov, P. N., Filatova, T. V., Orekhova, N. P., Kulik, V., Weil, I., & Brailov, A. (2018). The impact of the central bank key rate and commercial banks credit rates on creating and maintaining of a favorable investment climate in the country. *Journal of Reviews on Global Economics*, (7), 360–376. https://doi.org/10.6000/1929-7092.2018.07.31
24. Brusov, P. N., Filatova, T. V., & Orekhova, N. P. (2018). *Modern corporate finance and investments*. Monograph. Knorus Publishing House, 517 p.
25. Filatova, T., Orehova, N., & Brusova, A. (2008). Weighted average cost of capital in the theory of Modigliani–Miller, modified for a finite life–time company. *Bull FU, 48*, 68–77.
26. Modigliani, F., & Miller, M. (1966). Some estimates of the cost of capital to the electric utility industry 1954–1957. *The American Economic Review, 56*, 333–391.
27. Myers, S. (2001). Capital structure. *The Journal of Economic Perspectives, 15*(2), 81–102.
28. Sethi, S. P., Derzko, N. A., & Lehoczky, J. P. (1991). A stochastic extension of the Miller-Modigliani framework. *Mathematical Finance, 1*(4), 57–76.
29. Peter, M., & DeMarzo. (1988). An extension of the Modigliani–Miller theorem to stochastic economies with incomplete markets and interdependent securities. *Journal of Economic Theory, 45*(2), 353–369.
30. Arzac, E. R., & Glosten, L. R. (2005). A reconsideration of tax shield valuation. *European Financial Management, 11*(4), 453–461 (INTERNATIONAL JOURNAL OF BUSINESS, 12(3), 2007 411).

Chapter 7
Modification of the Modigliani–Miller Theory for the Case of Advance Tax on Profit Payments

7.1 Introduction

The first serious study (and first quantitative study) of influence of capital structure of the company on its indicators of activities was the work by Nobel Prize Winners Modigliani and Miller. Their theory has a lot of limitations. One of the most important and serious assumptions of the Modigliani–Miller theory is that all financial flows as well as all companies are perpetuity. This limitation was lift out by Brusov–Filatova–Orekhova in 2008 [1], who have created BFO theory—modern theory of capital cost and capital structure for companies of arbitrary age [2–14].

Despite the fact that the Modigliani–Miller theory [2–17] is currently a particular case of the general theory of capital cost and capital structure—Brusov–Filatova–Orekhova (BFO) theory—it is still widely used at the West.

In current chapter we discuss one more limitation of Modigliani–Miller theory: a method of tax on profit payments. Modigliani–Miller theory accounts these payments as annuity–immediate while in practice these payments are making in advance and thus should be accounting as annuity–due.

We generalize the Modigliani–Miller theory for the case of advance tax on profit payments, which is widely used in practice, and show that this leads to some important consequences, which change seriously all the main statements by Modigliani and Miller. These consequences are as following: WACC starts depend on debt cost k_d, WACC turns out to be lower than in case of classical Modigliani–Miller theory and thus company capitalization becomes higher than in ordinary Modigliani–Miller theory. We show that equity dependence on leverage level L is still linear, but the tilt angle with respect to L-axis turns out to be smaller: this could lead to modification of the divident policy of the company.

Correct account of a method of tax on profit payments demonstrates that shortcomings of Modigliani–Miller theory are dipper, than everybody suggested: the underestimation of WACC really turns out to be bigger, as well as overestimation of the capitalization of the company. This means that systematic risks arising from the

P. Brusov, T. Filatova, *Business Valuation and Company Value*, Contributions to Finance and Accounting, https://doi.org/10.1007/978-3-031-90847-7_7

use of modified Modigliani–Miller theory (which is more correct than "classical' one) in practice is higher than it was suggested by the "classical" version of this theory.

7.2 Modified Modigliani–Miller Theory in Case of Advance Tax Payments

7.2.1 Tax Shield in Case of Advance Tax Payments

Modigliani–Miller theory accounts these payments as annuity–immediate while in practice these payments are making in advance and thus should be accounting as annuity–due.

To calculate tax shield *TS* in case of advance tax payments one should use annuity–due (Fig. 7.1).

$$\mathrm{TS} = k_{\mathrm{d}}Dt + \frac{k_{\mathrm{d}}Dt}{(1+k_{\mathrm{d}})} + \frac{k_{\mathrm{d}}Dt}{(1+k_{\mathrm{d}})^2} + \ldots = \frac{k_{\mathrm{d}}Dt}{\left(1-(1+k_{\mathrm{d}})^{-1}\right)} = Dt(1+k_{\mathrm{d}}) \quad (7.1)$$

This expression is different from the case of classical Modigliani–Miller theory (which used annuity–immediate) (Fig. 7.2).

$$\mathrm{TS} = \frac{k_{\mathrm{d}}Dt}{(1+k_{\mathrm{d}})} + \frac{k_{\mathrm{d}}Dt}{(1+k_{\mathrm{d}})^2} + \ldots = \frac{k_{\mathrm{d}}Dt}{(1+k_{\mathrm{d}})\left(1-(1+k_{\mathrm{d}})^{-1}\right)} = Dt \quad (7.2)$$

Thus in the former case tax shield TS is bigger by multiplier (1 + k_{d}). This is connected with the time value of money: money today is more expensive than money tomorrow due to the possibility of their alternative investment.

7.2.2 Capitalization of the Company

Modigliani–Miller theorem for capitalization of the company

Fig. 7.1 Annuity–Due

Fig. 7.2 Annuity–Immediate

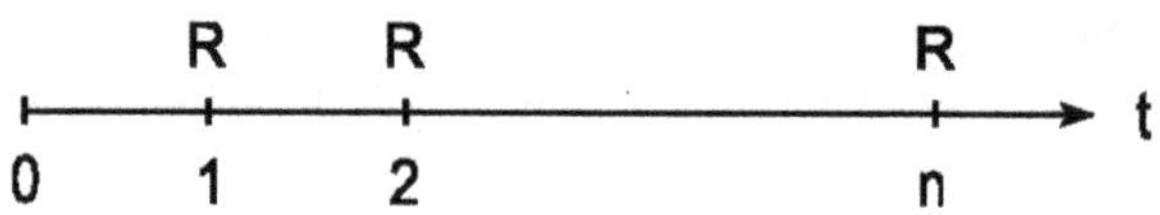

$$V = V_0 + \mathrm{TS} \tag{7.3}$$

takes the following form:

$$V = V_0 + Dt(1 + k_\mathrm{d}). \tag{7.4}$$

Thus, we arrive to the following statement, which modifies the original Modigliani and Miller.

The value of financially dependent company making tax on profit payments in advance is equal to the value of the company of the same risk group used no leverage, increased by the value of tax shield arising from financial leverage, and equal to the product of rate of corporate income tax t, the value of debt D and multiplier (1 + k_d*).*

Substituting $D = w_\mathrm{d} \cdot V$
one has

$$V\left(1 - w_\mathrm{d} t(1 + k_\mathrm{d})\right) = V_0$$

$$\frac{\mathrm{CF}}{\mathrm{WACC}} \cdot \left(1 - w_d t(1 + k_d)\right) = \frac{\mathrm{CF}}{k_0}$$

And for WACC we have the following formula:

$$\mathrm{WACC} = k_0\left(1 - w_\mathrm{d} t(1 + k_\mathrm{d})\right). \tag{7.5}$$

$$\text{At } L \rightarrow \qquad \mathrm{WACC} = k_0(1 - t(1 + k_\mathrm{d})). \tag{7.6}$$

This expression is different from the similar one in classical Modigliani–Miller theory

$$\mathrm{WACC} = k_0\left(1 - w_\mathrm{d} t\right).$$

$$\text{At } L \rightarrow \qquad \mathrm{WACC} = k_0(1 - t).$$

From these expressions it is seen that WACC decreases with L, achieving lower value WACC = $k_0(1 - t(1 + k_\mathrm{d}))$ at $L \rightarrow \infty$ in considering case comparing with the classical Modigliani–Miller theory $WACC = k_0(1 - t)$.

This means also, that company capitalization becomes higher than in ordinary Modigliani–Miller theory (Fig. 7.3).

7.2.3 Equity Cost

Let us find equity cost

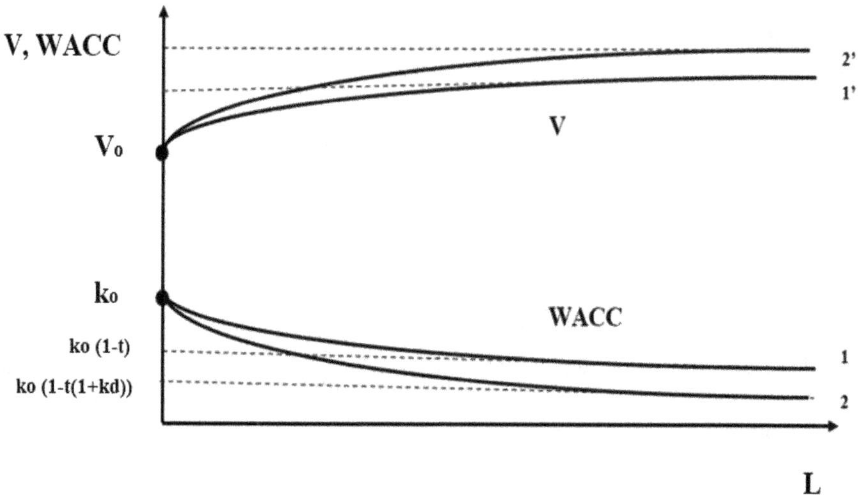

Fig. 7.3 Dependence of company capitalization, V and WACC on leverage level in "classical" Modigliani–Miller theory (curve 1′ and curve 1) and in "modified" Modigliani–Miller theory (curve 2′ and curve 2)

$$\mathrm{WACC} = k_e w_e + k_d w_d (1 - t). \tag{7.7}$$

Equating (7.5) and (7.7), we obtain:

$$k_0 (1 - w_d t (1 + k_d)) = k_e w_e + k_d w_d (1 - t), \tag{7.8}$$

whence we get the following expression for the equity cost:

$$\begin{aligned} k_e &= k_0 \frac{(1 - w_d t(1 + k_d))}{w_e} - k_d \frac{w_d}{w_e}(1 - t) = \\ &= k_0 \frac{1}{w_e} - k_0 \frac{w_d}{w_e}(1 + k_d) \cdot t - k_d \frac{D}{S}(1 - t) = \\ &= k_0 \frac{D + S}{S} - k_0 \frac{D}{S}(1 + k_d) \cdot t - k_d \frac{D}{S}(1 - t) = \\ &= k_0 + L[(1 - t)(k_0 - k_d) - k_0 k_d t]. \end{aligned} \tag{7.9}$$

Finally, we have for the equity cost:

$$k_e = k_0 + L[(1 - t)(k_0 - k_d) - k_0 k_d t]. \tag{7.10}$$

Thus, we arrive to the following statement, which modifies the original Modigliani and Miller.

Equity cost of leverage company k_e making tax on profit payments in advance could be found as equity cost of financially independent company k_0 of the same

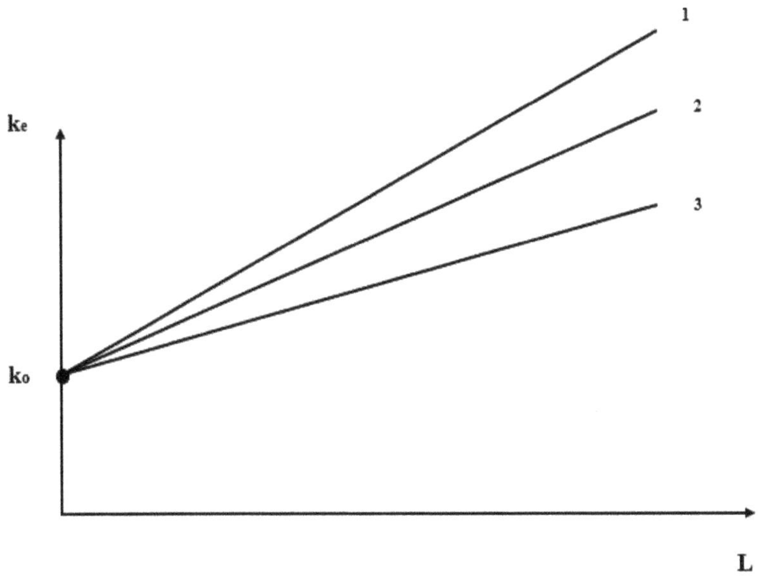

Fig. 7.4 Dependence of equity cost of the company, ke, on leverage level in "classical" Modigliani–Miller theory (curve 2), in "modified" Modigliani–Miller theory (curve 3) and for financially independent company ($t = 0$) (curve 1)

group of risk, plus premium for risk, the value which is equal to production of leverage level L on production of difference $(k_0 - k_d)$ *and tax shield (1 − t), decreasing by the value:* $k_0 k_d t$ (Fig. 7.4).

This means, that equity cost dependence on leverage level L is still linear, but the tilt angle with respect to L-axis turns out to be smaller: $tg\alpha = (k_0 - k_d) \cdot (1 - t) - k_0 k_d t$.

This could lead to modification of the dividend policy of the company, because the equity cost represents itself economically sound value of dividends. Thus, company could decreases the value of dividends.

7.3 The Dependence of the Weighted Average Cost of Capital, WACC, on Leverage Level in "Classical" Modigliani–Miller Theory (MM Theory) and in Modified Modigliani–Miller Theory (MMM Theory)

Let us compare the dependence of the weighted average cost of capital, WACC, on leverage level in "classical" Modigliani–Miller theory (MM theory) and in modified Modigliani–Miller theory (MMM theory). Study of such dependence is very

Table 7.1 Dependence of WACC on leverage level in "classical" Modigliani–Miller theory

k_o	L	t	WACC
0.2	0	0.2	0.2000
0.2	1	0.2	0.1800
0.2	2	0.2	0.1733
0.2	3	0.2	0.1700
0.2	4	0.2	0.1680
0.2	5	0.2	0.1667
0.2	6	0.2	0.1657
0.2	7	0.2	0.1650
0.2	8	0.2	0.1644
0.2	9	0.2	0.1640
0.2	10	0.2	0.1636

Table 7.2 Dependence of WACC on leverage level in modified Modigliani–Miller theory (MMM theory) at $k_d = 0.18$

k_o	L	t	k_d	WACC
0.2	0	0.2	0.18	0.2000
0.2	1	0.2	0.18	0.1764
0.2	2	0.2	0.18	0.1685
0.2	3	0.2	0.18	0.1646
0.2	4	0.2	0.18	0.1622
0.2	5	0.2	0.18	0.1607
0.2	6	0.2	0.18	0.1595
0.2	7	0.2	0.18	0.1587
0.2	8	0.2	0.18	0.1580
0.2	9	0.2	0.18	0.1575
0.2	10	0.2	0.18	0.1571

important, because, the weighted average cost of capital, WACC, plays the role of discount rate in operating financial flows discounting as well as of financial flows in rating methodologies. WACC's value determines as well the capitalization of the company $V = \text{CF/WACC}$. We use Microsoft Excel for calculations.

From the Tables 7.1, 7.2, 7.3 and 7.4 and from Fig. 7.5 it is seen that WACC in MMM theory turns out to be lower than in case of classical Modigliani–Miller theory and thus company capitalization becomes higher than in ordinary Modigliani–Miller theory. It is seen, that WACC decreases with debt cost k_d.

Correct account of a method of tax on profit payments demonstrates that shortcomings of Modigliani–Miller theory are dipper, than everybody suggested: the underestimation of WACC really turns out to be bigger, as well as overestimation of the capitalization of the company. This means that systematic risks arising from the use of modified Modigliani–Miller theory (MMM theory) (which is more correct than "classical" one) in practice are higher than it was suggested by the "classical" version of this theory.

Table 7.3 Dependence of WACC on leverage level in modified Modigliani–Miller theory (MMM theory) at $k_d = 0.14$

k_o	L	t	k_d	WACC
0.2	0	0.2	0.14	0.2000
0.2	1	0.2	0.14	0.1772
0.2	2	0.2	0.14	0.1696
0.2	3	0.2	0.14	0.1658
0.2	4	0.2	0.14	0.1635
0.2	5	0.2	0.14	0.1620
0.2	6	0.2	0.14	0.1609
0.2	7	0.2	0.14	0.1601
0.2	8	0.2	0.14	0.1595
0.2	9	0.2	0.14	0.1590
0.2	10	0.2	0.14	0.1585

Table 7.4 Dependence of WACC on leverage level in modified Modigliani–Miller theory (MMM theory) at $k_d = 0.1$

k_o	L	t	k_d	WACC
0.2	0	0.2	0.1	0.2000
0.2	1	0.2	0.1	0.1780
0.2	2	0.2	0.1	0.1707
0.2	3	0.2	0.1	0.1670
0.2	4	0.2	0.1	0.1648
0.2	5	0.2	0.1	0.1633
0.2	6	0.2	0.1	0.1623
0.2	7	0.2	0.1	0.1615
0.2	8	0.2	0.1	0.1609
0.2	9	0.2	0.1	0.1604
0.2	10	0.2	0.1	0.1600

7.4 Conclusions

Despite the fact that the Modigliani–Miller theory is currently a particular case of the general theory of capital cost and capital structure—Brusov–Filatova–Orekhova (BFO) theory—it is still widely used at the West. We generalize the Modigliani–Miller theory for the case of advance tax on profit payments, which is widely used in practice, and show that this leads to some important consequences, which change seriously all the main statements by Modigliani and Miller. These consequences are as following: WACC starts depend on debt cost k_d, WACC turns out to be lower than in case of classical Modigliani–Miller theory and thus company capitalization becomes higher than in ordinary Modigliani–Miller theory.

We study as well the equity cost dependence on leverage level L and show that it is still linear, but the tilt angle with respect to L-axis turns out to be smaller in MMM theory $tg\alpha = (k_0 - k_d) \cdot (1 - t) - k_0 k_d t$, than in classical Modigliani–Miller theory $tg\alpha = (k_0 - k_d) \cdot (1 - t)$. This could lead to modification of the dividend policy of the company, because the equity cost represents itself economically sound value of

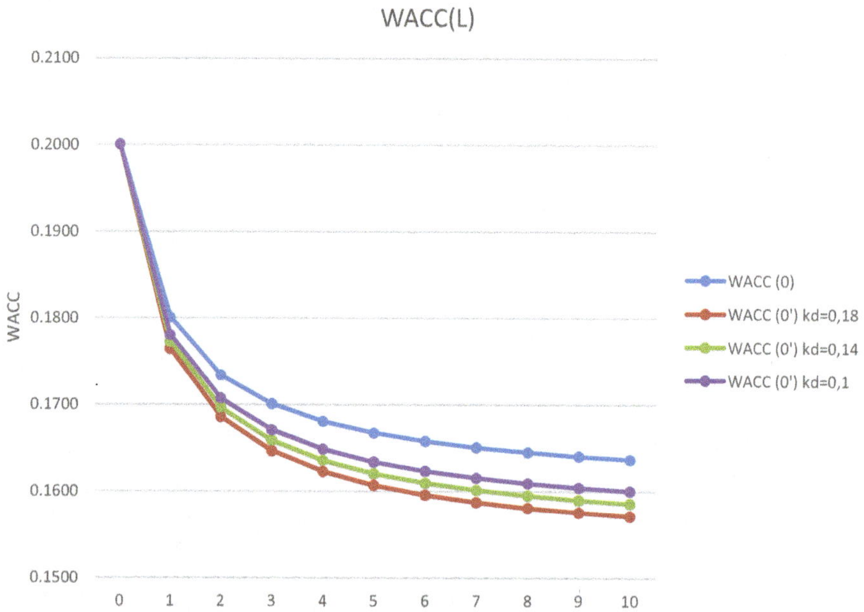

Fig. 7.5 Dependence of WACC on leverage level L—in "classical" Modigliani–Miller theory (curve WACC(0)) and—in modified Modigliani–Miller theory (MMM theory) at different values of debt cost: $k_d = 0.18$; $k_d = 0.14$; $k_d = 0.1$ (curves WACC(0′))

dividends. Thus, company could decreases the value of dividends, which they should pay to shareholders.

Correct account of a method of tax on profit payments demonstrates that shortcomings of Modigliani–Miller theory are dipper, than everybody suggested: the underestimation of WACC really turns out to be bigger, as well as overestimation of the capitalization of the company. This means that systematic risks arising from the use of modified Modigliani–Miller theory (which is more correct than "classical' one) in practice is higher than it was suggested by the "classical" version of this theory.

References

1. Filatova, T., Orehova, N., & Brusova, A. (2008). Weighted average cost of capital in the theory of Modigliani–Miller, modified for a finite life-time company. *Bull FU, 48*, 68–77.
2. Brusov, P., Filatova, T., Orehova, N., & Eskindarov, M. (2015). *Modern corporate finance, investments and taxation*. Springer International Publishing, 373 p. monograph. https://www.springer.com/gp/book/9783319147314
3. Brusov, P., Filatova, T., Orehova, N., & Eskindarov, M. (2018). *Modern corporate finance, investments, taxation and ratings*. Springer Nature Publishing, 571 p. monograph.

4. Brusov, P. N., Filatova, T. V., Orekhova, N. P., Kulik, V. L., Chang, S.-I., & Lin, Y. C. G. (2020). Modification of the Modigliani–Miller theory for the case of advance payments of tax on profit. *Journal of Reviews on Global Economics, 9*, 257–268.
5. Brusov, P. N., Filatova, T. V., Orekhova, N. P., Kulik, V. L., & Weil, I. (2019). Ratings of the investment projects of arbitrary durations: New methodology. *Journal of Reviews on Global Economics, 8*, 437–448. https://doi.org/10.6000/1929-7092.2019.08.37
6. Brusov, P. (2018). Editorial: Introduction on special issue on the Banking System and financial markets of Russia and other countries: Problems and prospects. *Journal of Reviews on Global Economics, 7*, i–vi.
7. Filatova, T. V., Brusov, P. N., Orekhova, N. P., & Kulik, V. L. (2018). Ratings of the long-term projects: New approach. *Journal of Reviews on Global Economics, 7*, 645–661. https://doi.org/10.6000/1929-7092.2018.07.59
8. Brusov, P. N., Filatova, T. V., Orekhova, N. P., & Kulik, V. L. (2018). Rating: New approach. *Journal of Reviews on Global Economics, 7*, 37–62. https://doi.org/10.6000/1929-7092.2018.07.05
9. Brusov, P. N., Filatova, T. V., Orekhova, N. P., & Kulik, V. L. (2018). A "golden age" of the companies: Conditions of its existence. *Journal of Reviews on Global Economics, 7*, 88–103. https://doi.org/10.6000/1929-7092.2018.07.07
10. Brusov, P. N., Filatova, T. V., Orekhova, N. P., & Kulik, V. L. (2018). Rating methodology: New look and new horizons. *Journal of Reviews on Global Economics, 7*, 63–87. https://doi.org/10.6000/1929-7092.2018.07.06
11. Brusov, P. N., Filatova, T. V., Orekhova, N. P., Kulik, V. L., & Weil, I. (2018). New meaningful effects in modern capital structure theory. *Journal of Reviews on Global Economics, 7*, 104–122. https://doi.org/10.6000/1929-7092.2018.07.08
12. Brusov, P. (2018). Editorial. *Journal of Reviews on Global Economics, 7*, i–v.
13. Brusov, P., Filatova, T., Orekhova, N., Kulik, V., Weil, I., & Brailov, A. (2018). The impact of the central bank key rate and commercial banks credit rates on creating and maintaining of a favorable investment climate in the country. *Journal of Reviews on Global Economics, 7*, 360–376. https://doi.org/10.6000/1929-7092.2018.07.31
14. Brusov, P. N., Filatova, T. V., & Orekhova, N. P. (2018). *Modern corporate finance and investments, monograph*. Knorus Publishing House, 517 p.
15. Modigliani, F., & Miller, M. (1958). The cost of capital, corporate finance, and the theory of investment. *The American Economic Review, 48*, 261–297.
16. Modigliani, F., & Miller, M. (1963). Corporate income taxes and the cost of capital: A correction. *The American Economic Review, 53*, 147–175.
17. Modigliani, F., & Miller, M. (1966). Some estimates of the cost of capital to the electric utility industry 1954–1957. *The American Economic Review, 56*, 333–391.

Chapter 8
How Frequently Should Companies Pay Tax on Profit

8.1 Introduction

Brusov and Filatova [1] have modified the Modigliani–Miller theory for the case of arbitrary frequency of payment of tax on profit and have shown that the account of frequency of payment of tax on profit within the Modigliani–Miller theory turns out to be very important and leads to significant consequences:

1. All Modigliani–Miller theorems, statements and all formulas change;
2. All main financial indicators, such as the weighted average cost of capital, WACC, company value, V and equity cost, k_e depend on the frequency of tax on profit payments. This allows to company manage WACC, V, k_e etc. by choosing the number of payments of tax of profit p per year;
3. Obtained results allow to company choose the number of payments of tax of profit per year, as many, as it is profitable to it (of course, within actual tax legislation);
4. In case of income tax payments more than once per year (at $p \neq 1$), as it takes place in practice, the weighted average cost of capital, WACC, company value, V and equity cost, k_e start depend on k_d, while in ordinary (classical) Modigliani–Miller theory all these values *DO NOT depend on* k_d.
5. The tilt angle of the curve of equity cost, k_e (L), decreases with the number of payments of tax of profit p, this modifies the dividend policy of the company, because the economically justified value of dividends is equal to equity cost.

As it was shown in that paper [1], more frequent payments of income tax are beneficial for both parties: for the company and for the tax regulator: for the company, this leads to decrease of the cost of attracting capital and thus to an increase in the value of the company, and for the tax regulator, earlier payments are beneficial due to the time value of money.

But companies could pay tax of profit in advance and as we show in current Chapter this fact changes drastically the main conclusion: more frequent payments

P. Brusov, T. Filatova, *Business Valuation and Company Value*, Contributions to Finance and Accounting, https://doi.org/10.1007/978-3-031-90847-7_8

of income tax are *NOT* beneficial for the company: this leads to increase of the cost of attracting capital and thus to decrease in the value of the company, but for the tax regulator it remains beneficial: earlier payments are beneficial due to the time value of money.

The structure of the chapter is as following:

1. We give a literature review for Modigliani–Miller theory and for its modifications.
2. We modify the Modigliani–Miller theory for the case of frequent payment of tax on profit (in case of advanced payments) and get modified Modigliani–Miller theorems as well as new formulas for the weighted average cost of capital, WACC, for the equity cost, k_e, and for the capitalization of the company, V.
3. Within new Modified Modigliani–Miller theory (MMM theory) we study the dependence of the main financial indicators of the company (WACC, k_e, V) on leverage level L, on debt cost k_d and on tax on profit t at different frequency of payments of tax on profit (at the end of period and in advance).
4. Based on obtained results we come to some very important conclusions, which allow company choose the method of payments of tax of profit (at the end of period or in advance) and number of payments of tax of profit per year, as many, as it is profitable to it (of course, within actual tax legislation).

8.2 Literature Review

In their first paper Modigliani and Miller [2] considered the problem of capital cost and capital structure for perpetuity companies without any taxes. In the second article, Modigliani and Miller [3] considered taxation of corporate profits, but did not take into account the presence in the economy of individual taxes of investors. Merton Miller [4] has introduced the model, demonstrating impact of leverage on the company value with account of the corporate and individual taxes. To describe his model, we will enter the following legends: T_C—tax on corporate profits rate, T_S—the tax rate on income of an individual investor from his ownership by stock of corporation, T_D—tax rate on interest income from the provision of investor–individuals of credits to other investors and companies. Income from shares partly comes in the form of a dividend and, in part, as capital profits, so that T_S is a weighted average value of effective rates of tax on dividends and capital profits on shares, while the income from the provision of loans usually comes in the form of the interests. The last are usually taxed at a higher rate. In the light of the individual taxes, and with the same assumptions that have been made for Modigliani–Miller models previously, the financially independent company value can be determined as follows:

$$V_U = \frac{\text{EBIT}(1 - T_C)(1 - T_S)}{k_0}. \tag{8.1}$$

A term $(1 - T_S)$ allows to take into account the individual taxes in the formula. In this way, numerator indicates which part of the operating company's profit remains in the possession of the investors, after the company will pay taxes on their profits, and its shareholders then will pay individual taxes on income from stock ownership. Since individual taxes reduce profits, remaining in the disposal of investors, the last, at other things being equal circumstances, also reduce and an overall assessment of the financially independent company value.

Robert Hamada [5] unites Capital Asset Pricing Model (CAPM) with Modigliani–Miller model taxation. As a result, he derived the following formula for calculation of the equity cost of financially dependent company, including both financial and business risk of company:

$$k_e = k_F + (k_M - k_F)b_U + (k_M - k_F)b_U \frac{D}{S}(1 - T), \tag{8.2}$$

where b_U is the β-coefficient of the company of the same group of business risk, that the company under consideration, but with zero financial leverage. The formula (8.2) represents the desired profitability of equity capital k_e as a sum of three components: risk-free profitability k_F, compensating to shareholders a temporary value of their money, premium for business risk $(k_M - k_F)b_U$, and premium for financial risk $(k_M - k_F)b_U \frac{D}{S}(1 - T)$.

If the company does not have borrowing ($D = 0$), the financial risk factor will be equal to zero (the third term is drawn to zero), and its owners will only receive the premium for business risk.

More general formula for the WACC, then famous Modigliani–Miller (MM) one has been derived and discussed by a few authors ([6, 7], and J.R. Ezzel, Peter MDeMarzo et al). It takes the following form (Eq. (18) in FGS, 2006):

$$\text{WACC} = k_0(1 - w_d T) - k_d t w_d + k_{TS} t w_d \tag{8.3}$$

where k_0, k_d and k_{TS} are the expected returns respectively on the unlevered firm, the debt and the tax shield, V is the value of the levered firm, VTS is the value of the tax shield, D is the value of the debt and TC is the corporate tax rate.

This formula is derived from the balance sheet identity and the definition of the weighted average cost of capital (for a similar presentation, see [8, 9]). It should therefore be verified at any point in time, whatever returns are annually or continuously compounded.

Although Eq. (8.3) is fully general, its practical applicability requires additional conditions. Indeed, when the WACC is constant over time, the value of a levered firm can be computed by discounting with the WACC the unlevered free cash-flows. Therefore, we pay a special attention to the special cases that make the WACC constant. The resulting particular formulas can also be found in textbooks [9].

First, Modigliani and Miller [3] assume that the level of debt D is constant. Then, as the expected after-tax cash-flow of the unlevered firm is fixed, V_0 is also constant. By assumption, $k_{TS} = k_D$ and the value of the tax shield is TS = tD. Therefore, the value of the firm V is a constant and the general WACC formula (8.15) simplifies to a constant WACC:

$$\mathrm{WACC} = k_0(1 - w_d T) \tag{8.4}$$

But our opinion is that "classical" Modigliani–Miller (MM) theory, which coming from the suggestion that the expected returns on the debt k_d and the tax shield k_{TS} are equals (because both of them have debt nature), is much more reasonable and in our paper we modify namely "classical" Modigliani–Miller (MM) theory, which is still widely used in practice.

Recently the Modigliani–Miller theory has been modified by Brusov et al. [10] into two ways: it was applied for rating methodologies needs and it was generalized for the case of advance payments of tax on profit, which are widely used in practice (MMM theory). As well Brusov et al. [10] used the Modified Modigliani–Miller theory (MMM theory) and apply it for rating methodologies needs. The financial "ratios" (main rating parameters) were introduced into MMM theory. The dependence of the weighted average cost of capital (WACC), which plays the role of discount rate in financial flows discounting in rating methodologies, on coverage and leverage ratios is analyzed. Obtained results will help improve the existing rating methodologies.

One of the most serious limitations of the Modigliani–Miller theory is the suggestion about perpetuity of the companies. In 2008, Brusov–Filatova–Orekhova have lifted up this limitation and have shown that the accounting of the finite life-time (or finite age) of the company leads to significant changes of all Modigliani–Miller results [2, 3, 11]: capitalization of the company is changed, as well as the equity cost, k_e, and the weighted average cost of capital, WACC, in the presence of corporative taxes. Besides, a number of qualitatively new effects in corporate finance, obtained within Brusov–Filatova–Orekhova theory (BFO theory) [10, 12–28]), are absent in Modigliani–Miller theory [2, 3, 29–33].

The famous Brusov–Filatova–Orekhova formula for weighted average cost of capital, WACC, has the following form:

$$\frac{1-(1+\mathrm{WACC})^{-n}}{\mathrm{WACC}} = \frac{1-(1+k_0)^{-n}}{k_0[1-\omega_d T(1-(1+k_d)^{-n})]}. \tag{8.5}$$

Here, S—the value of own (equity) capital of the company, $w_d = \frac{D}{D+S}$—the share of debt capital; $k_e, w_e = \frac{S}{D+S}$—the cost and the share of the equity of the company, $L = D/S$—financial leverage.

A lot of meaningful effects has been discovered within BFO theory: these effects are absent within Modigliani–Miller theory. BFO theory has destroyed some main existing principles of financial management: among them trade off theory, which was considered as keystone of formation of optimal capital structure of the company during many decades and *bankruptcy of which has been proven* by Brusov–Filatova–Orekhova.

8.3 Focus and Objective of the Chapter

The tax shield plays a crucial role in both main capital structure theories: Brusov–Filatova–Orekhova (BFO theory) and its perpetuity limit Modigliani–Miller theory. How it is formed influences the results of both theories. Both of them describe the case of annual payments of tax of profit, but in practice these payments are made more frequent: semiannually, quarterly, monthly. While a return is not required more than once a year, businesses may be responsible for filing estimated taxes based on profits earned. This requirement is dependent on showing a profit. For example, sole proprietors must file estimated taxes on profits quarterly, on the 15th day of April, June, September, and January. In Russia tax on profit payments could be made annually, quarterly, monthly. Brusov and Filatova [1] have shown that the account of frequency of payment of tax on profit within the Modigliani–Miller theory turns out to be very important and leads to significant consequences.

To study this problem they have modified the Modigliani–Miller theory for the case of arbitrary frequency of payment of tax on profit and show that:

1. All Modigliani–Miller theorems, statements and all formulas change;
2. All main financial indicators, such as the weighted average cost of capital, WACC, company value, V and equity cost, k_e depend on the frequency of tax on profit payments. This allows to company manage WACC, V, k_e etc. by choosing the number of payments of tax of profit p per year;
3. Increase the number of payments of tax of profit per year leads to decrease of the cost of attracting capital and increase of the company value;
4. Obtained results allow to company choose the number of payments of tax of profit per year, as many, as it is profitable to it (of course, within actual tax legislation);
5. In case of income tax payments more than once per year (at $p \neq 1$), as it takes place in practice, the weighted average cost of capital, WACC, company value, V, and equity cost, ke start depend on k_d, while in ordinary (classical) Modigliani–Miller theory all these values DO NOT depend on k_d.
6. The tilt angle of the curve of equity cost, k_e (L), decreases with the number of payments of tax of profit p, this modifies the dividend policy of the company, because the economically justified value of dividends is equal to equity cost.

As it was shown in that paper [1], more frequent payments of income tax are beneficial for both parties: for the company and for the tax regulator: for the company, this leads to decrease of the cost of attracting capital and thus to an increase in the value of the company, and for the tax regulator, earlier payments are beneficial due to the time value of money.

But companies could pay tax of profit in advance and as we show in current paper this fact changes drastically the main conclusion: more frequent payments of income tax are *NOT* beneficial for the company: this leads to increase of the cost of attracting capital and thus to decrease in the value of the company, but for the tax regulator it remains beneficial: earlier payments are beneficial due to the time value of money.

8.4 Modification of the Modigliani–Miller Theory for the Case of Arbitrary Frequency of Advanced Payments of Tax on Profit

8.4.1 Tax Shield

Let us calculate first the tax shield TS in the Modigliani–Miller theory in the case of arbitrary frequency (p per year) of advanced payments of tax on profit:

$$\mathrm{TS} = \frac{k_\mathrm{d}Dt}{p} + \frac{k_\mathrm{d}Dt}{p(1+k_\mathrm{d})^{1/p}} + \frac{k_\mathrm{d}Dt}{p(1+k_\mathrm{d})^{2/p}} + \dots \tag{8.6}$$

(TS) represents a geometric progression with denominator $q = \frac{1}{(1+k_\mathrm{d})^{1/p}}$.

Here D is debt value, k_d is debt cost, t is tax on profit rate, p is the frequency of payments of tax on profit.

Summing the progression, one gets:

$$\mathrm{TS} = \frac{k_\mathrm{d}Dt}{p\left(1-(1+k_\mathrm{d})^{-1/p}\right)} = \frac{k_\mathrm{d}Dt\cdot(1+k_\mathrm{d})^{1/p}}{p\left((1+k_\mathrm{d})^{1/p}-1\right)}. \tag{8.7}$$

Note, that in case of classical Modigliani–Miller theory, modified by us for the case of one advanced payment of tax on profit per year:

$$\mathrm{TS} = Dt(1+k_\mathrm{d}). \tag{8.8}$$

It is easy to get this result from (8.7), putting $p = 1$.

8.4.2 *The Weighted Average Cost of Capital, WACC*

Let us calculate now the weighted average cost of capital, WACC in the Modigliani–Miller theory in the case of arbitrary frequency (p per year) of advanced payments of tax on profit.

From the Modigliani–Miller theorem we have for the company value V:

$$V = V_0 + \mathrm{TS}. \tag{8.9}$$

Here V_0 is the financially independent company value.

Putting expression for tax shield TS, we get:

$$V = V_0 + k_d D t \cdot \frac{(1 + k_d)^{1/p}}{p\left((1 + k_d)^{1/p - 1}\right).} \tag{8.10}$$

Substituting $D = w_d V$, one gets:

$$V\left(1 - \frac{k_d w t d \cdot (1 + k_d)^{1/p}}{p\left((1 + k_d)^{1/p} - 1\right)}\right) = V_0. \tag{8.11}$$

Accounting, that:

$$V = \frac{\mathrm{CF}}{\mathrm{WACC}}; V_0 = \frac{\mathrm{CF}}{k_0}, \tag{8.12}$$

we have:

$$\frac{\mathrm{CF}}{\mathrm{WACC}} \cdot \left(1 - \frac{k_d w t d \cdot (1 + k_d)^{1/p}}{p\left((1 + k_d)^{1/p} - 1\right)}\right) = \frac{\mathrm{CF}}{k_0}. \tag{8.13}$$

From here we arrive to the expression for the weighted average cost of capital, WACC,

$$\mathrm{WACC} = k_0 \cdot \left(1 - \frac{k_d w_d t \cdot (1 + k_d)^{1/p}}{p\left((1 + k_d)^{1/p} - 1\right)}\right). \tag{8.14}$$

Note, that in case of classical Modigliani–Miller theory, modified by us for the case of one advanced payment of tax on profit per year:

$$\mathrm{WACC} = k_0 \cdot (1 - w_d t \cdot (1 + k_d)). \tag{8.15}$$

It is easy to get this result from (8.14), putting $p = 1$.

8.4.3 Company Value, V

Let us calculate now the company value, V, in the Modigliani–Miller theory in the case of arbitrary frequency (p per year) of advanced payments of tax on profit.

Putting WACC value into first formula of (8.12) one gets for company value, V:

$$V = \frac{\mathrm{CF}}{\mathrm{WACC}} = \frac{\mathrm{CF}}{k_0 \cdot \left(1 - \frac{k_d w_d t \cdot (1+k_d)^{1/p}}{p\left((1+k_d)^{1/p} - 1\right)}\right)}. \tag{8.16}$$

Here CF is the profit of the company per year, w_d is the debt share.

In case of classical Modigliani–Miller theory, modified by us for the case of one advanced payment of tax on profit per year [12–14], one has:

$$V = \frac{\mathrm{CF}}{\mathrm{WACC}} = \frac{\mathrm{CF}}{k_0 \cdot (1 - w_d t (1 + k_d))}. \tag{8.17}$$

It is easy to get this result from (8.16), putting $p = 1$.

8.4.4 Equity Cost

Let us calculate now the equity cost, k_d, in the Modigliani–Miller theory in the case of arbitrary frequency (p per year) of advanced payments of tax on profit.

For the weighted average cost of capital, WACC, we have from formula (8.14)

$$\mathrm{WACC} = k_0 \cdot \left(1 - \frac{k_d w_d t \cdot (1 + k_d)^{1/p}}{p\left((1 + k_d)^{1/p} - 1\right)}\right). \tag{8.18}$$

From other side, upon definition one has:

$$\mathrm{WACC} = k_e w_e + k_d w_d (1 - t). \tag{8.19}$$

From (8.19) we get:

$$k_e = \frac{\text{WACC}}{w_e} - \frac{k_{\text{d}} w_{\text{d}}(1-t)}{w_e} \tag{8.20}$$

Putting (8.18) into (8.20), we arrive to expression for equity cost k_e

$$\begin{aligned} k_e &= \frac{k_0}{w_e}\cdot\left(1 - \frac{k_{\text{d}} w_{\text{d}} t \cdot 1 + k_{\text{d}}{}^{1/p}}{p1 + k_{\text{d}}{}^{1/p} - 1}\right) - \frac{k_{\text{d}} w_{\text{d}} 1 - t}{w_e} = \\ &= k_0\cdot\left(1 + L - \frac{k_{\text{d}} L t \cdot 1 + k_{\text{d}}{}^{1/p}}{p1 + k_{\text{d}}{}^{1/p} - 1}\right) - k_{\text{d}} L 1 - t = \\ &= k_0 + L\left(k_0 - k_{\text{d}} 1 - t - \frac{k_0 k_{\text{d}} t \cdot 1 + k_{\text{d}}{}^{1/p}}{p1 + k_{\text{d}}{}^{1/p} - 1}\right) \end{aligned} \tag{8.21}$$

Finally, one has:

$$k_e = k_0 + L\left(k_0 - k_{\text{d}}(1-t) - \frac{k_0 k_{\text{d}} t \cdot (1 + k_{\text{d}})^{1/p}}{p\left((1 + k_{\text{d}})^{1/p} - 1\right)}\right) \tag{8.22}$$

In case of classical Modigliani–Miller theory, modified by us for the case of one advanced payment of tax on profit per year [12–14], one has:

$$k_e = k_0 + L(k_0 - k_{\text{d}} \cdot (1-t) - k_0 t(1 + k_{\text{d}})). \tag{8.23}$$

Putting in formula (8.22) $p = 1$, it is easy to get formula (8.23).

8.5 Dependence of the Weighted Average Cost of Capital, WACC, on Leverage Level *L*

8.5.1 Dependence of the Weighted Average Cost of Capital, WACC, on Leverage Level L *at Different* p *and Fixed* k_d

Let us start from the study numerically with use of Microsoft Excel of the dependence of the weighted average cost of capital, WACC, on leverage level L at different frequency of payments p and fixed debt cost k_{d}, using obtained by us formulas:

$\text{WACC} = k_0\cdot\left(1 - \frac{k_{\text{d}} w_{\text{d}} t}{p\left((1+k_{\text{d}})^{1/p} - 1\right)}\right)$ and $\text{WACC} = k_0\cdot\left(1 - \frac{k_{\text{d}} w_{\text{d}} t \cdot (1+k_{\text{d}})^{1/p}}{p\left((1+k_{\text{d}})^{1/p} - 1\right)}\right)$

Table 8.1 Dependence of WACC on leverage level L at $k_0 = 0.2$, $k_d = 0.16$ and $p = 1$

k_0	k_d	w_d	L	t	p	i	WACC
0.2	0.16	0	0	0.2	1	1.16	0.2
0.2	0.16	0.5	1	0.2	1	1.16	0.18
0.2	0.16	0.666667	2	0.2	1	1.16	0.173333
0.2	0.16	0.75	3	0.2	1	1.16	0.17
0.2	0.16	0.8	4	0.2	1	1.16	0.168
0.2	0.16	0.833333	5	0.2	1	1.16	0.166667
0.2	0.16	0.857143	6	0.2	1	1.16	0.165714
0.2	0.16	0.875	7	0.2	1	1.16	0.165
0.2	0.16	0.888889	8	0.2	1	1.16	0.164444
0.2	0.16	0.9	9	0.2	1	1.16	0.164
0.2	0.16	0.909091	10	0.2	1	1.16	0.163636

Table 8.2 Dependence of WACC on leverage level L at $k_0 = 0.2$, $k_d = 0.16$ and $p = 12$

k_0	k_d	w_d	L	t	p	i	WACC
0.2	0.16	0	0	0.2	12	1.012445	0.2
0.2	0.16	0.5	1	0.2	12	1.012445	0.178573
0.2	0.16	0.666667	2	0.2	12	1.012445	0.17143
0.2	0.16	0.75	3	0.2	12	1.012445	0.167859
0.2	0.16	0.8	4	0.2	12	1.012445	0.165716
0.2	0.16	0.833333	5	0.2	12	1.012445	0.164288
0.2	0.16	0.857143	6	0.2	12	1.012445	0.163267
0.2	0.16	0.875	7	0.2	12	1.012445	0.162502
0.2	0.16	0.888889	8	0.2	12	1.012445	0.161907
0.2	0.16	0.9	9	0.2	12	1.012445	0.161431
0.2	0.16	0.909091	10	0.2	12	1.012445	0.161041

Table 8.3 Dependence of WACC on leverage level L at $k_0 = 0.2$, $k_d = 0.16$ and $p = 1$ (advanced payment of tax of profit)

k_0	k_d	w_d	L	t	p	i	WACC
0.2	0.16	0	0	0.2	1	1.16	0.2
0.2	0.16	0.5	1	0.2	1	1.16	0.1768
0.2	0.16	0.666667	2	0.2	1	1.16	0.169067
0.2	0.16	0.75	3	0.2	1	1.16	0.1652
0.2	0.16	0.8	4	0.2	1	1.16	0.16288
0.2	0.16	0.833333	5	0.2	1	1.16	0.161333
0.2	0.16	0.857143	6	0.2	1	1.16	0.160229
0.2	0.16	0.875	7	0.2	1	1.16	0.1594
0.2	0.16	0.888889	8	0.2	1	1.16	0.158756
0.2	0.16	0.9	9	0.2	1	1.16	0.15824
0.2	0.16	0.909091	10	0.2	1	1.16	0.157818

From Tables 8.1, 8.2, 8.3, 8.4 and Figs. 8.1, 8.2 it is seen, that the weighted average cost of capital, WACC, decreases with leverage level L at all values of frequency of payments of tax on profit p. With increase of the frequency of payments

Table 8.4 Dependence of WACC on leverage level L at $k_0 = 0.2$, $k_d = 0.16$ and $p = 12$ (advanced payment of tax of profit)

k_0	k_d	w_d	L	t	p	i	WACC
0.2	0.16	0	0	0.2	12	1.012445	0.2
0.2	0.16	0.5	1	0.2	12	1.012445	0.178306
0.2	0.16	0.666667	2	0.2	12	1.012445	0.171075
0.2	0.16	0.75	3	0.2	12	1.012445	0.167459
0.2	0.16	0.8	4	0.2	12	1.012445	0.16529
0.2	0.16	0.833333	5	0.2	12	1.012445	0.163843
0.2	0.16	0.857143	6	0.2	12	1.012445	0.16281
0.2	0.16	0.875	7	0.2	12	1.012445	0.162035
0.2	0.16	0.888889	8	0.2	12	1.012445	0.161433
0.2	0.16	0.9	9	0.2	12	1.012445	0.160951
0.2	0.16	0.909091	10	0.2	12	1.012445	0.160556

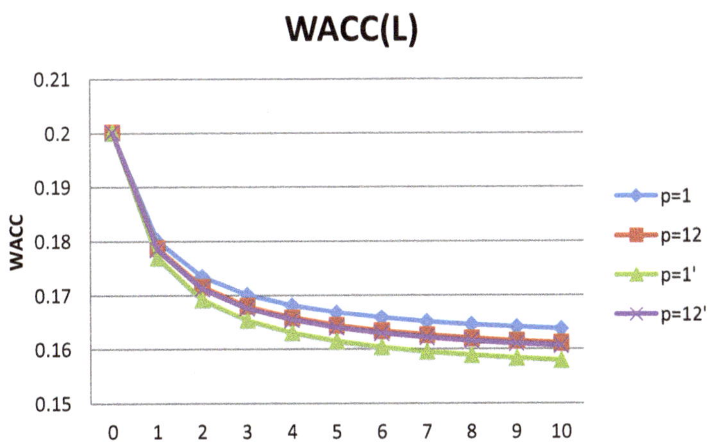

Fig. 8.1 Dependence of the weighted average cost of capital, WACC, on leverage level L at different p and fixed $k_d = 0.16$

of tax on profit per year p WACC (1) decreases in the case of payments at the ends of the periods and WACC (2) increases in the case of advanced payments of tax on profit. As we'll see below, this will lead to increase of company value, V, with increase of p in the case (1) and to decrease of company value, V, with increase of p in the case (2).

Note, that the weighted average cost of capital, WACC, in the case of advanced payments of tax on profit for all values of frequency of payments turns out to be low than for any values of frequency of payments in the case of payments at the ends of the periods.

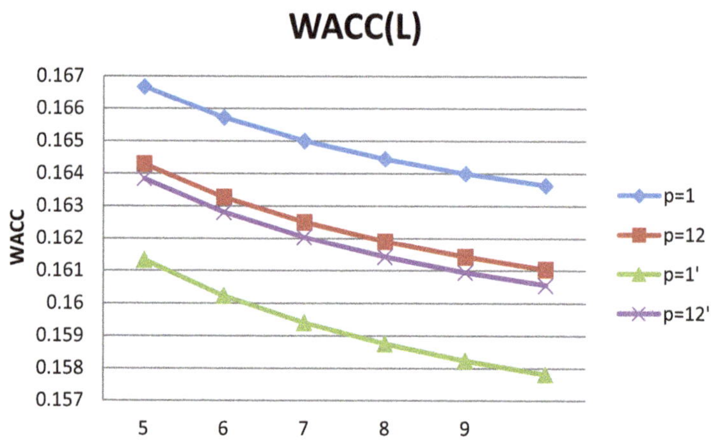

Fig. 8.2 Dependence of the Weighted Average Cost of Capital, WACC, on Leverage Level L at Different p and Fixed $k_d = 0.16$ (Larger Scale)

Table 8.5 Dependence of WACC on debt cost value k_d at $k_0 = 0.2$, at $L = 1$ and $p = 1$

k_0	k_d	w_d	L	t	p	i	WACC	CF	V	k_e
0.2	0.08	0.5	1	0.2	1	1.08	0.18	1000	5555.556	0.296
0.2	0.09	0.5	1	0.2	1	1.09	0.18	1000	5555.556	0.288
0.2	0.1	0.5	1	0.2	1	1.1	0.18	1000	5555.556	0.28
0.2	0.11	0.5	1	0.2	1	1.11	0.18	1000	5555.556	0.272
0.2	0.12	0.5	1	0.2	1	1.12	0.18	1000	5555.556	0.264
0.2	0.13	0.5	1	0.2	1	1.13	0.18	1000	5555.556	0.256
0.2	0.14	0.5	1	0.2	1	1.14	0.18	1000	5555.556	0.248
0.2	0.15	0.5	1	0.2	1	1.15	0.18	1000	5555.556	0.24
0.2	0.16	0.5	1	0.2	1	1.16	0.18	1000	5555.556	0.232
0.2	0.17	0.5	1	0.2	1	1.17	0.18	1000	5555.556	0.224
0.2	0.18	0.5	1	0.2	1	1.18	0.18	1000	5555.556	0.216

8.5.2 Dependence of the Weighted Average Cost of Capital, WACC, and of Company Value, V*, on Debt Cost Value* k_d *at Fixed Value of Leverage Level* L *and Fixed* p

In this paragraph we investigate the dependence of the weighted average cost of capital, WACC, and of company value, V on debt cost value k_d at fixed value of leverage level L and fixed p.

From Tables 8.5, 8.6, 8.7, 8.8, and Fig. 8.3 it is seen, that the weighted average cost of capital, WACC, decreases with increase of debt cost value k_d, except the classical MM case $p = 1$ and payment at the end of the period. It supports obtained by us result that account of each of two effects (either advanced payments of tax on profit, or more frequent than once per year payments) leads to dependence of all

Table 8.6 Dependence of WACC on debt cost value k_d at $k_0 = 0.2$, at $L = 1$ and $p = 12$

k_0	k_d	w_d	L	t	p	i	WACC	CF	V	k_e
0.2	0.08	0.5	1	0.2	12	1.006434	0.179277	1000	5577.965	0.294554
0.2	0.09	0.5	1	0.2	12	1.007207	0.179188	1000	5580.736	0.286376
0.2	0.1	0.5	1	0.2	12	1.007974	0.179099	1000	5583.501	0.278198
0.2	0.11	0.5	1	0.2	12	1.008735	0.179011	1000	5586.259	0.270021
0.2	0.12	0.5	1	0.2	12	1.009489	0.178923	1000	5589.012	0.261845
0.2	0.13	0.5	1	0.2	12	1.010237	0.178835	1000	5591.758	0.253669
0.2	0.14	0.5	1	0.2	12	1.010979	0.178747	1000	5594.499	0.245494
0.2	0.15	0.5	1	0.2	12	1.011715	0.17866	1000	5597.234	0.237319
0.2	0.16	0.5	1	0.2	12	1.012445	0.178573	1000	5599.963	0.229145
0.2	0.17	0.5	1	0.2	12	1.01317	0.178486	1000	5602.686	0.220972
0.2	0.18	0.5	1	0.2	12	1.013888	0.178399	1000	5605.404	0.212799

Table 8.7 Dependence of WACC on debt cost value k_d at $k_0 = 0.2$, at $L = 1$ and $p = 1$ (for advanced payments of tax of profit)

k_0	k_d	w_d	L	t	p	i	WACC	CF	V	k_e
0.2	0.08	0.5	1	0.2	1	1.08	0.1784	1000	5605.381	0.2928
0.2	0.09	0.5	1	0.2	1	1.09	0.1782	1000	5611.672	0.2844
0.2	0.1	0.5	1	0.2	1	1.1	0.178	1000	5617.978	0.276
0.2	0.11	0.5	1	0.2	1	1.11	0.1778	1000	5624.297	0.2676
0.2	0.12	0.5	1	0.2	1	1.12	0.1776	1000	5630.631	0.2592
0.2	0.13	0.5	1	0.2	1	1.13	0.1774	1000	5636.979	0.2508
0.2	0.14	0.5	1	0.2	1	1.14	0.1772	1000	5643.341	0.2424
0.2	0.15	0.5	1	0.2	1	1.15	0.177	1000	5649.718	0.234
0.2	0.16	0.5	1	0.2	1	1.16	0.1768	1000	5656.109	0.2256
0.2	0.17	0.5	1	0.2	1	1.17	0.1766	1000	5662.514	0.2172
0.2	0.18	0.5	1	0.2	1	1.18	0.1764	1000	5668.934	0.2088

Table 8.8 Dependence of WACC on debt cost value k_d at $k_0 = 0.2$, at $L = 1$ and $p = 12$ (for advanced payments of tax of profit)

k_0	k_d	w_d	L	t	p	i	WACC	CF	V	k_e
0.2	0.08	0.5	1	0.2	12	1.006434	0.179144	1000	5582.116	0.294287
0.2	0.09	0.5	1	0.2	12	1.007207	0.179038	1000	5585.412	0.286076
0.2	0.1	0.5	1	0.2	12	1.007974	0.178932	1000	5588.702	0.277865
0.2	0.11	0.5	1	0.2	12	1.008735	0.178827	1000	5591.986	0.269655
0.2	0.12	0.5	1	0.2	12	1.009489	0.178723	1000	5595.266	0.261445
0.2	0.13	0.5	1	0.2	12	1.010237	0.178618	1000	5598.541	0.253236
0.2	0.14	0.5	1	0.2	12	1.010979	0.178514	1000	5601.811	0.245027
0.2	0.15	0.5	1	0.2	12	1.011715	0.17841	1000	5605.077	0.236819
0.2	0.16	0.5	1	0.2	12	1.012445	0.178306	1000	5608.338	0.228612
0.2	0.17	0.5	1	0.2	12	1.01317	0.178202	1000	5611.594	0.220405
0.2	0.18	0.5	1	0.2	12	1.013888	0.178099	1000	5614.846	0.212199

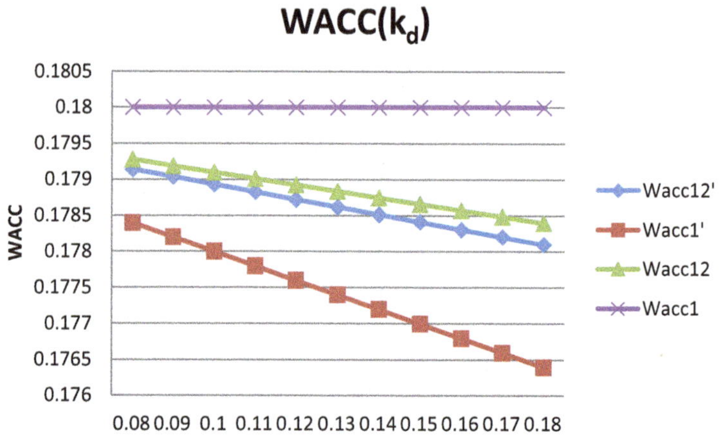

Fig. 8.3 Dependence of WACC on debt cost value k_d at $k_0 = 0.2$, at $L = 1$ and $p = 1$ and $p = 12$ ('means advanced payments of tax of profit)

main financial indicators, such as the weighted average cost of capital, WACC, company value, V, equity cost, k_e, etc. on debt cost value k_d, while in the classical MM case these financial indicators DO NOT depend on debt cost value k_d.

Similar to previous paragraph it is seen, that with increase of the frequency of payments of tax on profit per year p WACC (1) decreases in the case of payments at the ends of the periods and WACC (2) increases in the case of advanced payments of tax on profit. As we'll see below, this will lead to increase of company value, V, with increase of p in the case (1) and to decrease of company value, V, with increase of p in the case (2).

Note, that the weighted average cost of capital, WACC, in the case of advanced payments of tax on profit for all values of frequency of payments turns out to be low than for any values of frequency of payments in the case of payments at the ends of the periods.

From Tables 8.5, 8.6, 8.7, 8.8 and Fig. 8.4 it is seen, that the company value, V, increases with increase of debt cost value k_d, except the classical Modigliani–Miller theory case $p = 1$ and payment at the end of the period, when V remains constant. It supports obtained by us result that account of each of two effects (either advanced payments of tax on profit, or more frequent than once per year) leads to dependence of all main financial indicators, such as the weighted average cost of capital, WACC, company value, V, equity cost, k_e, etc. on debt cost value k_d, while in the classical Modigliani–Miller theory case these financial indicators DO NOT depend on debt cost value k_d.

Similar to previous paragraph it is seen, that with increase of the frequency of payments of tax on profit per year p the company value, V, (1) increases in the case of payments at the ends of the periods and WACC (2) decreases in the case of advanced payments of tax on profit.

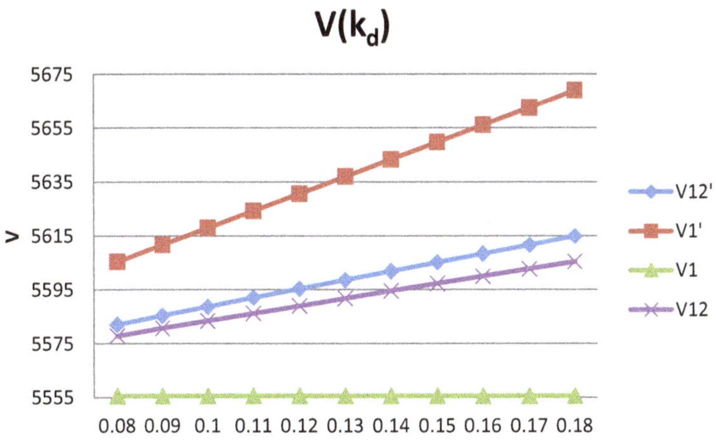

Fig. 8.4 Dependence of company value V on debt cost value k_d at $k_0 = 0.2$, at $L = 1$ and $p = 1$ and $p = 12$ ('means advanced payments of tax of profit)

Note, that the company value, V, in the case of advanced payments of tax on profit for all values of frequency of payments turns out to be higher than for any values of frequency of payments in the case of payments at the ends of the periods.

8.6 Dependence of the Company Capitalization, V, on Leverage Level L

8.6.1 Dependence of the Company Capitalization, V, *on Leverage Level* L *at Different* p *and Fixed* k_d

Let us now calculate the dependence of the company capitalization, V, on leverage level L at different p and fixed k_d.

From Tables 8.9, 8.10, 8.11, 8.12 and Figs. 8.5 and 8.6 it is seen, that the company value, V, increases with the increase of leverage level L for all frequencies of payment of tax on profit per year p.

While in case of payments at the end of the periods the company value, V, increases with the increase of the frequency of payment of tax on profit per year p, in case of advanced payments of tax on profit the company value, V, decreases with the increase of the frequency of payment of tax on profit per year p. These facts consistent with the dependence of the weighted average cost of capital, WACC, on the leverage level L.

All the company values, V, in case of advanced payments of tax on profit turns out to be bigger than in case of payments at the end of the periods. Thus if the companies tend to increase their capitalization, they should pay tax on profit (1) in advance; (2) as seldom as tax legislation allows.

Table 8.9 Dependence of the company capitalization on leverage level L (at $k_d = 0.16$ and at $p = 1$)

k_0	k_d	w_d	L	t	p	i	WACC	CF	V
0.2	0.16	0	0	0.2	1	1.16	0.2	1000	5000
0.2	0.16	0.5	1	0.2	1	1.16	0.18	1000	5555.556
0.2	0.16	0.666667	2	0.2	1	1.16	0.173333	1000	5769.231
0.2	0.16	0.75	3	0.2	1	1.16	0.17	1000	5882.353
0.2	0.16	0.8	4	0.2	1	1.16	0.168	1000	5952.381
0.2	0.16	0.833333	5	0.2	1	1.16	0.166667	1000	6000
0.2	0.16	0.857143	6	0.2	1	1.16	0.165714	1000	6034.483
0.2	0.16	0.875	7	0.2	1	1.16	0.165	1000	6060.606
0.2	0.16	0.888889	8	0.2	1	1.16	0.164444	1000	6081.081
0.2	0.16	0.9	9	0.2	1	1.16	0.164	1000	6097.561
0.2	0.16	0.909091	10	0.2	1	1.16	0.163636	1000	6111.111

Table 8.10 Dependence of the company capitalization on leverage level L (at $k_d = 0.16$ and at $p = 12$)

k_0	k_d	w_d	L	t	p	i	WACC	CF	V
0.2	0.16	0	0	0.2	12	1.012445	0.2	1000	5000
0.2	0.16	0.5	1	0.2	12	1.012445	0.178573	1000	5599.963
0.2	0.16	0.666667	2	0.2	12	1.012445	0.17143	1000	5833.279
0.2	0.16	0.75	3	0.2	12	1.012445	0.167859	1000	5957.383
0.2	0.16	0.8	4	0.2	12	1.012445	0.165716	1000	6034.413
0.2	0.16	0.833333	5	0.2	12	1.012445	0.164288	1000	6086.883
0.2	0.16	0.857143	6	0.2	12	1.012445	0.163267	1000	6124.923
0.2	0.16	0.875	7	0.2	12	1.012445	0.162502	1000	6153.767
0.2	0.16	0.888889	8	0.2	12	1.012445	0.161907	1000	6176.39
0.2	0.16	0.9	9	0.2	12	1.012445	0.161431	1000	6194.608
0.2	0.16	0.909091	10	0.2	12	1.012445	0.161041	1000	6209.594

Table 8.11 Dependence of the company capitalization on leverage level L (at $k_d = 0.16$ and at $p = 1$) (for advanced payments of tax of profit)

k_0	k_d	w_d	L	t	p	i	WACC	CF	V
0.2	0.16	0	0	0.2	1	1.16	0.2	1000	5000
0.2	0.16	0.5	1	0.2	1	1.16	0.1768	1000	5656.109
0.2	0.16	0.666667	2	0.2	1	1.16	0.169067	1000	5914.826
0.2	0.16	0.75	3	0.2	1	1.16	0.1652	1000	6053.269
0.2	0.16	0.8	4	0.2	1	1.16	0.16288	1000	6139.489
0.2	0.16	0.833333	5	0.2	1	1.16	0.161333	1000	6198.347
0.2	0.16	0.857143	6	0.2	1	1.16	0.160229	1000	6241.084
0.2	0.16	0.875	7	0.2	1	1.16	0.1594	1000	6273.526
0.2	0.16	0.888889	8	0.2	1	1.16	0.158756	1000	6298.992
0.2	0.16	0.9	9	0.2	1	1.16	0.15824	1000	6319.515
0.2	0.16	0.909091	10	0.2	1	1.16	0.157818	1000	6336.406

Table 8.12 Dependence of the company capitalization on leverage level L (at $k_d = 0.16$ and at $p = 12$) (for advanced payments of tax of profit)

k_0	k_d	w_d	L	t	p	i	WACC	CF	V
0.2	0.16	0	0	0.2	12	1.012445	0.2	1000	5000
0.2	0.16	0.5	1	0.2	12	1.012445	0.178306	1000	5608.338
0.2	0.16	0.666667	2	0.2	12	1.012445	0.171075	1000	5845.403
0.2	0.16	0.75	3	0.2	12	1.012445	0.167459	1000	5971.613
0.2	0.16	0.8	4	0.2	12	1.012445	0.16529	1000	6049.99
0.2	0.16	0.833333	5	0.2	12	1.012445	0.163843	1000	6103.394
0.2	0.16	0.857143	6	0.2	12	1.012445	0.16281	1000	6142.121
0.2	0.16	0.875	7	0.2	12	1.012445	0.162035	1000	6171.49
0.2	0.16	0.888889	8	0.2	12	1.012445	0.161433	1000	6194.528
0.2	0.16	0.9	9	0.2	12	1.012445	0.160951	1000	6213.082
0.2	0.16	0.909091	10	0.2	12	1.012445	0.160556	1000	6228.345

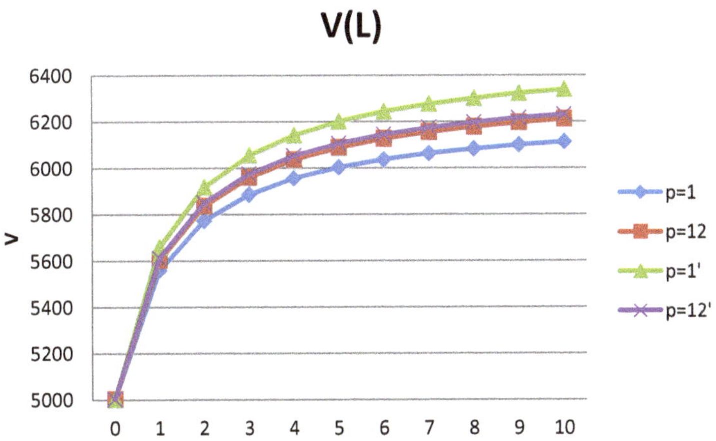

Fig. 8.5 Dependence of the company capitalization, V, on leverage level L at $k_d = 0.16$ and $p = 1$ and 12 ('means advanced payment of tax of profit)

8.7 Dependence of the Equity Cost, k_e, on Leverage Level L

Let us now calculate the dependence of the equity cost, ke, on leverage level L at different p and fixed k_d.

From Tables 8.13, 8.14, 8.15, 8.16 and Figs. 8.7, 8.8 and 8.9 it is seen, that the equity cost, k_e, linearly increases with the increase of leverage level L for all frequencies of payment of tax on profit per year p.

While in the case of tax on profit payments at the end of the periods the tilt of the curve of the equity cost, k_e, decreases with the increase of the frequency of payment of tax on profit per year p, in case of the advanced payments of tax on profit, the tilt of the curve of the equity cost, k_e, increases with the increase of the frequency of

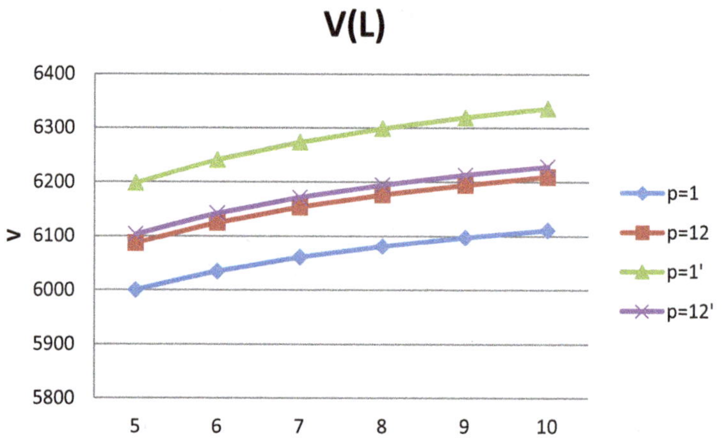

Fig. 8.6 Dependence of the company capitalization, *V*, on leverage level *L* at $k_d = 0.16$ and $p = 1$ and 12 ('means advanced payment of tax of profit) (larger scale)

Table 8.13 Dependence of the equity cost, k_e, on leverage level *L* (at $k_d = 0.16$ and at $p = 1$)

k_0	k_d	w_d	L	t	p	i	WACC	CF	V	k_e
0.2	0.16	0	0	0.2	1	1.16	0.2	1000	5000	0.2
0.2	0.16	0.5	1	0.2	1	1.16	0.18	1000	5555.556	0.232
0.2	0.16	0.666667	2	0.2	1	1.16	0.173333	1000	5769.231	0.264
0.2	0.16	0.75	3	0.2	1	1.16	0.17	1000	5882.353	0.296
0.2	0.16	0.8	4	0.2	1	1.16	0.168	1000	5952.381	0.328
0.2	0.16	0.833333	5	0.2	1	1.16	0.166667	1000	6000	0.36
0.2	0.16	0.857143	6	0.2	1	1.16	0.165714	1000	6034.483	0.392
0.2	0.16	0.875	7	0.2	1	1.16	0.165	1000	6060.606	0.424
0.2	0.16	0.888889	8	0.2	1	1.16	0.164444	1000	6081.081	0.456
0.2	0.16	0.9	9	0.2	1	1.16	0.164	1000	6097.561	0.488
0.2	0.16	0.909091	10	0.2	1	1.16	0.163636	1000	6111.111	0.52

payment of tax on profit per year *p*. These facts should be accounted when developing the company's dividend policy, because the value of the equity cost, k_e, is the economically justified amount of dividends.

The company's dividend policy should be as following:

(1) *If the company would like* ***to minimize the dividend payments:***

- in case of payments of tax on profit at the ends of periods they should be as frequently as tax legislation allows.
- in case of advanced payments of tax on profit they should be as seldom as tax legislation allows.

(2) *If the company would like* ***to maximize the dividend payments:***

Table 8.14 Dependence of the equity cost, k_e, on leverage level L (at $k_d = 0.16$ and at $p = 12$)

k_0	k_d	w_d	L	t	p	i	WACC	CF	V	k_e
0.2	0.16	0	0	0.2	12	1.012445	0.2	1000	5000	0.2
0.2	0.16	0.5	1	0.2	12	1.012445	0.178573	1000	5599.963	0.229145
0.2	0.16	0.666667	2	0.2	12	1.012445	0.17143	1000	5833.279	0.25829
0.2	0.16	0.75	3	0.2	12	1.012445	0.167859	1000	5957.383	0.287436
0.2	0.16	0.8	4	0.2	12	1.012445	0.165716	1000	6034.413	0.316581
0.2	0.16	0.833333	5	0.2	12	1.012445	0.164288	1000	6086.883	0.345726
0.2	0.16	0.857143	6	0.2	12	1.012445	0.163267	1000	6124.923	0.374871
0.2	0.16	0.875	7	0.2	12	1.012445	0.162502	1000	6153.767	0.404017
0.2	0.16	0.888889	8	0.2	12	1.012445	0.161907	1000	6176.39	0.433162
0.2	0.16	0.9	9	0.2	12	1.012445	0.161431	1000	6194.608	0.462307
0.2	0.16	0.909091	10	0.2	12	1.012445	0.161041	1000	6209.594	0.491452

Table 8.15 Dependence of the equity cost, k_e, on leverage level L (at $k_d = 0.16$ and at $p = 1$) (for advanced payments of tax of profit)

k_0	k_d	w_d	L	t	p	i	WACC	CF	V	k_e
0.2	0.16	0	0	0.2	1	1.16	0.2	1000	5000	0.2
0.2	0.16	0.5	1	0.2	1	1.16	0.1768	1000	5656.109	0.2256
0.2	0.16	0.666667	2	0.2	1	1.16	0.169067	1000	5914.826	0.2512
0.2	0.16	0.75	3	0.2	1	1.16	0.1652	1000	6053.269	0.2768
0.2	0.16	0.8	4	0.2	1	1.16	0.16288	1000	6139.489	0.3024
0.2	0.16	0.833333	5	0.2	1	1.16	0.161333	1000	6198.347	0.328
0.2	0.16	0.857143	6	0.2	1	1.16	0.160229	1000	6241.084	0.3536
0.2	0.16	0.875	7	0.2	1	1.16	0.1594	1000	6273.526	0.3792
0.2	0.16	0.888889	8	0.2	1	1.16	0.158756	1000	6298.992	0.4048
0.2	0.16	0.9	9	0.2	1	1.16	0.15824	1000	6319.515	0.4304
0.2	0.16	0.909091	10	0.2	1	1.16	0.157818	1000	6336.406	0.456

Table 8.16 Dependence of the equity cost, k_e, on leverage level L (at $k_d = 0.16$ and at $p = 12$) (for advanced payments of tax of profit)

k_0	k_d	w_d	L	t	p	i	WACC	CF	V	k_e
0.2	0.16	0	0	0.2	12	1.012445	0.2	1000	5000	0.2
0.2	0.16	0.5	1	0.2	12	1.012445	0.178306	1000	5608.338	0.228612
0.2	0.16	0.666667	2	0.2	12	1.012445	0.171075	1000	5845.403	0.257224
0.2	0.16	0.75	3	0.2	12	1.012445	0.167459	1000	5971.613	0.285836
0.2	0.16	0.8	4	0.2	12	1.012445	0.16529	1000	6049.99	0.314448
0.2	0.16	0.833333	5	0.2	12	1.012445	0.163843	1000	6103.394	0.34306
0.2	0.16	0.857143	6	0.2	12	1.012445	0.16281	1000	6142.121	0.371671
0.2	0.16	0.875	7	0.2	12	1.012445	0.162035	1000	6171.49	0.400283
0.2	0.16	0.888889	8	0.2	12	1.012445	0.161433	1000	6194.528	0.428895
0.2	0.16	0.9	9	0.2	12	1.012445	0.160951	1000	6213.082	0.457507
0.2	0.16	0.909091	10	0.2	12	1.012445	0.160556	1000	6228.345	0.486119

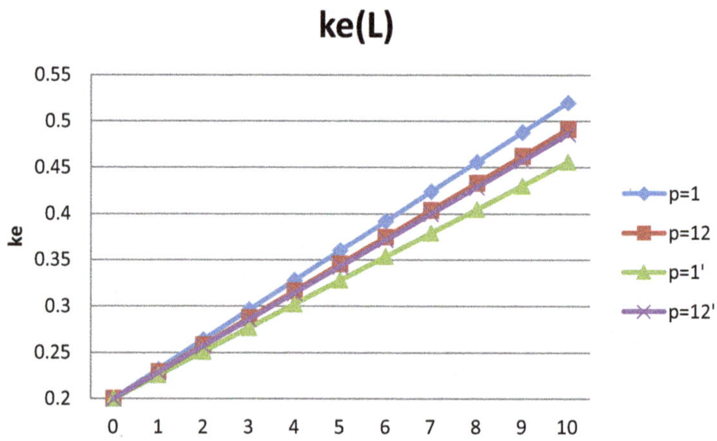

Fig. 8.7 Dependence of the dependence of the equity cost, k_e, on leverage level L at different p and fixed $k_d = 0.16$ ('means advanced payment of tax of profit)

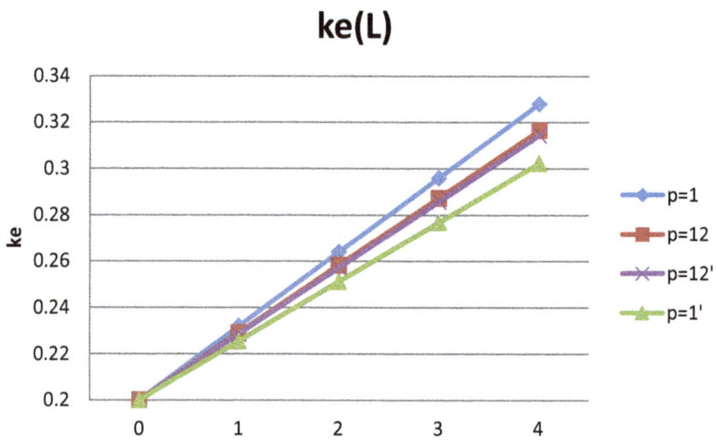

Fig. 8.8 Dependence of the dependence of the equity cost, k_e, on leverage level L at different p and fixed $k_d = 0.16$ ('means advanced payment of tax of profit) (larger scale)

- in case of payments of tax on profit at the ends of periods they should be as seldom as tax legislation allows.
- in case of advanced payments of tax on profit they should be as frequently as tax legislation allows.

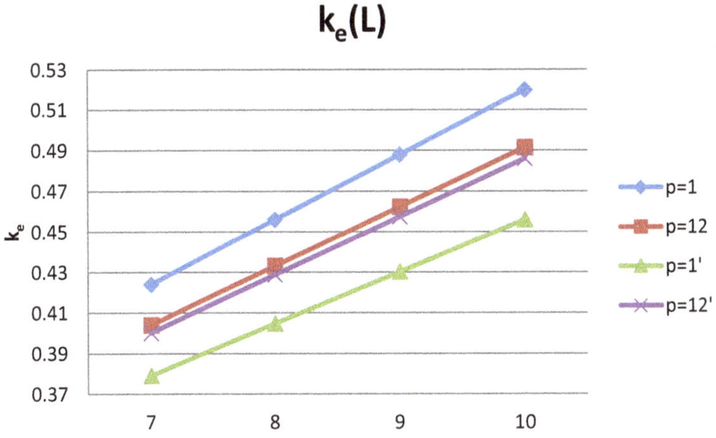

Fig. 8.9 Dependence of the dependence of the equity cost, k_e, on leverage level L at different p and fixed $k_d = 0.16$ ('means advanced payment of tax of profit) (larger scale)

Table 8.17 Dependence of the equity cost, k_e, and the weighted average cost of capital, WACC, on tax on profit value at $p = 1$

k_0	k_d	w_d	L	t	p	i	WACC	CF	V	k_e
0.2	0.16	0.5	1	0	1	1.16	0.2	1000	5000	0.24
0.2	0.16	0.5	1	0.1	1	1.16	0.19	1000	5263.158	0.236
0.2	0.16	0.5	1	0.2	1	1.16	0.18	1000	5555.556	0.232
0.2	0.16	0.5	1	0.3	1	1.16	0.17	1000	5882.353	0.228
0.2	0.16	0.5	1	0.4	1	1.16	0.16	1000	6250	0.224
0.2	0.16	0.5	1	0.5	1	1.16	0.15	1000	6666.667	0.22
0.2	0.16	0.5	1	0.6	1	1.16	0.14	1000	7142.857	0.216
0.2	0.16	0.5	1	0.7	1	1.16	0.13	1000	7692.308	0.212
0.2	0.16	0.5	1	0.8	1	1.16	0.12	1000	8333.333	0.208
0.2	0.16	0.5	1	0.9	1	1.16	0.11	1000	9090.909	0.204
0.2	0.16	0.5	1	1	1	1.16	0.1	1000	10,000	0.2

8.8 Dependence of the Equity Cost, k_e, and the Weighted Average Cost of Capital, WACC, on Tax on Profit Value t

In this paragraph we study the dependence of the equity cost, k_e, and the weighted average cost of capital, WACC, on tax on profit value.

Let us first analyze the dependence of the equity cost, k_e, on tax on profit value t at different p and fixed debt cost k_d.

From Tables 8.17, 8.18, 8.19, 8.20 and Figs. 8.10 and 8.11 it is seen, that the equity cost, k_e, decreases with the increase of on tax on profit value t for all frequencies of payment of tax on profit per year p.

Table 8.18 Dependence of the equity cost, k_e, and the weighted average cost of capital, WACC, on tax on profit value at $p = 12$

k_0	k_d	w_d	L	t	p	i	WACC	CF	V	k_e
0.2	0.16	0.5	1	0	12	1.012445	0.2	1000	5000	0.24
0.2	0.16	0.5	1	0.1	12	1.012445	0.189286	1000	5283.002	0.234573
0.2	0.16	0.5	1	0.2	12	1.012445	0.178573	1000	5599.963	0.229145
0.2	0.16	0.5	1	0.3	12	1.012445	0.167859	1000	5957.383	0.223718
0.2	0.16	0.5	1	0.4	12	1.012445	0.157145	1000	6363.54	0.21829
0.2	0.16	0.5	1	0.5	12	1.012445	0.146432	1000	6829.129	0.212863
0.2	0.16	0.5	1	0.6	12	1.012445	0.135718	1000	7368.227	0.207436
0.2	0.16	0.5	1	0.7	12	1.012445	0.125004	1000	7999.733	0.202008
0.2	0.16	0.5	1	0.8	12	1.012445	0.11429	1000	8749.634	0.196581
0.2	0.16	0.5	1	0.9	12	1.012445	0.103577	1000	9654.672	0.191154
0.2	0.16	0.5	1	1	12	1.012445	0.092863	1000	10,768.54	0.185726

Table 8.19 Dependence of the equity cost, k_e, and the weighted average cost of capital, WACC, on tax on profit value at $p = 1$ (for advanced payments of tax of profit)

k_0	k_d	w_d	L	t	p	i	WACC	CF	V	k_e
0.2	0.16	0.5	1	0	1	1.16	0.2	1000	5000	0.24
0.2	0.16	0.5	1	0.1	1	1.16	0.1884	1000	5307.856	0.2328
0.2	0.16	0.5	1	0.2	1	1.16	0.1768	1000	5656.109	0.2256
0.2	0.16	0.5	1	0.3	1	1.16	0.1652	1000	6053.269	0.2184
0.2	0.16	0.5	1	0.4	1	1.16	0.1536	1000	6510.417	0.2112
0.2	0.16	0.5	1	0.5	1	1.16	0.142	1000	7042.254	0.204
0.2	0.16	0.5	1	0.6	1	1.16	0.1304	1000	7668.712	0.1968
0.2	0.16	0.5	1	0.7	1	1.16	0.1188	1000	8417.508	0.1896
0.2	0.16	0.5	1	0.8	1	1.16	0.1072	1000	9328.358	0.1824
0.2	0.16	0.5	1	0.9	1	1.16	0.0956	1000	10,460.25	0.1752
0.2	0.16	0.5	1	1	1	1.16	0.084	1000	11,904.76	0.168

Table 8.20 Dependence of the equity cost, k_e, and the weighted average cost of capital, WACC, on tax on profit value at $p = 12$ (for advanced payments of tax of profit)

k_0	k_d	w_d	L	t	p	i	WACC	CF	V	k_e
0.2	0.16	0.5	1	0	12	1.012445	0.2	1000	5000	0.24
0.2	0.16	0.5	1	0.1	12	1.012445	0.189153	1000	5286.726	0.234306
0.2	0.16	0.5	1	0.2	12	1.012445	0.178306	1000	5608.338	0.228612
0.2	0.16	0.5	1	0.3	12	1.012445	0.167459	1000	5971.613	0.222918
0.2	0.16	0.5	1	0.4	12	1.012445	0.156612	1000	6385.21	0.217224
0.2	0.16	0.5	1	0.5	12	1.012445	0.145765	1000	6860.363	0.21153
0.2	0.16	0.5	1	0.6	12	1.012445	0.134918	1000	7411.917	0.205836
0.2	0.16	0.5	1	0.7	12	1.012445	0.124071	1000	8059.911	0.200142
0.2	0.16	0.5	1	0.8	12	1.012445	0.113224	1000	8832.063	0.194448
0.2	0.16	0.5	1	0.9	12	1.012445	0.102377	1000	9767.838	0.188754
0.2	0.16	0.5	1	1	12	1.012445	0.09153	1000	10,925.41	0.18306

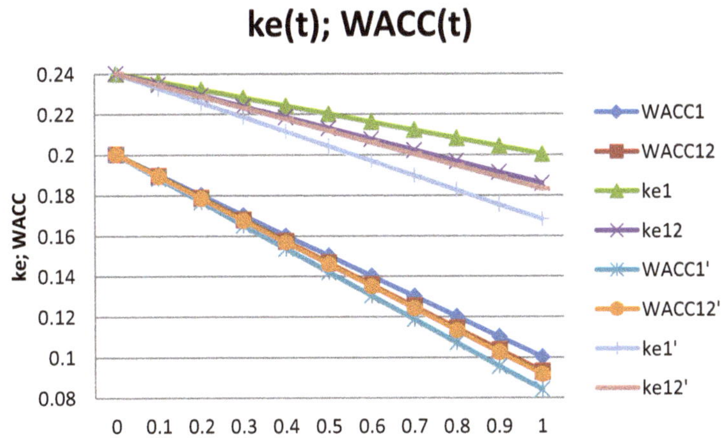

Fig. 8.10 Dependence of the equity cost, k_e, and the weighted average cost of capital, WACC, on tax on profit value at different p and fixed $k_d = 0.16$ ('means advanced payment of tax of profit)

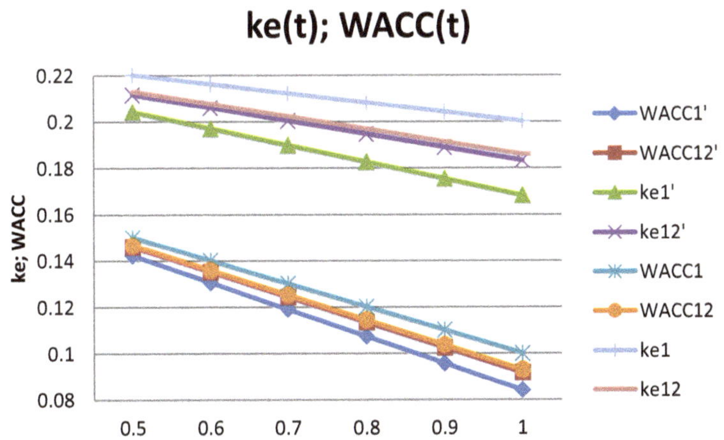

Fig. 8.11 Dependence of the dependence of the equity cost, k_e, and the weighted average cost of capital, WACC, on tax on profit value at different p and fixed $k_d = 0.16$ ('means advanced payment of tax of profit) (larger scale)

While in case of payments at the end of the periods the equity cost, k_e, decreases with the increase of the frequency of payments of tax on profit per year p (negative tilt of the curve k_e(t) increases on modulus), in case of advanced payments of tax on profit t the equity cost, k_e, increases with the increase of the frequency of payment of tax on profit per year p (negative tilt of the curve k_e(t) increases on modulus).

All the equity cost, k_e, values in case of advanced payments of tax on profit turns out to be lower than in case of payments at the end of the periods.

Let us now analyze the dependence of the weighted average cost of capital, WACC, on tax on profit value at different p and fixed debt cost k_d.

From Tables 8.17, 8.18, 8.19, 8.20 and Figs. 8.10 and 8.11 it is seen, that the weighted average cost of capital, WACC, decreases with the increase of on tax on profit value t for all frequencies of payment of tax on profit per year p.

While in case of payments at the end of the periods the weighted average cost of capital, WACC, decreases with the increase of the frequency of payments of tax on profit per year p (negative tilt of the curve WACC(t) increases on modulus), in case of advanced payments of tax on profit t the weighted average cost of capital, WACC, increases with the increase of the frequency of payment of tax on profit per year p (negative tilt of the curve WACC(t) increases on modulus).

All the weighted average cost of capital, WACC, values in case of advanced payments of tax on profit turns out to be lower than in case of payments at the end of the periods.

8.9 Discussions

Let us discuss the main obtained results.

1. The weighted average cost of capital, WACC, decreases with leverage level L at all values of frequency of payments of tax on profit p. With increase of the frequency of payments of tax on profit per year p the weighted average cost of capital, WACC, (1) decreases in the case of payments at the ends of the periods and WACC (2) increases in the case of advanced payments of tax on profit. Note, that the weighted average cost of capital, WACC, in the case of advanced payments of tax on profit for all values of frequency of payments turns out to be low than for any values of frequency of payments in the case of payments at the ends of the periods.
2. The weighted average cost of capital, WACC, decreases with increase of debt cost value k_d, except the classical Modigliani–Miller case $p = 1$ and payment at the end of the period. It supports obtained by us result that account of each of two effects (either advanced payments of tax on profit, or more frequent than once per year) leads to dependence of all main financial indicators, such as the weighted average cost of capital, WACC, company value, V, equity cost, k_e, etc. on debt cost value k_d, while in the classical Modigliani–Miller case these financial indicators DO NOT depend on debt cost value k_d.

   Similar to previous paragraph it is seen, that with increase of the frequency of payments of tax on profit per year p WACC (1) decreases in the case of payments at the ends of the periods and WACC (2) increases in the case of advanced payments of tax on profit. As we'll see below, this will lead to increase of company value, V, with increase of p in the case (1) and to decrease of company value, V, with increase of p in the case (2).

   Note, that the weighted average cost of capital, WACC, in the case of advanced payments of tax on profit for all values of frequency of payments turns out to be low than for any values of frequency of payments in the case of payments at the ends of the periods.

3. The company value, V, increases with the increase of leverage level L for all frequencies of payment of tax on profit per year p.

   While in case of payments at the end of the periods the company value, V, increases with the increase of the frequency of payment of tax on profit per year p, in case of advanced payments of tax on profit the company value, V, decreases with the increase of the frequency of payment of tax on profit per year p. These facts consistent with the dependence of the weighted average cost of capital, WACC, on the leverage level L.

   All the company values, V, in case of advanced payments of tax on profit turns out to be bigger than in case of payments at the end of the periods. Thus if the companies tend to increase their capitalization, they should pay tax on profit (1) in advance; (2) as seldom as tax legislation allows.

4. The equity cost, k_e, linearly increases with the increase of leverage level L for all frequencies of payment of tax on profit per year p.

   While in the case of tax on profit payments at the end of the periods the tilt of the curve of the equity cost, k_e, decreases with the increase of the frequency of payment of tax on profit per year p, in case of the advanced payments of tax on profit, the tilt of the curve of the equity cost, k_e, increases with the increase of the frequency of payment of tax on profit per year p. These facts should be accounted when developing the company's dividend policy, because the value of the equity cost, k_e, is the economically justified amount of dividends.

   The company's dividend policy should be as following:

(1) *If the company would like* ***to minimize the dividend payments****:*

   - in case of payments of tax on profit at the ends of periods they should be as frequently as tax legislation allows.
   - in case of advanced payments of tax on profit they should be as seldom as tax legislation allows.

(2) *If the company would like* ***to maximize the dividend payments****:*

   - in case of payments of tax on profit at the ends of periods they should be as seldom as tax legislation allows.
   - in case of advanced payments of tax on profit they should be as frequently as tax legislation allows.

5. The equity cost, k_e, decreases with the increase of on tax on profit value t for all frequencies of payment of tax on profit per year p.

While in case of payments at the end of the periods the equity cost, k_e, decreases with the increase of the frequency of payments of tax on profit per year p (negative tilt of the curve $k_e(t)$ increases on modulus), in case of advanced payments of tax on profit t the equity cost, k_e, increases with the increase of the frequency of payment of tax on profit per year p (negative tilt of the curve $k_e(t)$ increases on modulus).

All the equity cost, k_e, values in case of advanced payments of tax on profit turns out to be lower than in case of payments at the end of the periods.

The weighted average cost of capital, WACC, decreases with the increase of on tax on profit value t for all frequencies of payment of tax on profit per year p.

While in case of payments at the end of the periods the weighted average cost of capital, WACC, decreases with the increase of the frequency of payments of tax on profit per year p (negative tilt of the curve WACC(t) increases on modulus), in case of advanced payments of tax on profit t the weighted average cost of capital, WACC, increases with the increase of the frequency of payment of tax on profit per year p (negative tilt of the curve WACC(t) increases on modulus).

All the weighted average cost of capital, WACC, values in case of advanced payments of tax on profit turns out to be lower than in case of payments at the end of the periods.

8.10 Conclusions

Brusov and Filatova [1] have modified the Modigliani–Miller theory for the case of arbitrary frequency of payment of tax on profit t and show that the account of frequency of payment of tax on profit within the Modigliani–Miller theory turns out to be very important and leads to significant consequences: (1) all Modigliani–Miller theorems, statements and all formulas change; (2) all main financial indicators, such as the weighted average cost of capital, WACC, company value, V and equity cost, k_e depend on the frequency of tax on profit payments p; (3) this allows to company choose the number of payments of tax of profit per year, as many, as it is profitable to it (of course, within actual tax legislation).

As it was shown in that paper [1], more frequent payments of income tax are beneficial for both parties: for the company and for the tax regulator: for the company, this leads to decrease of the cost of attracting capital and thus to an increase in the value of the company, and for the tax regulator, earlier payments are beneficial due to the time value of money.

But companies could (and sometimes should) pay tax of profit in advance. In current paper we use two modifications of the Modigliani–Miller theory simultaneously: (1) frequent payments of income tax; (2) advance payments of income tax and show that this fact changes drastically the main conclusions of the paper [1]: more frequent payments of income tax are *NOT* beneficial for the company: this leads to increase of the cost of attracting capital (the weighted average cost of capital, WACC, and the equity cost, k_e,) and thus to decrease in the value of the company, but for the tax regulator it remains beneficial: earlier payments are beneficial due to the time value of money.

Thus, the main conclusions of current Chapter are as follows:

1. For payments of tax of profit at the end of periods more frequent payments of income tax are beneficial for both parties: for the company and for the tax

regulator: (1) for the company, this leads to decrease of the cost of attracting capital, WACC, and thus to an increase in the value of the company, *V*, and (2) for the tax regulator, more frequent payments are beneficial due to the time value of money.

2. For payments of tax of profit in advance more frequent payments of income tax are NOT beneficial for the company: this leads to increase of the cost of attracting capital, WACC, and thus to decrease in the value of the company, *V*, but for the tax regulator it remains beneficial: earlier payments are beneficial due to the time value of money.

References

1. Brusov, P., & Filatova, T. (2021). The Modigliani—Miller theory with arbitrary frequency of payment of tax on profit. *Mathematics, 9*(11), 1198. https://doi.org/10.3390/math9111198
2. Modigliani, F., & Miller, M. (1958). The cost of capital, corporate finance, and the theory of investment. *The American Economic Review, 48*, 261–297.
3. Modigliani, F., & Miller, M. (1963). Corporate income taxes and the cost of capital: A correction. *The American Economic Review, 53*, 147–175.
4. Miller, M. (1977). Debt and taxes. *The Journal of Finance, 32*(2), 261–275. https://doi.org/10.1111/j.1540-6261.1977.tb03267.x
5. Hamada, R. (1969). Portfolio analysis, market equilibrium, and corporate finance. *Journal of Finance, 24*, 13–31.
6. Farber, A., Gillet, R., & Szafarz, A. (2007). A general formula for the WACC. *International Journal of Business, 11*(2), 211–218.
7. Harris, R., & Pringle, J. (1985). Risk–Adjusted Discount Rates—Extension form the Average–Risk Case. *The Journal of Financial Research, 8*(3), 237–244.
8. Berk, J., & DeMarzo, P. (2007). *Corporate finance*. Pearson-Addison Wesley.
9. Ross, S., Westerfiel, R., & Jaffee, D. (2005). *Corporate finance* (7th ed.). McGraw Hill.
10. Brusov, P. N., Filatova, T. V., Orekhova, N. P., Kulik, V. L., & Weil, I. (2019). Ratings of the investment projects of arbitrary durations: New methodology. *Journal of Reviews on Global Economics, 8*, 437–448. https://doi.org/10.6000/1929-7092.2019.08.37
11. Modigliani, F., & Miller, M. (1966). Some estimates of the cost of capital to the electric utility industry 1954–1957. *The American Economic Review, 56*, 333–391.
12. Brusov, P., Filatova, T., & Orehova, N. (2020). *Ratings: Critical analysis and new approaches of quantitative and qualitative methodology*. Springer Nature Publishing. Monograph. https://www.springer.com/de/book/9783030562427
13. Brusov, P. N., Filatova, T. V., Orekhova, N. P., Kulik, V. L., Chang, S.-I., & Lin, Y. C. G. (2020). Application of the Modigliani–Miller theory, modified for the case of advance payments of tax on profit, in rating methodologies. *Journal of Reviews on Global Economics, 9*, 282–292. https://doi.org/10.6000/1929-7092.2020.09.28
14. Brusov, P. N., Filatova, T. V., Orekhova, N. P., Kulik, V. L., Chang, S.-I., & Lin, Y. C. G. (2020). Modification of the Modigliani–Miller theory for the case of advance payments of tax on profit. *Journal of Reviews on Global Economics, 9*, 257–267. https://doi.org/10.6000/1929-7092.2020.09.25
15. Arzac, E. R., & Glosten, L. R. (2005). A reconsideration of tax shield valuation. *European Financial Management,* 11(4), 453–461 (2007. *International Journal of Business*, 12(3), 411).
16. Brusov, P., Filatova, T., Orehova, N., & Eskindarov, M. (2015). *Modern corporate finance, investments and taxation*. Springer International Publishing. 373 p. monograph. https://www.springer.com/gp/book/9783319147314

17. Brusov, P., Filatova, T., Orehova, N., & Eskindarov, M. (2018). *Modern corporate finance, investments, taxation and ratings*. Springer Nature Publishing, 571 p. monograph.
18. Brusov, P. (2018). Editorial: Introduction on special issue on the banking system and financial markets of Russia and other countries: Problems and prospects. *Journal of Reviews on Global Economics, 7*, i–vi.
19. Brusov, P. N., Filatova, T. V., Orekhova, N. P., & Kulik, V. L. (2018). Rating: New approach. *Journal of Reviews on Global Economics, 7*, 37–62. https://doi.org/10.6000/1929-7092.2018.07.05
20. Brusov, P. N., Filatova, T. V., Orekhova, N. P., & Kulik, V. L. (2018). A "golden age" of the companies: Conditions of its existence. *Journal of Reviews on Global Economics, 7*, 88–103. https://doi.org/10.6000/1929-7092.2018.07.07
21. Brusov, P. N., Filatova, T. V., Orekhova, N. P., & Kulik, V. L. (2018). Rating methodology: New look and new horizons. *Journal of Reviews on Global Economics, 7*, 63–87. https://doi.org/10.6000/1929-7092.2018.07.06
22. Brusov, P. N., Filatova, T. V., Orekhova, N. P., Kulik, V. L., & Weil, I. (2018). New meaningful effects in modern capital structure theory. *Journal of Reviews on Global Economics, 7*, 104–122. https://doi.org/10.6000/1929-7092.2018.07.08
23. Brusov, P. (2018). Editorial. *Journal of Reviews on Global Economics, 7*, i–v.
24. Brusov, P., Filatova, T., Orekhova, N., Kulik, V., Weil, I., & Brailov, A. (2018). The impact of the central bank key rate and commercial banks credit rates on creating and maintaining of a favorable investment climate in the country. *Journal of Reviews on Global Economics, 7*, 360–376. https://doi.org/10.6000/1929-7092.2018.07.31
25. Brusov, P. N., Filatova, T. V., & Orekhova, N. P. (2018). *Modern corporate finance and investments*. Monograph. Knorus Publishing house, 517 p.
26. Fernandez, P. (2007). A general formula for the WACC: A comment. *International Journal of Business* (this issue).
27. Filatova, T. V., Brusov, P. N., Orekhova, N. P., & Kulik, V. L. (2018). Ratings of the long–term projects: New approach. *Journal of Reviews on Global Economics, 7*, 645–661. https://doi.org/10.6000/1929-7092.2018.07.59
28. Filatova, T., Orehova, N., & Brusova, A. (2008). Weighted average cost of capital in the theory of Modigliani–Miller, modified for a finite life–time company. *Bull FU, 48*, 68–77.
29. De Marzo, P. M. (1988). An extension of the Modigliani–Miller theorem to stochastic economies with incomplete markets and interdependent securities. *Journal of Economic Theory, 45*(2), 353–369.
30. Miles, J. A., & Ezzel, J. R. (1980). The weighted average cost of capital, perfect capital markets and project life: A clarification. *Journal of Financial and Quantitative Analysis, 15*(3), 719–730.
31. Myers, S. (2001). Capital structure. *The Journal of Economic Perspectives, 15*(2), 81–102.
32. Sethi, S. P., Derzko, N. A., & Lehoczky, J. P. A stochastic extension of the Miller-Modigliani framework. *Mathematical Finance, 1, 4*, 57–76.
33. Taggart, R. A. (1991). Consistent valuation and cost of capital expressions with corporate and personal taxes. *Financial Management, 20*, 8–20.

Chapter 9
Generalization of the Modigliani–Miller Theory for the Case of Variable Profit

9.1 Introduction

The original theory by Nobel Prize Winners Modigliani and Miller [1–3] has been modified by many authors and we shortly discuss some of these. A few important modifications have been done by the authors of this paper [4–6], who created the general theory of capital cost and capital structure, the Brusov–Filatova–Orekhova (BFO) theory, which generalized the Modigliani–Miller theory for the case of companies of arbitrary age (and arbitrary lifetime), as well as for the case of advance payments of tax on profit [6], for rating needs [5, 6] as well as for variable debt cost. Note that a stochastic extension of the Miller–Modigliani theory has been created by some authors [7, 8].

In the current paper, for the first time we have generalized the world-famous theory by Nobel Prize winners Modigliani and Miller for the case of variable profit, which significantly extends the application of the theory in practice, specifically in business valuation, ratings, investments and in other areas of the economy and of finance. We consider the case of growing profit as well as decreasing profit and show that all theorems, all statements by Modigliani and Miller (and all their main formulas) are changed significantly.

Within the new Generalized Modigliani–Miller theory (GMM theory), we study the dependence of the weighted average cost of capital, WACC, the equity cost, k_e, the discount rate, i, and the capitalization of the company, V, on leverage level L. Some important results have been obtained, which allows for the development of a new approach to financial policy and financial strategy for the company. Some of these are as follows:

- The discount rate for a leveraged company changes from the weighted average cost of capital, WACC, to WACC $-$ g (where g is growth rate), for an unleveraged company from k_0 to $k_0 - g$. This means that WACC and k_0 are no longer the discount rates they are in the case of classical Modigliani–Miller theory with constant profit.

P. Brusov, T. Filatova, *Business Valuation and Company Value*, Contributions to Finance and Accounting, https://doi.org/10.1007/978-3-031-90847-7_9

- All curves WACC(L) for different values of g start from one point, k_0. They decrease with L at $g < k_0$ and increase at $g > k_0$. It turns out that WACC grows with g, while real discount rates WACC $- g$ and $k_0 - g$ decrease with g. This leads to an increase of company capitalization along with g: $V =$ CF/WACC $- g$. Knowing the correct value of the discount rate allows for management of companies' financial flows.
- The equity cost, k_e, which grows linearly with the leverage level, increases with g: the tilt angle $k_e(L)$ grows along with the growth rate g.

A qualitatively new effect in corporate finance has been discovered: at rate $g < g^*$ the slope of the curve $k_e(L)$ turns out to be negative. Two these effects, which are absent in classical Modigliani–Miller theory, could significantly alter the principles behind the company's dividend policy, because the economically justified value of dividends is equal to equity cost.

The final effect is similar to the qualitatively new effect in corporate finance that was discovered by Brusov–Filatova–Orekhova within the BFO theory: the abnormal dependence of equity cost on leverage level at tax on profit rate T, which exceeds rate value T^*. This discovery also significantly alters the principles behind the company's dividend policy.

The current paper, contrary to papers on (stochastically) growing cash flows, can be directly applied for calculation of all the main indicators for companies. The results obtained will have applications in corporate finance, business valuation, ratings, etc.

The structure of the Chapter is as follows:

1. We give an introduction to the traditional approach of the Modigliani–Miller theory and to its modifications.

   - We generalize the Modigliani–Miller theory to the case of variable profit and obtain generalized Modigliani–Miller theorems, as well as new formulae for the weighted average cost of capital, WACC, equity cost, k_e, discount rate, i, and capitalization of the company, V.
   - Within the new Generalized Modigliani–Miller theory (GMM theory), we numerically study (with MS Excel) the dependence of the main financial indicators of the company (WACC, k_e, i, V) on leverage level L.
   - We discuss the results obtained and based on these, we arrive at some important conclusions.

   Below we discuss the problem of capital structure.

   Capital structure is the relationship between the debt and the equity capital of the company. Does capital structure impact the main financial indicators of the company, such as the cost of capital, profit, value of the company, etc., and, if so, how? The choice of an optimal capital structure, i.e., a capital structure that maximizes the company's capitalization, V, and minimizes the weighted average cost of capital, WACC, is one of the most important problems to be solved by the financial manager and senior management of a company. The first quantitative study of the impact of company capital structure on its financial indicators was performed by Modigliani and Miller [1]. Before 1958, the traditional approach, based on empirical data analysis, was used.

9.1.1 The Traditional Approach

The traditional approach supposes that the weighted average cost of capital, WACC, and the associated company capitalization value, $V = \mathrm{CF/WACC}$, depend on the capital structure (the level of leverage (L)). The reason for this is that the debt cost always turns out to be lower than the equity cost because the first one has lower risk, due to the fact that, in the event of bankruptcy, creditor claims are met prior to shareholders claims.

Thus, if the firm increases the share of lower–cost debt capital in the overall capital structure, this will lead to a lower weighted average cost of capital and WACC up to the limit, which does not cause violation of financial sustainability or result in increase of bankruptcy risk.

Investor profitability is required and the cost of equity grows with the leverage level; however, its growth has not lead to compensation of benefits from the increasing use of lower–cost debt capital. Therefore, at a low leverage level WACC decreases with the increase of leverage $L = D/S$ and company capitalization increases. At a high leverage level when the risk of bankruptcy becomes higher, WACC may increase with the increase of leverage L and company capitalization decreases. Thus, the tradeoff between advantages of debt financing at a low leverage level and its shortcomings at a high leverage level forms an optimal capital structure, which maximizes the company capitalization, V, and minimizes the weighted average cost of capital, WACC.

The traditional approach has existed up to 1958, when the first quantitative theory by Modigliani and Miller has appeared [1].

9.1.2 Modigliani–Miller Theory Without Taxes

In their first paper, [1] Modigliani and Miller (MM) under a lot of assumptions (there are no taxes, no bankruptcy costs, no transaction costs, perfect financial markets exist with symmetry information, equivalence in borrowing costs for both companies and investors, etc.), came to the conclusions that choosing of the ratio between the debt and equity capital does not affect company value as well as capital costs (Fig. 9.1). These conclusions were fundamentally different from the conclusions of the traditional approach.

Modigliani and Miller, under the above assumptions, have analyzed the impact of financial leverage, assuming the absence of any taxes (on corporate profit as well as individual one). They have formulated and proven two following theorems.

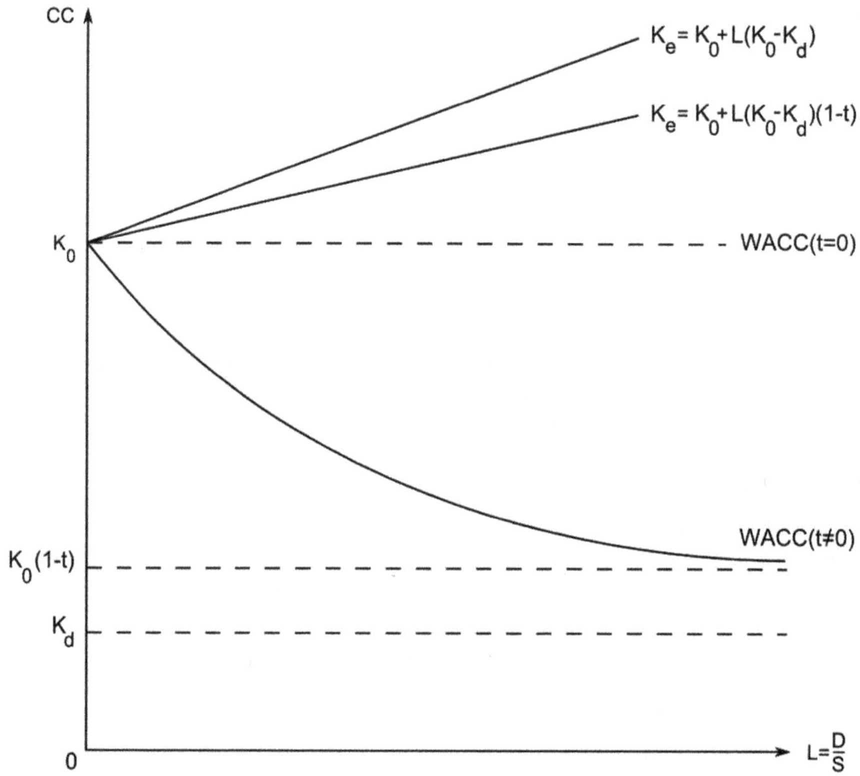

Fig. 9.1 Dependence of equity capital cost, debt cost and WACC on leverage in Modigliani–Miller theory without taxes ($t = 0$) and with taxes ($t \neq 0$)

Without taxes, the total cost of any company is determined by the value of its EBIT (Earnings Before Interest and Taxes), discounted with fixed rate k_0, corresponding to group of business risk of this company:

$$V = V_0 = \frac{\text{EBIT}}{k_0} \tag{9.1}$$

This leads to the following expressions for WACC:

$$\text{WACC} = k_0 \tag{9.2}$$

Note, that k_0 here and below is the weighted average cost of capital, WACC for an unleveraged company. For a leveraged company, k_0 is the equity cost (and weighted average cost of capital, WACC) at zero leverage level ($L = 0$).

From the first Modigliani–Miller theorem [1], it is easy to derive an expression for the equity capital cost:

$$\mathrm{WACC} = k_0 = k_e w_e + k_d w_d. \tag{9.3}$$

Finding from here k_e, one gets

$$k_e = \frac{k_0}{w_e} - k_d \frac{w_d}{w_e} = \frac{k_0(S + D)}{S} - k_d \frac{D}{S} = k_0 + (k_0 - k_d) \frac{D}{S} = k_0 + (k_0 - k_d) L \tag{9.4}$$

Here,

D	value of debt capital of the company;
S	value of equity capital of the company;
$k_d, w_d = \frac{D}{D+S}$	cost and fraction of debt capital of the company;
$k_e, w_e = \frac{S}{D+S}$	cost and fraction of equity capital of the company;
$L = D/S$	financial leverage;
WACC	weighted average cost of capital.

Thus, we come to second theorem of Modigliani–Miller theory regarding the equity cost of a leveraged company [1].

The equity cost of a leveraged company (k_e) could be found as k_0 (the cost of equity of an unleveraged company with the same asset risk), plus premium for risk, which is equal to the product of difference $(k_0 - k_d)$ and leverage level L:

$$k_e = k_0 + L(k_0 - k_d) \tag{9.5}$$

Formula (9.5) shows that equity cost of the company increases linearly with is leverage level.

9.2 Some Modifications of Modigliani–Miller Theory

9.2.1 Modigliani–Miller Theory with Taxes

In 1963 Modigliani and Miller [2] accounted for the effect of corporate taxes and obtained the following result for the capitalization value of a leveraged company, V:

$$V = V_0 + DT, \tag{9.6}$$

where V_0 is the value of unleveraged company, D is the debt value and T is the corporate taxes on profit rate.

The value of a leveraged company is equal to the value of the unleveraged company with the same asset risk, plus the value of tax shield arising from

financial leverage, which is equal to the product of corporate income tax rate (*T*) and the debt value (*D*).

The following section presents the formal derivation result for WACC and the equity capital cost (k_e) of the company with consideration of corporate taxes.

9.2.1.1 Weighted Average Cost of Capital, WACC

$$V = V_0 + DT$$

$$V(1 - w_d T) = V_0 \quad (9.7)$$

$$\frac{\mathrm{CF}}{\mathrm{WACC}}(1 - w_d T) = \frac{\mathrm{CF}}{k_0} \quad (9.8)$$

$$\mathrm{WACC} = k_0 \cdot (1 - w_d t) \quad (9.9)$$

The above is a very important formula for weighted average cost of capital (WACC) and is one of the main results of the Modigliani–Miller theory with taxes.

9.2.1.2 Equity Cost

Let us derive formula for equity cost.

On definition of the weighted average cost of capital with "tax shield", we have:

$$\mathrm{WACC} = k_0 w_e + k_d w_d (1 - T). \quad (9.10)$$

Equating Eqs. (9.9) and (9.10), one gets:

$$k_0(1 - w_d T) = k_0 w_e + k_d w_d (1 - T) \quad (9.11)$$

and from here, for equity cost, we develop the following expression:

$$\begin{aligned} k_e &= k_0 \frac{(1 - w_d T)}{w_e} - k_d \frac{w_d}{w_e}(1 - T) = k_0 \frac{1}{w_e} - k_0 \frac{w_d}{w_e} T - k_d \frac{D}{S}(1 - T) \\ &= k_0 \frac{D + S}{S} - k_0 \frac{D}{S} T - k_d \frac{D}{S}(1 - T) = k_0 + L(1 - T)(k_0 - k_d). \end{aligned} \quad (9.12)$$

Therefore, we see the following theorem obtained by Modigliani and Miller in 1963 [2]:

Equity cost of leveraged company (k_e) could be found as equity cost of unleveraged company (k_0) with the same business risk, plus premium for risk, the value of which is equal to the triple product of difference between the cost of

capital for an unleveraged firm and cost of debt $(k_0 - k_d)$**, leverage level** $(L = \frac{D}{S})$ **and tax corrector** $(1 - T)$**.**

It should be noted that Formula (9.12) is different from Formula (9.5) without tax only by the multiplier $(1 - T)$, suggesting the tax benefit of debt will lower the cost of equity as well.

Analysis of Formulas (9.5), (9.9) and (9.12) leads to the following conclusions. When leverage grows:

1. The value of company increases.
   - The weighted average cost of capital WACC decreases from k_0 (at $L = 0$) up to $k_0(1 - T)$ (at $L = \infty$) (when the company is funded solely by borrowed funds).
   - Equity cost increases linearly from k_0 (at $L = 0$) up to ∞ (at $L = \infty$).

   Within their theory, Modigliani and Miller [2] came to the following conclusions regarding the growth of financial leverage (Fig. 9.1).

9.2.2 *Taking into Account Market Risk: Hamada Model*

In 1969 Hamada [9] united CAPM (Capital Asset Pricing Model) and Modigliani–Miller theory with taxes. For the equity cost of a leveraged company, the following formula has been derived, which includes financial and business risk of company:

$$k_e = k_F + (k_M - k_F)b_U + (k_M - k_F)b_U\frac{D}{S}(1 - T), \tag{9.13}$$

here b_U is the CAPM βeta of the unleveraged company with the same asset risk as the leveraged company under consideration. Formula (9.13) represents the cost of equity (k_e) for a leveraged firm as a sum of three components: risk-free rate (k_F), risk premium for business/asset risk $(k_M - k_F)b_U$ and risk premium for financial risk $(k_M - k_F)b_U\frac{D}{S}(1 - T)$.

If the company does not have borrowing ($D = 0$), the financial risk factor will be equal to zero (the third term is drawn to zero), and equity holder will only require the premium for business/asset risk.

9.2.3 *The Account of Corporate and Individual Taxes (Miller Model)*

In their second article, Modigliani and Miller [2] only considered corporate taxes benefit of debt but did not take into account the effect of individual investors' income taxes.

Merton Miller [10], in 1977, developed such a model, showing the influence of financial leverage on the capitalization of the company when the effects of both corporate and individual taxes are accounted for, i.e., there is double taxation on corporate earnings. We will use the following definitions: T_C—corporate taxes rate and T_S—a weighted average value of effective taxes rates on dividends and capital gains on shares. With the same other assumptions that have been made for Modigliani–Miller models previously, the unleveraged company value can be determined as follows:

$$V_U = \frac{\text{EBIT}(1 - T_C)(1 - T_S)}{k_0}.$$

A term $(1 - T_S)$ accounts for the individual taxes. The numerator in V_U indicates which part of the operating company's profit remains in the possession of the investors, after the company earnings are taxed twice: by corporate taxes and individual taxes. Since individual taxes reduce investors' residual income, it also reduces the overall assessment of the unleveraged company value.

9.2.4 Brusov–Filatova–Orekhova (BFO) Theory

One of the most serious limitations of the Modigliani–Miller theory is the suggestion about the perpetuity of the companies. In 2008, Brusov–Filatova–Orekhova [4] considered this limitation and showed that accounting for the finite lifetime length (or arbitrary age n) of the company leads to significant changes of all Modigliani–Miller results [1–3]: capitalization of the company V is changed, as well as the equity cost, k_e, and the weighted average cost of capital, WACC, in the presence of corporative taxes. Moreover, a number of new findings for corporate finance, in the Brusov–Filatova–Orekhova theory [4], are absent in Modigliani–Miller theory.

The formula for weighted average cost of capital (WACC) for the company with arbitrary age n, as derived by Brusov–Filatova–Orekhova, has the following form [4]:

$$\frac{1 - (1 + \text{WACC})^{-n}}{\text{WACC}} = \frac{1 - (1 + k_0)^{-n}}{k_0[1 - w_d T(1 - (1 + k_d)^{-n})]} \tag{9.14}$$

Here, $w_d \left(= \frac{D}{D+S}\right)$ is the debt share; k_0 is the cost of capital for an unleveraged firm with the same asset risk; k_d the cost of debt; T denotes the corporate taxes rate; and n stands for the firm's lifetime length (age).

A perpetuity (Modigliani–Miller) limit could be easily obtained from (9.14), substituting $n \to \infty$.

A lot of meaningful effects have been discovered within the BFO theory: these effects are absent within MM theory. BFO theory has challenged some main existing principles of financial management: among them the trade-off theory, which has been prevailing for many decades and established the foundation to claim the existence of an optimal capital structure for firms. However, BFO theory has proven the bankruptcy of trade-off theory [4].

9.2.5 The General WACC Formula

A more general formula for WACC, than in the famous Modigliani–Miller theory (MM) has been derived and discussed by a few authors in 2006–2007 [11–14]. It takes the following form (Eq. (9.18) in [11]).

$$\text{WACC} = k_0(1 - w_d t) - k_d t w_d + k_{TS} t w_d \tag{9.15}$$

where k_0 is the required return on unleveraged company, k_d is the required return on its debt, k_{TS} is the expected return on the tax shield and t is the corporate taxes rate.

This formula is derived from the definition of the weighted average cost of capital and the balance sheet identity (for a similar presentation, see [13]). At any point in time, it should therefore be verified, regardless of whether returns are annually or continuously compounded.

Practical applicability of Eq. (9.15) (while it is fairly general) requires additional conditions. Indeed, when the WACC is constant over time, the value of a leveraged company can be computed by discounting with the WACC of the unlevered free cash–flows. Therefore, it is interesting to consider the special cases when WACC is constant. The resulting formulas can also be found in textbooks [15, 16].

It was assumed by Modigliani and Miller [2] that the debt value D is constant. As the expected after-tax cash-flow of the unleveraged company is fixed, V_0 is also constant. By this assumption, $k_{TS} = k_D$ and the value of the tax shield is $\text{TS} = tD$. Therefore, the capitalization of the leveraged company V is a constant and the general WACC Formula (9.15) simplifies to a constant WACC:

$$\text{WACC} = k_0(1 - w_d T)$$

However, our opinion is that "classical" Modigliani–Miller (MM) theory, which suggests that the expected returns on the debt k_d and the tax shield k_{TS} are equals (because both of them have debt nature), is much more reasonable and in our paper we modify the "classical" Modigliani–Miller (MM) theory, which is still widely used in practice.

9.2.6 *Trade-Off Theory*

The world-famous **trade-off theory** has been considered the cornerstone in the solution of the problem of optimal capital structure for a company for many decades and is still used today for decision analysis on capital structure. Below we give two examples.

Frank, M., and Goyal, V., in their 2009 paper [17], "examines the relative importance of many factors in the capital structure decisions of publicly traded American firms from 1950 to 2003. The most reliable factors for explaining market leverage are: median industry leverage (+ effect on leverage), market-to-book assets ratio (−), tangibility (+), profits (−), log of assets (+), and expected inflation (+)". In addition, the authors have found that "dividend-paying firms tend to have lower leverage. When considering book leverage, somewhat similar effects are found. However, for book leverage, the impact of firm size, the market-to-book ratio, and the effect of inflation are not reliable." The empirical evidence seems to be reasonably consistent with some versions of the trade-off theory of capital structure.

Serrasqueiro, Z., and Caetano, A., in 2015 [18], analyzed "to what extent decisions on the capital structure of small and medium–sized enterprises (SMEs) are closer to the assumptions of trade–off theory or to the assumptions of hierarchy theory. They used a sample of small and medium–sized enterprises located in the Portuguese hinterland, using dynamic LSDVC as the valuation method, and the empirical evidence suggests that the most profitable and oldest SMEs are less leveraged, confirming Pecking Order Theory's forecasts. Larger SMEs are leveraging more borrowing, confirming the predictions of trade–off theory and hierarchy theory. In addition, SMEs are significantly adjusting their current debt levels towards the optimal debt ratio, which is consistent with the predictions of the compromise theory. It was concluded that theories of compromise and hierarchy are not mutually exclusive in explaining capital structure decisions of small and medium–sized enterprises".

However, the bankruptcy of trade-off theory has been proven by Brusov et al. in 2013 [4]. They have shown that risky debt financing (and growing credit rate near the bankruptcy) in contrast to waiting results does not lead to growing of weighted average cost of capital, WACC, which still decreases with leverage. This means the absence of minimum in the dependence of WACC on leverage level as well as the absence of maximum in the dependence of company capitalization on leverage. Thus, the well-known trade-off theory lacks an optimal capital structure. The explanation for this fact was made by Brusov et al. in 2013 [4] by analyzing the dependence of the cost of equity capital on the leverage level on the assumption that debt capital is risky.

Modigliani–Miller have considered tax shields from the interest on debt can increase the value of companies. In 1980, De Angelo and Masulis [19] moved further in the theoretical examination of tax shields. They have noted that there are tax deductibles for companies other than debt to reduce their corporate tax burden and debt and non-debt tax shields should be accounted for. Depreciation, investment tax credits, or net-loss carry forwards could represents examples of such kind of non-debt tax shields. The first to test for these tax effects (suggested by DeAngelo and Masulis in 1980 [19]) has been carried out by Bradley, Jarrell and Kim in 1984 [20]. In contrast to the prediction in De Angelo and Masulis [19], by regressing company-specific debt-to-value ratios on non-debt tax shields, they have shown that debt is positively related to non-debt tax shields as measured by depreciation and investment tax credits. Titman and Wessels in 1988 [18] found that "their results do not provide support for an effect on debt ratios arising from non-debt tax shields..." It was pointed out in 2003 by Graham [21], if a company invests heavily and uses debt financing to invest, a positive relation between such proxies for non-debt tax shield and debt may result. A mechanical positive relation of this type overwhelms and renders any substitution effects between debt and non-debt tax shields.

The original theory by Nobel Prize Winners Modigliani and Miller [1–3] has been modified by many authors and above we shortly discussed some of them. In next paragraph we will generalize for the first time the Modigliani–Miller theory for the case of variable profit.

9.3 Generalization of the Modigliani–Miller Theory for the Case of Variable Profit

9.3.1 *Modigliani–Miller Theory Without Taxes*

9.3.1.1 Company Value, *V*

In this part we generalize for the first time the world-famous theory by Nobel Prize winners Modigliani and Miller for the case of variable profit. Let us calculate capitalization for leverage company, assuming profit per period grows with growth rate g.

$$V = \frac{\mathrm{CF}}{1+\mathrm{WACC}} + \frac{\mathrm{CF}(1+g)}{(1+\mathrm{WACC})^2} + \frac{\mathrm{CF}(1+g)^2}{(1+\mathrm{WACC})^3} + \tag{9.16}$$

Here, CF is the annual profit of company, WACC is the weighted average cost of capital.

This is geometric progression with denominator:

$$\widehat{g} = \frac{1+g}{1+\mathrm{WACC}} \tag{9.17}$$

Summing (9.16), one gets

$$V = \frac{\mathrm{CF}}{1+\mathrm{WACC}} \cdot \frac{1}{1-\frac{1+g}{1+\mathrm{WACC}}} = \frac{\mathrm{CF}}{\mathrm{WACC}-g} \tag{9.18}$$

Let us calculate the capitalization for a financially independent company:

$$V_0 = \frac{\mathrm{CF}}{1+k_0} + \frac{\mathrm{CF}(1+g)}{(1+k_0)^2} + \frac{\mathrm{CF}(1+g)^2}{(1+k_0)^3} + \tag{9.19}$$

This is geometric progression with denominator:

$$\widehat{g} = \frac{1+g}{1+k_0} \tag{9.20}$$

Summing (9.19), one gets:

$$V_0 = \frac{\mathrm{CF}}{1+k_0} \cdot \frac{1}{1-\frac{1+g}{1+k_0}} = \frac{\mathrm{CF}}{k_0-g} \tag{9.21}$$

The original first theorem by Modigliani–Miller changes by the following one:

Without taxes, the total cost of any company with growing profit with rate g is determined by the value of its EBIT—Earnings Before Interest and Taxes—discounted with fixed rate $k_0 - g$, corresponding to group of business risk of this company:

$$V = V_0 = \frac{\mathrm{EBIT}}{k_0-g} \tag{9.22}$$

Note, that this formula for company capitalization is similar to one of the Gordon Growth Model (GGM), which assumes a company is perpetuity and pays dividends per share that increase at a constant rate. To estimate the value of a stock, the model takes the infinite series of dividends per share and discounts them back into the present using the required rate of return.

$$P = \frac{\mathrm{DPS}_1}{k_e-g} \tag{9.23}$$

where p—value of stock;

DPS_1—expected dividends one year from now (next period);

k_e—required rate of return for equity investors (equity cost);

g—growth rate in dividends forever.

Note that, despite some formal similarities, these are not the same model. In our model, CF is the company's annual income (not dividends), and we study all the financial indicators of the company (WACC, V, k_e, i, etc.), all of them (except V) are absent in Gordon's growth model.

9.3.1.2 The Weighted Average Cost of Capital, WACC

Below we calculate the WACC value:

$$V = V_0$$
$$\frac{\mathrm{CF}}{\mathrm{WACC} - g} = \frac{\mathrm{CF}}{k_0 - g}$$
$$\mathrm{WACC} = k_0 \tag{9.24}$$

It is seen, that the WACC value turns out to be the same as in case of classical Modigliani–Miller theory with constant profit.

9.3.1.3 Equity Cost, k_e

It is easy to show that the expression for equity cost turns out to be the same as in case of classical Modigliani–Miller theory with constant profit:

$$k_e = k_0 + (k_0 - k_d)L. \tag{9.25}$$

Therefore, as we show in the case of the Modigliani–Miller theory without taxes accounting of variable profit changes the first of three Modigliani–Miller theorems only. As we show below, in the case of the Modigliani–Miller theory with taxes (practically interesting case) all three Modigliani–Miller theorems (concerning value of the company, V, its WACC and its equity cost, k_e) change significantly.

9.3.2 Modigliani–Miller Theory with Taxes

9.3.2.1 The Weighted Average Cost of Capital, WACC

The following result is obtained by Modigliani and Miller [2]:

The value of leveraged company is equal to the value of the company of the same risk group used no leverage, increased by the value of tax shield arising from financial leverage, and equal to the product of rate of corporate income tax *T* and the value of debt *D*.

$$V = V_0 + DT \tag{9.26}$$

From here it is easy to get expression for WACC (see above).

$$V(1 - w_d T) = V_0 \tag{9.27}$$

$$\frac{\text{CF}}{\text{WACC}}(1 - w_\text{d} T) = \frac{\text{CF}}{k_0} \tag{9.28}$$

$$\text{WACC} = k_0 \cdot (1 - w_\text{d} t) \tag{9.29}$$

Under variable profit, as we have seen above in the case of Modigliani–Miller theory without taxes, one has the same statement for capitalization, but discount rates are different:

$$\text{WACC} \to \text{WACC} - g$$

$$k_0 \to k_0 - g \tag{9.30}$$

$$\frac{\text{CF}}{\text{WACC} - g}(1 - w_\text{d} T) = \frac{\text{CF}}{k_0 - g} \tag{9.31}$$

$$\text{WACC} - g = (k_0 - g) \cdot (1 - w_\text{d} T) \tag{9.32}$$

$$\text{WACC} = (k_0 - g) \cdot (1 - w_\text{d} T) + g \tag{9.33}$$

This expression for WACC is different from the original result for WACC by Modigliani and Miller (9.29):

$$\text{WACC} = k_0 \cdot (1 - w_\text{d} T)$$

In order to understand the character of dependence of WACC(L) let us calculate the derivative of WACC with respect to L using Eq. (9.33):

$$\text{WACC} = (k_0 - g) \cdot (1 - w_\text{d} T) + g = k_0 - (k_0 - g) \cdot T \cdot \frac{L}{1 + L} \tag{9.34}$$

$$\text{WACC}'_L = -\frac{(k_0 - g) \cdot T}{(1 + L)^2} \tag{9.35}$$

At $g < k_0$ the derivative WACC'_L is negative, thus WACC(L) decreases with L, while at the $g > k_0$ derivative WACC'_L is positive and WACC(L) increases with L. We will see these effects in part 4 under numerical calculations.

9.3.2.2 The Company Value, *V*

For the company value from the Formulas (9.30) and (9.32) we get the following formula:

$$V = \frac{\mathrm{CF}}{\mathrm{WACC} - g} = \frac{\mathrm{CF}}{(k_0 - g)\cdot(1 - w_d t)} \tag{9.36}$$

We should note that in case of growing rate g equals to equity cost (at $L = 0$) k_0, the company value V becomes infinite. This is the limitation of the perpetuity Modigliani–Miller theory. As we will see in our future publications such kind of restriction is absent for companies of finite ages when we consider the generalization of BFO theory for the case of variable profit.

9.3.2.3 Equity Cost, k_e

Let us calculate the equity cost k_e. From (9.33) we have for the weighted average cost of capital, WACC,

$$\mathrm{WACC} = (k_0 - g)\cdot(1 - w_d T) + g \tag{9.37}$$

As well according to the definition of the weighted average cost of capital, considering the "tax shield" we have:

$$\mathrm{WACC} = k_e w_e + k_d w_d (1 - T) \tag{9.38}$$

Equating (9.37) and (9.38), we obtain:

$$(k_0 - g)\cdot(1 - w_d T) + g = k_e w_e + k_d w_d (1 - T) \tag{9.39}$$

whence we get the following expression for the cost of equity:

$$\begin{aligned} k_e &= \mathrm{WACC}(1 + L) - L k_d (1 - T) \\ &= k_0 + L\cdot[(k_0 - k_d)\cdot(1 - T) + gT] \end{aligned} \tag{9.40}$$

This expression is different from Formula (9.12) in case of the classical Modigliani–Miller theory with tax. Thus, we have the following modified statement for equity cost within Generalized Modigliani–Miller theory.

Equity cost of leverage company k_e could be found as equity cost of unleveraged company k_0 of the same group of risk, plus premium for risk, the value which is equal to production of leverage level L on sum of production of difference $(k_0 - k_d)$ on tax corrector and gT:

$$k_e = k_0 + L[(k_0 - k_d)(1 - T) + gT] \tag{9.41}$$

The Qualitatively New Effect in Corporate Finance

From the Formula (9.41) it is seen that with decrease of g tilt angle of straight $k_e(L)$ decreases and at some value of $g = g^*$

$$k_e = k_0 \tag{9.42}$$

Let us find g^* value. Equating k_e to k_0 in Formula (9.41), one has:

$$L[(k_0 - k_d)(1 - T) + gT] = 0$$

and from here we find g^*:

$$g^* = -\frac{(k_0 - k_d)(1 - T)}{T} \tag{9.43}$$

It is clear that at with growth rate $g < g^*$ the slope of the straight $k_e(L)$ turns out to be negative.

Thus, the qualitatively new effect in corporate finance has been discovered: at growth rate $g < g^*$ the slope of the straight $k_e(L)$ turns out to be negative. This effect, which is absent in the classical Modigliani–Miller theory, could significantly alter the principles of the company's dividend policy. Rating methodology takes into account the company's dividend policy, thus the results obtained could change the credit rating of an issuer.

We have shown in Sect. 9.3, that in case of the Modigliani–Miller theory with taxes, generalized for the case of variable profit all three Modigliani–Miller statements (concerning value of the company, V, the weighted average cost of capital, WACC, and its equity cost, k_e) change significantly. The consequences of such changes will be investigated numerically in next part.

9.4 Results and Discussions

In this section we study numerically (within Microsoft Excel) the dependence of the weighted average cost of capital, WACC, discount rate, i, company value, V, and equity cost, k_e, on leverage level L in the Generalized Modigliani–Miller theory (GMM theory) at two values of equity cost k_0 (0.2 and 0.3) and different values of growth rate, g. The obtained results are discussed.

9.4.1 Dependence of WACC on Leverage Level L in Generalized Modigliani–Miller Theory (GMM Theory) at $k_0 = 0.2$ and Different Values of g (0.4; 0.3; 0.2; 0.0; −0.2; −0.3; −0.4)

We study first the dependence of the weighted average cost of capital, WACC, on leverage level L in Generalized Modigliani–Miller theory (GMM theory) at $k_0 = 0.2$ and different values of g (0.4; 0.3; 0.2; 0.0; −0.2; −0.3; −0.4).

The results of Table 9.1 are shown in Fig. 9.2.

From Fig. 9.2 it is seen that all curves WACC(L) for different g start from one point k_0, in this case from point (0; 0.2). They decrease with leverage level L at $g <$ 0.2 (at $g = 0$; ±0.2; −0.3; −0.4) and increase at $g > 0.2$ (at $g = 0.3$; 0.4). The curves WACC(L) increase with growth rate, g. Note, that cut–off value of g, which separate increasing curves WACC(L) from decreasing ones is equal to $k_0 = 0.2$, and WACC is constant at $g = k_0$ and equal to k_0. Below we check this observation at different value of k_0 ($k_0 = 0.3$).

9.4.2 Dependence of the Weighted Average Cost of Capital WACC on Leverage Level L in Generalized Modigliani–Miller Theory (GMM Theory) at $k_0 = 0.3$ and Different Values of g

Let us study the dependence of WACC on leverage level L in Generalized Modigliani–Miller theory (GMM theory) at $k_0 = 0.3$ and different values of g.

The results of Table 9.2 are shown in Fig. 9.3.

Table 9.1 Dependence of WACC on leverage level L in generalized Modigliani–Miller theory (GMM theory) at $k_0 = 0.2$ and $g = 0.4$; 0.3; 0.2; 0.0; −0.2; −0.3; −0.4

			WACC						
L	T	k_0	$g = 0.4$	$g = 0.3$	$g = 0.2$	$g = 0$	$g = -0.2$	$g = -0.3$	$g = -0.4$
0	0.2	0.2	0.20000	0.20000	0.20000	0.20000	0.20000	0.20000	0.20000
1	0.2	0.2	0.22000	0.21000	0.20000	0.18000	0.16000	0.15000	0.14000
2	0.2	0.2	0.22667	0.21333	0.20000	0.17333	0.14667	0.13333	0.12000
3	0.2	0.2	0.23000	0.21500	0.20000	0.17000	0.14000	0.12500	0.11000
4	0.2	0.2	0.23200	0.21600	0.20000	0.16800	0.13600	0.12000	0.10400
5	0.2	0.2	0.23333	0.21667	0.20000	0.16667	0.13333	0.11667	0.10000
6	0.2	0.2	0.23429	0.21714	0.20000	0.16571	0.13143	0.11429	0.09714
7	0.2	0.2	0.23500	0.21750	0.20000	0.16500	0.13000	0.11250	0.09500
8	0.2	0.2	0.23556	0.21778	0.20000	0.16444	0.12889	0.11111	0.09333
9	0.2	0.2	0.23600	0.21800	0.20000	0.16400	0.12800	0.11000	0.09200
10	0.2	0.2	0.23636	0.21818	0.20000	0.16364	0.12727	0.10909	0.09091

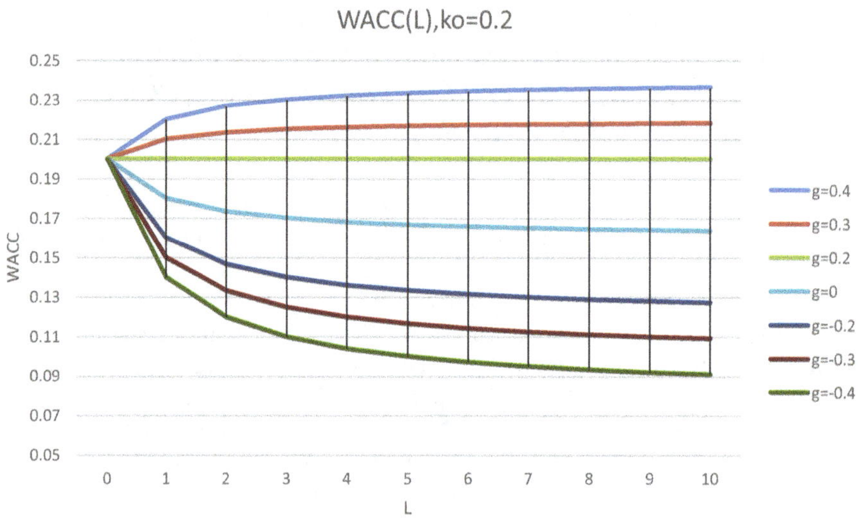

Fig. 9.2 Dependence of WACC on leverage level L in Generalized Modigliani–Miller theory (GMM theory) at $k_0 = 0.2$ and $g = 0; \pm0.2; \pm0.3; \pm0.4$

Table 9.2 Dependence of WACC on leverage level L in generalized Modigliani–Miller theory (GMM theory) at $k_0 = 0.3$ and $g = 0; \pm0.2; -0.3; -0.4$

			WACC						
L	T	k_0	$g = 0.4$	$g = 0.3$	$g = 0.2$	$g = 0$	$g = -0.2$	$g = -0.3$	$g = -0.4$
0	0.2	0.3	0.300	0.300	0.300	0.300	0.300	0.300	0.300
1	0.2	0.3	0.310	0.300	0.290	0.270	0.250	0.240	0.230
2	0.2	0.3	0.313	0.300	0.287	0.260	0.233	0.220	0.207
3	0.2	0.3	0.315	0.300	0.285	0.255	0.225	0.210	0.195
4	0.2	0.3	0.316	0.300	0.284	0.252	0.220	0.204	0.188
5	0.2	0.3	0.317	0.300	0.283	0.250	0.217	0.200	0.183
6	0.2	0.3	0.317	0.300	0.283	0.249	0.214	0.197	0.180
7	0.2	0.3	0.318	0.300	0.283	0.248	0.213	0.195	0.178
8	0.2	0.3	0.318	0.300	0.282	0.247	0.211	0.193	0.176
9	0.2	0.3	0.318	0.300	0.282	0.246	0.210	0.192	0.174
10	0.2	0.3	0.318	0.300	0.282	0.245	0.209	0.191	0.173

From Fig. 9.3 it is seen that all curves WACC(L) for different g start from one point, k_0, in this case from (0; 0.3). They decrease with L at $g < 0.3$ (at $g = 0; \pm0.2; \pm0.3; -0.4$) and increase at $g > 0.3$ (at $g = 0.4$). Note, that thr cut-off value of g, which separate increasing curves WACC(L) from decreasing ones as in above case is equal to k_0 ($k_0 = 0.3$), and WACC is constant at $g = k_0$ and equal to k_0. Therefore, our observation that this conclusion is valid at different value of k_0 is right.

From Figs. 9.2 and 9.3 it is seen that all curves WACC(L) for different g start from one point, k_0. They decrease with L at $g < k_0$ and increase at $g > k_0$. It turns out

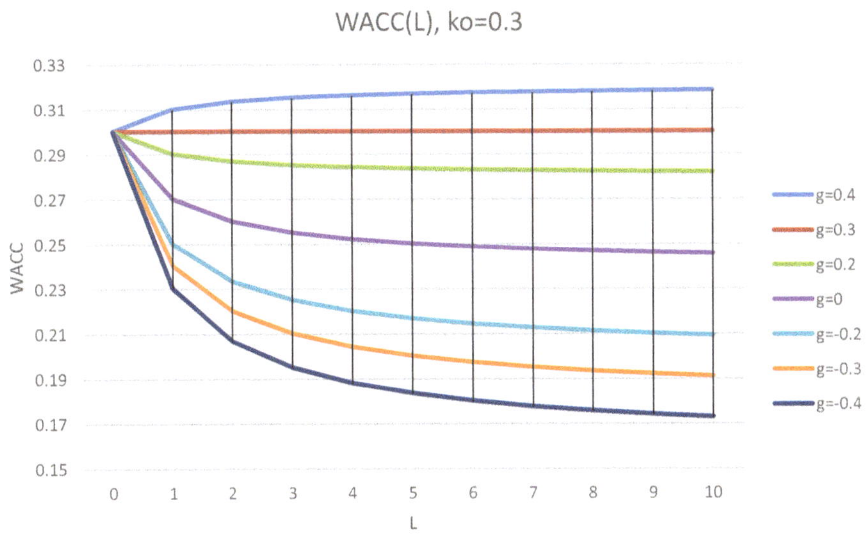

Fig. 9.3 Dependence of WACC on leverage level L in generalized Modigliani–Miller theory (GMM theory) at $k_0 = 0.3$ and $g = 0; \pm0.2; \pm0.3; \pm0.4$

that WACC grows with g, while, as we will see below, real discount rates WACC – g and $k_0 - g$ decrease with g. This leads to increase of company capitalization with g: $V = \text{CF/WACC} - g$.

9.4.3 *Dependence of Discount Rate* i *on Leverage Level* L *in Generalized Modigliani–Miller Theory (GMM Theory) at* $k_0 = 0.2$ *and Different Values of* g

Discount rate for the leverage company change from the weighted average cost of capital, WACC, to WACC – g (where g is growth rate), for a financially independent company from k_0 to $k_0 - g$. This means that WACC and k_0 are no longer the discount rates as it takes place in case of the classical Modigliani–Miller theory with constant profit. Below we study the dependence of discount rate i on leverage level L in Generalized Modigliani–Miller theory (GMM theory) at two values of k_0 (0.2; 0.3) and different values of g (–0.4; –0.3; –0.2; 0.0; 0.1; 0.15; 0.2; 0.3; 0.4). Let us start from $k_0 = 0.2$.

The results of Table 9.3 are shown in Fig. 9.4.

From Fig. 9.4 it is seen that discount rate i decreases with leverage level L at growth values $g < k_0$ (at $g = 0; 0.1; 0.15; \pm0.2; -0.3; -0.4$). Discount rate i in contrast to WACC decreases with g: this provides the increase of company value V with g.

Table 9.3 Dependence of discount rate i on leverage level L in generalized Modigliani–Miller theory (GMM theory) at $k_0 = 0.2$ and $g = -0.4$; -0.3; -0.2; 0.0; 0.1; 0.15; 0.2; 0.3; 0.4

			i								
L	t	k_0	$g = 0.4$	$g = 0.3$	$g =$ 0.2	$g =$ 0.15	$g =$ 0.1	$g = 0$	$g = -$ 0.2	$g = -$ 0.3	$g = -$ 0.4
0	0.2	0.2	−0.200	−0.100	0.000	0.050	0.100	0.200	0.400	0.500	0.600
1	0.2	0.2	−0.180	−0.090	0.000	0.045	0.090	0.180	0.360	0.450	0.540
2	0.2	0.2	−0.173	−0.087	0.000	0.043	0.087	0.173	0.347	0.433	0.520
3	0.2	0.2	−0.170	−0.085	0.000	0.043	0.085	0.170	0.340	0.425	0.510
4	0.2	0.2	−0.168	−0.084	0.000	0.042	0.084	0.168	0.336	0.420	0.504
5	0.2	0.2	−0.167	−0.083	0.000	0.042	0.083	0.167	0.333	0.417	0.500
6	0.2	0.2	−0.166	−0.083	0.000	0.041	0.083	0.166	0.331	0.414	0.497
7	0.2	0.2	−0.165	−0.083	0.000	0.041	0.083	0.165	0.330	0.413	0.495
8	0.2	0.2	−0.164	−0.082	0.000	0.041	0.082	0.164	0.329	0.411	0.493
9	0.2	0.2	−0.164	−0.082	0.000	0.041	0.082	0.164	0.328	0.410	0.492
10	0.2	0.2	−0.164	−0.082	0.000	0.041	0.082	0.164	0.327	0.409	0.491

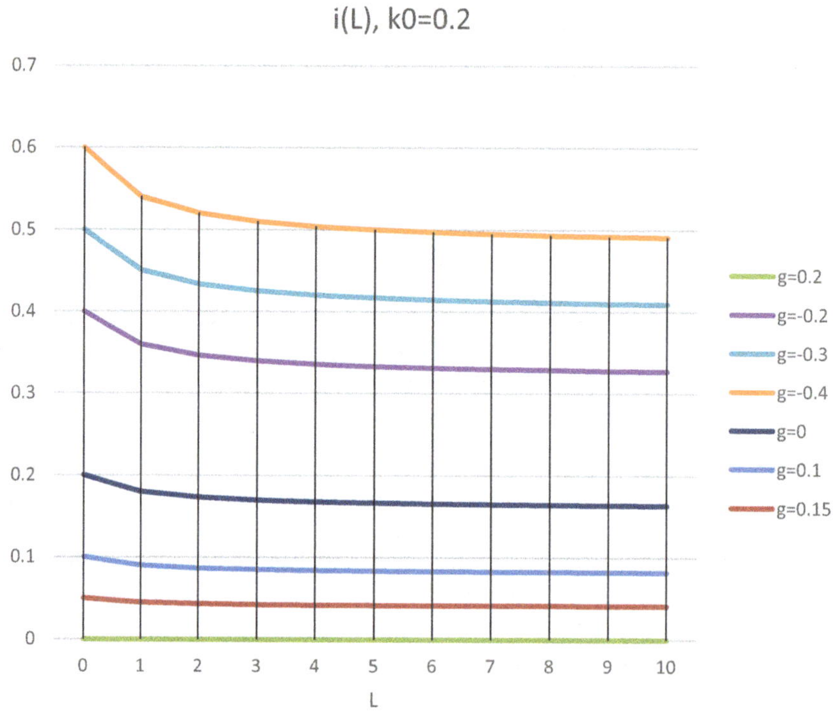

Fig. 9.4 Dependence of discount rate i on leverage level L in Generalized Modigliani–Miller theory (GMM theory) at $k_0 = 0.2$ and $g = 0$; 0.1; 0.15; ±0.2; −0.3; −0.4

At $g > k_0$ discount rate i increases with L, but being negative it is not shown in Fig. 9.4.

9.4.4 Dependence of Discount Rate i on Leverage Level L in Generalized Modigliani–Miller Theory (GMM Theory) at $k_0 = 0.3$ and Different Values of g

Let us study the dependence of discount rate i on leverage level L in Generalized Modigliani–Miller theory (GMM theory) at $k_0 = 0.3$ and different values of g.

The results of Table 9.4 are shown in Fig. 9.5.

From Fig. 9.5 it is seen that discount rate i decreases with leverage level L at growth values at $g < k_0$ (at $g = 0$; 0.1; 0.15; ±0.2; −0.3; −0.4). Discount rate i in opposite to WACC decreases with g: this provides the increase of company value V with g.

At $g > k_0$ discount rate i increases with L, but being negative it is not shown in Fig. 9.5.

From Figs. 9.4 and 9.5 it is seen that for $k_0 = 0.2$ and 0.3 discount rate i decreases with g: this provides the increase of company value V with g. Discount rate i decreases with L at $g < k_0$ (at $g = 0$; 0.1; 0.15; ±0.2; −0.3; −0.4). At $g > k_0$ discount rate i increases with L, but being negative it is not shown in Figs. 9.4 and 9.5.

Table 9.4 Dependence of discount rate i on leverage level L in generalized Modigliani–Miller theory (GMM theory) at $k_0 = 0.3$ and $g = 0$; 0.1; 0.15; ±0.2; ±0.3; ±0.4

			i								
L	T	k_0	$g = 0.4$	$g =$ 0.3	$g =$ 0.2	$g =$ 0.15	$g =$ 0.1	$g = 0$	$g = -$ 0.2	$g = -$ 0.3	$g = -$ 0.4
0	0.2	0.3	−0.100	0.000	0.100	0.150	0.200	0.300	0.500	0.600	0.700
1	0.2	0.3	−0.090	0.000	0.090	0.135	0.180	0.270	0.450	0.540	0.630
2	0.2	0.3	−0.087	0.000	0.087	0.130	0.173	0.260	0.433	0.520	0.607
3	0.2	0.3	−0.085	0.000	0.085	0.128	0.170	0.255	0.425	0.510	0.595
4	0.2	0.3	−0.084	0.000	0.084	0.126	0.168	0.252	0.420	0.504	0.588
5	0.2	0.3	−0.083	0.000	0.083	0.125	0.167	0.250	0.417	0.500	0.583
6	0.2	0.3	−0.083	0.000	0.083	0.124	0.166	0.249	0.414	0.497	0.580
7	0.2	0.3	−0.083	0.000	0.083	0.124	0.165	0.248	0.413	0.495	0.578
8	0.2	0.3	−0.082	0.000	0.082	0.123	0.164	0.247	0.411	0.493	0.576
9	0.2	0.3	−0.082	0.000	0.082	0.123	0.164	0.246	0.410	0.492	0.574
10	0.2	0.3	−0.082	0.000	0.082	0.123	0.164	0.245	0.409	0.491	0.573

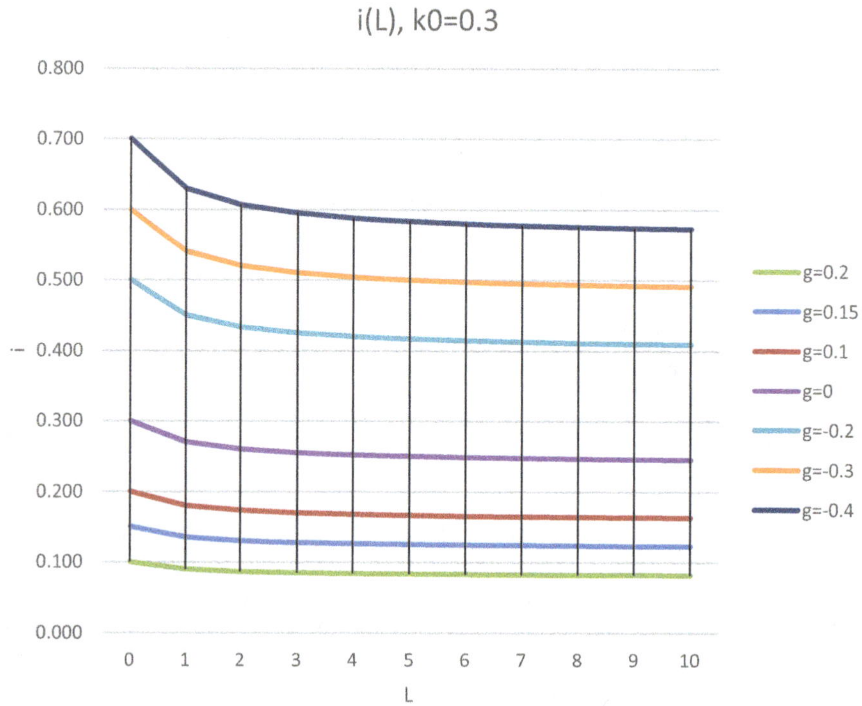

Fig. 9.5 Dependence of discount rate i on leverage level L in Generalized Modigliani–Miller theory (GMM theory) at $k_0 = 0.3$ and $g = 0$; 0.1; 0.15; ±0.2; −0.3; −0.4

9.4.5 Dependence of Company Value V on Leverage Level L in Generalized Modigliani–Miller Theory (GMM Theory) at $k_0 = 0.2$ and Different Values of g

Let us study the dependence of company value V on leverage level L in Generalized Modigliani–Miller theory (GMM theory) at $k_0 = 0.2$ and different values of g.

The results of Table 9.5 are shown in Fig. 9.6.

From Fig. 9.6 it is seen, that at $k_0 = 0.2$ and $g = 0$; 0.1; 0.15; −0.2; −0.3; −0.4 the company value V at fixed growth rate g increases with leverage level L in Generalized Modigliani–Miller theory (GMM theory). The company value V also increases with growth rate g.

Table 9.5 Dependence of company value V on leverage level L in generalized Modigliani–Miller theory (GMM theory) at $k_0 = 0.2$ and different values of $g = 0.0; 0.1; 0.15; -0.2; -0.3; -0.4$

					V					
L	t	k_0	k_d	CF	$g = 0.1$	$g = 0.15$	$g = 0$	$g = -0.2$	$g = -0.3$	$g = -0.4$
0	0.2	0.2	0.16	100	1000	2000	500	250	200	166.6667
1	0.2	0.2	0.16	100	1111.111	2222.222	555.5556	277.7778	222.2222	185.1852
2	0.2	0.2	0.16	100	1153.846	2307.692	576.9231	288.4615	230.7692	192.3077
3	0.2	0.2	0.16	100	1176.471	2352.941	588.2353	294.1176	235.2941	196.0784
4	0.2	0.2	0.16	100	1190.476	2380.952	595.2381	297.619	238.0952	198.4127
5	0.2	0.2	0.16	100	1200	2400	600	300	240	200
6	0.2	0.2	0.16	100	1206.897	2413.793	603.4483	301.7241	241.3793	201.1494
7	0.2	0.2	0.16	100	1212.121	2424.242	606.0606	303.0303	242.4242	202.0202
8	0.2	0.2	0.16	100	1216.216	2432.432	608.1081	304.0541	243.2432	202.7027
9	0.2	0.2	0.16	100	1219.512	2439.024	609.7561	304.878	243.9024	203.252
10	0.2	0.2	0.16	100	1222.222	2444.444	611.1111	305.5556	244.4444	203.7037

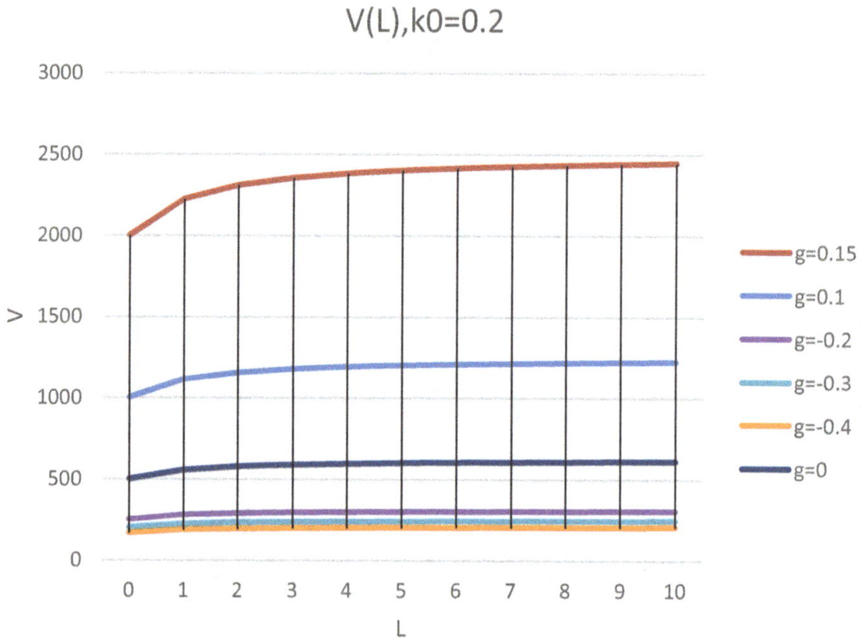

Fig. 9.6 Dependence of company value V on leverage level L in Generalized Modigliani–Miller theory (GMM theory) at $k_0 = 0.2$ and $g = 0; 0.1; 0.15; -0.2; -0.3; -0.4$

9.4.6 Dependence of Company Value V *on Leverage Level* L *in Generalized Modigliani–Miller Theory (GMM Theory) at* $k_0 = 0.3$ *and Different Values of* g

Let us study the dependence of company value V on leverage level L in generalized Modigliani–Miller theory (GMM theory) at $k_0 = 0.3$ and different values of g.

The results of Table 9.6 are shown in Fig. 9.7.

From Fig. 9.7 it is seen, that at $k_0 = 0.3$ and $g = 0; 0.1; 0.15; -0.2; -0.3; -0.4$ the company value V at fixed growth rate g increases with leverage level L in Generalized Modigliani–Miller theory (GMM theory). The company value V as well increases with growth rate g.

From Figs. 9.6 and 9.7 it is seen, that at two values of k_0 (0.2; 0.3) and $g = 0; 0.1; 0.15; -0.2; -0.3; -0.4$ the company value V at fixed growth rate g increases with leverage level L in Generalized Modigliani–Miller theory (GMM theory). The company value V as well increases with growth rate g.

We should note that in case of growing rate g equals to equity cost (at $L = 0$) k_0, the company value V becomes infinite. This is the limitation of the perpetuity Modigliani–Miller theory. As one can see in our future publications where we will consider the generalization of BFO theory for the case of variable profit such kind of restriction is absent for companies of finite ages.

Table 9.6 Dependence of company value V on leverage level L in generalized Modigliani–Miller theory (GMM theory) at $k_0 = 0.3$ and $g = 0.25$

L	t	k_0	k_d	CF	V						
					$g = 0.1$	$g = 0.15$	$g = 0.25$	$g = 0$	$g = -0.2$	$g = -0.3$	$g = -0.4$
0	0.2	0.3	0.16	100	500	666.6667	2000	333.3333	200	166.6667	142.8571
1	0.2	0.3	0.16	100	555.5556	740.7407	2222.222	370.3704	222.2222	185.1852	158.7302
2	0.2	0.3	0.16	100	576.9231	769.2308	2307.692	384.6154	230.7692	192.3077	164.8352
3	0.2	0.3	0.16	100	588.2353	784.3137	2352.941	392.1569	235.2941	196.0784	168.0672
4	0.2	0.3	0.16	100	595.2381	793.6508	2380.952	396.8254	238.0952	198.4127	170.068
5	0.2	0.3	0.16	100	600	800	2400	400	240	200	171.4286
6	0.2	0.3	0.16	100	603.4483	804.5977	2413.793	402.2989	241.3793	201.1494	172.4138
7	0.2	0.3	0.16	100	606.0606	808.0808	2424.242	404.0404	242.4242	202.0202	173.1602
8	0.2	0.3	0.16	100	608.1081	810.8108	2432.432	405.4054	243.2432	202.7027	173.7452
9	0.2	0.3	0.16	100	609.7561	813.0081	2439.024	406.5041	243.9024	203.252	174.216
10	0.2	0.3	0.16	100	611.1111	814.8148	2444.444	407.4074	244.4444	203.7037	174.6032

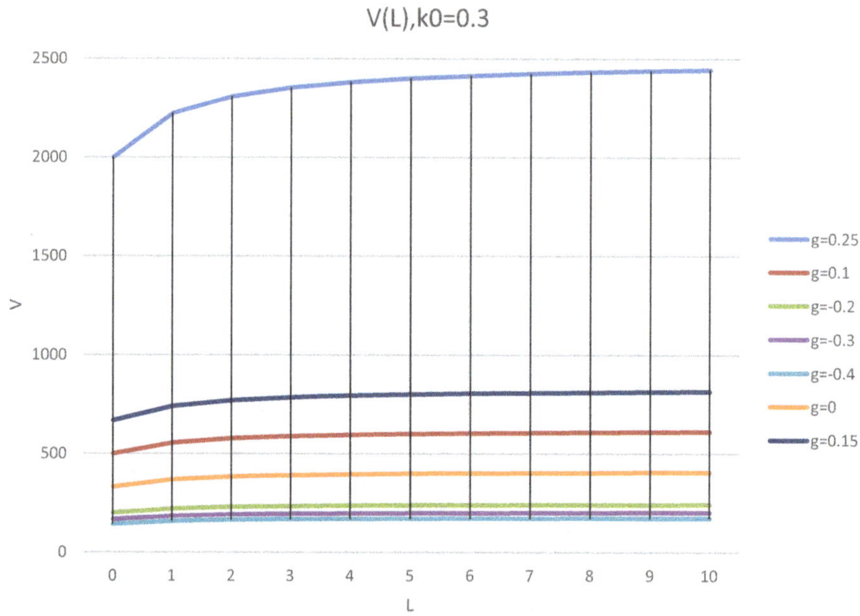

Fig. 9.7 Dependence of company value V on leverage level L in Generalized Modigliani–Miller theory (GMM theory) at $k_0 = 0.3$ and $g = 0; 0.1; 0.15; 0.25; -0.2; -0.3; -0.4$

9.4.7 Dependence of Equity Cost k_e on Leverage Level L in Generalized Modigliani–Miller Theory (GMM Theory) at $k_0 = 0.2$ and Different Values of g (0; ±0.2; ±0.3; ±0.4)

The economically justified value of dividends is equal to equity cost. Thus, knowing the dependence of equity cost of company on leverage level L, on growth rate g is very important, because it could impact the dividend policy of the company and help to company management to develop reasonable dividend policy.

Below we study the dependence of equity cost k_e on leverage level L in Generalized Modigliani–Miller theory (GMM theory) at $k_0 = 0.2$ and different values of g (0; ±0.2; ±0.3; ±0.4).

The results of Table 9.7 are shown in Fig. 9.8.

We have investigated the dependence of equity cost k_e on leverage level L in the Generalized Modigliani–Miller theory (GMM theory) at $k_0 = 0.2$ and $g = 0; \pm 0.2; \pm 0.3; \pm 0.4$. From Fig. 9.8 it is seen that the equity cost, k_e, which linearly grows with leverage level L increases with g: the tilt angle $k_e(L)$ grows with g. It is interesting, that at $k_0 = 0.2$; $k_d = 0.16$ and at $g^* = -0.16$ in accordance with formula $g^* = -\frac{(k_0 - k_d)\cdot(1 - T)}{T}$ the equity cost k_e turns out to be equal to k_0 and does not change with leverage level L.

Table 9.7 Dependence of equity cost k_e on leverage level L in generalized Modigliani–Miller theory (GMM theory) at $k_0 = 0.2$ and different values of g (0; ±0.2; ±0.3; ±0.4)

				k_e						
L	T	k_0	k_d	$g = 0.4$	$g = 0.3$	$g = 0.2$	$g = 0$	$g = -0.2$	$g = -0.3$	$g = -0.4$
0	0.2	0.2	0.16	0.2	0.2	0.2	0.2	0.2	0.2	0.2
1	0.2	0.2	0.16	0.312	0.292	0.272	0.232	0.192	0.172	0.152
2	0.2	0.2	0.16	0.424	0.384	0.344	0.264	0.184	0.144	0.104
3	0.2	0.2	0.16	0.536	0.476	0.416	0.296	0.176	0.116	0.056
4	0.2	0.2	0.16	0.648	0.568	0.488	0.328	0.168	0.088	0.008
5	0.2	0.2	0.16	0.76	0.66	0.560	0.360	0.16	0.060	−0.040
6	0.2	0.2	0.16	0.872	0.752	0.632	0.392	0.152	0.032	−0.088
7	0.2	0.2	0.16	0.984	0.844	0.704	0.424	0.144	0.004	−0.136
8	0.2	0.2	0.16	1.096	0.936	0.776	0.456	0.136	−0.024	−0.184
9	0.2	0.2	0.16	1.208	1.028	0.848	0.488	0.128	−0.052	−0.232
10	0.2	0.2	0.16	1.320	1.120	0.920	0.520	0.120	−0.080	−0.280

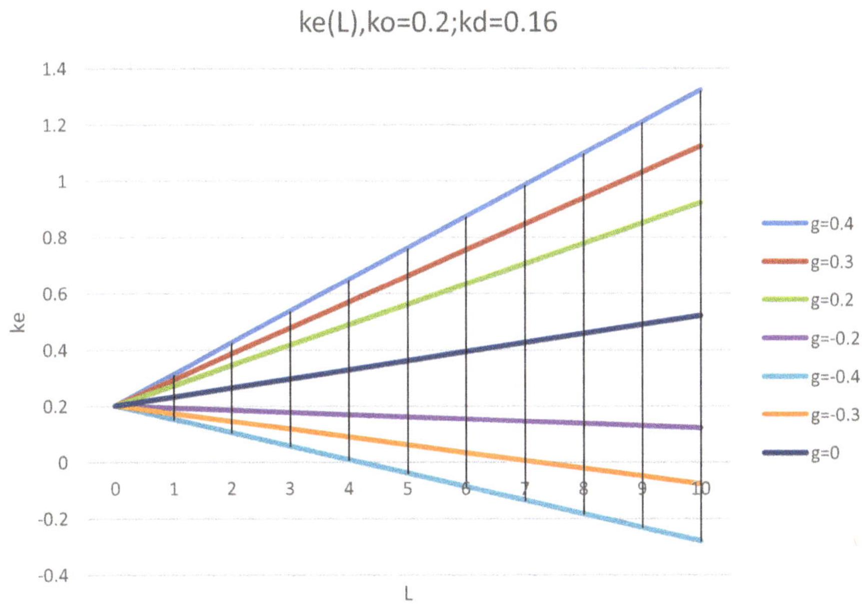

Fig. 9.8 Dependence of equity cost k_e on leverage level L in Generalized Modigliani–Miller theory (GMM theory) at $k_0 = 0.2$; $k_d = 0.16$ and $g = 0$; ±0.2; ±0.3; ±0.4

This should change the dividend policy of the company, because the economically justified value of dividends is equal to equity cost.

9.4.8 Dependence of Equity Cost k_e on Leverage Level L in Generalized Modigliani–Miller Theory (GMM Theory) at $k_0 = 0.3$ and Different Values of g

Presented below is the dependence of equity cost k_e on leverage level L in the Generalized Modigliani–Miller theory (GMM theory) at $k_0 = 0.3$ and different values of g (0; ±0.2; ±0.3; ±0.4).

The results of Table 9.8 are shown in Fig. 9.9.

We have investigated the dependence of equity cost k_e on leverage level L in the Generalized Modigliani–Miller theory (GMM theory) at $k_0 = 0.3$ and $g = 0$; ±0.2; ±0.3; ±0.4. From Fig. 9.9 it is seen that the equity cost, k_e, which linearly grows with leverage level increases with g: the tilt angle $k_e(L)$ grows with g. It is interesting, that at $k_0 = 0.3$; $k_d = 0.22$ and at $g^* = -0.32$ in accordance with Formula (9.43) the equity cost k_e turns out to be equal to k_0 and does not change with leverage level L.

Here, we have investigated the dependence of equity cost k_e on leverage level L in Generalized Modigliani–Miller theory (GMM theory) at two values of k_0 (0.2 and 0.3) and different values of g. The equity cost, k_e, which linearly grows with leverage level increases with g: the tilt angle $k_e(L)$ grows with g. This should change the dividend policy of the company, because the economically justified value of dividends is equal to equity cost.

It is interesting, that at $k_0 = 0.2$; $k_d = 0.16$ and at $g^* = -0.16$ and at $k_0 = 0.3$; $k_d = 0.22$ and at $g^* = -0.32$ in accordance with Formula (9.43) the equity cost k_e turns out to be equal to k_0 and does not change with leverage level L.

The qualitatively new effect in corporate finance has been discovered: at rate $g < g^*$ the slope of the curve $k_e(L)$ turns out to be negative. This effect, which is absent in classical Modigliani–Miller theory, could significantly alter the principles of the company's dividend policy.

Table 9.8 Dependence of equity cost k_e on leverage level L in generalized Modigliani–Miller theory (GMM theory) at $k_0 = 0.3$ and different g (0; ± 0.2; ±0.3; ±0.4)

				k_e						
L	T	k_0	k_d	$g = 0.4$	$g = 0.3$	$g = 0.2$	$g = 0$	$g = -0.2$	$g = -0.3$	$g = -0.4$
0	0.2	0.3	0.16	0.3	0.3	0.3	0.3	0.3	0.3	0.3
1	0.2	0.3	0.16	0.444	0.424	0.404	0.364	0.324	0.304	0.284
2	0.2	0.3	0.16	0.588	0.548	0.508	0.428	0.348	0.308	0.268
3	0.2	0.3	0.16	0.732	0.672	0.612	0.492	0.372	0.312	0.252
4	0.2	0.3	0.16	0.876	0.796	0.716	0.556	0.396	0.316	0.236
5	0.2	0.3	0.16	1.020	0.920	0.820	0.620	0.420	0.32	0.220
6	0.2	0.3	0.16	1.164	1.044	0.924	0.684	0.444	0.324	0.204
7	0.2	0.3	0.16	1.308	1.168	1.028	0.748	0.468	0.328	0.188
8	0.2	0.3	0.16	1.452	1.292	1.132	0.812	0.492	0.332	0.172
9	0.2	0.3	0.16	1.596	1.416	1.236	0.876	0.516	0.336	0.156
10	0.2	0.3	0.16	1.740	1.540	1.340	0.940	0.540	0.340	0.140

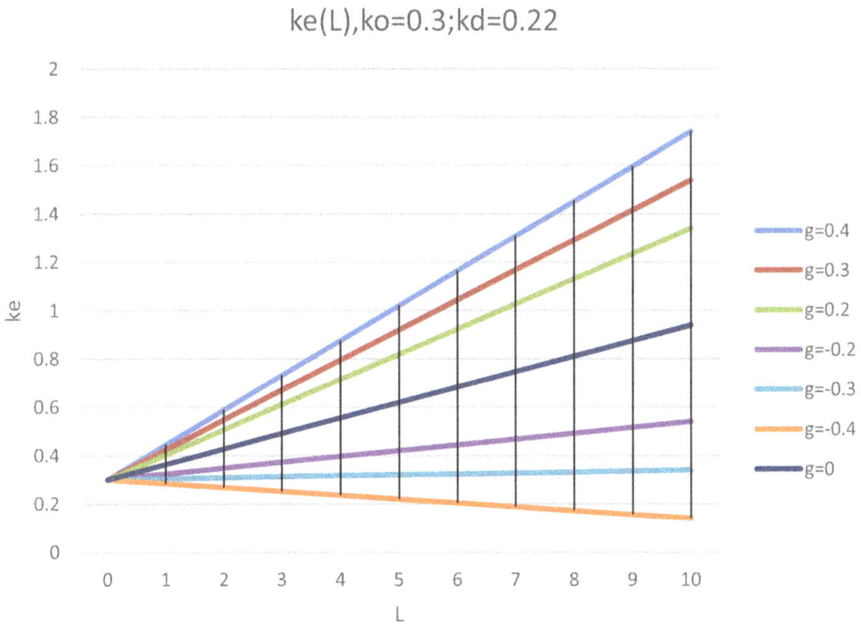

Fig. 9.9 Dependence of equity cost k_e on leverage level L in generalized Modigliani–Miller theory (GMM theory) at $k_0 = 0.3$; $k_d = 0.22$ and $g = 0$; ±0.2; ±0.3; ±0.4

The last effect is similar to the qualitatively new effect in corporate finance, which has been discovered by Brusov–Filatova–Orekhova within the BFO theory [4]: abnormal dependence of equity cost k_e on leverage level at tax on profit rate T, which exceed some rate value T^*: this discovery also significantly alters the principles of the company's dividend policy.

9.5 Conclusions

The main purpose of the current study is to generalize the Modigliani–Miller theory by taking into account one of the most important conditions of companies real functioning: variable profit. We use analytical and numerical methods: we derive all main formulas of the Generalized Modigliani–Miller theory theoretically and then use them to obtain all the main financial indicators of company and their dependences on different parameters by MS Excel. The generalized Modigliani–Miller theory significantly extends its application in practice, especially in business valuation, in ratings and in other areas of economy and finance. We consider the case of growing profit as well as decreasing of profit and show, that all the statements by Modigliani and Miller change significantly.

Within the new Generalized Modigliani–Miller theory (GMM theory), we study the dependence of the weighted average cost of capital, WACC, the cost of equity, k_e, the discount rate i and the capitalization of the company, V, on leverage level L at different values of growth rate g and obtained the following results:

1. Discount rate for leverage company changes from the weighted average cost of capital, WACC, to WACC $- g$ (where g is growing rate), for a financially independent company from k_0 to $k_0 - g$. This means that WACC and k0 are no longer the discount rates as it takes place in case of classical Modigliani–Miller theory with constant profit.
2. All curves WACC(L) for different g start from one point, k_0. They decrease with L at $g < k_0$ and increase at $g > k_0$. It turns out that WACC grows with g, while real discount rates WACC $- g$ and $k_0 - g$ decrease with g. This leads to increase of company capitalization with g: $V =$ CF/WACC $- g$.
3. We have investigated the dependence of equity cost k_e on leverage level L in the Generalized Modigliani–Miller theory (GMM theory) at two values of k_0 (0.2 and 0.3) and different values of g. The equity cost, k_e, which linearly grows with leverage level L increases with g: the tilt angle $k_e(L)$ grows with g. This should change the dividend policy of the company, because the economically justified value of dividends is equal to equity cost.

It is interesting, that at $k_0 = 0.2$; $k_d = 0.16$ and at $g^* = -0.16$ and at $k_0 = 0.3$; $k_d = 0.22$ and at $g^* = -0.32$ in accordance with Formula (9.43) the equity cost k_e turns out to be equal to k_0 and does not change with leverage level L.

A new qualitative finding related to corporate finance has been discovered in this study: at rate $g < g^*$ the slope of the curve $k_e(L)$ turns out to be negative. This effect, which is absent in the classical Modigliani–Miller theory, could significantly alter the principles of the company's dividend policy. Rating methodology should take into account the company's dividend policy [5], thus such results obtained could change the credit rating of an issuer.

The last effect found in this study is similar to the qualitatively new effect in corporate finance, discussed above, which has been discovered by Brusov–Filatova–Orekhova within BFO theory: abnormal dependence of equity cost on leverage level when corporate taxes rate T exceeds some rate value T^*; this discovery also significantly alters the principles of the company's dividend policy.

Thus, within the new Generalized Modigliani–Miller theory, a lot of significant implications are obtained, allowing us to develop new approach to financial policy and financial strategy of the company.

Note that the limitations of the Generalized Modigliani–Miller theory are similar to the limitations of the classical Modigliani–Miller theory and are connected with the numerous assumptions of classical Modigliani–Miller theory [1–3]. Therefore, the direction of future research is clear: we will consider the generalization of BFO theory, which describes the companies of arbitrary age, for the case of variable profit. The limitations of the Generalized Modigliani–Miller theory will be absent in the generalized BFO theory, which is valid for companies of an arbitrary age.

The Generalized Modigliani–Miller theory could and should be used in corporate finance and corporate management, in investments, business valuation, taxation, ratings, etc., for more correct assessments and more qualified management decisions.

Authors planning further development of the Modigliani–Miller theory should account for practical conditions of business and its application in all mentioned above areas. In particular, more frequent payments of tax of profit will be considered, combination of this case with advance payments of tax of profit, modification of BFO theory (which is valid for the companies of arbitrary ages) for the case of variable profit.

References

1. Modigliani, F., & Miller, M. (1958). The cost of capital, corporate finance, and the theory of investment. *The American Economic Review, 48*, 261–297.
2. Modigliani, F., & Miller, M. (1963). Corporate income taxes and the cost of capital: A correction. *The American Economic Review, 53*, 147–175.
3. Modigliani, F., & Miller, M. (1966). Some estimates of the cost of capital to the electric utility industry 1954–1957. *The American Economic Review, 56*, 333–391.
4. Brusov, P., Filatova, T., Orehova, N., & Eskindarov, M. (2018). *Modern corporate finance, investments, taxation and ratings* (2nd ed., pp. 1–571). Springer Nature Publishing.
5. Brusov, P., Filatova, T., & Orehova, N. (2020). *Ratings: Critical analysis and new approaches of quantitative and qualitative methodology* (pp. 1–369). Springer Nature Publishing. Retrieved from https://www.springer.com/de/book/9783030562427
6. Brusov, P. N., Filatova, T. V., Orekhova, N. P., Kulik, V. L., Chang, S.-I., & Lin, Y. C. G. (2020). Application of the Modigliani–Miller theory, modified for the case of advance payments of tax on profit, in rating methodologies. *Journal of Reviews on Global Economics, 9*, 282–292.
7. Sethi, S. P., Derzko, N. A., & Lehoczky, J. P. (1991). A stochastic extension of the Miller-Modigliani framework. *Mathematical Finance, 1*(4), 57–76.
8. DeMarzo, P. M. (1988). An extension of the Modigliani–Miller theorem to stochastic economies with incomplete markets and interdependent securities. *Journal of Economic Theory, 45*(2), 353–369.
9. Hamada, R. (1969). Portfolio analysis, market equilibrium, and corporate finance. *Journal of Finance, 24*(1), 13–31.
10. Miller, M. (1977). Debt and taxes. *The Journal of Finance, 32*(2), 261–275. https://doi.org/10.1111/j.1540-6261.1977.tb03267.x
11. Farber, A., Gillet, R., & Szafarz, A. (2006). A general formula for the WACC. *International Journal of Business, 11*(2), 211–218.
12. Fernandez, P. (2006). A general formula for the WACC: A comment. *International Journal of Business, 11*(2), 219.
13. Berk, J., & DeMarzo, P. (2007). *Corporate finance*. Pearson-Addison Wesley.
14. Harris, R., & Pringle, J. (1985). Risk-adjusted discount rates-extension form the average-risk case. *The Journal of Financial Research, 8*(3), 237–244.
15. Brealey, R., Myers, S., & Allen, F. (2005). *Principle of corporate finance* (7th ed.). McGraw Hill.
16. Ross, S., Westerfiel, R., & Jaffee, D. (2005). *Corporate finance* (7th ed.). McGraw Hill.
17. Frank, M., & Goyal, V. (2009). Capital structure decisions: Which factors are reliably important? *Financial Management, 38*(1), 1–37. https://doi.org/10.1111/j.1755-053X.2009.01026.x

18. Sheridan, T., & Wessels, R. (1988). The determinants of capital structure choice. *Journal of Finance, 43*, 1–19.
19. Harry, D. A., & Masulis, R. W. (1980). Optimal capital structure under corporate and personal taxation. *Journal of Financial Economics, 8*, 3–29.
20. Bradley, M., Jarrell, G. A., & Kim, E. H. (1984). On the existence of an optimal capital structure: Theory and evidence. *Journal of Finance, 39*, 857–878.
21. Graham, J. R. (2003). Taxes and corporate finance: A review. *Review of Financial Studies, 16*, 1075–1129.

Chapter 10
Modern Theory of Capital Cost and Capital Structure: Brusov-Filatova-Orekhova Theory (BFO Theory)

One of the serious limitations of the Modigliani–Miller theory is the suggestion about perpetuity of the companies. In 2008 Brusov–Filatova–Orehova [1] have lifted up this limitation and shown, that the accounting of the finite lifetime of the company leads to significant changes of all Modigliani–Miller results [2–4]: capitalization of the company is changed, as well as the equity cost, k_e, and the weighted average cost of capital, WACC, in the presence of corporative taxes. Besides a number of qualitatively new effects in corporate finance, obtained in Brusov–Filatova–Orehova theory [5–15], is absent in Modigliani–Miller theory.

Only in the absence of corporative taxes we give a rigorous proof of the Brusov–Filatova–Orekhova theorem, that cost of company equity, k_e, as well as its weighted average cost, WACC, do not depend on the lifetime or age of the company, so the Modigliani–Miller theory could be generalized for arbitrary life-time companies.

Until recently (before 2008, when the first paper by Brusov–Filatova–Orekhova [1] has appeared) the basic theory (and the first quantitative one) of the cost of capital and capital structure of companies was the theory of Nobel Prize winners Modigliani and Miller [2–4]. One of the serious limitations of the Modigliani–Miller theory is the suggestion about perpetuity of the companies. We lift up this limitation and show, that the accounting of the finite lifetime of the company leads to change of the equity cost, k_e, as well as of the weighted average cost of capital WACC in the presence of corporative taxes. The effect of leverage on the cost of equity capital of the company, k_e, with an arbitrary lifetime, and its weighted average cost of WACC is investigated. We give a rigorous proof of the Brusov–Filatova–Orekhova theorem, that in the absence of corporative taxes cost of company equity, k_e, as well as its weighted average cost, WACC, do not depend on the lifetime of the company.

P. Brusov, T. Filatova, *Business Valuation and Company Value*, Contributions to Finance and Accounting, https://doi.org/10.1007/978-3-031-90847-7_10

10.1 Companies with Arbitrary Lifetime: Brusov-Filatova-Orekhova Equation

Let us consider the situation with finite lifetime companies. First of all we will find the value of tax shields, TS, of the company for n years

$$\mathrm{TS} = k_d DT \sum_{t=1}^{n} (1+k_d)^{-t} = DT[1-(1+k_d)^{-n}]. \tag{10.1}$$

(We used the formula for the sum of n terms of a geometric progression).

Here, D is the value of debt capital; k_d—the cost of debt capital, T—income tax rate.

Next, we use the Modigliani–Miller theorem [2–4]:

The value of financially dependent company is equal to the value of the company of the same risk group used no leverage, increased by the value of tax shield arising from financial leverage and equal to the product of rate of corporate income tax T and the value of debt D.

$$V = V_0 + DT \tag{10.2}$$

This theorem was formulated by Modigliani and Miller for perpetuity companies, but we modify it for a company with a finite lifetime.

$$\begin{aligned} V = V_0 + \mathrm{TS} &= V_0 + k_d DT \sum_{t=1}^{\infty} (1+k_d)^{-t} \\ &= V_0 + w_d VT[1-(1+k_d)^{-n}] \end{aligned} \tag{10.3}$$

$$V(1 - w_d VT[1-(1+k_d)^{-n}]) = V_0 \tag{10.4}$$

There is a common use of the following two formulas for the cost of the financially independent and financially dependent companies [2–4]:

$$V_0 = \mathrm{CF}/k_0 \text{ and } V = \mathrm{CF}/\mathrm{WACC} \tag{10.5}$$

However, these almost always used formulas were derived for perpetuity company and in case of a company with a finite lifetime they must be modified in the same manner as the value of tax shields [1, 5–15]:

$$V_0 = \mathrm{CF}[1-(1+k_0)^{-n}]/k_0; \quad V = \mathrm{CF}[1-(1+\mathrm{WACC})^{-n}]/\mathrm{WACC}. \tag{10.6}$$

From formula (Eq. 10.4) we get Brusov–Filatova–Orekhova equation for WACC [1, 5–15]:

$$\frac{1-(1+\mathrm{WACC})^{-n}}{\mathrm{WACC}}=\frac{1-(1+k_0)^{-n}}{k_0[1-\omega_\mathrm{d}T(1-(1+k_\mathrm{d})^{-n})]}. \tag{10.7}$$

Here, S—the value of own (equity) capital of the company, $w_\mathrm{d}=\frac{D}{D+S}$—the share of debt capital; $k_\mathrm{e}, w_\mathrm{e}=\frac{S}{D+S}$—the cost and the share of the equity of the company, $L=D/S$—financial leverage.

At $n=1$ we get Myers [16] formula for one-year company

$$\mathrm{WACC}=k_0-\frac{(1+k_0)k_\mathrm{d}}{1+k_\mathrm{d}}w_\mathrm{d}T \tag{10.8}$$

For $n=2$ one has

$$\frac{1-(1+\mathrm{WACC})^{-2}}{\mathrm{WACC}}=\frac{1-(1+k_0)^{-2}}{k_0\left[1-\omega_\mathrm{d}T\left(1-(1+k_\mathrm{d})^{-2}\right)\right]}. \tag{10.9}$$

This equation can be solved for WACC analytically:

$$\mathrm{WACC}=\frac{1-2\alpha\pm\sqrt{4\alpha+1}}{2\alpha}, \tag{10.10}$$

where

$$\alpha=\frac{2+k_0}{(1+k_0)^2\left[1-\omega_\mathrm{d}T\frac{2k_\mathrm{d}+k_\mathrm{d}^2}{(1+k_\mathrm{d})^2}\right]}. \tag{10.11}$$

For $n=3$ and $n=4$ equation for the WACC becomes more complicate, but it still can be solved analytically, while for $n>4$ it can be solved only numerically.

We would like to make **an important methodological notice**: taking into account the finite life-time of the company, all formulas, without exception, should be received with use formulas (Eq. 10.6) instead of their perpetuity limits (Eq. 10.5).

Below, we will describe the algorithm for the numerical solution of the equation (Eq. 10.7).

Algorithm for Finding of WACC in Case of Arbitrary Life-Time of the Project
Let us return back to n-year project (n-year company). We have the following equation for WACC in n-year case:

$$\frac{1-(1+\mathrm{WACC})^{-n}}{\mathrm{WACC}}-A(n)=0, \tag{10.12}$$

where

$$A(n)=\frac{1-(1+k_0)^{-n}}{k_0[1-\omega_d T(1-(1+k_d)^{-n})]}. \tag{10.13}$$

The algorithm of the solving of the Eq. (10.12) should be as following:

(1) Putting the values of parameters k_0, ω_d, T and given n, we calculate $A(n)$;
(2) We determine two WACC values, for which the left part of the Eq. (10.12) has opposite signs. It is obviously that as these two values we can use WACC_1 and WACC_∞, because $\text{WACC}_1 > \text{WACC}_n > \text{WACC}_\infty$ for finite $n \geq 2$.
(3) Using, for example, the bisection method, we can solve the Eq. (10.12) numerically.

10.2 Comparison of Modigliani–Miller Results (Perpetuity Company) with Myers Results (One Year Company) and Brusov-Filatova-Orekhova Ones (Company with Arbitrary Lifetime)

Myers [16] has compared his result for one-year project (Eq. 10.8) with Modigliani and Miller's result for perpetuity limits (Eq. 2.8). He has used the following values of parameters:

$$k_0=8\%\div 24\%; k_d=7\%; T=50\%; w_d=0\%\div 60\%$$

and estimated the difference in the WACC values following from the formulas Eqs. (10.8) and (2.8). We did make the similar calculations for two-, three-,five- and ten-year project for the same set of parameters and we have gotten the following results, shown in Tables (Table 10.1 (second line (bulk)), Table 10.2 (second line (bulk)) and Table 10.3) and corresponding Figures (Figs. 10.1, 10.2, and 10.3).

Note, that data for equity cost $k_0 = 8\%$ turn out to be a little bit uncertain: this could be relate to the fact that this value of equity cost is quite close to value of interest rate of the debt $k_d = 7\%$. For all other values of equity cost the results are reproducible and very informative and are discussed below.

For a graphic illustration of the results, we use data for $n = 1, 2, \infty$, that adequately reflect the results we obtained.

Discussion of Results

1. From Table 10.1 and Fig. 10.2 it is obviously that WACC has a maximum for one year project and decreases with the life-time of the project, reaching the minimum in the Modigliani–Miller perpetuity case. Dependence of all WACC values on debt share w_d turn out to be linear at any equity cost k_0 for all considered durations of the project (life-time values of the companies). It is natural for one-year project,

Table 10.1 WACC dependence on debt share w_d for different values of equity cost k_0 for companies with different lifetime n

k_0	n	$w_d = 10\%$	20%	30%	40%	50%	60%
$k_0 = 8\%$	$n = 1$	7.6%	7.3	6.9	6.6	6.2	5.9
	$n = 2$	**7.52**	**7.08**	**6.6**	**6.17**	**5.67**	**5.21**
	$n = \infty$	7.6	7.2	6.8	6.4	6.0	5.6
$k_0 = 10\%$	$n = 1$	9.7	9.3	8.9	8.6	8.2	7.8
	$n = 2$	**9.51**	**9.05**	**8.59**	**8.13**	**7.64**	**7.16**
	$n = \infty$	9.5	9.0	8.5	8.0	7.5	7.0
$k_0 = 12\%$	$n = 1$	11.6	11.3	10.9	10.5	10.2	9.8
	$n = 2$	**11.51**	**11.02**	**10.54**	**10.07**	**9.6**	**9.09**
	$n = 3$	**11.46**	**10.93**	**10.39**	**9.85**	**9.31**	**8.77**
	$n = 5$	**11.42**	**10.83**	**10.25**	**9.66**	**9.06**	**8.46**
	$n = 10$	**11.396**	**10.786**	**10.1695**	**9.5455**	**8.914**	**8.2745**
	$n = \infty$	11.4	10.8	10.2	9.6	9.0	8.4
$k_0 = 16\%$	$n = 1$	15.62	15.2	14.9	14.5	14.1	13.7
	$n = 2$	**15.52**	**14.99**	**14.5**	**13.98**	**13.47**	**12.96**
	$n = 3$	**15.44**	**14.88**	**14.31**	**13.75**	**13.18**	**12.61**
	$n = 5$	**15.38**	**14.76**	**14.14**	**13.51**	**12.88**	**12.24**
	$n = 10$	**15.34**	**14.67**	**13.99**	**13.31**	**12.62**	**11.92**
	$n = \infty$	15.2	14.4	13.6	12.8	12.0	11.2
$k_0 = 20\%$	$n = 1$	19.6	19.2	18.8	18.4	18.1	17.7
	$n = 2$	**19.45**	**18.97**	**18.45**	**17.93**	**17.37**	**16.86**
	$n = 3$	**19.41**	**18.82**	**18.23**	**17.64**	**17.05**	**16.45**
	$n = 5$	**19.35**	**18.69**	**18.03**	**17.36**	**16.70**	**16.03**
	$n = 10$	**19.27**	**18.54**	**17.80**	**17.05**	**16.30**	**15.54**
	$n = \infty$	19.0	18.0	17.0	16.0	15.0	14.0
$k_0 = 24\%$	$n = 1$	23.6	23.2	22.8	22.4	22.0	21.6
	$n = 2$	**23.46**	**22.94**	**22.37**	**21.80**	**21.30**	**20.75**
	$n = 3$	**23.39**	**22.77**	**22.15**	**21.54**	**20.91**	**20.29**
	$n = 5$	**23.31**	**22.61**	**21.91**	**21.21**	**20.51**	**19.80**
	$n = 10$	**23.21**	**22.40**	**21.60**	**20.78**	**19.96**	**19.13**
	$n = \infty$	22.8	21.6	20.4	19.2	18.0	16.8

because it is described by Myers linear formula (10.8) as well as in the Modigliani–Miller perpetuity case, describing by the formula:

$$\mathrm{WACC} = k_0(1 - \omega_d T),$$

which is linear too, but it is surprised for two-year project, where formula for WACC (10.7) is obviously nonlinear.

The negative slope in WACC increases with the equity cost k_0.

2. As it follows from the Table 10.2 and Fig. 10.3 the dependence of the average ratios $\overline{r} = <\Delta_1/\Delta_2>$ on debt share w_d is quite weak and can be considered as almost constant. The value of this constant increases practically linear with the equity cost k_0 from 1.22 at $k_0 = 10\%$ up to 5.69 at $k_0 = 24\%$ (see Fig. 10.4).
3. The relative difference between one-year and two-year projects increases when the equity cost k_0 decreases. This means that the error in one-year project WACC

Table 10.2 Dependence of the differences $\Delta_1 = \text{WACC}_1 - \text{WACC}_\infty$ (first line), $\Delta_2 = \text{WACC}_1 - \text{WACC}_2$ (second line (bulk)) and their ratio $r = \Delta_1/\Delta_2$ (third line) on debt share w_d for different values of equity cost k_0

	$w_d = 10\%$	20%	30%	40%	50%	60%
$k_0 = 10\%$	0.20	0.30	0.4	0.60	0.7	0.8
	0.19	**0.25**	**0.31**	**0.47**	**0.56**	**0.64**
	1.05	1.2	1.29	1.28	1.25	1.25
$k_0 = 12\%$	0.2	0.5	0.7	0.9	1.2	1.4
	0.09	**0.28**	**0.36**	**0.43**	**0.6**	**0.71**
	2.22	1.76	1.94	2.09	2	1.97
$k_0 = 16\%$	0.4	0.8	1.3	1.7	2.1	2.5
	0.08	**0.21**	**0.4**	**0.52**	**0.63**	**0.74**
	5.0	3.81	3.25	3.27	3.33	3.38
$k_0 = 20\%$	0.6	1.2	1.8	2.4	3.1	3.7
	0.15	**0.23**	**0.35**	**0.47**	**0.73**	**0.84**
	4.0	5.22	5.14	5.11	4.25	4.4
$k_0 = 24\%$	0.8	1.6	2.4	3.2	4.0	4.8
	0.14	**0.26**	**0.43**	**0.6**	**0.7**	**0.85**
	5.7	6.15	5.58	5.33	5.71	5.65

Table 10.3 Average (by debt share w_d) values of ratios $\bar{r} = <\Delta_1/\Delta_2>$ for $k_0 = 10\ \%\ ;\ 12\ \%\ ;\ 16\ \%\ ;\ 20\%$ and 24%

k_0	10%	12%	16%	20%	24%
$\bar{r} = <\Delta_1/\Delta_2>$	1.22	2.00	3.67	4.69	5.69

for two-year project increases when the equity cost k_0 decreases as well. At the same time the relative difference between two-year project and perpetuity MM project increases with the equity cost k_0.

10.3 Brusov–Filatova–Orekhova Theorem

10.3.1 Case of Absence of Corporate Taxes

Modigliani–Miller theory in case of absence of corporate taxes gives the following results for dependence of WACC and equity cost k_e on leverage:

$$(1)\ V_0 = V_L; \text{CF}/k_0 = \text{CF}/\text{WACC}, \text{and thus WACC} = k_0. \tag{10.14}$$

(2) $\text{WACC} = w_e \cdot k_e + w_d \cdot k_d$; and thus

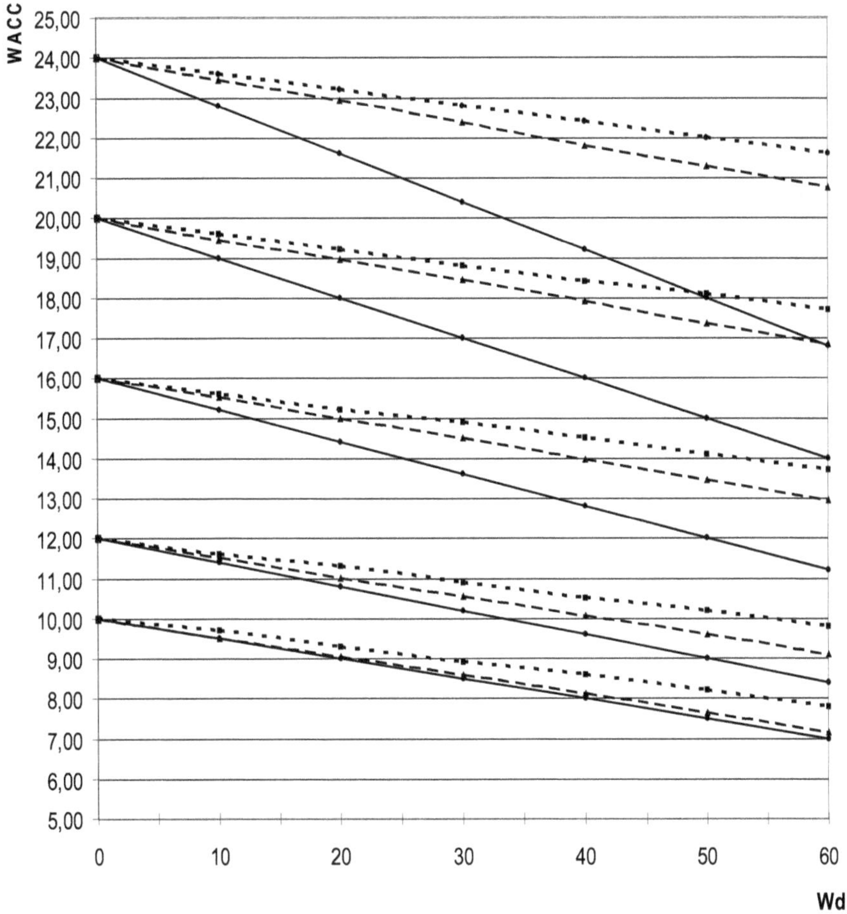

Fig. 10.1 The dependence of the WACC on debt share w_d for companies with different lifetimes for different cost of equity, k_0 (from Table 10.1)

$$k_e = \frac{\text{WACC} - w_d \cdot k_d}{w_e} = \frac{k_0 - \dfrac{L}{1+L} k_d}{\dfrac{1}{1+L}}. \qquad (10.15)$$
$$= k_0 + L(k_0 - k_d)$$

For the finite life-time companies Modigliani–Miller theorem about equality of value of financially independent and financially dependent companies ($V_0 = V_L$) has the following view [5–15]:

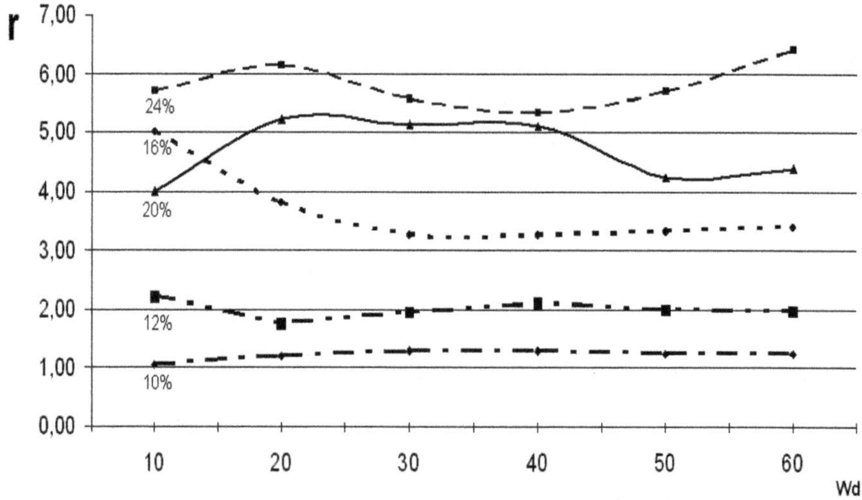

Fig. 10.2 Dependence of the ratio $r = \Delta_1/\Delta_2$ of differences $\Delta_1 = \mathrm{WACC}_1 - \mathrm{WACC}_\infty$, $\Delta_2 = \mathrm{WACC}_1 - \mathrm{WACC}_2$ on debt share w_d for different values of equity cost k_0 (from Table 10.2)

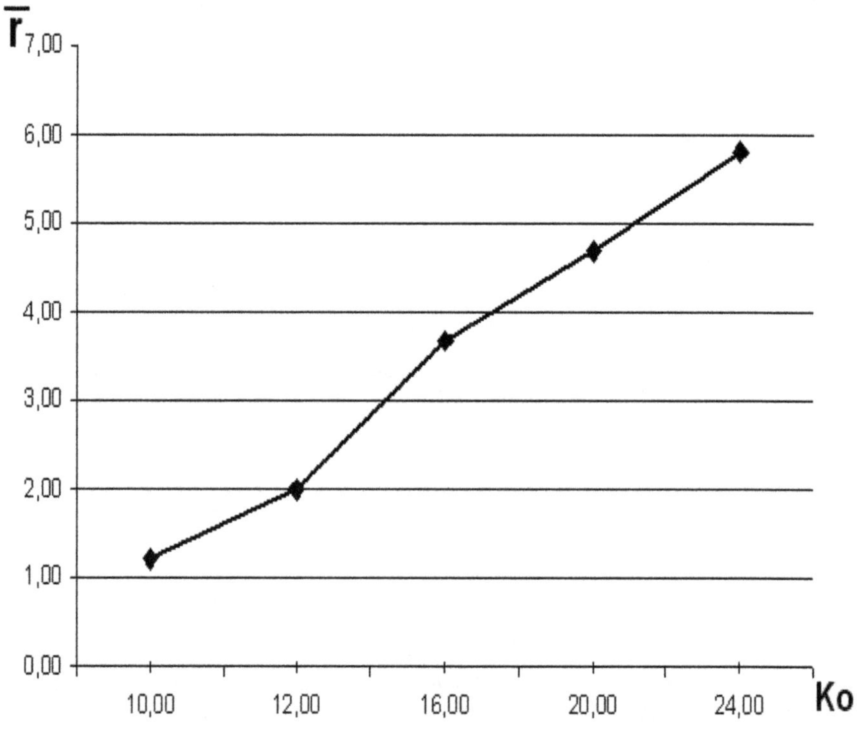

Fig. 10.3 Dependence of the average values of ratio $\bar{r} = <\Delta_1/\Delta_2>$ on the equity cost k_0

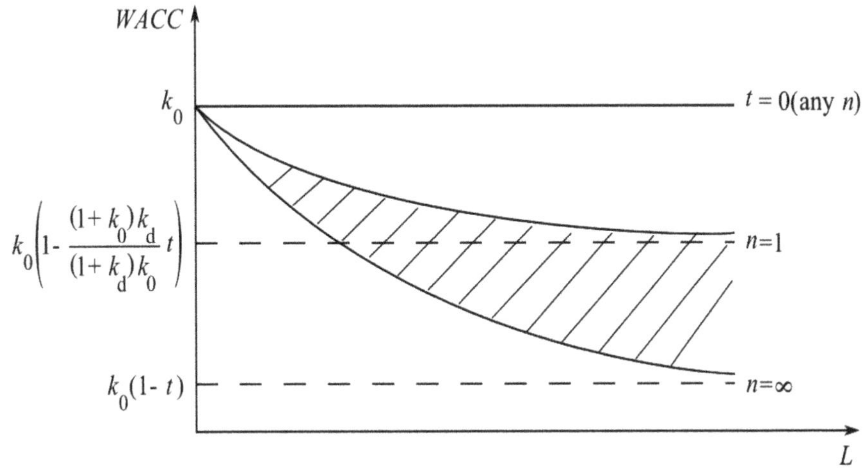

Fig. 10.4 The dependence of the WACC on leverage in the absence of corporate taxes (the horizontal line ($t = 0$)), as well as in the presence of corporate taxes (for one-year ($n = 1$) and perpetuity companies ($n = \infty$)). Curves for the WACC of companies with an intermediate lifetime ($1 < n < \infty$) lie within the shaded region

$$V_0 = V_L;$$
$$\mathrm{CF}\cdot\frac{[1-(1+k_0)^{-n}]}{k_0} = \mathrm{CF}\cdot\frac{[1-(1+\mathrm{WACC})^{-n}]}{\mathrm{WACC}}. \tag{10.16}$$

Using this relation, we prove an important Brusov–Filatova–Orekhova theorem:

Under absence of corporate taxes the equity cost of the company, k_e, as well as its weighted average cost of capital, WACC, do not depend on the lifetime of the company and are equal respectively to:

$$k_e = k_0 + L(k_0 - k_d);\ \mathrm{WACC} = k_0. \tag{10.17}$$

Let us consider first the one- and two-year companies

(a) For one-year company one has from (10.16)

$$\frac{1-(1+k_0)^{-1}}{k_0} = \frac{1-(1+\mathrm{WACC})^{-1}}{\mathrm{WACC}}, \tag{10.18}$$

and thus:

$$\frac{1}{1+k_0} = \frac{1}{1+\mathrm{WACC}}. \tag{10.19}$$

Hence

$$\text{WACC} = k_0. \tag{10.20}$$

Formula for equity cost $k_e = k_0 + L(k_0 - k_d)$ now obtained by substituting WACC $= k_0$ into (10.15).

(b) For two-years company one has from (10.16)

$$\frac{\left[1 - (1 + k_0)^{-2}\right]}{k_0} = \frac{\left[1 - (1 + \text{WACC})^{-2}\right]}{\text{WACC}},$$

and thus:

$$\frac{2 + k_0}{(1 + k_0)^2} = \frac{2 + \text{WACC}}{(1 + \text{WACC})^2}. \tag{10.21}$$

Denoting $\alpha = \frac{2+k_0}{(1+k_0)^2}$, we get the following quadratic equation for WACC:

$$\alpha \cdot \text{WACC}^2 + (2\alpha - 1) \cdot \text{WACC} + (\alpha - 2) = 0. \tag{10.22}$$

It has two solutions:

$$\text{WACC}_{1,2} = \frac{1 - 2\alpha \pm \sqrt{4\alpha + 1}}{2\alpha}. \tag{10.23}$$

Substituting $\alpha = \frac{2+k_0}{(1+k_0)^2}$, we get:

$$\text{WACC}_{1,2} = \frac{\left(k_0^2 - 3\right) \pm (k_0 + 3)(1 + k_0)}{2(2 + k_0)}. \tag{10.24}$$

$$\text{WACC}_1 = k_0; \text{WACC}_2 = -\frac{2k_0 + 3}{k_0 + 2} < 0. \tag{10.25}$$

The second root is negative, but the weighted average cost of capital can only be positive, so only one value remains:

$$\text{WACC}_1 = k_0.$$

(c) For company with arbitrary lifetime n Brusov–Filatova–Orekhova –Orekhova formula (10.16) gives:

$$\frac{1 - (1 + k_0)^{-n}}{k_0} = \frac{1 - (1 + \text{WACC})^{-n}}{\text{WACC}}. \tag{10.26}$$

For a fixed k_0 (10.26) is an equation of n-degree relative to WACC. It has n roots (in general complex). One of the roots, as shown by direct substitution, is always WACC $= k_0$. Investigation of the remaining roots is difficult and not part of our problem.

Formula for equity cost $k_e = k_0 + L(k_0 - k_d)$ now obtained by substituting WACC $= k_0$ into (10.15).

Thus we have proved the Brusov–Filatova–Orekhova theorem.

10.3.2 Case of the Presence of Corporate Taxes

Modigliani–Miller theory in case of presence of corporate taxes gives the following results for dependence of WACC and equity cost k_e on leverage:

(1) WACC

$$V_L = V_0 + Dt; D = w_d V_L; \tag{10.27}$$

$$\mathrm{CF}/\mathrm{WACC} = \mathrm{CF}/k_0 + Dt = \mathrm{CF}/k_0 + w_d t \mathrm{CF}/\mathrm{WACC}; \tag{10.28}$$

$$\frac{1 - w_d t}{\mathrm{WACC}} = \frac{1}{k_0}; \tag{10.29}$$

$$\mathrm{WACC} = k_0(1 - w_d t) = k_0\left(1 - \frac{L}{1+L}t\right). \tag{10.30}$$

Thus, WACC decreases with leverage from k_0 (in the absence of debt financing ($L = 0$)) up to $k_0(1 - t)$ (at $L = \infty$).

(2) The equity cost k_e:

$$\mathrm{WACC} = k_0(1 - w_d t) = w_e \cdot k_e + w_d \cdot k_d(1 - t);$$

and thus:

$$k_e = \frac{\mathrm{WACC} - w_d \cdot k_d \cdot (1 - t)}{w_e}$$
$$= \frac{k_0(1 - w_d t) - \frac{L}{1+L}k_d(1 - t)}{\frac{1}{1+L}} = k_0 + L(k_0 - k_d)(1 - t). \tag{10.31}$$

10.4 From Modigliani–Miller to General Theory of Capital Cost and Capital Structure

Let us consider how the weighted average cost of capital, WACC, and the cost of equity capital, k_e, will be changed under taking into account the finite lifetime of the company.

(a) One-year company

From (10.7) one has:

$$\frac{1-(1+\mathrm{WACC})^{-n}}{\mathrm{WACC}}=\frac{1-(1+k_0)^{-n}}{k_0[1-w_d t(1-(1+k_d)^{-n})]}. \tag{10.32}$$

For one-year company we get:

$$\frac{1-(1+\mathrm{WACC})^{-1}}{\mathrm{WACC}}=\frac{1-(1+k_0)^{-1}}{k_0\left[1-w_d t\left(1-(1+k_d)^{-1}\right)\right]} \tag{10.33}$$

From (Eq. 10.33) we obtain the well-known Myers formula (Eq. 10.8), which is the particular case of Brusov–Filatova–Orekhova formula (Eq. 10.7).

$$\mathrm{WACC}=k_0-\frac{1+k_0}{1+k_d}k_d w_d t.$$

Thus:

$$\mathrm{WACC}=k_0\left(1-\frac{(1+k_0)\cdot k_d}{(1+k_d)\cdot k_0}\cdot\frac{L}{1+L}t\right). \tag{10.34}$$

Thus, WACC decreases with leverage from k_0 (in the absence of debt financing ($L=0$)) up to $k_0\left(1-\frac{(1+k_0)\cdot k_d}{(1+k_d)\cdot k_0}t\right)$ (at $L=\infty$).

Equating the right part to general expression for WACC

$$\mathrm{WACC}=w_e\cdot k_e+w_d\cdot k_d(1-t), \tag{10.35}$$

one gets:

$$k_0-\frac{1+k_0}{1+k_d}k_d w_d t=w_e\cdot k_e+w_d\cdot k_d(1-t). \tag{10.36}$$

Thus:

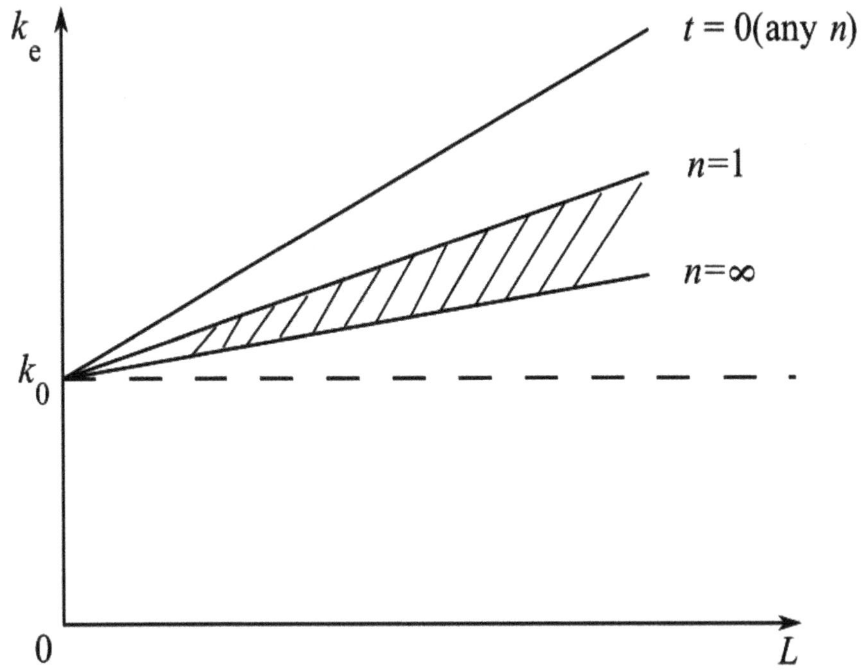

Fig. 10.5 Dependence of the equity cost, k_e, on leverage in the absence of corporate taxes (the upper line ($t = 0$)), as well as in the presence of corporate taxes (for one-year ($n = 1$) and perpetuity companies ($n = \infty$)). Dependences of the cost of equity capital of companies, k_e, with an intermediate lifetime ($1 < n < \infty$) lie within the shaded region

$$
\begin{aligned}
k_e &= \frac{1}{w_e}\left[k_0 - \frac{1+k_0}{1+k_d}k_d w_d t - k_d w_d(1-t)\right] \\
&= (1+L)k_0 - L\frac{k_d}{1+k_d}[(1+k_0)t + (1+k_d)(1-t)] \\
&= k_0 + L(k_0 - k_d)\left(1 - \frac{k_d}{1+k_d}t\right).
\end{aligned}
$$

$$
k_e = k_0 + L(k_0 - k_d)\left(1 - \frac{k_d}{1+k_d}t\right). \tag{10.37}
$$

So, we see, that in case of one-year company the perpetuity limit $k_e = k_0 + L(k_0 - k_d)(1 - t)$ is replaced by (Eq. 10.37).

Difference is due to different values of the tax shield for a one-year company and perpetuity one (Fig. 10.5).

Let us investigate the question of the tax shields value for companies with different life-time in more detail.

10.5 Tax Shield

General expression for the tax shield has the form (Brusov–Filatova–Orekhova):

$$\mathrm{TS}=\sum_{i=1}^{n}\frac{k_{\mathrm{d}}Dt}{(1+k_{\mathrm{d}})^{i}}=\frac{k_{\mathrm{d}}Dt[1-(1+k_{\mathrm{d}})^{-n}]}{(1+k_{\mathrm{d}})\left(1-(1+k_{\mathrm{d}})^{-1}\right)}=Dt[1-(1+k_{\mathrm{d}})^{-n}]. \quad (10.38)$$

(1) In perpetuity limit ($n \to \infty$) tax shield is equal to $\mathrm{TS}_{\infty}=Dt$, which leads to the so-called effect of the tax shield associated with the appearance of a factor $(1-t)$ in the equity cost: $k_{\mathrm{e}}=k_0+L(k_0-k_{\mathrm{d}})(1-t)$.
(2) For the one-year company tax shield value is equal to:

$$\mathrm{TS}_1=Dt\left(1-(1+k_{\mathrm{d}})^{-1}\right)=Dtk_{\mathrm{d}}/(1+k_{\mathrm{d}}) \quad (10.39)$$

This leads to appearance of a factor $\left(1-\frac{k_{\mathrm{d}}}{1+k_{\mathrm{d}}}t\right)$ in the equity cost (Eq. 10.37)

$$k_{\mathrm{e}}=k_0+L(k_0-k_{\mathrm{d}})\left(1-\frac{k_{\mathrm{d}}}{1+k_{\mathrm{d}}}t\right).$$

(3) Tax shield for a two-year company is equal to:

$$\mathrm{TS}_2=Dt\left(1-(1+k_{\mathrm{d}})^{-2}\right)=Dtk_{\mathrm{d}}(2+k_{\mathrm{d}})/(1+k_{\mathrm{d}})^2 \quad (10.40)$$

and if the analogy with one-year company will keep, then factor $(1-t)$ in the Modigliani–Miller theory would be replaced by the factor:

$$\left(1-\frac{k_{\mathrm{d}}(2+k_{\mathrm{d}})}{(1+k_{\mathrm{d}})^2}t\right). \quad (10.41)$$

However, due to a nonlinear relation between WACC and k_0 and k_{d} in Brusov–Filatova–Orekhova formula (Eqs. 10.9–10.11) for two-year company (and companies with longer life-time), such a simple analogy is no longer observed, and the calculations become more complex.

10.6 Conclusions

In this Chapter an important step towards a general theory of capital cost and capital structure of the company has been done. For this perpetuity theory of Nobel Prize winners Modigliani and Miller, which was until recently the basic theory of capital

Table 10.4 Comparison of results, obtained within Modigliani–Miller theory and general Brusov–Filatova–Orehova theory

Financial parameter	Modigliani–Miller (MM) results	Brusov–Filatova–Orehova (BFO) results
Capitalization of financially independent company	$V_0 = \mathrm{CF}/k_0$	$V_0 = \frac{\mathrm{CF}}{k_0}\left[1-(1+k_0)^{-n}\right]$
Capitalization of leverage (financially dependent) company	$V = \mathrm{CF}/\mathrm{WACC}$	$V = \frac{\mathrm{CF}}{\mathrm{WACC}}\left[1-(1+\mathrm{WACC})^{-n}\right]$
Tax shield	$(\mathrm{TS})_\infty = DT$	$(\mathrm{TS})_n = DT[1-(1+k_d)^{-n}]$
Modigliani–Miller theorem with taxes	$V = V_0 + DT$	$V = V_0 + DT[1-(1+k_d)^{-n}]$
Weighted average cost of capital, WACC	$\mathrm{WACC} = k_0(1-w_d t)$	$\frac{1-(1+\mathrm{WACC})^{-n}}{\mathrm{WACC}} = \frac{1-(1+k_0)^{-n}}{k_0[1-\omega_d T(1-(1+k_d)^{-n})]}$
Equity cost, k_e	$k_e = k_0 + L(k_0-k_d)(1-t)$	$k_e = (1+L)\cdot\mathrm{WACC} - k_d L(1-T)$

cost and capital structure of companies, extended to the case of companies with an arbitrary life-time, as well as for projects of arbitrary duration.

We show that taking into account the finite lifetime of the company in the presence of corporate taxes leads to a change in the equity cost of the company, k_e, as well as in its weighted average cost, WACC.

Thus, we have removed one of the most serious limitations of the theory of Modigliani–Miller connected with the assumption of perpetuity of the companies. The effect of leverage on the cost of equity capital of the company with an arbitrary lifetime, k_e, and its weighted average cost, WACC, is investigated. We give a rigorous proof of an important Brusov–Filatova–Orekhova theorem, that in the absence of corporate tax equity cost of companies, k_e, as well as its weighted average cost, WACC, do not depend on the lifetime of the company.

We summarize the difference in results obtained within modern Brusov–Filatova–Orehova theory and within classical Modigliani–Miller one in Table 10.4.

The first four formulas from the right-hand column are sometimes used in practice, but there are several significant nuances. The first—these formulas do not take account of the residual value of the company, and only take into account the operating flows and this must be borne in mind. The second—these formulas contain the weighted average cost of capital of the company, WACC. If it is estimated within the traditional approach or the theory of Modigliani-Miller, it gives a lower WACC value, than the real, and, therefore, overestimates the capitalization of both financially-dependent and financially-independent companies. Therefore, in order to assess of company capitalization by the first two formulas one needs to use Brusov–Filatova–Orehova formulas for weighted average cost of capital, WACC, and equity cost, k_e.

To calculate the equity cost in BFO approximation (last line in Table 10.4) one needs to first use the Brusov–Filatova–Orehova formula for weighted average cost of capital, WACC [1, 5–15].

References

1. Filatova, T., Orehova, N., & Brusova, A. (2008). Weighted average cost of capital in the theory of Modigliani–Miller, modified for a finite life–time company. *Bulletin of the FU, 48*, 68–77.
2. Modigliani, F., & Miller, M. (1958). The cost of capital, corporate finance, and the theory of investment. *American Economic Review, 48*, 261–297.
3. Modigliani, F., & Miller, M. (1963). Corporate income taxes and the cost of capital: A correction. *American Economic Review, 53*, 147–175.
4. Modigliani, F., & Miller, M. (1966). Some estimates of the cost of capital to the electric utility industry 1954–1957. *American Economic Review, 56*, 333–391.
5. Brusov, P., Filatova, T., Orehova, N., & Brusova, A. (2011). Weighted average cost of capital in the theory of Modigliani–Miller, modified for a finite life–time company. *Applied Financial Economics, 21*(11), 815–824.
6. Brusov, P., Filatova, P., & Orekhova, N. (2013). Absence of an optimal capital structure in the famous tradeoff theory! *Journal of Reviews on Global Economics, 2*, 94–116.
7. Brusov, P., Filatova, P., & Orekhova, N. (2014). Mechanism of formation of the company optimal capital structure, different from suggested by trade off theory. *Cogent Economics & Finance, 2*, 1–13. https://doi.org/10.1080/23322039.2014.946150
8. Brusov, P., Filatova, T., Orehova, N., et al. (2011). From Modigliani–Miller to general theory of capital cost and capital structure of the company. *Research Journal of Economics, Business and ICT, 2*, 16–21.
9. Brusov, P., Filatova, T., Eskindarov, M., & Orehova, N. (2012). Influence of debt financing on the effectiveness of the finite duration investment project. *Applied Financial Economics, 22*(13), 1043–1052.
10. Brusov, P., Filatova, T., Orehova, N., et al. (2011). Influence of debt financing on the effectiveness of the investment project within the Modigliani–Miller theory. *Research Journal of Economics, Business and ICT (UK), 2*, 11–15.
11. Brusov, P., Filatova, T., Eskindarov, M., & Orehova, N. (2012). Hidden global causes of the global financial crisis. *Journal of Reviews on Global Economics, 1*, 106–111.
12. Brusov, P. N., & Filatova, T. V. (2011). From Modigliani–Miller to general theory of capital cost and capital structure of the company. *Finance and Credit, 435*, 2–8.
13. Brusov, P., Filatova, T., Orehova, N., Brusov, P. P., & Brusova, N. (2011). From Modigliani–Miller to general theory of capital cost and capital structure of the company. *Research Journal of Economics, Business and ICT, 2*, 16–21.
14. Brusov, P., Filatova, T., & Orehova, N. (2014). Inflation in Brusov–Filatova–Orekhova theory and in its perpetuity limit – Modigliani–Miller Theory. *Journal of Reviews on Global Economics, 3*, 175–185.
15. Brusov, P., Filatova, T., & Orehova, N. (2013). A qualitatively new effect in corporative finance: Abnormal dependence of cost of equity of company on leverage. *Journal of Reviews on Global Economics, 2*, 183–193.
16. Myers, S. (2001). Capital structure. *Journal of Economic Perspectives, 15*(2), 81–102.

Chapter 11
Inflation in Brusov–Filatova–Orekhova Theory and in Its Perpetuity Limit: Modigliani–Miller Theory

In this chapter the influence of inflation on capital cost and capitalization of the company within modern theory of capital cost and capital structure—Brusov–Filatova–Orekhova theory (BFO theory) [1–12] and within its perpetuity limit—Modigliani–Miller theory [13–15] is investigated. By direct incorporation of inflation into both theories, it is shown for the first time, that inflation not only increases the equity cost and the weighted average cost of capital, but as well it changes their dependence on leverage. In particular, it increases growing rate of equity cost with leverage. Capitalization of the company is decreased under accounting of inflation.

Introduction Created more than half a century ago by Nobel Prize winners Modigliani and Miller theory of capital cost and capital structure [13–15] did not account a lot of factors of a real economy, such as taxing, bankruptcy, unperfected capital markets, inflation, and many others. But while taxing has been included into consideration by authors themselves and some other limitations have been taken off by their followers, direct incorporation of inflation to Modigliani–Miller theory is absent still now.

The influence of inflation on valuation of capital cost of company and its capitalization is investigated within Modigliani–Miller theory (MM) [13–15], which is now outdate, but still widely used at the West, as well as within modern theory of capital cost and capital structure—Brusov–Filatova–Orekhova theory (BFO theory) [1–12], which should replace Modigliani–Miller theory [13–15]. It is shown, that inflation not only increases the equity cost and the weighted average cost of capital, but as well it changes their dependence on leverage. In particular, it increases growing rate of equity cost with leverage. Capitalization of the company is decreased under accounting of inflation.

We start from the study of inflation within Modigliani–Miller theory without taxing [13], than with taxing [14] and finally within modern theory of capital cost and capital structure– Brusov–Filatova–Orekhova theory (BFO theory) [1–12, 16].

P. Brusov, T. Filatova, *Business Valuation and Company Value*, Contributions to Finance and Accounting, https://doi.org/10.1007/978-3-031-90847-7_11

11.1 Accounting of Inflation in Modigliani–Miller Theory Without Taxes

Note, that any modification of Modigliani–Miller theory, as well as of any other one, requires going behind the frame of modified theory. Thus, in current case we should go behind the frame of perpetuity of the company (remind to reader, that Modigliani–Miller theory describes only perpetuity companies—companies with infinite lifetime), come to the companies with finite lifetime, make necessary calculations and then use the perpetuity limit.

As known, in profit approach capitalization of the company is equal to discounted sum of profits of the company. Suppose that profit is constant for all periods and equal to CF, one gets for capitalization of the financially independent company V_0, existing n years at market:

$$V_0 = \frac{\mathrm{CF}}{1+k_0} + \frac{\mathrm{CF}}{(1+k_0)^2} + \cdots + \frac{\mathrm{CF}}{(1+k_0)^n}. \tag{11.1}$$

Here k_0 capital cost of the financially independent company.

Under inflation with rate α the capitalization of the financially independent company V_0^* becomes equal to:

$$V_0^* = \frac{\mathrm{CF}}{(1+k_0)(1+\alpha)} + \frac{\mathrm{CF}}{[(1+k_0)(1+\alpha)]^2} + \cdots + \frac{\mathrm{CF}}{[(1+k_0)(1+\alpha)]^n}. \tag{11.2}$$

Using the formula for sum of the terms of indefinitely diminishing geometrical progression with the first term:

$$a_1 = \frac{\mathrm{CF}}{(1+k_0)(1+\alpha)} \tag{11.3}$$

and denominator:

$$q = \frac{1}{(1+k_0)(1+\alpha)} \tag{11.4}$$

one gets for capitalization of the financially independent company V_0^* the following expression:

$$V_0^* = \frac{a_1}{1-q} = \frac{\mathrm{CF}}{(1+k_0)(1+\alpha)\left[1-((1+k_0)(1+\alpha))^{-1}\right]}$$
$$= \frac{\mathrm{CF}}{(1+k_0)(1+\alpha)-1} = \frac{\mathrm{CF}}{k_0(1+\alpha)+\alpha}.$$
$$V_0^* = \frac{\mathrm{CF}}{k_0(1+\alpha)+\alpha}. \tag{11.5}$$

It is seen, that under accounting of inflation the capitalization of the company decreases.

At discount rate $k_0 = 10\%$ and inflation rate $\alpha = 3\%$ the decrease is equal to 5.7%, and at discount rate $k_0 = 15\%$ and inflation rate $\alpha = 7\%$ the decrease is equal to 35%. One can see that influence of inflation on the company capitalization could be significant enough and negative.

For leverage company, using debt capital one has without inflation

$$V_L = \frac{\mathrm{CF}}{1+\mathrm{WACC}} + \frac{\mathrm{CF}}{(1+\mathrm{WACC})^2} + \cdots + \frac{\mathrm{CF}}{(1+\mathrm{WACC})^n}, \tag{11.6}$$

And in perpetuity limit:

$$V_L = \frac{\mathrm{CF}}{\mathrm{WACC}}. \tag{11.7}$$

Under accounting of inflation the capitalization of the company is equal to:

$$V_L^* = \frac{\mathrm{CF}}{(1+\mathrm{WACC})(1+\alpha)} + \frac{\mathrm{CF}}{[(1+\mathrm{WACC})(1+\alpha)]^2} + \cdots + \frac{\mathrm{CF}}{[(1+\mathrm{WACC})(1+\alpha)]^n}. \tag{11.8}$$

Summing the infinite set, we get for leverage company capitalization under accounting of inflation in Modigliani–Miller limit:

$$V_L^* = \frac{a_1}{1-q} = \frac{\mathrm{CF}}{(1+\mathrm{WACC})(1+\alpha)\left[1-((1+\mathrm{WACC})(1+\alpha))^{-1}\right]}$$
$$= \frac{\mathrm{CF}}{(1+\mathrm{WACC})(1+\alpha)-1} = \frac{\mathrm{CF}}{\mathrm{WACC}(1+\alpha)+\alpha}$$
$$V_L^* = \frac{\mathrm{CF}}{\mathrm{WACC}(1+\alpha)+\alpha}. \tag{11.9}$$

It is seen, that similar to the case of the financially independent company inflation decreases the company capitalization and the decrease could be significant. From the formulas Eqs. (11.7) and (11.9) it follows that effective values of capital costs (equity and WACC) are equal to:

$$k_0^* = k_0(1+\alpha)+\alpha \tag{11.10}$$

$$\mathrm{WACC}^* = \mathrm{WACC}\cdot(1+\alpha)+\alpha \tag{11.11}$$

Note, that both capital costs increases under inflation.

We can compare obtained results with Fisher formula for inflation.

$$i^* = \frac{i-\alpha}{1+\alpha} \tag{11.12}$$

Solving this equation with respect to nominal rate i, one gets equation, similar Eqs. (11.10) and (11.11):

$$i = i^*\cdot(1+\alpha)+\alpha. \tag{11.13}$$

Thus, effective capital costs in our case have meaning of nominal ones, accounting inflation.

From the Modigliani–Miller theorem, that the weighted average cost of capital WACC does not depends on leverage level (without taxing), formulating under accounting of inflation, it is easy to get expression for the equity cost:

$$\mathrm{WACC}^* = k_0^* = k_e^* w_e + k_d^* w_d. \tag{11.14}$$

Finding from here k_e^*, one gets:

$$k_e^* = \frac{k_0^*}{w_e} - k_d^*\frac{w_d}{w_e} = \frac{k_0^*(S+D)}{S} - k_d^*\frac{D}{S} = k_0^* + \left(k_0^* - k_d^*\right)\frac{D}{S} = k_0^* + \left(k_0^* - k_d^*\right)L \tag{11.15}$$

Putting instead of k_0^*, k_d^* their expressions, one gets finally:

$$k_e^* = k_0^* + \left(k_0^* - k_d^*\right)L = k_0(1+\alpha)+\alpha + L(k_0 - k_d)(1+\alpha) = (1+\alpha)[k_0 + \alpha + L(k_0 - k_d)]$$

$$k_e^* = k_0(1+\alpha)+\alpha + L(k_0 - k_d)(1+\alpha). \tag{11.16}$$

It is seen, that inflation not only increases the equity cost, but as well it changes its dependence on leverage. In particular, it increases growing rate of equity cost with leverage by multiplier $(1 + \alpha)$. The growing rate of equity cost with leverage, which is equal to $(k_0 - k_d)$ without inflation becomes equal to $(k_0 - k_d)(1 + \alpha)$ under accounting of inflation (Fig. 11.1).

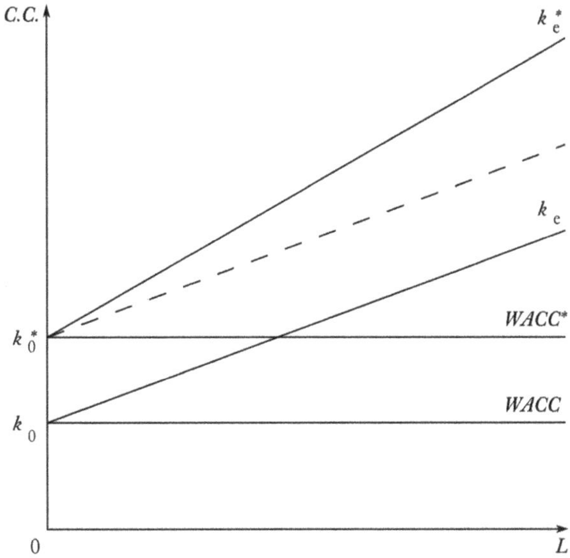

Fig. 11.1 Dependence of the equity cost and the weighted average cost of capital on leverage in the Modigliani–Miller theory without taxing under accounting of inflation. It is seen, that growing rate of equity cost increases with leverage. Axis *y* means capital costs—C.C

Thus, we come to conclusion, that it is necessary to modified the second statement of the Modigliani–Miller theory [13] concerning the equity cost of leverage company.

Second Original MM Statement

Equity cost of leverage company k_e could be found as equity cost of financially independent company k_0 of the same group of risk, plus premium for risk, which value is equal to production of difference $(k_0 - k_d)$ on leverage level *L*.

Second Modified MM-BFO statement

under existing of inflation with rate α equity cost of leverage company k_e could be found as equity cost of financially independent company k_0 of the same group of risk, multiplied by $(1 + \alpha)$, plus inflation rate α and plus premium for risk, which value is equal to production of difference $(k_0 - k_d)$ on leverage level *L* and on multiplier $(1 + \alpha)$.

11.2 Accounting of Inflation in Modigliani–Miller Theory with Corporate Taxes

Let us calculate first the tax shield for perpetuity company under accounting of inflation:

$$(\mathrm{PV})_{\mathrm{TS}} = k_{\mathrm{d}}^{*} DT \sum_{t=1}^{\infty} \left(1 + k_{\mathrm{d}}^{*}\right)^{-t} = DT \tag{11.17}$$

It is interesting to note, that in spite of dependence of each term of set on effective credit rate:

k_{d}^{*} tax shield turns out to be independent of it and equal to "inflationless" value DT and Modigliani–Miller theorem under accounting of inflation takes the form [14]:

$$V_{\mathrm{L}}^{*} = V_{0}^{*} + DT. \tag{11.18}$$

Substituting $D = w_{\mathrm{d}} V_{\mathrm{L}}^{*}$,
one gets:

$$V_{\mathrm{L}}^{*} = \mathrm{CF}/k_{0}^{*} + w_{\mathrm{d}} V_{\mathrm{L}}^{*} T \tag{11.19}$$

Or:

$$V_{\mathrm{L}}^{*}(1 - w_{\mathrm{d}} T) = \mathrm{CF}/k_{0}^{*}. \tag{11.20}$$

Because leverage company capitalization is equal to $V_{\mathrm{L}}^{*} = \mathrm{CF}/\mathrm{WACC}^{*}$ for the weighted average cost of capital one has:

$$\mathrm{WACC}^{*} = k_{0}^{*}(1 - w_{\mathrm{d}} T). \tag{11.21}$$

From (Eq. 11.21) we get the dependence of WACC* on leverage level $L = D/S$:

$$\mathrm{WACC}^{*} = k_{0}^{*}(1 - LT/(1 + L)).$$

$$\mathrm{WACC}^{*} = [k_{0}(1 + \alpha) + \alpha] \cdot (1 - w_{\mathrm{d}} T) \tag{11.22}$$

On definition of the weighted average cost of capital with accounting of the tax shield one has:

$$\mathrm{WACC}^{*} = k_{0}^{*} w_{\mathrm{e}} + k_{\mathrm{d}}^{*} w_{\mathrm{d}}(1 - T). \tag{11.23}$$

Equating right hand parts of expressions (Eq. 11.21) and (Eq. 11.23), we get:

$$k_{0}^{*}(1 - w_{\mathrm{d}} T) = k_{0}^{*} w_{\mathrm{e}} + k_{\mathrm{d}}^{*} w_{\mathrm{d}}(1 - T) \tag{11.24}$$

from where one obtains the following expression for equity cost:

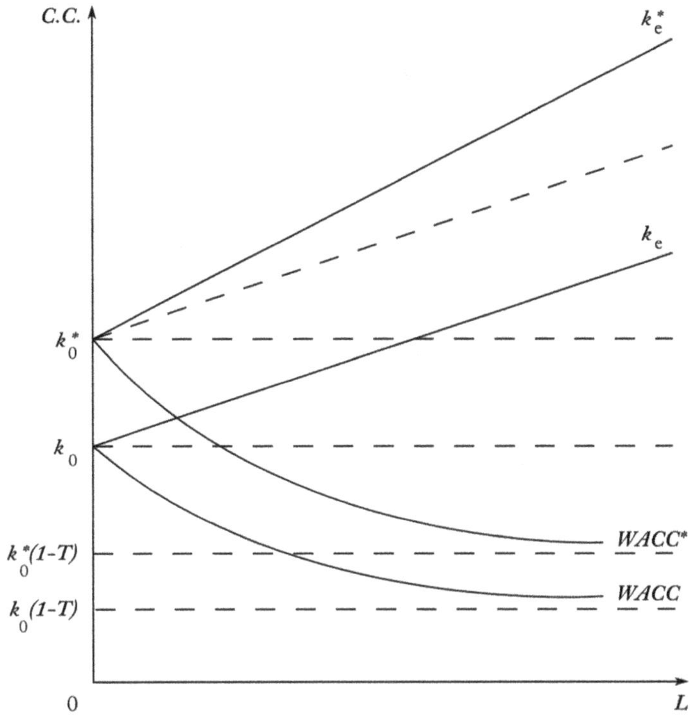

Fig. 11.2 Dependence of the equity cost and the weighted average cost of capital on leverage in the Modigliani–Miller theory with taxing under accounting of inflation. It is seen, that growing rate of equity cost increases with leverage. Axis *y* means capital costs–C.C

$$
\begin{aligned}
k_e^* &= k_0^* \frac{(1 - w_d T)}{w_e} - k_d^* \frac{w_d}{w_e}(1 - T) = k_e^* \frac{1}{w_e} - k_0^* \frac{w_d}{w_e} T - k_d^* \frac{D}{S}(1 - T) \\
&= k_0^* \frac{D + S}{S} - k_0^* \frac{D}{S} T - k_d^* \frac{D}{S}(1 - T) = k_0^* + L(1 - T)\left(k_0^* - k_d^*\right).
\end{aligned}
$$

$$
k_e^* = k_0^* + L(1 - T)\left(k_0^* - k_d^*\right) = [k_0(1 + \alpha) + \alpha] + L(1 - T)(k_0 - k_d)(1 + \alpha) \quad (11.25)
$$

It is seen, that similar to the case without taxes inflation not only increases the equity cost, but as well it changes its dependence on leverage (Fig. 11.2). In particular, it increases growing rate of equity cost with leverage by multiplier $(1 + \alpha)$. The growing rate of equity cost with leverage, which is equal to $(k_0 - k_d)(1 - T)$ without inflation becomes equal to $(k_0 - k_d)(1 + \alpha)(1 - T)$ under accounting of inflation.

We can now reformulate the fourth statement of the Modigliani–Miller theory [14] concerning the equity cost of leverage company for case of accounting of inflation.

Fourth Original MM Statement
equity cost of leverage company k_e paying tax on profit could be found as equity cost of financially independent company k_0 of the same group of risk, plus premium for risk, which value is equal to production of difference $(k_0 - k_d)$ on leverage level L and on tax shield $(1 - T)$ and on multiplier $(1 + \alpha)$.

Fourth Modified MM-BFO Statement
equity cost of leverage company k_e paying tax on profit under existing of inflation with rate α could be found as equity cost of financially independent company k_0 of the same group of risk, multiplied by $(1 + \alpha)$, plus inflation rate α and plus premium for risk, which value is equal to production of difference $(k_0 - k_d)$ on leverage level L, on tax shield $(1 - T)$ and on multiplier $(1 + \alpha)$.

11.3 Accounting of Inflation in Brusov–Filatova–Orekhova Theory with Corporate Taxes

11.3.1 Generalized Brusov–Filatova–Orekhova Theorem

Brusov–Filatova–Orekhova, generalized the Modigliani–Miller theory for the case of the companies with arbitrary lifetime [1–12], have proved the following important theorem in case of absence of corporate taxing:

Without corporate taxing the equity cost k_0, as well as the weighted average cost of capital WACC do not depend on company lifetime and are equal to

$$k_e = k_0 + L(k_0 - k_d) \text{ and } \mathrm{WACC} = k_0. \tag{11.26}$$

consequently.

Thus, the theorem has been proved, that without corporate taxes (say, in offshore zones) the Modigliani–Miller results for capital costs, in spite of the fact, that they have been obtained in perpetuity limit, remain in force for companies with arbitrary lifetime, describing by Brusov–Filatova–Orekhova theory (BFO theory). To prove this theorem Brusov, Filatova and Orekhova, of course, had to go behind Modigliani–Miller approximation.

Under accounting of inflation we can generalize this theorem [1–12]:

Generalized Brusov–Filatova–Orekhova Theorem
Under accounting of inflation without corporate taxing the equity cost k_e^*, as well as the weighted average cost of capital WACC^* do not depend on company lifetime and are equal to

$$k_e^* = k_0^* + L(k_0^* - k_d^*) = k_0(1+\alpha) + \alpha + L(k_0 - k_d)(1+\alpha)$$
$$\text{and WACC}^* = k_0^* = k_0(1+\alpha) + \alpha \tag{11.27}$$

consequently.

Following to Brusov–Filatova–Orekhova [1–12], let us consider the situation for arbitrary lifetime companies with account of corporate taxing. They have derived the famous formula for weighted average cost of capital of companies with arbitrary lifetime:

$$\frac{1-(1+\text{WACC})^{-n}}{\text{WACC}} = \frac{1-(1+k_0)^{-n}}{k_0[1-\omega_d T(1-(1+k_d)^{-n})]} \tag{11.28}$$

The application of BFO formula (11.28) is very wide: authors have applied it in corporate finance, in investments, in taxing, in business valuation, in banking and some other areas [1, 2, 4]***. Using this formula (11.28), one can study the dependence of the weighted average cost of capital, WACC, as well as the equity cost k_e on leverage level L, on tax on profit rate t, on lifetime of the company n and on relation between equity and debt cost. The qualitatively new effect in corporate finance has been discovered: decrease of the equity cost k_e on leverage level L, which is quite important for corporate finance in general and, in particular, for creating the adequate dividend policy.***

Below we generalize formula (11.28) under existing of inflation.

11.4 Generalized Brusov–Filatova–Orekhova Formula Under Existing of Inflation

In the presence of inflation it is necessary to replace all capital costs: the equity, the debt and the weighted average cost of capital k_0, k_d, WACC by effective ones $k_0^*, k_d^*, \text{WACC}^*$,

where

$$k_0^* = k_0(1+\alpha) + \alpha$$
$$k_d^* = k_d(1+\alpha) + \alpha$$
$$\text{WACC}^* = \text{WACC}\cdot(1+\alpha) + \alpha$$

Rewriting the equations for tax shield $(\text{TS})_n$, capitalization of financially independent company V_0^* as well as for financially dependent company V_L^* for the case of existing of inflation, one gets:

$$(\mathrm{PV})_{\mathrm{TS}} = (\mathrm{TS})_n = k_d^* DT \sum_{t=1}^{n} \left(1 + k_d^*\right)^{-t} = DT\left[1 - \left(1 + k_d^*\right)^{-n}\right] \tag{11.29}$$

$$V_0^* = \mathrm{CF} \sum_{t=1}^{n} \left(1 + k_0^*\right)^{-t} = \mathrm{CF}\left[1 - \left(1 + k_0^*\right)^{-n}\right] / k_0^* \tag{11.30}$$

$$V_L^* = \mathrm{CF} \sum_{t=1}^{n} (1 + \mathrm{WACC}^*)^{-t} = \mathrm{CF}[1 - (1 + \mathrm{WACC}^*)^{-n}] / \mathrm{WACC}^* \tag{11.31}$$

$$V_L^* = V_0^* + (\mathrm{TS})_n, \tag{11.32}$$

After substitution $D = w_d V_L^*$ we have:

$$V_L^* = \mathrm{CF}/k_0^* + w_d V_L^* T \tag{11.33}$$

From here after some transformations we get generalized Brusov-Filatova-Orekhova formula under existing risk of inflation:

$$\frac{1 - (1 + \mathrm{WACC}^*)^{-n}}{\mathrm{WACC}^*} = \frac{1 - \left(1 + k_0^*\right)^{-n}}{k_0^*\left[1 - \omega_d T\left(1 - \left(1 + k_d^*\right)^{-n}\right)\right]}, \tag{11.34}$$

or after substitutions:

$$k_0^* = k_0(1 + \alpha) + \alpha; \quad k_d^* = k_d(1 + \alpha) + \alpha$$

one gets finally:

$$\frac{1 - (1 + \mathrm{WACC}^*)^{-n}}{\mathrm{WACC}^*} = \frac{1 - [(1 + k_0)(1 + \alpha)]^{-n}}{(k_0(1 + \alpha) + \alpha) \cdot [1 - \omega_d T(1 - ((1 + k_0)(1 + \alpha))^{-n})]} \tag{11.35}$$

Formula (11.35) is the generalized Brusov-Filatova-Orekhova formula under existing of inflation.

Let us show some figures, illustrating obtained results.

At Figs. 11.3 and 11.4 the dependence of the weighted average cost of capital WACC on debt ratio w_d at different inflation rate α ($1 - \alpha = 3\%$; $2 - \alpha = 5\%$; $3 - \alpha = 7\%$; $4 - \alpha = 9\%$) for five-year company as well as for two-year company. It is seen that with increase of inflation rate lines, showing the dependence WACC (w_d) shift practically homogeneously to higher values.

It is seen, that difference in results for two-year company and five-year company is very small. More obviously it could be observed from below tables (Tables 11.1 and 11.2).

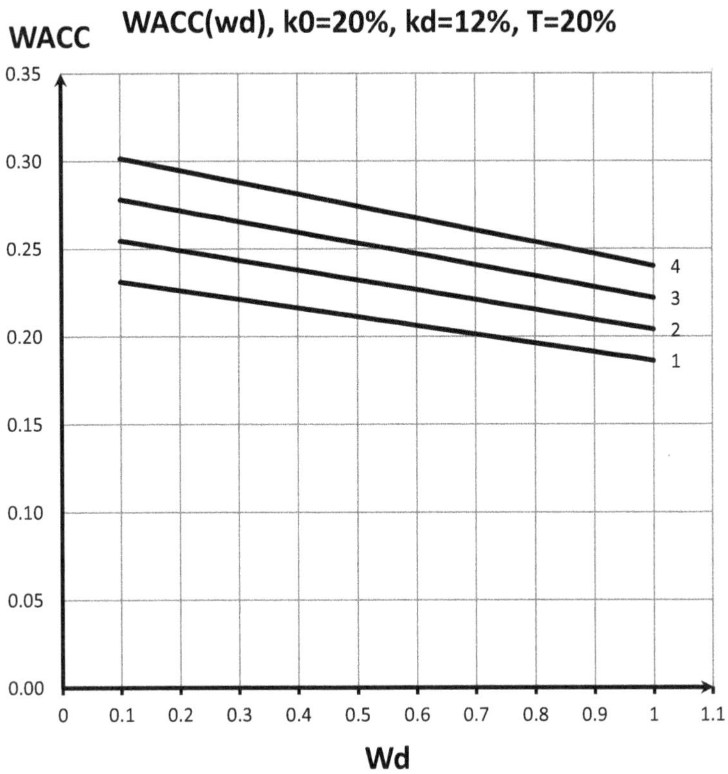

Fig. 11.3 Dependence of the weighted average cost of capital WACC on debt ratio w_d at different inflation rate α ($1 - \alpha = 3\%$; $2 - \alpha = 5\%$; $3 - \alpha = 7\%$; $4 - \alpha = 9\%$) for five-year company

Below we show the dependences of the weighted average cost of capital WACC on debt ratio w_d at different tax on profit rate from $T = 10\%$ up to $T = 100\%$ at different inflation rate $\alpha = 3\%, 5\%, 7\%, 9\%$ for five-year company (Figs. 11.5, 11.6, 11.7, and 11.8) as well as for two-year company (Figs. 11.9, 11.10, 11.11, and 11.12). Tax on profit rate increases from $T = 0.1$ (upper line) up to $T = 1$ (lowest line) with the step 0.1.

The analysis of Figs. 11.5, 11.6, 11.7, 11.8, 11.9, 11.10, 11.11, and 11.12 shows, that the weighted average cost of capital WACC decreases with debt ratio w_d faster with increase of tax on profit rate. The space between lines, corresponding to different tax on profit rates, increases with inflation rate. The variation range of WACC increases with inflation rate as well as with lifetime of the company.

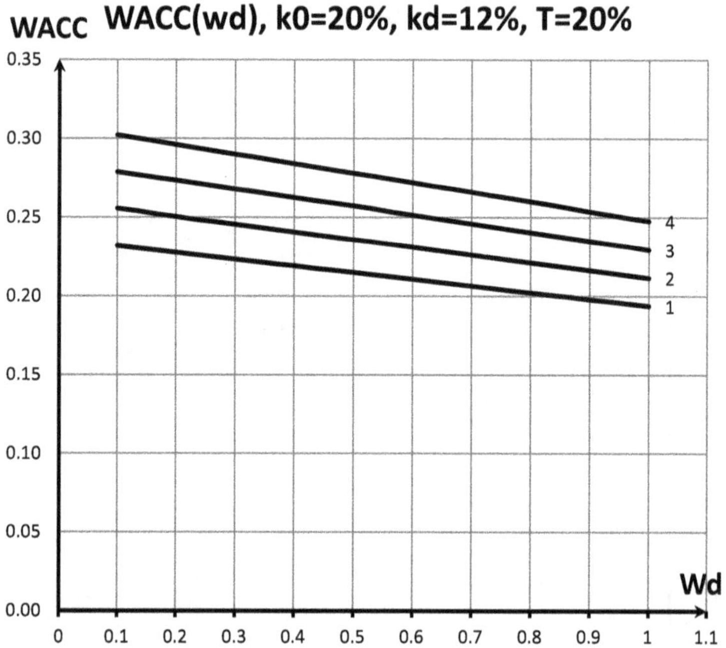

Fig. 11.4 Dependence of the weighted average cost of capital WACC on debt ratio w_d at different inflation rate α ($1 - \alpha = 3\%$; $2 - \alpha = 5\%$; $3 - \alpha = 7\%$; $4 - \alpha = 9\%$) for two-year company

11.5 Irregular Inflation

Above we considered inflation rate as constant. Really, as a rule, the inflation rate is a variable. It is possible to generalize all above consideration for the case of non-homogeneous inflation, introducing effective inflation for a few periods.

The effective inflation rate for a few periods $t = t_1 + t_2 + \cdots + t_n$ is equal to

$$\alpha = (1 + \alpha_1)(1 + \alpha_2)\ldots(1 + \alpha_n) - 1, \tag{11.36}$$

where $\alpha_1, \alpha_2, \ldots, \alpha_n$ are inflation rates for periods $t_1, t_2, \ldots, t_n$.

The proof of the formula (11.36) will be done below in Sect. 11.6.

In the case of non-homogeneous inflation it could be accounted in both theories: Modigliani–Miller and Brusov–Filatova–Orekhova theory (BFO theory) either through effective inflation rate, or directly upon discounting of financial flow.

Conclusions

In this Chapter the influence of inflation on capital cost and capitalization of the company within modern theory of capital cost and capital structure—Brusov–Filatova–Orekhova theory (BFO theory) [1–12] and in its perpetuity limit—Modigliani–Miller theory [13–15], which is now outdate, but still widely used at

Table 11.1 Dependence of the weighted average cost of capital WACC on debt ratio w_d at different inflation rate $\alpha = 3\%; 5\%; 7\%; 9\%$ for two-year company

α/w_d	0.1	0.2	0.3	0.4	0.5	0.6	0.7	0.8	0.9	1
0.03	0.2318	0.2276	0.2233	0.2191	0.2149	0.2106	0.2064	0.2021	0.1979	0.1937
0.05	0.2557	0.2503	0.2455	0.2406	0.2358	0.2309	0.2261	0.2212	0.2164	0.2115
0.07	0.2786	0.2733	0.2679	0.2626	0.2573	0.2514	0.2459	0.2404	0.2350	0.2295
0.09	0.3020	0.2960	0.2900	0.2839	0.2779	0.2720	0.2661	0.2602	0.2537	0.2476

Table 11.2 Dependence of the weighted average cost of capital WACC on debt ratio w_d at different inflation $\rho\alpha\tau\varepsilon$ α =3%; 5%; 7%; 9% for five-year company

α/w_d	0.1	0.2	0.3	0.4	0.5	0.6	0.7	0.8	0.9	1
0.03	0.2311	0.2262	0.2213	0.2163	0.2113	0.2064	0.2013	0.1963	0.1912	0.1863
0.05	0.2546	0.2491	0.2434	0.2379	0.2323	0.2267	0.2210	0.2154	0.2097	0.2040
0.07	0.2781	0.2718	0.2657	0.2595	0.2534	0.2472	0.2408	0.2346	0.2283	0.2219
0.09	0.3015	0.2947	0.2879	0.2812	0.2744	0.2676	0.2608	0.2539	0.2471	0.2400

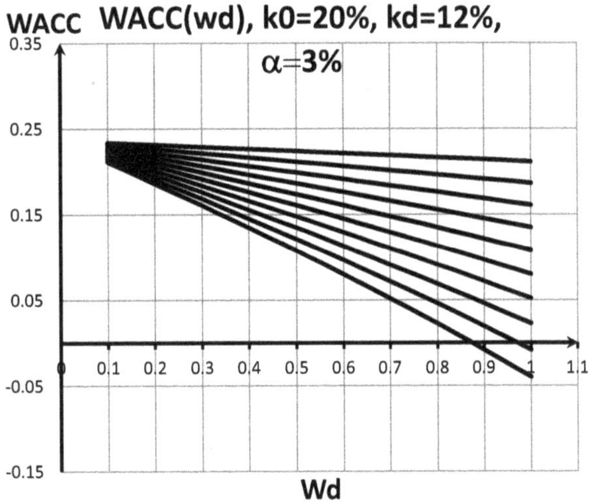

Fig. 11.5 Dependence of the weighted average cost of capital WACC on debt ratio w_d at different tax on profit rate at inflation rate $\alpha = 3\%$ for five-year company. Tax on profit rate increases from $T = 0.1$ (upper line) up to $T = 1$ (lowest line) with step 0.1

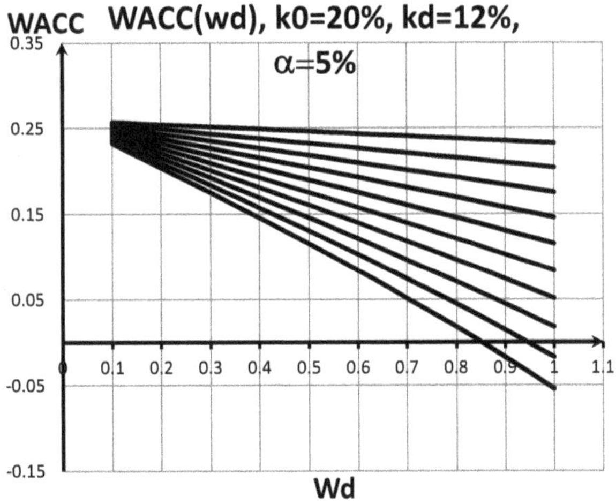

Fig. 11.6 Dependence of the weighted average cost of capital WACC on debt ratio w_{d} at different tax on profit rate at inflation rate $\alpha = 5\%$ for five-year company. Tax on profit rate increases from $T = 0.1$ (upper line) up to $T = 1$ (lowest line) with step 0.1

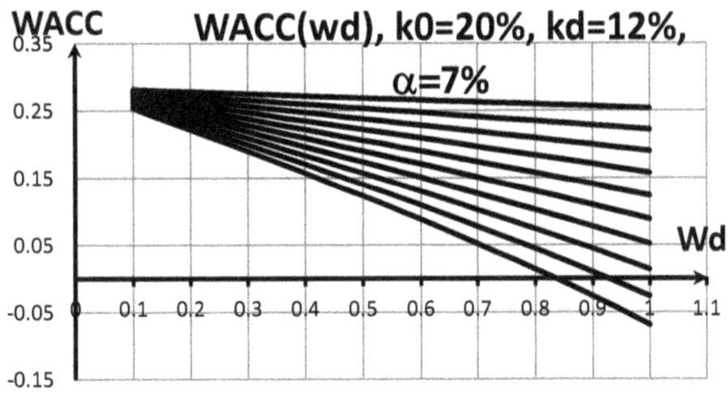

Fig. 11.7 Dependence of the weighted average cost of capital WACC on debt ratio w_{d} at different tax on profit rate at inflation rate $\alpha = 7\%$ for five-year company. Tax on profit rate increases from $T = 0.1$ (upper line) up to $T = 1$ (lowest line) with step 0.1

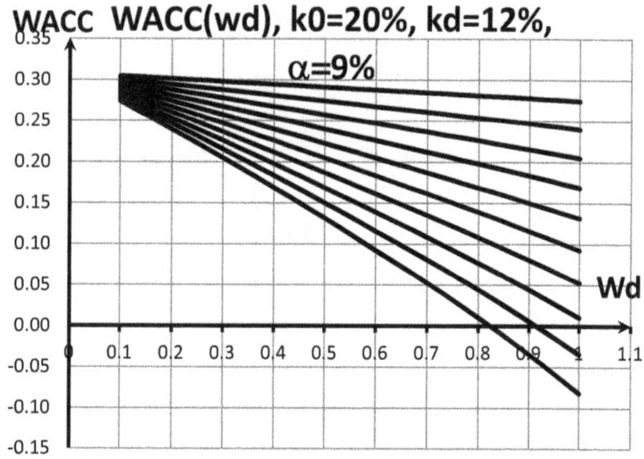

Fig. 11.8 Dependence of the weighted average cost of capital WACC on debt ratio w_d at different tax on profit rate at inflation rate $\alpha = 9\%$ for five-year company. Tax on profit rate increases from $T = 0.1$ (upper line) up to $T = 1$ (lowest line) with step 0.1

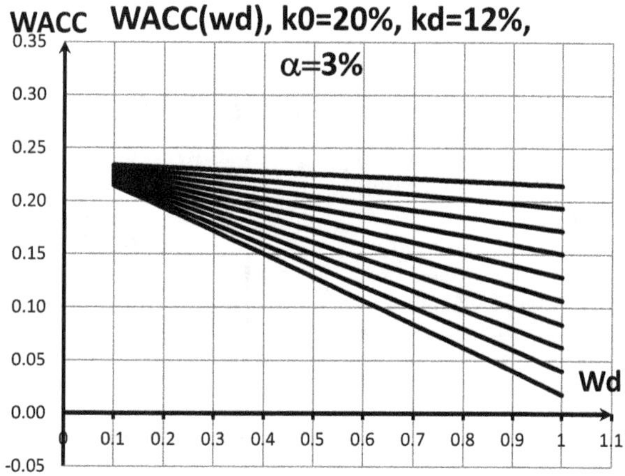

Fig. 11.9 Dependence of the weighted average cost of capital WACC on debt ratio w_d at different tax on profit rate at inflation rate $\alpha = 3\%$ for two-year company. Tax on profit rate increases from $T = 0.1$ (upper line) up to $T = 1$ (lowest line) with step 0.1

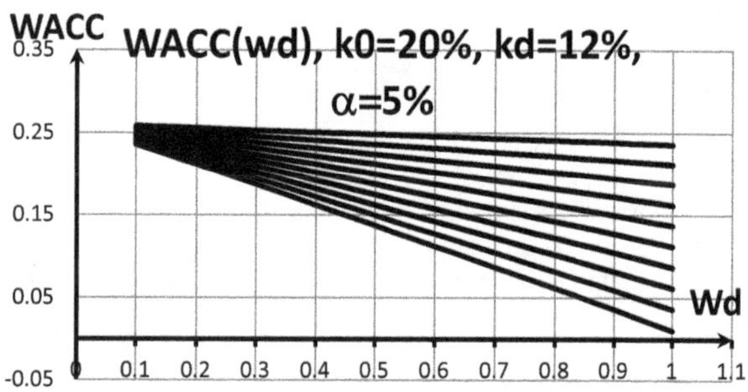

Fig. 11.10 Dependence of the weighted average cost of capital WACC on debt ratio w_d at different tax on profit rate at inflation rate $\alpha = 5\%$ for two-year company. Tax on profit rate increases from $T = 0.1$ (upper line) up to $T = 1$ (lowest line) with step 0.1

the West, is investigated. All basic results of Modigliani–Miller theory were modified. It is shown, that inflation not only increases the equity cost and the weighted average cost of capital, but as well it changes their dependence on leverage. In

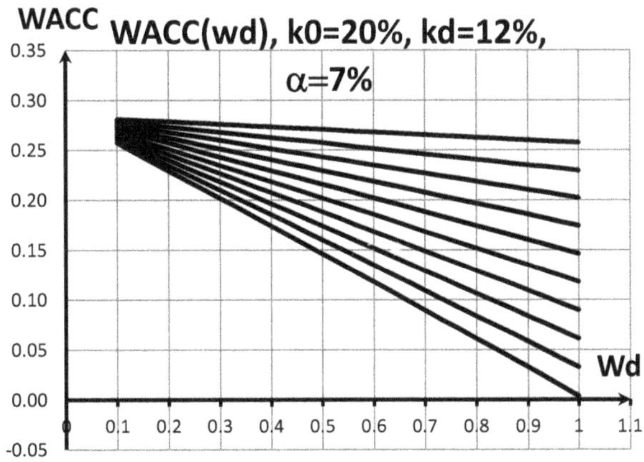

Fig. 11.11 Dependence of the weighted average cost of capital WACC on debt ratio w_d at different tax on profit rate at inflation rate $\alpha = 7\%$ for two-year company. Tax on profit rate increases from $T = 0.1$ (upper line) up to $T = 1$ (lowest line) with step 0.1

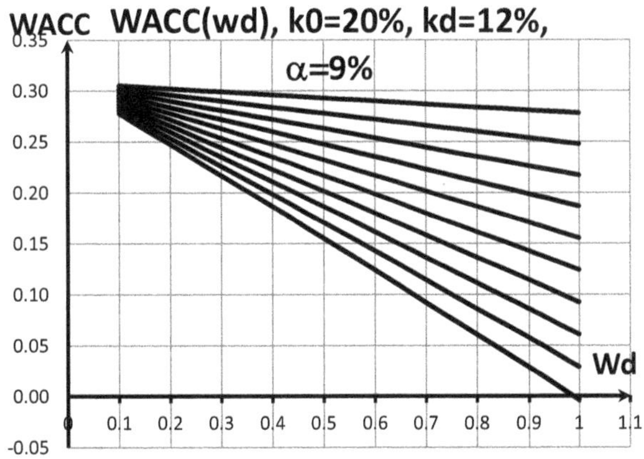

Fig. 11.12 Dependence of the weighted average cost of capital WACC on debt ratio w_d at different tax on profit rate at inflation rate $\alpha = 9\%$ for two-year company. Tax on profit rate increases from $T = 0.1$ (upper line) up to $T = 1$ (lowest line) with step 0.1

particular, it increases growing rate of equity cost with leverage. Capitalization of the company is decreased under accounting of inflation.

Within modern theory of capital cost and capital structure—Brusov–Filatova–Orekhova theory (BFO theory) the modified equation for the weighted average cost of capital, WACC, applicable to companies with arbitrary lifetime under accounting of inflation has been derived. Modified BFO equation allow to investigate the dependence of the weighted average cost of capital, WACC, and equity cost, k_e, on leverage level L, on tax on profit rate t, on lifetime of the company n, on equity cost of financially independent company, k_0, and debt cost, k_d, as well as on inflation rate α.

Using modified BFO equation the analysis of the dependence of the weighted average cost of capital, WACC, on debt ratio, w_d, at different tax on profit rate t, as well as inflation rate α has been done.

It has been shown, that WACC decreases with debt ratio, w_d, faster at bigger tax on profit rate t. The space between lines, corresponding to different values of tax on profit rate at the same step (10%), increases with inflation rate α. The variation region (with change of tax on profit rate t) of the weighted average cost of capital, WACC, increases with inflation rate α, as well as with lifetime of the company n.

11.6 Inflation Rate for a Few Periods

Suppose that the inflation rate for the consistent time periods $t_1, t_2, \ldots, t_n$ are equal to $\alpha_1, \alpha_2, \ldots, \alpha_n$ consequently. Let us find the inflation rate α for total time period $t = t_1 + t_2 + \cdots + t_n$.

Common sense dictates that inflation rate is an additive value, so that α, at least approximately, is equal to the sum of the inflation rates: $\alpha_1, \alpha_2, \ldots, \alpha_n$

$$\alpha \approx \alpha_1 + \alpha_2 + \cdots + \alpha_n. \tag{11.37}$$

Below we will get an exact expression for inflation rate for the total period of time t, and will see how it is different from an intuitive result (11.37).

At the end of the first commitment period the gained sum will be equal to the amount $S_1 = S_0(1 + i)$, and with the account of inflation $S_{1\alpha} = S_0(1+i)^{t_1}/(1+\alpha_1)$. At the end of the second commitment period the gained sum will be equal to the amount $S_2 = S_0(1+i)^{t_1+t_2}$, and with the account of inflation $S_{2\alpha} = S_0(1+i)^{t_1+t_2}/(1+\alpha_1)(1+\alpha_2)$. At the end of the n-th commitment period the gained sum will be equal to the amount $S_n = S_0(1+i)^{t_1+t_2+\cdots+t_n}$, and with the account of inflation:

$$S_{n\alpha} = S_0(1+i)^{t_1+t_2+\cdots+t_n}/(1+\alpha_1)(1+\alpha_2)\ldots(1+\alpha_n). \tag{11.38}$$

On the other hand, at inflation rate α for the total period at $t = t_1 + t_2 + \cdots + t_n$ at the end of this period t gained sum will be equal to:

$$S_{n\alpha} = S_0(1+i)^t/(1+\alpha). \tag{11.39}$$

Equating the right-hand part of (11.38) and (11.39), we get:

$$(1+\alpha_1)(1+\alpha_2)\ldots(1+\alpha_n) = 1+\alpha. \tag{11.40}$$

From where:

$$\alpha = (1+\alpha_1)(1+\alpha_2)\ldots(1+\alpha_n) - 1. \tag{11.41}$$

It is easy to get a strict proof of this formula by the method of mathematical induction. Note that inflation rate for the n-periods does not depend on both of the length of constituting periods and on the period t.

For equal inflation rates $\alpha_1 = \alpha_2 = \cdots = \alpha_n$ (it is interesting to note, that herewith the time intervals $t_1, t_2, \ldots, t_n$ can be arbitrary and do not equal each other) one has:

$$\alpha = (1 + \alpha_1)^n - 1. \tag{11.42}$$

References

1. Brusov, P., Filatova, T., Orehova, N., & Brusova, A. (2011). Weighted average cost of capital in the theory of Modigliani–Miller, modified for a finite life–time company. *Applied Financial Economics, 21*(11), 815–824.
2. Brusov, P., Filatova, P., & Orekhova, N. (2013). Absence of an optimal capital structure in the famous tradeoff theory! *Journal of Reviews on Global Economics, 2*, 94–116.
3. Brusov, P., Filatova, P., & Orekhova, N. (2014). Mechanism of formation of the company optimal capital structure, different from suggested by trade off theory. *Cogent Economics & Finance, 2*, 1–13. https://doi.org/10.1080/23322039.2014.946150
4. Brusov, P., Filatova, T., Orehova, N., et al. (2011). From Modigliani–Miller to general theory of capital cost and capital structure of the company. *Research Journal of Economics, Business and ICT, 2*, 16–21.
5. Brusov, P., Filatova, T., Eskindarov, M., & Orehova, N. (2012). Influence of debt financing on the effectiveness of the finite duration investment project. *Applied Financial Economics, 22*(13), 1043–1052.
6. Brusov, P., Filatova, T., Orehova, N., et al. (2011). Influence of debt financing on the effectiveness of the investment project within the Modigliani–Miller theory. *Research Journal of Economics, Business and ICT (UK), 2*, 11–15.
7. Brusov, P., Filatova, T., Eskindarov, M., & Orehova, N. (2012). Hidden global causes of the global financial crisis. *Journal of Reviews on Global Economics, 1*, 106–111.
8. Brusov, P. N., & Filatova, T. V. (2011). From Modigliani–Miller to general theory of capital cost and capital structure of the company. *Finance and Credit, 435*, 2–8.
9. Brusov, P., Filatova, T., Orehova, N., Brusov, P. P., & Brusova, N. (2011). From Modigliani–Miller to general theory of capital cost and capital structure of the company. *Research Journal of Economics, Business and ICT, 2*, 16–21.
10. Brusov, P., Filatova, T., & Orehova, N. (2014). Inflation in Brusov–Filatova–Orekhova theory and in its perpetuity limit – Modigliani – Miller Theory. *Journal of Reviews on Global Economics, 3*, 175–185.
11. Brusov, P., Filatova, T., & Orehova, N. (2013). A qualitatively new effect in corporative finance: Abnormal dependence of cost of equity of company on leverage. *Journal of Reviews on Global Economics, 2*, 183–193.
12. Filatova, T., Orehova, N., & Brusova, A. (2008). Weighted average cost of capital in the theory of Modigliani–Miller, modified for a finite life–time company. *Bulletin of the FU, 48*, 68–77.
13. Modigliani, F., & Miller, M. (1958). The cost of capital, corporate finance, and the theory of investment. *American Economic Review, 48*, 261–297.
14. Modigliani, F., & Miller, M. (1963). Corporate income taxes and the cost of capital: A correction. *American Economic Review, 53*, 147–175.
15. Modigliani, F., & Miller, M. (1966). Some estimates of the cost of capital to the electric utility industry 1954–1957. *American Economic Review, 56*, 333–391.
16. Brusova, A. (2011). A Comparison of the three methods of estimation of weighted average cost of capital and equity cost of company. *Financial Analysis: Problems and Solutions, 34*(76), 36–42.

Chapter 12
Benefits of Advance Payments of Tax on Profit: Consideration Within Brusov–Filatova–Orekhova (BFO) Theory

12.1 Introduction

The importance of accounting for upfront payments rests on two points: (1) companies can pay tax payments upfront with subsequent adjustments if they earn a stable and predictable income; (2) the development of a methodology for advance payments allows the regulator to expand this practice, which ensures an increase in the stability of budget revenues. As results of this Chapter show the advance payments of tax on income are quite important for both parties: for companies and for regulator.

We give below the short review of evolution of the theory of capital cost and capital structure.

(1) The Modigliani-Miller (MM) theory, created by Nobel Prize winners in 1958 [1], was the first quantitative theory of capital structure. It was based on many restrictions, the main of which were the absence of taxes and the perpetual nature of all financial flows and companies. The first limitation was removed by the MMs themselves [2, 3] and the following formulas were obtained: For tax shield TS:

$$\mathrm{TS} = DT; \tag{12.1}$$

For company value, V, financially independent company V_0, leverage company, V:

$$V_0 = \mathrm{CF}/k_0 \text{ and } \quad V = \mathrm{CF}/\mathrm{WACC}. \tag{12.2}$$

For weighted average cost of capital, WACC:

P. Brusov, T. Filatova, *Business Valuation and Company Value*, Contributions to Finance and Accounting, https://doi.org/10.1007/978-3-031-90847-7_12

$$\text{WACC} = k_0 \cdot (1 - w_d t); \tag{12.3}$$

For equity cost, k_e:

$$k_e = k_0 + L(k_e - k_d) \cdot (1 - t). \tag{12.4}$$

(2) Hamada [4] has derived the following formula for the leveraged company equity cost , accounting both financial and business risk of company, united the Modigliani–Miller theory with Capital Asset Pricing Model (CAPM):

$$k_e = k_F + (k_M - k_F) b_U + (k_M - k_F) b_U \frac{D}{S}(1 - T), \tag{12.5}$$

where b_U is the β–coefficient of the company of the same group of business risk, that the considering company, but with $L = 0$.

(3) Merton Miller [5] took into account the corporate and individual taxes he got for the capitalization of the financially independent company the following formula:

$$V = V_0 + \left[1 - \frac{(1 - T_C)(1 - T_S)}{(1 - T_D)}\right] \cdot D \tag{12.6}$$

Here T_C–the tax on corporate income rate, T_S–the tax rate on income of an individual investor from his ownership by corporation stock, T_D–tax rate on interest income from the provision of investor–individuals of credits to other investors and companies.

(4) In 2008, Brusov–Filatova–Orekhova [6, 7] have lifted up the limitation concerning the perpetuity of flows and companies and created the modern theory of capital cost and capital structure—BFO theory—which describes the companies of arbitrary age (and arbitrary life–time). The generalization of MM theory for the companies of arbitrary age (and arbitrary life–time) required the modification of the valuation of the tax shield TS, as well as of the valuation of the company capitalization: financially independent, V_0 as well as financially dependent, V.

The following formulas were obtained: for the tax shield, TS, company value, V, and for weighted average cost of capital WACC:

$$\text{TS} = k_d D T \sum_{t=1}^{n} (1 + k_d)^{-t} = DT[1 - (1 + k_d)^{-n}]. \tag{12.7}$$

$$V_0 = \text{CF}[1 - (1 + k_0)^{-n}]/k_0; \quad V = \text{CF}[1 - (1 + \text{WACC})^{-n}]/\text{WACC}. \tag{12.8}$$

$$\frac{1-(1+\text{WACC})^{-n}}{\text{WACC}}=\frac{1-(1+k_0)^{-n}}{k_0[1-\omega_\text{d}T(1-(1+k_\text{d})^{-n})]}. \tag{12.9}$$

Here, S—the value of equity capital of the company, $w_\text{d}=\frac{D}{D+S}$—the share of debt capital, $k_\text{e}, w_\text{e}=\frac{S}{D+S}$—the cost and the share of the equity capital of the company, and $L=D/S$—financial leverage, D—the value of debt capital.

At $n=1$, from BFO formula for WACC we get Myers [8] formula for 1-year company

$$\text{WACC}=k_0-\frac{(1+k_0)k_\text{d}}{1+k_\text{d}}w_\text{d}T; \tag{12.10}$$

and at $n=\infty$ we get Modigliani–Miller formula for WACC [2]:

$$\text{WACC}=k_0\cdot(1-w_\text{d}t) \tag{12.11}$$

Note, that before BFO discovery in 2008 there were two results only: Modigliani–Miller for perpetuity companies [1–3] and Myers one for a year companies [8]. The accounting of the company's finite age, as it was shown by BFO authors leads to significant changes of all Modigliani–Miller results [1–3]. Besides, a number of qualitatively new effects in corporate finance, obtained in Brusov–Filatova–Orekhova theory [6, 7] are absent in Modigliani–Miller theory.

(5) Recently we have modified the Modigliani–Miller theory for the case of advanced payments of income tax [7, 9] and have shown that obtained results are quite different from ones in "classical" Modigliani–Miller theory.

12.2 Modification of the Brusov–Filatova–Orekhova (BFO) Theory for Companies with Frequent Payments of Tax on Income

12.2.1 Calculation of the Tax Shield

Let us calculate the tax shield within Brusov–Filatova–Orekhova theory for the case of advance payments of tax on profit. The tax shield, TS, for a period of n-years is equal to sum of discounted values of benefits from the use of tax incentives:

$$(\text{TS})_n=k_\text{d}Dt+\frac{k_\text{d}Dt}{(1+k_\text{d})}+\frac{k_\text{d}Dt}{(1+k_\text{d})^2}+\cdots+\frac{k_\text{d}Dt}{(1+k_\text{d})^{n-1}} \tag{12.12}$$

This expression represents a geometric progression with denominator:

$$q=\frac{1}{(1+k_{\mathrm{d}})} \tag{12.13}$$

After summing the progression, one obtains:

$$(\mathrm{TS})_n=\frac{k_{\mathrm{d}}Dt(1-(1+k_{\mathrm{d}})^{-n})}{\left(1-(1+k_{\mathrm{d}})^{-1}\right)}=Dt(1-(1+k_{\mathrm{d}})^{-n})(1+k_{\mathrm{d}}) \tag{12.14}$$

In classical Brusov–Filatova–Orekhova theory under payments at the end of periods one has:

$$(\mathrm{TS})_n=Dt(1-(1+k_{\mathrm{d}})^{-n}) \tag{12.15}$$

12.2.2 Company Value

For the financially dependent company value V we have:

$$V=V_0+(\mathrm{TS})_n \tag{12.16}$$

$$V=V_0+Dt(1-(1+k_{\mathrm{d}})^{-n})\cdot(1+k_{\mathrm{d}}). \tag{12.17}$$

This formula differs from the equation for company value V in case of payments of income tax at the end of the year:

$$V=V_0+Dt(1-(1+k_{\mathrm{d}})^{-n}). \tag{12.18}$$

12.2.3 The Weighted Average Cost of Capital, WACC

Substituting

$$D=w_{\mathrm{d}}V \tag{12.19}$$

into (12.17), we get:

$$V\left(1-w_{\mathrm{d}}t(1-(1+k_{\mathrm{d}})^{-n})\cdot(1+k_{\mathrm{d}})\right)=V_0 \tag{12.20}$$

Using the values of financially dependent company, V, and financially independent company, V_0,

$$V=\frac{\mathrm{CF}(1-(1+\mathrm{WACC})^{-n})}{\mathrm{WACC}};\quad V_0=\frac{\mathrm{CF}}{k_0}\cdot(1-(1+k_0)^{-n}),\tag{12.21}$$

we come to the following intermediate equation:

$$\frac{\mathrm{CF}(1-(1+\mathrm{WACC})^{-n})}{\mathrm{WACC}}\cdot(1-w_d t\cdot(1-(1+k_d)^{-n}))\cdot(1+k_d)$$
$$=\frac{\mathrm{CF}}{k_0}\cdot(1-(1+k_0)^{-n})\tag{12.22}$$

and then to the final Brusov–Filatova–Orekhova equation for WACC for the case of advanced payments of income tax

$$\frac{(1-(1+\mathrm{WACC})^{-n})}{\mathrm{WACC}}=\frac{(1-(1+k_0)^{-n})}{k_0\left(1-w_d t\cdot(1-(1+k_d)^{-n})\right)\cdot(1+k_d)}.\tag{12.23}$$

This formula differs from the classical Brusov–Filatova–Orekhova equation by the factor $(1 + k_d)$ in the left denominator

$$\frac{(1-(1+\mathrm{WACC})^{-n})}{\mathrm{WACC}}=\frac{(1-(1+k_0)^{-n})}{k_0\left(1-w_d t\cdot(1-(1+k_d)^{-n})\right)}\tag{12.24}$$

12.2.3.1 Calculation of the Equity Cost

To calculate the equity cost k_e one should use the equation of WACC definition:

$$\mathrm{WACC}=k_e w_e+k_d w_d(1-t)\tag{12.25}$$

from where one gets the expression for k_e:

$$k_e=\frac{\mathrm{WACC}}{w_e}-\frac{k_d(1-t)w_d}{w_e}=\mathrm{WACC}(1+L)-Lk_d(1-t)\tag{12.26}$$

One should substitute WACC from the formula (12.12), calculating the equity cost k_e for the case of advanced payments of income tax and from the formula (12.13), calculating the equity cost k_e for the case of payments of income tax at the end of the year.

12.3 Results

Here the dependence of the weighted average cost of capital, WACC, capital value, V, equity cost, k_e, on leverage level L for three-year and six-year companies, using Microsoft Excel is being studied. We consider two types of payments of income tax: (1) at the end of the year and (2) in advance. As we mentioned above, for WACC we use formulas (12.12) and (12.13), for capital value, V, we use formulas (12.6) and (12.7) and for equity cost, k_e, we use formula (12.15).

We use the following parameters: $k_0 = 0.2$; $k_d = 0.18$; $t = 0.2$; $n = 3$; 6; CF $= 100$.

12.3.1 Dependence of the Weighted Average Cost of Capital, WACC, Capital Value, V*, Equity Cost,* k$_e$*, on Leverage Level* L *for Three-year Company*

Table 12.1 Dependence of the weighted average cost of capital, WACC, equity cost, k_e and company value, V, on leverage level L in the cases of payments of tax on profit at the end of the year (1) and in the beginning of the year (2) for three-year company

$n = 3$	WACC		k_e		V	
L	1	2	1	2	1	2
1	0.1749	0.1703	0.2057	0.1966	219.2281	220.8472
2	0.1664	0.1603	0.2113	0.1930	222.2455	224.4700
3	0.1622	0.1553	0.2168	0.1892	223.7856	226.3263
4	0.1596	0.1523	0.2222	0.1855	224.7199	227.4549
5	0.1579	0.1503	0.2277	0.1817	225.3471	228.2136
6	0.1567	0.1489	0.2331	0.1780	225.7973	228.7586
7	0.1558	0.1478	0.2386	0.1742	226.1361	229.1691
8	0.1551	0.1469	0.2440	0.1704	226.4003	229.4893
9	0.1545	0.1463	0.2495	0.1666	226.6122	229.7462
10	0.1541	0.1457	0.2549	0.1628	226.7858	229.9568

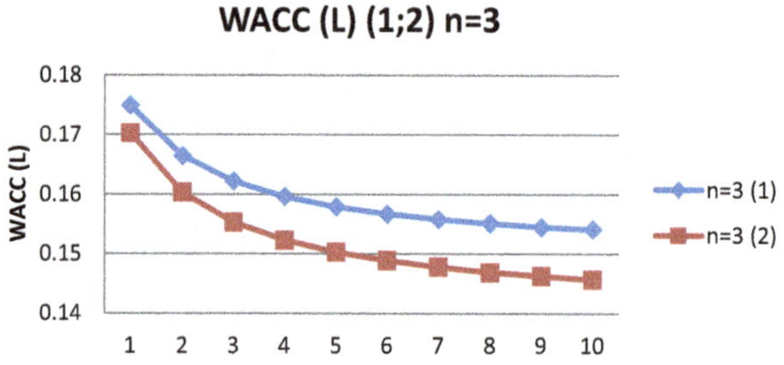

Fig. 12.1 Dependence of the weighted average cost of capital, WACC, on leverage level L in the cases of payments of tax on profit at the end of the year (1) and in the beginning of the year (2) for three-year company

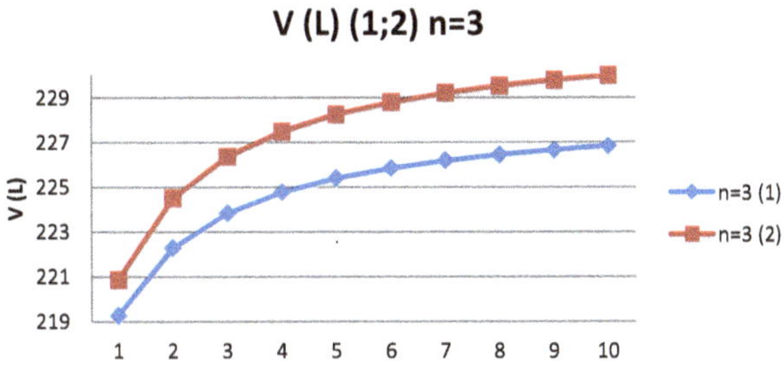

Fig. 12.2 Dependence of the company value, V, on leverage level L in the cases of payments of tax on profit at the end of the year (1) and in the beginning of the year (2) for three-year company

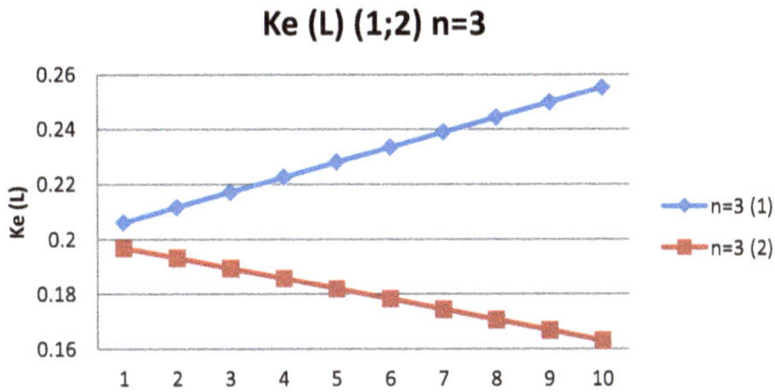

Fig. 12.3 Dependence of the equity cost, k_e on leverage level L in the cases of payments of tax on profit at the end of the year (1) and in the beginning of the year (2) for three-year company

12.3.2 Dependence of the Weighted Average Cost of Capital, WACC, Capital Value, V*, Equity Cost,* k_e*, on Leverage Level* L *for Six-Year Company*

Table 12.2 Dependence of the weighted average cost of capital, WACC, equity cost, k_e and company value, V, on leverage level L in the cases of payments of tax on profit at the end of the year (1) and in the beginning of the year (2) for six-year company

$n = 6$	WACC		k_e		V	
L	1	2	1	2	1	2
1	0.1743	0.1697	0.2047	0.1953	354.8940	359.2385
2	0.1657	0.1593	0.2090	0.1900	363.0242	369.1124
3	0.1613	0.1542	0.2131	0.1846	367.2305	374.2557
4	0.1586	0.1510	0.2172	0.1791	369.8015	377.4110
5	0.1569	0.1489	0.2213	0.1736	371.5355	379.5443
6	0.1556	0.1474	0.2254	0.1680	372.7841	381.0830
7	0.1547	0.1463	0.2294	0.1624	373.7261	382.2451
8	0.1539	0.1454	0.2335	0.1569	374.4620	383.1540
9	0.1534	0.1447	0.2376	0.1513	375.0529	383.8841
10	0.1529	0.1442	0.2416	0.1457	375.5377	384.4836

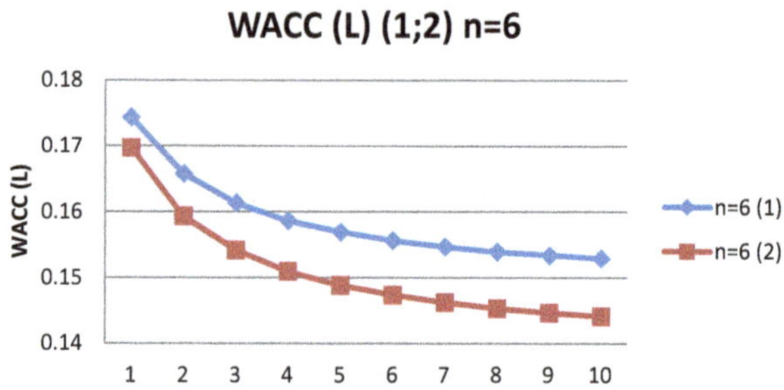

Fig. 12.4 Dependence of the weighted average cost of capital, WACC, on leverage level L in the cases of payments of tax on profit at the end of the year (1) and in the beginning of the year (2) for six-year company

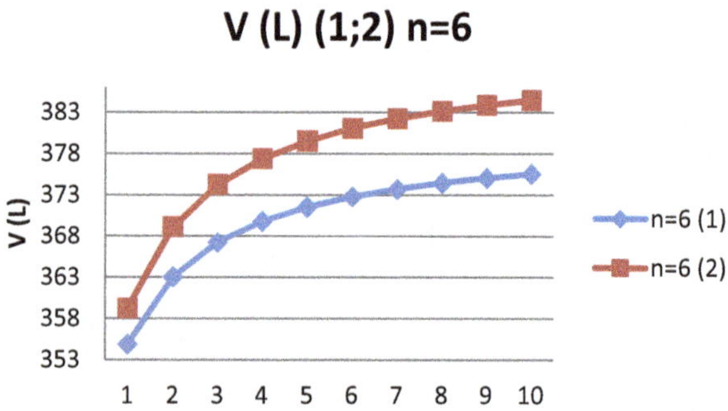

Fig. 12.5 Dependence of the company value, V, on leverage level L in the cases of payments of tax on profit at the end of the year (1) and in the beginning of the year (2) for six-year company

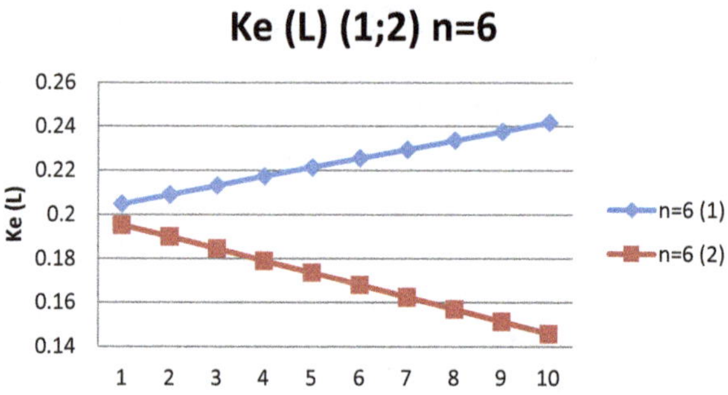

Fig. 12.6 Dependence of the equity cost, k_e on leverage level L in the cases of payments of tax on profit at the end of the year (1) and in the beginning of the year (2) for three-year company

12.4 Comparison of Results for Three-Year and Six-Year Companies

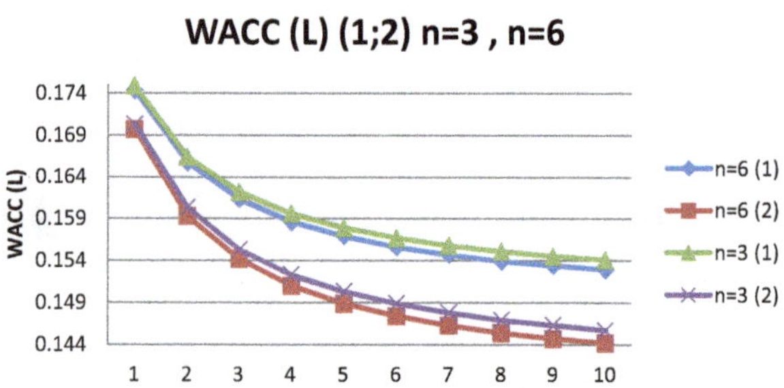

Fig. 12.7 Dependence of the weighted average cost of capital, WACC, on leverage level L in the cases of payments of tax on profit at the end of the year (1) and in the beginning of the year (2) for three-year and six-year companies

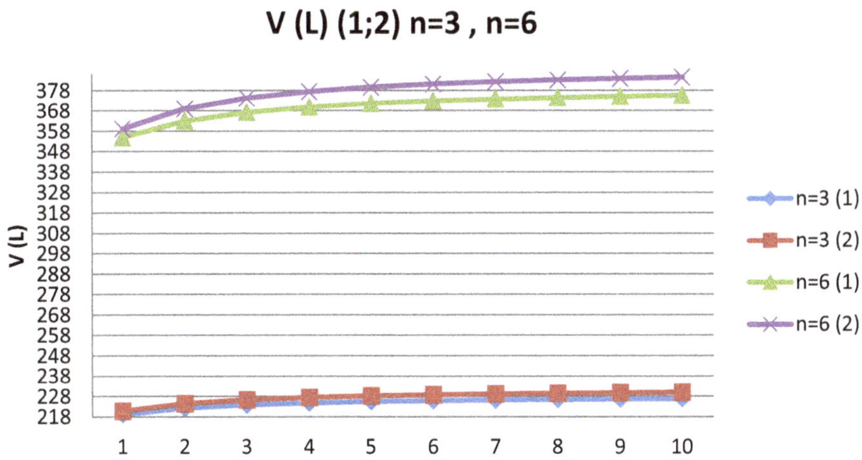

Fig. 12.8 Dependence of the company value, *V*, on leverage level *L* in the cases of payments of tax on profit at the end of the year (1) and in the beginning of the year (2) for three-year and six-year companies

Fig. 12.9 Dependence of the equity cost, k_e, on leverage level *L* in the cases of payments of tax on profit at the end of the year (1) and in the beginning of the year (2) for three-year and six-year companies

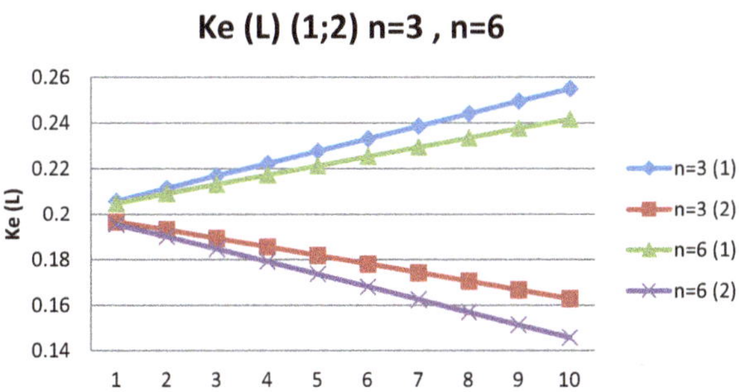

12.5 Discussion

We derive BFO formulas for WACC, *V*, k_e for the case of advanced payments of income tax. Making the calculations using these formulas within Microsoft Excel we get the following results (see Tables 12.1 and 12.2 and Figs. 12.1, 12.2, 12.3, 12.4, 12.5, 12.6, 12.7, 12.8, and 12.9).

1. WACC(*L*) decreases with *L* in both cases: advanced payments of income tax and payments at the ends of years. This means that debt financing is important and should be used by company—it leads to decrease of attracting capital cost with *L*.
2. WACC turns out to be lower in the case of advanced payments of income tax, this tells about importance of the use of advanced payments of income tax for companies.

3. WACC decreases with company age: this is one of the important results of classical BFO theory.
4. Company value, V, increases with L in both cases, this follows from the decrease of attracting capital cost with L.
5. Company value, V, turns out to be bigger in the case of advanced payments of income tax: this tells about importance of the use of advanced payments of income tax for companies.
6. Company value, V, increases with company age.
7. Equity cost k_e decreases with company age in both cases: this is one of the important results of classical BFO theory.
8. Equity cost k_e increases with leverage level L in the case of payments of income tax at the end of the year.
9. Equity cost k_e decreases with leverage level L in the case of advanced payments of income tax. This means the appearance of a qualitatively new effect that can greatly change the company's dividend policy, because the economically justified amount of dividends is equal to the equity cost.

12.6 Summary and Conclusions

Advance income tax payments are beneficial to both parties: to companies, because they lead to decrease of cost of attracting capital and increase of company values; to regulator, because earlier replenishment of the budget ensures an increase in the stability of budget revenues. Thus the regulator should extend the practice of tax payments in advance by the companies.

An important conclusion drawn in this Chapter is that the tax shield is very important and the way it is formed (payments at the end of the year or in advance) leads to very important consequences, changing all the financial indicators of the company, such as the cost of raising capital and company value and radically changing the company's dividend policy. If with payments at the end of the year, the amount of dividends should increase with the increase in the use of debt financing, with advance payments of income tax the amount of dividends should decrease with the increase in the use of debt financing: this is a pioneering result that radically changes the company's dividend policy.

References

1. Modigliani, F., & Miller, M. (1958). The cost of capital, corporate finance, and the theory of investment. *American Economic Review, 48*, 261–297.
2. Modigliani, F., & Miller, M. (1963). Corporate income taxes and the cost of capital: A correction. *American Economic Review, 53*, 147–175.
3. Modigliani, F., & Miller, M. (1966). Some estimates of the cost of capital to the electric utility industry 1954–1957. *American Economic Review, 56*, 333–391.

4. Hamada, R. (1969). Portfolio analysis, market equilibrium, and corporate finance. *Journal of Finance*, 13–31.
5. Miller, M. (1977). Debt and taxes. *The Journal of Finance, 32*(2), 261–275. https://doi.org/10.1111/j.1540-6261.1977.tb03267.x
6. Brusov, P., Filatova, T., Orehova, N., & Eskindarov, M. (2018). *Modern corporate finance, investments, taxation and ratings* (2nd ed., pp. 1–571). Springer Nature Publishing.
7. Brusov, P., Filatova, T., & Orehova, N. (2022). *Generalized Modigliani–Miller theory. Applications in corporate finance, investments, taxation and ratings* (pp. 1–362). Springer Nature Publishing. Retrieved from https://rd.springer.com/book/12.1007/978-3-030-93893-2
8. Myers, S. (2001). Capital structure. *Journal of Economic Perspectives, 15*(2), 81–102.
9. Brusov, P. N., Filatova, T. V., Orekhova, N. P., Kulik, V. L., Chang, S.-I., & Lin, Y. C. G. (2020). Modification of the Modigliani–Miller theory for the case of advance payments of tax on profit. *Journal of Reviews on Global Economics, 2020*(9), 257–267.

Chapter 13
The Generalization of the Brusov–Filatova–Orekhova Theory for the Case of Payments of Tax on Profit with Arbitrary Frequency

13.1 Introduction

The Chapter has the following structure:

In Introduction we describe some approaches to the capital structure theory: traditional approach, Modigliani–Miller theory and some of its modifications.

We generalize the Brusov–Filatova–Orekhova (BFO) theory for the case of arbitrary frequency of payments of income tax and get generalized Brusov–Filatova–Orekhova (BFO) formulas for the weighted average cost of capital, WACC, for the equity cost, k_e, and for the capitalization of the company, V.

Within the generalized Brusov–Filatova–Orekhova (BFO) theory the dependence of the main financial parameters of the company (the weighted average cost of capital, WACC, equity cost, k_e, company capitalization, V) on leverage level L at different frequency of payments of income tax p is studied.

Obtained results allow come to some very important conclusions and give recommendations how frequently company should pay income tax in order to decrease the cost of attracting capital and to increase its capitalization.

13.1.1 Capital Structure of the Company

This Chapter is devoted to one of the most important problem of financial management—problem of capital cost and capital structure of the company. The management by capital structure of the company (the relationship between equity and debt capital of the company) allows to company's management solve the main task—increase the company value. This relates as well with the problem of the optimal capital structure, i.e., with the capital structure, which minimizes the weighted average cost of capital, WACC, and maximizes the value of the company, V, which is one of the most important problem in financial management. The first

P. Brusov, T. Filatova, *Business Valuation and Company Value*, Contributions to Finance and Accounting, https://doi.org/10.1007/978-3-031-90847-7_13

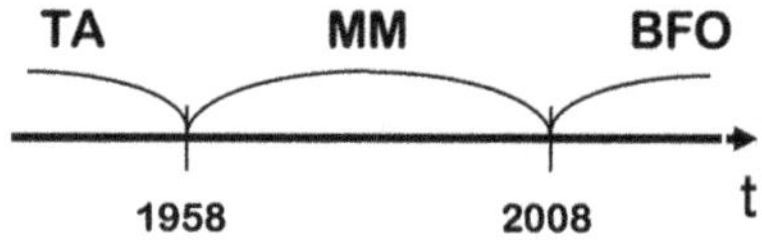

Fig. 13.1 Historical aspects of capital cost and capital structure theory: here empirical traditional approach—TA, Modigliani–Miller theory—MM, Brusov–Filatova–Orekhova theory—BFO

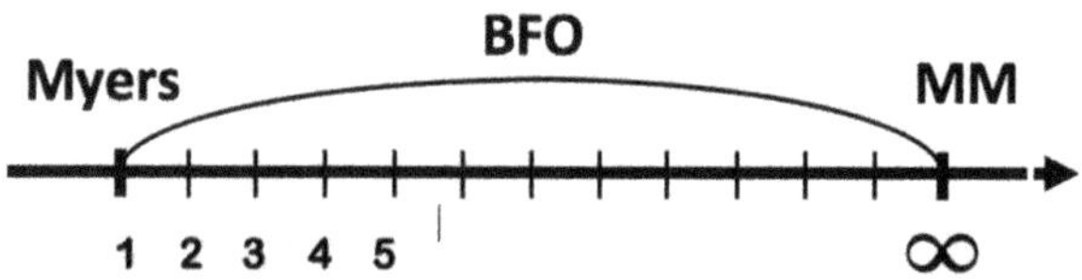

Fig. 13.2 Myers result describes one-year company, Modigliani–Miller theory (MM) corresponds to perpetuity limit, while Brusov–Filatova–Orekhova (BFO) theory fills the whole numeric axis (from $n = 1$ up to perpetuity limit $n = \infty$)

quantitative study of influence of capital structure of the company on its (company) financial indicators was the work by Nobel Prize winners Modigliani and Miller [1]. Before their work, the traditional approach existed, based on empirical data analysis. In 2008 the modern capital cost and capital structure theory—Brusov–Filatova–Orekhova (BFO) theory has been developed [2], which made Modigliani–Miller theory its particular case. Within BFO theory many qualitatively new effects has been discovered, which are absent within Modigliani–Miller theory. Some main existing principles of financial management have been destroyed by BFO theory, among them the world famous trade off theory [3–5], which was considered as keystone of formation of optimal capital structure of the company during many decades. Brusov–Filatova–Orekhova have proven the bankruptcy of trade off theory and have found the cause of this (Chap. 4 in monograph [6]). Below we briefly describe the historical development of the theory of the capital cost and the capital structure.

Figure 13.1 shows the historical development of capital cost and capital structure theory (from the empirical traditional approach, through perpetuity Modigliani–Miller theory to general capital cost and capital structure theory—Brusov–Filatova–Orekhova (BFO) theory [6].

Myers [7] in 2001 has considered the one-year company case and show that the weighted average cost of capital, WACC, in this case is higher than in Modigliani–Miller case, and the capitalization of the company, V, is less than in Modigliani–Miller case.

So, before 2008 only two results for capital structure of the company were available: Modigliani–Miller for perpetuity company [1, 8, 9] and Myers [7] for one-year company (see Fig. 13.2). Wherein Brusov–Filatova–Orekhova theory has filled out whole interval between $t = 1$ and $t = \infty$. Brusov–Filatova–Orekhova theory allows calculate the company value, V, the weighted average cost of capital, $WACC$, the equity cost k_e and other financial parameters for arbitrary age companies and for arbitrary life–time companies. BFO theory has led to a lot of new meaningful effects in modern capital structure theory, absent in Modigliani–Miller theory [6].

The main purpose of this Chapter is to generalize and further develop the Brusov–Filatova–Orekhova (BFO) theory, taking into account the case of paying income tax with an arbitrary frequency (monthly, quarterly, semi–annual or annual payments), which takes place in real economic practice. Note that both major theories on the cost of capital and capital structure of the Brusov–Filatova–Orekhova (BFO) theory and its perpetuity limit, the Modigliani–Miller theory, take into account annual income tax payments. The latter was recently generalized by us to the case of an arbitrary periodicity in the payment of income tax [10], which showed a significant change in the results of the classical Modigliani–Miller theory. Here we make this generalization for the Brusov–Filatova–Orekhova (BFO) theory.

Below we discuss the traditional approach, the Modigliani–Miller theory and its different modifications, the Brusov–Filatova–Orekhova (BFO) theory.

13.1.2 The Modigliani–Miller Theory

Before the Modigliani–Miller work, the traditional approach existed, based on empirical data analysis. In the traditional approach weighted average cost of capital, WACC, and the company value, *V*, depend on the capital structure, on the leverage level, *L*. The cost of debt is lower than cost of equity via the fact that first one has lower risk, because creditor claims are met prior to shareholders claims in the event of bankruptcy. Thus the increase of the share of lower–cost debt capital up to the values which does not cause violation of financial sustainability and growth of risk of bankruptcy leads to lower weighted average cost of capital, WACC. The required by investors profitability, which is equal to the equity cost, is growing; but its growth does not compensate the benefits from use of lower–cost debt capital. Therefore, in the traditional approach the increase of leverage level (at low leverage level) leads to decrease of WACC and to the associated increase of the value of company *V*. At high leverage level, financial difficulties and the risk of bankruptcy appear. The debt capital cost increases as well as the equity cost (which always increases with leverage level) and these lead to increase of WACC and decrease of company value *V*. Thus, competition between the advantages of debt financing at a low level of leverage and its disadvantages at a high level of leverage forms the optimal capital structure, when WACC reaches minimum and capital value *V* reaches maximum. This is the main point of the traditional approach and as well as trade off theory. This empirical approach has existed up to creation of the first quantitative theory by Nobel Prize Winners Modigliani and Miller (1958) [1].

Modigliani and Miller (MM) in their first Chapter [1] have come to the conclusions which were fundamentally different from the conclusions of traditional approach. Under assumptions that there are no taxes, no transaction costs, no bankruptcy costs, perfect financial markets exist with symmetry information, equivalence in borrowing costs for both companies and investors, etc., they have showed that choosing of the ratio between the debt and equity capital does not affect company value V, as well as weighted average cost of capital, WACC.

Modigliani and Miller analyzing the influence of financial leverage on V and V_0, weighted average cost of capital, WACC and k_e, get the following results: (here, EBIT–earnings before interest and taxes, k_0 is discount rate, V and V_0 are the values of levered and unlevered companies, k_e is the equity cost)

$$V = V_0 = \frac{\text{EBIT}}{k_0} \tag{13.1}$$

This leads to the following expressions for weighted average cost of capital, WACC:

$$\text{WACC} = k_0. \tag{13.2}$$

Note, that k_0 here and below is the equity cost for financially independent company. For financially dependent company k_0 is the equity cost at zero leverage level ($L = 0$).

From (13.1), and formula for WACC definition

$$\text{WACC} = k_0 = k_e w_e + k_d w_d \tag{13.3}$$

it is easy to derive an expression for the equity capital cost.

Finding k_e, one gets:

$$k_e = \frac{k_0}{w_e} - k_d \frac{w_d}{w_e} = \frac{k_0(S+D)}{S} - k_d \frac{D}{S} = k_0 + (k_0 - k_d)\frac{D}{S} = k_0 + (k_0 - k_d)L \tag{13.4}$$

Here,

D	value of debt capital of the company
S	value of equity capital of the company
$k_d, w_d = \frac{D}{D+S}$	cost and fraction of debt capital of the company
$k_e, w_e = \frac{S}{D+S}$	cost and fraction of equity capital of the company
$L = D/S$	financial leverage
WACC	weighted average cost of capital.

Thus, we have the following formula for equity cost:

$$k_e = k_0 + L(k_0 - k_d) \tag{13.5}$$

Formula (13.5) shows that equity cost of the company increases linearly with leverage level.

In 1963 Modigliani and Miller [8] have accounted the tax on profit and get the following formula for the cost of financially dependent company, V,

$$V = V_0 + DT, \tag{13.6}$$

where V_0 is the value of financially independent company, D is debt value and T is the tax on profit.

Below we get the expression for WACC and the equity capital cost k_e of the company under the existence of corporate taxes.

Weighted Average Cost of Capital, WACC

Substituting into (13.6) $D = w_d V$, one gets:

$$V(1 - w_d T) = V_0. \tag{13.7}$$

Putting $V = \frac{\mathrm{CF}}{\mathrm{WACC}}; V_0 = \frac{\mathrm{CF}}{k_0}$ into formula (13.7), we get:

$$\frac{\mathrm{CF}}{\mathrm{WACC}}(1 - w_d T) = \frac{\mathrm{CF}}{k_0}. \tag{13.8}$$

Here CF is the income of the company for one period.

And from (13.8) we arrive to the expression for weighted average cost of capital, WACC

$$\mathrm{WACC} = k_0 \cdot (1 - w_d t) \tag{13.9}$$

This formula for WACC is one of the main result of Modigliani–Miller theory with taxes.

The Equity Cost

Let us derive formula for equity cost.

On definition of the weighted average cost of capital WACC in the presence of corporate taxes, we have

$$\mathrm{WACC} = k_0 w_e + k_d w_d (1 - T). \tag{13.10}$$

Equating Eqs. (13.9) and (13.10), one gets:

$$k_0(1 - w_d T) = k_0 w_e + k_d w_d (1 - T) \tag{13.11}$$

and from here, for equity cost, k_e we get the following expression [8]:

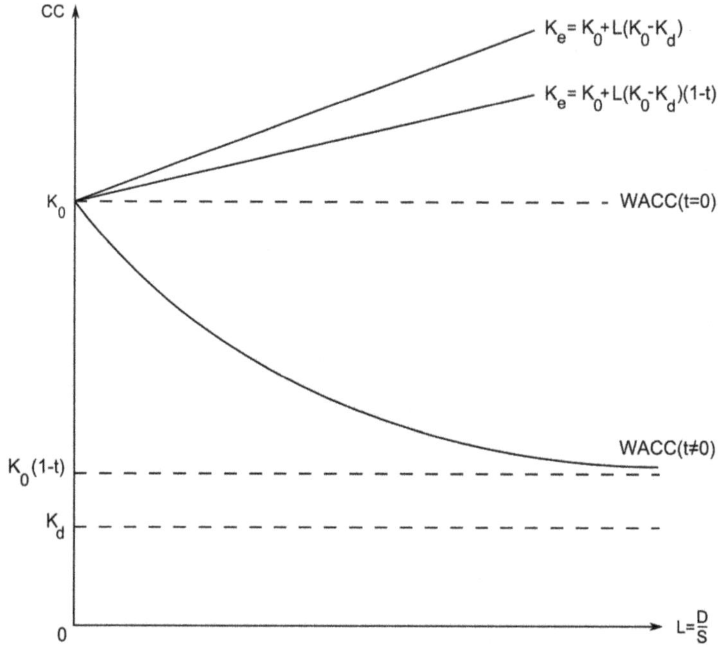

Fig. 13.3 Impact of leverage level on equity capital cost k_e, and WACC within the Modigliani–Miller theory without taxes ($t = 0$) and with taxes ($t \neq 0$)

$$\begin{aligned} k_e &= k_0 \frac{(1 - w_d T)}{w_e} - k_d \frac{w_d}{w_e}(1 - T) = k_0 \frac{1}{w_e} - k_0 \frac{w_d}{w_e} T - k_d \frac{D}{S}(1 - T) \\ &= k_0 \frac{D + S}{S} - k_0 \frac{D}{S} T - k_d \frac{D}{S}(1 - T) = k_0 + L(1 - T)(k_0 - k_d). \end{aligned} \tag{13.12}$$

The formula (13.12) is different from the formula (13.5) without tax only by the multiplier $(1 - T)$ in term, indicating a premium for risk. As the multiplier is less than unit, the corporate income tax leads to decrease of the slope of the curve $k_e(L)$ with respect to L–axis.

From formulas (13.5), (13.9) and (13.12) we get the following conclusions (see Fig. 13.3).

When leverage level grows:

(1) Value of company, V, increases;
(2) Weighted average cost of capital, WACC decreases from k_0 up to $k_0(1 - T)$ (at $L = \infty$);
(3) Equity cost increases linearly from k_0 up to infinity.

13.2 Some Modifications of Modigliani–Miller Theory

13.2.1 Hamada Model: Accounting Market Risk

The Modigliani–Miller theory were united with Capital Asset Pricing Model (CAPM) in 1969 [11] by R. Hamada, who has derived the following formula for the leveraged company equity cost , accounting both financial and business risk of company:

$$k_e = k_F + (k_M - k_F)b_U + (k_M - k_F)b_U \frac{D}{S}(1 - T), \qquad (13.13)$$

where b_U is the β–coefficient of the same group of business risk company, that the considering company, but with $L = 0$. The formula (13.13) represents the desired equity capital k_e profitability as a sum of three components: risk–free profitability k_F, compensating to shareholders a temporary value of their money, premium for business risk $(k_M - k_F)b_U$, and premium for financial risk $(k_M - k_F)b_U \frac{D}{S}(1 - T)$. If the company does not have debt financing, the financial risk factor will be equal to zero (the third term is drawn to zero), and only the premium for business risk will be received by shareholders.

13.2.2 The Account of Corporate and Individual Taxes (Miller Model)

In 1963 Modigliani and Miller [8] have accounted the taxation of corporate income, but the individual taxes of investors were not take into account.

The account of the corporate and individual taxes has been done by Merton Miller in 1977 [12], who study the influence of debt financing on the company capitalization. In Miller model, the following definitions have been used: TC–the tax on corporate income rate, TS–the tax rate on income of an individual investor from his ownership by corporation stock, TD–tax rate on interest income from the provision of investor–individuals of credits to other investors and companies. Because the income from shares partly comes in the form of a dividend and, in part, as capital profits, so TS is a weighted average value of the rates of tax on capital profits on shares and on dividends, while the profit from the provision of loans usually comes in the form of the interests. The former are usually taxed at a higher rate. For the capitalization of the financially independent company one has:

$$V_U = \frac{\text{EBIT}(1 - T_C)(1 - T_S)}{k_0}. \qquad (13.14)$$

A term $(1 - T_S)$ allows to take into account the individual taxes in the formula. In this way, numerator indicates which part of the operating company's profit remains in the possession of the investors, after the company will pay taxes on their profits, and its shareholders then will pay individual taxes on income from stock ownership. Since individual taxes reduce profits, remaining in the disposal of investors, the last, at other things being equal circumstances, also reduce and an overall assessment of the financially independent company value.

13.2.3 More General Case for WACC Formula

In [13–16] more general than Modigliani–Miller (MM) formula for the WACC has been derived: it is described by the following formula [13]

$$\mathrm{WACC} = k_0(1 - w_d T) - k_d t w_d + k_{TS} t w_d. \tag{13.15}$$

Here k_0, k_d and k_{TS} are the expected returns respectively on the financially independent company, the debt and the tax shield, V is the capitalization of the financially dependent company, VTS is the tax shield value, D is the debt value and TC is the rate of corporate tax.

Formula (13.15) is derived from the WACC definition and the balance identity (see as well Berk and De Marzo, 2007 [15]). Although Eq. (13.15) is general enough, some additional conditions are required its practical applicability. If the WACC is constant over time, as it stated in [13] the levered company value can be found by discounting with the WACC for the unlevered free cash–flows. The resulting formulas, which describe the special cases, where the WACC is constant, can be found in textbooks [15, 16].

First, Modigliani and Miller in 1963 [8] assume that the level of debt D is constant. Then, as the expected after–tax cash–flow of the unlevered firm is fixed, V_0 is also constant. By assumption, $k_{TS} = k_D$ and the value of the tax shield is TS = tD. Therefore, the company value V is a constant and the general WACC formula (13.15) simplifies to a constant WACC:

$$\mathrm{WACC} = k_0(1 - w_d T)$$

We think that "classical" Modigliani–Miller (MM) theory, which suggests that the expected returns on the debt k_d and the tax shield k_{TS} are equals (because both of them have debt nature), is much more reasonable and namely the "classical" Modigliani–Miller (MM) theory, which is still widely used in practice, has been modified by us in [10].

13.2.4 Fiscal Pressure, Financial Liquidity, Financial Solvency, and Financial Leverage

The problem of fiscal pressure is more current than ever in most countries. The article by Batrancea [17] was the first empirical study of the impact of fiscal pressure on the financial equilibrium of energy companies listed on the New York Stock Exchange. Empirical results obtained by econometric models showed that fiscal pressure had a stronger impact on the short–term and long–term equilibrium of electricity and oil companies than on the equilibrium of gas companies. The research could be useful for the managers of energy companies interested in estimating the evolution of company equilibrium state when considering different possible financial crises.

How financial liquidity and financial solvency influenced the performance of healthcare companies that are publicly traded on the New York Stock Exchange has been studied by Batrancea [18], who used econometric models with two–stage least squares (2SLS) panel and panel generalized method of moments (GMM). Empirical evidence showed that the financial indicators current liquidity ratio, quick liquidity ratio, and financial leverage significantly influenced company performance measured by return on assets, gross margin ratio, operating margin ratio, earnings before interest, tax, depreciation, and amortization. Strategies intended to improve business performance based on liquidity and solvency insights were also addressed.

13.2.5 Brusov–Filatova–Orekhova (BFO) Theory

The suggestion about perpetuity of the companies and of all financial flows is the most serious limitations of the Modigliani–Miller theory. This limitation have been lifted up in 2008 by Brusov, Filatova and Orekhova [4], who have created the modern theory of capital cost and capital structure, generalizing the Modigliani–Miller theory for the companies of arbitrary age (arbitrary life–time). They have shown that in this case all Modigliani–Miller results [1–3] significant changes: in the presence of corporative taxes capitalization of the company V is changed, as well as the equity cost, k_e, and the weighted average cost of capital, WACC. A number of qualitatively new effects in corporate finance, obtained in Brusov–Filatova–Orekhova theory [4], are absent in Modigliani–Miller theory [1, 8, 9].

The Brusov–Filatova–Orekhova formula for weighted average cost of capital, WACC for the company of arbitrary age n, takes the following form [4]

$$\frac{1-(1+\mathrm{WACC})^{-n}}{\mathrm{WACC}}=\frac{1-(1+k_0)^{-n}}{k_0[1-w_d T(1-(1+k_d)^{-n})]}. \qquad (13.16)$$

Table 13.1 The similarities and differences between BFO and MM results

	BFO	MM
Time parameter, n	Arbitrary age of companies, arbitrary life–time of companies	Perpetuity companies, perpetuity financial flows; $n=\infty$
Company age and life–time	Are different	Are the same and equal infinity
Dependence of all main financial indicators: WACC, V, k_e on time parameter, n	Allow study this dependences	Time parameter, n is absent
all main financial indicators: WACC, V, k_e depend on debt cost k_d	Yes	None
WACC	Correct estimation; decreases with leverage level L	Underestimation; decreases with leverage level L
V	Correct estimation; increases with leverage level L	Overestimation; increases with leverage level L
Equity cost k_e	increases linearly with leverage level L	Increases linearly with leverage level L
Equity cost k_e: the slope of the curve $k_e(L)$	Less than in MM theory	Bigger than in BFO theory
k_e: the slope of the curve k_e (L)	**could be negative at $T > T^*$ (qualitatively new effect)**	Always is positive **New effect is absent**
Golden and silver age effects [6]	Exist	Absent

Here, S—the equity capital of the company value, $w_d = \frac{D}{D+S}$—the share of debt capital; $k_e, w_e = \frac{S}{D+S}$—the cost and the share of the equity capital of the company, $L = D/S$—leverage level.

Substituting in (13.14) $n \to \infty$, we easy arrive to the perpetuity (Modigliani–Miller) limit formula for weighted average cost of capital, WACC.

$$\mathrm{WACC} = k_0(1 - w_d T). \tag{13.17}$$

Results of BFO theory are well known in the world literature (for example, see references [19–25]). Some Chapters [25] use the BFO theory in practical calculations.

The similarities and differences between BFO and MM results are shown in Table 13.1.

13.2.6 Trade–Off Theory

As we mentioned above, the main theory of optimal capital structure of the company during many decades was the world famous trade–off theory [3–5], which is still used now for decision on capital structure making. But, in 2013 Brusov et al. [6] have proven the trade–off theory's bankruptcy. In opposite to waiting result, it was shown that suggestion of risky debt financing which causes the growing of the credit rate near the bankruptcy, does not lead to the growing of weighted average cost of capital, WACC, which still decreases with leverage level. Thus, the minimum in the dependence of WACC on leverage level is absent as well. This means, that the world famous trade–off theory lacks an optimal capital structure. Brusov et al. in 2013 [6] have given the explanation of this fact by analyzing the equity capital cost dependence on the leverage level on the risky debt capital assumption.

13.3 Modification of the Brusov–Filatova–Orekhova (BFO) Theory for Companies with Frequent Payments of Tax on Income

Below we use the following definitions:

D	value of debt capital of the company
S	value of equity capital of the company
$k_d, w_d = \frac{D}{D+S}$	cost and fraction of debt capital of the company
$k_e, w_e = \frac{S}{D+S}$	cost and fraction of equity capital of the company
$L = D/S$	financial leverage
WACC	weighted average cost of capital

k_0 is the equity cost for financially independent company;
p is the number of payments of tax on profit per year;
T is tax on profit.

13.3.1 Calculation of the Tax Shield

We start from the calculation of the tax shield within Brusov–Filatova–Orekhova theory for the case of p payments of tax on profit per year (payments are made at the end of periods). The tax shield, TS, for period of n-years is equal to sum of discounted values of benefits from the use of tax incentives:

$$(\mathrm{TS})_n = \frac{k_{\mathrm{d}}Dt}{p(1+k_{\mathrm{d}})^{1/p}} + \frac{k_{\mathrm{d}}Dt}{p(1+k_{\mathrm{d}})^{2/p}} + \cdots + \frac{k_{\mathrm{d}}Dt}{p(1+k_{\mathrm{d}})^{np/p}} \tag{13.18}$$

We have a geometric progression with denominator $q = \frac{1}{(1+k_{\mathrm{d}})^{1/p}}$.

After summing the progression, one obtains:

$$(\mathrm{TS})_n = \frac{k_{\mathrm{d}}Dt(1-(1+k_{\mathrm{d}})^{-n})}{p(1+k_{\mathrm{d}})^{1/p}\left(1-(1+k_{\mathrm{d}})^{-1/p}\right)} = \frac{k_{\mathrm{d}}Dt(1-(1+k_{\mathrm{d}})^{-n})}{p\left((1+k_{\mathrm{d}})^{1/p}-1\right)} \tag{13.19}$$

In classical Brusov–Filatova–Orekhova theory (at $p = 1$):

$$(\mathrm{TS})_n = Dt(1-(1+k_{\mathrm{d}})^{-n}) \tag{13.20}$$

It is easy to obtain this result from (13.19), putting the frequency of payments of tax on profit $p = 1$.

13.3.2 Derivation of the Modified BFO Formula for Weighted Average Cost of Capital (WACC)

We derive now the modified formula BFO for weighted average cost of capital (WACC) for the case of p payments of tax on profit per year (payments at the end of periods).

One has for the financially dependent company value V:

$$V = V_0 + (\mathrm{TS})_n \tag{13.21}$$

Here V_0 is the value of a financially independent company. Putting expression (13.19) for TS, one has:

$$V = V_0 + \frac{k_{\mathrm{d}}Dt(1-(1+k_{\mathrm{d}})^{-n})}{p\left((1+k_{\mathrm{d}})^{1/p-1}\right)} \tag{13.22}$$

After substituting $D = w_{\mathrm{d}}V$, we get:

$$V\left(1-\frac{k_{\mathrm{d}}w_dt(1-(1+k_{\mathrm{d}})^{-n})}{p\left((1+k_{\mathrm{d}})^{1/p}-1\right)}\right)=V_0 \tag{13.23}$$

Accounting, that the values of financially dependent company, V, and financially independent company, V_0, are respectively equal to

$$V=\frac{\mathrm{CF}(1-(1+\mathrm{WACC})^{-n})}{\mathrm{WACC}};V_0=\frac{\mathrm{CF}(1-(1+k_0)^{-n})}{k_0} \tag{13.24}$$

one gets:

$$\frac{\mathrm{CF}(1-(1+\mathrm{WACC})^{-n})}{\mathrm{WACC}}\cdot\left(1-\frac{k_{\mathrm{d}}w_dt(1-(1+k_{\mathrm{d}})^{-n})}{p\left((1+k_{\mathrm{d}})^{1/p}-1\right)}\right)$$
$$=\frac{\mathrm{CF}}{k_0}\cdot(1-(1+k_0)^{-n}) \tag{13.25}$$

From here we derive now the modified formula BFO for weighted average cost of capital (WACC) for company of age n years for the case of p payments of tax on profit per year (payments at the end of periods)

$$\frac{1-(1+\mathrm{WACC})^{-n}}{\mathrm{WACC}}=\frac{\mathrm{CF}(1-(1+k_0)^{-n})}{k_0\left(1-\frac{k_{\mathrm{d}}w_dt(1-(1+k_{\mathrm{d}})^{-n})}{p\left((1+k_{\mathrm{d}})^{1/p}-1\right)}\right)} \tag{13.26}$$

At $p=1$ we get **classical** BFO formula

$$\frac{1-(1+\mathrm{WACC})^{-n}}{\mathrm{WACC}}=\frac{1-(1+k_0)^{-n}}{k_0\cdot(1-w_{\mathrm{d}}t[1-(1+k_{\mathrm{d}})^{-n}])} \tag{13.27}$$

13.3.3 Formulas for Capital Value, V, and Equity Cost, k_e

Below in Sect. 13.4 we will investigate the dependence of the weighted average cost of capital, WACC, capital value, V, equity cost, k_{e}, on leverage level L at different frequency of payment of tax on profit p for three-year and six-year companies, using Microsoft Excel. For WACC we will use formula (13.10) and for capital value, V, and equity cost, k_{e}, we will use formulas (13.28) and (13.29) respectively (see below).

Company of age n capitalization could be calculated by the following formula:

$$V=\frac{\mathrm{CF}\cdot\left(1-(1+\mathrm{WACC})^{-n}\right)}{\mathrm{WACC}}. \tag{13.28}$$

k_e should be found from the equation

$$\mathrm{WACC}=k_e w_e+k_d w_d(1-t)$$

$$k_e=\frac{\mathrm{WACC}}{w_e}-\frac{k_d(1-t)w_d}{w_e}=\mathrm{WACC}(1+L)-Lk_d(1-t) \tag{13.29}$$

where one should substitute WACC from the formula (13.26).

13.4 Results

In this section we study the dependence of the weighted average cost of capital, WACC, capital value, V, equity cost, k_e, on leverage level L at different frequency of payment of tax on profit p for three-year and six-year companies, using Microsoft Excel. As we mentioned above, for WACC we use formula (13.10) and for capital value, V, and equity cost, k_e, we use formulas (13.11) and (13.13) respectively.

We use the following parameters: $k_0 = 0.2$; $k_d = 0.18$; $t = 0.2$; $n = 3$; 6; $\mathrm{CF} = 100$.

13.4.1 Dependence of the Weighted Average Cost of Capital, WACC, Capital Value, V, Equity Cost, k$_e$, on Leverage Level L at Different Frequency of Payment of Tax on Profit p for Three-Year Company

13.4.1.1 Dependence of the Weighted Average Cost of Capital, WACC, on Leverage Level L at Different Frequency of Payment of Tax on Profit p for Three-Year Company

Dependence of the weighted average cost of capital, WACC, on leverage level L at different frequency of payments of tax on profit p **for three-year company** is shown **in** Table 13.2 and at Fig. 13.4.

The weighted average cost of capital, WACC, decreases with leverage level L at any frequency of payments of tax on profit p. The difference between the WACC(L) curves is maximum when moving from annual ($p = 1$) to semi–annual ($p = 2$) income tax payments and decreases when moving from semi–annual ($p = 2$) to quarterly ($p = 4$) payments and from quarterly ($p = 4$) to monthly ($p = 12$) payments.

Table 13.2 Dependence of the weighted average cost of capital, WACC, on leverage level L at different frequency of payments of tax on profit p for three-year company

		WACC				
L	w_d	$p = 1$	$p = 2$	$p = 4$	$p = 6$	$p = 12$
1	0.5	0.1749	0.1738	0.1732	0.1730	0.1728
2	0.666667	0.1664	0.1650	0.1642	0.1640	0.1637
3	0.75	0.1622	0.1605	0.1597	0.1594	0.1591
4	0.8	0.1596	0.1579	0.1570	0.1567	0.1564
5	0.833333	0.1579	0.1561	0.1552	0.1549	0.1545
6	0.857143	0.1567	0.1548	0.1539	0.1536	0.1532
7	0.875	0.1558	0.1539	0.1529	0.1526	0.1522
8	0.888889	0.1551	0.1532	0.1522	0.1518	0.1515
9	0.9	0.1545	0.1526	0.1515	0.1512	0.1509
10	0.909091	0.1541	0.1521	0.1511	0.1507	0.1504

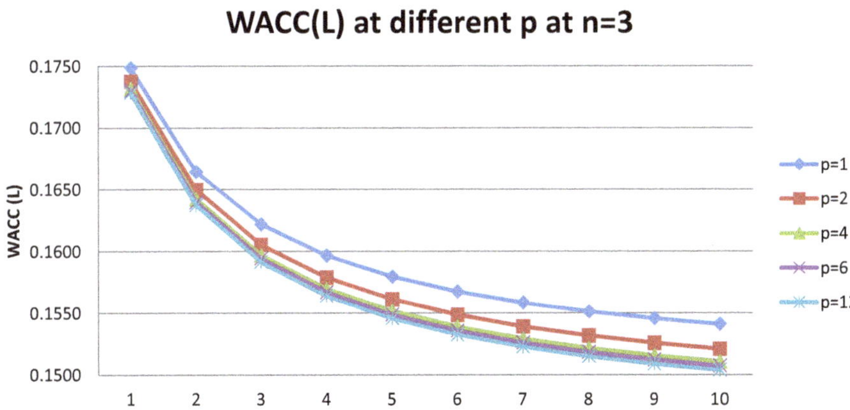

Fig. 13.4 Dependence of the weighted average cost of capital, WACC, on leverage level L at different frequency of payments of tax on profit p for three-year company

13.4.1.2 Dependence of the Company Value, on Leverage Level L at Different Frequency of Payment of Tax on Profit p for Three-Year Company

Dependence of the company value, V, on leverage level L at different frequency of payments of tax on profit p for three-year company is shown **in** Table 13.3 and at Fig. 13.5.

The value of company V increases with the frequency p. The largest increase occurs when moving from annual ($p = 1$) to semi–annual ($p = 2$) income tax payments and decreases when moving from semi–annual ($p = 2$) to quarterly ($p = 4$) payments and from quarterly ($p = 4$) to monthly ($p = 12$) payments.

Table 13.3 Dependence of the company value, V, on leverage level L at different frequency of payments of tax on profit p for three-year company

		V				
L	w_d	$p = 1$	$p = 2$	$p = 4$	$p = 6$	$p = 12$
1	0.5	219.2281	219.6139	219.8116	219.8781	219.9451
2	0.666667	222.2455	222.7746	223.0459	223.1373	223.2292
3	0.75	223.7856	224.3893	224.6989	224.8033	224.9082
4	0.8	224.7199	225.3694	225.7025	225.8149	225.9278
5	0.833333	225.3471	226.0275	226.3766	226.4944	226.6127
6	0.857143	225.7973	226.5000	226.8606	226.9822	227.1045
7	0.875	226.1361	226.8557	227.2250	227.3495	227.4747
8	0.888889	226.4003	227.1331	227.5091	227.6360	227.7635
9	0.9	226.6122	227.3555	227.7370	227.8657	227.9951
10	0.909091	226.7858	227.5378	227.9238	228.0540	228.1849

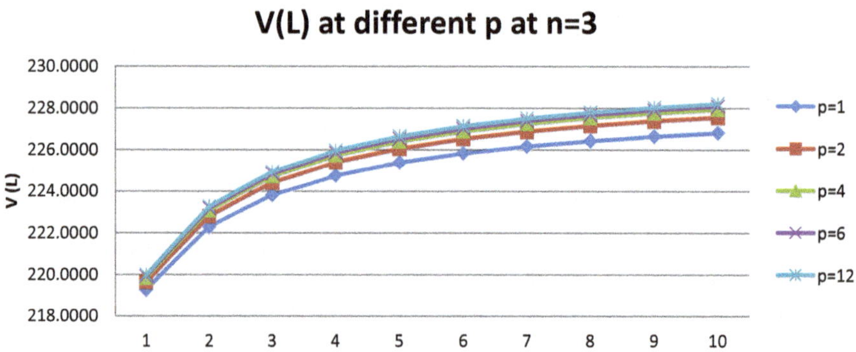

Fig. 13.5 Dependence of the company value, V, on leverage level L at different frequency of payments of tax on profit p for three-year company

13.4.1.3 Dependence of the Equity Cost, k_e, on Leverage Level L at Different Frequency of Payment of Tax on Profit p for Three-Year Company

Dependence of the equity cost, k_e, on leverage level L at different frequency of payments of tax on profit p for three-year company is shown in Table 13.4 and at Fig. 13.6.

The cost of equity k_e increases linearly with the level of leverage L. The slope of the curve $k_e(L)$ depends on the frequency of paying income tax: it decreases with increasing p, most rapidly when moving from annual ($p = 1$) to semi–annual ($p = 2$) payments of income tax and slower in the transition from semi–annual ($p = 2$) to quarterly ($p = 4$) payments and from quarterly ($p = 4$) to monthly ($p = 12$) payments.

Table 13.4 Dependence of the equity cost, k_e, on leverage level L at different frequency of payments of tax on profit p for three-year company

		k_e				
L	w_d	$p = 1$	$p = 2$	$p = 4$	$p = 6$	$p = 12$
1	0.5	0.2057	0.2036	0.2024	0.2021	0.2017
2	0.666667	0.2113	0.2069	0.2046	0.2039	0.2031
3	0.75	0.2168	0.2102	0.2068	0.2057	0.2045
4	0.8	0.2222	0.2134	0.2089	0.2074	0.2059
5	0.833333	0.2277	0.2167	0.2111	0.2092	0.2073
6	0.857143	0.2331	0.2199	0.2132	0.2109	0.2086
7	0.875	0.2386	0.2232	0.2153	0.2126	0.2100
8	0.888889	0.2440	0.2264	0.2174	0.2144	0.2113
9	0.9	0.2495	0.2296	0.2195	0.2161	0.2127
10	0.909091	0.2549	0.2329	0.2216	0.2178	0.2140

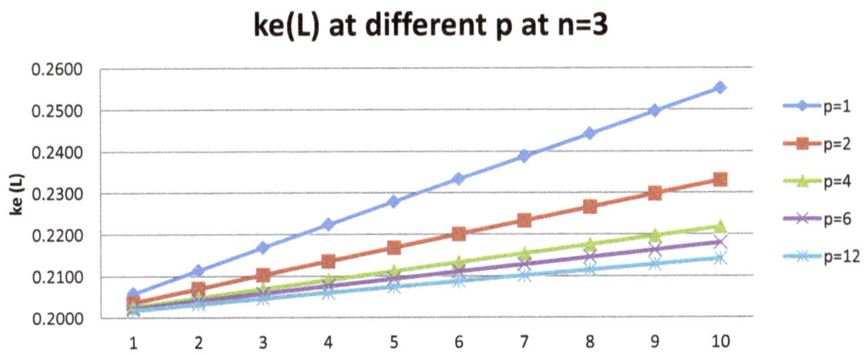

Fig. 13.6 Dependence of the equity cost, k_e, on leverage level L at different frequency of payments of tax on profit p for three-year company

13.4.2 *Dependence of the Weighted Average Cost of Capital, WACC, Capital Value,* V*, Equity Cost,* k_e*, on Leverage Level* L *at Different Frequency of Payment of Tax on Profit* p *for Six-Year Company*

13.4.2.1 Dependence of the Weighted Average Cost of Capital, WACC, on Leverage Level L at Different Frequency of Payment of Tax on Profit p for Six-Year Company

Dependence of the weighted average cost of capital, WACC, on leverage level L at different frequency of payments of tax on profit p for six-year company is shown **in Table 13.5** and at **Fig. 13.7.**

We use the following parameters: $k_0 = 0.2$; $k_d = 0.18$; $t = 0.2$; $n = 6$;

Table 13.5 Dependence of the weighted average cost of capital, WACC, on leverage level L at different frequency of payments of tax on profit p for six-year company

		WACC				
L	w_d	$p = 1$	$p = 2$	$p = 4$	$p = 6$	$p = 12$
1	0.5	0.1743	0.1732	0.1726	0.1725	0.1723
2	0.666667	0.1657	0.1641	0.1634	0.1631	0.1629
3	0.75	0.1613	0.1596	0.1587	0.1584	0.1581
4	0.8	0.1586	0.1568	0.1559	0.1556	0.1553
5	0.833333	0.1569	0.1550	0.1540	0.1537	0.1534
6	0.857143	0.1556	0.1537	0.1527	0.1523	0.1520
7	0.875	0.1547	0.1527	0.1517	0.1513	0.1510
8	0.888889	0.1539	0.1519	0.1509	0.1505	0.1502
9	0.9	0.1534	0.1513	0.1502	0.1499	0.1495
10	0.909091	0.1529	0.1508	0.1497	0.1494	0.1490

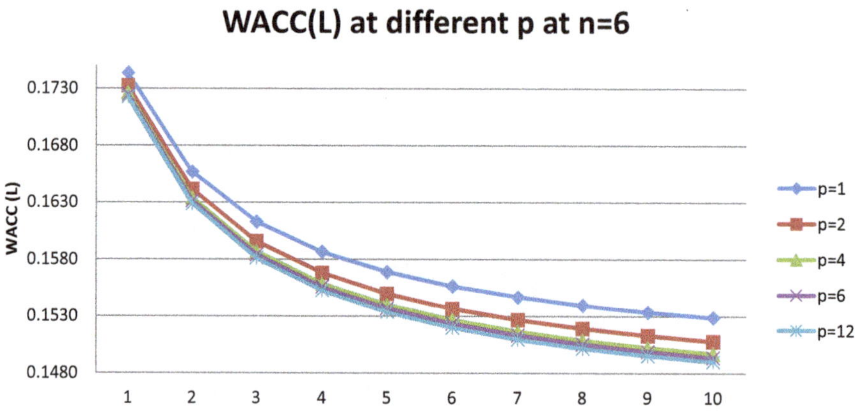

Fig. 13.7 Dependence of the weighted average cost of capital, WACC, on leverage level L at different frequency of payments of tax on profit p for six-year company

The weighted average cost of capital, WACC, decreases with leverage level L at any frequency of payments of tax on profit p. The difference between the WACC (L) curves is maximum when moving from annual ($p = 1$) to semi–annual ($p = 2$) income tax payments and decreases when moving from semi–annual ($p = 2$) to quarterly ($p = 4$) payments and from quarterly ($p = 4$) to monthly ($p = 12$) payments.

13.4.2.2 Dependence of the Company Value, *V*, on Leverage Level *L* at Different Frequency of Payment of Tax on Profit *p* for Six-Year Company

Dependence of the company value, *V*, on leverage level *L* at different frequency of payments of tax on profit *p* for six-year company is shown **in Table 13.6** and at **Fig. 13.8.**

The value of company V increases with the frequency *p*. The largest increase occurs when moving from annual ($p = 1$) to semi–annual ($p = 2$) income tax payments and decreases when moving from semi–annual ($p = 2$) to quarterly ($p = 4$) payments and from quarterly ($p = 4$) to monthly ($p = 12$) payments.

Table 13.6 Dependence of the company value, *V*, on leverage level *L* at different frequency of payments of tax on profit *p* for six-year company

		V				
L	w_d	$p = 1$	$p = 2$	$p = 4$	$p = 6$	$p = 12$
1	0.5	354.8940	355.9256	356.4549	356.6333	356.8127
2	0.666667	363.0242	364.4649	365.2052	365.4550	365.7063
3	0.75	367.2305	368.8900	369.7435	370.0315	370.3214
4	0.8	369.8015	371.5971	372.5210	372.8329	373.1468
5	0.833333	371.5355	373.4240	374.3960	374.7242	375.0545
6	0.857143	372.7841	374.7399	375.7468	376.0869	376.4291
7	0.875	373.7261	375.7330	376.7664	377.1154	377.4667
8	0.888889	374.4620	376.5090	377.5632	377.9193	378.2776
9	0.9	375.0529	377.1322	378.2031	378.5648	378.9289
10	0.909091	375.5377	377.6435	378.7283	379.0947	379.4635

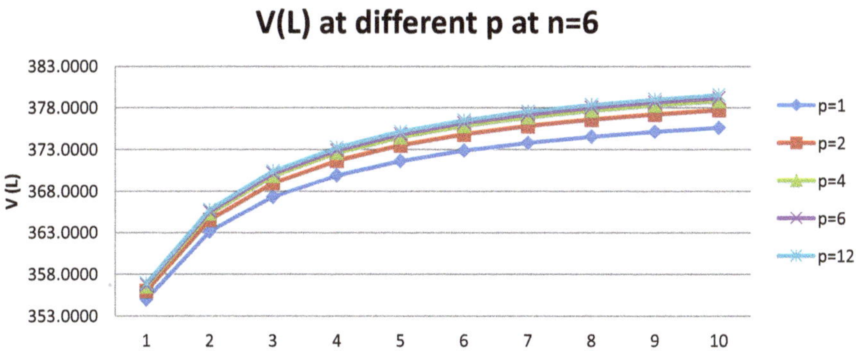

Fig. 13.8 Dependence of the company value, *V*, on leverage level *L* at different frequency of payments of tax on profit *p* for six-year company

13.4.2.3 Dependence of the Equity Cost, k_e, on Leverage Level L at Different Frequency of Payment of Tax on Profit p for Six-Year Company

Dependence of the equity cost, k_e, on leverage level L at different frequency of payments of tax on profit p for six-year company is shown **in** Table 13.7 and at Fig. 13.9.

The cost of equity k_e increases linearly with the level of leverage L. The slope of the curve $k_e(L)$ depends on the frequency of paying income tax: it decreases with increasing p, most rapidly when moving from annual ($p = 1$) to semi–annual ($p = 2$) payments of income tax and slower in the transition from semi–annual ($p = 2$) to quarterly ($p = 4$) payments and from quarterly ($p = 4$) to monthly ($p = 12$) payments.

One can see, that for a six-year–old company with monthly income tax payments **a qualitatively new anomalous effect takes place**: $k_e(L)$ decreases with an increase

Table 13.7 Dependence of the equity cost, k_e, on leverage level L at different frequency of payments of tax on profit p for six-year company

		k_e				
L	w_d	$p = 1$	$p = 2$	$p = 4$	$p = 6$	$p = 12$
1	0.5	0.2047	0.2024	0.2013	0.2009	0.2005
2	0.666667	0.2090	0.2044	0.2021	0.2013	0.2006
3	0.75	0.2131	0.2063	0.2028	0.2016	0.2005
4	0.8	0.2172	0.2081	0.2035	0.2019	0.2003
5	0.833333	0.2213	0.2099	0.2041	0.2021	0.2001
6	0.857143	0.2254	0.2117	0.2047	0.2023	0.1999
7	0.875	0.2294	0.2134	0.2052	0.2025	0.1997
8	0.888889	0.2335	0.2152	0.2058	0.2027	0.1995
9	0.9	0.2376	0.2169	0.2064	0.2029	0.1993
10	0.909091	0.2416	0.2187	0.2070	0.2030	0.1991

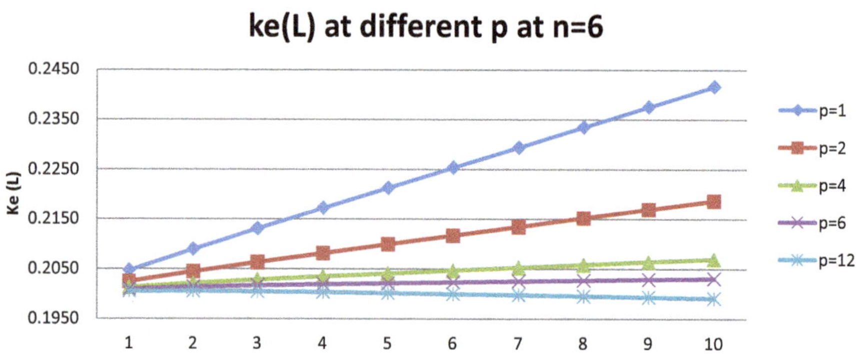

Fig. 13.9 Dependence of the equity cost, k_e, on leverage level L at different frequency of payments of tax on profit p for six-year company

in the level of leverage L. This radically changes the company's dividend policy, since the economically justified amount of the dividends is equal to the cost of equity.

13.5 The Discussion and Conclusions

The main purpose of the Chapter is bring the Brusov–Filatova–Orekhova (BFO) theory closer to economic practice, taking into account one of the features of the real functioning of companies—the frequent payments of tax on profit. We generalize for the first time the Brusov–Filatova–Orekhova (BFO) theory for the case of tax on profit payments with an arbitrary frequency. We derive modified BFO formulas and show that: (1) all BFO formulas change; (2) all main financial parameters of the company, such as company value, V, the weighted average cost of capital, WACC, and equity cost, k_e, depend on the frequency of tax on profit payments. It turns out that the increase of the number of payments of tax of profit per year leads to decrease of the cost of attracting capital and increase of the company value. More frequent payments of income tax are beneficial for both parties—for the company and for the tax regulator: for the company, this leads to an increase in the value of the company, and for the tax regulator, earlier payments are beneficial due to the time value of money.

We numerically studied the dependence of the weighted average cost of capital WACC, value of company, V, cost of equity k_e on the level of leverage L at different frequency of income tax payments p for three- and six-year companies and obtained the following results. The calculation of financial indicators are made for some typical parameters of enterprises (equity and debt costs etc.).

The weighted average cost of capital, WACC, decreases with leverage level L at any frequency of payments of tax on profit p. The difference between the WACC(L) curves is maximum when moving from annual ($p = 1$) to semi–annual ($p = 2$) income tax payments and decreases when moving from semi–annual ($p = 2$) to quarterly ($p = 4$) payments and from quarterly ($p = 4$) to monthly ($p = 12$) payments.

The value of company V increases with the frequency p. The largest increase occurs when moving from annual ($p = 1$) to semi–annual ($p = 2$) income tax payments and decreases when moving from semi–annual ($p = 2$) to quarterly ($p = 4$) payments and from quarterly ($p = 4$) to monthly ($p = 12$) payments.

The cost of equity k_e increases linearly with the level of leverage L. The slope of the curve $k_e(L)$ depends on the frequency of paying income tax: it decreases with increasing p, most rapidly when moving from annual ($p = 1$) to semi–annual ($p = 2$) payments of income tax and slower in the transition from semi–annual ($p = 2$) to quarterly ($p = 4$) payments and from quarterly ($p = 4$) to monthly ($p = 12$) payments.

At a certain age of the company, **a qualitatively new anomalous effect** takes place: $k_e(L)$ decreases with an increase in the level of leverage L. We obtain this

Table 13.8 Comparison of the impact of the frequency of income tax payments in BFO theory and in MM theory

	BFO	MM
Theorems, statements and formulas	Change	Change
All main financial indicators: WACC, V, k_e	Depend on the frequency of tax on profit payments p	Depend on the frequency of tax on profit payments p
All main financial indicators: WACC, V, k_e	Value of the effect of p on all indicators is higher, than in MM	Value of the effect of p on all indicators is smaller, than in BFO
All main financial indicators: WACC, V, k_e depend on debt cost k_d	At all p	At $p \neq 1$ only
WACC	Decreases with p	Decreases with p
V	Increases with p	Increases with p
k_e: the slope of the curve $k_e(L)$	Decreases with increasing p	Decreases with increasing p
k_e: the slope of the curve $k_e(L)$	**Could be negative (qualitatively new effect)**	Always is positive

result for a six-year-old company with monthly income tax payments, but it is clear that this effect should take place for the companies of different age and at different frequency of income tax payment. This radically changes the company's dividend policy, since the economically justified amount of the dividends is equal to the cost of equity.

Note, that this result remind discovery, made by us earlier (Chap. 8 in monograph [6]): the abnormal dependence of equity cost, k_e, on leverage level L: above some value of tax on income, T^*, equity cost k_e decreases with the leverage level L.

Thus, it is interesting to note, that **increase of frequency of income tax payments p and increase of tax on profit T lead to the similar abnormal effect: decrease of equity cost k_e with leverage level L.** If in the first case there are threshold values for the frequency of paying income tax, p^*, and the age of the company, n^*, then in the second case there is a threshold value for income tax T^*.

In Table 13.8 we compare the impact of the frequency of income tax payments in BFO theory and in MM theory.

Recommendations

From the analysis of the results of current Chapter the following recommendations could be done:

For company: company should pay tax on profit as frequently as it is possible, because in this case company value increases.

For regulator: should encourage more frequent payments of tax on profit, because earlier payments are beneficial for budget due to the time value of money.

References

1. Modigliani, F., & Miller, M. (1958). The cost of capital, corporate finance, and the theory of investment. *American Economic Review, 48*, 261–297.
2. Filatova, T., Orehova, N., & Brusova, A. (2008). Weighted average cost of capital in the theory of Modigliani–Miller, modified for a finite life–time company. *Bulletin of the FU, 48*, 68–77.
3. Brennan, M., & Schwartz, E. (1978). Corporate income taxes, valuation, and the problem of optimal capital structure. *The Journal of Business, 51*, 103–114.
4. Brennan, M., & Schwartz, E. (1984). Optimal financial policy and firm valuation. *Journal of Finance, 39*, 593–607.
5. Leland, H. (1994). Corporate debt value, bond covenants, and optimal capital structure. *Journal of Finance, 49*(4), 1213–1252.
6. Brusov, P., Filatova, T., Orehova, N., & Eskindarov, M. (2018). *Modern corporate finance, investments, taxation and ratings*. Springer Nature Publishing.
7. Myers, S. (2001). Capital structure. *Journal of Economic Perspectives, 15*(2), 81–102.
8. Modigliani, F., & Miller, M. (1963). Corporate income taxes and the cost of capital: A correction. *American Economic Review, 53*, 147–175.
9. Modigliani, F., & Miller, M. (1966). Some estimates of the cost of capital to the electric utility industry 1954–1957. *American Economic Review, 56*, 333–391.
10. Tatiana, F., Brusov, P., & Orekhova, N. (2022). Impact of advance payments of tax on profit on effectiveness of investments. *Mathematics, 10*(4), 666. https://doi.org/10.3390/math10040666
11. Hamada, R. (1969). Portfolio analysis, market equilibrium, and corporate finance. *J Financ, 24*(1), 13–31.
12. Miller, M. (1977). Debt and taxes. *The Journal of Finance, 32*(2), 261–275. https://doi.org/10.1111/j.1540-6261.1977.tb03267.x
13. Farber, A., Gillet, R., & Szafarz, A. (2006). A general formula for the WACC. *International Journal of Business, 11*(2), 211–218.
14. Fernandez, P. (2006). A general formula for the WACC: A comment. *International Journal of Business, 11*(2), 219.
15. Berk, J., & DeMarzo, P. (2007). *Corporate finance*. Pearson-Addison Wesley.
16. Harris, R., & Pringle, J. (1985). Risk-adjusted discount rates – Extension from the average–risk case. *The Journal of Financial Research, 8*(3), 237–244.
17. Batrancea, L. (2021). An econometric approach regarding the impact of fiscal pressure on equilibrium: Evidence from electricity, gas and oil companies listed on the New York Stock Exchange. *Mathematics, 9*(6), 630. https://doi.org/10.3390/math906063
18. Batrancea, L. (2021). The influence of liquidity and solvency on performance within the healthcare industry: Evidence from publicly listed companies. *Mathematics, 9*, 2231. https://doi.org/10.3390/math9182231
19. Dimitropoulos, P. (2014). Capital structure and corporate governance of soccer clubs: European evidence. *Management Research Review, 37*(7), 658–678. https://doi.org/10.1108/MRR-09-2012-0207
20. Machado, L. K. C., et al. (2015). The relevance of capital structure on firm performance: A multivariate analysis of publicly traded Brazilian companies. *REPeC Brasília, 9*(4), 384–401. https://doi.org/10.17524/repec.v9i4.1313
21. Barbi, M. (2011). On the risk–neutral value of debt tax shields. *Applied Financial Economics, 22*(3), 251–258. https://doi.org/10.1080/09603107.2013.613754
22. Pavel, Z. (2018). The impact of cash flows and weighted average cost of capital to enterprise value in the oil and gas sector. *Journal of Reviews on Global Economics, 7*, 138–145.
23. Justyna, F.-D., et al. (2021). Energy sector risk and cost of capital assessment—Companies and investors perspective. *Energies, 14*(6), 1613. https://doi.org/10.3390/en14061613

24. Angotti, M., de Lacerda Moreira, R., do Nascimento, J. H. B., & de Almeida Bispo, O. N. (2018). Analysis of an equity investment strategy based on accounting and financial reports in Latin American markets. *Reficont, 5*(2).
25. Vergara-Novoa, C., Sepúlveda-Rojas, J. P., Alfaro, M. D., & Nicolás, R. (2018). Cost of capital estimation for highway concessionaires in Chile. *Journal of Advanced Transportation*. https://doi.org/10.1155/2018/2153536

Chapter 14
Influence of Method and Frequency of Profit Tax Payments on Company Financial Indicators

14.1 Introduction

In practice, profit tax payments are (1) made more frequently than annually and (2) can be made in advance. Recently, the Brusov–Filatova–Orekhova (BFO) theory [1] was generalized for the case of profit tax payments with an arbitrary frequency. Using modified BFO formulae, it was shown that all financial indicators of a company depend on the frequency of profit tax payments. The weighted average cost of capital (WACC) decreases with the payments and the company value increases with the payments. A new effect was thus discovered: the decrease in equity cost with an increase in the level of leverage (L). Due to the fact that the economically justified amount of dividends is equal to the equity cost, this can significantly change the dividend policy of a company.

More frequent profit tax payments are beneficial for both parties: the company and the tax regulator. For the company, this practice increases the company value and it helps the tax regulator due to the time value of money.

In current Chapter, for the first time, the Brusov–Filatova–Orekhova (BFO) theory was generalized for the case of advance profit tax payments with an arbitrary frequency. We compared the obtained results to those that we described recently for profit tax payments at the end of the financial period [1] and found them to be totally different.

We showed that in spite the fact that the WACC decreases with the payments and the company value increases with the payments in this case [1], the WACC values turned out to be bigger and the company value turned out to be smaller than in case of advance profit tax payments of any frequency. This underlined the importance of advance profit tax payments.

The novelty and originality of this study lie in the fact that the simultaneous influence of two effects was investigated for the first time: (1) advance profit tax payments and (2) frequent profit tax payments.

P. Brusov, T. Filatova, *Business Valuation and Company Value*, Contributions to Finance and Accounting, https://doi.org/10.1007/978-3-031-90847-7_14

Before this study, we generalized the Brusov–Filatova–Orekhova (BFO) theory to (1) the case of frequent profit tax payments at the end of the financial period and (2) the case of advance annual profit tax payments.

In this article, we also compared the obtained results to the results of consideration (1) and found a large difference between the results. Thus, in this Chapter, an in-depth study of frequent profit tax payments was carried out for both possible cases of payments: advance and at the end of the financial period.

The robustness of the outcomes was provided by the use of the modern Brusov–Filatova–Orekhova (BFO) theory of capital cost and capital structure, which has been tested in real economic practice and its correct generalization for the case of frequent advance profit tax payments has been proven.

The structure of the Chapter is as follows.

In the introduction, we present a literature review of the development of the capital structure and capital cost theory (Sect. 14.1.1). The Brusov–Filatova–Orekhova (BFO) theory is generalized for frequent advance profit tax payments for the first time in Sect. 14.2, The modified formulae are derived for the tax shield in Sect. 14.2.1, company value in Sect. 14.2.2 and the equity cost and WACC in Sect. 14.2.3.

In Sect. 14.3, using the generalized BFO formulae that were obtained in Sect. 14.2 and Microsoft Excel, we numerically calculate the dependence of a company's financial parameters, such as WACC, company value and equity cost, on the level of leverage for an arbitrary frequency of profit tax payment for three- and six-year-old companies (Sects. 14.3.1 and 14.3.2, respectively).

We considered frequent advance profit tax payments. In Sect. 14.3, we also compare the obtained results to those that were found previously for frequent profit tax payments tax at the end of the financial period [1]. This section also compares the results for companies of different ages (three- and six-year-old companies). In Sect. 14.4, the discussion and conclusions are presented.

14.1.1 A Literature Review on the Development of the Capital Cost and Capital Structure Theory

In this section, we present an overview of the development of the capital cost and capital structure theory. Following the traditional empirical approaches, the first quantitative theory of capital cost and capital structure was proposed the Nobel Prize winners Modigliani and Miller (the MM theory) in 1958 [2]. Among the many limitations of this theory, two main shortcomings can be distinguished: the non-accounting of taxes and the infinite lifetime of companies. The first constraint was removed by Modigliani and Miller themselves [3, 4], who obtained the following formulae.

The tax shield (TS) had the following form:

$$\mathrm{TS} = DT \tag{14.1}$$

Company value (V) and financially independent companies (V_0) were expressed as:

$$V_0 = \mathrm{CF}/k_0 \text{ and } \quad V = \mathrm{CF}/\mathrm{WACC}. \tag{14.2}$$

For WACC, Modigliani and Miller obtained the following formula:

$$\mathrm{WACC} = k_0 \cdot (1 - w_d t) \tag{14.3}$$

For equity cost (k_e), they had:

$$k_e = k_0 + L(k_e - k_d) \cdot (1 - t) \tag{14.4}$$

In 1969, the capital asset pricing model (CAPM) was combined with the Modigliani–Miller theory by Hamada [5], which accounted for both the business and financial risk of companies and used the below formula for the equity cost (k_e) of financially dependent companies:

$$k_e = k_F + (k_M - k_F)b_U + (k_M - k_F)b_U \frac{D}{S}(1 - T), \tag{14.5}$$

where b_U is the β–coefficient of a company with $L = 0$.

Corporate and individual taxes were then taken into account by Miller in 1977 [6]. The following formula was obtained for unleveraged company value:

$$V = V_0 + \left[1 - \frac{(1 - T_C)(1 - T_S)}{(1 - T_D)}\right] \cdot D \tag{14.6}$$

where T_S is the profit tax rate for an individual investor for their ownership of corporation stock, T_C is the tax rate for corporate profit and T_D is the profit tax interest rate for the provision of credit to other investors or companies.

In [7–10], a more general expression for WACC than the Modigliani–Miller (MM) version was obtained [7]:

$$\mathrm{WACC} = k_0(1 - w_d t) - k_d w_d t + k_{TS} t w_d \tag{14.7}$$

where k_0 is the equity cost of the unleveraged company, k_d is the debt cost, k_{TS} is the expected return on the tax shield, V is the leveraged company value, V_{TS} is the value of the tax shield and D is the debt value.

Derived in [9], Formula (14.7) is more general than the Modigliani–Miller (MM) expression, so some additional conditions are necessary for its use in practice. When the WACC value is constant over time, the leveraged company value can be found by discounting the WACC from the unleveraged company cash flows. For this specific case, the formulae can be found in textbooks [9, 10].

In 1963, Modigliani and Miller [3] supposed that the level of debt was constant. The expected after-tax cash flow of a financially independent company was also fixed, thus V_0 was constant. By that assumption, $k_{TS} = k_d$ and the tax shield value was $TS = tD$. Thus, for the financially dependent company value, we used the ordinary Modigliani–Miller (MM) formula for WACC instead of Formula (14.7):

$$\mathrm{WACC} = k_0(1 - w_d t) \tag{14.8}$$

From our point of view, the "classical" Modigliani–Miller (MM) theory supposed that the debt cost and expected return on the tax shield were equal (due to the fact that both of them are debt-like in nature), which is much more reasonable and so we modified the "classical" Modigliani–Miller (MM) in [11].

A study on the influence of tax pressure on the financial balance of energy companies was carried out in [12]. The results showed that tax pressure has a stronger influence on the financial equilibrium (both in the short term and long term) of oil and electricity companies than gas companies. These results are useful for the managers of energy companies who underestimate the evolution of the financial equilibrium of the company because they take into account various possible financial crises.

A study on the influence of financial liquidity and financial solvency on the performance of healthcare companies was conducted in [13] using econometric models. From the empirical evidence, it follows that certain financial parameters such as current liquidity ratio, quick liquidity ratio and financial leverage significantly influence company performance, as measured by the return on assets, gross margin ratio, operating margin ratio, taxes, earnings before interest, amortization and depreciation.

The limitation that was related to the infinite lifetime of companies was removed in 2008 by Brusov et al. [14], who developed the modern theory of capital cost and capital structure (the BFO theory), which is applicable to companies of all ages.

Modifications to the calculations for the tax shield and company value (unleveraged, V_0; leveraged, V) were required for the generalization of the MM theory (see the formulae below):

$$\mathrm{TS} = k_d DT \sum_{t=1}^{n} (1 + k_d)^{-t} = DT[1 - (1 + k_d)^{-n}]. \tag{14.9}$$

$$V_0 = \mathrm{CF}[1 - (1 + k_0)^{-n}]/k_0; \quad V = \mathrm{CF}[1 - (1 + \mathrm{WACC})^{-n}]/\mathrm{WACC}. \tag{14.10}$$

$$\frac{1 - (1 + \mathrm{WACC})^{-n}}{\mathrm{WACC}} = \frac{1 - (1 + k_0)^{-n}}{k_0[1 - \omega_d T(1 - (1 + k_d)^{-n})]}. \tag{14.11}$$

where S is the equity capital value, $w_d = \frac{D}{D+S}$ is the debt capital, k_e, $w_e = \frac{S}{D+S}$ is the equity capital, $L = D/S$ is the value of financial leverage and D is the debt value.

The Myers formula for a one-year company [15] can be easily obtained from Formula (14.11) by substituting $n = 1$:

$$\mathrm{WACC} = k_0 - \frac{(1 + k_0)k_\mathrm{d}}{1 + k_\mathrm{d}} w_\mathrm{d} T \tag{14.12}$$

Then, by substituting $n = \infty$, we arrive at the Modigliani–Miller formula for WACC [2]:

$$\mathrm{WACC} = k_0 \cdot (1 - w_\mathrm{d} t) \tag{14.13}$$

Until 2008, when the BFO theory was created, there were only two options: the Modigliani–Miller theory for perpetual companies [2–4] and the Myers theory for one-year companies [15]. As the authors of the BFO showed, taking the finite age of a company into account leads to significant changes in all Modigliani–Miller results [2–4]. In addition, a number of the new effects that are observed within corporate finance using the Brusov–Filatov–Orekhova theory [14] are absent in the Modigliani–Miller theory.

In [11], the Modigliani-Miller theory was modified for the case of advance profit tax payments and significant differences were found between the modified theory and the "classical" Modigliani-Miller theory.

In the literature, the methodology and results of the Brusov–Filatova–Orekhova (BFO) theory are well known (for example, see references [16–27]). A number of scientists (see, for example, [22–24]) have used the BFO theory in practical calculations.

The authors of [20] valuated the capital costs of energy companies by including an investor and market risk approach. The study also concerned a WACC intra-industry analysis the companies, which was related to the sector characteristics of revenue, total assets, company age and market capitalization.

In [23], the influence of the overconfidence of finance managers on the capital structure decisions of family-run businesses in India was studied. It was shown that the capital structure decisions of managers could be explained by measurable managerial characteristics. In [24], the correlation between company risk and capital structure was explored using datasets from sugar and cement companies in Pakistan. It was shown that the role of risk assessment and capital structure is vital for the sustainable growth of companies and the increase in the wealth of shareholders.

The authors of [25] considered the adjusted present value method, the free cash flow (FCF) method, the flow-to-equity method and the relationships between these methods. The authors departed from Modigliani and Miller method and used a stationary FCF method and the Miles and Ezzell method to consider FCF as a first-order autoregressive possession. The authors derived DCF valuation formulae for annuities. The authors found (1) the correct discount rate for the tax shield when the free cash flow is a first-order autoregressive annuity and (2) the tax shield valuation of a first-order autoregressive free cash flow annuity.

The author of [26] studied the impacts of internal and external corporate governance mechanisms on the financial performance of banks in the MENA region during the COVID-19 pandemic. The results showed that the corporate governance measures of the presence of independent members on the board of directors, strong legal protection, a high concentration of ownership and the lack of political pressure on board members had positive effects on the financial indicators of banks.

In [27], the influence of leverage level on the expected equity returns of listed companies was studied. The existence of an optimal capital structure to maximize the expected equity returns was discussed. The authors thought that this problem should be discussed from the perspective of maximizing shareholder wealth and not from the perspective of maximizing company value due to the existence of financial distress costs and agency costs. Using Gebhardt–Lee–Swaminathan (GLS) model to measure the implied capital costs and to measure the expected equity returns, the authors empirically found that the leverage level had a positive correlation with the equity returns of a listed company. The relationship between the leverage level and the expected equity returns appeared to an inverse "U":

- When the leverage level fell below the optimal level, increasing the leverage could increase the expected equity returns;
- When the leverage level was higher than the optimal level, deleveraging was beneficial to increase the expected equity returns.

Within a standard capital structure model, the authors of [28] investigated the leverage decisions of over 1500 companies that were listed in the ASX for 2000–2012 and divided the sample into mining and non-mining industries. Applying a dummy variable approach, the authors showed that there were fundamental differences between two these types of companies when making leverage decisions. It turned out that mining companies were more sensitive to profitability and asset tangibility, whereas neither profitability nor asset tangibility were significant for non-mining companies. The obtained results suggested that industry type matters for companies when making leverage decisions.

In [18], the authors discussed the disagreement in the financial literature about the meaning of the "value of tax shields." Although it is accepted that the tax deductibility of interest increases the value of a firm, the correct valuation of this added value is controversial. Using a risk-neutral approach, the authors derived a general formula for the valuation of tax shields, which shows that this value equals the summation of the discounted future tax savings. As soon as the authors specified a leverage level policy and cash flow dynamics, they obtained some well-known formulae for the calculation of tax shields.

14.2 The Modified Brusov–Filatova–Orekhova (BFO) Theory for the Case of Frequent Advance Profit Tax Payments

In this section, we generalize the Brusov–Filatova–Orekhova (BFO) theory for the case of frequent advance profit tax payments. The expressions for the tax shield, company value, WACC and equity cost are presented in Sects. 14.2.1, 14.2.2 and 14.2.3, respectively.

In this section, we use the following definitions.

D	Debt Capital Value
S	Equity Capital Value
$k_d, w_d = \frac{D}{D+S}$	Debt Capital Cost and Share
$k_e, w_e = \frac{S}{D+S}$	Equity Capital Cost and Share
$L = D/S$ WACC	Leverage Level Weighted Average Cost of Capital

k_0, the equity cost for a financially independent company; p, the number of payments per year; T, profit tax.

14.2.1 The Tax Shield Calculation

We started with the calculation of the tax shield for the case of frequent advance profit tax payments using the Brusov–Filatova–Orekhova theory (where p is the number of payments per year when payments are made at the beginning of the financial period). For a period of n years, the tax shield (TS) was equal to the sum of the discounted values of benefits from the use of tax incentives:

$$(\mathrm{TS})_n = \frac{k_d Dt}{p} + \frac{k_d Dt}{p(1+k_d)^{1/p}} + \frac{k_d Dt}{p(1+k_d)^{2/p}} + \cdots + \frac{k_d Dt}{p(1+k_d)^{np/p}} \quad (14.14)$$

We used a geometric progression with the denominator $q = \frac{1}{(1+k_d)^{1/p}}$.

After summation, we obtained:

$$(\mathrm{TS})_n = \frac{k_d Dt(1-(1+k_d)^{-n})}{p\left(1-(1+k_d)^{-1/p}\right)} = \frac{k_d Dt(1-(1+k_d)^{-n})\cdot(1+k_d)^{1/p}}{p\left((1+k_d)^{1/p}-1\right)} \quad (14.15)$$

Using the classical Brusov–Filatova–Orekhova theory at $p = 1$, we obtained:

$$(\mathrm{TS})_n = Dt(1-(1+k_\mathrm{d})^{-n})\cdot(1+k_\mathrm{d}) \tag{14.16}$$

It was easy to obtain this result from Formula (14.15) by inputting the value of $p = 1$ for the number of profit tax payments.

14.2.2 Derivation of the Modified BFO Formula for the Weighted Average Cost of Capital (WACC)

We now present the modified BFO formula for the weighted average cost of capital (WACC) for the case of frequent advance profit tax payments (where p is the number of payments per year when the payments are made at the beginning of the financial period).

For the value of a leveraged company (V), we obtained:

$$V = V_0 + (\mathrm{TS})_n \tag{14.17}$$

where V_0 is the value of an unleveraged company. By substituting Formula (14.19) for TS, we obtained:

$$V = V_0 + \frac{k_\mathrm{d}Dt(1-(1+k_\mathrm{d})^{-n})\cdot(1+k_\mathrm{d})^{1/p}}{p((1+k_\mathrm{d})^{1/p}-1)} \tag{14.18}$$

After substituting $D = w_\mathrm{d}V$, we obtained:

$$V\left(1-\frac{k_\mathrm{d}w_\mathrm{d}t(1-(1+k_\mathrm{d})^{-n})\cdot(1+k_\mathrm{d})^{1/p}}{p((1+k_\mathrm{d})^{1/p}-1)}\right) = V_0 \tag{14.19}$$

Assuming that the values of a leveraged company (V) and an unleveraged company (V_0) were equal to:

$$V = \frac{\mathrm{CF}(1-(1+\mathrm{WACC})^{-n})}{\mathrm{WACC}}; V_0 = \frac{\mathrm{CF}(1-(1+k_0)^{-n})}{k_0} \tag{14.20}$$

respectively, we obtained:

$$\frac{\mathrm{CF}(1-(1+\mathrm{WACC})^{-n})}{\mathrm{WACC}} \cdot \left(1 - \frac{k_d w_d t(1-(1+k_d)^{-n}) \cdot (1+k_d)^{1/p}}{p((1+k_d)^{1/p}-1)}\right) = \frac{\mathrm{CF}}{k_0} \cdot (1-(1+k_0)^{-n}) \tag{14.21}$$

From there, we could derive the modified BFO formula for the WACC of a company of age n years for the case of p profit tax payments per year (payments at the beginning of the financial period):

$$\frac{1-(1+\mathrm{WACC})^{-n}}{\mathrm{WACC}} = \frac{\mathrm{CF}(1-(1+k_0)^{-n})}{k_0\left(1 - \frac{k_d w_d t(1-(1+k_d)^{-n}) \cdot (1+k_d)^{1/p}}{p\left((1+k_d)^{1/p}-1\right)}\right)} \tag{14.22}$$

For $p = 1$, we obtained the modified BFO formula for advance annual profit tax payments:

$$\frac{1-(1+\mathrm{WACC})^{-n}}{\mathrm{WACC}} = \frac{1-(1+k_0)^{-n}}{k_0 \cdot (1 - w_d t[1-(1+k_d)^{-n}] \cdot (1+k_d))} \tag{14.23}$$

14.2.3 Formulae for the Capital Value and Equity Cost

In Sect. 14.3, we present our investigation into the impact of the frequency of profit tax payments on the dependence of the WACC, capital value and equity cost on the leverage level of companies of different ages (3 or 6 years) using Microsoft Excel. For the WACC, we used Formula (14.10) and for capital value and equity cost, we used Formulae (14.24) and (14.25), respectively (see below).

The value of a company of age n could be calculated using the following formula:

$$V = \frac{\mathrm{CF} \cdot (1-(1+\mathrm{WACC})^{-n})}{\mathrm{WACC}} \tag{14.24}$$

Then, the equity cost could be found using the following equation:

$$\begin{aligned} \mathrm{WACC} &= k_e w_e + k_d w_d (1-t) \\ k_e &= \frac{\mathrm{WACC}}{w_e} - \frac{k_d(1-t) w_d}{w_e} = \mathrm{WACC}(1+L) - L k_d (1-t) \end{aligned} \tag{14.25}$$

where the WACC value is found using Formula (14.22).

14.3 Results and Discussions

We studied the impact of the frequency of advance profit tax payments on the dependence of the WACC, capital value and equity cost on the leverage level of companies of different ages (three or six years) using Microsoft Excel. We considered the advance profit tax payments using the formulae that were obtained above: Formula (14.22) was used for WACC, Formula (14.24) was used for capital value and Formula (14.25) was used for equity cost. By using the BFO theory, we could study companies of any age. Three- and six-year-old companies were considered as examples to estimate the impact of age on the financial parameters of a company. We used a large database, which can be made available to interested readers upon request. We used data that showed the influence of advance profit tax payments with an arbitrary frequency on the financial parameters of the selected companies. Then, we compared the results of the calculations of two types of profit tax payments: (1) payments at the end of the financial period [1] and (2) at the beginning of the financial period (i.e., in advance).

The following parameters were used: $k_0 = 0.2$; $k_d = 0.18$; $t = 0.2$; $n = 3$ or 6; CF = 100. The chosen parameters were typical for companies and the results that were obtained using other parameters were similar and did not show any qualitative differences.

14.3.1 The Impact of the Frequency of Profit Tax Payments on the Dependence of the Weighted Average Cost of Capital, Capital Value and Equity Cost on the Leverage Level of a Three-Year-Old Company

In Tables 14.1, 14.2, and 14.3, we present the results of our study on the impact of the frequency of profit tax payments on the dependence of the WACC, capital value and equity cost on the leverage level of a three-year-old company with payment frequencies of $p = 1, 2, 4, 6$, and 12.

As can be seen from Fig. 14.1, the WACC decreased with L for all frequencies (annual, semi-annual, quarterly, monthly, etc.) of advance profit tax payments. So, the debt financing led to a decrease in the corresponding capital cost with L and thus, its use by the company was important. The WACC values increased with the frequency of payments, which meant that more infrequent payments were beneficial for the company. The influence of the payment frequency was stronger for annual, semi-annual and quarterly payments and decreased for monthly payments.

As can be seen from Fig. 14.2, the company value (V) increased with L for all frequencies (annual, semi-annual, quarterly, monthly, etc.) of advance profit tax payments. So, the debt financing led to an increase in the company value with L and thus, its use by the company was important. This was consistent with the decrease in WACC with the leverage level, as shown in the previous section. The company

Table 14.1 The dependence of the WACC on the leverage level (L) of a three-year-old company for advance profit tax payments with frequencies of $p = 1, 2, 4, 6$, and 12

		WACC, $n = 3$ (2)				
L	w_d	$p = 1$	$p = 2$	$p = 4$	$p = 6$	$p = 12$
1	0.5	0.1703	0.1715	0.1721	0.1723	0.1725
2	0.666667	0.1603	0.1619	0.1627	0.1629	0.1632
3	0.75	0.1553	0.1571	0.1580	0.1583	0.1586
4	0.8	0.1523	0.1542	0.1552	0.1555	0.1558
5	0.833333	0.1503	0.1523	0.1533	0.1536	0.1539
6	0.857143	0.1489	0.1509	0.1519	0.1522	0.1526
7	0.875	0.1478	0.1499	0.1509	0.1512	0.1516
8	0.888889	0.1469	0.1491	0.1501	0.1505	0.1508
9	0.9	0.1463	0.1484	0.1495	0.1498	0.1502
10	0.909091	0.1457	0.1479	0.1490	0.1493	0.1497

Table 14.2 The dependence of the company value (V) on the leverage level (L) of a three-year-old company for advance profit tax payments with frequencies of $p = 1, 2, 4, 6$, and 12

		V				
L	w_d	$p = 1$	$p = 2$	$p = 4$	$p = 6$	$p = 12$
1	0.5	220.8472	220.4234	220.2163	220.1479	220.0800
2	0.666667	224.4700	223.8866	223.6018	223.5079	223.4144
3	0.75	226.3263	225.6593	225.3339	225.2266	225.1199
4	0.8	227.4549	226.7365	226.3861	226.2705	226.1557
5	0.833333	228.2136	227.4603	227.0930	226.9719	226.8515
6	0.857143	228.7586	227.9802	227.6007	227.4756	227.3512
7	0.875	229.1691	228.3717	227.9829	227.8548	227.7274
8	0.888889	229.4893	228.6771	228.2811	228.1506	228.0208
9	0.9	229.7462	228.9220	228.5202	228.3878	228.2561
10	0.909091	229.9568	229.1228	228.7162	228.5823	228.4490

Table 14.3 The dependence of the equity cost (k_e) on the leverage level (L) of a three-year-old company for advance profit tax payments with frequencies of $p = 1, 2, 4, 6$, and 12

		k_e				
L	w_d	$p = 1$	$p = 2$	$p = 4$	$p = 6$	$p = 12$
1	0.5	0.1966	0.1990	0.2002	0.2005	0.2009
2	0.666667	0.1930	0.1977	0.2001	0.2008	0.2016
3	0.75	0.1892	0.1964	0.1999	0.2011	0.2022
4	0.8	0.1855	0.1951	0.1998	0.2013	0.2028
5	0.833333	0.1817	0.1937	0.1996	0.2015	0.2034
6	0.857143	0.1780	0.1923	0.1994	0.2017	0.2040
7	0.875	0.1742	0.1910	0.1992	0.2019	0.2046
8	0.888889	0.1704	0.1896	0.1990	0.2021	0.2052
9	0.9	0.1666	0.1882	0.1988	0.2023	0.2058
10	0.909091	0.1628	0.1868	0.1986	0.2025	0.2063

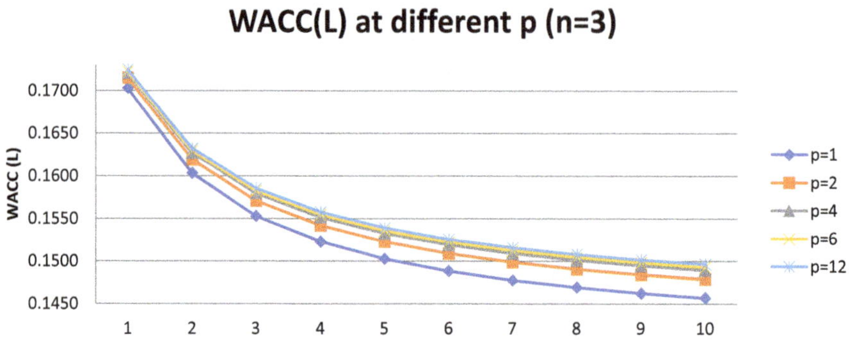

Fig. 14.1 The dependence of the WACC on the leverage level (L) of a three-year-old company for advance profit tax payments with frequencies of $p = 1, 2, 4, 6$, and 12

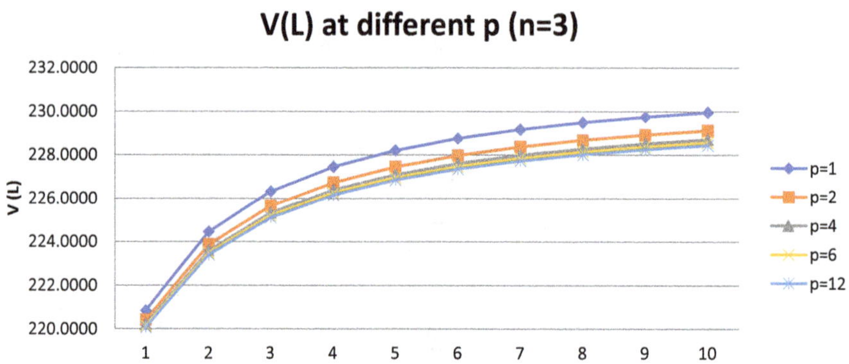

Fig. 14.2 The dependence of the company value (V) on the leverage level (L) of a three-year-old company for advance profit tax payments with frequencies of $p = 1, 2, 4, 6$, and 12

value decreased with the frequency of payments, which meant that more infrequent payments were beneficial for the company. The influence of the payment frequency was stronger for annual, semi-annual, and quarterly payments and decreased for monthly payments. This was consistent with the WACC behavior that was shown in the previous section.

As can be seen from Fig. 14.3, the equity cost (k_e) depended on the leverage level in a linear manner. It was not expected for $k_e(L)$ to be linear because the BFO equation for the WACC was an $n + 1$ power equation. However, as can be seen from the figures and tables for the equity cost results, this occurred with a high accuracy. We called the relationship quasi-linear (or practically linear). Our entire extensive database showed that this was the case. The equity cost decreased for annual ($p = 1$), semi-annual ($p = 2$) and quarterly ($p = 4$) payments and increased for monthly ($p = 12$) and bi-monthly payments ($p = 6$). The tilt angle of k_e(L) increased with p and the transition from k_e increasing with L to k_e decreasing with L occurred at $p = 4$.

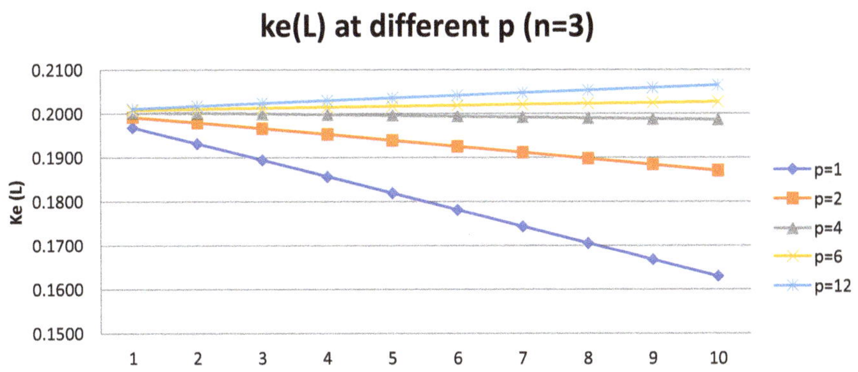

Fig. 14.3 The dependence of the equity cost (k_e) on the leverage level (L) of a three-year-old company for advance profit tax payments with frequencies of $p = 1, 2, 4, 6$, and 12

The economically justified amount of dividends was equal to the equity cost; therefore, the decrease in the equity cost with the leverage level signaled the discovery of a new effect that could significantly change the company's dividend policy.

Below, we compare the dependence of the WACC, company value and equity cost on the leverage level of a three-year-old company for profit tax payments that were (1) made at the end of the financial period (data from Ref. [1]) and (2) made at the beginning of the financial period (i.e., advance payments) with frequencies of $p = 2, 4, 6$, and 14.

The comparison of the dependence of the WACC on the leverage level (L) of a three-year-old company for profit tax payments that were (1) made at the end of the financial period and (2) made at the beginning of the financial period (i.e., advance payments) with frequencies of $p = 2, 4, 6$, and 12 is shown in Fig. 14.4.

The WACC(L) decreased with L in both cases: advance profit tax payments and payments at the end of the financial year. This meant that debt financing was important and should be used by the company as it led to a decrease in the corresponding capital cost with L.

In the case of profit tax payments that were made at the end of the financial year, the WACC(L) decreased with p, while in the case of advance profit tax payments, the WACC(L) increased with p. However, in spite of this, the WACC(L) always turned out to be lower in the case of advance profit tax payments. From Fig. 14.4, it can be seen, for example, that WACC($L = 10$) $= 0.1504$ at $p = 12$ for profit tax payments that were made at the end of the financial year while WACC($L = 10$) $=$ 0.1497 at $p = 12$ for advance profit tax payments. The WACC(L) curves for two these cases never crossed, i.e., the WACC values for advance profit tax payments always turned out to be lower than in the case of profit tax payments that were made at the end of the financial year. This demonstrated the importance of the use of advance profit tax payments for companies because the lower values of WACC corresponded to higher company values.

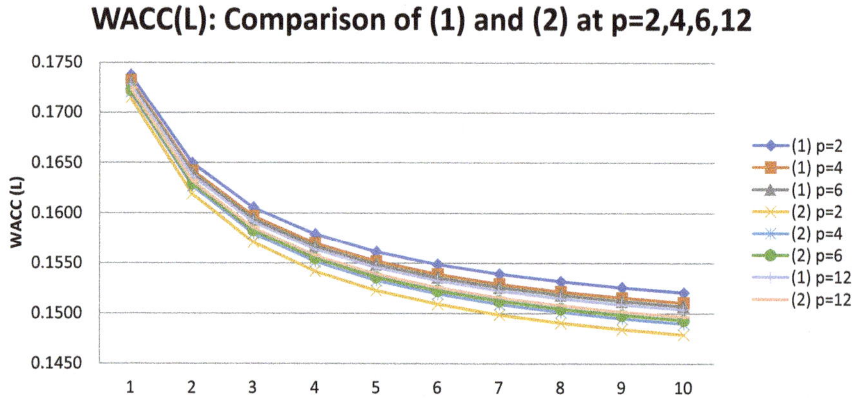

Fig. 14.4 The comparison of the dependence of the WACC on the leverage level (L) of a three-year-old company for profit tax payments that were (1) made at the end of the financial period and (2) made at the beginning of the financial period (i.e., advance payments) with frequencies of $p =$ 2, 4, 6, and 12

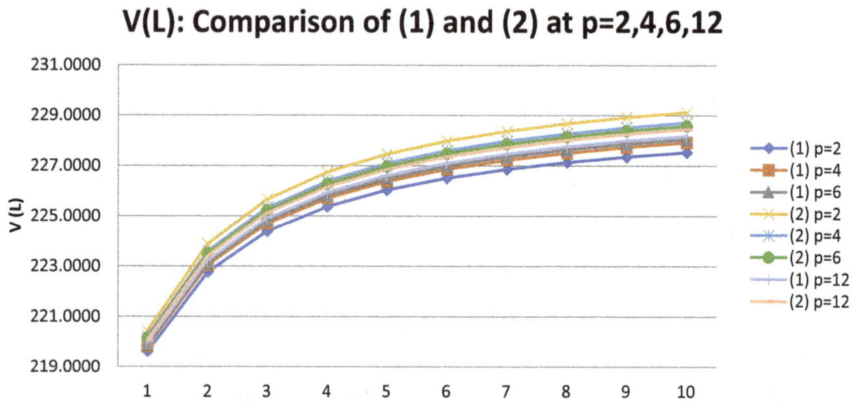

Fig. 14.5 The comparison of the dependence of the company value (V) on the leverage level (L) of a three-year-old company for profit tax payments that were (1) made at the end of the financial period and (2) made at the beginning of the financial period (i.e., advance payments) with frequencies of $p = 2$, 4, 6, and 12 is shown in Fig. 14.4

The comparison of the dependence of the company value (V) on the leverage level (L) of a three-year-old company for profit tax payments that were (1) made at the end of the financial period and (2) made at the beginning of the financial period (i.e., advance payments) with frequencies of $p = 2$, 4, 6, and 12 is shown in Fig. 14.5.

Figure 14.5 shows that the company value increased with L for all payment frequencies (annual, semi-annual, quarterly, monthly, etc.) in both cases: advance profit tax payments and payments the end of the financial period. So, the debt financing led to an increase in the company value with L and thus, it was important

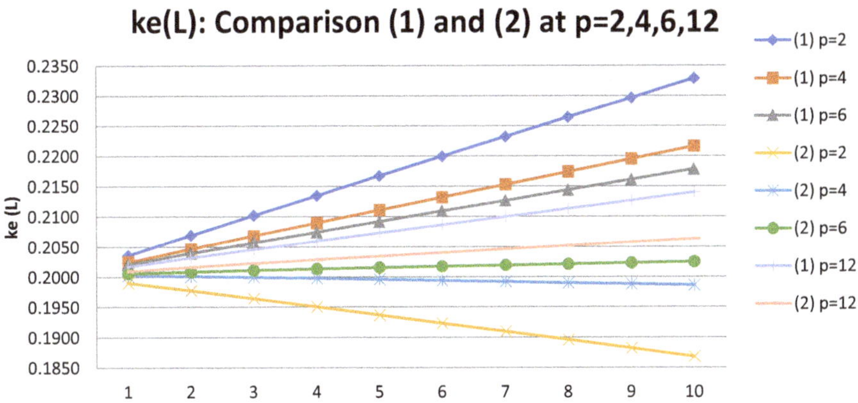

Fig. 14.6 The comparison of the dependence of the equity cost (k_e) on the leverage level (L) of a three-year-old company for profit tax payments that were (1) made at the end of the financial period and (2) made at the beginning of the financial period (i.e., advance payments) with frequencies of p = 2, 4, 6, and 12

and should be used by the company. This was consistent with the decrease in the WACC with L, as shown in the previous section. The company value decreased with the frequency of payments in the case of profit tax payments being made at the beginning of the financial period (i.e., advance payments), which meant that more infrequent payments were beneficial for the company. The company value increased with the frequency of payments in the case of profit tax payments being made at the end of the financial period. However, in spite of this, $V(L)$ always turned out to be bigger in the case of advance profit tax payments. From Fig. 14.5, it can be seen, for example, that $V(L = 10) = 228.1849$ at $p = 12$ for profit tax payments that were made at the end of the financial year, while $V(L = 10) = 228.4490$ at $p = 12$ for advance profit tax payments. Note that the $V(L)$ curves that corresponded to these two cases never crossed. The influence of the payment frequency was stronger for annual, semi-annual and quarterly payments and decreased for monthly payments. This was consistent with the WACC behavior that was shown in the previous section.

The comparison of the dependence of the equity cost (k_e) on the leverage level (L) of a three-year-old company for profit tax payments that were (1) made at the end of the financial period and (2) made at the beginning of the financial period (i.e., advance payments) with frequencies of $p = 2$, 4, 6 and 12 is shown in Fig. 14.6.

From Fig. 14.6, it can be seen that the equity cost (k_e) depended on the leverage level (L) in a linear manner in both cases ((1) and (2)) for all values of p. In the case of profit tax payments being made at the end of the financial period, the tilt angle of $k_e(L)$ decreased with p but k_e(L) continued to increase with L for all values of p. In the case of advance profit tax payments, the tilt angle of $k_e(L)$ increased with p and the transition from k_e increasing with L to k_e decreasing with L occurred at $p =$ 4. Additionally, $k_e(L)$ decreased with L for annual ($p = 1$), semi-annual ($p = 2$) and quarterly ($p = 4$) payments and increased for monthly ($p = 12$) and bi-monthly payments ($p = 6$). The economically justified amount of dividends was equal to the equity cost; therefore, the decrease in the equity cost with the leverage level signaled the discovery of a new effect that could significantly change the dividend policy of a company. This effect only occurred in the case of advance profit tax payments. Note that with the increase in p, the $k_e(L)$ curves for the two cases ((1) and (2)) approached each other but never crossed.

14.3.2 *The Impact of the Frequency of Profit Tax Payments on the Dependence of the Weighted Average Cost of Capital, Capital Value and Equity Cost on the Leverage Level of a Six-Year-Old Company*

In Tables 14.4, 14.5, and 14.6, we present the results of our study on the impact of the frequency of profit tax payments on the dependence of the WACC, capital value and equity cost on the leverage level of a six-year-old company with payment frequencies of $p = 1, 2, 4, 6$, and 12.

As can be seen from Fig. 14.7, the WACC decreased with L in the case of advance profit tax payments for all payment frequencies (annual, semi-annual, quarterly, monthly, etc.). So, debt financing led to a decrease in the corresponding capital cost with L and thus, should be used by the company. The WACC values increased with the frequency of payments, which meant that more infrequent payments were beneficial for the company. The influence of the payment frequency was stronger for annual, semi-annual and quarterly payments and decreased for monthly payments.

As can be seen from Fig. 14.8, the company value (V) increased with L in the case of advance profit tax payments for all payment frequencies (annual, semi-annual, quarterly, monthly, etc.). This was consistent with the decrease in the WACC with

Table 14.4 The dependence of the WACC on the leverage level (L) of a six-year-old company for advance profit tax payments with frequencies of $p = 1, 2, 4, 6$, and 12

		WACC, $n = 6$ (2)				
L	w_d	$p = 1$	$p = 2$	$p = 4$	$p = 6$	$p = 12$
1	0.5	0.1697	0.1709	0.1715	0.1717	0.1719
2	0.666667	0.1593	0.1610	0.1618	0.1621	0.1623
3	0.75	0.1542	0.1560	0.1569	0.1572	0.1575
4	0.8	0.1510	0.1530	0.1540	0.1543	0.1546
5	0.833333	0.1489	0.1510	0.1520	0.1524	0.1527
6	0.857143	0.1474	0.1496	0.1506	0.1510	0.1513
7	0.875	0.1463	0.1485	0.1496	0.1499	0.1503
8	0.888889	0.1454	0.1477	0.1487	0.1491	0.1495
9	0.9	0.1447	0.1470	0.1481	0.1484	0.1488
10	0.909091	0.1442	0.1464	0.1475	0.1479	0.1483

Table 14.5 The dependence of the company value (V) on the leverage level (L) of a six-year-old company for advance profit tax payments with frequencies of $p = 1, 2, 4, 6$, and 12

		V				
L	w_d	$p = 1$	$p = 2$	$p = 4$	$p = 6$	$p = 12$
1	0.5	359.2385	358.0973	357.5406	357.3571	357.1746
2	0.666667	369.1124	367.5078	366.7265	366.4692	366.2133
3	0.75	374.2557	372.4010	371.4988	371.2017	370.9064
4	0.8	377.4110	375.4000	374.4222	374.1003	373.7805
5	0.833333	379.5443	377.4263	376.3969	376.0581	375.7214
6	0.857143	381.0830	378.8871	377.8201	377.4690	377.1201
7	0.875	382.2451	379.9901	378.8947	378.5342	378.1761
8	0.888889	383.1540	380.8525	379.7347	379.3669	379.0014
9	0.9	383.8841	381.5452	380.4093	380.0356	379.6643
10	0.909091	384.4836	382.1139	380.9631	380.5845	380.2084

Table 14.6 The dependence of the equity cost (k_e) on the leverage level (L) of a six-year-old company for advance profit tax payments with frequencies of $p = 1, 2, 4, 6$, and 12

		k_e				
L	w_d	$p = 1$	$p = 2$	$p = 4$	$p = 6$	$p = 12$
1	0.5	0.1953	0.1978	0.1990	0.1994	0.1997
2	0.666667	0.1900	0.1950	0.1974	0.1982	0.1990
3	0.75	0.1846	0.1920	0.1957	0.1969	0.1981
4	0.8	0.1791	0.1891	0.1939	0.1955	0.1971
5	0.833333	0.1736	0.1860	0.1921	0.1941	0.1961
6	0.857143	0.1680	0.1830	0.1903	0.1927	0.1951
7	0.875	0.1624	0.1799	0.1885	0.1913	0.1941
8	0.888889	0.1569	0.1769	0.1867	0.1899	0.1931
9	0.9	0.1513	0.1738	0.1848	0.1885	0.1921
10	0.909091	0.1457	0.1707	0.1830	0.1870	0.1911

the leverage level, as shown in the previous section. The company value decreased with the frequency of payments, which meant that more infrequent payments were beneficial for the company. The influence of the payment frequency was stronger for annual, semi-annual and quarterly payments and decreased for monthly payments. This was consistent with the WACC behavior that was shown in the previous section.

As can be seen from Fig. 14.9, the equity cost (k_e) depended on the leverage level (L) in a linear manner. It decreased for annual ($p = 1$), semi-annual ($p = 2$) and quarterly ($p = 4$) payments and increased for monthly ($p = 12$) and bi-monthly payments ($p = 6$). The economically justified amount of dividends was equal to the equity cost; therefore, the decrease in the equity cost with the leverage level signaled the discovery of a new effect that could significantly change the dividend policy of the company. Note that for the three-year-old company, k_e increased with L for

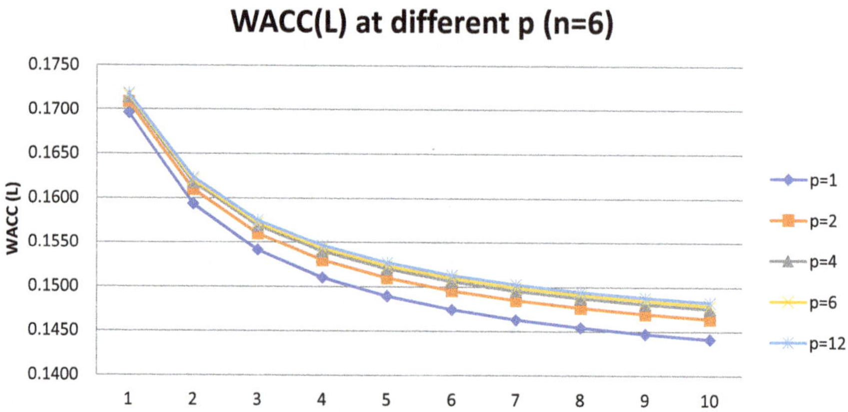

Fig. 14.7 The dependence of the WACC on the leverage level (L) of a six-year-old company for advance profit tax payments with frequencies of $p = 1, 2, 4, 6$, and 12

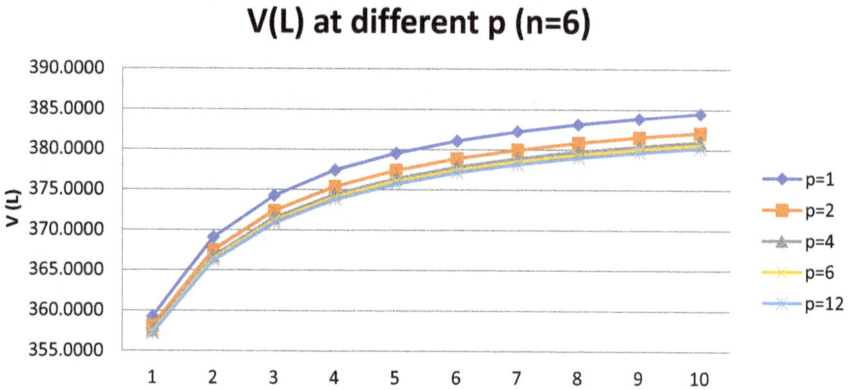

Fig. 14.8 The dependence of the company value (V) on the leverage level (L) of a six-year-old company for advance profit tax payments with frequencies of $p = 1, 2, 4, 6$, and 12

monthly ($p = 12$) and bi-monthly payments ($p = 6$) and the transition from decreasing to increasing behavior occurred at $p = 4$.

The comparison of the dependence of the WACC on the leverage level (L) of a six-year-old company for profit tax payments that were (1) made at the end of the financial period and (2) made at the beginning of the financial period (i.e., advance payments) with frequencies of $p = 2, 4, 6$, and 12 is shown in Fig. 14.10.

The WACC(L) decreased with L in both cases: profit tax payments that were made at the beginning of the financial period and payments that were made at the end of the financial period. This meant that the debt financing led to a decrease in the corresponding capital cost with L and should be used by the company.

The WACC(L) decreased with p in the case of profit tax payments being made at the end of the financial period, while in the case of profit tax payments being made at

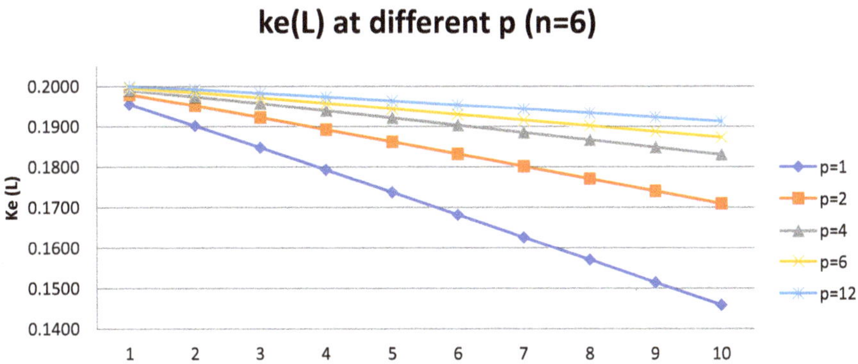

Fig. 14.9 The dependence of the equity cost (k_e) on the leverage level (L) of a six-year-old company for advance profit tax payments with frequencies of $p = 1, 2, 4, 6,$ and 12

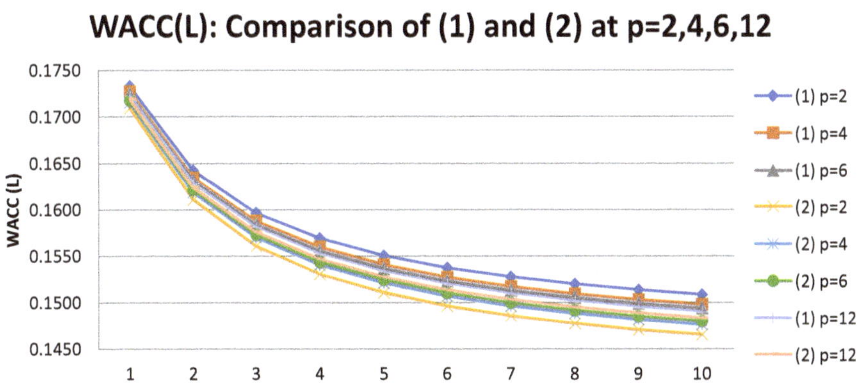

Fig. 14.10 The comparison of the dependence of the WACC on the leverage level (L) of a six-year-old company for profit tax payments that were (1) made at the end of the financial period and (2) made at the beginning of the financial period (i.e., advance payments) with frequencies of $p =$ 2, 4, 6, and 12

the beginning of the financial period (i.e., advance payments), the WACC(L) increased with p. However, in spite of this, the WACC(L) always turned out to be lower in the latter case. From Fig. 14.10, it can be seen, for example, that WACC ($L = 10$) = 0.1490 at $p = 12$ for profit tax payments that were made at the end of the financial year while WACC($L = 10$) = 0.1483 at $p = 12$ for advance profit tax payments. The WACC(L) curves for two these cases never crossed, i.e., the WACC values for advance profit tax payments always turned out to be lower than in the case of profit tax payments that were made at the end of the financial year. Thus, the use of advance profit tax payments was quite important for the company because the lower values of WACC corresponded to higher company values. This conclusion is valid for companies of any age.

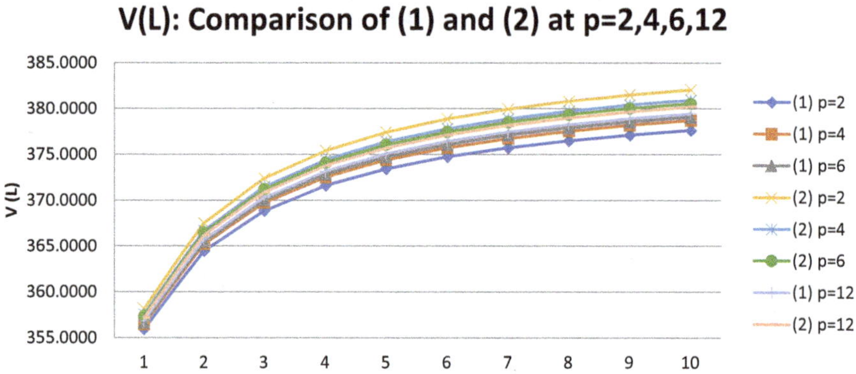

Fig. 14.11 The comparison of the dependence of the company value (V) on the leverage level (L) of a six-year-old company for profit tax payments that were (1) made at the end of the financial period and (2) made at the beginning of the financial period (i.e., advance payments) with frequencies of $p = 2$, 4, 6, and 12

The comparison of the dependence of the company cost (V) on the leverage level (L) of a six-year-old company for profit tax payments that were (1) made at the end of the financial period and (2) made at the beginning of the financial period (i.e., advance payments) with frequencies of $p = 2$, 4, 6, and 12 is shown in Fig. 14.11.

As can be seen from Fig. 14.11, the company value (V) increased with L for all payment frequencies (annual, semi-annual, quarterly, monthly, etc.) in both cases: profit tax payments that were made at the beginning of the financial period (i.e., advance payments) and profit tax payments that were made at the end of the financial period. So, the debt financing led to an increase in the company value with the leverage level and thus, should be used by the company. This was consistent with the decrease in the WACC with the leverage level, as shown in the previous section. The company value decreased with the frequency of payments in the case of advance profit tax payments, which meant that more infrequent payments were beneficial for the company, and increased with the frequency of payments in the case of profit tax payments that were made at the end of the financial period. However, in spite of this, the $V(L)$ always turned out to be bigger in the case of advance profit tax payments. From Fig. 14.11, it can be seen, for example, that $V(L = 10) = 379.4635$ at $p = 12$ for profit tax payments that were made at the end of the financial year while $V(L = 10) = 380.2084$ at $p = 12$ for advance profit tax payments. Note that the $V(L)$ curves that corresponded to these two cases never crossed. The influence of the payment frequency was stronger for annual, semi-annual and quarterly payments and decreased for monthly payments. This was consistent with the WACC behavior that was shown in the previous section.

The comparison of the dependence of the equity cost (k_e) on the leverage level (L) of a six-year-old company for profit tax payments that were (1) made at the end of the financial period and (2) made at the beginning of the financial period (i.e., advance payments) with frequencies of $p = 2$, 4, 6, and 12 is shown in Fig. 14.12.

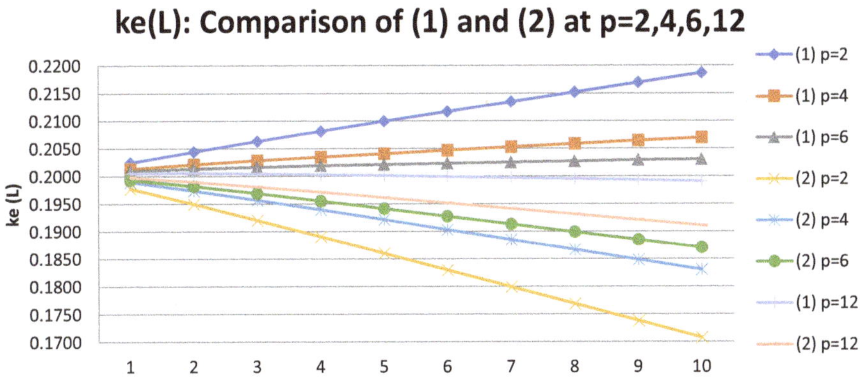

Fig. 14.12 The comparison of the dependence of the equity cost (k_e) on the leverage level (L) of a six-year-old company for profit tax payments that were (1) made at the end of the financial period and (2) made at the beginning of the financial period (i.e., advance payments) with frequencies of $p = 2$, 4, 6, and 12

From Fig. 14.12, it can be seen that the equity cost (k_e) depended on the leverage level (L) in a linear manner in both cases ((1) and (2)) for all values of p (similar to the case of the three-year-old company). In the case of profit tax payments that were made at the end of the financial period, the tilt angle of $k_e(L)$ decreased with p, but $k_e(L)$ continued to increase with L at $p = 2$, 4 and 6 and we only observed the transition to a slight decrease with leverage level at $p = 14$. Note that for the three-year-old company, this transition was absent and $k_e(L)$ increased with L for all values of p.

In the case of profit tax payments that were made at the beginning of the financial period (i.e., advance payments), the tilt angle of $k_e(L)$ increased with p, but $k_e(L)$ decreased with L for all payment frequencies. The economically justified amount of dividends was equal to the equity cost; thus, the discovery of the decrease in equity cost with the leverage level represented a new effect that could greatly change the company's dividend policy. This effect only occurred in the case of advance profit tax payments. Note that with the increase in p, the $k_e(L)$ curves for the two cases ((1) and (2)) approached each other but never crossed.

14.4 Conclusions

In this Chapter, we studied the influence of frequent advance profit tax payments on the main financial indicators of a company. For this, we generalized the Brusov–Filatova–Orekhova (BFO) theory for the case of advance profit tax payments with an arbitrary frequency for the first time. We derived modified BFO formulae and showed that (1) all BFO formulae could change and (2) all of the financial parameters of a company, such as WACC, company value and equity cost, depended on the frequency of profit tax payments. The weighted average cost of capital (WACC) increased with the payments and the company value decreased with the payments. This meant that more infrequent payments were beneficial for the companies. The tilt angle of the equity cost ($k_e(L)$) also increased with the payments. Depending on the company age, $k_e(L)$ either decreased with L for all payment frequencies or a transition from the decrease to an increase in k_e with L occurred for some frequencies.

We compared the obtained results to those that we recently obtained for frequent profit tax payments that were made at the end of the financial period [1] and found them to be totally different. We showed that in spite the fact that the WACC decreased with the payments and the company value increased with the payments in this case [1], the WACC values turned out to be bigger and the company values turned out to be smaller than in the case of frequent advance profit tax payments for all payment frequencies. This underlined the importance of advance profit tax payments. Regulators should extend the practice of advance income tax payments to profit tax payments.

The economically justified amount of dividends was equal to the equity cost; thus, the discovered decrease in the equity cost with the leverage level represented a new effect that could greatly change a company's dividend policy. In the case of profit tax payments that were made at the end of the financial period, the tilt angle of $k_e(L)$ decreased with the payments, but $k_e(L)$ continued to increase with L at $p = 2, 4$ and 6 and we only observed the transition to a slight decrease with leverage level at $p = 14$. Note that for the three-year-old company, this transition was absent and $k_e(L)$ increased with L for all values of p.

In the case of advance profit tax payments, the tilt angle of $k_e(L)$ increased with the payments, but $k_e(L)$ decreased with L for all payment frequencies for the six-year-old company. For the three-year-old company, $k_e(L)$ decreased at $p = 1$, 2 and 4 and increased at $p = 6$ and 14. Thus, for the three-year-old company, the transition from the decreasing $k_e(L)$ behavior to the $k_e(L)$ increasing behavior occurred at $p = 6$, while for the six-year-old company, this transition was absent.

The novelty of the Chapter is as follows:

1. We generalized the BFO theory for the case of advance profit tax payments with an arbitrary frequency for the first time;
2. By comparing the obtained results to those for frequent profit tax payments at the end of the financial period, we showed that in spite the fact that the WACC decreased with the payments and the company value increased with the payments, the WACC value in this case turned out to be bigger and the company value turned out to be smaller than in case of frequent advance profit tax payments for all payment frequencies, which underlined the importance of advance profit tax payments;
3. A new effect of the dependence of equity cost on leverage level was discovered and this pioneering result could radically change the company dividend policies. In the case of advance profit tax payments, the tilt angle of $k_e(L)$ increased with the payments, but $k_e(L)$ decreased with L for all payment frequencies for the

six-year-old company and at $p = 1$, 2 and 4 for the three-year-old company, although it increased at $p = 6$ and 14. Thus, for the three-year-old company, the transition from the decreasing $k_e(L)$ behavior to the increasing $k_e(L)$ behavior occurred at $p = 6$, while this transition was absent for the six-year-old company;

4. Regulator recommendations were developed to expand the practice of advance profit tax payments, which would be beneficial to both parties, the companies and the state, (1) this practice would lead to a decrease in the cost of raising capital and an increase in the company value for companies and (2) ensure an increase in the stability of budget revenues for the regulators.

There was one limitation of the current consideration: the BFO theory was used with the assumption of a constant income. We will generalize the BFO theory for the case of variable incomes in the next Chapter.

References

1. Brusov, P. N., Filatova, T. V., Orekhova, N. P., Kulik, V. L., Chang, S. I., & Lin, Y. C. G. (2022). The generalization of the Brusov–Filatova–Orekhova theory for the case of payments of tax on profit with arbitrary frequency. *Mathematics, 10*, 1343. https://doi.org/10.3390/math10081343
2. Modigliani, F., & Miller, M. (1958). The cost of capital, corporate finance, and the theory of investment. *American Economic Review, 48*, 261–297.
3. Modigliani, F., & Miller, M. (1963). Corporate income taxes and the cost of capital: A correction. *American Economic Review, 53*, 147–175.
4. Modigliani, F., & Miller, M. (1966). Some estimates of the cost of capital to the electric utility industry 1954–1957. *American Economic Review, 56*, 333–391.
5. Hamada, R. (1969). Portfolio analysis, market equilibrium, and corporate finance. *Journal of Finance, 24*, 13–31.
6. Miller, M. (1977). Debt and taxes. *Journal of Finance, 32*, 261–275. https://doi.org/10.1111/j.1540-6261.1977.tb03267.x
7. Farber, A., Gillet, R., & Szafarz, A. A. (2006). General formula for the WACC. *International Journal of Business, 11*, 211–218.
8. Fernandez, P. A. (2006). General formula for the WACC: A comment. *International Journal of Business, 11*, 219.
9. Berk, J., & De Marzo, P. (2007). *Corporate finance*. Pearson-Addison Wesley.
10. Harris, R., & Pringle, J. (1985). Risk-adjusted discount rates–extension from the average-risk case. *Journal of Financial Research, 8*, 237–244.
11. Brusov, P. N., Filatova, T. V., Orekhova, N. P., Kulik, V. L., Chang, S. I., & Lin, Y. C. G. (2020). Modification of the Modigliani–Miller theory for the case of advance payments of tax on profit. *Journal of Reviews on Global Economics, 9*, 257–267.
12. Batrancea, L. (2021). An econometric approach regarding the impact of fiscal pressure on equilibrium: Evidence from electricity, gas and oil companies listed on the New York Stock Exchange. *Mathematics, 9*, 630.
13. Batrancea, L. (2021). The influence of liquidity and solvency on performance within the healthcare industry: Evidence from publicly listed companies. *Mathematics, 9*, 2231.
14. Brusov, P., Filatova, T., Orehova, N., & Eskindarov, M. (2018). *Modern corporate finance, investments, taxation and ratings* (2nd ed., pp. 1–571). Springer Nature Publishing.
15. Myers, S. (2001). Capital structure. *Journal of Economic Perspectives, 15*, 81–102.
16. Dimitropoulos, P. (2014). Capital structure and corporate governance of soccer clubs: European evidence. *Management Research Review, 37*, 658–678.

17. Luiz, K., & Cruz, M. (2015). The relevance of capital structure on firm performance: A multivariate analysis of publicly traded Brazilian companies. *REPeC Brasília, 9*, 384–401.
18. Barbi, M. (2011). On the risk-neutral value of debt tax shields. *Applied Financial Economics, 22*, 251–258.
19. Zhukov, P. (2018). The impact of cash flows and weighted average cost of capital to enterprise value in the oil and gas sector. *Journal of Reviews on Global Economics, 7*, 138–145.
20. Franc-Dąbrowska, J., Mądra-Sawicka, M., & Milewska, A. (2021). Energy sector risk and cost of capital assessment—Companies and investors perspective. *Energies, 14*, 1613.
21. Angotti, M., de Lacerda Moreira, R., Bernardes, H., do Nascimento, J., & Neto de Almeida Bispo, O. (2018). Analysis of an equity investment strategy based on accounting and financial reports in Latin American markets. *Reficont, 5*, 22–40.
22. Vergara-Novoa, C., Sepúlveda-Rojas, J. P., Miguel, D. A., & Nicolás, R. (2018). Cost of capital estimation for highway concessionaires in Chile. *Journal of Advanced Transportation, 2018*, 2153536.
23. Mundi, H. S., Kaur, P., & Murty, R. L. N. (2022). A qualitative inquiry into the capital structure decisions of overconfident finance managers of family-owned businesses in India. *Qualitative Research in Financial Markets, 14*, 357–379.
24. Sadiq, M., Alajlani, S., Hussain, M. S., Bashir, F., & Chupradit, S. (2022). Impact of credit, liquidity, and systematic risk on financial structure: Comparative investigation from sustainable production. *Environmental Science and Pollution Research International, 29*, 20963–20975.
25. Becker, D. M. (2022). Getting the valuation formulas right when it comes to annuities. *Managerial Finance, 48*, 470.
26. El-Chaarani, H., Abraham, R., & Skaf, Y. (2022). The impact of corporate governance on the financial performance of the banking sector in the MENA (Middle Eastern and North African) region: An immunity test of banks for COVID-19. *Journal of Risk and Financial Management, 15*, 82.
27. Huang, S., Sun, H., Zhao, H., & Zhang, Y. (2020). Influence of leverage on the return on equity. *Systems Engineering - Theory & Practice, 40*, 355.
28. Islam, S. Z., & Khandaker, S. (2015). Firm leverage decisions: Does industry matter? *The North-American Journal of Economics and Finance, 31*, 94.

Chapter 15
Generalization of the Brusov–Filatova–Orekhova Theory for the Case of Variable Income

15.1 Introduction

Among the theories of capital structure, the two main ones are the Brusov–Filatova–Orekhova (BFO) theory and the Modigliani–Miller (MM) theory, which is the eternal limit of the BFO theory. Both of them consider the case of constant income, although in practice, a company's income is, of course, variable. The generalization of these two theories of capital structure for the case of variable income is very important, since it allows them to expand their applicability in practice. Recently, we have generalized the case of variable income the Modigliani-Miller theory [1], and here we have generalized for the first time the case of variable income of the Brusov–Filatova–Orekhova theory. This generalization significantly expands the applicability of the modern capital structure theory, which is valid for companies of any age, and in practice, for investments, corporate finance, business valuation, banking, ratings, etc.

15.1.1 Literature Review

The Brusov–Filatova–Orekhova (BFO) theory and its perpetual limit—the theory of Nobel laureates Modigliani and Miller—study capital structure problems. Capital structure is the ratio between the company's own and borrowed capital, and asks whether capital structure affects key company metrics, such as weighted average cost of capital, WACC, cost of equity, k_e, earnings, company value, and others, and if so, how? The determination of the optimal capital structure, that is, the determination of capital structure in which WACC is minimal and the company value is maximized, is one of the main problems, solved by the company's management. The Modigliani and Miller Chapter [2] was the first quantitative study of the influence of

P. Brusov, T. Filatova, *Business Valuation and Company Value*, Contributions to Finance and Accounting, https://doi.org/10.1007/978-3-031-90847-7_15

capital structure on a company's financial parameters. Prior to this, there was a traditional approach based on empirical data analysis.

15.1.2 *Before the Modigliani and Miller Work*

In the traditional approach, the WACC and the value of the company, *V*, depend on the level of leverage, *L* (and, therefore, on the capital structure). Debt is always cheaper than equity, since the former has less risk, and because in the event of bankruptcy, the claims of creditors are satisfied earlier than the claims of shareholders.

Thus, an increase in the share of cheaper borrowed capital of the total capital structure to the limit that does not violate financial stability and does not increase the risk of bankruptcy leads to a decrease in the WACC and an increase in the value of the company, *V*.

Further increase of debt financing could lead to financial stability violation and an increase in the bankruptcy risk. WACC increases and the company value, *V*, decreases. The competition of advantages of debt financing and its shortcomings at low and at high leverage levels forms the optimal capital structure where WACC is minimal and the company value, *V*, is maximum. These traditional approach results have been used in the trade-off theory.

15.1.3 *Modigliani–Miller Theory*

15.1.3.1 Modigliani–Miller Theory Without Taxes

Modigliani and Miller (MM) [2], under a lot of assumptions, including the absence of corporate and individual taxes, the perpetuity of all companies and all cash flows, etc. obtained results that are completely different from the results of the traditional approach: capital structure does not affect capital cost and company value.

Under the above restrictions, Modigliani and Miller have shown that without taxes, the company value, *V*, is equal

$$V = V_0 = \frac{\text{EBIT}}{k_0} \tag{15.1}$$

Here, EBIT is Earnings Before Interest and Taxes, k_0 is discount rate, and V_0 stands for the unlevered company value.

From (15.1), it is easy to obtain WACC:

$$\mathrm{WACC} = k_0 \tag{15.2}$$

k_0 is the cost of equity for a company without borrowed funds, and for a company with borrowed capital, k_0 is the cost of equity with a zero level of borrowed funds ($L = 0$).

From (15.1) and the expression for WACC

$$\mathrm{WACC} = k_0 = k_\mathrm{e} w_\mathrm{e} + k_\mathrm{d} w_\mathrm{d}. \tag{15.3}$$

One can obtain the cost of equity, k_e

$$\begin{aligned} k_\mathrm{e} &= \frac{k_0}{w_\mathrm{e}} - k_\mathrm{d}\frac{w_\mathrm{d}}{w_\mathrm{e}} = \frac{k_0(S+D)}{S} - k_\mathrm{d}\frac{D}{S} = k_0 + (k_0 - k_\mathrm{d})\frac{D}{S} \\ &= k_0 + (k_0 - k_\mathrm{d})L \end{aligned} \tag{15.4}$$

Here, WACC is weighted average cost of capital; L is leverage level; D is debt capital value; S is equity capital value; k_d and w_d are the cost and share of the company's debt capital; and k_e and w_e are the equity capital cost and share. From (15.4), it follows that the equity cost increases linearly with the leverage level.

15.1.3.2 Modigliani–Miller Theory with Taxes

Taking into account income tax, in 1963, Modigliani and Miller [3, 4] obtained the following result for the value of the levered company, V,

$$V = V_0 + D \cdot t \tag{15.5}$$

Here, V_0 stands for the unlevered company value, t is the tax on income, and D is debt value.

From (15.5), it is easy to derive the expression for the WACC

$$\mathrm{WACC} = k_0 \cdot (1 - w_d t) \tag{15.6}$$

From (15.6), one can obtain the formula for equity cost, k_e, within the Modigliani–Miller theory with taxes

$$k_e = k_0 + L \cdot (k_0 - k_d)(1 - t) \tag{15.7}$$

Formula (15.7) differs from formula (15.4) (MM without taxes) by the factor ($1-t$), which is called the tax corrector. This is less than one, so the tilt of the $k_e(L)$ curve decreases with taxes.

15.1.4 Unification of Capital Asset Pricing Model (CAPM) with Modigliani–Miller Model

Unification of Capital Asset Pricing Model (CAPM) with the Modigliani–Miller model was done in 1969 [5]. Hamada derived the below formula for the cost of equity of a levered company, which included the financial and business risks of a company:

$$k_e = k_F + (k_M - k_F)b_U + (k_M - k_F)b_U \frac{D}{S}(1 - T), \tag{15.8}$$

where b_U is the β–coefficient of the company of the same group of business risk, that the company under consideration, but with $L = 0$. The formula (15.8) consists of three terms: risk-free income ability k_F, compensating shareholders a time value of their money, business risk premium$(k_M - k_F)b_U$, and financial risk premium $(k_M - k_F)b_U \frac{D}{S}(1 - T)$.

For a financially independent company, the financial risk is equal to zero (the third term disappears), and its owners will only receive the business risk premium.

Miller Model

In [6], Miller has accounted for corporate and individual taxes to obtain the following formula for unlevered company value, V_U,

$$V_U = \frac{\mathrm{EBIT}(1 - T_C)(1 - T_S)}{k_0}. \tag{15.9}$$

Here, T_C—tax rate on corporate income, T_S—the tax rate on incomes of an individual investor from his ownership through corporation stocks.

15.1.5 Brusov–Filatova–Orekhova (BFO) Theory

The perpetuity of all company cash flow and of a company's lifetime was one of the main restrictions of the Modigliani–Miller (MM) theory, which has been lifted up in 2008 by Brusov–Filatova–Orekhova [7, 8]. They generalized the MM theory for the case of the company of any age, n, and derived the following Brusov–Filatova–Orekhova formula for the WACC

$$\frac{1 - (1 + \mathrm{WACC})^{-n}}{\mathrm{WACC}} = \frac{1 - (1 + k_0)^{-n}}{k_0[1 - w_d T(1 - (1 + k_d)^{-n})]} \tag{15.10}$$

To obtain the Modigliani–Miller formula for the WACC from (15.10), one should substitute $n \to \infty$.

It was shown [7, 8] that a number of innovative effects, discovered in the BFO theory, are absent in the MM theory [2–4].

Some main existing principles of financial management spanning many decades have been destroyed by the BFO theory; among them is the keystone of optimal capital structure formation—trade-off theory, and the bankruptcy of this theory has been proven within the BFO theory [7, 8].

15.1.6 Alternate WACC Formula

An alternate formula for the WACC has been suggested [9–12]. It has the form below (Eq. (18) in [9])

$$\mathrm{WACC} = k_0(1 - w_d t) - k_d t w_d + k_{\mathrm{TS}} t w_d \tag{15.11}$$

Here, k_0, k_d, and k_{TS} are the returns on the financially independent company, the debt, and the tax shield, respectively, t is the corporate tax rate, and w_d is the debt share.

Although Eq. (15.11) is quite general, additional conditions are needed for practical applicability. When the WACC remains constant over time, the value of a leveraged company can be found by discounting the unleveraged free cash flows using the WACC. In this case, specific formulas can be found in textbook [11].

In the Modigliani-Miller theory [3], the debt value D is constant. V_0 is also constant, as the expected after-tax cash-flow of the financially independent company is fixed. By assumption, $k_{\mathrm{TS}} = k_d$ and the tax shield value is $\mathrm{TS} = tD$. Therefore, the company value V is a constant and the alternate WACC formula (15.11) simplifies the MM formula:

$$\mathrm{WACC} = k_0(1 - w_d t)$$

The "classical" MM theory, suggesting that the returns on the debt k_d and the tax shield k_{TS} are equals (both these values have debt nature), is much more reasonable, so this is why in [1], we modify, namely, the "classical" MM theory.

15.1.7 Trade-Off Theory

In the study of the problem of optimal capital structure of the company during many decades, the cornerstone was the world-famous trade-off theory. It is still widely used now for decisions on capital structure. In [13], the relative importance of different factors in capital structure decisions of publicly traded American companies has been studied. The most important factors to explain leverage level are: median industry leverage (+ effect on leverage), log of assets (+), market-to-book assets ratio (−), inflation (+), tangibility (+), and incomes (−). It was noted that companies that pay dividends tend to have lower leverage levels. The related effects have been found under considering book leverage. Authors found empirical data consistent with some trade-off theory versions.

In [14], authors compare the applicability of the trade-off theory and pecking order theory for small and medium-sized companies' decisions about capital structure. It was found that the most lucrative and oldest companies have smaller leverage levels, which confirms the forecasts of the pecking order theory. Larger companies have a higher level of leverage, which is consistent with the predictions of the trade-off theory and pecking order theory. It is concluded that the trade-off theory and pecking order theory for small and medium-sized companies are not mutually exclusive when explaining capital structure decisions.

However, in 2013, Brusov et al. [7, 8] proved the inconsistency of the trade-off theory in the framework of the BFO theory they created. It is shown that the assumption of risky debt financing does not lead to an increase in the WACC, which still decreases with increasing leverage. Thus, there is no minimum depending on the level of WACC leverage and no maximum depending on the value of the company from the level of leverage. Therefore, in the world-famous theory of trade-offs, there is no optimal capital structure. Brusov et al., in 2013 [7, 8], having analyzed the equity cost dependence on the level of leverage under the assumption that debt capital is risky, gave an explanation for this fact.

The Modigliani–Miller theory proved that tax shields provided substantial gains to the company. In [15], the theoretical study of tax shields was continued. It was noted that companies may have tax deductibles other than debt. Such non-debt tax shields are investment tax credits, depreciation, and net-loss carry forwards. In [16], the tax effects suggested in [15] have been tested. In contrast to the prediction in [15], it was shown that debt is positively related to non-debt tax shields, as measured by investment tax credits and depreciation. The results of [14] do not provide support for an effect on debt ratios arising from nondebt tax shields. In [17], it was pointed out that a positive relationship between such proxies for non-debt tax shield and debt may result if a company invests heavily and borrows to invest. Any substitution effects between debt and non-debt tax shields could be suppressed by a mechanical positive relation of this type.

The Brusov–Filatova–Orekhova (BFO) theory methodology and results are well known in the literature (for example, see references [18–28]). Chapters [22–24, 28] use the BFO theory in practice. The impact on capital structure decisions of the overconfidence of finance managers of family-run businesses in India has been studied in [23]. The study concludes that manager decisions about capital structure could be explained by measurable managerial characteristics. In [24], the correlation between capital structure and company risk was studied using datasets from Pakistani companies. It was shown that the role of capital structure and risk valuation is vital for the increase in the wealth of shareholders and the sustainable growth of companies. In [25], the adjusted present value method, the free cash flow (FCF) method, the flow-to-equity method, and the relationships between these methods have been considered. The authors used a stationary FCF method and the Miles and

Ezzell method instead of the Modigliani–Miller method to derive DCF valuation formulas for annuities. In [26], the influence of internal and external corporate governance mechanisms on the financial performance of banks in the MENA region is studied. It was shown that the corporate governance had positive effects on the financial indicators of banks. The energy companies' capital costs, by including an investor and market risk approach, have been evaluated in [27]. The WACC intra-industry analysis of the companies has been done. The connection of capitalization and income ability in the BRICS banking sector has been examined in [28] under the signaling theory, the bankruptcy cost theory, the agency theory, the pecking order theory, the Modigliani and Miller theory, and the general theory of the cost of capital and capital structure—the Brusov–Filatova–Orekhova (BFO) theory. Over the past 2 years, the theory of capital structure has received a new impetus. A large-scale modification of both main theories of the capital structure, BFO and MM, has been carried out and continues in order to better take into account the conditions for the real functioning of companies, such as variable income, advance income tax payments, frequent income tax payments, their combinations, etc. [1, 29, 30]. A study of different aspects of emerging markets was carried out in [31–38]. In [34], the impact of intellectual capital on firm performance within a modified and extended VAIC model has been studied.

In the near future, the authors plan to publish a large review, which will examine in detail the problems of capital structure.

15.1.8 Materials and Methods

We combine analytical and numerical methods. First, we derive formulas for the company's leveraged value, V; the leverage less value of the company, V_0; the tax shield TS, and finally the WACC in the case of variable income.

Then, using Microsoft Excel, we study the dependences of the following values: weighted average cost of capital, WACC, discount rate, WACC-g, company value, V, and the cost of equity, k_e, on the level of leverage, L, at different values of the growth rate, g. We have created a large database for sets of cost of equity k_0 and cost of debt k_d, which is available upon request.

15.2 Modification of the BFO Theory for the Case of Companies with Variable Incomes

15.2.1 The Levered Company Value, V

Below, for the first time, we generalize the modern theory of the capital cost and capital structure—the Brusov–Filatova–Orekhova theory—to the case of variable income. We start by deriving the capitalization formula for a financially dependent company, assuming that income for the period grows with the growth rate g.

When accounting for the cost of any asset being equal to the sum of discounted values of incomes generated by this asset, one could write the capitalization for a financially dependent company's V of age, n, as the following expression

$$V = \frac{\mathrm{CF}}{1+\mathrm{WACC}} + \frac{\mathrm{CF}(1+g)}{(1+\mathrm{WACC})^2} + \frac{\mathrm{CF}(1+g)^2}{(1+\mathrm{WACC})^3} + \cdots + \frac{\mathrm{CF}(1+g)^{n-1}}{(1+\mathrm{WACC})^n} \tag{15.12}$$

Here, WACC is the weighted average cost of capital, CF is an annual income of company, and (15.12) is geometric progression with denominator

$$g = \frac{(1+g)}{(1+\mathrm{WACC})} \tag{15.13}$$

Summarizing (15.12), we get the expression for the capitalization, V, of the levered company of age n

$$V = \frac{\mathrm{CF}}{1+\mathrm{WACC}} \cdot \frac{1-\left(\frac{1+g}{1+\mathrm{WACC}}\right)^n}{1-\frac{1+g}{1+\mathrm{WACC}}} = \frac{\mathrm{CF}}{\mathrm{WACC}-g} \cdot \left(1-\left(\frac{1+g}{1+\mathrm{WACC}}\right)^n\right) \tag{15.14}$$

In the perpetuity limit ($n \to \infty$), we get the following formula for the levered company value, V,

$$V = \frac{\mathrm{CF}}{\mathrm{WACC}-g} \tag{15.15}$$

This formula shows that the discount rate is WACC–g, and not WACC.

15.2.2 The Unlevered Company Value, V_0

Let us now derive the capitalization formula for a financially independent company, assuming that income for the period grows with the growth rate g.

$$V_0 = \frac{\mathrm{CF}}{1+k_0} + \frac{\mathrm{CF}(1+g)}{(1+k_0)^2} + \frac{\mathrm{CF}(1+g)^2}{(1+k_0)^3} + \cdots + \frac{\mathrm{CF}(1+g)^{n-1}}{(1+k_0)^n} \tag{15.16}$$

(15.16) is geometric progression with denominator

$$g = \frac{(1+g)}{(1+k_0)} \tag{15.17}$$

Summarizing (15.16), we get the expression for the value V_0 of unlevered company of age n

$$V_0 = \frac{\mathrm{CF}}{1+k_0} \cdot \frac{1-\left(\frac{1+g}{1+k_0}\right)^n}{1-\frac{1+g}{1+k_0}} = \frac{\mathrm{CF}}{k_0-g} \cdot \left(1-\left(\frac{1+g}{1+k_0}\right)^n\right) \tag{15.18}$$

In the perpetuity limit ($n \to \infty$), we get the following formula for the unlevered company value, V_0,

$$V_0 = \frac{\mathrm{CF}}{k_0-g} \tag{15.19}$$

This formula shows that the discount rate is $k_0 - g$, and not k_0.

15.2.3 The Tax Shield Value

The tax shield for n–years is equal to

$$(\mathrm{TS})_n = \frac{tk_dD}{1+k_d} + \frac{tk_dD}{(1+k_d)^2} + \cdots + \frac{tk_dD}{(1+k_d)^n} \tag{15.20}$$

(15.20) is geometric progression with denominator

$$g = \frac{1}{(1+k_d)} \tag{15.21}$$

Summarizing (15.20), for the tax shield, we have

$$(\mathrm{TS})_n = \frac{tk_dD}{1+k_d} \cdot \frac{1-(1+k_d)^{-n}}{1-\frac{1}{1+k_d}} = Dt(1-(1+k_d)^{-n}) \tag{15.22}$$

$$(\mathrm{TS})_n = Dt(1-(1+k_d)^{-n}) \tag{15.23}$$

Using an analog of the first theorem by Modigliani–Miller for finite time, one gets

$$V = V_0 + (\mathrm{TS})_n \tag{15.24}$$

Substituting

$$D = w_d V \tag{15.25}$$

we arrive to the following expression

$$V(1-w_d t(1-(1+k_d)^{-n})) = V_0 \tag{15.26}$$

Substituting into this equation the values of an unlevered company, V_0 (15.18) and of levered company, V, (15.14) one gets the following expression

$$\frac{\mathrm{CF}\cdot\left(1-\left(\frac{1+g}{1+\mathrm{WACC}}\right)^n\right)\cdot(1-w_d t[1-(1+k_d)^{-n}])}{\mathrm{WACC}-g} = \frac{\mathrm{CF}\cdot\left(1-\left(\frac{1+g}{1+k_0}\right)^n\right)}{(k_0-g)} \tag{15.27}$$

Dividing both parts by $(1 - w_d t(1 - (1 + k_d)^{-n}))$, we get the BFO equation for the WACC of the company with variable income

$$\frac{1-\left(\frac{1+g}{1+\mathrm{WACC}}\right)^n}{\mathrm{WACC}-g} = \frac{1-\left(\frac{1+g}{1+k_0}\right)^n}{(k_0-g)\cdot(1-w_d t[1-(1+k_d)^{-n}])} \tag{15.28}$$

This is the main theoretical result of the current chapter.

In the perpetuity limit ($n \to \infty$), we get the following equation for WACC in the case of variable income [1]

$$\mathrm{WACC}-g = (k_0-g)\cdot(1-w_d t) \tag{15.29}$$

$$\mathrm{WACC} = (k_0-g)\cdot(1-w_d t) + g \tag{15.30}$$

15.3 Results and Discussions

In order to understand the impact of a variable growth rate g on the main financial indicators of a company within the framework of the generalized theory of Brusov–Filatova–Orekhova (GBFO), we study the dependences of the following values: weighted average cost of capital, WACC, discount rate, WACC $-$ g, company value, V, and the cost of equity, k_e, on the level of leverage L at different values of the growth rate, g. We have created a large database for sets of cost of equity k_0 and cost of debt k_d, which is available upon request. To illustrate the results obtained and the conclusions drawn below, we present below the results for the following financial parameters of the company:

$k_0 = 0.18; k_d = 0.16; t = 0.2; \mathrm{CF} = 100; n = 2$ and $n = 4; g = 0.2; 0.1; 0.0; -0.1; -0.2$

Here, k_0 is the equity cost at zero leverage level; k_d is the debt cost; t is the tax on income; CF is income per period; n is the company age; L is the leverage level; and g is the growth rate.

Note that if the results for different parameters could be and are numerically different, then the qualitative effect of the variable growth rate g on the main financial indicators is similar.

15.3.1 Calculations for 2-Year Company

15.3.1.1 Calculations of Weighted Average Cost of Capital, WACC

As can be seen from Table 15.1 and Fig. 15.1, for different values of g, all curves WACC(L) start from one point (0; $k_0 = 0.18$). These curves WACC(L) demonstrate the decrease of WACC with leverage level L at all g values. The curves WACC(L) increase with growth rate, g.

The results for a 2-year company differ from the results for a perpetual limit—the theory of Modigliani and Miller [1]. In the latter case, the WACC(L) curves decrease

Table 15.1 The WACC depending on the level of leverage L at different growth rates g in Generalized Brusov–Filatova–Orekhova theory (GBFO theory)

	WACC, $n = 2$				
L	$g = 0.2$	$g = -0.1$	$g = 0$	$g = 0.1$	$g = 0.2$
0	0.1799864	0.1799858	0.179995	0.1799849	0.1799844
1	0.1583502	0.1583545	0.1591545	0.1594954	0.1597882
2	0.15084	0.1513935	0.1521805	0.1523026	0.1526805
3	0.1472744	0.1478992	0.148688	0.1489255	0.1493521
4	0.1451307	0.1457982	0.1465907	0.1468947	0.1473505
5	0.1436999	0.1443958	0.1451917	0.1455391	0.1460143
6	0.142677	0.1433932	0.1441921	0.1445699	0.145059
7	0.1419093	0.1426408	0.1434421	0.1438425	0.144342
8	0.141312	0.1420553	0.1428587	0.1432764	0.1437841
9	0.140834	0.1415867	0.1423919	0.1428234	0.1433376
10	0.1404427	0.1412033	0.1420099	0.1424527	0.1429721

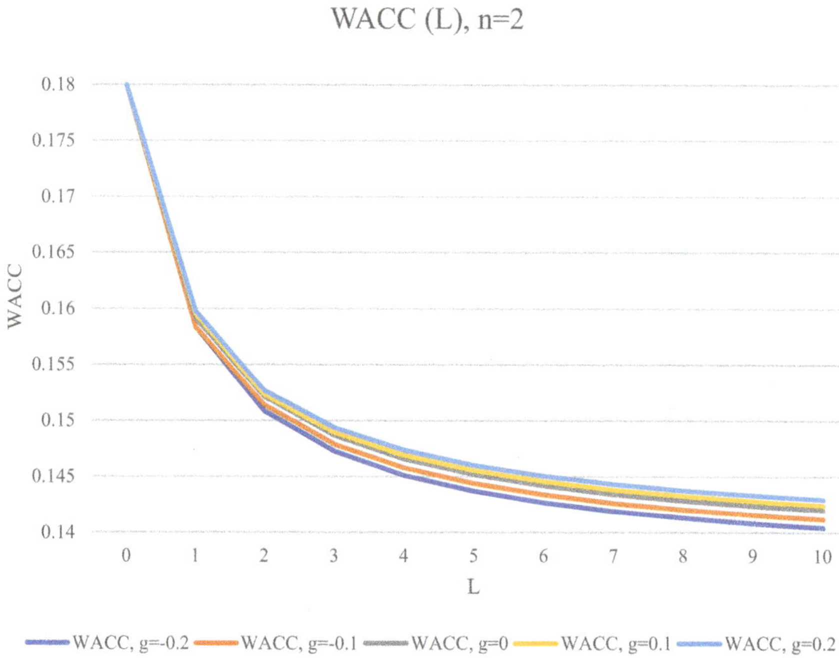

Fig. 15.1 The WACC depending on the level of leverage L at different growth rates g in Generalized Brusov–Filatova–Orekhova theory (GBFO theory)

with the level of leverage L at $g < k_0$ and increase at $g > k_0$. k_0 is the threshold value g separating the increasing WACC(L) curves from the decreasing ones, and for $g = k_0$ WACC $=$ const $= k_0$. In the first case (BFO theory), the WACC(L) curves decrease with increasing leverage L for all values of the growth rate g. The WACC(L) curves increase with the rate g both in the Brusov–Filatova–Orekhova theory and in the Modigliani and Miller theory. This is the first indication that WACC is no longer a discount rate, since it is intuitive that the discount rate must decrease in order to increase the value of company V. As we will see below, WACC–g and WACC–k_0 play the role of the discount rate.

The discount rate moves for a financially dependent company from the weighted average cost of capital, WACC, to WACC–g (where g is the growth rate), for a financially independent company from k_0 to k_0–g. As we intuitively understood above, this means that WACC and k_0 are no longer discount rates, as is the case for the classical Brusov–Filatova–Orekhova constant income theory. Below, we study the dependence of the discount rate WACC–g on the level of leverage L in the Generalized theory of Brusov–Filatova–Orekhova (the GBFO theory) at k_0=$_0$.18 and different values of g (−0.2; −0.1; 0.0; 0.1; 0.2) for a 2- and 4-year company.

Table 15.2 The discount rate WACC–g depending on the level of leverage L at different growth rates g in Generalized Brusov–Filatova–Orekhova theory (GBFO theory)

	(WACC–g), $n = 2$				
L	$g = -0.2$	$g = -0.1$	$g = 0$	$g = 0.1$	$g = 0.2$
0	0.3799864	0.2799858	0.179995	0.0799849	−0.020016
1	0.3583502	0.2583545	0.1591545	0.0594954	−0.040212
2	0.35084	0.2513935	0.1521805	0.0523026	−0.047319
3	0.3472744	0.2478992	0.148688	0.0489255	−0.050648
4	0.3451307	0.2457982	0.1465907	0.0468947	−0.052649
5	0.3436999	0.2443958	0.1451917	0.0455391	−0.053986
6	0.342677	0.2433932	0.1441921	0.0445699	−0.054941
7	0.3419093	0.2426408	0.1434421	0.0438425	−0.055658
8	0.341312	0.2420553	0.1428587	0.0432764	−0.056216
9	0.340834	0.2415867	0.1423919	0.0428234	−0.056662
10	0.3404427	0.2412033	0.1420099	0.0424527	−0.057028

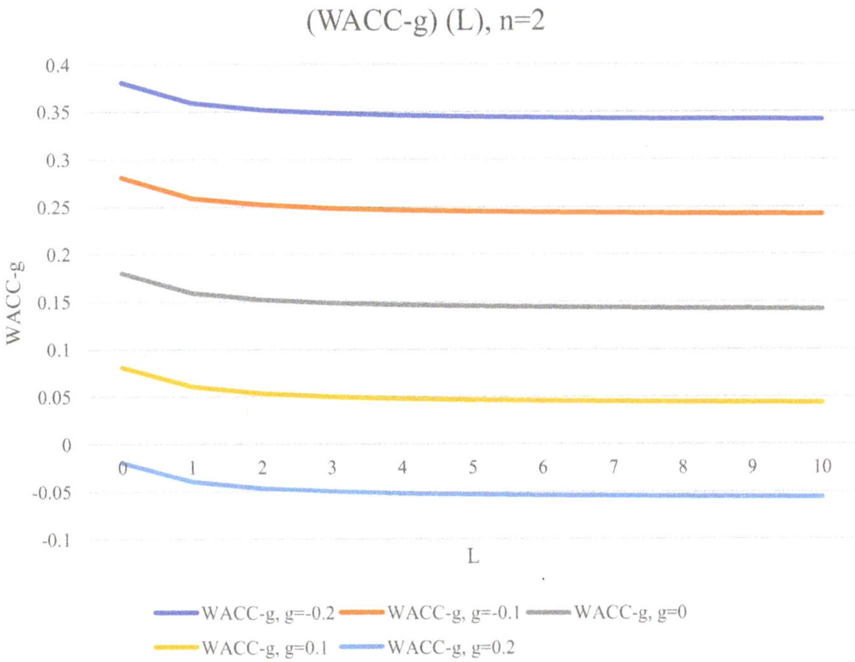

Fig. 15.2 The discount rate WACC–g depending on the level of leverage L at different growth rates g in Generalized Brusov–Filatova–Orekhova theory (GBFO theory)

15.3.1.2 Calculations of the Discount Rate, WACC–g

As can be seen from Table 15.2 and Fig. 15.2, the curves (WACC–g)(L) demonstrate the decrease of WACC–g with leverage level L at all g values. The curves (WACC–g)(L) decrease with growth rate, g.

This behavior of the (WACC–g)(L) curves can be explained as follows: all WACC(L) curves originate from the same point ($L = 0$; WACC $= 0.18$). The (WACC–g)(L) curves will be ordered as follows for $L = 0$: the larger g, the lower the starting point and hence the entire graph lies, since the curves do not intersect. As we will see below, the decrease of (WACC–g)(L) with growth rate, g, will lead to an increase of the company value, V, with g.

15.3.1.3 Calculations of the Company Value, V

It is seen from Table 15.3 and Fig. 15.3 that the company value V at fixed growth rate g increases with leverage level L in the Generalized Brusov–Filatova–Orekhova theory (GBFO theory). The company value V as well increases with growth rate g. This is a consequence of a decrease in the discount rate (WACC–g)(L) with an increase in the growth rate g.

Below, we have studied the dependence of cost of equity, k_e, on leverage level L and on growth rate, g, in the Generalized Brusov–Filatova–Orekhova theory (GBFO theory) at $k_0 = 0.18$; $k_d = 0.16$; and $g = 0;\pm0.1;\pm0.2$.

Table 15.3 The company value, V, depending on the level of leverage L at different growth rates g in Generalized Brusov–Filatova–Orekhova theory (GBFO theory)

	$V, n = 2$				
L	$g = -0.2$	$g = -0.1$	$g = 0$	$g = 0.1$	$g = 0.2$
0	142.20282	149.38493	156.56518	163.74916	170.93128
1	145.95218	153.40417	160.69451	168.06349	175.43476
2	147.29627	154.73958	162.12039	169.62647	177.07002
3	147.94237	155.41792	162.84303	170.36931	177.84511
4	148.3333	155.82837	163.27978	170.81878	178.31411
5	148.5953	156.10344	163.57228	171.12001	178.62842
6	148.78312	156.30063	163.78186	171.33596	178.85374
7	148.92435	156.44891	163.93941	171.49835	179.02318
8	149.03442	156.56448	164.06217	171.6249	179.15523
9	149.12261	156.65707	164.16051	171.7263	179.26104
10	149.19486	156.73293	164.24106	171.80938	179.34772

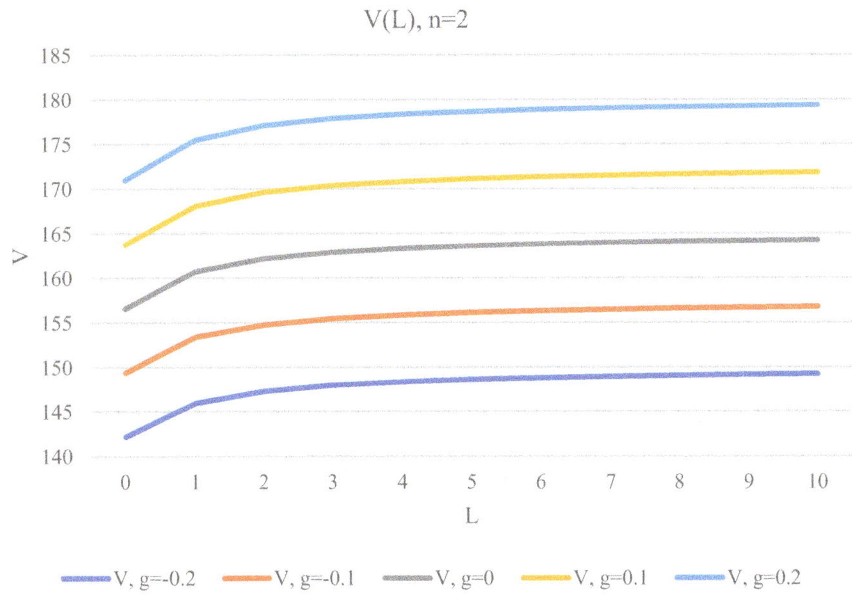

Fig. 15.3 The company value, *V*, depending on the level of leverage *L* at different growth rates *g* in Generalized Brusov–Filatova–Orekhova theory (GBFO theory)

Table 15.4 The equity cost, k_e, depending on the level of leverage *L* at different growth rates *g* in Generalized Brusov–Filatova–Orekhova theory (GBFO theory)

	k_e, $n = 2$				
L	$g = -0.2$	$g = -0.1$	$g = 0$	$g = 0.1$	$g = 0.2$
0	0.1799864	0.1799858	0.179995	0.1799849	0.1799844
1	0.1887005	0.1887091	0.1903091	0.1909909	0.1915764
2	0.19652	0.1981805	0.2005415	0.2009079	0.2020416
3	0.2050974	0.2075966	0.2107521	0.2117018	0.2134083
4	0.2136535	0.2169911	0.2209535	0.2224736	0.2247527
5	0.2221991	0.2263748	0.2311504	0.2332344	0.236086
6	0.2307387	0.2357525	0.2413446	0.243989	0.2474131
7	0.2392746	0.2451264	0.2515371	0.2547397	0.2587362
8	0.247808	0.2544978	0.2617285	0.2654878	0.2700568
9	0.2563397	0.2638674	0.271919	0.2762342	0.2813755
10	0.2648701	0.2732358	0.282109	0.2869792	0.292693

15.3.1.4 Calculations of the Equity Cost, k_e

As is seen from Table 15.4 and Fig. 15.4, the equity cost, k_e, practically linearly grows with leverage level L at all growth rate g values. The tilt angle $k_e(L)$ grows with g.

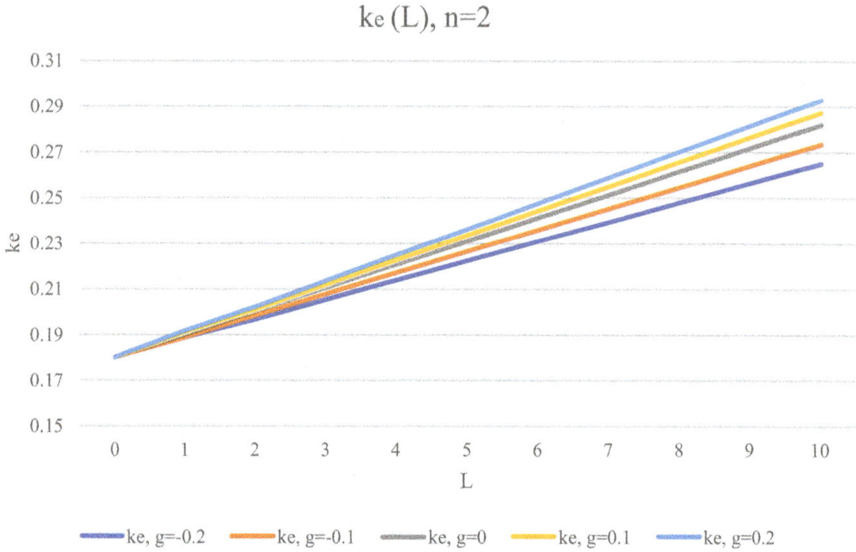

Fig. 15.4 The equity cost, k_e, depending on the level of leverage L at different growth rates g in Generalized Brusov–Filatova–Orekhova theory (GBFO theory)

15.3.2 Calculations for 4-Year Company

Below, we study the dependence of WACC, WACC–g, V, and k_e on leverage level L in the Generalized Brusov–Filatova–Orekhova theory (GBFO theory) at $k_0 = 0.18$; $k_d = 0.16$; $t = 0.2$; $g = 0.2; 0.1; 0.0; -0.1; -0.2$ for a 4-year company and compare obtained results with ones for a 2-year company (see above).

15.3.2.1 Calculations of Weighted Average Cost of Capital, WACC

From Table 15.5 and Fig. 15.5, it follows that for different values of g, all curves WACC(L) start from one point (0; $k_0 = 0.18$). These curves WACC(L) demonstrate the decrease of WACC with leverage level L at all g values. The curves WACC(L) increase with growth rate, g. In this part, the results for a 4-year company are similar to ones for a 2-year company. There is quantitative difference between them; the splitting of the curves WACC(L) Δ for $g = 0.2$ and $g = -0.2$ increases with g (at $L = 10$ $\Delta = 0.0018$ for $n = 2$, while for $n = 4$ $\Delta = 0.009$). In the perpetual case, ($n = \infty$) $\Delta = 0.073$) [1].

The results for a 4-year company, as it was seen for a 2-year company, differ from the results for the perpetual limit—the theory of Modigliani and Miller [1]; the WACC(L) curves decrease with increasing leverage L for all values of the growth rate g. As is known from [1], in the case of Modigliani and Miller, the WACC(L) curves decrease with the level of leverage L for $g < k_0$ and increase for $g > k_0$.

Table 15.5 The WACC depending on the level of leverage L at different growth rates g in Generalized Brusov–Filatova–Orekhova theory (GBFO theory)

	WACC				
L	$g = -0.2$	$g = -0.1$	$g = 0$	$g = 0.1$	$g = 0.2$
0	0.1799864	0.1799858	0.179995	0.1799849	0.1799844
1	0.1538597	0.1554667	0.1567983	0.1579005	0.1588493
2	0.1450677	0.1472039	0.1489781	0.1504635	0.1517081
3	0.1406528	0.1430555	0.1450509	0.1467206	0.1481196
4	0.1379983	0.1405609	0.1426891	0.1444691	0.145877
5	0.1362263	0.1388955	0.1411122	0.1429656	0.1445542
6	0.1349595	0.1377048	0.1399846	0.1418905	0.1434877
7	0.1340088	0.1368112	0.1391383	0.1410836	0.1427137
8	0.1332689	0.1361157	0.1384797	0.1404555	0.1421113
9	0.1326768	0.1355592	0.1379526	0.1399529	0.1416291
10	0.1321922	0.1351036	0.1375212	0.1395414	0.1412344

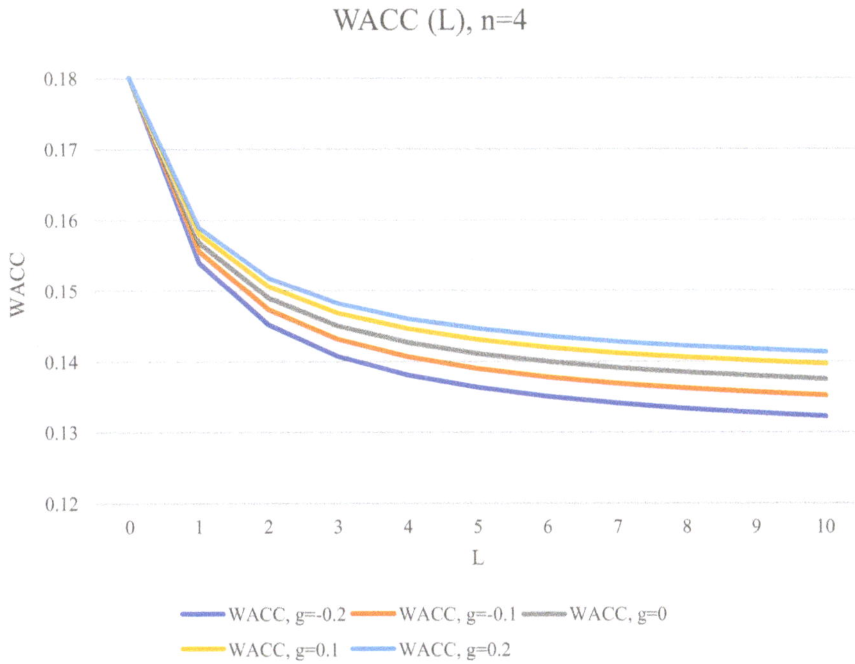

Fig. 15.5 The WACC depending on the level of leverage L at different growth rates g in Generalized Brusov–Filatova–Orekhova theory (GBFO theory)

Therefore, k_0 is the threshold value g separating the increasing WACC(L) curves from the decreasing ones, and for $g = k_0$ WACC = const = k_0. As we mentioned above, the increase in WACC(L) with the growth rate of g is the first indication that WACC is no longer a discount rate, since it is intuitive that the discount rate must decrease with growth rate g in order to increase the value of company V. As we will see below, WACC–g and WACC–k_0 play the role of the discount rates and satisfy the above condition.

15.3.2.2 Calculations of the Discount Rate, WACC–g

Table 15.6 and Fig. 15.6 show all the curves (WACC–g)(L) with leverage level L at all g values. The (WACC–g) values at fixed leverage level L decrease with growth rate g. This means that WACC–g is a suitable candidate for the discount rate.

As in the case of the 2-year company, the behavior of (WACC–g)(L) with g growth can be explained as follows: all WACC(L) curves originate from the same point ($L = 0$; WACC $= 0.18$). The (WACC–g)(L) curves will be ordered as follows for $L = 0$: the larger g, the lower the starting point and hence the entire graph lies, since the curves do not intersect. As we will see below, the decrease of (WACC–g)(L) with growth rate, g, will lead to an increase of the company value, V, with g.

15.3.2.3 Calculations of the Company Value, V

As can be seen from Table 15.7 and Fig. 15.7, the company value V at fixed growth rate g increases with leverage level L in Generalized Brusov–Filatova–Orekhova theory (GBFO theory). The company value V increases at fixed leverage level L with growth rate g at fixed as well. This is the consequence of a decrease in the discount rate (WACC–g)(L) with an increase in the growth rate g. Comparing with the results for the 2-year-old company, we see that the value of the company V increases with the age of the company: we have a range from 149 to 157 with $L = 1$ for g from –

Table 15.6 The discount rate, WACC–g, depending on the level of leverage L at different growth rates g in Generalized Brusov–Filatova–Orekhova theory (GBFO theory)

	(WACC–g), $n = 4$				
L	$g = -0.2$	$g = -0.1$	$g = 0$	$g = 0.1$	$g = 0.2$
0	0.3799864	0.2799858	0.179995	0.0799849	−0.020016
1	0.3538597	0.2554667	0.1567983	0.0579005	−0.041151
2	0.3450677	0.2472039	0.1489781	0.0504635	−0.048292
3	0.3406528	0.2430555	0.1450509	0.0467206	−0.05188
4	0.3379983	0.2405609	0.1426891	0.0444691	−0.054123
5	0.3362263	0.2388955	0.1411122	0.0429656	−0.055446
6	0.3349595	0.2377048	0.1399846	0.0418905	−0.056512
7	0.3340088	0.2368112	0.1391383	0.0410836	−0.057286
8	0.3332689	0.2361157	0.1384797	0.0404555	−0.057889
9	0.3326768	0.2355592	0.1379526	0.0399529	−0.058371
10	0.3321922	0.2351036	0.1375212	0.0395414	−0.058766

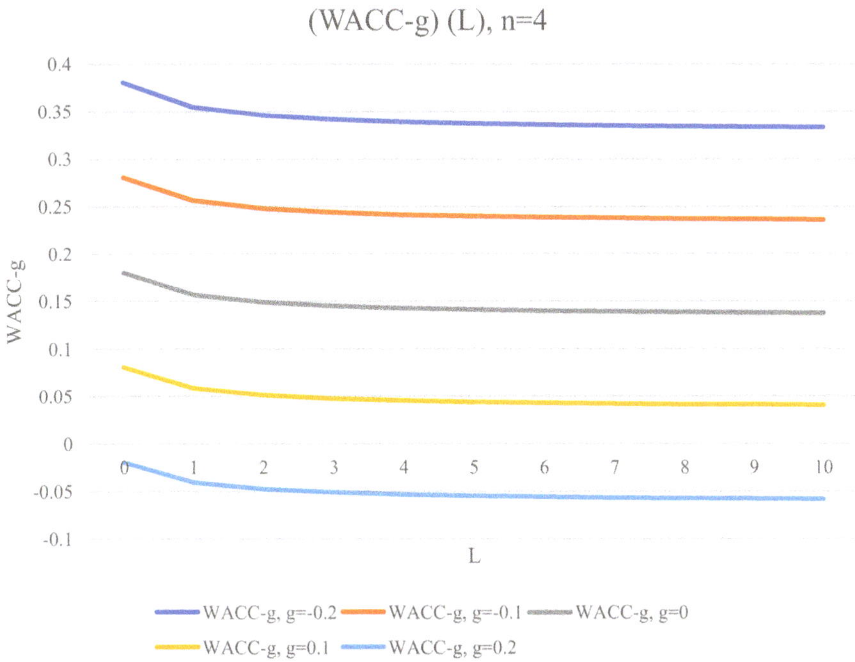

Fig. 15.6 The discount rate, WACC–g, depending on the level of leverage L at different growth rates g in Generalized Brusov–Filatova–Orekhova theory (GBFO theory)

Table 15.7 The company value, V, depending on the level of leverage L at different growth rates g in Generalized Brusov–Filatova–Orekhova theory (GBFO theory)

	$V, n = 4$				
L	$g = -0.2$	$g = -0.1$	$g = 0$	$g = 0.1$	$g = 0.2$
0	207.56615	236.28862	269.00881	306.05136	347.71061
1	217.29735	247.3597	281.61599	320.39442	363.9946
2	220.75339	251.2914	286.08901	325.47139	369.77167
3	222.5253	253.30558	288.37987	328.07604	372.72979
4	223.60263	254.53001	289.7723	329.65911	374.59742
5	224.32687	255.35305	290.70818	330.72309	375.70606
6	224.84716	255.94428	291.38042	331.48732	376.60369
7	225.23902	256.38954	291.88667	332.06283	377.25722
8	225.54478	256.73695	292.28165	332.51185	377.7671
9	225.79	257.01557	292.59842	332.87195	378.176
10	225.99105	257.244	292.85811	333.16716	378.51122

0.2 to 0.2 for the 2-year-old company and a range from 217 to 364 with $L = 1$ for g from -0.2 to 0.2 for a 4 year old company. This is the obvious conclusion, because it is well known that the value of any asset (company, stock, bond, etc.) is equal to the sum of the discounted returns generated by this asset. Since this value is

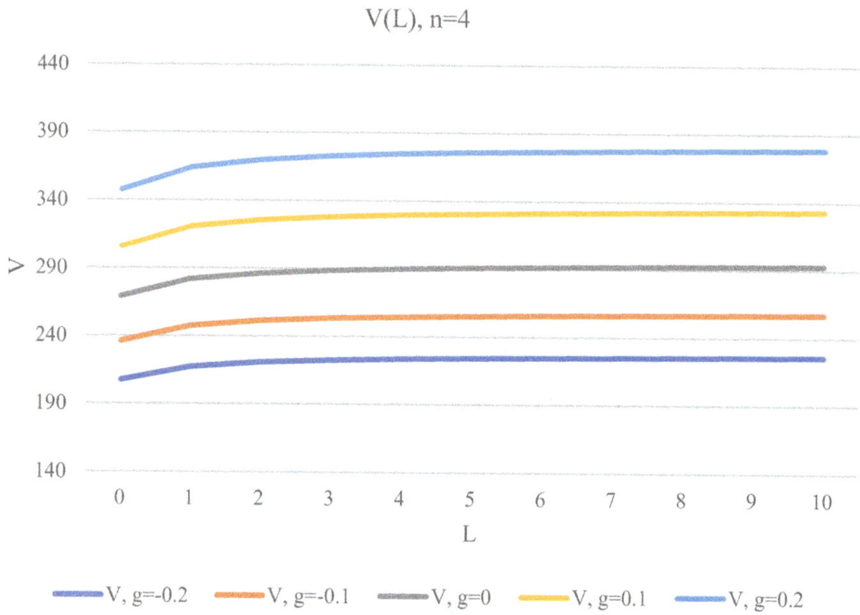

Fig. 15.7 The company value, V, depending on the level of leverage L at different growth rates g in Generalized Brusov–Filatova–Orekhova theory (GBFO theory)

proportional to the lifetime of this asset, the capitalization of the company will grow with its age.

The dependence of the cost of equity k_e on the level of leverage L in the Generalized theory of Brusov–Filatova–Orekhova (GBFO theory) was studied below for a 4-year company with growth rates $g = 0$; ± 0.1; and ± 0.2.

15.3.2.4 Calculations of the Cost of Equity k_e

From Table 15.8 and Fig. 15.8, it can be seen that the cost of equity k_e increases practically linearly with leverage level L at all growth rates g (except for $g = -0.2$ where we see a decrease in k_e with leverage level L). The slope angle $k_e(L)$ increases with g. This should change the dividend policy of an enterprise with variable income since, economically, the reasonable amount of dividends is equal to the cost of equity.

However, the biggest change in the company's dividend policy is related to the discovery of a qualitatively new effect in corporate finance: at a rate $g < g^*$, the slope of the $k_e(L)$ curve turns out to be negative (one can observe this effect here for $g = -0.2$ where a decrease in k_e with leverage level L takes place). This effect, which is absent in the classical Modigliani–Miller theory and the classical Brusov–Filatova–Orekhova theory with constant income, exists in the Modigliani–Miller theory with

Table 15.8 The equity cost, k_e, depending on the level of leverage L at different growth rates g in Generalized Brusov–Filatova–Orekhova theory (GBFO theory)

	k_e, $n = 4$				
L	$g = -0.2$	$g = -0.1$	$g = 0$	$g = 0.1$	$g = 0.2$
0	0.1799864	0.1799858	0.179995	0.1799849	0.1799844
1	0.1797193	0.1829333	0.1855967	0.187801	0.1896985
2	0.1792032	0.1856116	0.1909343	0.1953906	0.1991244
3	0.1786111	0.1882219	0.1962038	0.2028824	0.2084783
4	0.1779915	0.1908046	0.2014454	0.2103453	0.2173852
5	0.1773581	0.1933733	0.2066729	0.2177938	0.2273252
6	0.1767166	0.1959339	0.2118923	0.2252338	0.2364136
7	0.1760702	0.1984895	0.2171066	0.2326686	0.2457093
8	0.1754204	0.2010417	0.2223176	0.2400999	0.2550015
9	0.1747682	0.2035915	0.2275261	0.2475287	0.2642911
10	0.1741143	0.2061396	0.2327328	0.2549558	0.2735789

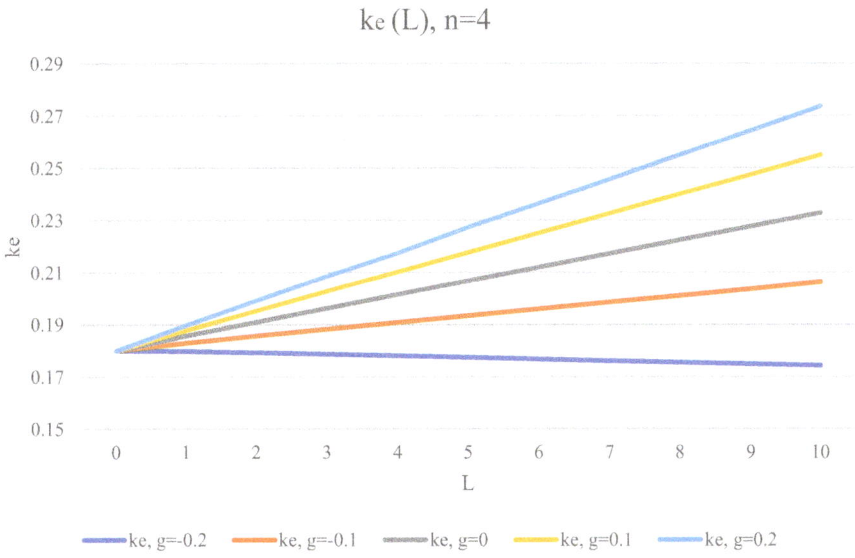

Fig. 15.8 The equity cost, k_e, depending on the level of leverage L at different growth rates g in Generalized Brusov–Filatova–Orekhova theory (GBFO theory)

variable income and in the Brusov–Filatova–Orekhova theory with variable income at a certain age of the company, n, which exceeds some cutoff age value, n^*.

The latter effect is similar to a qualitatively new effect in corporate finance, discovered by Brusov–Filatova–Orekhova within the framework of the BFO theory [7, 8]: anomalous dependences of the cost of equity k_e on the leverage level L when income tax T exceeds a certain value T^*. This discovery also significantly changes the principles of the company's dividend policy.

15.3.3 Comparison with the Theory of Modigliani and Miller with Variable Income

The main difference between the Modigliani–Miller (MM) theory and the Brusov–Filatova–Orekhova (BFO) theory is that the latter describes companies of an arbitrary age, whereas the former represents the eternal limit of the BFO theory and is valid only for perpetual companies (with an infinite lifetime). If there is no time factor in the MM theory, then within the framework of the modern BFO theory, it is possible to analyze the financial condition of a company of any age; in this chapter, this is done for two company ages—2 and 4 years.

Since the MM theory is the eternal BFO limit, it is important to note that some of the results, such as the change in the discount rate from WACC to WACC–g, etc., are general, and this emphasizes the correctness of both theories and the connection between them as a connection between a general theory and its limiting case.

Along with the fundamental differences between the results of the two theories indicated above, there are also qualitative and quantitative differences. One of them is the next. The BFO results for a 4-year company, as it was seen for a 2-year company as well, differ from the results for the perpetual limit—the theory of Modigliani and Miller [1], where the WACC(L) curves decrease with increasing leverage L for all values of the growth rate g. As is known from [1], in the case of Modigliani and Miller, the WACC(L) curves decrease with the level of leverage L for $g < k_0$ and increase for $g > k_0$.Therefore, k_0 is the threshold value g separating the increasing WACC(L) curves from the decreasing ones, and for $g = k_0$ WACC = const = k_0 (see Fig. 15.9 below).

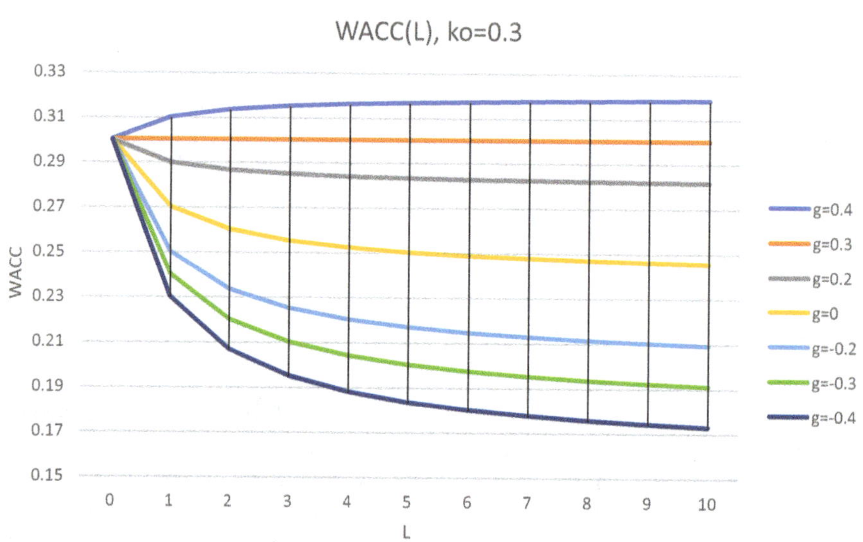

Fig. 15.9 WACC depending on leverage level L in Generalized Modigliani–Miller theory (GMM theory) at k_0=0.3 and $g = 0;\pm0.2;\pm0.3;\pm0.4$ (from [1])

We would like to emphasize an important observation. If in the classical versions of the Brusov–Filatova–Orekhova (BFO) theory and its perpetual limit, the theory of Nobel laureates Modigliani and Miller, where the case of constant income was considered, and where the gap between these two theories is huge (many qualitative effects that take place in the first theory, missing in the second), when taking into account the variable income, some effects of the BFO theory also take place in the Modigliani–Miller theory. This means that taking into account some effects that are present in economic practice (for example, variable income) brings both theories closer, and even the Modigliani–Miller theory, with all its many limitations, becomes more applicable in economic practice. However, it should be remembered that the Modigliani–Miller theory is only true for perpetual companies, whereas the BFO theory is valid for companies of any age, and from this point of view, they never coincide.

15.4 Conclusions

The Brusov–Filatova–Orekhova (BFO) theory of capital cost and capital structure as well as the theory by Nobel Prize winners Modigliani and Miller (perpetuity limit of BFO theory) consider the case of constant income, whereas in practice, the income of a company is, of course, variable. Recently, we have generalized the latter for the case of variable income, and here we have generalized for the first time the Brusov–Filatova–Orekhova theory for the case of variable income.

This generalization significantly expands the applicability of this modern capital structure theory, which is valid for companies of any age, and in practice for corporate finance, business valuation, investments, banking, ratings, etc.

We have derived the generalized BFO formula for WACC and this consists of a main theoretical result of a current chapter.

From this formula and as well from using this formula in MS Excel, we show that the role of the discount rate shifts from WACC to WACC–g (where g is the growth rate) for financially dependent companies and from k_0 to k_0–g (for financially independent companies). Whereas the WACC increases with g, the actual discount rates WACC–g and k_0–g decrease with g and, accordingly, the company value, V, increases with g. For the cost of equity k_e, the slope of the curve $k_e(L)$ increases with g. Since the cost of equity determines the economically justified amount of dividends, this should change the company's dividend policy. It turns out that at the rate $g < g^*$, the slope of the curve $k_e(L)$ becomes negative, which can significantly change the company's dividend policy principles. This means a qualitatively new effect on discoveries in corporate finance.

The novelty of the work is the generalization of the theory of Brusov–Filatova–Orekhova to the case of variable income, the derivation of generalized BFO formulas for the weighted average cost of capital, WACC, cost of equity, k_e, company value, V, and the use of these formulas to study the influence of the growth rate g on the dependence of the main financial performance of the company on debt financing.

The importance of the current consideration is due to a couple of points:

- Gives an idea of the behavior of the main financial indicators of the company with a change in the growth rate g for both cases—an increase in g and a decrease in g;
- Creates a developed methodology that allows the researcher to analyze the main financial indicators of the company (capital costs, company value, etc.) for the actual conditions of the company's functioning.

Limitations of the study: a generalization of the BFO theory was carried out for the case of paying income tax at the end of reporting periods. However, in practice, these payments may also be made in advance. This determines the further direction of the study: consideration of the case of variable income when income tax is paid in advance.

References

1. Brusov, P. N., Filatova, T. V., Orekhova, N. P., Kulik, V. L., Chang, S. I., & Lin, Y. C. G. (2021). Generalization of the Modigliani–Miller theory for the case of variable income. *Mathematics, 9*, 1286. https://doi.org/10.3390/math9111286
2. Modigliani, F., & Miller, M. (1958). The cost of capital, corporate finance, and the theory of investment. *The American Economic Review, 48*, 261–297.
3. Modigliani, F., & Miller, M. (1963). Corporate income taxes and the cost of capital: A correction. *The American Economic Review, 53*, 147–175.
4. Modigliani, F., & Miller, M. (1966). Some estimates of the cost of capital to the electric utility industry 1954–1957. *The American Economic Review, 56*, 333–391.
5. Hamada, R. (1969). Portfolio analysis, market equilibrium, and corporate finance. *The Journal of Finance, 24*, 13–31.
6. Miller, M. (1977). Debt and taxes. *The Journal of Finance, 32*, 261–275. https://doi.org/10.1111/j.1540-6261.1977.tb03267.x
7. Brusov, P., Filatova, T., Orehova, N., & Eskindarov, M. (2018). *Modern corporate finance, investments, taxation and ratings* (2nd ed., pp. 1–571). Springer Nature Publishing.
8. Brusov, P., Filatova, T., & Orehova, N. (2021). *Ratings: Critical analysis and new approaches of quantitative and qualitative methodology* (pp. 1–369). Springer Nature Publishing. https://www.springer.com/de/book/9783030562427
9. Farber, A., Gillet, R., & Szafarz, A. (2006). A general formula for the WACC. *International Journal of Business, 11*, 211–218.
10. Fernandez, P. (2006). A general formula for the WACC: A comment. *International Journal of Business, 11*, 219.
11. Berk, J., & DeMarzo, P. (2007). *Corporate finance*. Pearson–Addison Wesley.
12. Harris, R., & Pringle, J. (1985). Risk–Adjusted discount rates—Extension form the average–risk case. *Journal of Financial Research, 8*, 237–244.
13. Frank, M., & Goyal, V. (2009). Capital structure decisions: Which factors are reliably important? *Financial Management, 38*, 1–37. https://doi.org/10.1111/j.1755-053X.2009.01026.x
14. Sheridan, T., & Wessels, R. (1988). The determinants of capital structure choice. *The Journal of Finance, 43*, 1–19.
15. Harry, D. A., & Masulis, R. W. (1980). Optimal capital structure under corporate and personal taxation. *Journal of Financial Economics, 8*, 3–29.
16. Bradley, M., Jarrell, G. A., & Kim, E. H. (1984). On the existence of an optimal capital structure: Theory and evidence. *The Journal of Finance, 39*, 857–878.

17. Graham, J. R. (2003). Taxes and corporate finance: A review. *Review of Financial Studies, 16*, 1075–1129.
18. Dimitropoulos, P. (2014). Capital structure and corporate governance of soccer clubs: European evidence. *Management Research Review, 37*, 658–678.
19. Luiz, K., & Cruz, M. (2015). The relevance of capital structure on firm performance: A multivariate analysis of publicly traded Brazilian companies. *REPeC Brasília, 9*, 384–401.
20. Barbi, M. (2011). On the risk–neutral value of debt tax shields. *Applied Financial Economics, 22*, 251–258.
21. Franc-Dąbrowska, J., Mądra-Sawicka, M., & Milewska, A. (2021). Energy sector risk and cost of capital assessment—Companies and investors perspective. *Energies, 14*, 1615.
22. Angotti, M., de Lacerda Moreira, R., Bernardes, H., do Nascimento, J., & Neto de Almeida Bispo, O. (2018). Analysis of an equity investment strategy based on accounting and financial reports in Latin American markets. *Reficont, 5*, 22–40.
23. Vergara-Novoa, C., Sepúlveda-Rojas, J. P., Miguel, D. A., & Nicolás, R. (2018). Cost of capital estimation for highway concessionaires in Chile. *Journal of Advanced Transportation, 2018*, 2153536.
24. Mundi, H. S., Kaur, P., & Murty, R. L. N. (2022). A qualitative inquiry into the capital structure decisions of overconfident finance managers of family–owned businesses in India. *Qualitative Research in Financial Markets, 14*, 357–379.
25. Becker, D. M. (2022). Getting the valuation formulas right when it comes to annuities. *Managerial Finance, 48*, 470.
26. El-Chaarani, H., Abraham, R., & Skaf, Y. (2022). The impact of corporate governance on the financial performance of the banking sector in the MENA (Middle Eastern and North African) region: An immunity test of banks for COVID–19. *Journal of Risk and Financial Management, 15*, 82.
27. Sadiq, M., Alajlani, S., Hussain, M. S., Bashir, F., & Chupradit, S. (2022). Impact of credit, liquidity, and systematic risk on financial structure: Comparative investigation from sustainable production. *Environmental Science and Pollution Research International, 29*, 20963–20975.
28. Singhal, N., Goyal, S., Sharma, D., Kumari, S., & Nagar, S. (2022). Capitalization and profit ability: Applicability of capital theories in BRICS banking sector. *Future Business Journal, 8*, 30. https://doi.org/10.1186/s43093-022-00140-w
29. Brusov, P., & Filatova, T. (2022). Influence of method and frequency of profit tax payments on company financial indicators. *Mathematics, 10*, 2479. https://doi.org/10.3390/math10142479
30. Brusov, P., & Filatova, T. (2022). Veniamin Kulik benefits of advance payments of tax on profit: Consideration within the Brusov–Filatova–Orekhova (BFO) theory. *Mathematics, 10*, 2015. https://doi.org/10.3390/math10122013
31. Durana, P., Valaskova, K., Blazek, R., & Palo, J. (2022). Metamorphoses of earnings in the transport sector of the V4 region. *Mathematics, 10*, 1204. https://doi.org/10.3390/math10081204
32. Virglerova, Z., Ivanova, E., Dvorsky, J., Belas, J., & Krulický, T. (2021). Selected factors of internationalisation and their impact on the SME perception of the market risk. *Oeconomia Copernicana, 12*, 1011–1032. https://doi.org/10.24136/oc.2021.033
33. Durana, P., Ginevicius, R., Urbanski, M., Podhorska, I., & Tumpach, M. (2021). Parallels and differences in earnings management of the Visegrad Four and the Baltics. *Journal of Competitiveness, 13*, 39–55. https://doi.org/10.7441/joc.2021.03.03
34. Xu, J., & Liu, F. (2020). The impact of intellectual capital on firm performance: A modified and extended VAIC model. *Journal of Competitiveness, 12*, 161–176. https://doi.org/10.7441/joc.2010.01.10
35. Valaskova, K., Androniceanu, A.-M., Zvarikova, K., & Olah, J. (2021). Bonds between earnings management and corporate financial stability in the context of the competitive ability of enterprises. *Journal of Competitiveness, 13*, 167–184. https://doi.org/10.7441/joc.2021.04.10

36. Białek-Jaworska, A., & Klapkiv, L. (2021). Does withholding tax on interest limit international profit–shifting by FDI? *Equilibrium. Quarterly Journal of Economics and Economic Policy, 16*, 11–44. https://doi.org/10.24136/eq.2021.001
37. Valaskova, K., Kliestik, T., & Gajdosikova, D. (2021). Distinctive determinants of financial indebtedness: Evidence from Slovak and Czech enterprises. *Equilibrium. Quarterly Journal of Economics and Economic Policy, 16*, 639–659. https://doi.org/10.24136/eq.2021.023
38. Valaskova, K., Gavurova, B., Durana, P., & Kovacova, M. (2020). Alter ego only four times? The case study of business profits in the Visegrad Group. *E a M: Ekonomie a Management, 23*, 101–119. https://doi.org/10.15240/tul/001/2020-3-007

Chapter 16
BFO Theory with Variable Profit in Case of Advance Payments of Tax on Profit

16.1 Introduction

Two main capital structure theories—Brusov–Filatova–Orekhova (BFO) and its perpetuity limit—theory by Nobel Prize winners Modigliani and Miller—consider the companies with constant profit, while in practice, profit of the company is, of cause, variable. Recently the Modigliani–Miller (MM) theory has been generalized for the case of variable profit [1], and in the current chapter we have generalized for the first time BFO theory for the companies with variable profit and advance tax on profit payments. After such a generalization, the applicability of the BFO theory, which is valid for companies of any age, significantly expands in practice, in particular, in corporate finance, in investments, in business valuation, in banking, in ratings, etc.

16.1.1 Review of Literature

The capital structure is the ratio between the equity and debt value. Whether the capital structure affects main financial parameters and, if so, how?

The main tasks solved by the company's management is to determine the optimal capital structure, in which the cost of raising capital is minimal, and the value of the company V is maximum.

The first quantitative work on this topic was that of Modigliani and Miller (1958) [2]. Before this, there was the so-called traditional (empirical) approach.

P. Brusov, T. Filatova, *Business Valuation and Company Value*, Contributions to Finance and Accounting, https://doi.org/10.1007/978-3-031-90847-7_16

16.1.2 The Basis of the Traditional Approach (TA)

In TA, WACC and company value V depend on the level of leverage L (capital structure). Debt is cheaper than equity, because the former has less risk, since in the event of bankruptcy, the claims of creditors are satisfied before those of the shareholders.

Thus, increasing the share of cheaper debt capital in the total capital structure to a limit that does not violate financial stability and does not increase the risk of bankruptcy reduces WACC and increases the value of the company, V.

A further increase in debt financing leads to a violation of financial stability and to the risk of bankruptcy increase. WACC goes up, while company capitalization, V, goes down. The optimal capital structure is formed as a result of competition between the advantages of debt financing with a low level of leverage and its disadvantages with a high level of leverage. The trade-off theory has come to similar conclusions.

16.1.3 Modigliani–Miller Theory

16.1.3.1 Modigliani–Miller Theory Without Taxes

Modigliani and Miller [2], under numerous restrictions, came to the conclusion that, without taxes, the value of a company is constant and equal to

$$V = V_0 = \frac{\text{EBIT}}{k_0}, \tag{16.1}$$

where EBIT is earnings before interest and taxes, k_0 is the discount rate. This contradicts the findings of TA.

From (16.1) one obtains for WACC:

$$\text{WACC} = k_0. \tag{16.2}$$

k_0 is the equity cost at $L = 0$.

Using

$$\text{WACC} = k_0 = k_e w_e + k_d w_d. \tag{16.3}$$

and (16.1) we get for equity cost, k_e

$$k_e = \frac{k_0}{w_e} - k_d \frac{w_d}{w_e} = \frac{k_0(S+D)}{S} - k_d \frac{D}{S} = k_0 + (k_0 - k_d)\frac{D}{S} = k_0 + (k_0 - k_d)L, \tag{16.4}$$

where, L stands for leverage level, D is debt capital value of the company; WACC stands for weighted average cost of capital; S stands for equity value; k_d and wd stands for the debt cost and share; k_e and w_e stands for the equity capital cost and share. From (16.4), it follows that the equity cost increases linearly with the level of leverage.

16.1.3.2 Modigliani–Miller Theory with Taxes

Taking into account the income tax

In 1963, Modigliani and Miller [3, 4], accounting the tax on profit, have obtained for the levered company value, V,

$$V = V_0 + D \cdot t \tag{16.5}$$

where D stands for debt value, V_0 stands for the unlevered company value, and T stands for the tax on income rate.

From (16.5), one gets

$$\text{WACC} = k_0 \cdot (1 - w_d t) \tag{16.6}$$

One gets from (16.6) for equity cost k_e

$$k_e = k_0 + L \cdot (k_0 - k_d)(1 - t) \tag{16.7}$$

Formula (16.7) (MM with taxes) differs from formula (16.4) (MM without taxes) by the factor (1–t) (tax corrector). It is less than unity, so the $k_e(L)$ curve slope decreases when taxes are included.

16.1.4 Unification of Capital Asset Pricing Model (CAPM) with Modigliani–Miller Model

The Modigliani–Miller theory, with accounting taxes has been united with CAPM (Capital Asset Pricing Model) in 1961 by Hamada [5]. For the cost of equity of leveraged company, the below formula has been derived

$$k_e = k_F + (k_M - k_F)b_U + (k_M - k_F)b_U\frac{D}{S}(1 - T), \tag{16.8}$$

Here b_U is the β–coefficient of the unlevered company. First term represents risk-free profitability k_F, second term—business risk premium, $(k_M - k_F)b_U$, and third term—financial risk premium $(k_M - k_F)b_U\frac{D}{S}(1 - T)$.

In the case of an unlevered company ($D = 0$), the financial risk (the third term) is zero, and its shareholders receive only a business risk premium.

16.1.5 Miller Model

Corporate and individual taxes were taken into account by Miller [6], and the following formula was obtained for the value of a company without borrowed funds, V_U,

$$V_U = \frac{\text{EBIT}(1 - T_C)(1 - T_S)}{k_0}. \tag{16.9}$$

Here T_C stands for the corporate tax on income rate, T_S stands for the tax rate on profits of an individual investor from his ownership by stock of corporation, T_D–tax rate on interest profits from the provision of investor–individuals of credits to other investors and companies. A factor $(1 - T_S)$ accounts the individual taxes.

16.1.6 Brusov–Filatova–Orekhova (BFO) Theory

One of the most important limitations of the Modigliani–Miller theory, removed in 2008 by Brusov–Filatova–Orekhova [7, 8], is the perpetual nature of the company. The authors generalized the Modigliani–Miller theory for companies of arbitrary age and obtained the famous BFO formula for WACC.

$$\frac{1 - (1 + \text{WACC})^{-n}}{\text{WACC}} = \frac{1 - (1 + k_0)^{-n}}{k_0[1 - w_d T(1 - (1 + k_d)^{-n})]} \tag{16.10}$$

To get Modigliani–Miller limit, one should substitute $n \to \infty$.

A number of new effects, obtained in Brusov–Filatova–Orekhova theory [7, 8], are absent in Modigliani–Miller theory [1–3].

The BFO theory destroyed some of the main principles of financial management that have existed for many decades: among them, one of the cornerstones of the formation of an optimal capital structure—trade-off theory— the failure of which was proved by Brusov–Filatova–Orekhova [7, 8].

16.1.7 Alternative Expression for WACC

An alternative formula for the WACC, different from that of Modigliani–Miller has been derived in [9–12] from the WACC definition and the balance identity (see [9]):

$$\mathrm{WACC} = k_0(1 - w_d T) - k_d t w_d + k_{\mathrm{TS}} t w_d, \tag{16.11}$$

where k_O, k_d and k_{TS} are the expected returns, respectively, on the unlevered company, the debt and the tax shield.

Some additional conditions are required in Eq. (16.11) for practical applicability. If the WACC is constant over time, as it stated in [9], the levered company capitalization is found by discounting with the WACC of the unlevered company.

In textbooks [11, 12], formulas for the special cases, where the WACC is constant, could be found.

In 1963, Modigliani and Miller assumed that the debt value D is constant. Then, as the expected after-tax cash-flow of the unlevered firm is fixed, V_0 is constant as well. By assumption, $k_{\mathrm{TS}} = k_D$ and the value of the tax shield is $\mathrm{TS} = tD$. Thus, the capitalization of the company V is a constant and the alternative formula (16.11) becomes a formula for a constant WACC:

$$\mathrm{WACC} = k_0(1 - w_d T)$$

Because the debt k_d and the tax shield k_{TS} have debt nature, it seems reasonable that the expected returns on them are equals as suggested by the "classical" Modigliani–Miller (MM) theory, which has been modified by Brusov et al. [1] for cases of practical meanings.

16.1.8 Trade-Off Theory

The world famous trade-off theory was the cornerstone for many decades and is popular now (see, for example [13, 14]).

However, in 2013, the trade–off theory's bankruptcy has been proved by Brusov et al. [7, 8]. The risky debt financing does not lead to the WACC growth, and WACC still decreases with leverage. Thus the minimum in the WACC(L) curve and of maximum in the $V(L)$ are absent. This means that within the world-famous trade-off theory, the optimal capital structure is absent. Brusov et al. [7, 8] gave an explanation for this fact.

Modigliani–Miller stressed the importance of the tax shield. In [14–17], the study of the tax shield was continued, including non-debt tax shield [17]. It was shown, that the effects of substitution between debt and non-debt tax shields are suppressed by a positive relation of this type.

The Brusov–Filatova–Orekhova (BFO) theory, its methodology, and results are widely known (see, for example, [18–27]). A lot of authors of [22–24] use the BFO theory in practice.

16.2 Modification of the Brusov–Filatova–Orekhova (BFO) Theory to the Case of Companies with Variable Incomes and Advance Payments of Tax on Profit

16.2.1 The Financially Dependent Company Value, V

Below, for the first time, we generalize the BFO theory for the case of variable profit. Assuming that income for the period grows with rate g, we derive the formula for a financially dependent company value.

$$V = \frac{\mathrm{CF}}{1 + \mathrm{WACC}} + \frac{\mathrm{CF}(1+g)}{(1+\mathrm{WACC})^2} + \frac{\mathrm{CF}(1+g)^2}{(1+\mathrm{WACC})^3} + \cdots + \frac{\mathrm{CF}(1+g)^{n-1}}{(1+\mathrm{WACC})^n} \tag{16.12}$$

where CF is an annual profit of company.

Equation (16.12) is geometric progression with denominator

$$g = \frac{(1+g)}{(1+\mathrm{WACC})}. \tag{16.13}$$

Summing (16.12), one gets for V

$$V = \frac{\mathrm{CF}}{1+\mathrm{WACC}} \cdot \frac{1 - \left(\frac{1+g}{1+\mathrm{WACC}}\right)^n}{1 - \frac{1+g}{1+\mathrm{WACC}}} = \frac{\mathrm{CF}}{\mathrm{WACC} - g} \cdot \left(1 - \left(\frac{1+g}{1+\mathrm{WACC}}\right)^n\right) \tag{16.14}$$

In perpetuity (MM) limit ($n \to \infty$) one gets for V,

$$V = \frac{\mathrm{CF}}{\mathrm{WACC} - g} \tag{16.15}$$

This means that WACC–g plays the role of the discount rate, not WACC.

16.2.2 The Value of a Financially Independent Company, V_0

Let us, assuming that profit for the period grows with the rate g, derive formula for a financially independent company value V_0,

$$V_0 = \frac{\mathrm{CF}}{1+k_0} + \frac{\mathrm{CF}(1+g)}{(1+k_0)^2} + \frac{\mathrm{CF}(1+g)^2}{(1+k_0)^3} + \cdots + \frac{\mathrm{CF}(1+g)^{n-1}}{(1+k_0)^n} \tag{16.16}$$

(16.16) is geometric progression with denominator

$$g = \frac{(1+g)}{(1+k_0)} \tag{16.17}$$

Summing (16.16), one gets for a financially independent company value, V

$$V_0 = \frac{\mathrm{CF}}{1+k_0} \cdot \frac{1-\left(\frac{1+g}{1+k_0}\right)^n}{1-\frac{1+g}{1+k_0}} = \frac{\mathrm{CF}}{k_0-g} \cdot \left(1-\left(\frac{1+g}{1+k_0}\right)^n\right). \tag{16.18}$$

In perpetuity (MM) limit ($n \to \infty$) one gets for a financially independent company value, V_0,

$$V_0 = \frac{\mathrm{CF}}{k_0-g}. \tag{16.19}$$

This formula shows, that discount rate is k_0–g, and not k_0.

16.2.3 The Tax Shield Value

The tax shield for n–year company for advance payments of tax on income is equal

$$(\mathrm{TS})_n = tk_dD + \frac{tk_dD}{1+k_d} + \cdots + \frac{tk_dD}{(1+k_d)^{n-1}} \tag{16.20}$$

(16.20) is geometric progression with denominator

$$g = \frac{1}{(1+k_d)} \tag{16.21}$$

Summing (16.20), one gets for tax shield

$$(\mathrm{TS})_n = tk_dD \cdot \frac{1-(1+k_d)^{-n}}{1-\frac{1}{1+k_d}} = Dt(1-(1+k_d)^{-n}) \cdot (1+k_d) \tag{16.22}$$

$$(\mathrm{TS})_n = Dt(1-(1+k_d)^{-n}) \cdot (1+k_d) \tag{16.23}$$

Generalizing the first Modigliani–Miller theorem for finite company age, we get

$$V = V_0 + (\mathrm{TS})_n \tag{16.24}$$

Substituting

$$D = w_d V, \tag{16.25}$$

we arrive at the following expression

$$V(1 - w_d t(1-(1+k_d)^{-n}) \cdot (1+k_d)) = V_0 \tag{16.26}$$

Substituting here the values of a financially independent company, V_0 (18) and of a financially dependent company, V, (16.14) one gets

$$\frac{\mathrm{CF} \cdot \left(1 - \left(\frac{1+g}{1+\mathrm{WACC}}\right)^n\right) \cdot (1 - w_d t[1-(1+k_d)^{-n}] \cdot (1+k_d))}{\mathrm{WACC} - g} = \frac{\mathrm{CF} \cdot \left(1 - \left(\frac{1+g}{1+k_0}\right)^n\right)}{(k_0 - g)} \tag{16.27}$$

From (16.27) we arrive to the BFO equation for the case of the company variable profit

$$\frac{1 - \left(\frac{1+g}{1+\mathrm{WACC}}\right)^n}{\mathrm{WACC} - g} = \frac{1 - \left(\frac{1+g}{1+k_0}\right)^n}{(k_0 - g) \cdot (1 - w_d t[1-(1+k_d)^{-n}] \cdot (1+k_d))}. \tag{16.28}$$

This formula is the main theoretical result of the article.

For the Modigliani–Miller theory with the variable profit [1], one gets

$$\mathrm{WACC} - g = (k_0 - g) \cdot (1 - w_d t \cdot (1+k_d)) \tag{16.29}$$

$$\mathrm{WACC} = (k_0 - g) \cdot (1 - w_d t \cdot (1+k_d)) + g \tag{16.30}$$

16.3 Results and Discussions

Below we study within Microsoft Excel the impact of a growth rate g on the company financial parameters (WACC; WACC-g; V; k_e) using the formula (16.28), investigating their dependences on the level of leverage L. We present the results for the following financial parameters of the company:

$$k_0 = 0.26; k_d = 0.22; t = 0.2; \mathrm{CF} = 100; n = 5$$

Here k_0 stands for the equity cost at zero leverage level; k_d stands for the debt cost; t stands for the tax on profit; CF stands for profit per period; n stands for the company age; L stands for the leverage level.

Note that the qualitative effect of the growth rate g on the main financial parameters is similar, while the results for different parameters are different numerically.

16.3.1 *Five-Year Company*

16.3.1.1 Weighted Average Cost of Capital, WACC

As it could be seen from Fig. 16.1. all curves WACC(L) for different values of g start from one point (0; $k_0 = 0.26$). All curves WACC(L) decrease with leverage level L at all g values. The curves WACC(L) increase with g.

The 5-year company results differ from the results for a perpetual limit—the theory of Modigliani and Miller [1]. In the latter case, the WACC(L) curves decrease with the level of leverage L at $g<k_0$ and increase at $g>k_0$. k_0 is the threshold value g separating the increasing WACC(L) curves from the decreasing ones, and for $g = k_0$ WACC $=$ const $= k_0$. In the first case (BFO theory), the WACC(L) curves decrease with increasing leverage L for all values of the growth rate g. The WACC(L) curves increase with the rate g both in the Brusov–Filatova–Orekhova theory and in the Modigliani and Miller theory. This means that WACC is no longer a discount rate. As it will be seen below, the role of the discount rate play WACC–g and WACC–k_0.

Below we study the dependence of WACC–g on L in the Generalized theory of Brusov–Filatova–Orekhova (the GBFO theory) at $k_0 = 0.18$ and different values of g (−0.2; −0.1; 0.0 ; 0.1; 0.2) for a 2- and 4-year company.

16.3.1.2 Calculations of the Discount Rate, WACC–*g*

As it is seen from Fig. 16.2. the curves (WACC–g)(L) decrease with leverage level L at all g values. The curves (WACC–g)(L) decrease with growth rate, g.

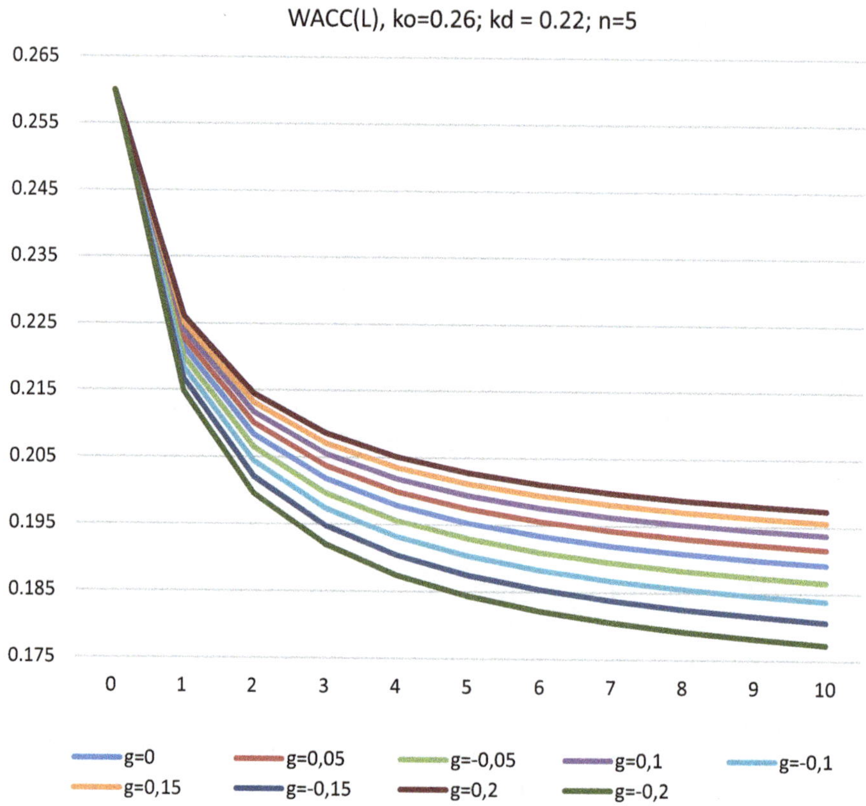

Fig. 16.1 The dependence of WACC on leverage level L in Generalized Brusov–Filatova–Orekhova theory (GBFO theory) with advance payments of tax on profit at $k_0 = 0.26$; $k_d = 0.22$; $t = 0.2$; $g = 0.2$; 0.15;0.1; 0.05; 0.0; −0.05; −0.1; −0.15; −0.2 for 5-year company

The explanation of the behavior of the (WACC–g)(L) curves is as follows: all WACC(L) curves originate from the same point ($L = 0$; WACC $= 0.26$). The (WACC–g)(L) curves will be ordered as follows for $L = 0$: the larger g, the lower the starting point and hence the entire graph lies, since the curves do not intersect. As we'll see below, the decrease of (WACC–g)(L) with growth rate, g, will lead to an increase of the company value, V, with g.

16.3.1.3 Calculations of the Company Value, *V*

It is seen from Fig. 16.3 that the company value V at fixed growth rate g increases with leverage level L. The company value V increases with growth rate g as well.

Below we study the equity cost, k_e, dependence on leverage level L and on growth rate, g, at $k_0 = 0.18$; $k_d = 0.16$ and $g = 0;\pm0.1;\pm0.2$.

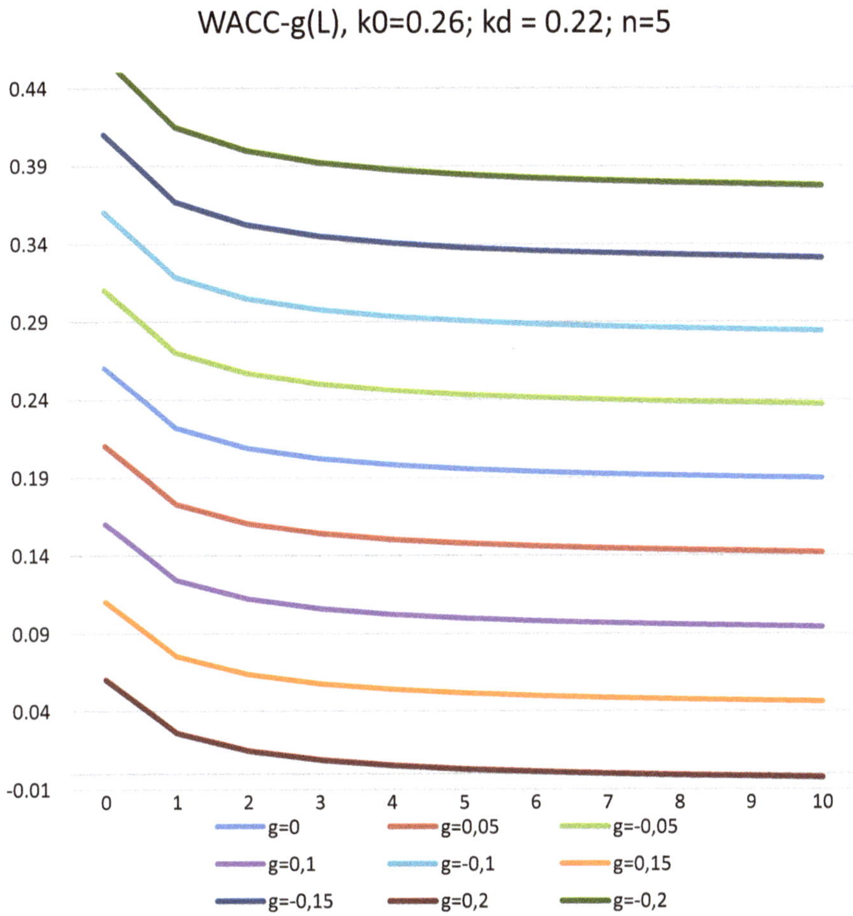

Fig. 16.2 The dependence of discount rate, WACC–g, on leverage level L in Generalized Brusov–Filatova–Orekhova theory (GBFO theory) with advance payments of tax on profit at $k_0 = 0.26$; $k_d = 0.22$; $t = 0.2$; $g = 0.2; 0.15;0.1; 0.05; 0.0; -0.05; -0.1; -0.15; -0.2$ for a 5-year company

16.3.1.4 Calculations of the Equity Cost, k_e

As it is seen from Fig. 16.4, the equity cost, k_e, practically linearly grows with leverage level L at all growth rate g values. The tilt angle $k_e(L)$ increases with g.

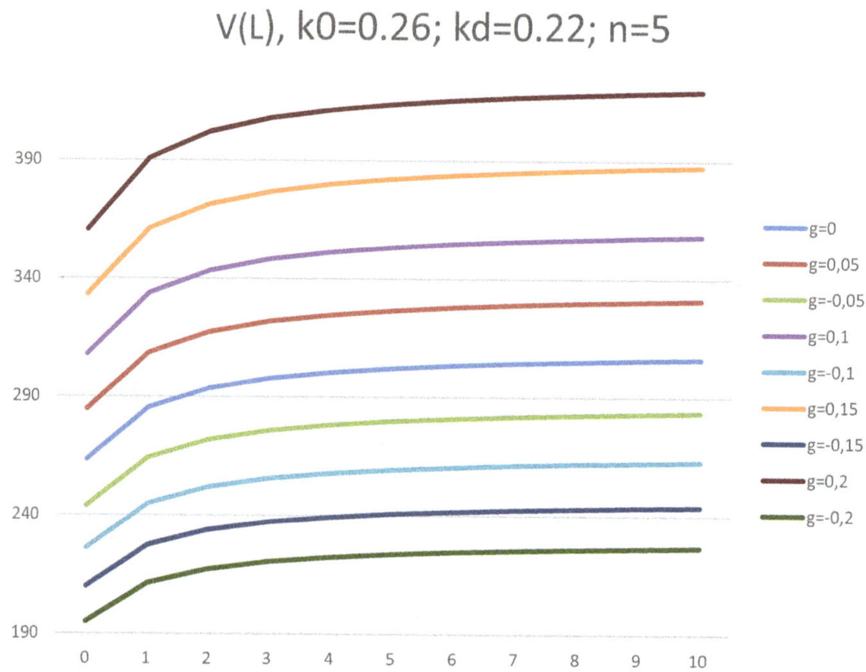

Fig. 16.3 The dependence of discount rate, WACC–g, on leverage level L in Generalized Brusov–Filatova–Orekhova theory (GBFO theory) with advance payments of tax on profit at $k_0 = 0.26$; $k_d = 0.22$; $t = 0.2$; $g = 0.2$; 0.15;0.1; 0.05; 0.0; −0.05; −0.1; −0.15; −0.2 for a 5-year company

16.3.2 *Study the Dependence of Financial Indicators on* $\mathbf{k_d}$

16.3.2.1 The Weighted Average Cost of Capital, WACC

It is seen from Fig. 16.5 that all curves WACC(L) at all values of k_d start from one point (0; 0.26). WACC(L) decrease with leverage level L at all values of k_d. WACC (L) decrease with the increase of debt cost k_d. This means that tax shield advantages the decrease of the cost of raising capital.

16.3.2.2 The Discount Rate, WACC–*g*

It is seen from Fig. 16.6 that all curves of discount rate (WACC–g) (L) at all values of k_d start from one point (0; 0.21). (WACC–g) (L) decrease with leverage level L at all values of k_d. (WACC–g)(L) decrease with the increase of debt cost k_d. This means that the tax shield tends to lower the value of the discount rate (WACC–g) and hence (as we will see below in 16.3.2.3) increase the value of the company, V.

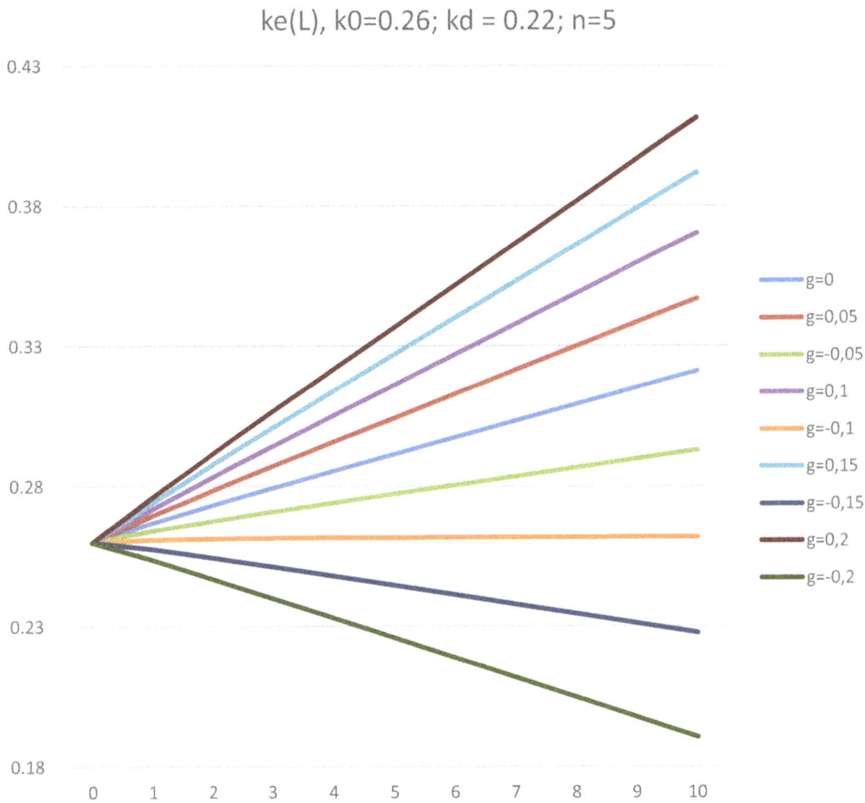

Fig. 16.4 The dependence of equity cost, k_e, on leverage level L in Generalized Brusov–Filatova–Orekhova theory (GBFO theory) with advance payments of tax on profit at $k_0 = 0.26$; $k_d = 0.22$; $t = 0.2$; $g = 0.2$; 0.15; 0.1; 0.05; 0.0; −0.05; −0.1; −0.15; −0.2 for a 5-year company

16.3.2.3 The Company Value, *V*

From Fig. 16.7 it follows that all curves of company value $V(L)$ at all values of k_d start from one point (0; 285). $V(L)$ increases with leverage level L at all values of k_d. $V(L)$ increases with the increase of debt cost k_d. This means that tax shield advantages the increase of the company value, V.

It follows from Fig. 16.8 that all curves of equity cost, $k_e(L)$, at all values of k_d start from one point (0; 0.26) and k_e increases with leverage level L at all values of k_d. The slope of the straight line $k_e(L)$ decreases with the cost of debt k_d. This means that debt cost k_d impacts the dividend policy of the company, because the equity cost k_e determines the economically justified amount of dividends.

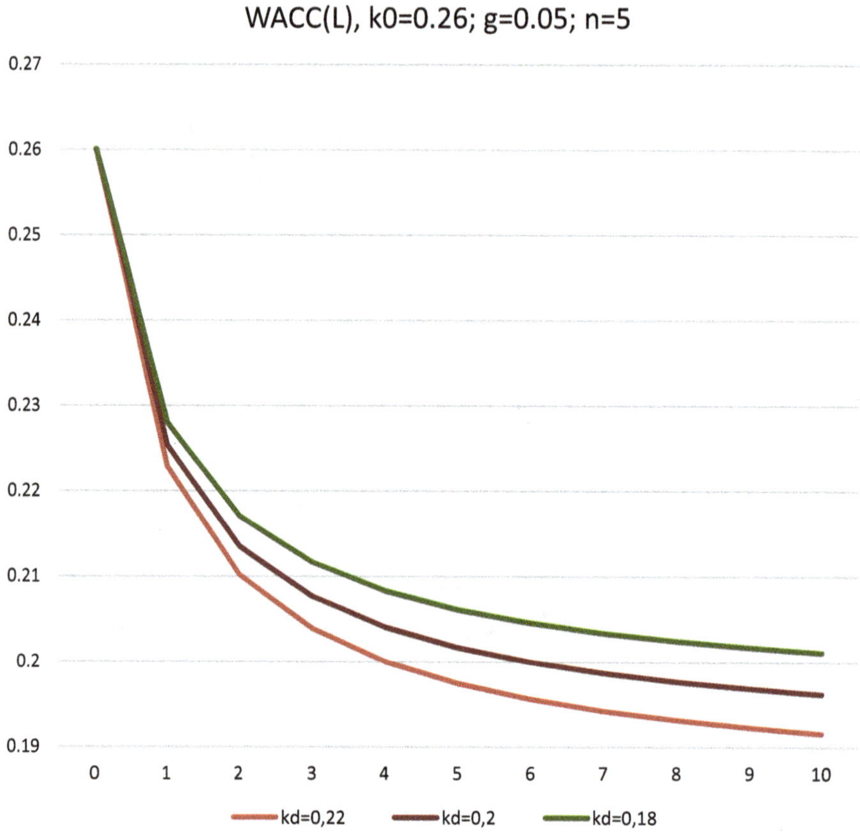

Fig. 16.5 The dependence of WACC on leverage level L in Generalized Brusov–Filatova–Orekhova theory (GBFO theory) with advance payments of tax on profit at $k_O = 0.26$; $k_d = 0.22$; 0.2; 0.18 $t = 0.2$; $g = 0.05$ for a 5-year company

16.3.3 *Impact of Company Age, n, on Main Financial Indicators of the Company*

Below we study the impact of company age, n, on main financial indicators of the company: WACC; WACC–g; V; k_e. We investigate the dependence of WACC; WACC–g; company value, V, and equity cost, k_e on leverage level L in Generalized Brusov–Filatova–Orekhova theory (GBFO theory) with advance payments of tax on profit at $k_O = 0.26$; $k_d = 0.22$; $t = 0.2$; $g = 0.2$ and $g = -0.2$ for 5-year and 10-year companies. We found a huge difference between the behavior of the main financial indicators of the company with a positive and negative growth rate g. This allows you to explore companies with growing profits and companies with decreasing profits. And also it allows study of the companies whose profits rise and fall in different periods.

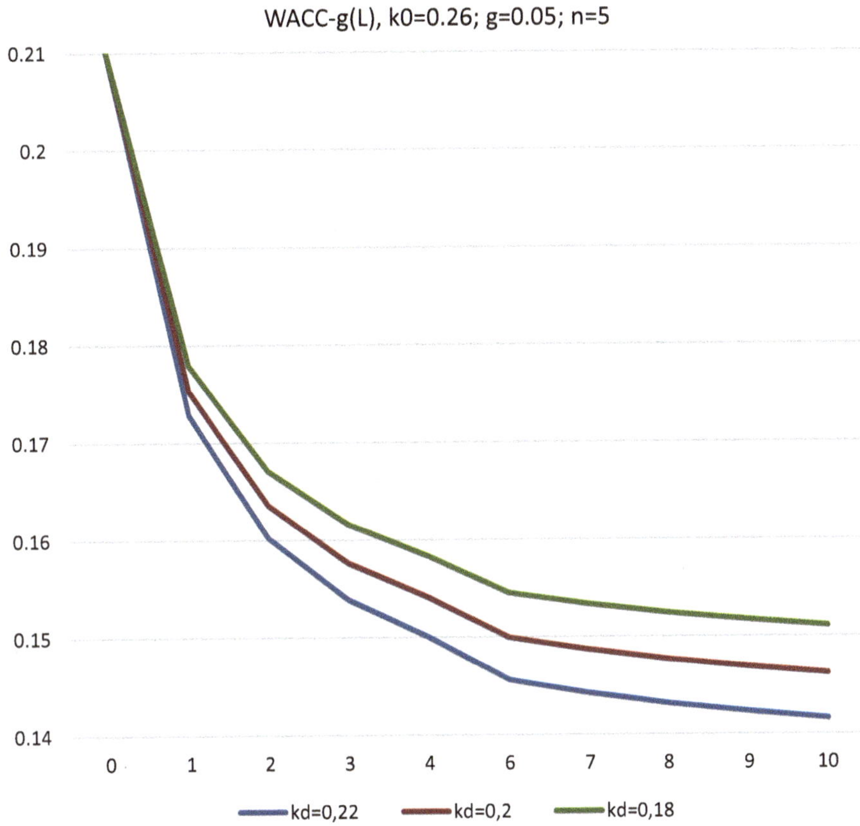

Fig. 16.6 The dependence of discount rate, WACC–g, on leverage level L in Generalized Brusov–Filatova–Orekhova theory (GBFO theory) with advance payments of tax on profit at $k_0 = 0.26$; $k_d = 0.22$; 0.2; 0.18; $t = 0.2$; $g = 0.05$ for a 5-year company

16.3.3.1 WACC(L)

We investigate the dependence of WACC on L for companies of two ages: 5 and 10 years old at two values of growth rates ($g = 0.2$ and $g = -0.2$) (see Figs. 16.9 and 16.10). In both cases, the two WACC(L) curves start from the point (0; 0.26) and decrease with L. For positive growth rate ($g = 0.2$) WACC is lower for the 5-year company, while for negative growth rate ($g = -0.2$), WACC is lower for the 10–year company. We are seeing this effect for the first time.

16.3.3.2 Discount Rate WACC–g

We investigate the dependence of discount rate WACC–g on L for companies of two ages: 5 and 10 years old at two values of growth rates ($g = 0.2$ and $g = -0.2$)

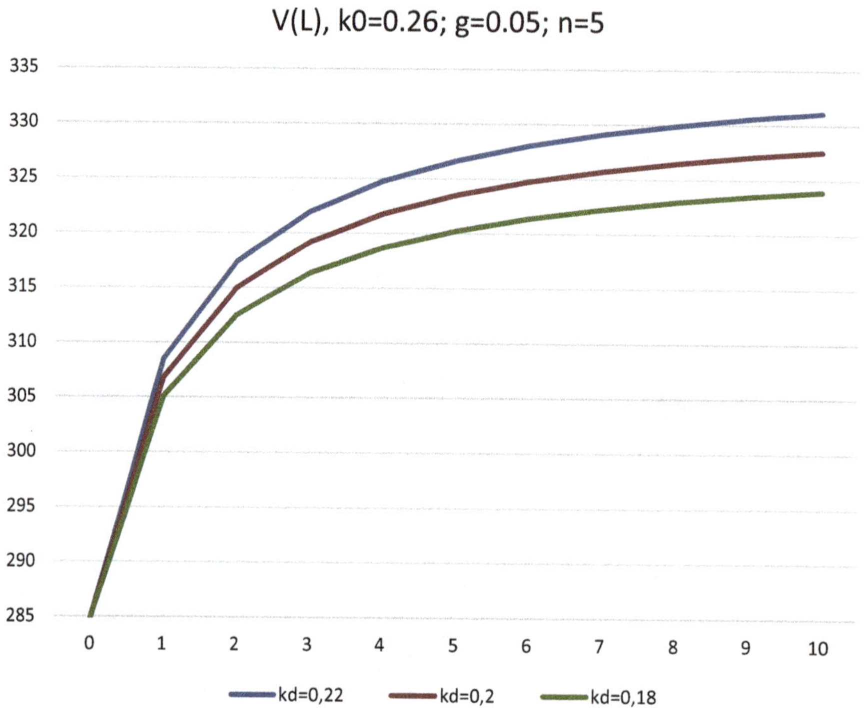

Fig. 16.7 The dependence of company value, V, on leverage level L in Generalized Brusov–Filatova–Orekhova theory (GBFO theory) with advance payments of tax on profit at $k_0 = 0.26$; $k_d = 0.22$; 0.2; 0.18; $t = 0.2$; $g = 0.05$ for a 5-year company

(see Figs. 16.11 and 16.12). In both cases, the two WACC(L) curves start from one point: (0; 0.06) for $g = 0.2$ and (0; 0.46) for $g = -0.2$ and decrease with L. For positive growth rate ($g = 0.2$) discount rate WACC–g is lower for the 5-year company, while for negative growth rate ($g = -0.2$), discount rate WACC–g is lower for the 10-year company. We are seeing this effect for the first time.

16.3.3.3 Company Value, V

Let us study the dependence of the value of a company V on L for companies of two ages: 5 and 10 years old at two values of growth rates ($g = 0.2$ and $g = -0.2$). In both cases, the value of the company V increases with the growth of L, and the greater the age of the company corresponds to, the greater the value of the company V. But if, with a positive growth rate ($g = 0.2$), the difference in the value of V for a 5-year company and a 10-year company is about 400, with negative growth rate ($g = -0.2$), this difference is 45 (ten times less) (see Figs. 16.13 and 16.14). So, in the competition between the age of the company and the size of the growth rate, the growth rate wins.

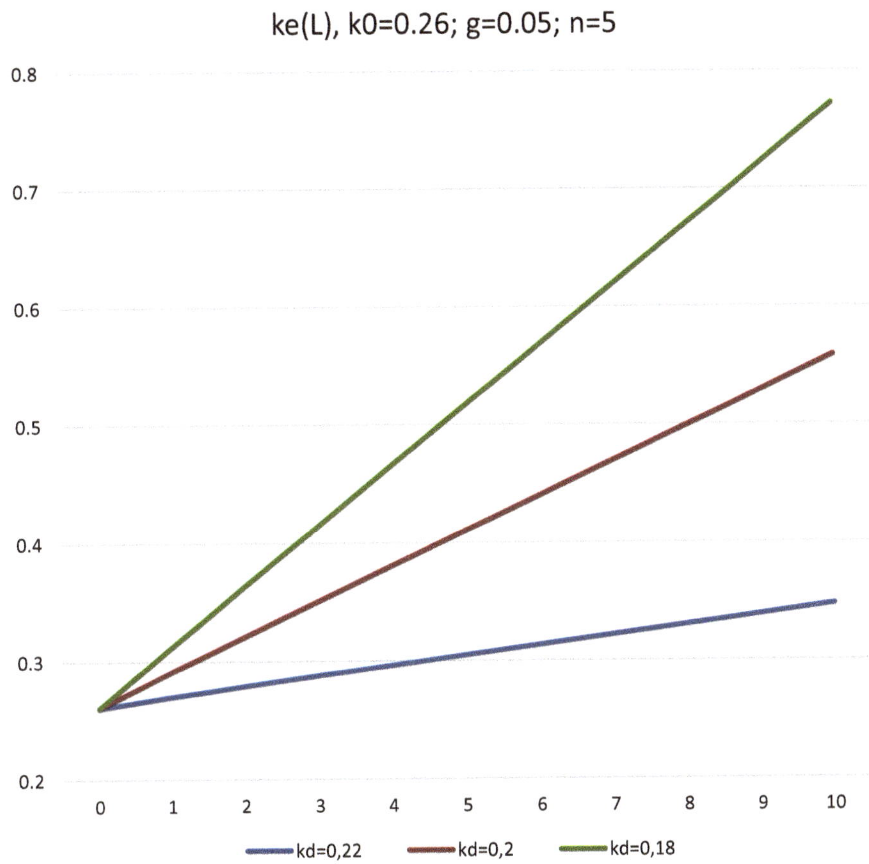

Fig. 16.8 The dependence of equity cost, k_e, on leverage level L in Generalized Brusov–Filatova–Orekhova theory (GBFO theory) with advance payments of tax on profit at $k_0 = 0.26$; $k_d = 0.22$; 0.2; 0.18 $t = 0.2$; $g = 0.05$ for a 5-year company

16.3.3.4 Equity Cost, k_e

Let us study the dependence of the equity cost, k_e, on L for companies of two ages: 5 and 10 years old at two values of growth rates ($g = 0.2$ and $g = -0.2$). In the case of a positive growth rate ($g = 0.2$), the cost of equity k_e increases linearly with L, and the slope $k_e(L)$ for a 10-year company is greater than for a 5-year one. With a negative growth rate ($g = -0.2$), the cost of equity k_e decreases linearly with increasing L, and the negative slope for a 10-year company is greater than for a 5-year one.

Since the cost of equity determines the economically justified amount of dividends, this means that the dividend policy of the company, when increasing profits and when decreasing profits, should be completely different. This is a qualitatively new effect, discovered by us for the first time (Figs. 16.15 and 16.16).

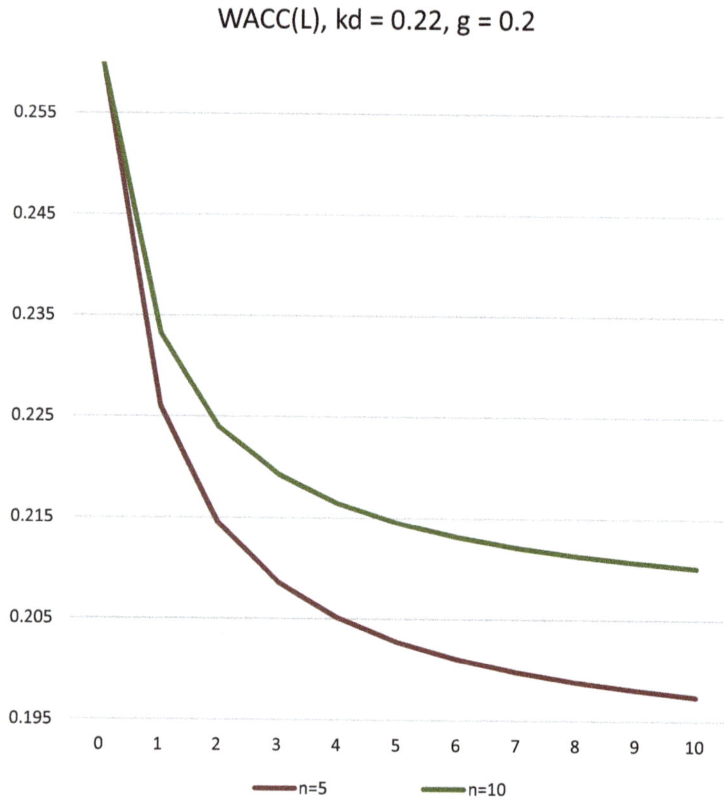

Fig. 16.9 The dependence of WACC on leverage level L in Generalized Brusov–Filatova–Orekhova theory (GBFO theory) with advance payments of tax on profit at $k_0 = 0.26$; $k_d = 0.22$; $t = 0.2$; $g = 0.2$ for 5-year and 10-year companies

16.3.3.5 Results Summary

The curves (WACC–g)(L) decrease with leverage level L at all g values. The curves (WACC–g)(L) decrease with growth rate, g.

The company value V increases with leverage level L at fixed growth rate g. The company value V increases with growth rate g as well. This is a consequence of a decrease of the discount rate (WACC–g)(L) with an increase of g.

The equity cost, k_e, practically linearly increases with leverage level L at all g values. The tilt angle $k_e(L)$ increases with g.

Studying the impact of the debt cost k_d on the main financial indicators, we found the following:

- All curves WACC(L) at all values of k_d start from one point (0; 0.26). WACC(L) decreases with leverage level L at all values of k_d. WACC(L) decreases with the

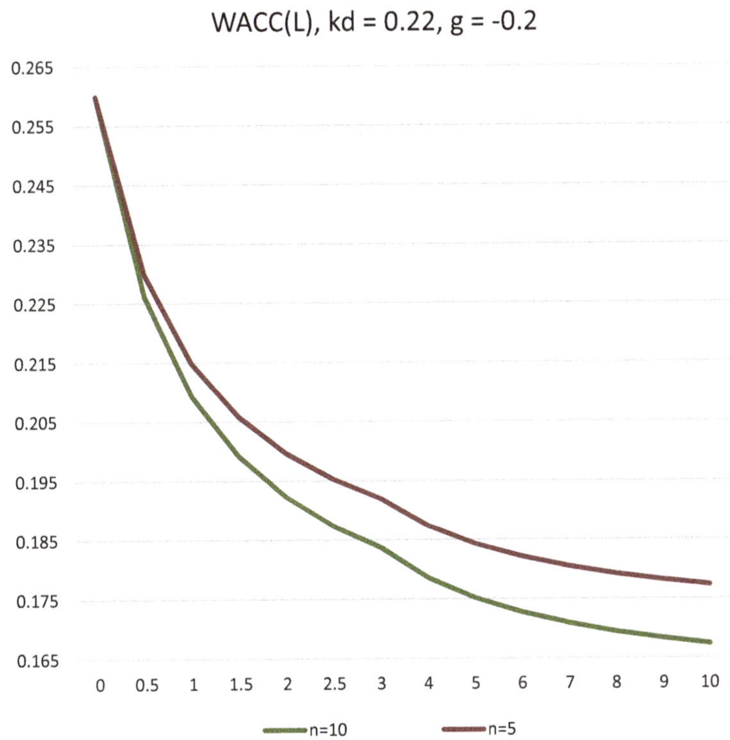

Fig. 16.10 The dependence of WACC on leverage level L in Generalized Brusov–Filatova–Orekhova theory (GBFO theory) with advance payments of tax on profit at $k_O = 0.26$; $k_d = 0.22$; $t = 0.2$; $g = -0.2$ for 5-year and 10-year companies

increase of debt cost k_d. This means that tax shield advantages the decrease of the cost of raising capital.

- All curves of discount rate (WACC–g) (L) at all values of k_d start from one point (0; 0.21). (WACC–g) (L) decrease with leverage level L at all values of k_d. (WACC–g)(L) decreases with the increase of debt cost k_d. This means that the tax shield tends to lower the value of the discount rate (WACC–g) and hence increase the value of the company, V.
- All curves of company value $V(L)$ at all values of k_d start from one point (0; 285). $V(L)$ increases with leverage level L at all values of k_d. $V(L)$ increases with the increase of debt cost k_d. This means that tax shield advantages the increase of the company value, V.
- All curves of equity cost, $k_e(L)$, at all values of k_d start from one point (0; 0.26) and k_e increases with leverage level L at all values of k_d. The slope of the straight line $k_e(L)$ decreases with the cost of debt k_d. This means that debt cost k_d impacts the dividend policy of the company, because the equity cost k_e determines the economically justified amount of dividends.

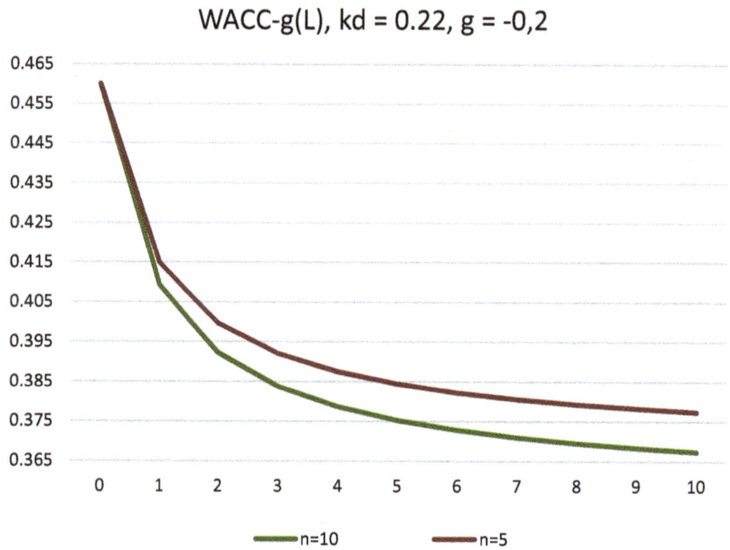

Fig. 16.11 The dependence of discount rate, WACC–g, on leverage level L in Generalized Brusov–Filatova–Orekhova theory (GBFO theory) with advance payments of tax on profit at $k_O = 0.26$; $k_d = 0.22$; $t = 0.2$; $g = -0.2$ for 5-year and 10-year companies

Studying the impact of company age, n, on main financial indicators of the company: WACC; WACC–g; V; k_e, we found a huge difference between the behavior of the main financial indicators of the company with a positive and negative growth rate g. This allows you to explore companies with growing profits and companies with decreasing profits. And also it allows study of the companies whose profits rise and fall in different periods.

Particular results here are as follows.

We investigate the dependence of WACC on L for companies of two ages: 5 and 10 years old at two values of growth rates ($g = 0.2$ and $g = -0.2$). In both cases, the two WACC(L) curves start from the point (0; 0.26) and decrease with L. For positive growth rate ($g = 0.2$) WACC is lower for 5-year company, while for negative growth rate ($g = -0.2$) WACC is lower for 10-year company. We are seeing this effect for the first time.

We investigate the dependence of discount rate WACC–g on L for companies of two ages: 5 and 10 years old at two values of growth rates ($g = 0.2$ and $g = -0.2$). In both cases, the two WACC(L) curves start from one point: (0; 0.06) for $g = 0.2$ and (0; 0.46) for $g = -0.2$ and decrease with L. For positive growth rate ($g = 0.2$) discount rate WACC–g is lower for 5-year company, while for negative growth rate ($g = -0.2$) discount rate WACC–g is lower for 10-year company. We are seeing this effect for the first time.

Let us analyze the dependence of the value of a company V on L for companies of two ages: 5 and 10 years old at two values of growth rates ($g = 0.2$ and $g = -0.2$). In both cases, the value of the company V increases with the growth of L, and the

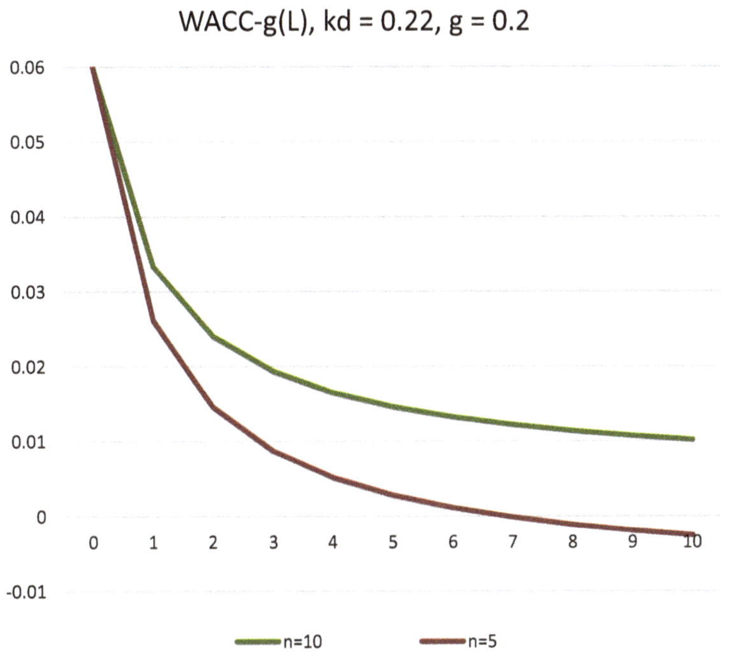

Fig. 16.12 The dependence of discount rate, WACC–g, on leverage level L in Generalized Brusov–Filatova–Orekhova theory (GBFO theory) with advance payments of tax on profit at $k_0 = 0.26$; $k_d = 0.22$; $t = 0.2$; $g = 0.2$ for 5-year and 10-year companies

greater the age of the company corresponds to the greater value of the company V. But if, with a positive growth rate ($g = 0.2$), the difference in the value of V for a 5-year company and a 10-year company is about 400, with negative growth rate ($g = -0.2$), this difference is 45 (ten times less). So, in the competition between the age of the company and the size of the growth rate, the growth rate wins.

We study the dependence of the equity cost, k_e, on L for companies of two ages: 5 and 10 years old at two values of growth rates ($g = 0.2$ and $g = -0.2$). In the case of a positive growth rate ($g = 0.2$), the cost of equity k_e increases linearly with L, and the slope $k_e(L)$ for a 10-year company is greater than for a 5-year one. With a negative growth rate ($g = -0.2$), the cost of equity k_e decreases linearly with increasing L, and the negative slope for a 10-year company is greater than for a 5-year one.

Since the cost of equity determines the economically justified amount of dividends, this means that the dividend policy of the company, when increasing profits and when decreasing profits, should be completely different. This is a qualitatively new effect, discovered by us for the first time.

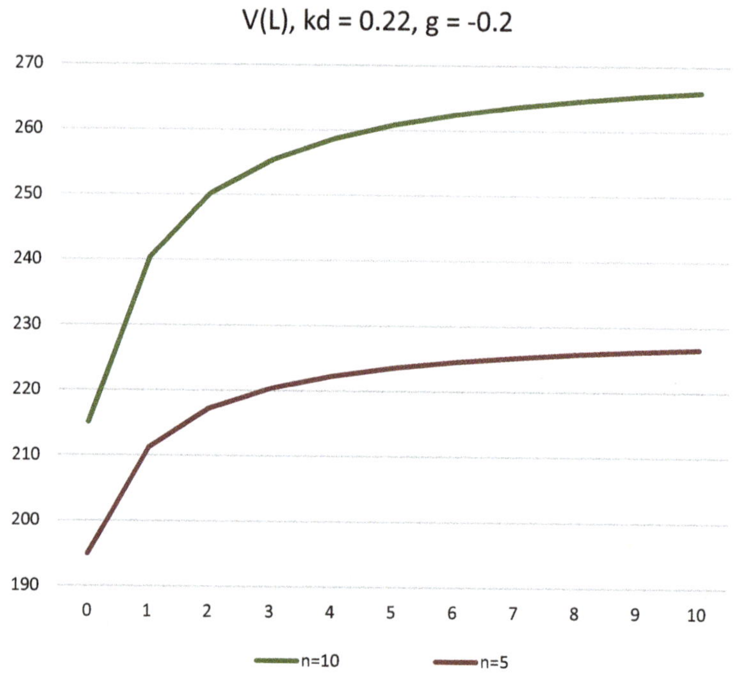

Fig. 16.13 The dependence of company value, V, on leverage level L in Generalized Brusov–Filatova–Orekhova theory (GBFO theory) with advance payments of tax on profit at $k_0 = 0.26$; $k_d = 0.22$; $t = 0.2$; $g = -0.2$ for 5-year and 10-year companies

16.4 Conclusions

The Brusov–Filatova–Orekhova (BFO) theory has been generalized for the case of variable income and advance payments of tax on profit. The generalized Brusov–Filatova–Orekhova formula for WACC has been derived. Using this formula the dependence of WACC, discount rate, WACC–g, company capitalization, V, the equity cost, k_e, on leverage level L at different values of the growth rate, g, on cost of debt capital, k_d, and on company age, n has been studied. It turns out that the WACC is no longer a discount rate. The role of the discount rate is played by WACC–g, which decreases with g, while the company's value increases with g. The slope of the curve $k_e(L)$ increases with g. It turns out that at the growth rate $g < g^*$ the slope of the curve $k_e(L)$ becomes negative, which can significantly change the principles of the company's dividend policy, since the economically justified amount of dividends is equal to the cost of equity. WACC(L) as well as the discount rate, WACC–g, decrease with the increase of debt cost k_d. $V(L)$ increases with leverage level L at all values of k_d. And $V(L)$ increases with the increase of debt cost k_d. This means that tax shield advantages the decrease of the cost of raising capital. Examining the behavior of the main financial parameters of the company with positive ($g = 0.2$) and negative ($g = -0.2$) growth rates, we found a huge difference in their

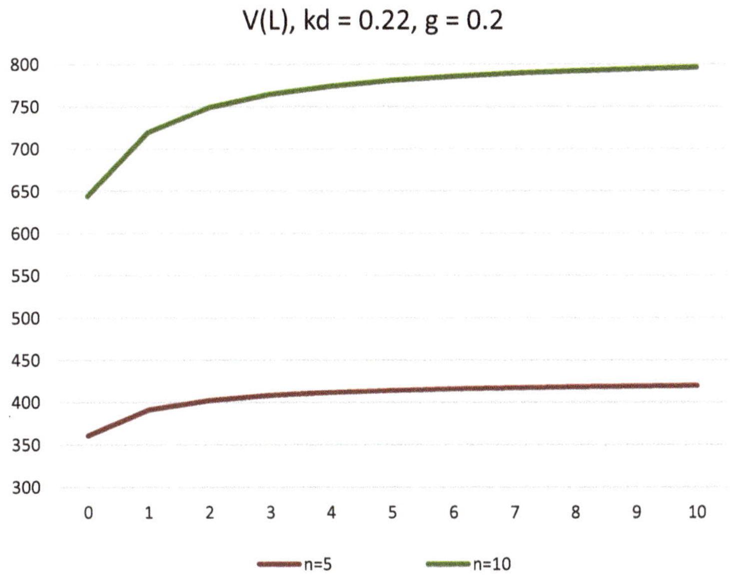

Fig. 16.14 The dependence of company value, V, on leverage level L in Generalized Brusov–Filatova–Orekhova theory (GBFO theory) with advance payments of tax on profit at $k_0 = 0.26$; $k_d = 0.22$; $t = 0.2$; $g = 0.2$ for 5-year and 10-year companies

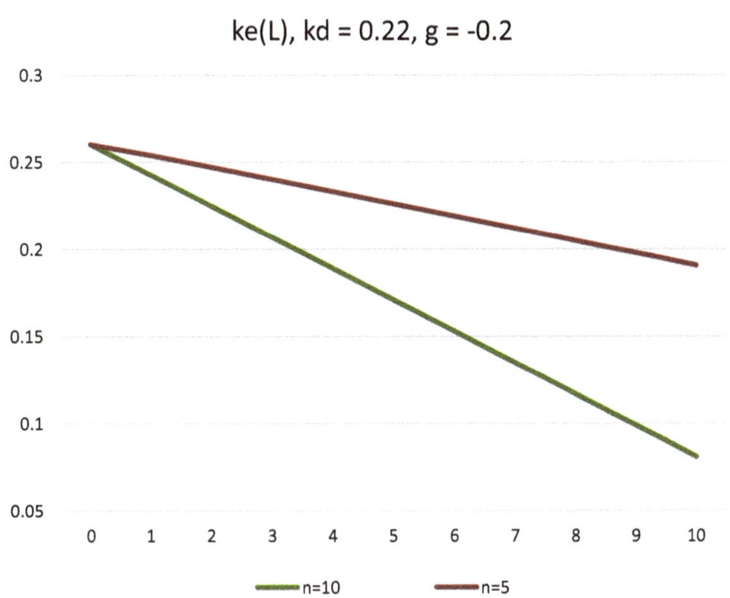

Fig. 16.15 The dependence of equity cost, k_e, on leverage level L in Generalized Brusov–Filatova–Orekhova theory (GBFO theory) with advance payments of tax on profit at $k_0 = 0.26$; $k_d = 0.22$; $t = 0.2$; $g = -0.2$ for 5-year and 10-year companies

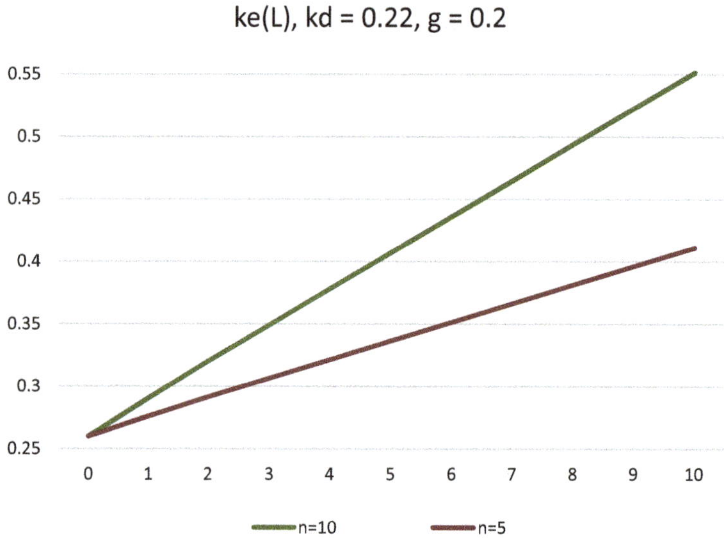

Fig. 16.16 The dependence of equity cost, k_e, on leverage level L in Generalized Brusov–Filatova–Orekhova theory (GBFO theory) with advance payments of tax on profit at $k_0 = 0.26$; $k_d = 0.22$; $t = 0.2$; $g = 0.2$ for 5-year and 10-year companies

behavior. This allows you to explore companies with growing profits and companies with decreasing profits. It also allows you to study companies whose profits rise and fall in different periods.

References

1. Brusov, P. N., Filatova, T. V., Orekhova, N. P., Kulik, V. L., Chang, S. I., & Lin, Y. C. G. (2021). Generalization of the Modigliani–Miller theory for the case of variable profit. *Mathematics, 9(11)*, 1286. https://doi.org/10.3390/math9111286
2. Modigliani, F., & Miller, M. (1958). The cost of capital, corporate finance, and the theory of investment. *The American Economic Review, 48*, 261–297.
3. Modigliani, F., & Miller, M. (1963). Corporate income taxes and the cost of capital: A correction. *The American Economic Review, 53*, 147–175.
4. Modigliani, F., & Miller, M. (1966). Some estimates of the cost of capital to the electric utility industry 1954–1957. *The American Economic Review, 56*, 333–391.
5. Hamada, R. (1969). Portfolio analysis, market equilibrium, and corporate finance. *Journal of Finance, 24*(1), 13–31.
6. Miller, M. (1977). Debt and taxes. *The Journal of Finance, 32*(2), 261–275. https://doi.org/10.1111/j.1540-6261.1977.tb03267.x
7. Brusov, P., Filatova, T., Orekhova, N., & Eskindarov, M. (2018). *Modern corporate finance, investments, taxation and ratings* (2nd ed., pp. 1–571). Springer Nature Publishing.
8. Brusov, P., Filatova, T., & Orehova, N. (2020). *Ratings: Critical analysis and new approaches of quantitative and qualitative methodology* (pp. 1–369). Springer Nature Publishing. https://www.springer.com/de/book/9783030562427

9. Farber, A., Gillet, R., & Szafarz, A. (2006). A general formula for the WACC. *International Journal of Business, 11*(2), 211–218.
10. Fernandez, P. (2006). A general formula for the WACC: A comment. *International Journal of Business, 11/2*, 219.
11. Berk, J., & DeMarzo, P. (2007). *Corporate finance*. Pearson–Addison Wesley.
12. Harris, R., & Pringle, J. (1985). Risk–adjusted discount rates – Extension form the average–risk case. *The Journal of Financial Research, 8*(3), 237–244.
13. Frank, M., & Goyal, V. (2009). Capital structure decisions: Which factors are reliably important? *Financial Management, 38*(1), 1–37. https://doi.org/10.1111/j.1755-053X.2009.01026.x
14. Sheridan, T., & Wessels, R. (1988). The determinants of capital structure choice. *Journal of Finance, 43*, 1–19.
15. Harry, D. A., & Masulis, R. W. (1980). Optimal capital structure under corporate and personal taxation. *Journal of Financial Economics, 8*, 3–29.
16. Bradley, M., Jarrell, G. A., & Kim, E. H. (1984). On the existence of an optimal capital structure: Theory and evidence. *Journal of Finance, 39*, 857–878.
17. Graham, J. R. (2003). Taxes and corporate finance: A review. *Review of Financial Studies, 16*, 1075–1129.
18. Dimitropoulos, P. (2014). Capital structure and corporate governance of soccer clubs: European evidence. *Management Research Review, 37*, 658–678.
19. Luiz, K., & Cruz, M. (2015). The relevance of capital structure on firm performance: A multivariate analysis of publicly traded Brazilian companies. *REPeC Brasília, 9*, 384–401.
20. Barbi, M. (2011). On the risk–neutral value of debt tax shields. *Applied Financial Economics, 22*, 251–258.
21. Franc-Dąbrowska, J., Mądra-Sawicka, M., & Milewska, A. (2021). Energy sector risk and cost of capital assessment—Companies and investors perspective. *Energies, 14*, 1613.
22. Angotti, M., de Lacerda Moreira, R., Hipólito Bernardes do Nascimento, J., & Neto de Almeida Bispo, O. (2018). Analysis of an equity investment strategy based on accounting and financial reports in Latin American markets. *Reficont, 5*, 22–40.
23. Vergara-Novoa, C., Sepúlveda-Rojas, J. P., Miguel, D. A., & Nicolás, R. (2018). Cost of capital estimation for highway concessionaires in Chile. *Journal of Advanced Transportation, 2018*, 2153536.
24. Mundi, H. S., Kaur, P., & Murty, R. L. N. (2022). A qualitative inquiry into the capital structure decisions of overconfident finance managers of family–owned businesses in India. *Qualitative Research in Financial Markets, 14*, 357–379.
25. Sadiq, M., Alajlani, S., Hussain, M. S., Bashir, F., & Chupradit, S. (2022). Impact of credit, liquidity, and systematic risk on financial structure: Comparative investigation from sustainable production. *Environmental Science and Pollution Research International, 29*, 20963–20975.
26. Becker, D. M. (2022). Getting the valuation formulas right when it comes to annuities. *Managerial Finance, 48*, 470.
27. El-Chaarani, H., Abraham, R., & Skaf, Y. (2022). The impact of corporate governance on the financial performance of the banking sector in the MENA (Middle Eastern and North African) Region: An immunity test of banks for COVID–19. *Journal of Risk and Financial Management, 15*, 82.

Chapter 17
BFO Theory with Variable Profit: Two Types of Payments of Tax on Profit: Advanced Payments and at the Ends of Periods

17.1 Introduction

The classical versions of the two main capital structure theories—Brusov–Filatova–Orekhova (BFO), which is valid for companies of arbitrary age, and the Modigliani–Miller theory, a perpetual version of the BFO theory, are limited to the case of constant profit. But in practice, the profit of the company, of course, is variable. Recently, both theories have been generalized to the case of variable income [1, 2] that expands significantly the applicability of both theories of capital structure in practice. In the Brusov–Filatova–Orekhova (BFO) theory, generalized to the case of variable income [2], income tax payments are made at the end of periods, while in practice, these payments can be made as well in advance. Here we consider two modifications of the Brusov–Filatova–Orekhova (BFO) theory with variable income: (1) with the payment of income tax at the end of periods and (2) with advance payments of income tax (this case is considered here for the first time). For these two cases, BFO formulas were derived for the weighted average cost of capital, WACC, for the value of the company, V, and within these formulas, a comprehensive analysis of the dependence of WACC, of the discount rate, WACC–g (here g is the growth rate), company capitalization, V, cost of equity, k_e, on debt financing at different values of the growth rate, g, at different values of the cost of debt capital, k_d, and at different values of the age of the company, n. The results for cases (1) and (2) are compared, which allows us to conclude that case (2) is always preferable for both the company and the regulator. This makes it possible to develop recommendations for both parties to expand the practice of advance income tax payments in the framework of real economic practice.

The structure of the chapter is as follows:

In 17.1 (Introduction), the formulation of the problem is discussed as well as the novelty of the study.

In 17.1.1, a review of the literature on capital structure is given.

P. Brusov, T. Filatova, *Business Valuation and Company Value*, Contributions to Finance and Accounting, https://doi.org/10.1007/978-3-031-90847-7_17

In 17.2 we consider the modification of the Brusov–Filatova–Orekhova (BFO) theory to the case of companies with variable incomes and advance payments of tax on profit. We derive here formulas for the value of a financially dependent company, V, for the value of a financially independent company, V_0, for the tax shield value, TS, for the weighted average cost of capital, WACC. The formulas are derived simultaneously for two cases: for income tax payments at the end of the periods and in advance in order to compare them. The latter are obtained here for the first time.

In 17.3, we describe the results of the study of the dependences of the following values: weighted average cost of capital, WACC, discount rate, WACC–g, company value, V, and the equity cost, k_e, on the debt financing L at different values of the growth rate, g, at different values of the debt cost, k_d, and at different ages of the companies.

17.1.1 A Literature Review

From a historical point of view, **five stages** in the development of the capital structure theory can be distinguished: **First** (before 1958)—the traditional approach that existed before the appearance of the first quantitative theory by Modigliani and Miller (**the second stage**) (1958–1963) [3–5]. **The third stage** (1964–2008) is the numerous attempts of scientists to modify the Modigliani—Miller theory. **The fourth stage** (2008–2019) is the appearance of Brusov–Filatova–Orekhova (BFO) theory, which removed the main restriction of the Modigliani—Miller (MM) theory associated with the infinite lifetime of the company [6]. And, finally, **the fifth stage** (2019 up to now), which began a couple years ago and is associated with the adaptation of the two main theories of the capital structure (Brusov–Filatova–Orekhova and Modigliani–Miller) to the established financial practice of the company's functioning by taking into account the real conditions of their work [1, 2].

The traditional approach is based on existing practical experience. From a theoretical point of view, the advantages of debt financing at a low level of leverage and its disadvantages at a high level of leverage are taken into account, the competition between which forms the optimal capital structure. The optimal capital structure is defined as the level of leverage at which WACC is lowest and company value, V, is highest. There are two versions of the Modigliani–Miller theory: without taxes and with taxes. Without taxes the following expressions for V, WACC and k_e

$$V = V_0 = \frac{\text{EBIT}}{k_0}, \tag{17.1}$$

were V_0 stands for the unlevered company value, EBIT stands for earnings before interest and taxes and k_0 stands for the equity cost at zero leverage level L.

From (17.1) one gets for the weighted average cost of capital WACC:

$$\text{WACC} = k_0. \tag{17.2}$$

From the expression for WACC

$$\text{WACC} = k_0 = k_e w_e + k_d w_d. \tag{17.3}$$

and accounting (17.1), one gets for the equity cost, k_e

$$\begin{aligned} k_e &= \frac{k_0}{w_e} - k_d \frac{w_d}{w_e} = \frac{k_0(S+D)}{S} - k_d \frac{D}{S} = k_0 + (k_0 - k_d)\frac{D}{S} \\ &= k_0 + (k_0 - k_d)L \end{aligned} \tag{17.4}$$

Here, D stands for debt capital value; S stands for equity capital value; k_d and $\mathrm{w_d}$ stand for the cost and share of the company's debt capital; k_e and w_e stand for the equity capital cost and share. It is seen from (17.4), that the equity increases linearly with the leverage level.

Within the framework of the Modigliani–Miller theory with taxes [4], the following expression was postulated for the value of a company using borrowed funds, V,

$$V = V_0 + D{\cdot}t. \tag{17.5}$$

The expression for WACC immediately follows from (17.5)

$$\text{WACC} = k_0{\cdot}(1 - w_d t) \tag{17.6}$$

The following formula for the cost of equity k_e can be obtained from (17.6) within the framework of the Modigliani–Miller theory with taxes

$$k_e = k_0 + L{\cdot}(k_0 - k_d)(1 - t) \tag{17.7}$$

Two formulas (17.7) (MM with taxes) and (17.4) (MM without taxes) differ by the multiplier (1–t), called the tax corrector. It is less than unit, thus the $k_e(L)$ curve slope decreases with accounting the taxes.

In [7], the Modigliani–Miller theory was combined with the CAPM (Capital Asset Pricing Model). This took into account both the financial and business risk of the company and was described by the formula below for the company's leveraged cost of equity, k_e

$$k_e = k_F + (k_M - k_F)b_U + (k_M - k_F)b_U \frac{D}{S}(1 - T), \tag{17.8}$$

where b_U is the β–coefficient of the unlevered company.

In [8], both corporate and individual taxes were taken into account. The following formula for the leveraged value of a company, V, was derived in 1977 by Miller [8]:

$$V = V_0 + \left[1 - \frac{(1 - T_C)(1 - T_S)}{(1 - T_D)}\right] \cdot D \tag{17.9}$$

where, T_S stands for the tax on income of an individual investor from his ownership by corporation stock rate, T_C stands for the tax rate on corporate income , T_D stands for the tax rate on interest income from the provision of investor–individuals of credits to other investors and companies.

Authors of [9–12] have supposed more general expression for WACC than the Modigliani–Miller (MM) one:

$$\mathrm{WACC} = k_0(1 - w_d t) - k_d w_d t + k_{\mathrm{TS}} t w_d \tag{17.10}$$

where, D stands for the debt value, k_0 stands for the equity cost at zero leverage level, k_d stands for the debt cost and k_{TS} stands for the tax shield returns , V stands for the levered company value, VTS stands for the tax shield value.

While formula (17.10), derived in [11], is more general than the Modigliani–Miller (MM) one, for its use in practice additional conditions are necessary. The levered company value can be found by discounting with the WACC for the unlevered company cash flows, in case when the WACC value is constant over time. For this special case, the formulas could be found in textbooks [11, 12].

Because, as it has been supposed by Modigliani and Miller in 1963 [4], the debt value, D, is constant, the unlevered company expected after-tax cash flow is fixed, thus V_0 is also constant. By assumption, $k_{\mathrm{TS}} = k_d$ and the tax shield value is $\mathrm{TS} = tD$. Thus, for the levered company value, V, one gets the ordinary Modigliani–Miller (MM) formula for WACC instead of formula (17.10):

$$\mathrm{WACC} = k_0(1 - w_d t)$$

We believe that the cost of debt k_d and the expected return of the tax shield k_{TS} are equal, since both of them are debt-based and thus the "classical" Modigliani–Miller (MM) theory is much more reasonable.

The influence of tax pressure on the financial balance of energy companies was studied in [13]. As it was shown, tax pressure has a stronger effect on the equilibrium of oil and electricity companies than of gas companies. In [14], by studying the impact of financial liquidity and solvency on the performance of healthcare companies within the framework of econometric models, it was shown that such financial indicators as the current liquidity ratio, quick ratio, and leverage level significantly affect the performance of the company.

The restriction associated with the infinite life of companies and the eternity of cash flows within the framework of the MM theory was removed in 2008 by Brusov, Filatova, and Orekhova [6], who created the modern theory of the cost of capital and capital structure—the BFO theory, which is valid both for companies of arbitrary age, as well as for companies with an arbitrary lifetime. A generalization of the assessment of the tax shield TS and the value of the company: without leverage V_0 and with leverage V was required to modify the theory of MM (see the formulas below):

$$\mathrm{TS} = k_\mathrm{d} DT \sum_{t=1}^{n} (1 + k_d)^{-t} = DT[1 - (1 + k_d)^{-n}].$$

$$V_0 = \mathrm{CF}[1 - (1 + k_0)^{-n}]/k_0; \quad V = \mathrm{CF}[1 - (1 + \mathrm{WACC})^{-n}]/\mathrm{WACC}.$$

$$\frac{1 - (1 + \mathrm{WACC})^{-n}}{\mathrm{WACC}} = \frac{1 - (1 + k_0)^{-n}}{k_0[1 - \omega_d T(1 - (1 + k_d)^{-n})]}. \quad (17.11)$$

Here, S stands for the equity capital value, $w_d = \frac{D}{D+S}$ stands for the debt capital share, $k_e, w_e = \frac{S}{D+S}$ stands for the equity cost and the equity capital share, $L = D/S$ stands for the value of financial leverage, and D stands for the debt capital value.

For a 1-year company, Steve Myers [15] has derived a formula, which could be obtained easily from formula (17.11) by substituting $n = 1$

$$\mathrm{WACC} = k_0 - \frac{(1 + k_0)k_d}{1 + k_d} w_d T$$

and by substituting $n = \infty$ we arrive to the Modigliani–Miller formula for WACC [4]:

$$\mathrm{WACC} = k_0 \cdot (1 - w_d t)$$

The methods and the conclusions of the Brusov–Filatova–Orekhova (BFO) theory are well known in the scientific literature (for example, see references [16–29]). In a number of works (see, for example, [22–24, 29]), the BFO theory is used in practical calculations.

17.1.2 Methods and Materials

We use both analytical and numerical methods. We derive analytically (from first principles) all the formulas for the main financial indicators of the company, and then using them in Microsoft Excel calculate the dependence of these indicators on debt financing, L, growth rate, g, cost of debt, k_d, age of the company, n, etc.

17.2 The Brusov–Filatova–Orekhova (BFO) Theory Modification to the Case of Companies with Variable Incomes and Advance Payments of Tax on Profit

17.2.1 The Value of a Financially Dependent Company, V

Below, for the first time, we generalize the modern theory of the capital cost and capital structure—the Brusov–Filatova–Orekhova theory—for the case of variable profit upon advance payments of tax on income. We simultaneously consider two modifications of the Brusov–Filatova–Orekhova (BFO) theory with variable income: (1) with the payment of income tax at the end of periods (formulas with ‘) and (2) with advance payments of income tax (this case is considered here) to see the difference in their derivation and in the resulting formulas.

Let's start by deriving the capitalization formula for a financially dependent company, assuming that income for the period grows with the growth rate g. Note that below (up to formula [17.20]), all derivations for the two cases (1) and (2) are similar; starting from formula (17.20), there will be big differences.

Accounting, that the cost of any asset is equal to sum of discounted values of incomes, generated by this asset, one could write for the capitalization for a financially dependent company V of age n the following expression

$$V=\frac{\mathrm{CF}}{1+\mathrm{WACC}}+\frac{\mathrm{CF}(1+g)}{(1+\mathrm{WACC})^2}+\frac{\mathrm{CF}(1+g)^2}{(1+\mathrm{WACC})^3}+\cdots+\frac{\mathrm{CF}(1+g)^{n-1}}{(1+\mathrm{WACC})^n} \tag{17.12}$$

Here WACC is the weighted average cost of capital, CF is an annual profit of company,

(17.12) is geometric progression with denominator

$$g=\frac{(1+g)}{(1+\mathrm{WACC})}. \tag{17.13}$$

Summing (17.12), we get the expression for the capitalization for a financially dependent company V of age n

$$V = \frac{\mathrm{CF}}{1+\mathrm{WACC}} \cdot \frac{1-\left(\frac{1+g}{1+\mathrm{WACC}}\right)^n}{1-\frac{1+g}{1+\mathrm{WACC}}} = \frac{\mathrm{CF}}{\mathrm{WACC}-g} \cdot \left(1-\left(\frac{1+g}{1+\mathrm{WACC}}\right)^n\right) \tag{17.14}$$

In perpetuity limit ($n \to \infty$), we get the following formula for a financially dependent company value, V,

$$V = \frac{\mathrm{CF}}{\mathrm{WACC}-g} \tag{17.15}$$

This formula shows, that discount rate is WACC–g, and not WACC.

17.2.2 The Value of a Financially Independent Company, V_0

Let us now derive the capitalization formula for a financially independent company, assuming that income for the period grows with the growth rate g.

$$V_0 = \frac{\mathrm{CF}}{1+k_0} + \frac{\mathrm{CF}(1+g)}{(1+k_0)^2} + \frac{\mathrm{CF}(1+g)^2}{(1+k_0)^3} + \cdots + \frac{\mathrm{CF}(1+g)^{n-1}}{(1+k_0)^n} \tag{17.16}$$

(17.16) is geometric progression with denominator

$$g = \frac{(1+g)}{(1+k_0)} \tag{17.17}$$

Summing (17.16), we get the expression for the capitalization for a financially independent company V of age n

$$V_0 = \frac{\mathrm{CF}}{1+k_0} \cdot \frac{1-\left(\frac{1+g}{1+k_0}\right)^n}{1-\frac{1+g}{1+k_0}} = \frac{\mathrm{CF}}{k_0-g} \cdot \left(1-\left(\frac{1+g}{1+k_0}\right)^n\right). \tag{17.18}$$

In perpetuity limit ($n \to \infty$), we get the following formula for a financially independent company value, V_0,

$$V_0 = \frac{\mathrm{CF}}{k_0-g}. \tag{17.19}$$

This formula shows, that discount rate is k_0–g, and not k_0.

17.2.3 The Tax Shield Value, TS

Starting from this place, formulas for cases (1) and (2) become different (formulas for the payments of income tax at the end of periods are marked with ').

The tax shield for *n*–years in case of advance payments of tax on profit is equal

$$(\mathrm{TS})_n = tk_dD + \frac{tk_dD}{1+k_d} + \cdots + \frac{tk_dD}{(1+k_d)^{n-1}} \tag{17.20}$$

The tax shield for *n*–years in case of payments of tax on profit in the ends of periods is equal

$$(\mathrm{TS})_n = \frac{tk_dD}{1+k_d} + \frac{tk_dD}{(1+k_d)^2} + \cdots + \frac{tk_dD}{(1+k_d)^n} \tag{17.20'}$$

(17.20) is geometric progression with denominator

$$g = \frac{1}{(1+k_d)} \tag{17.21}$$

Summing (17.20), one gets for tax shield

$$(\mathrm{TS})_n = tk_dD \cdot \frac{1-(1+k_d)^{-n}}{1-\frac{1}{1+k_d}} = Dt(1-(1+k_d)^{-n}) \cdot (1+k_d) \tag{17.22}$$

$$(\mathrm{TS})_n = \frac{tk_dD}{1+k_d} \cdot \frac{1-(1+k_d)^{-n}}{1-\frac{1}{1+k_d}} = Dt(1-(1+k_d)^{-n}) \tag{17.22'}$$

$$(\mathrm{TS})_n = Dt(1-(1+k_d)^{-n}) \cdot (1+k_d) \tag{17.23}$$

Using an analog of the first theorem by Modigliani–Miller for finite time, one gets

$$V = V_0 + (\mathrm{TS})_n \tag{17.24}$$

Substituting

$$D = w_dV, \tag{17.25}$$

we arrive to the following expression

$$V(1 - w_dt(1-(1+k_d)^{-n}) \cdot (1+k_d)) = V_0 \tag{17.26}$$

$$V(1 - w_dt(1-(1+k_d)^{-n})) = V_0 \tag{17.26'}$$

Substituting into this equation the values of a financially independent company, V_0 (17.18), and of a financially dependent company, V (17.14), one gets the following expression

$$\frac{\mathrm{CF}\cdot\left(1-\left(\frac{1+g}{1+\mathrm{WACC}}\right)^n\right)\cdot(1-w_d t[1-(1+k_d)^{-n}]\cdot(1+k_d))}{\mathrm{WACC}-g} = \frac{\mathrm{CF}\cdot\left(1-\left(\frac{1+g}{1+k_0}\right)^n\right)}{(k_0-g)} \tag{17.27}$$

$$\frac{CF\cdot\left(1-\left(\frac{1+g}{1+WACC}\right)^n\right)\cdot(1-w_d t[1-(1+k_d)^{-n}])}{WACC-g} = \frac{CF\cdot\left(1-\left(\frac{1+g}{1+k_0}\right)^n\right)}{(k_0-g)} \tag{17.27'}$$

Dividing both parts by $(1 - w_d t(1 - (1 + k_d)^{-n}) \cdot (1 + k_d))$,

$$(1-w_d t(1-(1+k_d)^{-n})),$$

we get the BFO equation for the case of variable profit of the company

$$\frac{1-\left(\frac{1+g}{1+\mathrm{WACC}}\right)^n}{\mathrm{WACC}-g} = \frac{1-\left(\frac{1+g}{1+k_0}\right)^n}{(k_0-g)\cdot(1-w_d t[1-(1+k_d)^{-n}]\cdot(1+k_d))}. \tag{17.28}$$

$$\frac{1-\left(\frac{1+g}{1+\mathrm{WACC}}\right)^n}{\mathrm{WACC}-g} = \frac{1-\left(\frac{1+g}{1+k_0}\right)^n}{(k_0-g)\cdot(1-w_d t[1-(1+k_d)^{-n}])}. \tag{17.28'}$$

These two formulas differ in the factor $(1 + k_d)$ in the denominator of the right side.

This is the main theoretical result of the current chapter.

In perpetuity limit ($n \to \infty$), we get the following equation for WACC in the Modigliani–Miller theory in the case of variable profit [1]

$$\mathrm{WACC}-g=(k_0-g)\cdot(1-w_d t\cdot(1+k_d)) \tag{17.29}$$

$$\mathrm{WACC}-g=(k_0-g)\cdot(1-w_d t) \tag{17.29'}$$

$$\mathrm{WACC}=(k_0-g)\cdot(1-w_d t\cdot(1+k_d))+g \tag{17.30}$$

$$\mathrm{WACC}=(k_0-g)\cdot(1-w_d t)+g \tag{17.30'}$$

$$k_e=\mathrm{WACC}(1+L)-Lk_d(1-t) \tag{17.31}$$

17.3 Results and Discussions

Below we describe the results of the study of the dependences of the following values: weighted average cost of capital, WACC, discount rate, WACC–g, company value, V, and the equity cost, k_e, on the debt financing L at different values of the growth rate, g, at different values of the debt cost, k_d and at different ages of the companies. A large database of results for different pairs (k_0, k_d) (k_0 is the equity cost, and k_d is the debt cost) has been created and is available upon request. To illustrate the obtained results and the conclusions, the results for the following typical financial indicators of the 5- and 10-year companies are presented: $k_0 = 0.26$; $k_d = 0.22$; $t = 0.2$ $g = 0.1; 0.05; 0.0; -0.05$; $CF = 100$, where k_0 stands for the equity cost at zero leverage level; k_d stands for the debt cost; t stands for the tax on income; CF stands for income per period; n stands for the company age; L stands for the leverage level; g stands for the growth rate.

Although the results for different values of the indicators may be numerically different, the detected qualitative effects are similar.

For both types of payments of tax on profit (at the ends of periods (1) and advance payments (2)), as it could be seen from Fig. 17.1, for different values of g, all curves WACC(L) start from one point ($L = 0$; $k_0 = 0.26$). These curves WACC(L) demonstrate the decrease of WACC with leverage level L at all g values. The curves WACC(L) increase with growth rate, g. This is the first manifestation of the fact that under variable income, the weighted average cost of capital, WACC, is no longer a discount rate. As we will see below, the role of the discount rate has been taken over by WACC–g (here g is the growth rate).

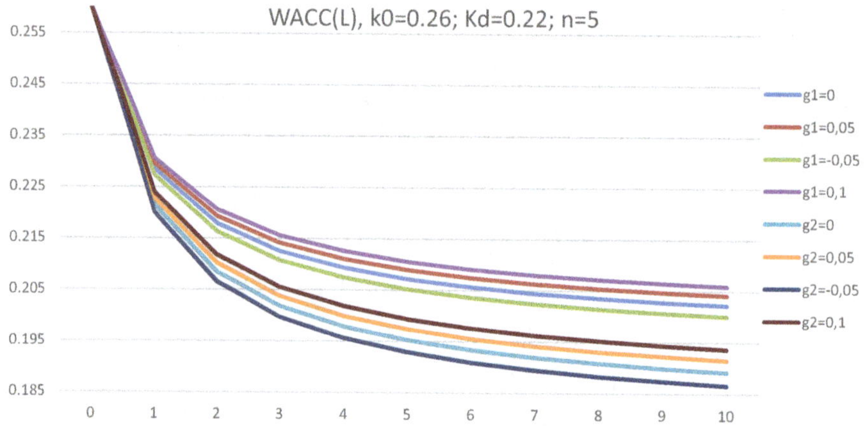

Fig. 17.1 The dependence of WACC on leverage level L in Generalized Brusov–Filatova–Orekhova theory (GBFO theory) with payments of tax on profit at the ends of periods (1) and advance payments (2) at $k_0 = 0.26$; $k_d = 0.22$; $t = 0.2$; $g = 0.1$; 0.05; 0.0; −0.05 for a 5-year company

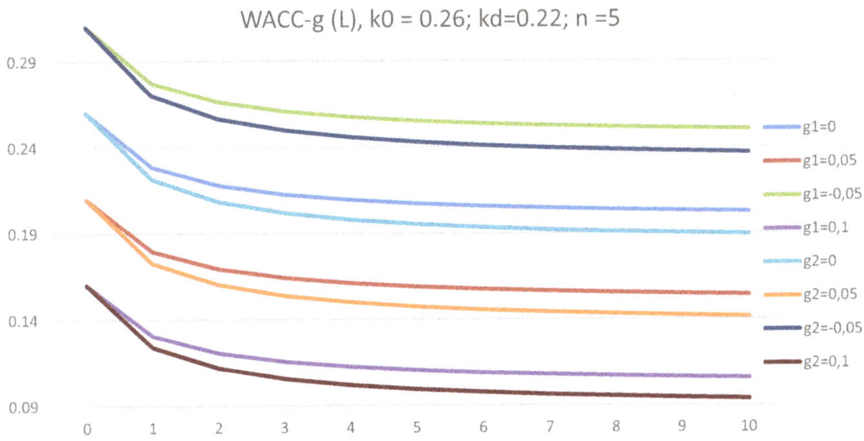

Fig. 17.2 The dependence of discount rate, WACC–g, on leverage level L in Generalized Brusov–Filatova–Orekhova theory (GBFO theory) with payments of tax on profit at the ends of periods (1) and advance payments (2) at $k_0 = 0.26$; $k_d = 0.22$; $t = 0.2$; $g = 0.1; 0.05; 0.0; -0.05$ for a 5-year company

The results for a 5-year company differ from the results for a perpetual limit—the theory of Modigliani and Miller with variable income [1]. In the latter case, the WACC(L) curves decrease with the level of leverage L at $g < k_0$ and increase at $g > k_0$. k_0 is the threshold value g separating the increasing WACC(L) curves from the decreasing ones, and for $g = k_0$ WACC = const = k_0. In the first case (BFO theory), the WACC(L) curves decrease with increasing leverage L for all values of the growth rate g. Note, that the WACC(L) curves increase with the rate g both in the Brusov–Filatova–Orekhova theory and in the Modigliani–Miller theory.

As it could be seen from Fig. 17.2, the curves (WACC–g)(L) demonstrate the decrease of the discount rate, WACC–g, with leverage level L at all g values. The curves (WACC–g)(L) decrease with growth rate, g. Thus, as it will be seen below, the company value, V, will increase with g.

This behavior of the (WACC–g)(L) curves can be explained as follows: all WACC(L) curves originate from the same point (L = 0; WACC = 0.26). The (WACC–g)(L) curves will be ordered as follows for L = 0: the larger the g, the lower the starting point and hence the entire graph lies, since the curves do not intersect. As we'll see below, the decrease of (WACC–g)(L) with growth rate, g, will lead to increase of the company value, V, with g.

Now, for each value of g, the two curves (WACC–g)(L) corresponding to two different types of income tax payments ((1) and (2)), start from the same point (k_0–g), and they are then split into two curves, and the income tax advances

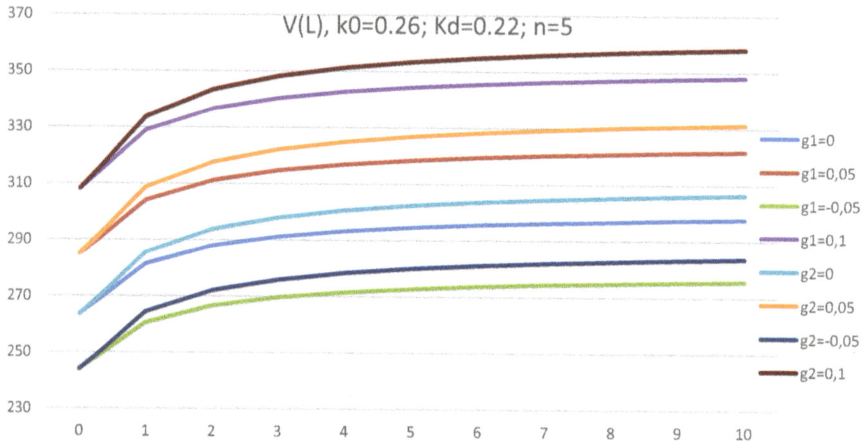

Fig. 17.3 The dependence of discount rate, WACC–g, on leverage level L in Generalized Brusov–Filatova–Orekhova theory (GBFO theory) with payments of tax on profit at the ends of periods (1) and advance payments (2) at $k_0 = 0.26$; $k_d = 0.22$; $t = 0.2$ $g = 0.1; 0.05; 0.0; -0.05$ for a 5-year company

correspond to the lower curve. This means that advance income tax payments result in lower WACC–g values for any growth rate g.

It could be seen from Fig. 17.3, that the company value V at fixed growth rate g increases with leverage level L in Generalized Brusov–Filatova–Orekhova theory (GBFO theory) for both type of payments of tax on profit: at the ends of periods (1) and advance payments (2). The company value V as well increases with growth rate g. This is a consequence of a decrease in the discount rate (WACC–g)(L) with an increase in the growth rate g.

Note that for each value of g, two $V(L)$ curves corresponding to two different types of income tax payments ((1) and (2)) start from the same point CF/(k_0–g), they are then split into two curves, and the income tax advances correspond to the upper curve. This means that advance payments of income tax result in a higher value for company V at any growth rate g.

Below we have studied the dependence of cost of equity, k_e, on leverage level L and on growth rate, g, in Generalized Brusov–Filatova–Orekhova theory (GBFO theory) at $k_0 = 0.26$; $k_d = 0.22$ and $g = 0.1; 0.05; 0.0; -0.05$ for a 5-year company for two types of payments of tax on profit: at the ends of periods (1) and advance payments (2).

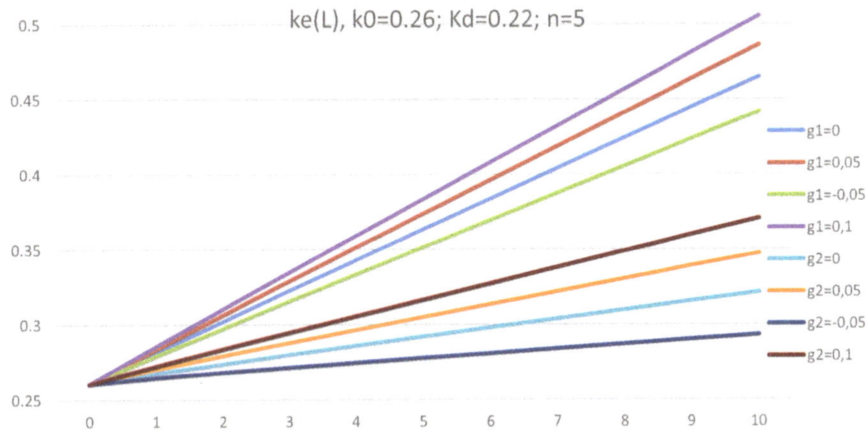

Fig. 17.4 The dependence of equity cost, k_e, on leverage level L in Generalized Brusov–Filatova–Orekhova theory (GBFO theory) with payments of tax on profit at the ends of periods (1) and advance payments (2) at $k_0 = 0.26$; $k_d = 0.22$; $t = 0.2$; $g = 0.1$; 0.05; 0.0; −0.05 for a 5-year company

17.3.1 Calculations of the Equity Cost, $\mathbf{k_e}$

As it is seen from Fig. 17.4 the equity cost, k_e, practically linearly grows with leverage level L at all growth rate g values. All curves $k_e(L)$ start from one point ($L = 0; k_e = k_0 = 0.26$).The tilt angle $k_e(L)$ grows with g. The slope of all curves $k_e(L)$ corresponding to the payment of income tax at the end of periods (1) is greater than that of the curves with advance payments of income tax.

Given that the cost of equity capital determines the economically justified amount of dividends, we can draw the following conclusions regarding the dividend policy of the company:

1. Higher growth rate g justifies higher dividends.
2. If a company pays income tax at the end of periods, it must pay large dividends.
3. If the company pays income tax in advance, dividends should be less.

One important remark can be made: when income tax is paid in advance, the dividends paid by the company should be less. In addition, as we saw above, the value of the company in this case is higher. So, advance payments on income tax are beneficial to the enterprise.

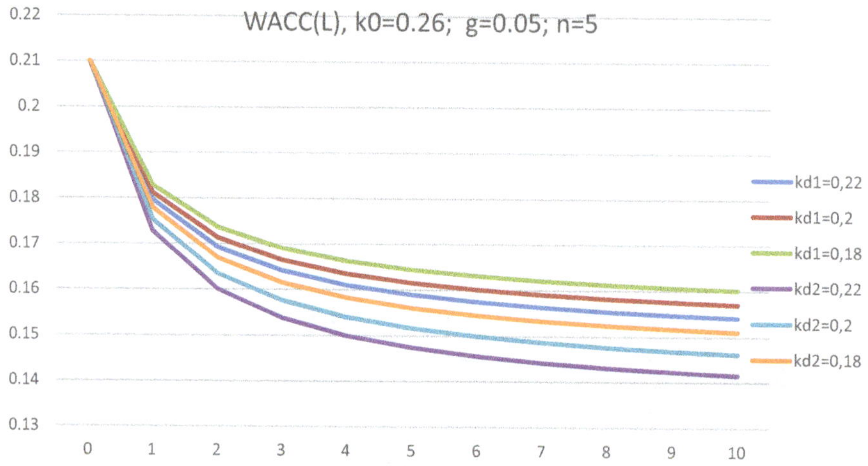

Fig. 17.5 The dependence of WACC on leverage level L in Generalized Brusov–Filatova–Orekhova theory (GBFO theory) with payments of tax on profit at the ends of periods (1) and advance payments (2) at $k_0 = 0.26$; $k_d = 0.22$; 0.2; 0.18 $t = 0.2$; $g = 0.05$ for a 5-year company

17.3.2 *Study the Dependence of Financial Indicators on* $\mathbf{k_d}$

17.3.2.1 The Weighted Average Cost of Capital, WACC

It is seen from Fig. 17.5 that all curves WACC(L) at all values of k_d start from one point (0; 0.26). WACC(L) decreases with leverage level L at all values of k_d. WACC (L) decreases with the increase of debt cost k_d. This means that tax shield advantages the decrease of the cost of raising capital. All curves WACC(L) in the case of advance income tax payments lie below the curves in the case of income tax payments at the end of periods.

17.3.2.2 The Discount Rate, WACC–g

It is seen from Fig. 17.6 that all curves of discount rate (WACC–g) (L) at all values of k_d start from one point ($L = 0$; $k_e = 0.21$). (WACC–g) (L) decreases with leverage level L at all values of k_d. (WACC–g)(L) decreases with the increase of debt cost k_d. This means that the tax shield tends to lower the value of the discount rate (WACC–g) and hence (as we will see below in 17.3.2.3) increase the value of the company, V. All curves (WACC–g) (L) in the case of advance income tax payments lie below the curves in the case of income tax payments at the end of periods.

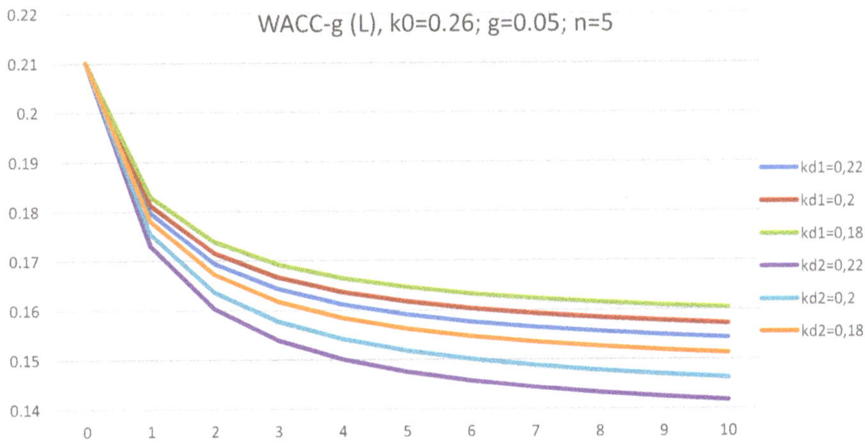

Fig. 17.6 The dependence of discount rate, WACC–g, on leverage level L in Generalized Brusov–Filatova–Orekhova theory (GBFO theory) with payments of tax on profit at the ends of periods (1) and advance payments (2) at $k_0 = 0.26$; $k_d = 0.22$; 0.2; 0.18; $t = 0.2$; $g = 0.05$ for a 5-year company

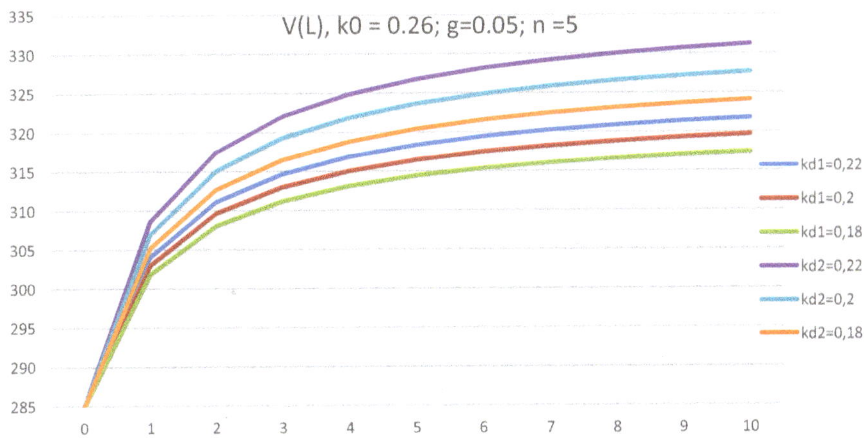

Fig. 17.7 The dependence of company value, V, on leverage level L in Generalized Brusov–Filatova–Orekhova theory (GBFO theory) with payments of tax on profit at the ends of periods (1) and advance payments (2) at $k_0 = 0.26$; $k_d = 0.22$; 0.2; 0.18; $t = 0.2$; $g = 0.05$ for a 5-year company

17.3.2.3 The Company Value, V

From Fig. 17.7 it follows that all curves of company value $V(L)$ at all values of k_d for a 5-year company start from one point (0; 285). $V(L)$ increases with leverage level

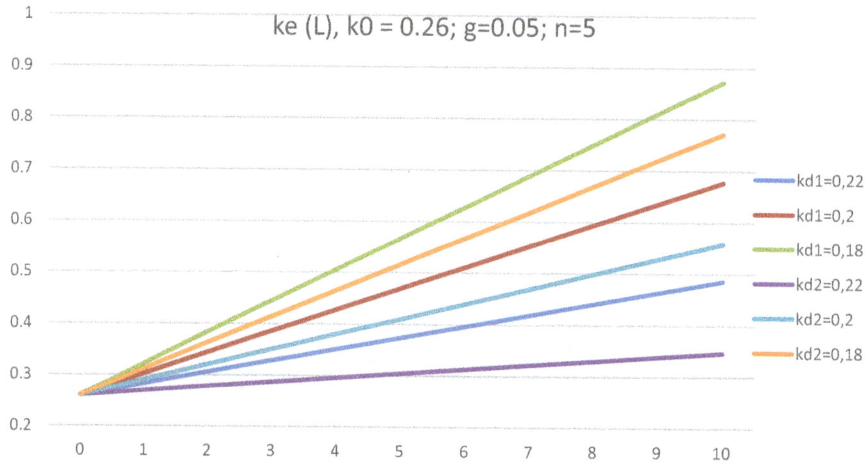

Fig. 17.8 The dependence of equity cost, k_e, on leverage level L in Generalized Brusov–Filatova–Orekhova theory (GBFO theory) with payments of tax on profit at the ends of periods (1) and advance payments (2) at $k_O = 0.26$; $k_d = 0.22$; 0.2; 0.18 $t = 0.2$; $g = 0.05$ for a 5-year company

L at all values of k_d. $V(L)$ increases with the increase of debt cost k_d. This means that tax shield advantages the increase of the company value, V. All curves $V(L)$ in the case of advance payments of income tax lie above the curves in the case of payment of income tax at the ends of periods. This means that advance income tax payments are beneficial for the company, increasing its capitalization.

It follows from Fig. 17.8 that all curves of equity cost, $k_e(L)$, at all values of k_d start from one point ($L = 0$; $k_e = k_O = 0.26$) and k_e increases with leverage level L at all values of k_d. The slope of the straight line $k_e(L)$ decreases with the cost of debt k_d. This means that debt cost k_d impacts the dividend policy of the company, because the equity cost k_e determines the economically justified amount of dividends. At a fixed cost of debt k_d, the slope of the straight line $k_e(L)$ in the case of paying income tax at the end of the periods is higher than in the case of advance payments.

17.3.3 *Impact of Company Age,* n, *on Main Financial Indicators of the Company*

Below we study the impact of company age, n, on main financial indicators of the company. We investigate the dependence of WACC; WACC–g; company value, V, and equity cost, k_e on leverage level L in Generalized Brusov–Filatova–

Orekhova theory (GBFO theory) with advance payments of tax on profit at $k_0 = 0.26$; $k_d = 0.22$; $t = 0.2$; $g = 0.2$ and $g = -0.2$ for 5-year and 10-year companies. We found a huge difference between the behavior of the main financial indicators of the company with a positive and negative growth rate g. This allows you to explore companies with growing profits and companies with decreasing profits. And also it allows study of the companies whose profits rise and fall in different periods.

17.3.3.1 WACC(*L*)

We investigate the dependence of WACC on L for companies of two ages: 5 and 10 years old at two values of growth rates ($g = 0.2$ and $g = -0.2$) (see Figs. 17.9 and 17.10). In both cases, the two WACC(L) curves start from the point (0; 0.26) and decrease with L. For positive growth rate ($g = 0.2$) WACC is lower for the 5-year company, while for negative growth rate ($g = -0.2$) WACC is lower for the 10-year company. We are seeing this effect for the first time.

17.3.3.2 Discount Rate WACC–*g*

We investigate the dependence of discount rate WACC–g on L for companies of two ages: 5 and 10 years old at two values of growth rates ($g = 0.2$ and $g = -0.2$)

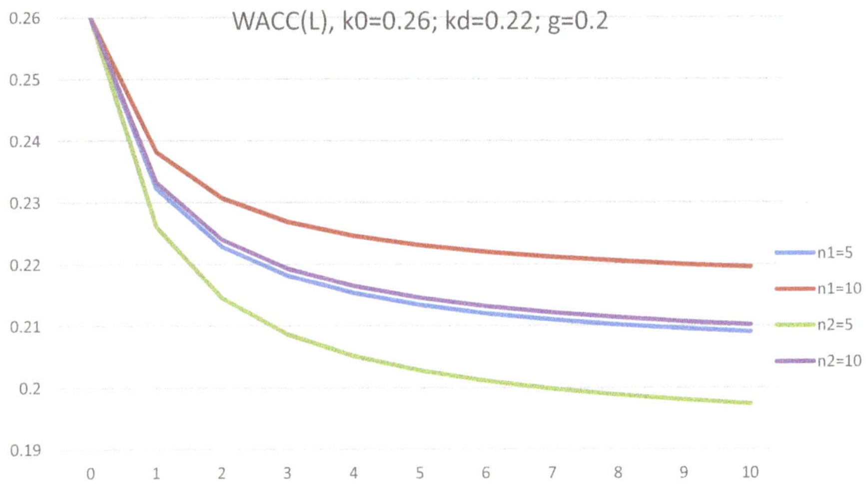

Fig. 17.9 The dependence of WACC on leverage level L in Generalized Brusov–Filatova–Orekhova theory (GBFO theory) with payments of tax on profit at the ends of periods (1) and advance payments (2) at $k_0 = 0.26$; $k_d = 0.22$; $t = 0.2$; $g = 0.2$ for 5-year and 10-year companies

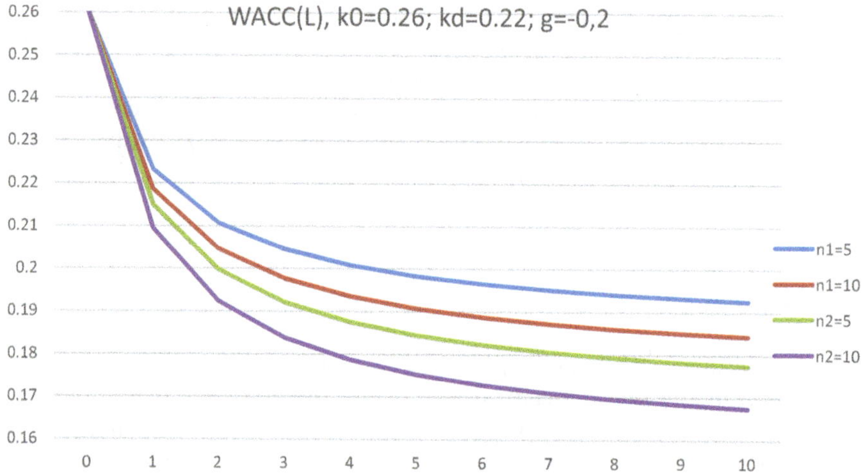

Fig. 17.10 The dependence of WACC on leverage level L in Generalized Brusov–Filatova–Orekhova theory (GBFO theory) with payments of tax on profit at the ends of periods (1) and advance payments (2) at $k_0 = 0.26$; $k_d = 0.22$; $t = 0.2$; $g = -0.2$ for 5-year and 10-year companies

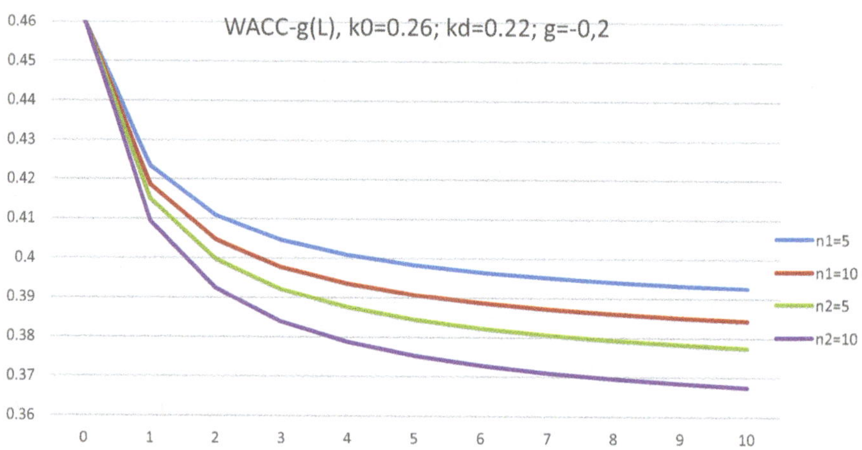

Fig. 17.11 The dependence of discount rate, WACC–g, on leverage level L in Generalized Brusov–Filatova–Orekhova theory (GBFO theory) with payments of tax on profit at the ends of periods (1) and advance payments (2) at $k_0 = 0.26$; $k_d = 0.22$; $t = 0.2$; $g = -0.2$ for 5-year and 10-year companies

(see Figs. 17.11 and 17.12). In both cases, the two WACC(L) curves start from one point: (0; 0.06) for $g = 0.2$ and (0; 0.46) for $g = -0.2$ and decrease with L. For positive growth rate ($g = 0.2$) discount rate WACC–g is lower for the 5-year

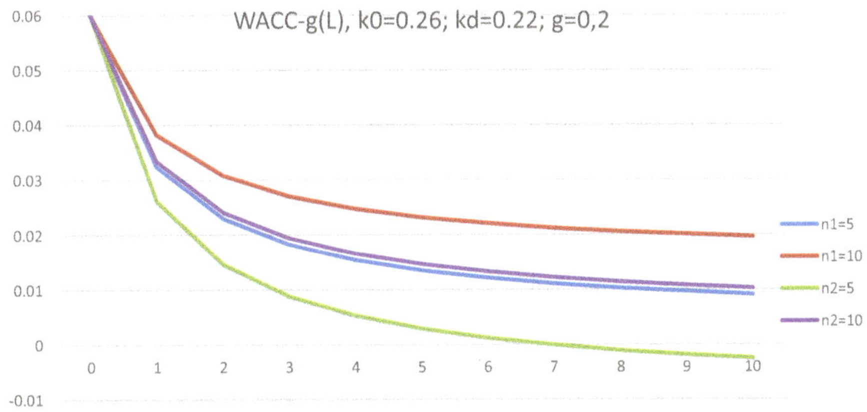

Fig. 17.12 The dependence of discount rate, WACC–g, on leverage level L in Generalized Brusov–Filatova–Orekhova theory (GBFO theory) with payments of tax on profit at the ends of periods (1) and advance payments (2) at $k_0 = 0.26$; $k_d = 0.22$; $t = 0.2$; $g = 0.2$ for 5-year and 10-year companies

company, while for negative growth rate ($g = -0.2$), discount rate WACC–g is lower for the 10-year company. This effect been observed here for the first time.

17.3.3.3 Company Value, V

Let us study the dependence of the value of a company V on L for companies of two ages: 5 and 10 years old at two values of growth rates ($g = 0.2$ and $g = -0.2$). In both cases, the value of the company V increases with the growth of L, and the greater the age of the company corresponds to, the greater the value of the company V. But if, with a positive growth rate ($g = 0.2$), the difference in the value of V for a 5-year company and a 10-year company is about 400, with negative growth rate ($g = -0.2$), this difference is 45 (ten times less) (see Figs. 17.13 and 17.14). So, in the competition between the age of the company and the size of the growth rate, the growth rate wins.

17.3.3.4 Equity Cost, k_e

Let us study the dependence of the equity cost, k_e, on L for companies of two ages: 5 and 10 years old at two values of growth rates ($g = 0.2$ and $g = -0.2$) (see Figs. 17.15 and 17.16). In the case of a positive growth rate ($g = 0.2$), the cost of equity k_e increases linearly with L, and the slope $k_e(L)$ for a 10-year company is greater than for a 5-year one. With a negative growth rate ($g = -0.2$), the cost of

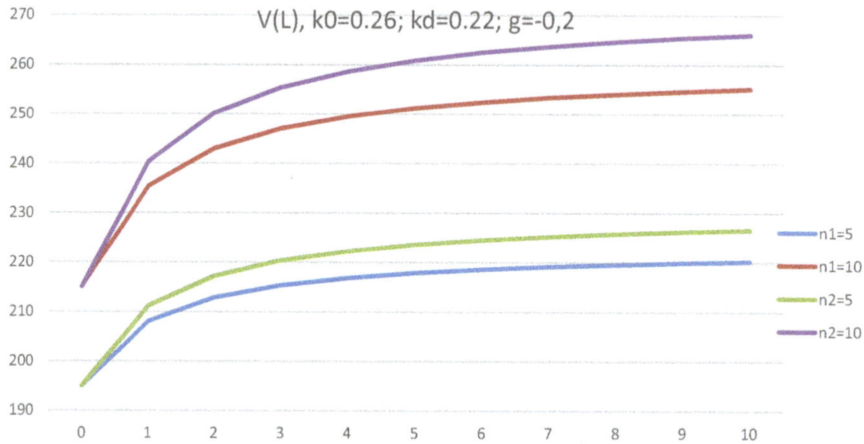

Fig. 17.13 The dependence of company value, V, on leverage level L in Generalized Brusov–Filatova–Orekhova theory (GBFO theory) with payments of tax on profit at the ends of periods (1) and advance payments (2) at $k_0 = 0.26$; $k_d = 0.22$; $t = 0.2$; $g = -0.2$ for 5-year and 10-year companies

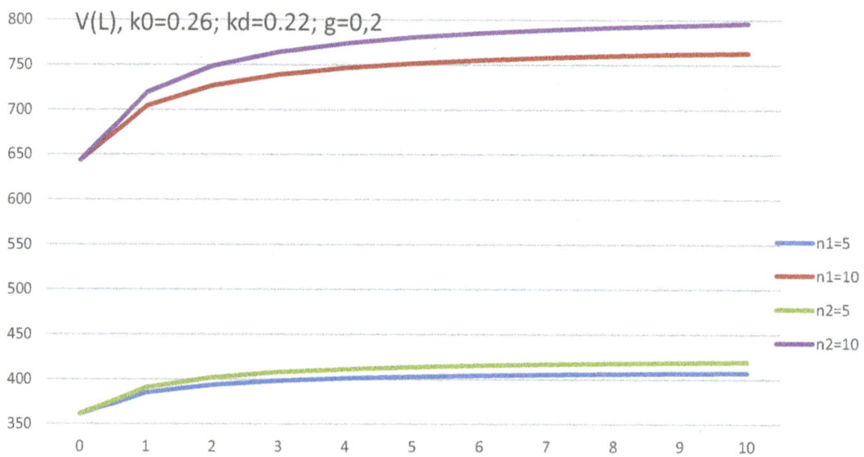

Fig. 17.14 The dependence of company value, V, on leverage level L in Generalized Brusov–Filatova–Orekhova theory (GBFO theory) with payments of tax on profit at the ends of periods (1) and advance payments (2) at $k_0 = 0.26$; $k_d = 0.22$; $t = 0.2$; $g = 0.2$ for 5-year and 10-year companies

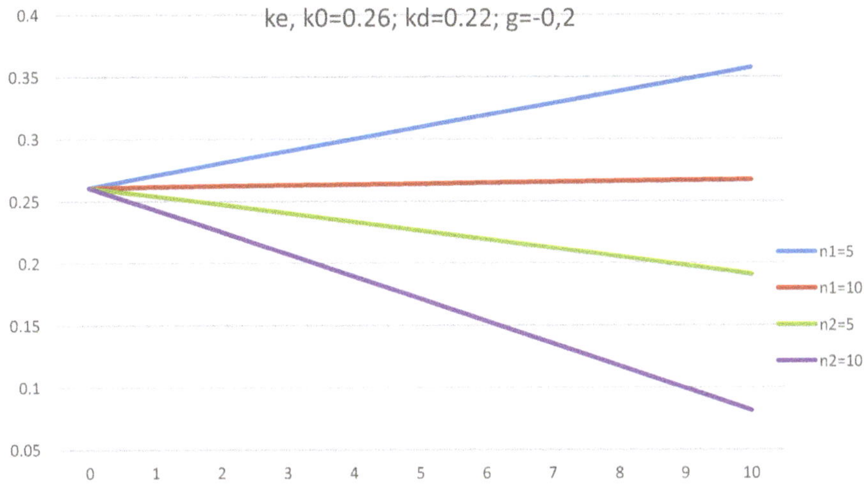

Fig. 17.15 The dependence of equity cost, k_e, on leverage level L in Generalized Brusov–Filatova–Orekhova theory (GBFO theory) with payments of tax on profit at the ends of periods (1) and advance payments (2) at $k_0 = 0.26$; $k_d = 0.22$; $t = 0.2$; $g = -0.2$ for 5-year and 10-year companies

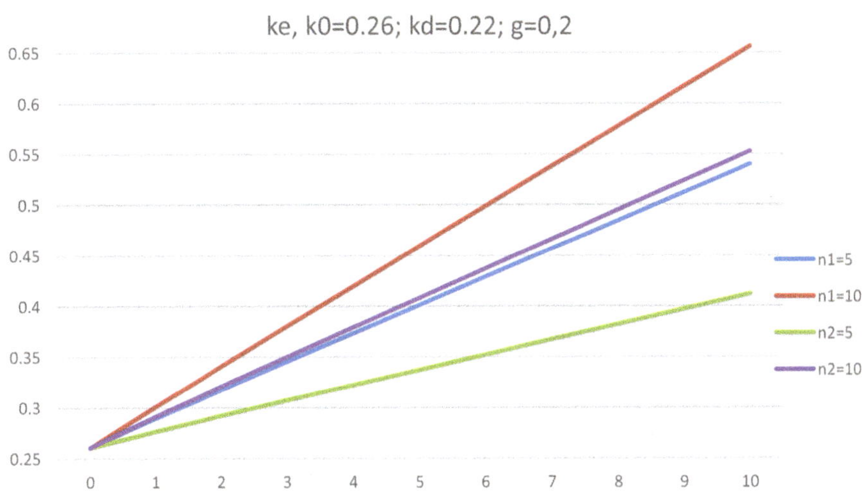

Fig. 17.16 The dependence of equity cost, k_e, on leverage level L in Generalized Brusov–Filatova–Orekhova theory (GBFO theory) with payments of tax on profit at the ends of periods (1) and advance payments (2) at $k_0 = 0.26$; $k_d = 0.22$; $t = 0.2$; $g = 0.2$ for 5-year and 10-year companies

equity k_e decreases linearly with increasing L, and the negative slope for a 10-year company is greater than for a 5-year one.

Since the cost of equity determines the economically justified amount of dividends, this means that the dividend policy of the company when increasing profits and when decreasing profits should be completely different. This is a qualitatively new effect, discovered by us for the first time.

17.3.3.5 Results Summary

The curves (WACC–g)(L) demonstrate the decrease of WACC–g with leverage level L at all g values. The curves (WACC–g)(L) decrease with growth rate, g.

The company value V at fixed growth rate g increases with leverage level L in Generalized Brusov–Filatova–Orekhova theory (GBFO theory). The company value V as well increases with growth rate g. This is a consequence of a decrease in the discount rate (WACC–g)(L) with an increase in the growth rate g.

The equity cost, k_e, practically linearly grows with leverage level L at all growth rate g values. The tilt angle $k_e(L)$ grows with g.

Studying the impact of the debt cost k_d on the main financial indicators, we found the following:

- All curves WACC(L) at all values of k_d start from one point (0; 0.26). WACC(L) decreases with leverage level L at all values of k_d. WACC(L) decreases with the increase of debt cost k_d. This means that tax shield advantages the decrease of the cost of raising capital.
- All curves of discount rate (WACC–g) (L) at all values of k_d start from one point (0; 0.21). (WACC–g) (L) decrease with leverage level L at all values of k_d. (WACC–g)(L) decreases with the increase of debt cost k_d. This means that the tax shield tends to lower the value of the discount rate (WACC–g) and hence increases the value of the company, V.
- All curves of company value $V(L)$ at all values of k_d start from one point (0; 285). $V(L)$ increases with leverage level L at all values of k_d. $V(L)$ increases with the increase of debt cost k_d. This means that tax shield advantages the increase of the company value, V.
- All curves of equity cost, $k_e(L)$, at all values of k_d start from one point (0; 0.26) and k_e increases with leverage level L at all values of k_d. The slope of the straight line $k_e(L)$ decreases with the cost of debt k_d. This means that debt cost k_d impacts the dividend policy of the company, because the equity cost k_e determines the economically justified amount of dividends.

Studying the impact of company age, n, on main financial indicators of the company: WACC; WACC–g; V; k_e, w_e found a huge difference between the behavior of the main financial indicators of the company with a positive and negative growth rate g. This allows you to explore companies with growing profits and companies with decreasing profits. And also it allows study of the companies whose profits rise and fall in different periods.

Particular Results Here Are as Follows

We investigate the dependence of WACC on L for companies of two ages: 5 and 10 years old at two values of growth rates ($g = 0.2$ and $g = -0.2$). In both cases, the two WACC(L) curves start from the point (0; 0.26) and decrease with L. For positive growth rate ($g = 0.2$) WACC is lower for a 5-year company, while for negative growth rate ($g = -0.2$) WACC is lower for a 10-year company. We are seeing this effect for the first time.

We investigate the dependence of discount rate WACC–g on L for companies of two ages: 5 and 10 years old at two values of growth rates ($g = 0.2$ and $g = -0.2$). In both cases, the two WACC(L) curves start from one point: (0; 0.06) for $g = 0.2$ and (0; 0.46) for $g = -0.2$ and decrease with L. For positive growth rate ($g = 0.2$) discount rate WACC–g is lower for a 5-year company, while for negative growth rate ($g = -0.2$) discount rate WACC–g is lower for a 10-year company. We are seeing this effect for the first time.

We study the dependence of the value of a company V on L for companies of two ages: 5 and 10 years old at two values of growth rates ($g = 0.2$ and $g = -0.2$). In both cases, the value of the company V increases with the growth of L, and the greater the age of the company corresponds to, the greater the value of the company V. But if, with a positive growth rate ($g = 0.2$), the difference in the value of V for a 5-year company and a 10-year company is about 400, with negative growth rate ($g = -0.2$), this difference is 45 (ten times less). So, in the competition between the age of the company and the size of the growth rate, the growth rate wins.

The dependence of the equity cost, k_e, on L is studied for companies of two ages: 5 and 10 years old at two values of growth rates ($g = 0.2$ and $g = -0.2$). In the case of a positive growth rate ($g = 0.2$), the cost of equity k_e increases linearly with L, and the slope $k_e(L)$ for a 10-year company is greater than for a 5-year one. With a negative growth rate ($g = -0.2$), the cost of equity k_e decreases linearly with increasing L, and the negative slope for a 10-year company is greater than for a 5-year one.

Since the cost of equity determines the economically justified amount of dividends, this means that the dividend policy of the company when increasing profits and when decreasing profits should be completely different. This is a qualitatively new effect, discovered by us for the first time.

As it is seen from the Table 17.1, the most significant qualitative differences between cases (1) and (2) occur for the dependence of financial indicators on the age of the company. For the cases of influence g and k_d, despite the fact that the financial parameters for cases (1) and (2) move in the same direction, there are significant differences between cases (1) and (2) from the quantitative point of view as well as with respect to interposition of the financial parameters.

Table 17.1 Comparison of the results in Generalized Brusov–Filatova–Orekhova theory (GBFO theory) with payments of tax on profit at the ends of periods (1) and advance payments (2) for parameters g (growth rate), k_d (debt cost), and n (company age [for 5-year and 10-year companies]) (↑ means an increase in the indicator with the current parameter; ↓ means a decrease in the indicator with the current parameter)

	g Growth rate		k_d Debt cost		n Company age			
	Payments at the ends of periods (1)	Advance payments (2)	Payments at the ends of periods (1)	Advance payments (2)		Payments at the ends of periods (1)	Advance payments (2)	Comparison of (1) and (2)
WACC	↑	↑	↓	↓	$g \geq 0$	↑	↑	1≥2
					$g \leq 0$	↓	↓	
WACC–g	↓	↓	↓	↓	$g \geq 0$	↑	↑	1≥2
					$g \leq 0$	↓	↓	
V	↑	↑	↑	↑	$g \geq 0$	↑	↑	1≤2
					$g \leq 0$	↑	↑	
k_e	↑	↑	↓	↓	$g \geq 0$	↑	↑	1≤2
					$g \leq 0$	↓	↓	

17.4 Conclusions

Two modifications of the Brusov–Filatova–Orekhova (BFO) theory with variable income have been considered: (1) with the payment of income tax at the end of periods and (2) with advance payments of income tax (for the first time). BFO formulas for the weighted average cost of capital, WACC, for company value, V, were derived for these two cases and within these formulas a comprehensive analysis of the dependence of WACC, of discount rate, WACC–g, company value, V, the equity cost, k_e, on the debt financing at different values of the growth rate, g, at different values of the cost of debt capital, k_d, and at different values of company age, n was carried out.

The results for cases (1) and (2) are compared, which allow us to conclude that case (2) is always preferable for both the company and the regulator. This allows to develop recommendations for both parties to expand the practice of advance payments of income tax.

Chapter novelties are as follows:

1. Generalization of the Brusov–Filatova–Orekhova theory to the case of variable income with advance payments of income tax.
2. Derivation of generalized BFO formulas for weighted average cost of capital, WACC, cost of equity, k_e, company value, V.
3. Using the obtained formulas, investigate the influence of the growth rate g, the cost of debt, k_d, the company's age, n, on the dependence of the company's financial performance on debt financing.
4. Comparison of the obtained results with those for the case of paying income tax at the end of the periods.
5. Developing the recommendations for both companies and regulators on how to pay income tax.
6. The developed methodology makes it possible to study companies with growing profits and companies with decreasing profits, that is quite important for practice. And also it allows study of the companies whose profits rise and fall in different periods.

Regarding the direction of future research: The proposed theory will be used in the future for a more detailed study of companies with both rising and falling profits.

References

1. Brusov, P., Filatova, T., Orekhova, N., Kulik, V., Chang, S.-I., & Lin, G. (2021). Generalization of the Modigliani–Miller theory for the case of variable profit. *Mathematics, 9*, 1286. https://doi.org/10.3390/math9111286
2. Brusov, P. (2022). Tatiana Filatova generalization of the Brusov–Filatova–Orekhova theory for the case of variable income. *Mathematics, 10*(19), 3661. https://doi.org/10.3390/math10193661

3. Modigliani, F., & Miller, M. (1958). The cost of capital, corporate finance, and the theory of investment. *The American Economic Review, 48*, 261–297.
4. Modigliani, F., & Miller, M. (1963). Corporate income taxes and the cost of capital: A correction. *The American Economic Review, 53*, 147–175.
5. Modigliani, F., & Miller, M. (1966). Some estimates of the cost of capital to the electric utility industry 1954–1957. *The American Economic Review, 56*, 333–391.
6. Brusov, P., Filatova, T., Orekhova, N., & Eskindarov, M. (2018). *Modern corporate finance, investments, taxation and ratings* (2nd ed., pp. 1–571). Springer Nature Publishing.
7. Hamada, R. (1969). Portfolio analysis, market equilibrium, and corporate finance. *The Journal of Finance, 24*, 13–31.
8. Miller, M. (1977). Debt and taxes. *The Journal of Finance, 32*, 261–275.
9. Farber, A., Gillet, R., & Szafarz, A. A. (2006). General formula for the WACC. *International Journal of Business, 11*, 211–218.
10. Fernandez, P. A. (2006). General formula for the WACC: A comment. *International Journal of Business, 11*, 219.
11. Berk, J., & De Marzo, P. (2007). *Corporate finance*. Pearson–Addison Wesley.
12. Harris, R., & Pringle, J. (1985). Risk–adjusted discount rates – Extension from the average–risk case. *Journal of Financial Research, 8*, 237–244.
13. Batrancea, L. (2021). An econometric approach regarding the impact of fiscal pressure on equilibrium: Evidence from electricity, gas and oil companies listed on the New York stock exchange. *Mathematics, 9*, 630.
14. Batrancea, L. (2021). The influence of liquidity and solvency on performance within the healthcare industry: Evidence from publicly listed companies. *Mathematics, 9*, 2231.
15. Myers, S. (2001). Capital structure. *The Journal of Economic Perspectives, 15*, 81–102.
16. Dimitropoulos, P. (2014). Capital structure and corporate governance of soccer clubs: European evidence. *Management Research Review, 37*, 658–678.
17. Luiz, K., & Cruz, M. (2015). The relevance of capital structure on firm performance: A multivariate analysis of publicly traded Brazilian companies. *REPeC Brasília, 9*, 384–401.
18. Barbi, M. (2011). On the risk–neutral value of debt tax shields. *Applied Financial Economics, 22*, 251–258.
19. Zhukov, P. (2018). The impact of cash flows and weighted average cost of capital to enterprise value in the oil and gas sector. *Journal of Reviews on Global Economics, 7*, 138–145.
20. Franc-Dąbrowska, J., Mądra-Sawicka, M., & Milewska, A. (2021). Energy sector risk and cost of capital assessment—Companies and investors perspective. *Energies, 14*, 1613.
21. Angotti, M., de Lacerda Moreira, R., Hipólito Bernardes do Nascimento, J., & Neto de Almeida Bispo, O. (2018). Analysis of an equity investment strategy based on accounting and financial reports in Latin American markets. *Reficont, 5*, 22–40.
22. Vergara-Novoa, C., Sepúlveda-Rojas, J. P., Miguel, D. A., & Nicolás, R. (2018). Cost of capital estimation for highway concessionaires in Chile. *Journal of Advanced Transportation*, 2153536.
23. Mundi, H. S., Kaur, P., & Murty, R. L. N. (2022). A qualitative inquiry into the capital structure decisions of overconfident finance managers of family–owned businesses in India. *Qualitative Research in Financial Markets, 14*, 357–379.
24. Sadiq, M., Alajlani, S., Hussain, M. S., Bashir, F., & Chupradit, S. (2022). Impact of credit, liquidity, and systematic risk on financial structure: Comparative investigation from sustainable production. *Environmental Science and Pollution Research International, 29*, 20963–20975.
25. Becker, D. M. (2022). Getting the valuation formulas right when it comes to annuities. *Managerial Finance, 48*, 470.
26. El-Chaarani, H., Abraham, R., & Skaf, Y. (2022). The impact of corporate governance on the financial performance of the banking sector in the MENA (Middle Eastern and North African) region: An immunity test of banks for COVID–19. *Journal of Risk and Financial Management, 15*, 82.

27. Huang, S., Sun, H., Zhao, H., & Zhang, Y. (2020). Influence of leverage on the return on equity. *Systems Engineering - Theory & Practice, 40*, 355.
28. Islam, S. Z., & Khandaker, S. (2015). Firm leverage decisions: Does industry matter? *North American Journal of Economics and Finance, 31*, 94.
29. Singhal, N., Goyal, S., Sharma, D., Kumari, S., & Nagar, S. (2022). Capitalization and profitability: Applicability of capital theories in BRICS banking sector. *Future Business Journal, 8*, 1. https://doi.org/10.1186/s43093-022-00140-w

Chapter 18
New Meaningful Effects in Modern Capital Structure Theory

18.1 Introduction

One of the main and the most important problems in corporate finance is the problem of cost of capital, the impact of capital structure on its cost and capitalization of the companies and problem of an optimal capital structure of the companies (at which the company capitalization is maximal, and weighted average cost of capital WACC is minimal). The importance of these problems is connected to the fact, that one can doing nothing, just by change the ratio between debt and equity (by change the capital structure) increase the capitalization of the company, i.e., solve the main task of any company.

However, to date, even the question of the existence of an optimal capital structure of the companies still remains open. Numerous theories and models, including the first and the only one until recently, the quantitative theory by Nobel laureates Modigliani and Miller (MM), not only do not solve the problem but also because of the large number of restrictions (such as, for example, theory of MM) have a weak relationship to the real economy. Herewith the qualitative theories and models, based on the empirical approaches, do not allow to carry out the necessary assessment.

This special issue is devoted to recent development of capital structure theory and its applications. Discussions will be made within both main theories: modern theory by Brusov, Filatova, and Orekhova (BFO theory) and its perpetuity limit—classical Modigliani–Miller (MM) theory, which will be compared in detail. From 2008, the BFO theory has replaced the famous theory of capital cost and capital structure by Nobel laureates Modigliani and Miller. The authors of BFO have moved from the assumption of Modigliani–Miller concerning the perpetuity (infinite time of life) of companies and further elaborated quantitative theory of valuation of core parameters of financial activities of companies of arbitrary age as well as of arbitrary time of life.

Results of modern BFO theory turn out to be quite different from those of the Modigliani–Miller theory. Brusov, Filatova, and Orekhova show that later, via its

P. Brusov, T. Filatova, *Business Valuation and Company Value*, Contributions to Finance and Accounting, https://doi.org/10.1007/978-3-031-90847-7_18

perpetuity, underestimates the assessment of weighted average cost of capital, the equity cost of the company and substantially overestimates the assessment of the capitalization of the company.

Such an incorrect assessment of key performance indicators of financial activities of companies has led to an underestimation of risks involved, and impossibility, or serious difficulties in adequate managerial decision-making, that was one of the implicit reasons of the global financial crisis of the year 2008.

Within the modern theory of capital cost and capital structure (BFO theory), a lot of qualitative new results, described in this paper, have been obtained, among them:

- **Bankruptcy of the famous trade-off theory has been proven.** BFO theory has destroyed some main existing principles of financial management: among them the trade-off theory, which was considered as keystone of formation of optimal capital structure of the company during many decades.

It would be a great pity if the optimal capital structure of the company does not exist in general, thus BFO authors have suggested the **Mechanism of formation of the company optimal capital structure**, different from that suggested by the trade-off theory.

- **The qualitative new effect in corporate finance** has been discovered by the BFO authors: abnormal dependence of equity cost on leverage, which significantly alters the principles of the company's **dividend policy**.
- Existence of **"A golden age" of the companies** has been discovered. It was shown for the first time that valuation of WACC in the Modigliani–Miller theory is not minimal and valuation of the company capitalization is not maximal, as all financiers have supposed up to now: at some age of the company, its WACC value turns out to be lower than in Modigliani–Miller theory (in perpetuity limit) and company capitalization V at some company age turns out to be greater, than company capitalization V in Modigliani–Miller theory.
- **The inflation in both Modigliani–Miller as well as in Brusov–Filatova–Orekhova theories** has been taken into account in explicit form, with the detected its non-trivial impact on the dependence of equity cost on leverage.
- **Study of the role of taxes and leverage** has been done, and the obtained results allow the Regulator to set the tax on profits rate, and to businesses to choose the optimal level of debt financing.
- **Investigation of the influence of tax on profit rate on effectiveness of investment projects** at different debt levels showed that increase of tax on profit rate from one side leads to decrease of project NPV, but from other side, it leads to decrease of sensitivity of NPV with respect to leverage level. At high leverage level L, the influence of tax on profit rate on effectiveness of investment projects becomes significantly less.
- **The influence of growth of tax on profit rate** on the efficiency of the investment as well has led to two qualitatively new effects in investments:

The growth of tax on profit rate changes the nature of the NPV dependence on leverage L at some value t^*: there is a transition from diminishing function NPV(L) at $t < t^*$, to growing function NPV(L) at $t > t^*$.

At high leverage levels, the growth of tax on profit rate leads to the growth of the efficiency of the investments.

Discovered effects in investments can be applied in a real economic practice for optimizing of the management of investments.

Established BFO theory allows to conduct a valid assessment of the core parameters of financial activities of companies, such as weighted average cost of capital and equity capital cost of the company, its capitalization. It allows the management of a company to make adequate decisions that improve the effectiveness of the company management. More generally, **the introduction of the new system of evaluation of the parameters of financial activities of companies into the systems of financial reporting (IFRS, GAAP etc.) would lead to lower risk of global financial crisis.**

Corporate management in the modern world is the management of financial flows. The proposed Brusov–Filatova–Orekhova theory allows correctly to identify a discount rate—basic parameters for discounting of financial flows to arbitrary time moment, compare financial flows with a view to adoption of literate managerial decisions. The discount rate is a key link of the existing financial system, by pulling on the which modern finance can be adequately built, and the BFO theory can assist in this.

In this paper, we discuss numerous new meaningful effects in modern capital structure theory.

18.2 Comparison of Modigliani–Miller (MM) and Brusov–Filatova–Orekhova (BFO) Results

18.2.1 The Traditional Approach

The traditional (empirical) approach informed businessmen that weighted average cost of capital, WACC, and the associated company capitalization, $V = \text{CF/WACC}$ depend on the capital structure, the level of leverage. Debt cost always turns out to be lower than equity cost because first one has lower risk, because in the event of bankruptcy, creditors' claims are met prior to shareholders' claims.

As a result, an increase in the proportion of lower-cost debt capital in the overall capital structure up to the limit which does not cause violation of financial sustainability and growth in risk of bankruptcy leads to lower weighted average cost of capital, WACC.

The required by investors profitability (the equity cost) is growing; however, its growth has not led to compensation benefits from use of more low-cost debt capital. Therefore, the traditional approach welcomes the increased leverage $L = D/S$, and

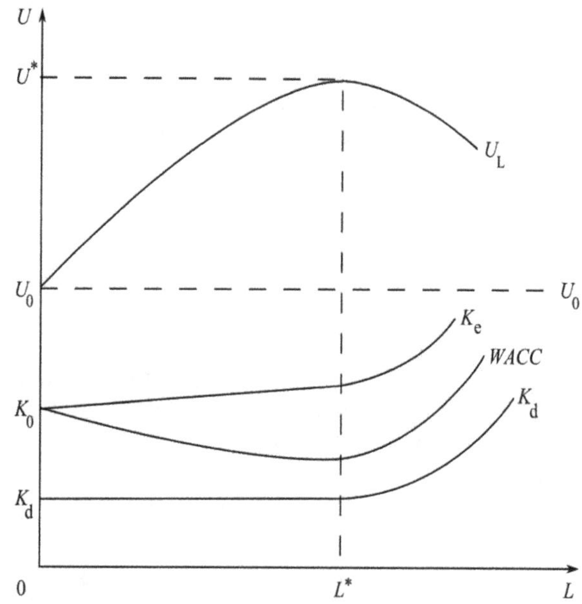

Fig. 18.1 Dependence of company capitalization, U_L, equity cost, k_e, debt cost, k_d, weighted average cost of capital, WACC, in traditional (empirical) approach

the associated increase of company capitalization. The traditional (empirical) approach has existed up to the appearance of the first quantitative theory by Modigliani and Miller in 1958 (Fig. 18.1) [1].

18.2.2 Modigliani–Miller Theory

The Modigliani–Miller theory with taxes is based on the following three formulae for capitalization: V, WACC, and equity cost k_e (Fig. 18.2).

$$V = V_0 + Dt,$$

$$\text{WACC} = k_0(1 - w_d T),$$

$$k_e = k_0 + L(1 - T)(k_0 - k_d).$$

One of the most important assumptions of the Modigliani–Miller theory is that all financial flows are in perpetuity.

This limitation was lifted out by Brusov–Filatova–Orekhova in 2008, who have created the BFO theory–modern theory of capital cost and capital structure for companies of arbitrary age (BFO–I) and for companies of arbitrary lifetime (BFO–II) (Fig. 18.3) [2].

Note that before 2008, only two results for capital structure of company were available:

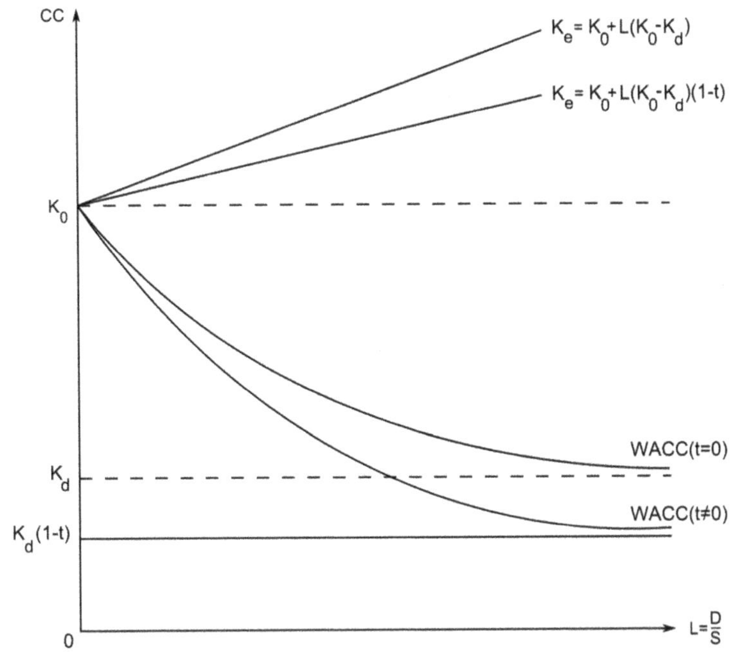

Fig. 18.2 Dependence of equity capital cost, debt cost, and WACC on leverage in Modigliani–Miller theory without taxes ($t = 0$) and with taxes ($t \neq 0$)

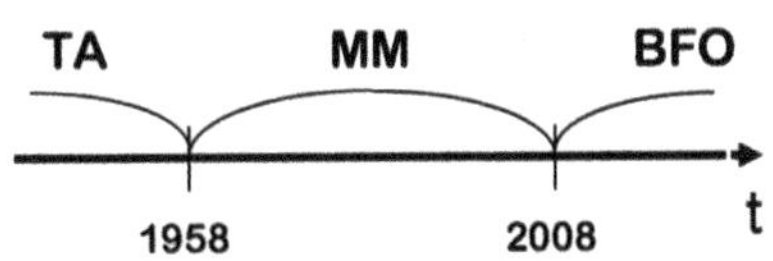

Fig. 18.3 Historical development of capital structure theory (here TA—traditional (empirical) approach, MM—Modigliani–Miller approach, and BFO—Brusov–Filatova–Orekhova theory)

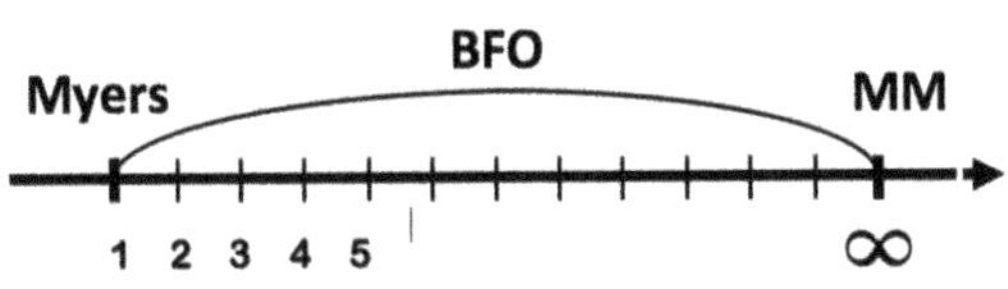

Fig. 18.4 The MM theory describes perpetuity limit, the Myers paper describes 1-year company, while the BFO theory fills the whole numeric axis (from $n = 1$ up to perpetuity limit $n = \infty$)

Modigliani–Miller for perpetuity company and Myers for 1-year company (see Fig. 18.4).

BFO theory has filled out whole interval between $t = 1$ and $t = \infty$. One got the possibility to calculate capitalization V, WACC, and equity cost k_e for companies of arbitrary age and for companies of arbitrary lifetime. The BFO theory has led to a lot of new meaningful effects in modern capital structure theory, as discussed in this paper.

The BFO theory is based on the famous formula

$$\frac{1-(1+\mathrm{WACC})^{-n}}{\mathrm{WACC}}=\frac{1-(1+k_0)^{-n}}{k_0[1-\omega_d T(1-(1+k_d)^{-n})]}. \tag{18.1}$$

Here, S—the value of own (equity) capital of the company, $w_d=\frac{D}{D+S}$—the share of debt capital; $k_e, w_e=\frac{S}{D+S}$— the cost and the share of the equity of the company, $L=D/S$— financial leverage.

18.3 Comparison of Modigliani–Miller Results (Perpetuity Company) with Myers Results (1-Year Company) and Brusov–Filatova–Orekhova Ones (Company of Arbitrary Age)

We could compare the Modigliani–Miller results (perpetuity company) with Myers results (1-year company) and Brusov–Filatova–Orekhova ones (company with arbitrary age) under valuation of WACC and equity cost.

From Tables 18.1, 18.2 and Fig. 18.5. it is obvious that WACC has a maximum for a 1-year company and decreases with the age (lifetime) of the company, reaching the minimum in the Modigliani–Miller perpetuity case. (Note, however, that this will not always be so via the effect of "golden age" of the company [see below]).

Results of modern the BFO theory turn out to be quite different from the ones of Modigliani–Miller theory. They show that later, via its perpetuity, it underestimates the assessment of weighted average cost of capital, the equity cost of the company, and substantially overestimates the assessment of the capitalization of the company.

Table 18.1 Dependence of WACC and k_e on leverage level for $n = 1$, and $n = \infty$

L	K_o	K_d	t	n	W_d	WACC	BFO	K_e	WACC MM	MM K_e
0	0.2	0.1	0.2	1	0.00	20.00%	0.000	0.2000	20.00%	0.2000
1	0.2	0.1	0.2	1	0.50	18.91%	0.000	0.2982	18.00%	0.2800
2	0.2	0.1	0.2	1	0.67	18.55%	0.000	0.3964	17.33%	0.3600
3	0.2	0.1	0.2	1	0.75	18.36%	0.000	0.4945	17.00%	0.4400
4	0.2	0.1	0.2	1	0.80	18.25%	0.000	0.5927	16.80%	0.5200
5	0.2	0.1	0.2	1	0.83	18.18%	0.000	0.6909	16.67%	0.6000
6	0.2	0.1	0.2	1	0.86	18.13%	0.000	0.7891	16.57%	0.6800
7	0.2	0.1	0.2	1	0.88	18.09%	0.000	0.8873	16.50%	0.7600
8	0.2	0.1	0.2	1	0.89	18.06%	0.000	0.9855	16.44%	0.8400
9	0.2	0.1	0.2	1	0.90	18.04%	0.000	1.0836	16.40%	0.9200
10	0.2	0.1	0.2	1	0.91	18.02%	0.000	1.1818	16.36%	1.0000

Table 18.2 Dependence of WACC and k_e on leverage level for $n = 3$, and $n = \infty$

L	K_o	K_d	t	n	W_d	WACC	BFO	K_e	WACC MM	MM K_e
0	0.2	0.1	0.2	3	0.00	20.00%	0.000	0.2000	20.00%	0.2000
1	0.2	0.1	0.2	3	0.50	18.41%	0.000	0.2881	18.00%	0.2800
2	0.2	0.1	0.2	3	0.67	17.87%	0.000	0.3762	17.33%	0.3600
3	0.2	0.1	0.2	3	0.75	17.61%	0.000	0.4642	17.00%	0.4400
4	0.2	0.1	0.2	3	0.80	17.44%	0.000	0.5522	16.80%	0.5200
5	0.2	0.1	0.2	3	0.83	17.34%	0.000	0.6402	16.67%	0.6000
6	0.2	0.1	0.2	3	0.86	17.26%	0.000	0.7283	16.57%	0.6800
7	0.2	0.1	0.2	3	0.88	17.20%	0.000	0.8163	16.50%	0.7600
8	0.2	0.1	0.2	3	0.89	17.16%	0.000	0.9043	16.44%	0.8400
9	0.2	0.1	0.2	3	0.90	17.12%	0.000	0.9923	16.40%	0.9200
10	0.2	0.1	0.2	3	0.91	17.09%	0.000	1.0803	16.36%	1.0000

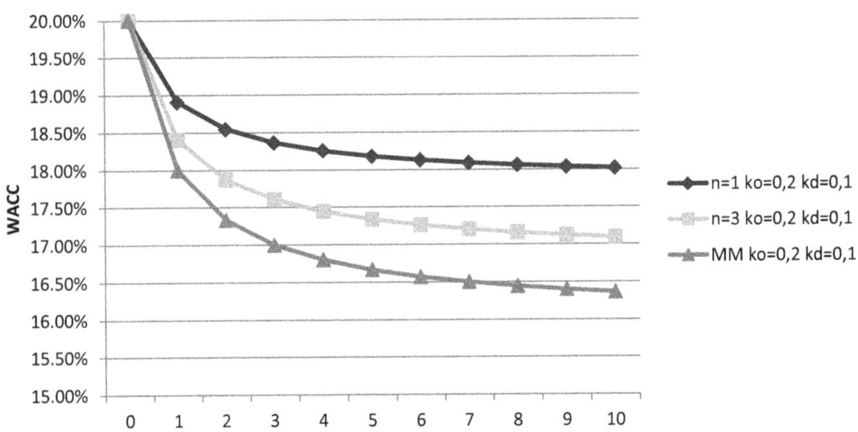

Fig. 18.5 Dependence of WACC on leverage level for $n = 1$, $n = 3$, and $n = \infty$

Such an incorrect assessment of key performance indicators of financial activities of companies has led to an underestimation of risks involved, and impossibility, or serious difficulties in adequate managerial decision-making, that was one of the implicit reasons for the global financial crisis in 2008.

The BFO theory allows to make a correct assessment of key parameters of financial activities of companies of arbitrary age (arbitrary lifetime) that leads accordingly to adequate managerial decision-making.

18.4 Bankruptcy of the Famous Trade-Off Theory

Within modern theory of capital structure and capital cost by Brusov–Filatova–Orekhova [3–13] analyses of widely known trade-off theory have been made. It is shown that suggestion of risky debt financing (and growing credit rate near the bankruptcy) as opposed to waiting result does not lead to growing of weighted average cost of capital, WACC, which still decreases with leverage. This means the absence of minimum in the dependence of WACC on leverage as well as the absence of maximum in the dependence of company capitalization on leverage. Thus, it means that the optimal capital structure is absent in the famous trade-off theory. The explanation to this fact has been done.

In modified Modigliani–Miller theory, we have proved the following theorem:

In modified Modigliani–Miller (MMM) theory (allowing riskiness debt capital), under arbitrary change of debt cost with leverage (growing, as well as decrease), weighted average cost of capital, WACC, always falls with leverage. This means the absence of the company optimal capital structure within MMM theory and proves insolvency well-known classical trade-off in its original formulation.

We consider linear and quadratic growth of debt cost $k_d$with leverage, starting from some value (with different coefficients), different values of k_0, and different terms of life of the companies. Let us find WACC values (Table 18.3, Fig. 18.6).

1. $n = 3$; $t = 20\ \%$; $L = 0, 1, 2, \ldots 10$

$$k_0 = 24\%; \overline{k}_d = \left\{ \begin{array}{c} 0,07; atL \le 2 \\ 0,07 + 0,01(L-2)^2; atL > 2 \end{array} \right\} \tag{18.2}$$

Let us see, how the growth of debt cost k_d with leverage affects the equity cost k_e dependence on leverage. We will consider the same cases as above for the calculations of dependences WACC(L) (Table 18.4, Fig. 18.7).

1.

$$n = 3; t = 20\%; L = 0, 1, 2, \ldots 10$$

$$k_0 = 24\%; \overline{k}_d = \left\{ \begin{array}{c} 0.07; atL \le 2 \\ 0.07 + 0.01(L-2)^2; atL > 2 \end{array} \right\} \tag{18.3}$$

The analysis of the well-known trade-off theory, conducted with the help of modern theory of capital structure and capital cost by Brusov–Filatova–Orekhova, has shown that that suggestion of risky debt financing (and growing credit rate near the bankruptcy) as opposed to waiting result does not lead to growing of WACC, which still decreases with leverage. This means the absence of minimum in the dependence of WACC on leverage as well as the absence of maximum in the

Table 18.3 Dependence of WACC on L

n	L	0	1	2	3	4	5	6	7	8	9	10
3	k_d	0.07	0.07	0.07	0.08	0.11	0.16	0.23	0.32	0.43	0.56	0.71
k_0	A	1.9813	2.0184	2.0311	2.0445	2.0703	2.1075	2.1520	2.1988	2.2438	2.2842	2.3186
0.24	WACC	0.2401	0.2279	0.2238	0.2195	0.2111	0.1997	0.1864	0.1730	0.1605	0.1496	0.1406

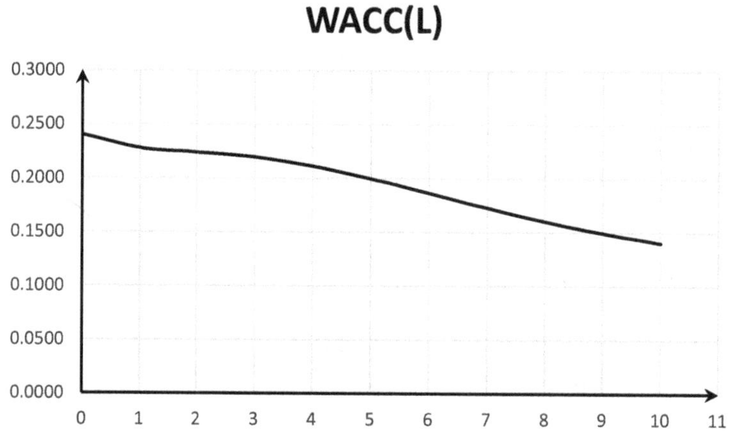

Fig. 18.6 Dependence of WACC on L

dependence of capitalization V on leverage. *Thus, it seems that the optimal capital structure is absent in famous trade-off theory.* The explanation to this fact has been done within the same Brusov–Filatova–Orekhova theory by studying the dependence of the equity cost k_e with leverage. It turned out that the growth of debt cost k_d with leverage leads to decrease of equity cost k_e with leverage, starting from some leverage level, which is higher than starting point of debt cost growth. This paradox conclusion gives the explanation of the absence of the optimal capital structure in the famous trade-off theory. This means, that competition of benefits from using of debt financing and of financial distress cost (or a bankruptcy cost) are NOT balanced and hopes that the trade-off theory gives us the optimal capital structure, unfortunately, do not realized.

The absence of the optimal capital structure in the trade-off theory questioned the existence of an optimal capital structure of the company (but as authors have shown, the optimal capital structure for the investment still exists [6, 8]). In the search for the "golden fleece," one needs to switch to study of other mechanisms for formation of the capital structure of the company, different from the ones considered in the trade-off theory.

18.5 The Qualitatively New Effect in Corporate Finance

Qualitatively new effect in corporative finance is discovered: decreasing of cost of equity k_e with leverage L. This effect, which is absent in perpetuity Modigliani–Miller limit, takes place under account of finite lifetime of the company at tax on profit rate, which exceeds some value T^*.

Table 18.4 Dependence of equity cost k_e on L

n	L	0	1	2	3	4	5	6	7	8	9	10
3	k_d	0.07	0.07	0.07	0.08	0.11	0.16	0.23	0.32	0.43	0.56	0.71
k_0	A	1.9813	2.0184	2.0311	2.0445	2.0703	2.1075	2.1520	2.1988	2.2438	2.2842	2.3186
0.24	Ke	0.2401	0.3997	0.5594	0.6861	0.7036	0.5581	0.2011	−0.4081	−1.3075	−2.5356	−4.133

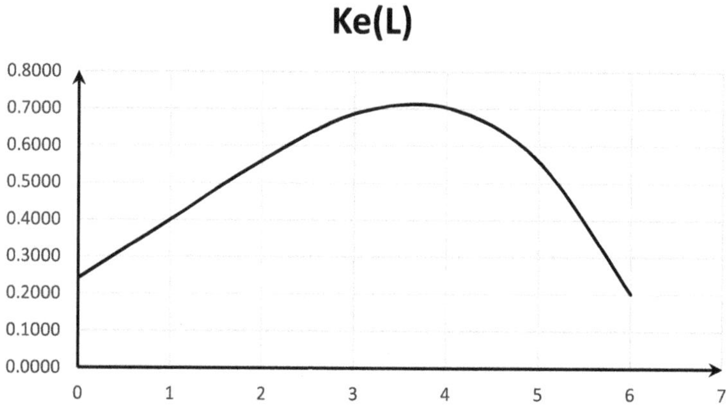

Fig. 18.7 Dependence of equity cost $k_e$on L

Fig. 18.8 Dependence of cost of equity on leverage L at different tax on profit rates T for the case $k_0 = 10\ \%$; $k_d = 8\%$ (1 – $T = 0$; 2 – $T = 0.1$; 3 – $T = 0.2$; 4 – $T = 0.3$; 5 – $T = 0.4$; 6 – $T = 0.5$; 7 – $T = 0.6$; 8 – $T = 0.7$; 9 – 10 – $T = 0.9$; 11 – $T = 1$)

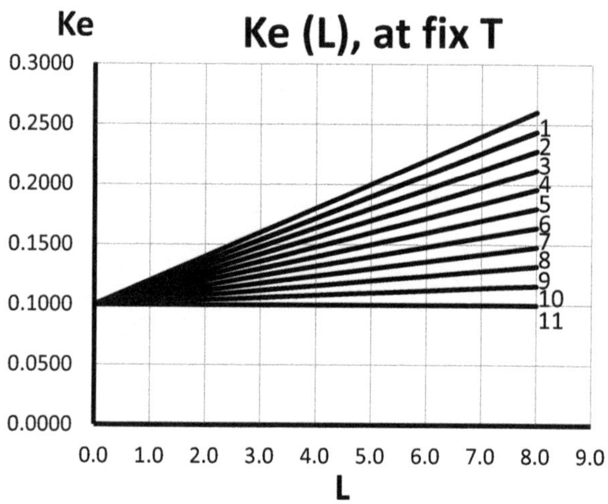

At some ratios between cost of debt and cost of equity, the discovered effect takes place at tax on profit rate, existing in western countries and Russia. This provides the practical meaning of discussed effect. Its accounting is important at modification of tax low and can change the dividend policy of the company.

18.5.1 Perpetuity Modigliani–Miller Limit

One sees in Fig. 18.8 that position limit of dependence of cost of equity on leverage L is horizontal line 11 at $T = 1$ (Fig. 18.9). Below we'll see that in BFO theory, the

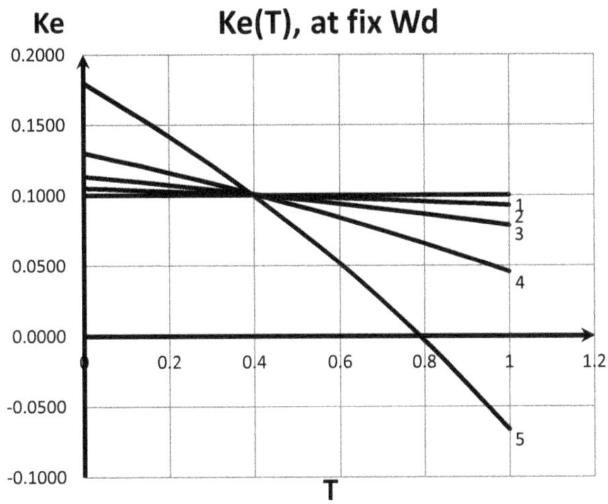

Fig. 18.9 Dependence of cost of equity k_e on tax on profit rate T at different fix leverage level L ($n = 10$, $k_0 = 10\ \%$, $k_d = 8\%$)(1 – $w_d = 0$; 2 – $w_d = 0.2$; 3 – $w_d = 0.4$; 4 – $w_d = 0.6$; 5 – $w_d = 0.8$)

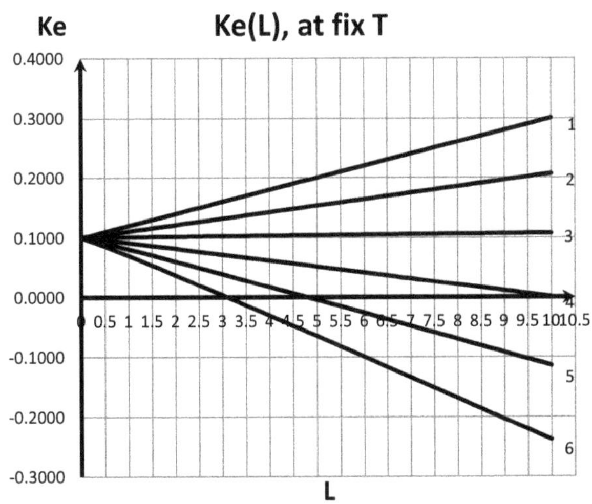

Fig. 18.10 Dependence of cost of equity k_e on leverage level L at different tax on profit rate T ($n = 5$, $k_0 = 10\ \%$, $k_d = 8\%$)(1 – $T = 0$; 2 – $T = 0.2$; 3 – $T = 0.4$; 4 – $T = 0.6$; 5 – $T = 0.8$; 6 – $T = 1$)

abnormal effect takes place (see Fig. 18.10) and dependence of cost of equity on leverage L line could have a negative slope.

18.5.2 BFO Theory

In Fig. 18.10 it is seen that dependence of cost of equity k_e on leverage level L with a good accuracy is linear. The tilt angle decreases with tax on profit rate like the perpetuity case.

However, for the finite lifetime of companies along with the behavior $k_e(L)$, similar to the perpetuity behavior of the Modigliani–Miller case (Fig. 18.8), for some sets of parameters n, k_0, k_d there is another behavior $k_e(L)$.

In the Fig. 18.10, it is seen that starting from some values of tax on profit rate T^* (in this case from $T^* = 40\%$, although at other sets of parameters n, k_0, k_d, critical values of tax on profit rate T^* could be lower), there is no rise in the cost of equity of the company with leverage but descending. Once again, the presence or the absence of such an effect depends on a set of parameters k_0, k_d, n.

This effect has been observed above in the dependence of cost of equity k_e on tax on profit rate T at fix leverage level, but it is more clearly visible, depending on value of cost of equity of the company on the leverage for various values of tax on profit rate T.

Note that this is a new effect, which may take place only for the finite lifetime company and which is not observed in perpetuity Modigliani–Miller limit.

It is easy to receive from the Modigliani–Miller formula for WACC

$$\mathrm{WACC} = k_e w_e + k_d w_d (1 - T)$$

formula for k_e

$$k_e = k_0 + L(1 - T)(k_0 - k_d),$$

from which one can see that at $T = 1(100\%)$ cost of equity k_e does not change with leverage: $k_e = k_0$, i.e., there is no decreasing of k_e with leverage at any tax on profit rate T.

Conclusions Qualitatively new effect in corporative finance is discovered: decreasing of cost of equity k_e with leverage L. This effect, which is absent in perpetuity Modigliani–Miller limit, takes place under account of finite lifetime of the company at tax on profit rate, which exceeds some value T^* [3–13].

At some ratios between debt cost and equity cost, the discovered effect takes place at tax on profit rate, existing in western countries and Russia. This provides the practical meaning of discussed effect. Its accounting is important at modification of tax low and can change the dividend policy of the company.

The complete and detailed investigation of discussed effect, discovered within the Brusov–Filatova–Orekhova (BFO) theory, has been done [3–13]. It has been shown, that the absence of the effect at some particular set of parameters is connected to the fact, that in these cases, T^* exceeds 100% (tax on profit rate is situated in a "non-financial" region).

In future, the papers and monographs will be devoted to discussion of discovered abnormal effect, but it is already now clear that we will have to abandon some established views in corporative finance.

Table 18.5 k_d, k_e and weighted average cost of capital, WACC, for companies with lifetimes $n =$ 1; 3;5;10

L	0	0.5	1	1.1	1.3	1.6	2	3	4
K_d	0.12	0.12	0.12	0.1188	0.1161	0.1107	0.1	0.04	−0.14
WACC ($n = 1$)	0.220	0.211	0.207	0.206	0.205	0.206	0.206	0.214	0.252
K_e ($n = 1$)	0.220	0.257	0.294	0.302	0.320	0.358	0.417	0.736	1.819
WACC ($n = 3$)	0.219	0.208	0.201	0.201	0.199	0.199	0.198	0.209	0.279
K_e ($n = 3$)	0.219	0.252	0.281	0.291	0.307	0.340	0.395	0.716	1.955
WACC ($n = 5$)	0.220	0.206	0.200	0.199	0.197	0.197	0.196	0.207	0.301
K_e ($n = 5$)	0.220	0.250	0.279	0.287	0.303	0.335	0.388	0.710	2.067
WACC ($n =$ 10)	0.220	0.206	0.199	0.198	0.196	0.196	0.194	0.205	0.383
K_e($n = 10$)	0.220	0.249	0.277	0.285	0.301	0.332	0.383	0.699	2.474

Table 18.6 k_d, k_e and weighted average cost of capital, WACC, for companies with lifetimes $n =$ 1; 3;5;10

L	0	0.5	1	1.1	1.3	1.6	2	3	4
K_d	0.12	0.12	0.12	0.1188	0.1161	0.1107	0.1	0.04	−0.14
WACC ($n =$ 1).	0.220	0.211	0.207	0.206	0.205	0.206	0.206	0.214	0.252
K_e ($n = 1$)	0.220	0.257	0.294	0.302	0.320	0.358	0.417	0.736	1.819
WACC ($n = 3$)	0.219	0.208	0.201	0.201	0.199	0.199	0.198	0.209	0.279
K_e ($n = 3$)	0.219	0.252	0.281	0.291	0.307	0.340	0.395	0.716	1.955
WACC ($n = 5$)	0.220	0.206	0.200	0.199	0.197	0.197	0.196	0.207	0.301
K_e ($n = 5$)	0.220	0.250	0.279	0.287	0.303	0.335	0.388	0.710	2.067
WACC ($n =$ 10)	0.220	0.206	0.199	0.198	0.196	0.196	0.194	0.205	0.383
K_e($n = 10$)	0.220	0.249	0.277	0.285	0.301	0.332	0.383	0.699	2.474

18.6 Mechanism of Formation of the Company Optimal Capital Structure

Under condition of proved by us insolvency of well–known classical trade-off theory, proved by us, question of finding of new mechanisms of formation of the company optimal capital structure, different from one, suggested by trade-off theory, becomes very important. One of such real mechanisms has been developed by us in this chapter. It is based on the decrease of debt cost with leverage, which is determined by growth of debt volume. This mechanism is absent in perpetuity Modigliani–Miller theory, even in modified version developed by us, and exists within more general modern theory of capital cost and capital structure in Brusov–Filatova–Orekhova (BFO theory).

Suggested mechanism of formation of the company optimal capital structure is based on the decrease of debt cost, which (in some range of leverage levels) is determined by the growth of the debt volume.

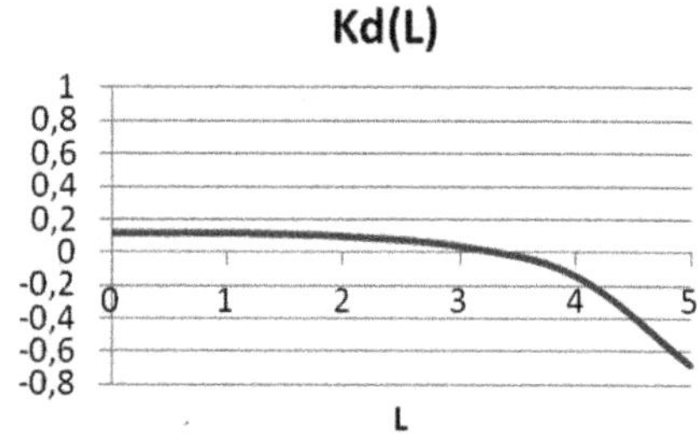

Fig. 18.11 Dependence of debt cost k_d on leverage level L in case of its exponential decrease at $\alpha = 0.01$

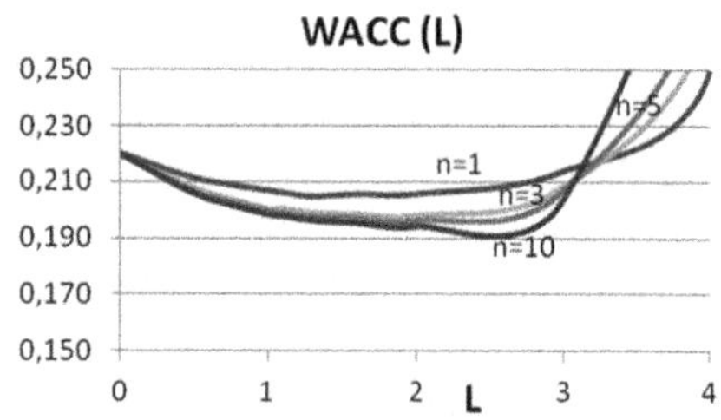

Fig. 18.12 Dependence of weighted average cost of capital, WACC on leverage level L in case of exponential decrease of debt cost at $\alpha = 0.01$

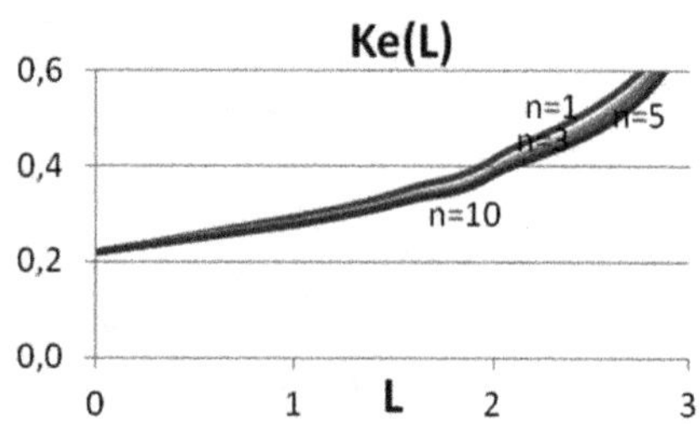

Fig. 18.13 Dependence of equity cost k_e on leverage level L in case of its exponential decrease at $\alpha = 0.01$

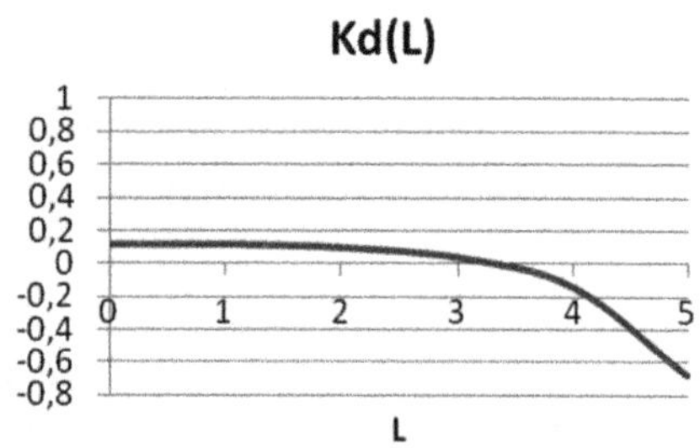

Fig. 18.14 Dependence of debt cost k_d on leverage level L in case of its exponential decrease at $\alpha = 0.01$

We will study below the dependence of equity cost $k_e$and weighted average cost of capital, WACC, on leverage level L in case of exponential decrease in debt cost k_d.

The case $\alpha = 0, 01$. Let us consider first the case $\alpha = 0, 01$ (Tables 18.5 and 18.6, Figs. 18.11, 18.12, and 18.13**).**

The case $\alpha = 0, 01$**.** Let us consider first the case $\alpha = 0, 01$**.**

We will study below the dependence of debt cost k_d, equity cost k_e, and weighted average cost of capital, WACC, on leverage level L in case of exponential decrease in k_d (Figs. 18.14, 18.15, and 18.16).

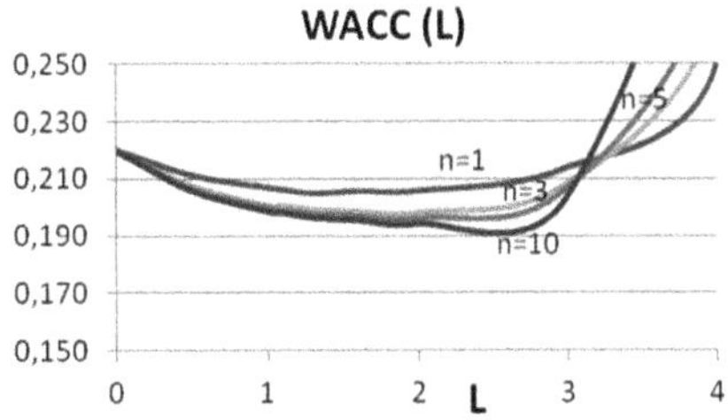

Fig. 18.15 Dependence of weighted average cost of capital, WACC, on leverage level L in case of exponential decrease of debt cost at $\alpha = 0.01$

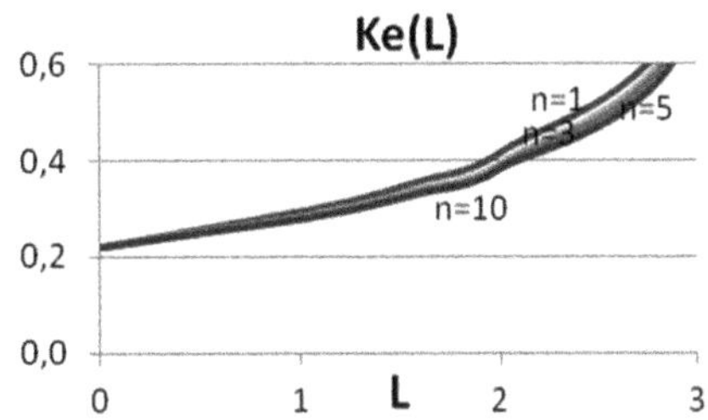

Fig. 18.16 Dependence of equity cost k_e on leverage level L in case of its exponential decrease at $\alpha = 0.01$

18.7 "A Golden Age" of the Company

Authors of the BFO theory have investigated the dependence of attracting capital cost on the time of life of company n at various leverage levels, at various values of capital costs with the aim of defining minimum cost of attracting capital. All calculations have been done within modern theory of capital cost and capital structure by Brusov–Filatova–Orekhova [2–13].

It was shown for the first time that valuation of WACC in the Modigliani–Miller theory [1, 14, 15] is not minimal and valuation of the company capitalization is not maximal, as all financiers have supposed up to now: at some age of the company, its WACC value turns out to be lower than in Modigliani–Miller theory and company capitalization V turns out to be greater than V in Modigliani–Miller theory.

It was shown that, from the point of view of cost of attracting capital, there are two types of dependences of weighted average cost of capital, WACC, on the time of life of company n: monotonic descending with n and descending with passage through minimum, followed by a limited growth. The first type takes place for the companies with low-cost capital, a characteristic of western companies. The second type takes place for higher costs capital costs of the company, a characteristic of Russian companies as well as of companies from other developing countries. This means that latter companies, in contrast to the western ones, can take advantage of the benefits given at a certain stage of development of company by discovered effect. Moreover, since the "golden age" of a company depends on the company's capital costs, by controlling them (for example, by modifying the value of dividend payments, that reflect the equity cost), a company may extend the "golden age" of the company, when the cost to attract capital becomes a minimal (less than perpetuity

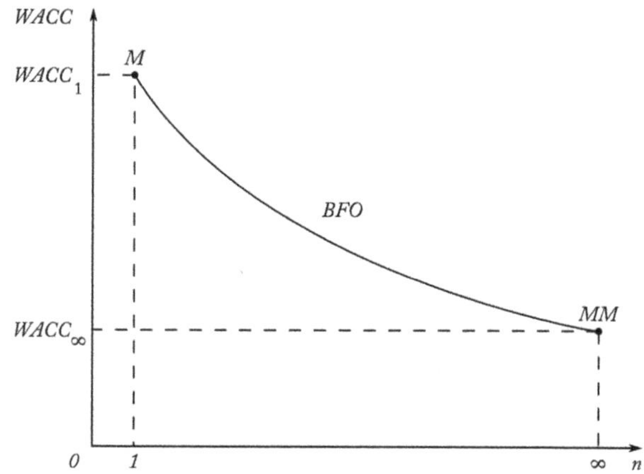

Fig. 18.17 Monotonic dependence of weighted average cost of capital, WACC, on lifetime of the company *n*

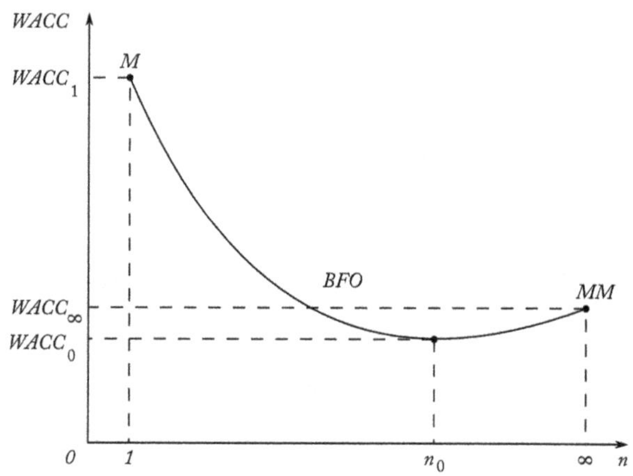

Fig. 18.18 Dependence of weighted average cost of capital, WACC, on lifetime of the company *n*, showing descending with *n*, and with the passage through a minimum and then a limited growth

limit), and capitalization of companies becomes maximal (above than perpetuity assessment) up to a specified time interval.

Concluded that existed up to the present conclusions of the results of the theory of Modigliani–Miller [1, 14, 15] in these aspects are incorrect. We discuss the use of opened effects in developing economics.

The conclusion made in this paper for the first time, that the assessment of weighted average cost of capital of the company, WACC, in the theory of Modigliani and Miller (MM) [1, 14, 15] is not the minimal, and capitalization is not maximal, seems to be very significant and important (Figs. 18.17, 18.18, and 18.19).

Below we show the dependence of weighted average cost of capital, WACC, on lifetime of the company *n* at fixed value of equity cost, $k_0 = 20\%$, and at four values of debt cost.

In Fig. 18.20, it is seen, that with increase of debt cost, k_d, the character of dependence of weighted average cost of capital, WACC, on lifetime of the company

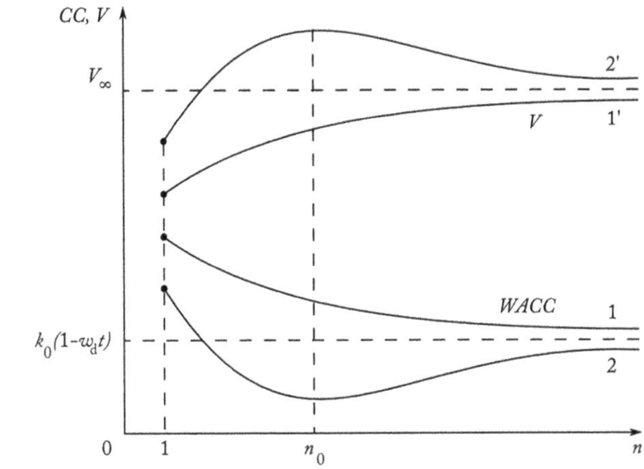

Fig. 18.19 Two kinds of dependences of weighted average cost of capital, WACC, and company capitalization, V, on lifetime of the company n: 1–1′—monotonic dependence of weighted average cost of capital, WACC, and company capitalization, V, on lifetime of the company n; 2–2′—showing descending of WACC with n, and with the passage through a minimum and then a limited growth and increase of V with the passage through a maximum (at n_0) and then a limited descending

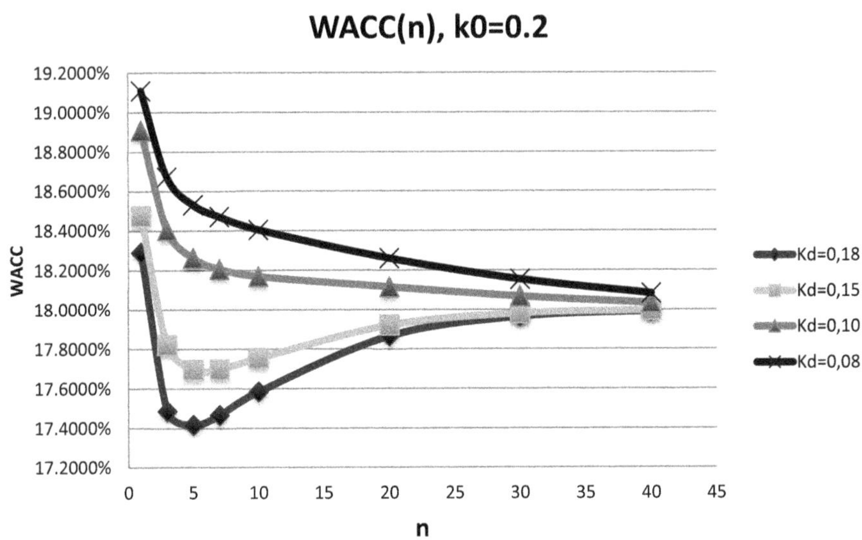

Fig. 18.20 Dependence of weighted average cost of capital, WACC, on lifetime of the company n at fixed value of equity cost, $k_0 = 20\%$, and at four values of debt cost, $k_d = 8\%;10\%;15\%$, and 18% at leverage level $L = 1$

n is changed from monotonic descending of WACC with n to descending of WACC with n with passage through minimum, followed by a limited growth.

Conclusions

Above it is shown for the first time within the BFO theory [3–13] that valuation of WACC in the Modigliani–Miller theory [1, 14, 15] is not minimal and valuation of the company capitalization is not maximal, as all financiers have supposed up to now: at some age of the company, its WACC value turns out to be lower than in the Modigliani–Miller theory and company capitalization V turns out to be greater than V in the Modigliani–Miller theory [1, 14, 15]. Thus, the existing up to the present presentations concerning the results of the Modigliani–Miller theory in this aspect [16] turns out to be incorrect.

It is shown that, from the point of view of cost of attracting capital, there are two types of dependences of weighted average cost of capital, WACC, on the time of life of company n: monotonic descending with n and descending with passage through minimum, followed by a limited growth (there is a third modification of dependences WACC(n), which leaves all conclusions valid). The first type takes place for the companies with low cost capital, a characteristic of western companies. The second type takes place for higher costs capital costs of the company, a characteristic of the Russian companies as well as of companies from other developing countries. This means that latter companies, in contrast to the western ones, can take advantage of the benefits given at a certain stage of development of company by discovered effect. **(For example, the capitalization of Russian oil company "Rosneft", which has been valued in 2014 by the Modigliani–Miller method, could be higher, accounting the discovered effect and the BFO theory).** Moreover, since the "golden age" of company depends on the company's capital costs, by controlling them (for example, by modifying the value of dividend payments, that reflect the equity cost), a company may extend the "golden age" of the company when the cost to attract capital becomes minimal (less than perpetuity limit), and capitalization of companies becomes maximal (higher than perpetuity assessment) up to a specified time interval.

It is important to note that the "golden age" of company effect changes the dependence of WACC on L: the curve WACC(L) for perpetuity company turns out to be not lowest for company with this effect—as it is seen in Tables 18.7, 18.8 and Fig. 18.21, below the curve WACC(L) for 3 years the company lies below the perpetuity curve.

18.8 Inflation in MM and BFO Theories

Here we describe the influence of inflation on capital cost and capitalization of the company within modern theory of capital cost and capital structure—Brusov–Filatova–Orekhova theory (BFO theory) [3–13] and within its perpetuity limit—Modigliani–Miller theory [1, 14, 15] . By direct incorporation of inflation into both

Table 18.7 Dependence of WACC and k_e on leverage level for $n = 1$, and $n = \infty$

L	K_o	K_d	t	n	W_d	WACC	BFO	K_e	WACC MM	MM K_e
0	0.2	0.15	0.2	1	0.00	20.00%	0.000	0.2000	20.00%	0.2000
1	0.2	0.15	0.2	1	0.50	18.43%	0.000	0.2487	18.00%	0.2400
2	0.2	0.15	0.2	1	0.67	17.91%	0.000	0.2974	17.33%	0.2800
3	0.2	0.15	0.2	1	0.75	17.65%	0.000	0.3461	17.00%	0.3200
4	0.2	0.15	0.2	1	0.80	17.50%	0.000	0.3948	16.80%	0.3600
5	0.2	0.15	0.2	1	0.83	17.39%	0.000	0.4435	16.67%	0.4000
6	0.2	0.15	0.2	1	0.86	17.32%	0.000	0.4922	16.57%	0.4400
7	0.2	0.15	0.2	1	0.88	17.26%	0.000	0.5409	16.50%	0.4800
8	0.2	0.15	0.2	1	0.89	17.22%	0.000	0.5896	16.44%	0.5200
9	0.2	0.15	0.2	1	0.90	17.18%	0.000	0.6383	16.40%	0.5600
10	0.2	0.15	0.2	1	0.91	17.15%	0.000	0.6870	16.36%	0.6000

Table 18.8 Dependence of WACC and k_e on leverage level for $n = 3$, and $n = \infty$

L	K_o	K_d	t	n	W_d	WACC	BFO	K_e	WACC MM	MM K_e
0	0.2	0.15	0.2	3	0.00	20.00%	0.000	0.2000	20.00%	0.2000
1	0.2	0.15	0.2	3	0.50	17.80%	0.000	0.2360	18.00%	0.2400
2	0.2	0.15	0.2	3	0.67	17.06%	0.000	0.2719	17.33%	0.2800
3	0.2	0.15	0.2	3	0.75	16.69%	0.000	0.3078	17.00%	0.3200
4	0.2	0.15	0.2	3	0.80	16.47%	0.000	0.3436	16.80%	0.3600
5	0.2	0.15	0.2	3	0.83	16.32%	0.000	0.3795	16.67%	0.4000
6	0.2	0.15	0.2	3	0.86	16.22%	0.000	0.4153	16.57%	0.4400
7	0.2	0.15	0.2	3	0.88	16.14%	0.000	0.4511	16.50%	0.4800
8	0.2	0.15	0.2	3	0.89	16.08%	0.000	0.4869	16.44%	0.5200
9	0.2	0.15	0.2	3	0.90	16.03%	0.000	0.5228	16.40%	0.5600
10	0.2	0.15	0.2	3	0.91	15.99%	0.000	0.5586	16.36%	0.6000

theories, Brusov–Filatova–Orekhova have shown for the first time that inflation not only increases the equity cost and the weighted average cost of capital but it changes their dependence on leverage as well. In particular, it increases growing rate of equity cost with leverage. Capitalization of the company is decreased under accounting of inflation.

Under accounting of inflation, all original MM (Modigliani–Miller) statements have been modified as it is done below.

2-nd Original MM Statement equity cost of leverage company k_e could be found as equity cost of financially independent company k_0 of the same group of risk, plus premium for risk, which value is equal to production of difference ($k_0 - k_d$) on leverage level L.

2-nd Modified MM–BFO Statement under existing of inflation with rate α equity cost of leverage company k_e could be found as equity cost of financially independent company k_0 of the same group of risk, multiplied by $(1+ \alpha)$, plus inflation rate α and

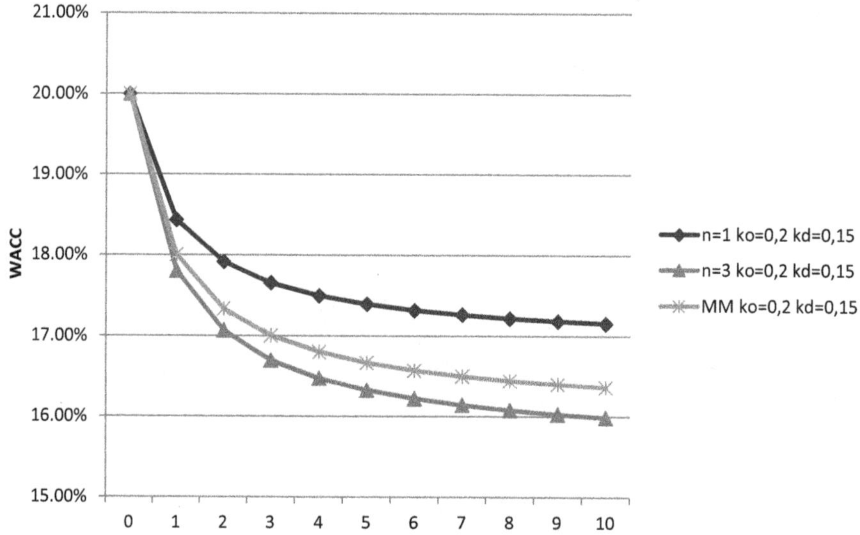

Fig. 18.21 The curve WACC(L) for perpetuity company turns out to be not lowest for company with the effect of "golden age": the curve WACC(L) for 3 years, the company lies below the perpetuity curve

plus premium for risk, which value is equal to production of difference (k_0 − k_d)on leverage level L and on multiplier (1 + α).

4-th Original MM Statement equity cost of leverage company k_e paying tax on profit could be found as equity cost of financially independent company k_0 of the same group of risk, plus premium for risk, which value is equal to production of difference (k_0 − k_d)on leverage level L and on tax shield *(1–T)* and on multiplier (1 + α).

4-th Modified MM–BFO Statement equity cost of leverage company k_e paying tax on profit under existing of inflation with rate α could be found as equity cost of financially independent company k_0 of the same group of risk, multiplied by (1 + α), plus inflation rate α and plus premium for risk, which value is equal to production of difference (k_0 − k_d)on leverage level L, on tax shield *(1–T)* and on multiplier (1 + α) (Fig. 18.22).

We generalized a very important Brusov–Filatova–Orekhova theorem under accounting of inflation.

Generalized Brusov–Filatova–Orekhova Theorem

Under accounting of inflation without corporate taxing, the equity cost k_0^*, *as well as the weighted average cost of capital* WACC* *do not depend on company lifetime and are equal to*

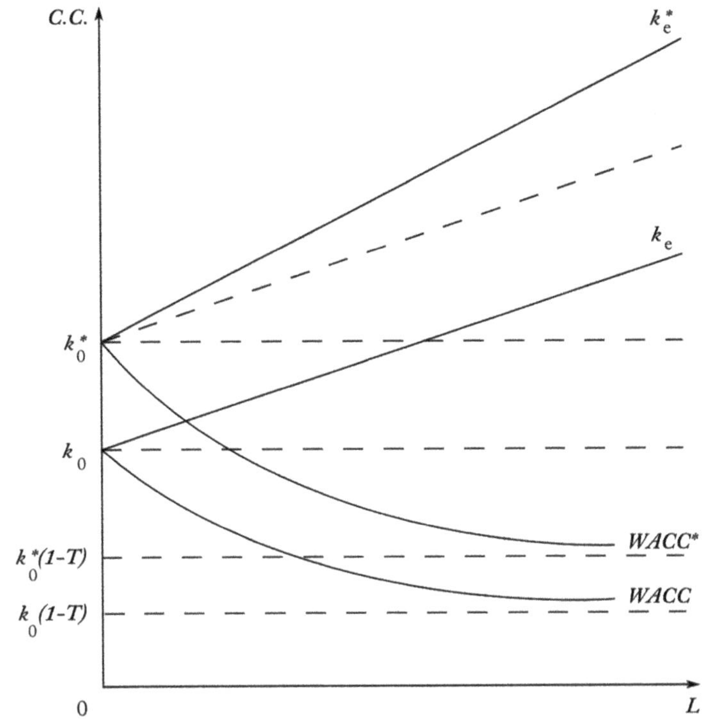

Fig. 18.22 Dependence of the equity cost and the weighted average cost of capital on leverage in the Modigliani–Miller theory with taxing under accounting of inflation. It is seen, that growing rate of equity cost increases with leverage. Axis y means capital costs –C.C

$$k_e^* = k_0^* + L(k_0^* - k_d^*) = k_0(1 + \alpha) + \alpha + L(k_0 - k_d)(1 + \alpha)$$
$$\text{and} \quad \text{WACC}^* = k_0^* = k_0(1 + \alpha) + \alpha. \tag{18.4}$$

consequently.

It is shown, that inflation not only increases the equity cost and the weighted average cost of capital but it changes their dependence on leverage as well. In particular, it increases growing rate of equity cost with leverage. Capitalization of the company is decreased under accounting of inflation.

Within modern theory of capital cost and capital structure—Brusov–Filatova–Orekhova theory (BFO theory)—the modified equation for the weighted average cost of capital, WACC, applicable to companies with arbitrary lifetime under accounting of inflation has been derived. The modified BFO equation allows to investigate the dependence of the weighted average cost of capital, WACC, and equity cost, k_e, on leverage level L, on tax on profit rate t, on lifetime of the company n, on equity cost of financially independent company, k_0, and debt cost, k_d, as well as on inflation rate α .

Using modified BFO equation, the analysis of the dependence of the weighted average cost of capital, WACC, on debt ratio, w_d, at different tax on profit rate t, as well as inflation rate α has been done.

It has been shown, that WACC decreases with debt ratio, w_d, faster at bigger tax on profit rate t. The space between lines, corresponding to different values of tax on profit rate at the same step (10%), increases with inflation rate α. The variation region (with change of tax on profit rate t) of the weighted average cost of capital, WACC, increases with inflation rate α, as well as with lifetime of the company n.

18.9 Effects Connected with Tax Shields, Taxes, and Leverage

The role of tax shields, taxes, and leverage is investigated within the theory of Modigliani–Miller as well as within the modern theory of corporate finance by Brusov–Filatova–Orekhova [3–13]. It is shown that equity cost of the company as well as weighted average cost of capital decrease with the growth of tax on profits rates. A detailed study of the dependence of weighted average cost of capital WACCand equity cost of the company $k_e$on tax on profits rates at fixed leverage level (fixed debt capital fraction w_d) as well as on leverage level (debt capital fraction w_d) at fixed tax on profits rate has been done. The dependences of weighted average cost of capital WACCand equity cost of the company k_e on company lifetime have been investigated as well.

The concept "tax operating lever" has been introduced. For companies with finite lifetime, a number of important qualitative effects that do not have analogs for perpetuity companies has been detected.

One such effect—decreasing of equity cost with leverage level at values of tax on profits rate T, which exceeds some critical value T^*—is described in detail in Chap. 10 (at certain ratios between the debt cost and equity capital discovered effect takes place at tax on profits rate, existing in the western countries and in Russia, that provide practical value effect.) Its accounting is important in improving tax legislation and may change dividend policy of the company.

18.10 Effects Connected with the Influence of Tax on Profit Rate on Effectiveness of Investment Projects

BFO authors have conducted the analysis of effectiveness of investment projects within the perpetuity (Modigliani–Miller) approximation [1, 14, 15] as well as within the BFO theory. They analyzed the effectiveness of investment projects for three cases:

1. At a constant difference between equity cost (at $L = 0$) and debt cost $\Delta k = k_0 - k_d$
2. At a constant equity cost (at $L = 0$) and varying debt cost k_d
3. At a constant debt cost k_d and varying equity cost (at $L = 0$) k_0

The dependence of NPV on investment value and/or equity value will be also analyzed. The results have been represented in the form of tables and graphs.

It should be noted that the obtained tables have played an important practical role in determining of the optimal, or acceptable debt level, at which the project remains effective. The optimal debt level there is for the situation, when in the dependence of NPV on leverage level L there is an optimum (leverage level value, at which NPV reaches a maximum value). An acceptable debt level there is for the situation, when NPV decreases with leverage. And, finally, it is possible that NPV is growing with leverage. In this case, an increase in borrowing leads to increased effectiveness of investment projects, and their limit is determined by financial sustainability of investing company.

18.11 Influence of Growth of Tax on Profit Rate

Within modern theory of capital cost and capital structure by Brusov–Filatova–Orekhova (BFO theory) [3–13] and created within this theory, modern investment models' influence of growth of tax on profit rate on the efficiency of the investment is investigated. It has been shown that for long-term investment projects as well as for arbitrary duration projects, the growth of tax on profit rate changes the nature of the NPV dependence on leverage at some value t^*: there is a transition from diminishing function NPV(L) when $t < t^*$ to growing function NPV(L). The t^* value depends on the duration of the project, cost of capital (equity and debt) values, and other parameters of the project.

Table 18.9 Dependence of NPV and ΔNPV on leverage level L at fixed levels of tax on profit rates t for 5-year project at $t = 0.3$

L	K_o	K_d	W_d	t	n	NOI	K_e	NPV	ΔNPV
0	0.18	0.14	0	0.3	5	800	0.18	751.22	4.922709
1	0.18	0.14	0.5	0.3	5	800	0.197488	756.14	−36.8599
2	0.18	0.14	0.66667	0.3	5	800	0.214367	719.28	−44.7663
3	0.18	0.14	0.75	0.3	5	800	0.231082	674.51	−46.126
4	0.18	0.14	0.8	0.3	5	800	0.24773	628.39	−45.4549
5	0.18	0.14	0.83333	0.3	5	800	0.264343	582.93	−44.027
6	0.18	0.14	0.85714	0.3	5	800	0.280937	538.90	−42.3084
7	0.18	0.14	0.875	0.3	5	800	0.297518	496.60	−40.4978
8	0.18	0.14	0.88889	0.3	5	800	0.314091	456.10	−38.6879
9	0.18	0.14	0.9	0.3	5	800	0.330658	417.41	−36.9239
10	0.18	0.14	0.90909	0.3	5	800	0.34722	380.49	

Table 18.10 Dependence of NPV and ΔNPV on leverage level L at fixed levels of tax on profit rates t for 5-year project at $t = 0.4$

L	K_o	K_d	W_d	t	n	NOI	K_e	NPV	ΔNPV
0	0.18	0.14	0	0.4	5	800	0.18	501.04	64.13345
1	0.18	0.14	0.5	0.4	5	800	0.189578	565.18	4.73089
2	0.18	0.14	0.66667	0.4	5	800	0.19803	569.91	−9.5017
3	0.18	0.14	0.75	0.4	5	800	0.206172	560.40	−14.7815
4	0.18	0.14	0.8	0.4	5	800	0.214184	545.62	−17.1025
5	0.18	0.14	0.83333	0.4	5	800	0.22213	528.52	−18.1709
6	0.18	0.14	0.85714	0.4	5	800	0.230037	510.35	−18.6246
7	0.18	0.14	0.875	0.4	5	800	0.23792	491.73	−18.7461
8	0.18	0.14	0.88889	0.4	5	800	0.245786	472.98	−18.6762
9	0.18	0.14	0.9	0.4	5	800	0.253642	454.30	−18.4911
10	0.18	0.14	0.90909	0.4	5	800	0.261488	435.81	

Table 18.11 Dependence of NPV and ΔNPV on leverage level L at fixed levels of tax on profit rates t for 5-year project at $t = 0.5$

L	K_o	K_d	W_d	t	n	NOI	K_e	NPV	ΔNPV
0	0.18	0.14	0	0.5	5	800	0.18	250.87	116.0669
1	0.18	0.14	0.5	0.5	5	800	0.181448	366.94	41.1323
2	0.18	0.14	0.66667	0.5	5	800	0.181065	408.07	22.57738
3	0.18	0.14	0.75	0.5	5	800	0.180162	430.65	15.19888
4	0.18	0.14	0.8	0.5	5	800	0.179041	445.84	11.52994
5	0.18	0.14	0.83333	0.5	5	800	0.177806	457.37	9.446706
6	0.18	0.14	0.85714	0.5	5	800	0.176505	466.82	8.154973
7	0.18	0.14	0.875	0.5	5	800	0.175162	474.98	7.302458
8	0.18	0.14	0.88889	0.5	5	800	0.173792	482.28	6.713275
9	0.18	0.14	0.9	0.5	5	800	0.172401	488.99	6.291579
10	0.18	0.14	0.90909	0.5	5	800	0.170996	495.28	

At high leverage levels, this leads to qualitatively new effect in investments: **growth of the efficiency of the investments with growth of tax on profit rate.** Discovered effects take place under consideration from the point of view of owners of equity capital as well as from the point of view of owners of equity and debt capital (Tables 18.9, 18.10, and 18.11).

One can see in the Figs. 18.23 and 18.24 that the nature of the NPV dependence on leverage at $t^* = 0.5$: there is a transition from diminishing function NPV(L) when $t < t^*$ to growing function NPV(L) at $t > t^*$.

Within modern theory of capital cost and capital structure by Brusov–Filatova–Orekhova (BFO theory) and created within this theory modern investment models' influence of growth of tax on profit rate on the efficiency of the investment is investigated. It has been shown that for arbitrary duration projects as well as for perpetuity projects, the growth of tax on profit rate change the nature of the NPV dependence on leverage at some value t^*: there is a transition from diminishing

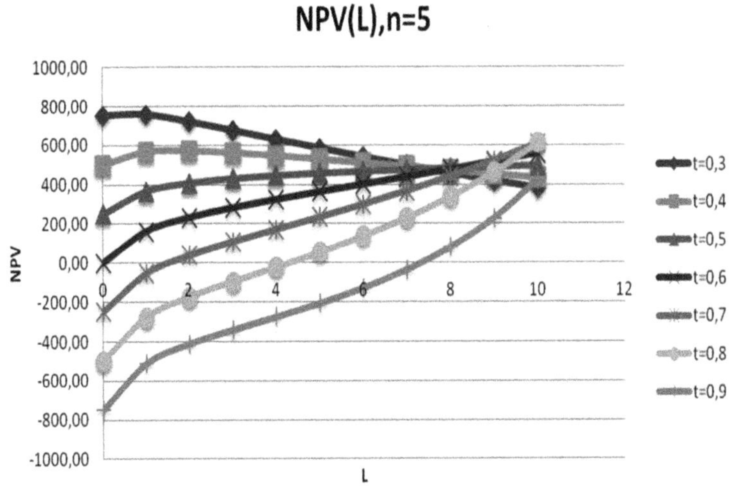

Fig. 18.23 Dependence of NPV on leverage level L at fixed levels of tax on profit rates t for 5-year project

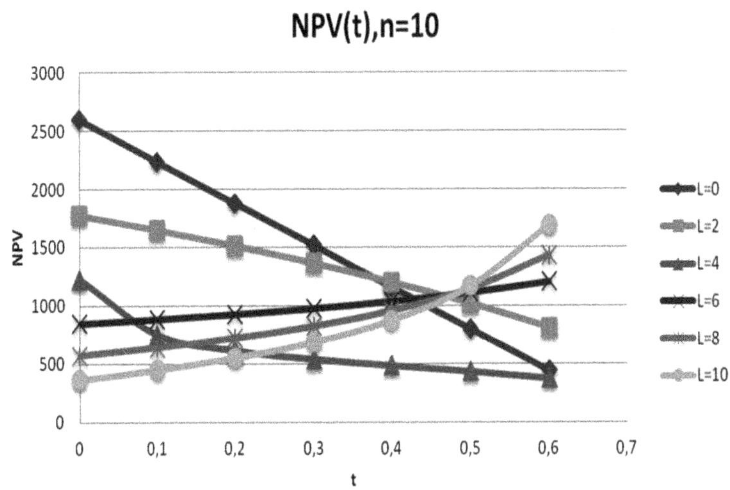

Fig. 18.24 Dependence of NPV on tax on profit rate t at fixed leverage level L for 10-year project

function NPV(L) when $t < t^*$ to growing function NPV(L). The t^* value depends on the duration of the project, cost of capital (equity and debt) values and other parameters of the project.

At high leverage levels, this leads to qualitatively new effect in investments: growth of the efficiency of the investments with growth of tax on profit rate. Discovered effects take place under consideration from the point of view of owners of equity capital as well as from the point of view of owners of equity and debt capital.

It is observed at high leverage levels' (starting from $L = 6$) increase of NPV with growth of the tax on profit rate t (Fig. 18.24) takes place at all values of t, which means that this is an entirely new effect in investments that can be applied in a real economic practice for optimizing of the management of investments.

So, *two very important qualitatively new effects in investments have been discovered:*

1. *Change the character of NPV dependence on leverage*
2. *Growth of the efficiency of the investments with growth of tax on profit rate*

Both effects could be used in practice to optimize the investments.

18.12 New Approach to Ratings

The application of the perpetuity limit of the BFO theory (MM theory) to the rating methodology is discussed in the monograph by Brusov et al. [17]: for the first time, the authors included the main parameters of ratings—rating "coefficients"—directly into the modern theory of capital structure. This allows to use the powerful methods and "toolkit" of this theory in rating and creates practically the new basis of a rating methodology, that allows to make more correct ratings.

A new approach to rating methodology has been suggested, key factors of which are: (1) The adequate use of discounting of financial flows virtually not used in existing rating methodologies; (2) The incorporation of rating parameters (financial "ratios") into the modern theory of capital structure (BFO theory) [2] (for beginning into its perpetuity limit). This, on the one hand, allows to use the powerful tools of this theory in the rating, and on the other, it ensures the correct discount rates when discounting of financial flows. We discuss also the interplay between rating ratios and leverage level which can be quite important in rating. All these create a new base for rating methodologies.

In conclusion, we remind again that the introduction of the new system of evaluation of the parameters of financial activities of companies into the systems of financial reporting (IFRS, GAAP, etc.) would lead to lower risk of global financial crisis.

References

1. Modigliani, F., & Miller, M. (1958). The cost of capital, corporate finance, and the theory of investment. *American Economic Review, 48*, 261–297.
2. Brusov, P., Filatova, T., Orehova, N., & Eskindarov, M. (2015). *Modern corporate finance, investment and taxation.* Springer International Publishing. 368 p.
3. Brusov, P., Filatova, T., Orehova, N., & Brusova, A. (2011a). Weighted average cost of capital in the theory of Modigliani–Miller, modified for a finite life-time company. *Applied Financial Economics, 21*(11), 815–824.

4. Brusov, P., Filatova, P., & Orekhova, N. (2013a). Absence of an optimal capital structure in the famous tradeoff theory! *Journal of Reviews on Global Economics, 2*, 94–116.
5. Brusov, P., Filatova, P., & Orekhova, N. (2014a). Mechanism of formation of the company optimal capital structure, different from suggested by trade off theory. *Cogent Economics & Finance, 2*, 1–13. https://doi.org/10.1080/23322039.2014.946150
6. Brusov, P., Filatova, T., Orehova, N., et al. (2011b). From Modigliani–Miller to general theory of capital cost and capital structure of the company. *Research Journal of Economics, Business and ICT, 2*, 16–21.
7. Brusov, P., Filatova, T., Eskindarov, M., & Orehova, N. (2012a). Influence of debt financing on the effectiveness of the finite duration investment project. *Applied Financial Economics, 22*(13), 1043–1052.
8. Brusov, P., Filatova, T., Orehova, N., et al. (2011c). Influence of debt financing on the effectiveness of the investment project within the Modigliani–Miller theory. *Research Journal of Economics, Business and ICT (UK), 2*, 11–15.
9. Brusov, P., Filatova, T., Eskindarov, M., & Orehova, N. (2012b). Hidden global causes of the global financial crisis. *Journal of Reviews on Global Economics, 1*, 106–111.
10. Brusov, P., Filatova, T., Orehova, N., Brusov, P. P., & Brusova, N. (2011e). From Modigliani–Miller to general theory of capital cost and capital structure of the company. *Research Journal of Economics, Business and ICT, 2*, 16–21.
11. Brusov, P., Filatova, T., & Orehova, N. (2014b). Inflation in Brusov–Filatova–Orekhova theory and in its Perpetuity Limit – Modigliani – Miller theory. *Journal of Reviews on Global Economics, 3*, 175–185.
12. Brusov, P., Filatova, T., & Orehova, N. (2013c). A qualitatively new effect in corporative finance: Abnormal dependence of cost of equity of company on leverage. *Journal of Reviews on Global Economics, 2*, 183–193.
13. Filatova, T., Orehova, N., & Brusova, A. (2008). Weighted average cost of capital in the theory of Modigliani–Miller, modified for a finite life–time company. *Bulletin of the FU, 48*, 68–77.
14. Modigliani, F., & Miller, M. (1963). Corporate income taxes and the cost of capital: A correction. *American Economic Review, 53*, 147–175.
15. Modigliani, F., & Miller, M. (1966). Some estimates of the cost of capital to the electric utility industry 1954–1957. *American Economic Review, 56*, 333–391.
16. Myers, S. (2001). Capital structure. *Journal of Economic Perspectives, 15*(2), 81–102.
17. Brusov, P., Filatova, T., & Orehova, N. (2021). *Ratings: Critical analysis and new approaches of quantitative and qualitative methodology (contributions to finance and accounting)*. Springer Nature. 365 p.
18. Brusov, P. N., & Filatova, T. V. (2011). From Modigliani–Miller to general theory of capital cost and capital structure of the company. *Finance and Credit, 435*, 2–8.
19. Brusova, A. (2011). A comparison of the three methods of estimation of weighted average cost of capital and equity cost of company. *Financial Analysis: Problems and Solutions, 34*(76), 36–42.

Part IV
Innovative Effects in Business Valuation

Chapter 19
Cost of Equity and Dividend Policy

Abbreviations

BFO	Brusov–Filatova–Orekhova theory
MM	the Modigliani–Miller theory
WACC	the weighted average cost of capital

19.1 Introduction

One of the main parameters of business valuation and company value is the company's income flow. Determining the appropriate income flows of a company is a complex and time-consuming task. Dividends paid to shareholders of a company represent a significant amount of income, but determining the adequate size of dividends is a complex problem. The economically justified value of dividends is the cost of the company's equity, the assessment of which is a non-trivial task that can be carried out correctly within the framework of the Brusov–Filatova–Orekhova (BFO) theory. This chapter examines in detail the problem of forming a company's dividend policy, determining the value of dividends and the impact on their size and on the size of the company's income, used in assessing the business and the company, the real conditions of the companies' functioning.

As is well known, valuation of equity cost, which determines the amount of dividends, is an extremely complex task. It is possible to do this within one of the two main capital structure theories: either the Modigliani–Miller (MM) theory or within the Brusov–Filatova–Orekhova (BFO) one. First one is valid for perpetual companies only, while the second describes the companies of arbitrary ages. Until recently, both of these theories did not take into account some real conditions for the functioning of companies, such as the variable income of companies, the frequent payments income tax, types of income tax payments (advance payments and

P. Brusov, T. Filatova, *Business Valuation and Company Value*, Contributions to Finance and Accounting, https://doi.org/10.1007/978-3-031-90847-7_19

payments at the end of reporting periods), etc. This was some gap in the study of the problem of the cost of equity. The challenge was to adapt both theories to these conditions and use the generalized Brusov–Filatova–Orekhova (BFO) and Modigliani–Miller (MM) theories in studying the problem of the cost of equity.

The correct dividend policy of the company, based on the correct assessment of the cost of equity, has both macroeconomic and microeconomic aspects.

From a macroeconomic point of view, it plays an important role in ensuring economic growth and economic efficiency, as well as the position in the financial and investment markets. At the macroeconomic level as well, an adequate dividend policy leads to increased investment and economic growth.

From a microeconomic point of view, the importance of a correct assessment of the cost of equity and the establishment of an adequate dividend policy of the company is determined by the following.

The rights of shareholders to receive adequate profit may be violated if the management of the company pursues an incorrect and inefficient dividend policy due to the inability of management to correctly determine the amount of dividends (the economically justified amount of which is the cost of equity), or as a deliberate violation of the rights of shareholders in this sphere.

After the recent generalization of both theories, taking into account the real conditions of the functioning of the companies, it becomes possible to estimate the cost of equity in these conditions. Dependence of the cost of equity on the level of leverage for different ages of the company, different values of equity and debt costs, for different frequency of income tax payment, for advance payments of income tax and payments at the end of periods, for variable income of companies, etc. is being studied. The behavior and properties of the cost of equity are explored within the framework of the BFO theory. Several very important innovative effects have been discovered, which significantly change the company's dividend policy. Among them the abnormal dependence of equity cost on leverage level at different growth rate of variable income, at different age of the company, at different frequency of income tax payment, at different debt cost, etc. The developed methodology and results will help the company's management to develop an adequate and effective dividend policy.

A comprehensive analysis was carried out of one of the most striking and significant phenomena in corporate finance in general and in the capital structure in particular—the anomalous dependence of the cost of equity on the leverage level, L. The first qualitative theory of the capital structure—the Modigliani–Miller theory [1–3]—had numerous limitations, the main of which were its perpetual nature and the absence of taxes: corporate and individual. Many scientists have tried to modify the Modigliani–Miller theory in different aspects [4–17].

In 2008 [18], the main limitation of the Modigliani–Miller theory—the eternity of companies—was removed by Brusov et al. and modern theories of the cost of capital and capital structure—Brusov–Filatova–Orekhova theory (BFO–theory) were created for companies of arbitrary age (BFO–1 theory) and for companies of arbitrary lifetime (BFO–2 theory) [19]. The well-known theory of the cost of capital and capital structure of Nobel laureates Modigliani and Miller has been replaced by the

Brusov–Filatova–Orekhova (BFO) theory. The authors departed from the Modigliani–Miller assumption about the eternity (infinity of life) of companies and developed a quantitative theory for estimating the main parameters of the financial activity of companies of arbitrary age. The results of the modern BFO theory turn out to be quite different from the results of the Modigliani–Miller theory. They show that the latter, due to its perpetuity, underestimates the cost of raising capital (WACC and equity) and significantly overestimates the company value. Such an incorrect assessment of key performance indicators of companies' financial activities led to an underestimation of the associated risks and the impossibility or serious difficulties in making adequate management decisions, which was one of the latent causes of the 2008 global financial crisis, which began with the crisis in the MBS (mortgage-backed securities) market.

In the Modigliani–Miller theory, there is no time factor (time is equal to infinity), which does not allow us to investigate the dependence of the company's financial performance on the time factor. But the theories of Brusov–Filatova–Orekhova (BFO–theories) are created for companies of arbitrary age and allow studying the dependence of a company's financial performance on the time factor.

Over the past couple of years, two main theories of the capital structure—Brusov–Filatova–Orekhova and Modigliani–Miller—have been adapted to the established financial practice of the functioning of companies, taking into account the real conditions of their work (see [20] and references there). This made it possible to investigate the impact of frequent income tax payments p with advance income tax payments and payments at the end of reporting periods, as well as the impact of the company's variable income on its main financial results. In [20], an analysis of all existing theories of the capital structure (with their advantages and disadvantages) was carried out in order to understand all aspects of the problem and make the right management decisions in practice. The role of the capital structure lies in the fact that the correct determination of the optimal capital structure allows the company's management to maximize the capitalization of the company and fulfill the long-term goal of the functioning of any company. In [20], the state of the theory of the structure of capital and the cost of capital is considered from the middle of the last century, when the first quantitative theory was created, to the present. The two main theories of Modigliani–Miller (MM) and Brusov–Filatova–Orekhova (BFO) are discussed and analyzed, as well as their numerous modifications and generalizations. The Brusov–Filatova–Orekhova (BFO) theory, its methodology, and results are widely known [21–30]. Many authors (see, for example, [25–27]) use the BFO theory in practice.

In addition to the approach of Modigliani and Miller, there are alternative approaches: Harris and Pringle [31], Miles and Ezzell [32].

Below we discuss them.

19.1.1 Alternate WACC Formula

An alternate formula for the *WACC* has been suggested [13, 32–34]. It has the form below

$$\text{WACC} = k_0(1 - w_d t) - k_d t w_d + k_{\text{TS}} t w_d \tag{19.1}$$

Here, k_0, k_d, and k_{TS} are the returns on the financially independent company, the debt, and the tax shield, respectively, t is the corporate tax rate, and wd is the debt share.

Although Eq. (19.1) is quite general, additional conditions are needed for practical applicability. When the *WACC* remains constant over time, the value of a leveraged company can be found by discounting the unleveraged free cash flows using the *WACC*. In this case, specific formulas can be found in textbooks [11].

In the Modigliani–Miller theory [1–3], the debt value D is constant. V_0 is also constant, as the expected after-tax cash flow of the financially independent company is fixed. By assumption, $k_{TS} = k_d$ and the tax shield value is $TS = tD$. Therefore, the company value V is a constant and the alternate *WACC* Formula (19.1) simplifies the MM formula:

$$\text{WACC} = k_0(1 - w_d t) \tag{19.2}$$

The "classical" MM theory, suggesting that the returns on the debt k_d and the tax shield k_{TS} are equals (both these values have debt nature), is much more reasonable, so this is why we modify, namely, the "classical" MM theory.

19.1.2 The Miles–Ezzell Model Versus the Modigliani–Miller Theory

Becker (2021) [12] discussed the differences between the Modigliani–Miller theory [1–3] and the Miles–Ezzell model [31], which deal with the stochasticity of free cash flows. The Modigliani–Miller theory considers a stationary process, while in the Miles–Ezzell model the process is stochastic. The author conducts a numerical experiment that allows you to determine the values and discount rates using a risk-neutral approach. He analyzes three formulas:

Modigliani–Miller theory [1–3],

$$\text{WACC} = k_0(1 - w_d t) \tag{19.3}$$

Miles–Ezzell model [32],

$$\mathrm{WACC} = k_0 - w_d t k_f \frac{1 + k_0}{1 + k_f} \tag{19.4}$$

Cooper and Nyborg (2006) [35]

$$\mathrm{WACC} = k_0 - w_d k_f t, \tag{19.5}$$

where k_f stands for the risk-free rate, which equals the required return of the debt holders.

The author shows that in the Miles–Ezzell model, all cash flows and the values depend on the path, in contrast to the Modigliani–Miller theory. Additionally, in the Miles–Ezzell model, all discount rates are time independent, with the exception of the discount rate used to discount tax shields, which depends on the duration of the cash flows. Conversely, in the Modigliani–Miller theory, all discount rates change over time except for the constant tax shield discount rate. This affects the applicability of the well-known formula for annuities and the development of models for estimating both finite and perpetual cash flows.

In this paper, Becker (2021) [12] raises the issue of paying the debt body together with the payment of interest on the debt. Regarding this issue, we would like to note that in both classical MM and BFO theories, the body of the debt is not paid. In the framework of the Modigliani–Miller theory, such an account is fundamentally impossible, while in the BFO theory it can be conducted and was conducted in the framework of the BFO-2 theory, where the amount of debt D decreases with time. This decrease in the value of debt D results in a decrease in the tax shield (see BFO-2 theory)[19, 36].

19.1.3 Dividend Policy Under Asymmetric Information

As we mentioned above, the economically justified amount of dividends is equal to the cost of equity. The task of calculating the cost of equity capital is one of the most difficult in financial management. Because of this problem, management decisions made by company managers are not always correct.

The rights of shareholders to receive adequate profits may be violated when the management of the company pursues an incorrect and ineffective dividend policy, due to the inability of management to correctly determine the amount of dividends (the economically justified amount of which is the cost of equity capital), or with a deliberate violation of the rights of shareholders in this area.

The asymmetric information means, that the company's managers know more than outside investors about the true state of the company's current earnings.

Miller and Rock [33] extended the standard financial model of dividend/investment/financing decisions of a company for conditions of asymmetric information. They showed that once stock trading along with asymmetric information is included in the model, Fisher's criterion for optimal investment becomes time-inconsistent: the market's belief that a company follows Fisher's rule creates incentives to violate the rule.

(Fisher's rule postulates that in the presence of efficient capital markets, a company's choice of investments is independent of the investment preferences of its owners, and therefore the company should be motivated only to maximize profits.)

Miller and Rock [33] showed that an informationally consistent signaling equilibrium exists under asymmetric information and stock trading, which restores the temporal consistency of investment policy but generally leads to lower levels of investment than the optimal achievable under perfect information and/or no trading. Contractual provisions that modify information asymmetry or the ability to profit from it could eliminate both temporary inconsistency and inefficiency in investment policies, but these contractual provisions are also likely to incur sunk costs.

In this article, we do not use an econometric model but rather a first principles approach. Both theories, MM and BFO, use this approach. However, there are empirical studies where scientists have considered the influence of various factors on the cost of equity capital using simpler models. One of them is the following.

[37] Examines the effect that environmental, social, and governance (ESG) disclosure generates on the cost of equity capital in the food and beverages (F&B) sector.

This study analyzes a sample of 171 internationally listed firms pertaining to the F&B sector and headquartered in North America, Western Europe, and the Asia Pacific (developed), forming an unbalanced panel of 1.316 observations, spanning the period 2010–2019. Authors run a fixed-effects panel regression model to test the relationship between ESG disclosure and the cost of equity capital. Authors' empirical outcomes suggest a significant negative relationship between ESG disclosure and the cost of equity capital. Authors find support for the notion that increased levels of ESG disclosure are linked to an improved access to financial resources for firms.

The hypotheses of the study are as follows: (1) the actual operating conditions of companies affect the dependence of the cost of equity on debt financing and the company's dividend policy; (2) the abnormal dependence of the cost of equity on the company's debt financing can radically change the company's dividend policy. The presence and magnitude of the anomalous effect can be changed and controlled by changing such parameters as the cost of equity, k_e and debt capital, k_d, the frequency and method of paying income tax, the growth of the income rate, the age of the company, etc. The aims of this paper is to study the influence of the conditions of the real functioning of companies on the dependence of the cost of equity on debt financing, as well as on the anomalous effect: on its existence and management. The following research methods are used: the generalized Brusov–Filatova–Orekhova (BFO theory) and calculations by Microsoft Excel the equity cost dependences on debt financing at different parameters, such as k_0 and k_d, the frequency and method of paying income tax, the growth of the income rate, the age of the company, etc. The anomalous effect affects the company's dividend policy and requires careful and detailed study. It's done here. The motivation of the study is to investigate one of the most striking effects in corporate finance—the anomalous dependence of the cost of equity on the company's debt financing. The significance of this study is determined by the importance of the anomalous effect that significantly affects the company's dividend policy. The limitation of the study is due to the fact that the case of a constant income growth rate is considered. In future publications, the case of a variable income growth rate will be considered. One more limitation of the consideration is related to the known limitations of the WACC approximation.

The structure of the paper is as follows:

We briefly discuss the literature review below. In 19.2, we will discuss the theoretical basis of our consideration and give the main formulas for WACC and k_d, which will be used to calculate WACC and k_e, and the influence of all considered factors on the dependence of equity cost k_e on the level of debt financing, L.

In 19.3, the influence of the following factors on the dependence of equity cost k_e on the level of debt financing, L are considered. In 19.3.1, increase of tax on profit is considered; in 19.3.2, variable income is considered; in 19.3.2.1, influence of growth rate g is considered for tax on income payments at the ends of periods. In 19.3.2.2, influence of growth rate g is considered with comparison of results with advance income tax payments and payments at the end of periods. In 19.3.3, frequent income tax payments is considered. In 19.3.3.1, income tax payments at the ends of periods is considered. In 19.3.3.2, frequent tax on income payments is considered compared with the results for advance payments of tax on income and payments at the ends of periods. In 19.3.4, simultaneous influence of the growth rate g and the frequency of income tax payment, p is considered. In 19.3.5, influence of company age, n is considered. In 19.4, the explanations of the observed effects have been done. In 19.5, conclusions are made.

19.2 Theoretical Basis

The following notation will be used in the text below.

k_d, $w_d = \frac{D}{D+S}$—the debt capital cost and debt capital share, $k_e, w_e = \frac{S}{D+S}$—the equity capital cost and the equity capital share, and $L = D/S$—the value of financial leverage, D—the debt capital value, S—the equity capital value, k_0—the equity capital cost at zero leverage level, g—income growth rate, p—frequency of tax on income payments, $WACC$—the weighted average cost of capital, t—tax on profit, n—company age.

The Brusov–Filatova–Orekhova (BFO) theory and its perpetual limit—the Modigliani–Miller theory—have recently been generalized to the established practice of the functioning of companies. This generalization took into account the real operating conditions of companies, such as variable income, frequent income tax payments, advance income tax payments, etc. This made it possible to investigate the impact of these conditions on its main financial performance [20]. In this paper, we use the generalized Brusov–Filatova–Orekhova (BFO) theory as well as the generalized Modigliani–Miller theory to study the influence of the conditions of the real functioning of companies on the dependence of the cost of equity on debt financing, as well as on the anomalous effect: on its existence and management.

Below we give a summary of the WACC formulas for the BFO-theory as well as for the MM-theory [20].

The classical BFO equation for *WACC*

$$\frac{1-(1+\mathrm{WACC})^{-n}}{\mathrm{WACC}-g}=\frac{1-(1+k_0)^{-n}}{(k_0-g)\cdot(1-w_dt[1-(1+k_d)^{-n}])} \tag{19.6}$$

and its limit for perpetuity companies (MM limit)

$$\mathrm{WACC}=k_0\cdot(1-w_dt) \tag{19.7}$$

The formula for the equity cost comes from the definition of WACC

$$\mathrm{WACC}=k_ew_e+k_dw_d(1-t), \tag{19.8}$$

accounting that

$$w_e=\frac{1}{1+L};w_d==\frac{L}{1+L} \tag{19.9}$$

and is as follows:

$$k_e=\mathrm{WACC}(1+L)-Lk_d(1-t) \tag{19.10}$$

The WACC formulas for the BFO-theory and for the MM-theory under the conditions of the real functioning of companies are presented below [20].

19.2.1 Variable Income Case

19.2.1.1 Income Tax Payments at the Ends of Periods

$$\mathrm{BFO}:\quad\frac{1-\left(\frac{1+g}{1+\mathrm{WACC}}\right)^n}{\mathrm{WACC}-g}=\frac{1-\left(\frac{1+g}{1+k_0}\right)^n}{(k_0-g)\cdot(1-w_dt[1-(1+k_d)^{-n}])} \tag{19.11}$$

$$\mathrm{MM}:\quad\mathrm{WACC}=(k_0-g)\cdot(1-w_dt)+g \tag{19.12}$$

19.2.1.2 Advance Income Tax Payments

$$\text{BFO}: \quad \frac{1-\left(\frac{1+g}{1+\text{WACC}}\right)^n}{\text{WACC}-g} = \frac{1-\left(\frac{1+g}{1+k_0}\right)^n}{(k_0-g)\cdot(1-w_dt[1-(1+k_d)^{-n}]\cdot(1+k_d))} \tag{19.13}$$

$$\text{MM}: \quad \text{WACC} = (k_0-g)\cdot(1-w_dt\cdot(1+k_d))+g \tag{19.14}$$

19.2.2 *Frequent Income Tax Payments*

19.2.2.1 Income Tax Payments at the Ends of Periods

$$\text{BFO}: \quad \frac{1-(1+\text{WACC})^{-n}}{\text{WACC}} = \frac{1-(1+k_0)^{-n}}{k_0\cdot\left(1-\frac{k_dw_dt}{p}\frac{[1-(1+k_d)^{-n}]}{(1+k_d)^{1/p}-1}\right)} \tag{19.15}$$

$$\text{MM}: \quad \text{WACC} = k_0\cdot\left(1-\frac{k_dw_dt}{p\cdot\left[(1+k_d)^{1/p}-1\right]}\right) \tag{19.16}$$

19.2.2.2 Advance Income Tax Payments

$$\text{BFO}: \quad \frac{1-(1+\text{WACC})^{-n}}{\text{WACC}} = \frac{1-(1+k_0)^{-n}}{k_0\cdot\left(1-\frac{k_dw_dt}{p}\frac{[1-(1+k_d)^{-n}]\cdot(1+k_d)^{1/p}}{(1+k_d)^{1/p}-1}\right)} \tag{19.17}$$

$$\text{MM}: \quad \text{WACC} = k_0\cdot\left(1-\frac{k_dw_dt\cdot(1+k_d)^{1/p}}{p\cdot\left[(1+k_d)^{1/p}-1\right]}\right) \tag{19.18}$$

19.2.3 *Simultaneous Accounting of Variable Income in Case of Frequent Income Tax Payments*

19.2.3.1 Income Tax Payments at the Ends of Periods

$$\text{BFO}: \quad \frac{1-\left(\frac{1+g}{1+\text{WACC}}\right)^n}{\text{WACC}-g}=\frac{1-\left(\frac{1+g}{1+k_0}\right)^n}{(k_0-g)\cdot\left(1-\frac{k_d w_d t}{p}\frac{\left[1-(1+k_d)^{-n}\right]}{(1+k_d)^{1/p}-1}\right)} \tag{19.19}$$

$$\text{MM}: \quad \text{WACC}-g=(k_0-g)\cdot\left(1-\frac{k_d w_d t}{p\cdot(1+k_d)^{1/p}-1}\right) \tag{19.20}$$

19.2.3.2 Advance Income Tax Payments

$$\text{BFO}: \quad \frac{1-\left(\frac{1+g}{1+\text{WACC}}\right)^n}{\text{WACC}-g}=\frac{1-\left(\frac{1+g}{1+k_0}\right)^n}{(k_0-g)\cdot\left(1-\frac{k_d w_d t}{p}\frac{\left[1-(1+k_d)^{-n}\right]\cdot(1+k_d)^{1/p}}{\left[(1+k_d)^{1/p}-1\right]}\right)} \tag{19.21}$$

$$\text{MM}: \quad \text{WACC}-g=(k_0-g)\cdot\left(1-\frac{k_d w_d t\cdot(1+k_d)^{1/p}}{p\cdot(1+k_d)^{1/p}-1}\right) \tag{19.22}$$

The general formula for equity cost, k_e, is as follows:

$$k_e=\text{WACC}(1+L)-Lk_d(1-t). \tag{19.23}$$

To study the dependence of the cost of equity capital, k_e, on various variables, it is first necessary to find the value of WACC and substitute it into formula (19.23).

19.3 Results and Discussions

Below we study the influence of the conditions of the real functioning of companies on the dependence of the cost of equity on debt financing, as well as the conditions for the existence of the "anomalous effect" and the possibility of controlling it in accordance with the program stated above.

Our consideration combines theoretical and numerical results: the latter are based on calculations in Microsoft Excel, using some typical data on the cost of equity and debt capital of companies, as well as on the actual data of a particular company (see paragraph 19.3.5 and Figs. 19.30, 19.31, 19.32, 19.33, 19.34, and 19.35, where calculations were made for PJSC GAZP). The data are taken from company reports; profitability, leverage level, cost of debt, etc. Then the data is processed to obtain the necessary calculation parameters, the most important of which is k_0, the cost of equity at zero leverage level L.

19.3.1 Increase of Tax on Profit

Within the Modigliani–Miller theory, formula for equity cost, k_e, has a view

$$k_e = k_0 + L(k_0 - k_d)(1 - t). \tag{19.24}$$

Using formula (19.19) we calculate the dependence of cost of equity, k_e, on leverage level, L at different tax on profit rates t.

It could be seen from Fig. 19.1, where the dependence of cost of equity, k_e, on leverage level, L at different tax on profit rates t is shown for the case $k_0 = 10\ \%$; $k_d = 8\%$ (line 1 corresponds to tax on profit rates $t = 0$; line 2 corresponds to tax on profit rates $t = 0.1$; line 3 corresponds to tax on profit rates $t = 0.2$; line 4 corresponds to tax on profit rates $t = 0.3$; line 5 corresponds to tax on profit rates $t = 0.4$; line 6 corresponds to tax on profit rates $t = 0.5$; line 7 corresponds to tax on profit rates $t = 0.6$; line 8 corresponds to tax on profit rates $t = 0.7$; line 9 corresponds to tax on profit rates t $= 0.8$; line 10 corresponds to tax on profit rates $t = 0.9$; line 11 corresponds to tax on profit rates $t = 1$), that the slope $k_e(L)$ decreases with t and becomes zero at $t = 1$. In this case (at $t = 1$), the line $k_e(L)$ is horizontal. This behavior can be explained by the presence of a multiplier (1–T) (tax corrector).

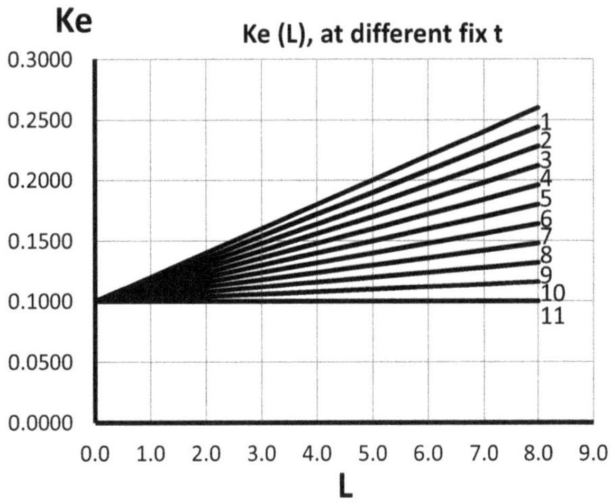

Fig. 19.1 Dependence of cost of equity on leverage level L at different tax on profit rates t

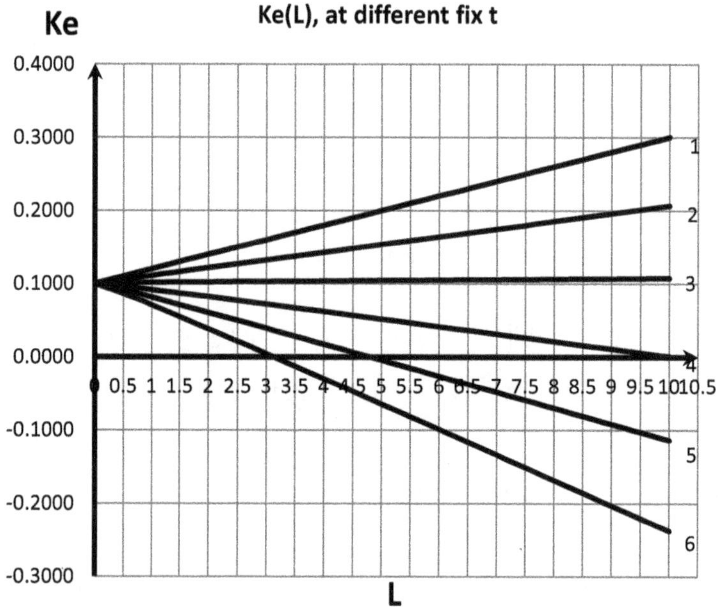

Fig. 19.2 Dependence of cost of equity k_e on leverage level L at different tax on profit rate t

The situation for the finite age company, described by the Brusov–Filatova–Orekhova (BFO) theory is completely different: there is an anomalous dependence of the cost of equity, k_e, on the level of leverage L: the cost of equity k_e decreases with the level of leverage L. This effect radically changes the principles of the company's dividend policy: shareholders receive smaller dividends with an increase in debt financing (which leads to financial difficulties and the risk of bankruptcy). In Fig. 19.2, the dependence of cost of equity k_e on leverage level L at different tax on profit rate t ($n = 5$, $k_0 = 10$ % , $k_d = 8$%)($1 - t = 0$; $2 - t = 0.2$; $3 - t = 0.4$; $4 - t = 0.6$; $5 - T = 0.8$; $6 - t = 1$) is shown.

From Fig. 19.2, it can be seen that starting from certain values of the income tax rate (in this case, from $t^* = 40\%$), the cost of equity of a company with leverage does not increase but decreases. Note that t^* value (as well as the presence or the absence of such an abnormal effect) depends on particular values of parameters n, k_0, k_d, and at other values of parameters n, k_0, k_d critical values of tax on profit rate t^* could be lower.

The anomalous effect could only take place for a finite age company and is not observed in the eternal Modigliani–Miller limit.

This effect exists within the framework of the BFO theory, NOT at all parameters values: below we give an example of the usual behavior similar to the Modigliani–Miller limit one (see Table 19.1 and Fig. 19.3).

Table 19.1 Dependence of cost of equity k_e on leverage level L at different fix tax on profit rates T for the case $n = 7$, $k_0 = 20$ %, $k_d = 10\%$

TL	0.0	1.0	2.0	3.0	4.0	5.0	6.0	7.0	8.0	9.0	10
0	0.2000	0.3000	0.4000	0.5000	0.6000	0.7000	0.8000	0.9000	1.0000	1.1000	1.2000
0.2	0.2000	0.2842	0.3682	0.4522	0.5362	0.6202	0.7042	0.7874	0.8713	0.9551	1.0389
0.4	0.2000	0.2677	0.3344	0.4008	0.4672	0.5335	0.5998	0.6661	0.7323	0.7986	0.8649
0.6	0.2000	0.2504	0.2984	0.3457	0.3928	0.4397	0.4865	0.5334	0.5802	0.6265	0.6731
0.8	0.2000	0.2323	0.2601	0.2861	0.3117	0.3369	0.3619	0.3867	0.4116	0.4364	0.4612
1	0.2000	0.2132	0.2185	0.2210	0.2223	0.2229	0.2231	0.2233	0.2231	0.2228	0.2224

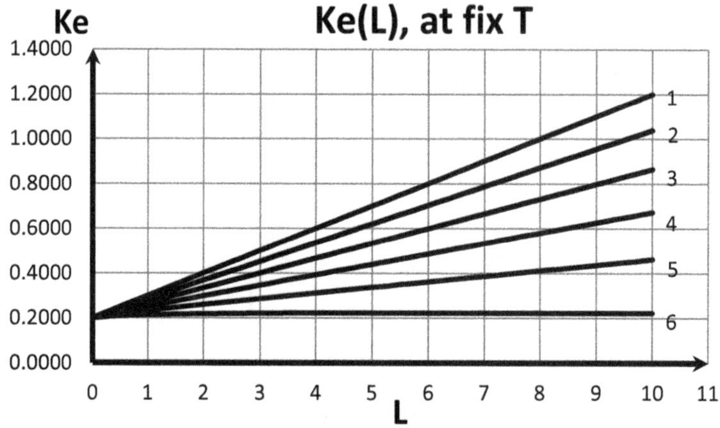

Fig. 19.3 Dependence of cost of equity k_e on leverage level L at different tax on profit rate T ($n = 7$, $k_0 = 20$ % , $k_d = 10\%$) ($1 - T = 0$; $2 - T = 0.2$; $3 - T = 0.4$; $4 - T = 0.6$; $5 - T = 0.8$; $6 - T = 1$)

Let us calculate the dependence of the critical value of the income tax rate, t^*, on the age of the company, n. To do this, we must first calculate the WACC value over a set of parameters, including the age of the company, n, by solving the classical BFO Eq. (19.1). Then obtained value of WACC at a fixed n should be substituted into Eq. (19.18) and the right side should be equated to k_0.

$$k_0 = \text{WACC}(1 + L) - Lk_d(1 - t) \tag{19.25}$$

By solving this equation, one can obtain the critical value of the income tax rate t^* for a fixed n. In Fig. 19.4, the results of such calculations are shown: the dependence of the critical value of tax on profit rate t^* on the age of the company, n, under variation of the difference between k_0 and k_d.

($\Delta k = k_0 - k_d = 2$ % ; 4 % ; 6 % ; 8%) ($1 - k_d = 6\%$, $k_0 = 8\%$; $2 - k_d = 6\%$, $k_0 =$ 10%; $3 - k_d = 6\%$, $k_0 = 12\%$; $4 - k_d = 6\%$, $k_0 = 14\%$)

From the Fig. 19.4, it follows that:

1. The critical value of tax on profit rate t^* increases with the difference $\Delta k = k_0 - k_d$, therefore a small difference between the value of cost of equity (at $L = 0$) k_0 of the company and the credit rate, k_d favor to existence of a new effect.
2. The critical value of tax on profit rate t^* decreases monotonically with the age of the company (only for 10 – years in case of $\Delta k = k_0 - k_d = 2\%$ it has a minimum). Therefore the probability of the anomaly effect is higher for "adult" companies.
3. Recapitulating 1 and 2, one can note that a small difference between the value of cost of equity (at $L = 0$) k_0 of the company and the credit rate k_d as well as old enough age of the company favor to existence of a new effect.

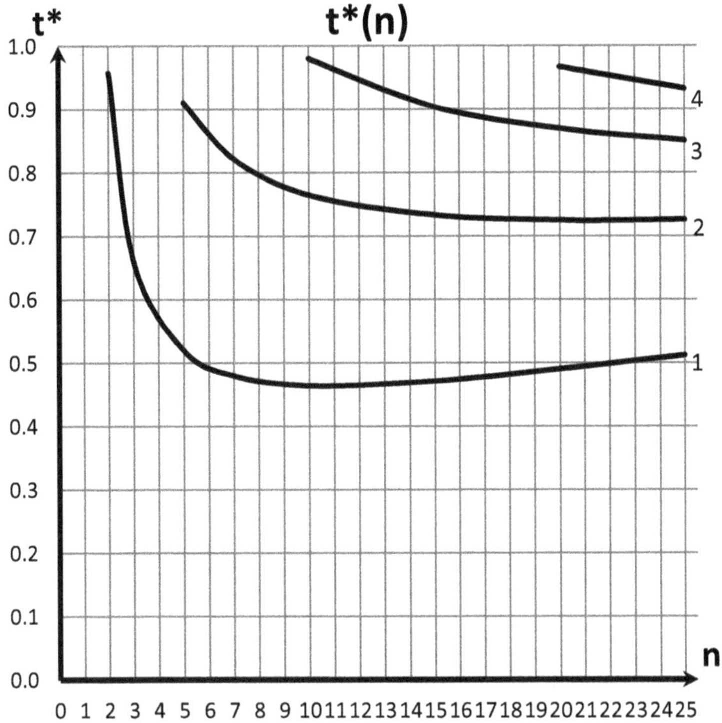

Fig. 19.4 The dependence of the critical value of tax on profit rate t^* on the age of the company

Table 19.2 The dependence of the critical value of tax on profit rate t^* on the age of the company under different values of $k_0$and k_d at constant difference between them $\Delta k = k_0 - k_d = 2\%$

$K_e(t)\backslash n$	2	3	5	7	10	15	20	25
$k_d = 6\%.k_0 = 8\%$	0.9575	0.6600	0.5200	0.4800	0.4640	0.4710	0.4903	0.5121
$k_d = 8\%.k_0 = 10\%$	0.7313	0.5125	0.4140	0.3905	0.3892	0.4138	0.4453	0.4803
$k_d = 10\%.k_0 = 12\%$	0.6000	0.4280	0.3510	0.3392	0.3467	0.3840	0.4285	0.4733
$k_d = 12\%.k_0 = 14\%$	0.5125	0.3687	0.3110	0.3043	0.3218	0.3697	0.4239	0.4788
$k_d = 14\%.k_0 = 16\%$	0.4437	0.3266	0.2810	0.2821	0.3043	0.3636	0.4277	0.4904
$k_d = 18\%.k_0 = 20\%$	0.3625	0.2710	0.2435	0.2549	0.2895	0.3677	0.4468	0.5221
$k_d = 22\%.k_0 = 24\%$	0.3100	0.2370	0.2220	0.2400	0.2875	0.3818	0.4759	0.5588

Below (in Table 19.2), the dependence of the critical value of tax on profit rate t^* on the company age, n, under different values of $k_0$and k_d at constant difference between them $\Delta k = k_0 - k_d = 2\%$ is studied.

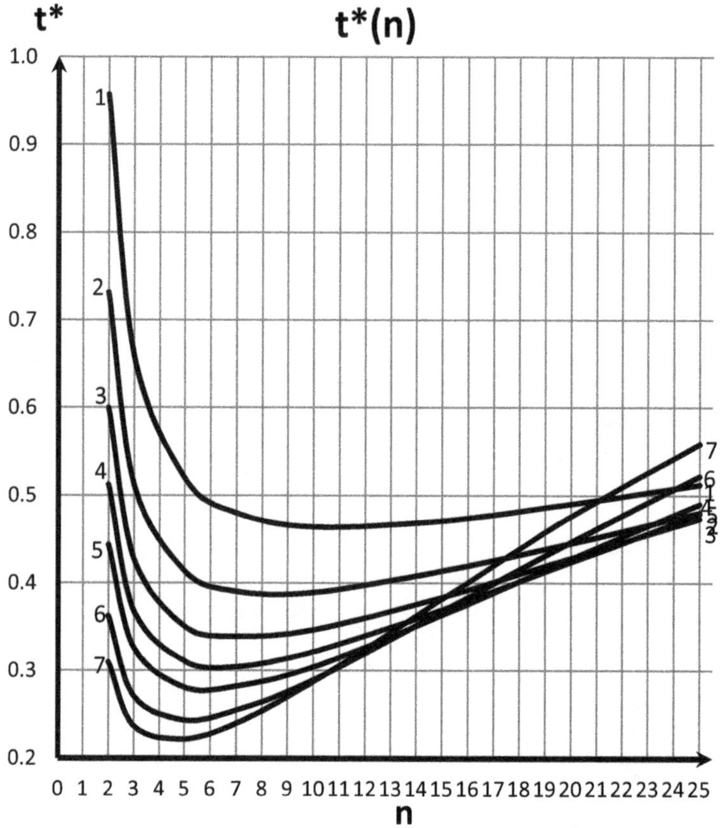

Fig. 19.5 The dependence of the critical value of tax on profit rate t^* on the company age, n

In Fig. 19.5, the dependence of the critical value of tax on profit rate, t^*, on the company age, n, is shown at different values of $k_0$and k_d at constant difference between them $\Delta k = k_0 - k_d = 2\%$ (1 – $k_0 = 8\%$; 2 – $k_0 = 10\%$; 3 – $k_0 = 12\%$; 4 – $k_0 = 14\%$; 5 – $k_0 = 16\%$; 6 – $k_0 = 20\%$; 7 – $k_0 = 24\%$)

The following conclusions could be made:

1. All curves are convex and the critical value of tax on profit rate t^* reaches minimum, which value decreases with k_0.

   $_{\text{Min}}\, t_* = {}_{22.2\%}$ at $k_0 = 24\%$, min $T^* = 24.35\%$ at $k_0 = 20\%$, min $t^* = 28.1\%$ at $k_0 = 16\%$, min $t^* = 30.43\%$ at $k_0 = 14\%$, min $t^* = 33.92\%$ at $k_0 = 12\%$, min $T^* = 38.92\%$ at $k_0 = 10\%$, min $t^* = 46.4\%$ at $k_0 = 8\%$. Therefore the higher value of $k_0$and the higher value of k_d at constant difference between them $\Delta k = k_0 - k_d = const$ favor for existence of a new effect.
2. The critical value of tax on profit rate t^* reaches minimum at company age, decreasing with k_0: $n = 4$, 5 years at $k_0 = 24\%$,$n = 5$, 5 years at $k_0 = 16\%$,$n = 6$, 5 years at $k_0 = 12\%$and $n = 10$, 5 years at $k_0 = 8\%$.

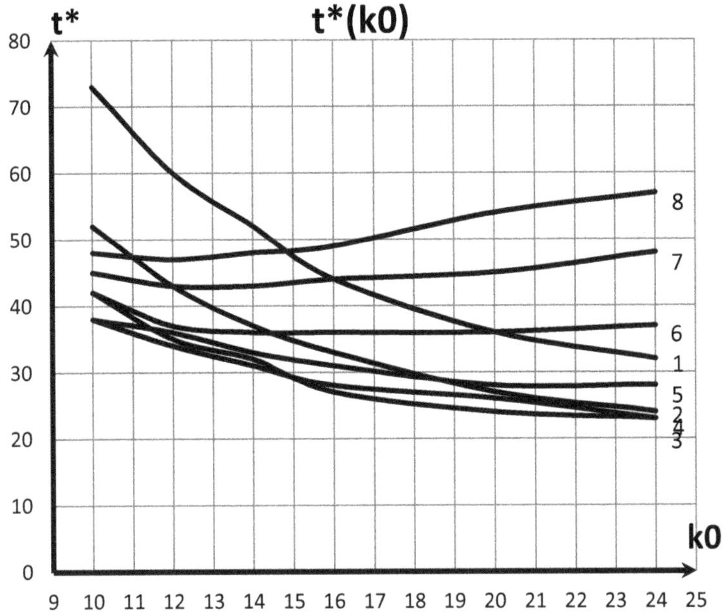

Fig. 19.6 The dependence of the critical value of tax on profit rate t^* on k_0 at constant difference between them $\Delta k = k_0 - k_d = 2\%$

3. Thus, a parallel shift up of rates $k_0$and k_d favor a for new effect, while company age, favorable for a new effect, decreases with k_0.

   The dependence of the critical value of tax on profit rate t^* on k_0(equity cost at zero leverage level, L) at constant difference between them $\Delta k = k_0 - k_d = 2\%$ is shown in Fig. 19.6, where ($1 - n = 2$; $2 - n = 3$; $3 - n = 5$; $4 - n = 7$; $5 - n = 10$; $6 - n = 15$; $7 - n = 20$; $8 - n = 25$)

What is the practical value of the effect? Does it exist in real life or is its discovery of purely theoretical interest? Since the new effect takes place at an income tax rate that is greater than a certain value of T^*, it is necessary to compare this value with the real income tax rates established in different countries. The biggest tax on profits of corporation rate is in the USA—39.2%. In Japan, it exceeds a little bit 38%. In France, tax on profits of corporation rate varies from 33.3% for small- and medium-sized companies up to 36% for the major ones. In England, tax on profits of corporation rate is in the range of 21% to 28%. In the Russian Federation, tax on profits of corporation rate is installed in the amount of 20%.

In the examples considered by us, the value T^* strongly depends on the ratio between k_0, k_d, n and reaches a minimal value of 22.2%, and it is quite likely even lower values of T^* with other ratios between k_0, k_d, n.

In this way, we come to the conclusion, that at some ratios between equity cost, debt cost, and company age k_0, k_d, n discovered by us, an effect takes place at tax on profits of corporation rate established in most developed countries, that provides the practical value of effect.

Its account is important in improving tax legislation and may change dividend policies of the company.

Opening the effect expands our view of the rules of the game in the economy.

If prior to that it was widely known that, with the rising of leverage, the cost of equity is always growing, that is associated with the decrease in financial sustainability of the companies, with an increase in the share of borrowing, when the shareholders require higher rate of return on the share.

But now it becomes clear that this is not always the case, and the dependence of cost of equity on leverage depends on the ratio between the parameters k_0, k_d, n, and, ultimately, on the tax on profit rate.

This effect has never been known, therefore, it was not taken into account by controls tax legislation, but tremendous possibilities are opening here.

The effect is also important for the development of the dividend policy of the company.

It turns out that the rule adopted by shareholders since time immemorial to demand a higher rate of return per share when increasing the share of debt capital does not always work now. This will allow the company's management to pursue a more realistic dividend policy, limiting shareholders' appetites to an economically justified amount of dividends.

As we mentioned in 19.2, the Brusov–Filatova–Orekhova (BFO) theory and its eternal limit—the Modigliani–Miller theory—have recently been generalized to the established practice of the functioning of companies, such as variable income, frequent income tax payments, advance payments on income tax, etc. Below, we use the generalized theory of Brusov–Filatova–Orekhova (BFO) to study the existence and behavior of an anomalous effect in a wide range of the financial parameters of the company. We start from the case of variable income.

19.3.2 Variable Income

19.3.2.1 Influence of Growth Rate g: Tax on Income Payments at the Ends of Periods

Below, we study the influence of growth rate of income g in case of tax on income payments at the ends of periods.

19.3.2.1.1 Six-Year Company

Figure 19.7 shows the dependence of equity cost, k_e, on leverage level, L, for a 6-year company at $p = 1$; $k_0 = 0.16$; $k_d = 0.14$; $t = 0.2$; at different $g = -0.2$; -0.15;

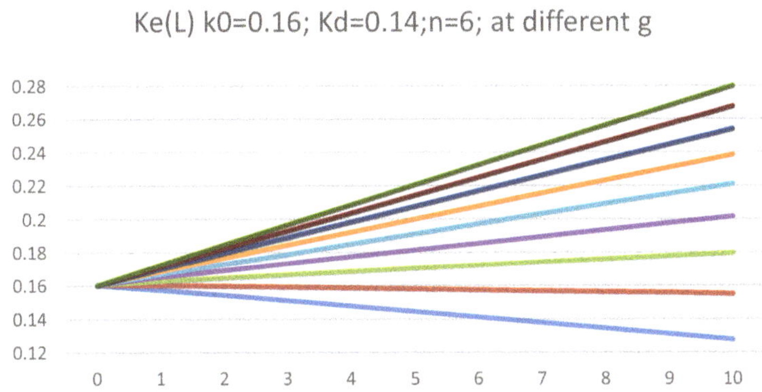

Fig. 19.7 The dependence of equity cost, k_e, on leverage level, L, at $p = 1$; $k_0 = 0.16$; $k_d = 0.14$; $t = 0.2$; $n = 6$ at different $g = -0.2; -0.15; -0.1; -0.05; 0; 0.05; 0.1; 0.15; 0.2$ (from the bottom to the top) with payments of tax on income at the ends of periods

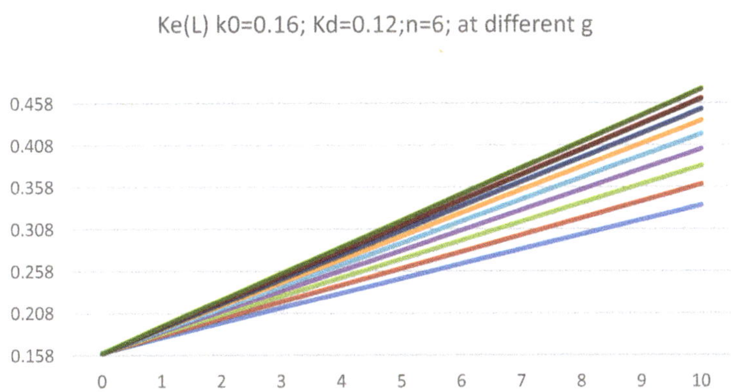

Fig. 19.8 The dependence of equity cost, k_e, on leverage level, L, at $p = 1$; $k_0 = 0.16$; $k_d = 0.12$; $t = 0.2$; $n = 6$ at different $g = -0.2; -0.15; -0.1; -0.05; 0; 0.05; 0.1; 0.15; 0.2$ (from the bottom to the top) with payments of tax on income at the ends of periods

$-0.1; -0.05; 0; 0.05; 0.1; 0.15; 0.2$ with payments of tax on income at the ends of periods.

From Fig. 19.7, it is seen that with an increase in the rate of income growth g, the slope of the curve $k_e(L)$ increases. The anomalous effect for a 6-year company at $k_0 = 0.16$; $k_d = 0.14$ takes place at $g < -1$: the slope of the curve $k_e(L)$ is negative.

Figure 19.8 shows the dependence of equity cost, k_e, on leverage level, L, for a 6-year company at $p = 1$; $k_0 = 0.16$; $k_d = 0.12$; $t = 0.2$; at different $g = -0.2; -0.15; -0.1; -0.05; 0; 0.05; 0.1; 0.15; 0.2$ with payments of tax on income at the ends of periods.

From Fig. 19.8, it is seen that, like the previous case, with an increase in the rate of income growth g, the slope of the curve $k_e(L)$ increases. The anomalous effect for a 6-year company at $k_0 = 0.16$; $k_d = 0.12$ 12 is absent at all g values: the slope of the curve $k_e(L)$ is positive.

19.3.2.1.2 Three-Year Company

Let us consider a 3-year company.

From Fig. 19.9, it is seen that, like the previous case, with an increase in the rate of income growth g, the slope of the curve $k_e(L)$ increases. The anomalous effect for a 3-year company at $k_0 = 0.16$; $k_d = 0.14$ is absent at all g values: the slope of the curve $k_e(L)$ is positive.

From Fig. 19.10, it is seen that, like the previous case, with an increase in the rate of income growth g, the slope of the curve $k_e(L)$ increases. The anomalous effect for a 3-year company at $k_0 = 0.16$; $k_d = 0.12$ is absent at all g values: the slope of the curve $k_e(L)$ is positive. The width of the bundle of straight lines $k_e\{L\}$ narrows with decreasing k_d.

From Figs. 19.7, 19.8, 19.9, 19.10, and 19.11, we could conclude the following:

1. With an increase in the rate of income growth g, the slope of the curve $k_e(L)$ increases, since in this case shareholders can count on larger dividends.
2. The anomalous effect exists for a small difference between k_0 and k_d $\Delta = k_0 - k_d$ and is absent for a larger value of this difference.

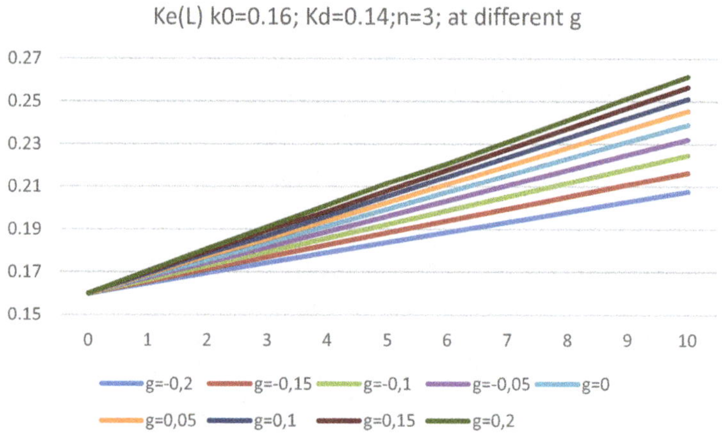

Fig. 19.9 The dependence of equity cost, k_e, on leverage level, L, at $p = 1$; $k_0 = 0.16$; $k_d = 0.14$; $t = 0.2$; $n = 3$ at different $g = -0.2; -0.15; -0.1; -0.05; 0; 0.05; 0.1; 0.15; 0.2$ (from the bottom to the top) with payments of tax on income at the ends of periods

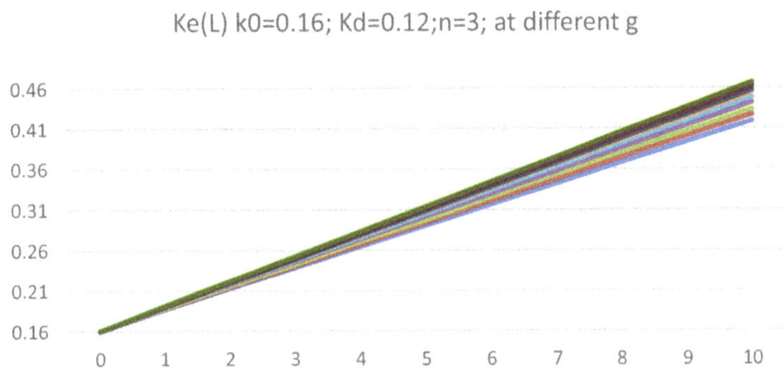

Fig. 19.10 The dependence of equity cost, k_e, on leverage level, L, at $p = 1$; $k_0 = 0.16$; $k_d = 0.12$; $t = 0.2$; $n = 3$ at different $g = -0.2; -0.15; -0.1; -0.05; 0; 0.05; 0.1; 0.15; 0.2$ (from the bottom to the top) with payments of tax on income at the ends of periods

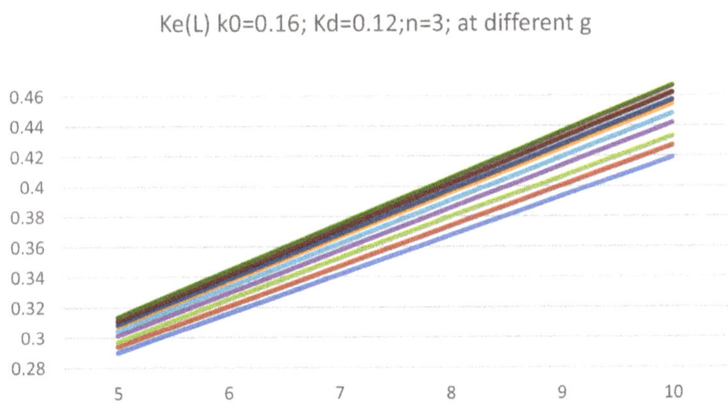

Fig. 19.11 The dependence of equity cost, k_e, on leverage level, L, at $p = 1$; $k_0 = 0.16$; $k_d = 0.12$; $t = 0.2$; $n = 3$ at different $g = -0.2; -0.15; -0.1; -0.05; 0; 0.05; 0.1; 0.15; 0.2$ (from the bottom to the top) with payments of tax on income at the ends of periods (lager scale)

19.3.2.2 Influence of Growth Rate *g*: Comparison of Results with Advance Income Tax Payments and Payments at the Ends of Periods

Below, we continue to study the influence of growth rate g. We consider the case of advance income tax payments for 6-year and 3-year companies. The results for advance income tax payments are shown in Figs. 19.12, 19.13, 19.14, 19.15, and 19.16.

Six-Year Company

From Fig. 19.12, it is seen that with an increase in the rate of income growth g, the slope of the curve $k_e(L)$ increases. The anomalous effect for a 6-year company with advance payments of tax on income at $k_0 = 0.16$; $k_d = 0.14$ takes place at $g < 0$: the slope of the curve $k_e(L)$ is negative for all negative g.

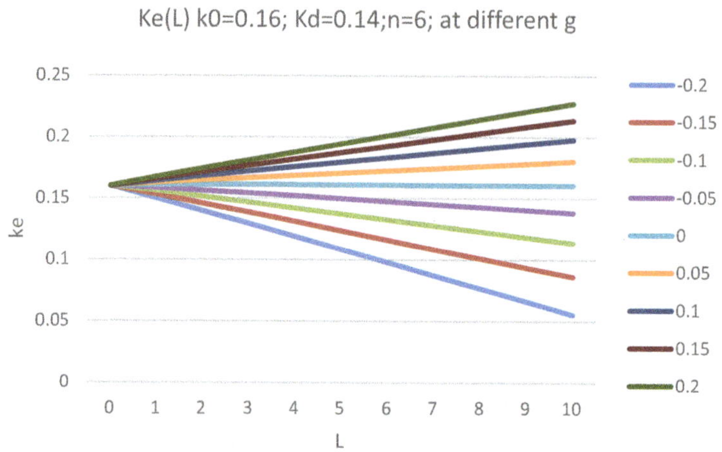

Fig. 19.12 The dependence of equity cost, k_e, on leverage level, L, at $p = 1$; $k_0 = 0.16$; $k_d = 0.14$; $t = 0.2$; $n = 6$ at different $g = -0.2; -0.15; -0.1; -0.05; 0; 0.05; 0.1; 0.15; 0.2$ (from the bottom to the top) with advance payments of tax on income

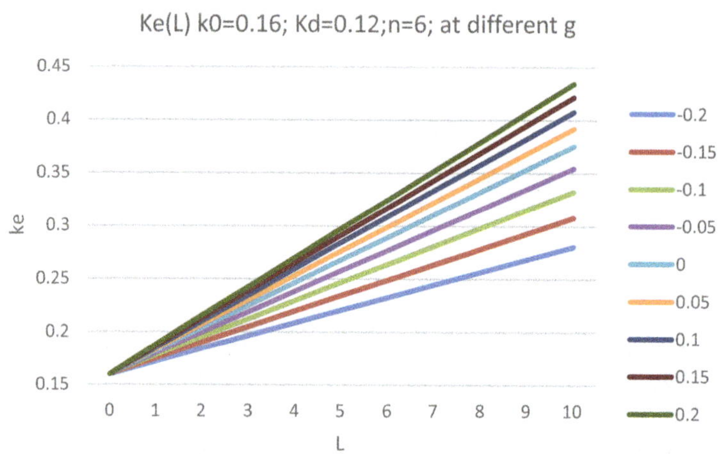

Fig. 19.13 The dependence of equity cost, k_e, on leverage level, L, at $p = 1$; $k_0 = 0.16$; $k_d = 0.12$; $t = 0.2$; $n = 6$ at different $g = -0.2; -0.15; -0.1; -0.05; 0; 0.05; 0.1; 0.15; 0.2$ (from the bottom to the top) with advance payments of tax on income

From Fig. 19.13, it is seen that with an increase in the rate of income growth g, the slope of the curve $k_e(L)$ increases. The anomalous effect for a 6-year company with advance payments of tax on income at $k_0 = 0.16$; $k_d = 0.12$ is absent: the slope of the curve $k_e(L)$ is positive for all g values.

Three-Year Company

Let us calculate for a 3-year company.

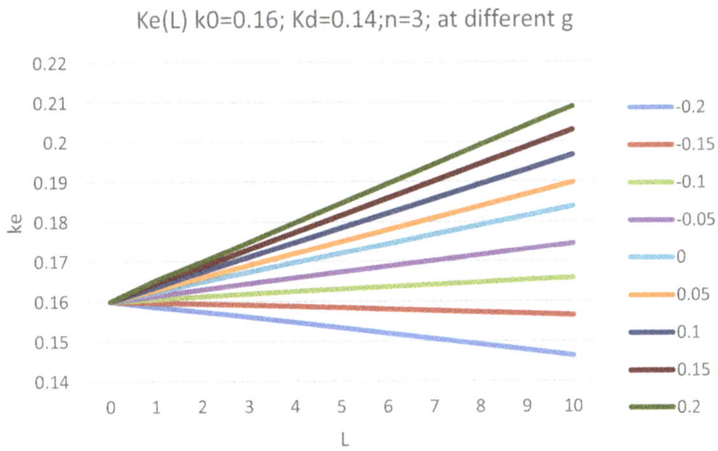

Fig. 19.14 The dependence of equity cost, k_e, on leverage level, L, at $p = 1$; $k_0 = 0.16$; $k_d = 0.14$; $t = 0.2$; $n = 3$ at different $g = -0.2; -0.15; -0.1; -0.05; 0; 0.05; 0.1; 0.15; 0.2$ (from the bottom to the top) with advance payments of tax on income

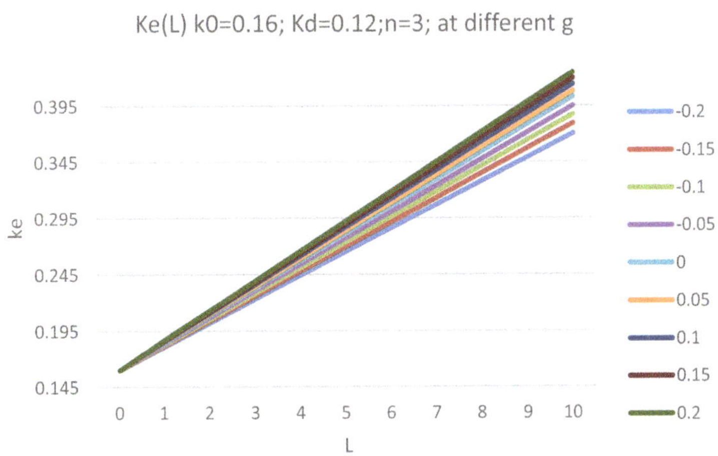

Fig. 19.15 The dependence of equity cost, k_e, on leverage level, L, at $p = 1$; $k_0 = 0.16$; $k_d = 0.12$; $t = 0.2$; $n = 3$ at different $g = -0.2; -0.15; -0.1; -0.05; 0; 0.05; 0.1; 0.15; 0.2$ (from the bottom to the top) with advance payments of tax on income

From Fig. 19.14, it is seen that with an increase in the rate of income growth g, the slope of the curve $k_e(L)$ increases. The anomalous effect for a 3-year company with advance payments of tax on income at $k_0 = 0.16$; $k_d = 0.14$ takes place at $g < -0.1$: the slope of the curve $k_e(L)$ is negative for $g < -0.1$. Note that for a 6-year company with advance payments of tax on income at $k_0 = 0.16$; $k_d = 0.14$, the anomalous effect takes place at bigger $g < 0$.

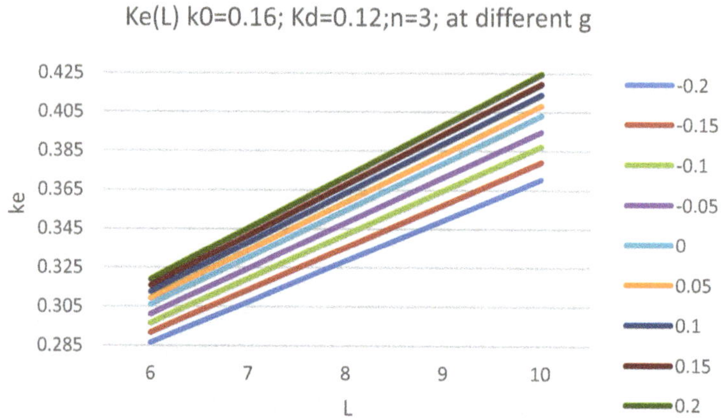

Fig. 19.16 The dependence of equity cost, k_e, on leverage level, L, at $p = 1$; $k_0 = 0.16$; $k_d = 0.12$; $t = 0.2$; $n = 3$ at different $g = -0.2; -0.15; -0.1; -0.05; 0; 0.05; 0.1; 0.15; 0.2$ (from the bottom to the top) with advance payments of tax on income (larger scale)

From Figs. 19.15 and 19.16, it is seen that with an increase in the rate of income growth g, the slope of the curve $k_e(L)$ increases. The anomalous effect for a 3-year company with advance payments of tax on income at $k_0 = 0.16$; $k_d = 0.12$ is absent: the slope of the curve $k_e(L)$ is positive for all g values. The width of the bundle of straight lines k_e narrows with decreasing k_d.

From Figs. 19.12, 19.13, 19.14, 19.15, and 19.16, the following conclusions follows:

1. With an increase in the rate of income growth g, the slope of the curve $k_e(L)$ increases, since, in this case, shareholders can count on larger dividends.
2. The anomalous effect exists for a small difference between k_0 and k_d $\Delta k = k_0 - k_d$ and is absent for a larger value of this difference.
3. The width of the bundle of straight lines k_e narrows with decreasing k_d.
4. Advance income tax payments are favorable for the existence of the anomalous effect.

The dependence of the existence of anomalous effect on the $\Delta k = k_0 – k_d$, company age, n, and value of g^*, below which ($g < g^*$) this effect exists is shown in Table 19.3.

From Table 19.3, it is seen that at company age $n = 6$, the effect exists at $\Delta k = 2\%$ only in both cases (for payments of tax on income at the ends [1] of periods as well as for advance payments of tax on income [2]) and that the g^*value increases under transition from (1) type of payments to (2) from – 0.125 to zero. At company age $n = 3$ the effect exists at $\Delta k = 2\%$ only in the case of advance payments of tax on income at the $g^* = - 0.125$.

Small value of Δk favorable to existence of effect, as well as **bigger age of company** and **advance payments of tax on income**.

Table 19.3 The dependence of the existence of anomalous effect on the Δk, company age, n, and value of g^*, below which ($g < g^*$) this effect exists

		Payments of tax on income at the ends of periods		Advance payments of tax on income	
	Δk	Existence of effect	g^* value	Existence of effect	g^* value
$n = 6$	2%	+	−0.125	+	0
	4%	−		−	
	6%	−		−	
$n = 3$	2%	−		+	−0.125
	4%	−		−	
	6%	−		−	

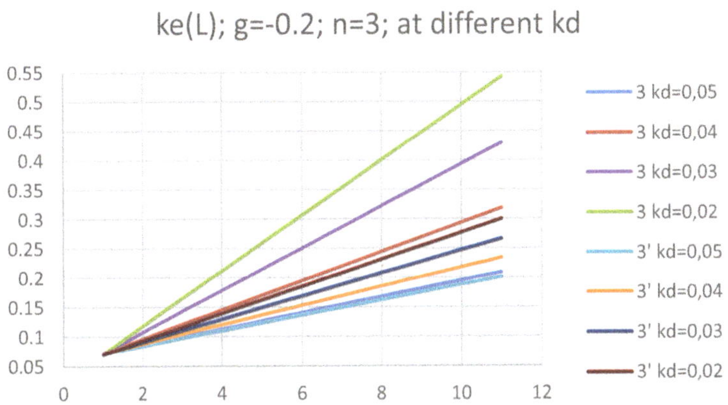

Fig. 19.17 The comparison of the dependences of equity cost, k_e, on leverage level, L, at $p = 1$; $k_0 = 0.07$; at different $k_d = 0.02$; 0.03; 0.04; 0.05; $t = 0.2$; $n = 3$ at negative $g = -0.2$ with advance payments of tax on income (3') and payments of tax on income at the ends of periods (3)

19.3.2.3 Influence of Debt Cost k_d

Below, we examine the impact of the cost of debt k_d on the existence of an anomalous effect for advance income tax payments and income tax payments at the ends of periods, both in the case of falling and rising profits.

It can be seen from Fig. 19.17 that the slope of the curve $k_e(L)$ decreases with the value of k_d for both cases of income tax payment (for payments at the ends of periods [1] and also for advance payments [2]). In this case, all curves (except curve at $k_d =$ 0.05) related to case (1) have a greater slope compared to the curves related to case (2).

It can be seen in Fig. 19.18 that the slope of the curve $k_e(L)$ decreases with the value of k_d for both cases of income tax payment (for payments at the ends of periods [1] and also for advance payments [2]). In this case, all curves (except curve at $k_d =$ 0.05) related to case (1) have a greater slope compared to the curves related to case (2).

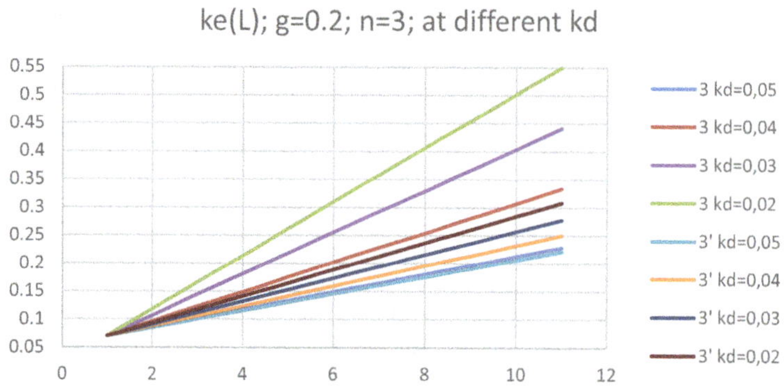

Fig. 19.18 The comparison of the dependences of equity cost, k_e, on leverage level, L, at $p = 1$; $k_0 = 0.07$; at different $k_d = 0.02$; 0.03; 0.04; 0.05; $t = 0.2$; $n = 3$ at positive $g = 0.2$ with advance payments of tax on income (3') and payments of tax on income at the ends of periods (3)

19.3.3 Frequent Income Tax Payments

19.3.3.1 Frequent Tax on Income Payments: Compare the Results for Advance Payments of Tax on Income and Payments at the Ends of Periods

Below, we investigate the impact of frequent income tax payments on existence and behavior of the anomalous effect for advance payments of tax on income and payments at the ends of periods for 3- and 6-year companies. For this, we calculate the dependence of equity cost, k_e, on leverage level, L, at different $p = 1; 2; 4; 6; 12$; $k_0 = 0.22$; at $k_d = 0.2$; $t = 0.2$; $n = 3$ with advance payments of tax on income (p') and payments of tax on income at the ends of periods (p). Results for a 3-year company are shown in Table 19.4 and Fig. 19.19, and for a 6-year company are shown in Table 19.5 and Fig. 19.20.

The slope $k_e(L)$ decreases with p when income tax is paid at the ends of periods and becomes negative at $p > 4$ and higher for $n = 3$ and at $p > 2$ and higher for $n = 6$ (Fig. 19.20). For advance income tax payments, the slope $k_e(L)$, which is negative for any income tax frequency payments, increases (decreases modulo) with a frequency p, but never intersects with the curves $k_e(L)$ in case of income tax payment at the ends of the periods. The anomalous effect disappears at $\Delta k = 4\%$ and above (see Fig. 19.21 below).

The slope of $k_e(L)$ is positive for all p and both payment methods: this means that there is no anomalous effect. The slope $k_e(L)$ decreases with p when income tax is paid at the end of the period and increases when income tax advances are paid. The order of the curves $k_e(L)$ corresponding to these two methods of payment turns out to be mixed. This can lead to very interesting effects, providing new opportunities and the ability to manage the payment of income tax in accordance with tax laws (Fig. 19.21).

Table 19.4 The comparison of the dependences of equity cost, k_e, on leverage level, L, at different $p = 1; 2; 4; 6; 12$; $k_0 = 0.22$; at $k_d = 0.2$; $t = 0.2$; $n = 3$ with advance payments of tax on income (p') and payments of tax on income at the ends of periods (p)

	0	1	2	3	4	5	6	7	8	9	10
P1	0.2202527	0.224593	0.228886	0.23312	0.237315	0.241494	0.245665	0.249829	0.25399	0.258148	0.262305
P2	0.2202527	0.221983	0.223697	0.225345	0.226963	0.228565	0.230162	0.231751	0.233333	0.234894	0.236496
P4	0.2202527	0.221029	0.220729	0.221286	0.221551	0.221801	0.222043	0.22228	0.222513	0.222744	0.222974
P6	0.2202527	0.220172	0.220075	0.219908	0.220674	0.220139	0.219594	0.219043	0.218488	0.217931	0.217371
P12	0.2202527	0.219711	0.219149	0.218517	0.217855	0.217179	0.216495	0.215805	0.215112	0.214416	0.213718
P'1	0.2199314	0.213774	0.207216	0.20056	0.193864	0.187147	0.180416	0.173609	0.166787	0.159962	0.153135
P'2	0.2199314	0.216589	0.212896	0.209117	0.205303	0.201472	0.19763	0.193783	0.189931	0.186075	0.182212
P'4	0.2199314	0.217965	0.21567	0.213294	0.210886	0.208461	0.206028	0.203588	0.201144	0.198697	0.196249
P'6	0.2199314	0.218419	0.216585	0.214672	0.212727	0.210766	0.208796	0.20682	0.204841	0.202858	0.200874
P'12	0.2199314	0.218871	0.217495	0.216041	0.214558	0.213058	0.211549	0.210035	0.208517	0.206996	0.205473

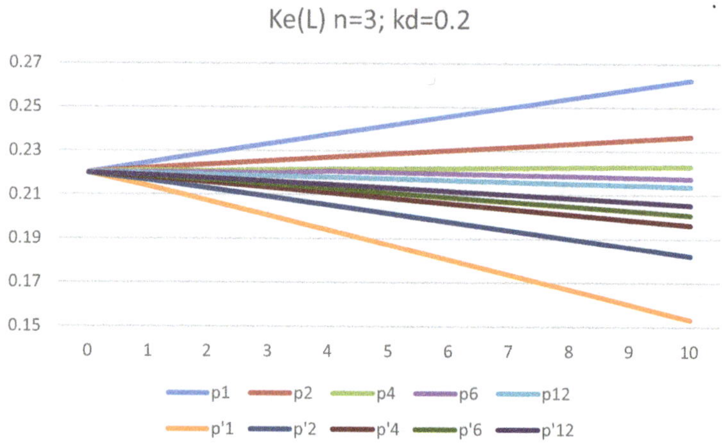

Fig. 19.19 The comparison of the dependences of equity cost, k_e, on leverage level, L, at different $p = 1; 2; 4; 6; 12$; $k_0 = 0.22$; at $k_d = 0.2$; $t = 0.2$; $n = 3$ with advance payments of tax on income (p') and payments of tax on income at the ends of periods (p)

19.3.4 Simultaneous Influence of the Growth Rate g and the Frequency of Income Tax Payment, p

19.3.4.1 Tax Payments at the Ends of periods

Below (see Figs. 19.22 and 19.23), we examine the impact of growth rate g and income tax frequency payments, p on the existence of an anomalous effect for income tax payments at the ends of periods, both in the case of falling and rising profits.

The slope $k_e(L)$ decreases with p and becomes negative at $p > 2$ ($p = 4; 6; 12$).

The slope $k_e(L)$ decreases with p and remains positive at all p.

The slope $k_e(L)$ decreases with p and remains positive at all p. Comparing to the case $g = 0$ the slope $k_e(L)$ is higher, this means that the slope $k_e(L)$ grows with g (Fig. 19.24).

19.3.4.2 Advance Tax Payments

Below, we examine the impact of the frequency of income tax payments on the existence of an anomalous effect for advance income tax payments, both in the case of falling and rising profits.

At negative growth rate g, the slope $k_e(L)$ increases with frequency of payments of tax on profit, p, being negative at $p = 1$ and about zero.

At $p = 1$, the slope $k_e(L)$ increases with growth rate g, being negative at $g = -0.2$.

At $p = 2$, the slope $k_e(L)$ increases with growth rate g, being positive at all g.

At $p = 6$, the slope $k_e(L)$ increases with growth rate g, being positive at all g.

Table 19.5 The comparison of the dependences of equity cost, k_e, on leverage level, L, at different $p = 1; 2; 4; 6; 12$; $k_0 = 0.22$; at $k_d = 0.2$; $t = 0.2$; $n = 6$ with advance payments of tax on income (p') and payments of tax on income at the ends of periods (p)

	0	1	2	3	4	5	6	7	8	9	10
P1	0.220001	0.223948	0.227332	0.230598	0.234363	0.237707	0.241035	0.244354	0.247667	0.250975	0.25428
P2	0.220001	0.221308	0.222079	0.222713	0.223284	0.223825	0.224352	0.224865	0.225367	0.225839	0.226362
P4	0.220001	0.219931	0.219315	0.218552	0.217729	0.216874	0.216737	0.215709	0.214677	0.213643	0.212606
P6	0.220001	0.219464	0.218373	0.217133	0.215831	0.214498	0.213147	0.211785	0.210415	0.209039	0.20766
P12	0.220001	0.218774	0.217249	0.215655	0.214025	0.21238	0.210725	0.209065	0.2074	0.205734	0.202836
P'1	0.219896	0.212774	0.20494	0.196401	0.188102	0.179762	0.1714	0.163023	0.154636	0.146243	0.137844
P'2	0.219896	0.215592	0.210457	0.205549	0.200371	0.195157	0.189921	0.184673	0.179415	0.174151	0.168883
P'4	0.219896	0.21697	0.213446	0.209762	0.20601	0.202225	0.19842	0.194603	0.190777	0.186946	0.18311
P'6	0.219896	0.217424	0.214366	0.21115	0.207869	0.204555	0.201221	0.197875	0.194521	0.191161	0.187798
P'12	0.219896	0.217876	0.215281	0.212531	0.209717	0.206871	0.204006	0.201128	0.198243	0.195353	0.192458

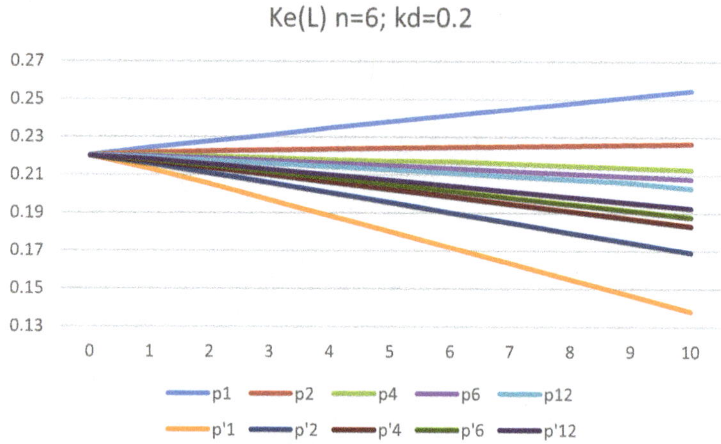

Fig. 19.20 The comparison of the dependences of equity cost, k_e, on leverage level, L, at different $p = 1; 2; 4; 6; 12$; $k_0 = 0.22$; at $k_d = 0.2$; $t = 0.2$; $n = 6$ with advance payments of tax on income (p') and payments of tax on income at the ends of periods (p)

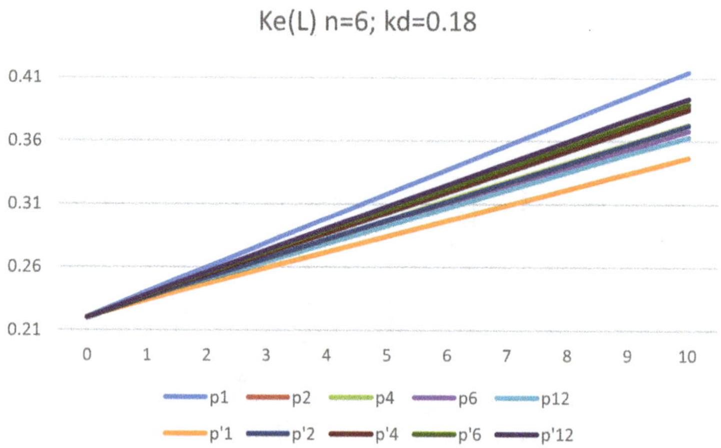

Fig. 19.21 The comparison of the dependences of equity cost, k_e, on leverage level, L, at different $p = 1; 2; 4; 6; 12$; $k_0 = 0.22$; at $k_d = 0.18$; $t = 0.2$; $n = 6$ with advance payments of tax on income (p') and payments of tax on income at the ends of periods (p)

At $p = 12$ the slope $k_e(L)$ increases with growth rate g, being positive at all g . From Figs. 19.25, 19.26, 19.27, 19.28, and 19.29, it follows that the slope of all curves $k_e(L)$ increase with frequency of advance payments of tax on income.

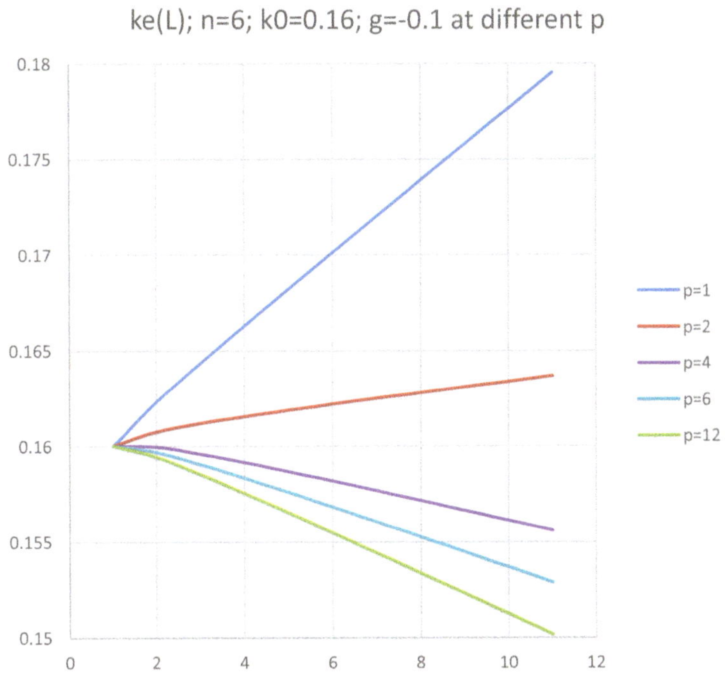

Fig. 19.22 The dependences of equity cost, k_e, on leverage level, L, at different $p = 1; 2; 4; 6; 12$; $k_0 = 0.16$; at $k_d = 0.14$; $t = 0.2$; $n = 6$, $g = -0.1$ with payments of tax on income at the ends of periods

19.3.5 Impact of Company Age

The dependence of the cost of equity, k_e, on the level of leverage, L, at different ages of the company, n, was studied. It was shown that there are two kind of dependences of slope $k_e(L)$ on company age, n: monotonic decrease of tilt angle $k_e(L)$ with n (see Fig. 19.30) and seemingly random addiction of tilt angle $k_e(L)$ with n (see Fig. 19.31).

In Fig. 19.30, at negative $g = -0.05$, the following ordering of curves $k_e(L)$ takes place: $n = 1$; $n = 3$; $n = 5$; $n = 7$; $n = 10$; $n = 15$; $n = 20$; $n = 25$; $n = 30$. This ordering is monotonic.

In Fig. 19.31, at positive $g = 0.05$, the following ordering of curves $k_e(L)$ takes place $n = 1$; $n = 30$; $n = 25$; $n = 20$; $n = 15$; $n = 3$; $n = 10$; $n = 5$; $n = 7$. This ordering is not monotonous, but seemingly random.

The reason for the existence of two types of dependences of the slope $k_e(L)$ on the age of the enterprise, n: a monotonous decrease in the slope angle $k_e(L)$ with n and a seemingly random dependence of the slope angle $k_e(L)$ with n, is the absence or presence of the so-called "golden age" company effect [20].

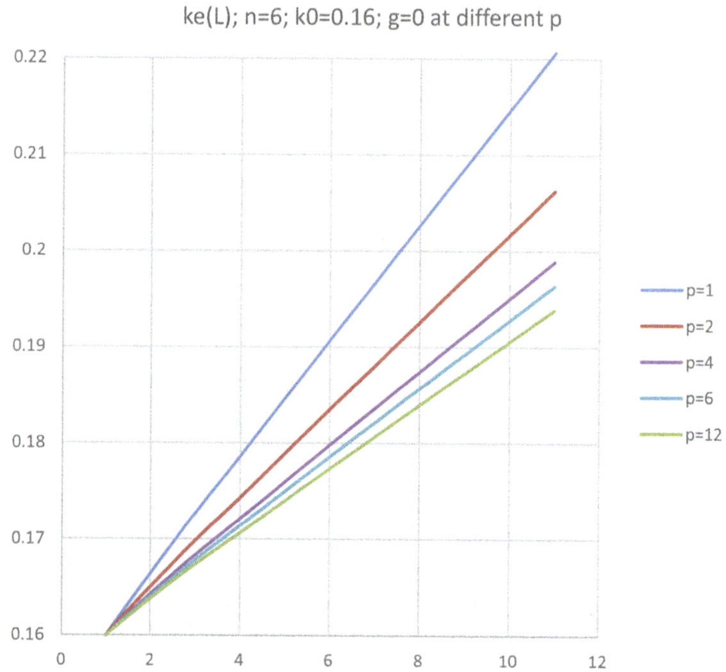

Fig. 19.23 The dependences of equity cost, k_e, on leverage level, L, at different $p = 1; 2; 4; 6; 12$; $k_0 = 0.16$; at $k_d = 0.14$; $t = 0.2$; $n = 6$, $g = 0$ with payments of tax on income at the ends of periods

The so-called "golden age" of the company effect is the situation, when the cost of capital raised (WACC) is below the perpetuity limit, and the company's value is higher. The absence or presence of the "golden age" company effect could explain the ordering of curves $k_e(L)$ (of slopes of curves $k_e(L)$) (see Figs. 19.32, 19.33, and 19.34):

- In case of the absence of the "golden age" company effect, the ordering of curves $k_e(L)$ is monotonic (Fig. 19.32).
- In case of the presence of the "golden age" company effect, the ordering of curves $k_e(L)$ is not monotonous (Fig. 19.33).

From Fig. 19.34, it is seen that the main factor affecting the slope $k_e(L)$ (between the cost of debt k_d and the growth rate g) is the cost of debt k_d. From Fig. 19.34, it can be seen that the slope $k_e(L)$ decreases as the cost of debt k_d increases. We deliberately choose a leverage level, $L = 10$, where the difference between the cost of equity values is the largest, in order to make this difference more noticeable. Of the doublet of curves corresponding to the fixed cost of debt k_d, the upper curve refers to the larger growth rate g. Note, that the shape of the $k_e(n)$ curve is similar to the shape of the WACC(n) curve.

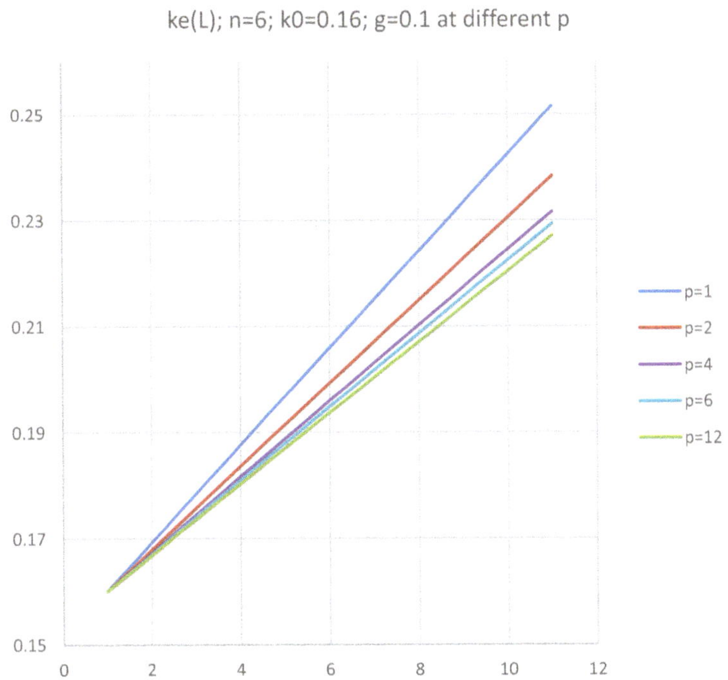

Fig. 19.24 The dependences of equity cost, k_e, on leverage level, L, at different $p = 1; 2; 4; 6; 12$; $k_0 = 0.16$; at $k_d = 0.14$; $t = 0.2$; $n = 6$, $g = 0.1$ with payments of tax on income at the ends of periods

19.3.6 Simultaneously Impact of Company Age, n*, and Growth Rate,* g

The conclusions from Fig. 19.35 are as follows:

- The slope of curves, $k_e(L)$, decreases with company age, n, and increases with growth rate, g.
- The curves corresponding to different company ages mix at different growth rates g. This can lead to interesting effects and create new options for the company's dividend policy.

19.4 The Explanations of the Observed Effects

19.4.1 Increase of k_e *with* L

$k_e(L)$ rises with L, because as L increases, financial distress and the risk of bankruptcy increase, shareholders demand higher returns per share.

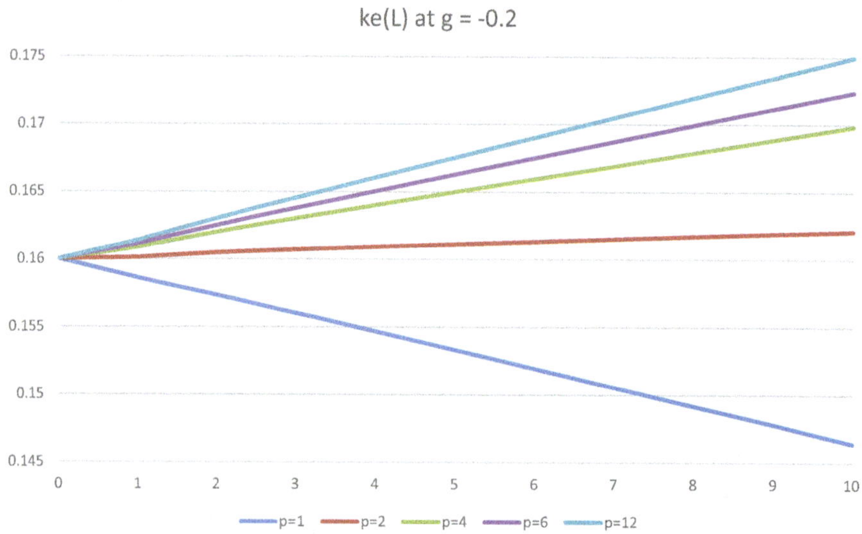

Fig. 19.25 The dependences of equity cost, k_e, on leverage level, L, at different $p = 1; 2; 4; 6; 12$; $k_0 = 0.16$; at $k_d = 0.14$; $t = 0.2$; $n = 6$, $g = -0.2$ with advance payments of tax on income

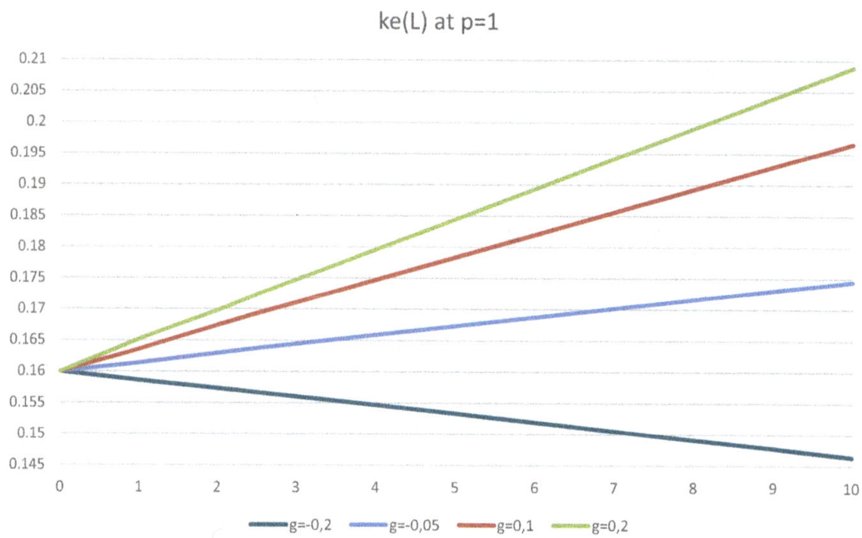

Fig. 19.26 The dependences of equity cost, k_e, on leverage level, L, at $p = 1$; $k_0 = 0.16$; at $k_d = 0.14$; $t = 0.2$; $n = 6$, at different $g = -0.2; -0.05; 0.1; 0.2$ with advance payments of tax on income

19.4.2 Tax Rate Increases

As taxes increase, the slope $k_e(L)$ decreases via tax corrector, which decreases with tax on profit rate.

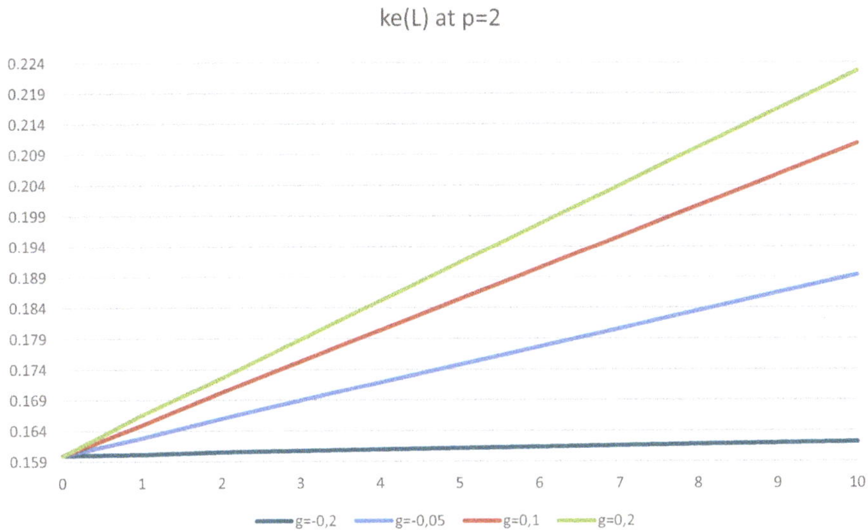

Fig. 19.27 The dependences of equity cost, k_e, on leverage level, L, at $p = 2$; $k_0 = 0.16$; at $k_d = 0.14$; $t = 0.2$; $n = 6$, at different $g = -0.2; -0.05; 0.1; 0.2$ with advance payments of tax on income

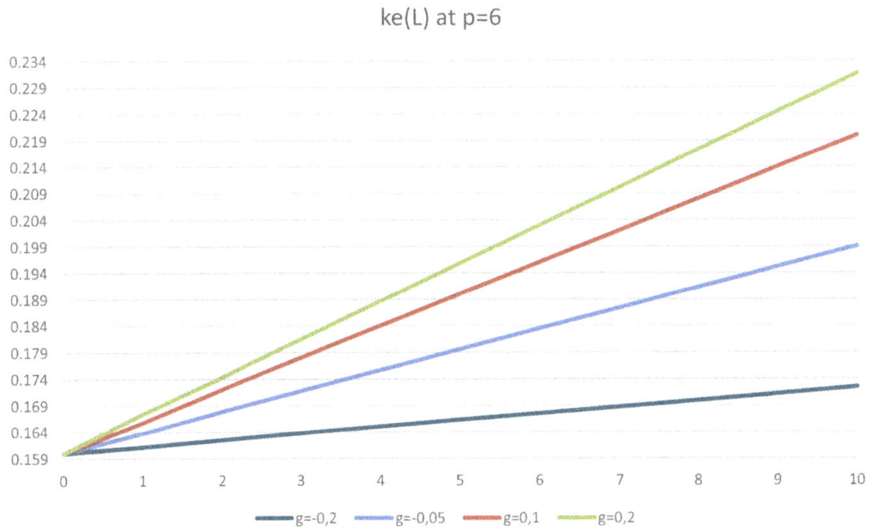

Fig. 19.28 The dependences of equity cost, k_e, on leverage level, L, at $p = 6$; $k_0 = 0.16$; at $k_d = 0.14$; $t = 0.2$; $n = 6$, at different $g = -0.2; -0.05; 0.1; 0.2$ with advance payments of tax on income

19.4.3 Influence of the Rate of Income Growth g

With an increase in the rate of income growth g, the slope of the curve $k_e(L)$ increases, since, in this case, shareholders can count on larger dividends.

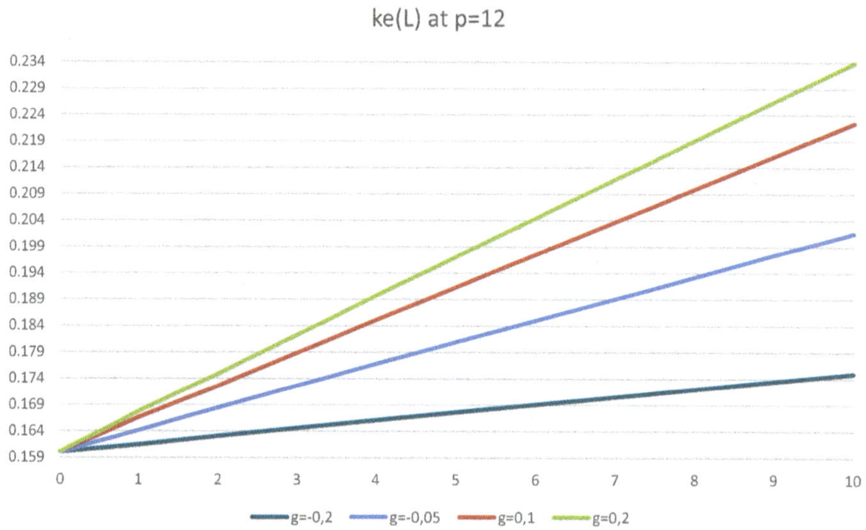

Fig. 19.29 The dependences of equity cost, k_e, on leverage level, L, at $p = 2$; $k_0 = 0.16$; at $k_d = 0.14$; $t = 0.2$; $n = 6$, at different $g = -0.2$; -0.05; 0.1; 0.2 with advance payments of tax on income

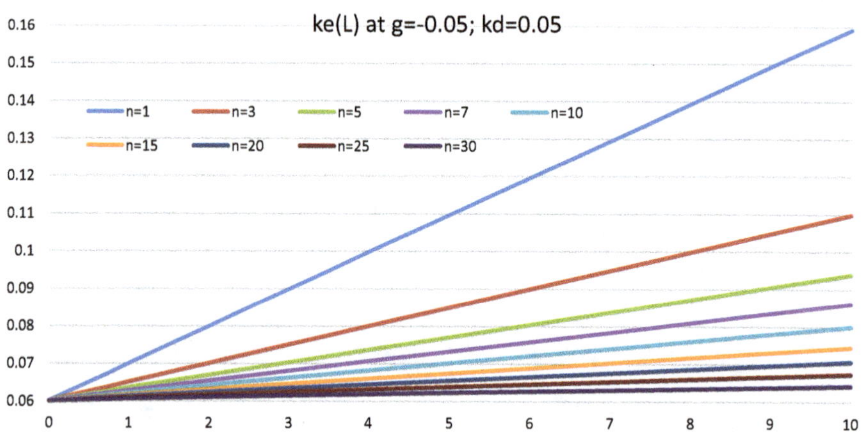

Fig. 19.30 The dependence of equity cost, k_e, on leverage level, L, at $p = 1$; $k_0 = 0.06$; $k_d = 0.05$; $t = 0.2$; at negative $g = -0.05$ and at different company age n with payments of tax on income at the ends of periods

– At $g < 0$, earnings per share decrease.

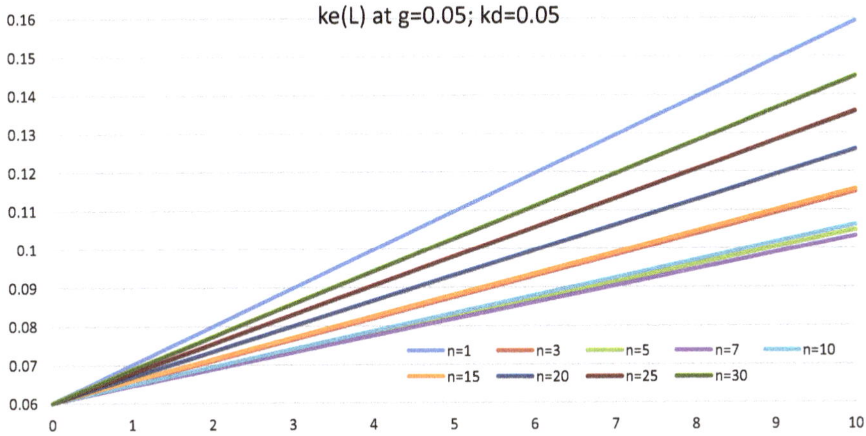

Fig. 19.31 The dependence of equity cost, k_e, on leverage level, L, at $p = 1$; $k_0 = 0.06$; $k_d = 0.05$; $t = 0.2$; at positive $g = 0.05$ and at different company age n with payments of tax on income at the ends of periods

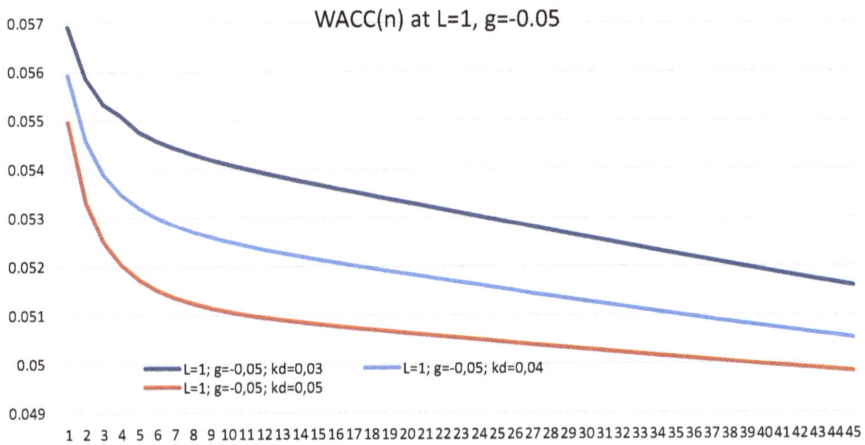

Fig. 19.32 The absence of the "golden age" company effect. The dependence of equity cost, k_e, on leverage level, L, at $p = 1$; $k_0 = 0.06$; $k_d = 0.03$; 0.04; 0.05; $t = 0.2$; at negative $g = -0.05$ and at different company age n with payments of tax on income at the ends of periods

19.4.4 Influence of the Frequency of Income Tax Payments

For income tax payments at the ends of periods, slope $k_e(L)$ decreases with p, thus earnings per share decrease (properties of ordinary annuity).

The increase in PV and FV of cash flows with the growth of p follows from their behavior in the ordinary p-term annuity with the growth of p. Both of them increase because an increase in p gives a percentage gain: their earlier accrual.

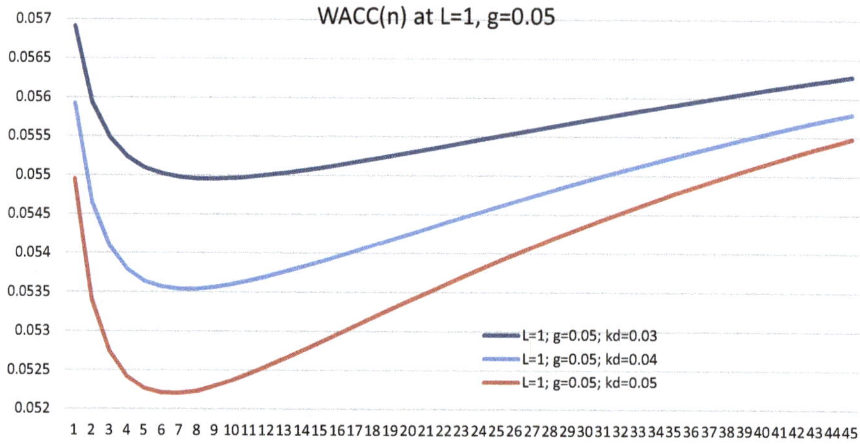

Fig. 19.33 The presence of the "golden age" company effect. The dependence of equity cost, k_e, on leverage level, L, at $p = 1$; $k_0 = 0.06$; $k_d = 0.03$; 0.04; 0.05; $t = 0.2$; at positive $g = 0.05$ and at different company age n with payments of tax on income at the ends of periods

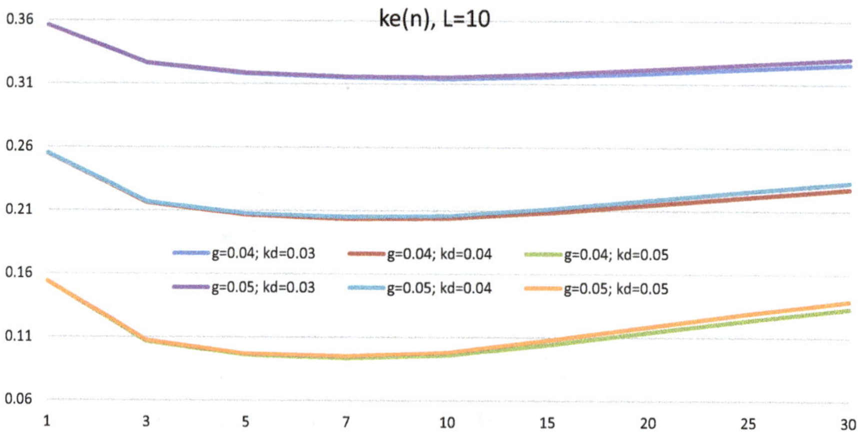

Fig. 19.34 The dependence of the cost of equity, k_e, on company age at different cost of debt k_d and growth rate g at leverage level $L = 10$ for advance payments of tax on income

For advance income tax payments, the slope $k_e(L)$ increases with p, thus earnings per share increase (properties of annuity due).

Advance income tax payments correspond to the case of p-term annuity due: the decrease in PV and FV of cash flows with increasing p follows from their behavior in the p-term annuity due with increasing p. They both decrease because an increase in p gives an interests decrease via charging interest on the smaller value of the annuity payment R/p instead of R.

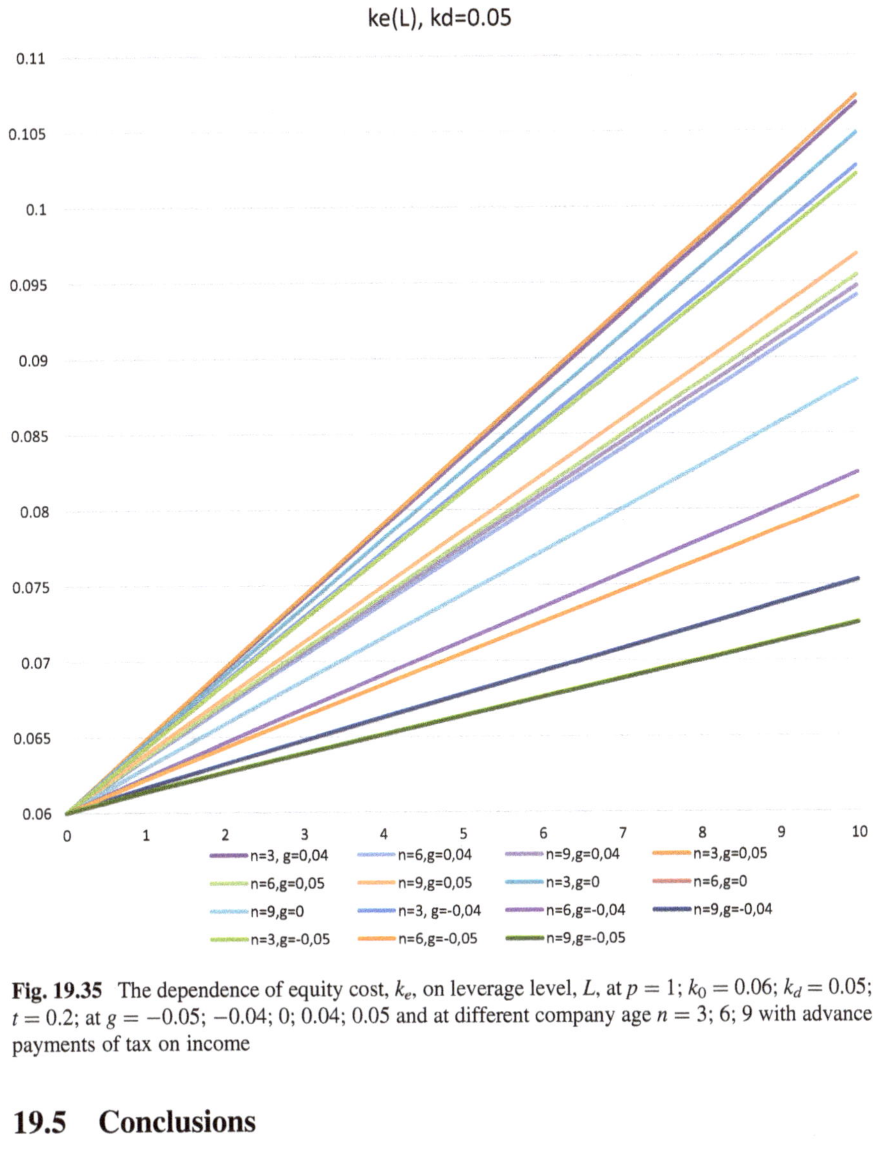

Fig. 19.35 The dependence of equity cost, k_e, on leverage level, L, at $p = 1$; $k_0 = 0.06$; $k_d = 0.05$; $t = 0.2$; at $g = -0.05$; -0.04; 0; 0.04; 0.05 and at different company age $n = 3$; 6; 9 with advance payments of tax on income

19.5 Conclusions

The contribution of the current results to the theory is related to the use of generalized theories of capital structure (both BFO and MM) to estimate the cost of equity capital in real-life operating conditions of companies. A study was conducted of the dependence of the cost of equity on the level of leverage for different ages of the company, different values of the leverage level, of cost of debt capital, for different frequencies of payment of income tax, for advance payments of income tax and payments at the ends of periods, for the company's variable income, etc. As a result, it becomes possible to estimate the cost of equity in these conditions. The behavior and properties of the cost of equity are explored within the framework of the BFO theory. Several very important innovative effects have been discovered, which significantly change the company's dividend policy. Among them the abnormal dependence of equity cost on leverage level at different growth rate of variable income, at different age of the company, at different frequency of income tax payment, at different debt cost, etc.

Developed approach can be applied to any country; you just need to use the tax rate for a specific company in a specific country. And the methods of paying income tax: advance payments or at the ends of periods and with what frequency: annually, semi-annually, quarterly, or monthly.

The following results have been obtained:

1. $k_e(L)$ rises with L, because as L increases, financial distress and the risk of bankruptcy increase, shareholders demand higher returns per share.
2. As taxes increase, the slope $k_e(L)$ decreases via tax corrector, which decreases with tax on profit rate.
3. With an increase in the rate of income growth g, the slope of the curve $k_e(L)$ increases, since, in this case, shareholders can count on larger dividends. At $g < 0$, earnings per share decrease.
4. For income tax payments at the ends of periods, slope $k_e(L)$ decreases with the frequency of income tax payments p, thus earnings per share decrease (properties of ordinary annuity). The increase in PV and FV of cash flows with the growth of p follows from their behavior in the ordinary p-term annuity with the growth of p. Both of them increase because an increase in p gives a percentage gain: their earlier accrual.

For advance income tax payments, the slope $k_e(L)$ increases with p, thus earnings per share increase (properties of annuity due).

5. Two types of dependences of the slope $k_e(L)$ on the age of the enterprise, n, were found: a monotonous decrease in the slope angle $k_e(L)$ with n and a seemingly random dependence of the slope angle $k_e(L)$ with n. The reason for this is the absence or presence of the so-called "golden age" company effect [20].

The so-called "golden age" of the company effect is the situation when the cost of capital raised (WACC) is below the perpetuity limit, and the company's value is higher. The absence or presence of the "golden age" company effect could explain the ordering of curves $k_e(L)$ (of slopes of curves $k_e[L]$) (see Figs. 19.32, 19.33, and 19.34):

- In the case of the absence of the "golden age" company effect, the ordering of curves $k_e(L)$ is monotonic (Fig. 19.32).
- In the case of the presence of the "golden age" company effect, the ordering of curves $k_e(L)$ is not monotonous (Fig. 19.33).

The developed methodology and results will help the company's management to develop an adequate and effective dividend policy, taking into account the real conditions for the functioning of companies, such as variable income of companies, frequent income tax payments, types of income tax payments. (advance payments and payments at the end of reporting periods), etc. This allows when developing a dividend policy the use of qualitatively new effects that we have discovered, such as the anomalous dependence of the cost of equity on the level of leverage at different growth rates of variable income, at different company ages (the "golden age" effect), at different frequency of income tax payments, at different cost of debt, etc.

The limitation of the study is due to the fact that the case of a constant income growth rate is considered. In future publications, the case of a variable income growth rate will be considered. One more limitation of the consideration is related to the known limitations of the WACC approximation.

Concerning the direction of further research: the authors plan to take into account both business and financial risks when assessing the financial performance of companies, including the cost of equity.

19.6 Direction for Further Research

When studying the cost of equity capital within the framework of BFO and MM theories, we take into account only the financial risks associated with the use of debt financing. There is the Capital Asset Pricing Model (CAPM) that also looks at return on equity, but only considers business risks that relate to the investment in the portfolio rather than in the risk-free assets. In practice, companies use debt financing and operate at non-zero levels of leverage. This means that it is necessary to take into account the financial risk associated with the use of debt financing along with the business one. A new approach to CAPM has been developed recently [34] that takes into account both business and financial risk. The authors combine the theory of CAPM and the Modigliani–Miller (MM) theory. The first is based on portfolio analysis and accounting for business risks in relation to the market (or industry). The second one (the Modigliani–Miller (MM) theory) describes a specific company and takes into account the financial risks associated with the use of debt financing. The combination of these two different approaches makes it possible to take into account both types of risks: business and financial ones. The authors [38–42] combine these two approaches analytically, while Hamada [4, 36] did it phenomenologically. Using the Modigliani–Miller (MM) theory [1–3, 40, 43, 44], it is shown that Hamada's model, first model, used for this purpose half a century ago, is incorrect. In addition to the renormalization of the beta-coefficient, obtained in the Hamada model, two additional terms are found: the renormalized risk-free return and the term dependent on the cost of debt k_d. A critical analysis of the Hamada model was carried out in [38, 39]. The vast majority of listing companies use debt financing and are levered, and the Hamada model is not applicable to them in contrast to a new approach applicable to leveraged companies. Two versions of CAPM (market or industry) have been considered.

References

1. Modigliani, F., & Miller, M. (1958). The cost of capital, corporate finance, and the theory of investment. *The American Economic Review, 48*, 261–297.
2. Modigliani, F., & Miller, M. (1963). Corporate income taxes and the cost of capital: A correction. *The American Economic Review, 53*, 147–175.
3. Modigliani, F., & Miller, M. (1966). Some estimates of the cost of capital to the electric utility industry 1954–1957. *The American Economic Review, 56*, 333–391.
4. Hamada, R. (1969). Portfolio analysis, market equilibrium, and corporate finance. *The Journal of Finance, 24*, 13–31.
5. Miller, M. (1977). Debt and taxes. *The Journal of Finance, 32*, 261–275. https://doi.org/10.1111/j.1540-6261.1977.tb03267.x
6. Farber, A., Gillet, R., & Szafarz, A. A. (2006). General formula for the WACC. *International Journal of Business, 11*, 211–218.
7. Fernandez, P. A. (2006). General formula for the WACC: A comment. *International Journal of Business, 11*, 219.
8. Berk, J., & De Marzo, P. (2007). *Corporate finance*. Pearson-Addison Wesley.
9. Harris, R., & Pringle, J. (1985). Risk–adjusted discount rates–Extension form the average–risk case. *Journal of Financial Research, 8*, 237–244.
10. Baumol, W., Panzar, J. G., & Willig, R. D. (1982). *Contestable markets and the theory of industry structure*. Harcourt College Publishers.
11. Black, F. N. (1986). *The Journal of Finance, 41*, 529–543.
12. Becker, D. M. (2021). The difference between Modigliani–Miller and Miles–Ezzell and its consequences for the valuation of annuities. *Cogent Economics & Finance, 9*, 1.
13. Berk, J., & DeMarzo, P. (2007). *Corporate finance*. Pearson-Addison Wesley.
14. Brealey, R. A., & Myers, S. C. (1981). *Principles of corporate finance* (1st ed.). McGraw-Hill.
15. Brealey, R. A., Myers, S. C., & Allen, F. (2014). *Principles of corporate finance* (11th ed.). McGraw-Hill.
16. Bhattacharya, S. (1978). Project valuation with mean–reacting cash flows. *The Journal of Finance, 33*, 1317–1331.
17. Myers, S. (2001). Capital structure. *The Journal of Economic Perspectives, 15*, 81–102.
18. Filatova, T., Orehova, N., & Brusova, A. (2008). Weighted average cost of capital in the theory of Modigliani–Miller, modified for a finite life–time company. *Bull FU, 48*, 68–77.
19. Brusov, P., Filatova, T., Orehova, N., & Eskindarov, M. (2018). *Modern corporate finance, investments, taxation and ratings* (2nd ed., pp. 1–571). Springer Nature Publishing.
20. Brusov, P., & Filatova, T. (2023). Capital structure theory: Past, present, future. *Mathematics, 11*(3), 616. https://doi.org/10.3390/math11030616
21. Dimitropoulos, P. (2014). Capital structure and corporate governance of soccer clubs: European evidence. *Management Research Review, 37*, 658–678.
22. Luiz, K., & Cruz, M. (2015). The relevance of capital structure on firm performance: A multivariate analysis of publicly traded Brazilian companies. *REPeC Brasília, 9*, 384–401.
23. Barbi, M. (2011). On the risk–neutral value of debt tax shields. *Applied Financial Economics, 22*, 251–258.
24. Franc-Dąbrowska, J., Mądra-Sawicka, M., & Milewska, A. (2021). Energy sector risk and cost of capital assessment—Companies and investors perspective. *Energies, 14*, 1613.
25. Angotti, M., de Lacerda Moreira, R., Hipólito Bernardes do Nascimento, J., & Neto de Almeida Bispo, O. (2018). Analysis of an equity investment strategy based on accounting and financial reports in Latin American markets. *Reficont, 5*, 22–40.
26. Vergara-Novoa, C., Sepúlveda-Rojas, J. P., Miguel, D. A., & Nicolás, R. (2018). Cost of capital estimation for highway concessionaires in Chile. *Journal of Advanced Transportation, 2018*, 2153536.

27. Mundi, H. S., Kaur, P., & Murty, R. L. N. (2022). A qualitative inquiry into the capital structure decisions of overconfident finance managers of family–owned businesses in India. *Qualitative Research in Financial Markets, 14*, 357–379.
28. Sadiq, M., Alajlani, S., Hussain, M. S., Bashir, F., & Chupradit, S. (2022). Impact of credit, liquidity, and systematic risk on financial structure: Comparative investigation from sustainable production. *Environmental Science and Pollution Research International, 29*, 20963–20975.
29. Becker, D. M. (2022). Getting the valuation formulas right when it comes to annuities. *Managerial Finance, 48*, 470.
30. El-Chaarani, H., Abraham, R., & Skaf, Y. (2022). The impact of corporate governance on the financial performance of the banking sector in the MENA (Middle Eastern and North African) region: An immunity test of banks for COVID–19. *Journal of Risk and Financial Management, 15*, 82.
31. Harris, R. S., & Pringle, J. J. (1985). Risk–adjusted discount rates–Extensions from the average–risk case. *Journal of Financial Research, 8*, 237–244.
32. Miles, J. A., & Ezzell, J. R. (1980). The weighted average cost of capital, perfect capital markets, and project life: A clarification. *Journal of Financial and Quantitative Analysis, 15*, 719–730.
33. Farber, A., Gillet, R., & Szafarz, A. (2006). A General Formula for the WACC. *International Journal of Business, 11*, 211–218.
34. Fernandez, P. (2006). A general formula for the WACC: A comment. *International Journal of Business, 11*, 219.
35. Cooper, I., & Nyborg, K. G. (2018). Consistent valuation of project finance and LBO's using the flow–to–equity method. *European Financial Management, 24*, 34–52.
36. Hamada, R. (1972). The effect of the firm's capital structure on the systematic risk of common stocks. *The Journal of Finance*, 435–452.
37. Brusov P, Filatova T, Orekhova N, (2023) *Theory of capital structure: Applications in corporate finance, investments, taxation and ratings*. 769 pp
38. Brusov, P., Filatova, T., & Kulik, V. (2023). Capital Asset Pricing Model 2.0: Account of business and financial risk. *Preprints*, 2023100347. https://doi.org/10.20944/preprints202310.0347.v1
39. Brusov, P. N., Filatova, T. V., & Kulik, V. L. (2024). Capital Asset Pricing Model (CAPM) 2.0: Account of business and financial risk. *Finance: Theory and Practice., 28*(2), 128–142. https://doi.org/10.26794/2587-5671-2024-28-2-128-142
40. Brusov, P., Filatova, T., & Orekhova, N. (2022). *Generalized Modigliani–Miller theory: Applications in corporate finance, investments, taxation and ratings* (362 pp). https://link.springer.com/book/10.1007/978–3–030–93893–2?page=2#toc
41. Brusov, P., Filatova, T., Orehova, N., & Kulik, V. (2015). The golden age of the company: (Three colors of company's time). *Journal of Reviews on Global Economics, 4*, 21–42.
42. Brusov, P., Filatova, T., Orehova, N., & Kulk, V. (2018). A "Golden Age" of the companies: Conditions of its existence. *Journal of Reviews on Global Economics, 7*, 88–103.
43. Miller, M. H., & Rock, K. (1985). Dividend policy under asymmetric information. *The Journal of Finance, 40*(4), 1031–1051. https://doi.org/10.1111/j.1540-6261.1985.tb02362.x
44. Raimo, N., de Nuccio, E., Giakoumelou, A., Petruzzella, F., & Vitolla, F. (2021). Non–financial information and cost of equity capital: An empirical analysis in the food and beverage industry. *British Food Journal, 123*(1), 49–65. https://doi.org/10.1108/BFJ-03-2020-0278

Chapter 20
The Golden Age of the Company (Three Colors of Company's Time)

20.1 Introduction

In business and company valuation, along with the company's income stream, the discount rate plays a key role. The main disadvantage of existing valuation methods is their inability to correctly estimate the discount rate. Currently, the only theory that allows doing this correctly is the modern theory of cost and capital structure BFO. This theory plays a fundamental role in the development of a qualitatively new approach to business valuation and company value created in this monograph. Within the framework of the BFO theory, the WACC(n) dependence is correctly estimated, which is the key point of the qualitatively new methodology for business valuation and company value developed in this monograph. Within the framework of the BFO theory, the authors of this monograph discovered the innovative effect of the company's golden age and conducted a detailed study of it. This effect, which exists for most companies, significantly 1and qualitatively changes the WACC(n) dependence, which leads to a significant change in the business valuation and company value.

It is well-known that the company goes through several stages in its development process: adolescence, maturity, and old age. Within the modern theory of capital cost and capital structure by Brusov-Filatova-Orekhova (BFO theory) [1–12], it is shown that the problem of the company development has an interpretation, which is absolutely different from the generally accepted one.

One of the most important problems in corporate finance is the problem of capital cost and capital structure. Before 2008, there were just two kinds of valuations of cost of capital: one of them was the first quantitative theory by Nobel Prize winners Modigliani and Miller [13–15], applicable to perpetuity (with infinite lifetime) companies, and the second one was the valuation applicable to one-year companies by Steve Myers [16].

So, before 2008, when the modern theory of capital cost and capital structure by Brusov-Filatova-Orekhova (BFO theory) had been created [1–12], only two points

P. Brusov, T. Filatova, *Business Valuation and Company Value*, Contributions to Finance and Accounting, https://doi.org/10.1007/978-3-031-90847-7_20

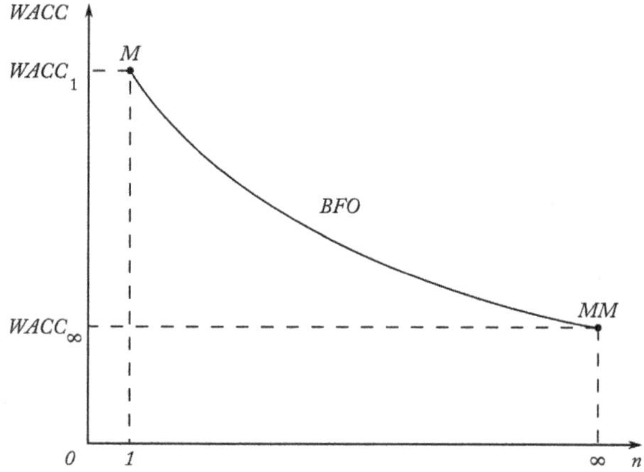

Fig. 20.1 Monotonic dependence of weighted average cost of capital, WACC, on life-time of the company *n*

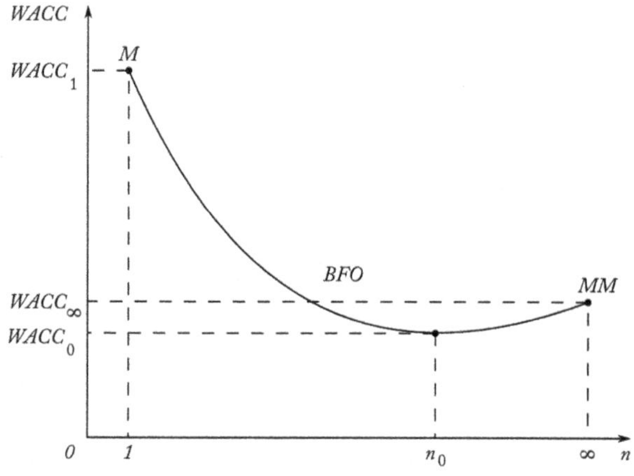

Fig. 20.2 Dependence of weighted average cost of capital, WACC, on life-time of the company *n*, showing descending with n, and with the passage through a minimum and then a limited growth

in time interval had been known: one-year and infinity. That time Steve Myers [16] has supposed that the Modigliani–Miller (MM) theorem [13–15] gives the lowest assessment for weighted average cost of capital, WACC, and consequently, the highest assessment for company capitalization. This mean that the weighted average cost of capital, WACC, monotonically descends with time of life of company, n, approaching to its perpetuity limit (Fig. 20.1), and, consequently, company capitalization monotonically increases approaching to its perpetuity limit (Fig. 20.2).

Created in 2008, the modern theory of capital cost and capital structure by Brusov-Filatova-Orekhova (BFO theory) [1–12] has turned out to be able to make valuation of capital cost and company capitalization for companies with arbitrary lifetime: this completes the whole time interval from $n = 1$ up to $n = \infty$. A lot of qualitative effects in corporate finance, investments, taxations, etc. has been made within BFO theory.

In this Chapter within BFO theory, it is shown that Steve Myers' suggestion [16] turns out to be wrong.

Choosing of optimal capital structure of the company, i.e., proportion of debt and equity, which minimizes weighted average cost of capital and maximizes the company capitalization, is one of the most important tasks of financial manager and the management of a company. The search for an optimal capital structure has attracted the attention of economists and financiers during many tens of years. And it is clear why: one can, nothing making, but only by changing the proportion between the values of equity capital and debt one of the company, significantly enhance the company capitalization, by other words to fulfill the primary task, to reach critical goal of the business management. Spend a little less of your own, loan slightly more (or vice versa), and company capitalization reaches a maximum.

Before now, the search for an optimal capital structure was made by study of the dependence of weighted average cost of capital, WACC, on leverage level in order to determine the optimal leverage level L_0, at which WACC is minimal and capitalization V is maximal. Here we apply an absolutely different method, studying the dependence of weighted average cost of capital, WACC, on the time of life of company n. Note that before the appearance of BFO theory, study of such kind of dependences was impossible via the absence of "time" parameter in perpetuity Modigliani–Miller theory [13–15].

As it is shown in this paper, from the point of view of cost of capital, there are two types of dependences of weighted average cost of capital, WACC, on the time of life of company n: monotonic descending of WACC with n and descending of WACC with passage through minimum, followed by a limited growth (Figs. 20.1 and 20.2).

The first type of behavior is linked with the comment by Myers [16], that the Modigliani–Miller (MM) theorem [13–15] gives the lowest assessment for weighted average cost of capital, WACC, that, as shown by us within the BFO theory, is, generally speaking, incorrect. In many cases, there is a second type of behavior of dependence of weighted average cost of capital, WACC, on the time of life of company n: descending with passage through minimum, followed by a limited growth.

Thus, in the general case, the comment by Myers [16] turns out to be wrong, and in the life of company there is a "golden age," or "the golden century," when the cost of attracting of the capital becomes a minimal, and capitalization companies—maximal (Figs. 20.2 and 20.3). In the life of a company, the same number of stages as usual can be allocated: youth, maturity, and old age. In youth, weighted average cost of capital, WACC, decreases with n, in the maturity value of attracting capital cost becomes minimal, and in the old-age this cost growth, approaches its perpetuity limit.

So, figuratively speaking, a current investigation transforms "black and white business world" (with monotonic descending of WACC with time of life of company, n) into "color business world" (with descending of WACC with n with passage through minimum, followed by a limited growth): there are really three colors of company's time.

The conclusion made in this paper for the first time, that the assessment of weighted average cost of capital of the company, WACC, in the theory of

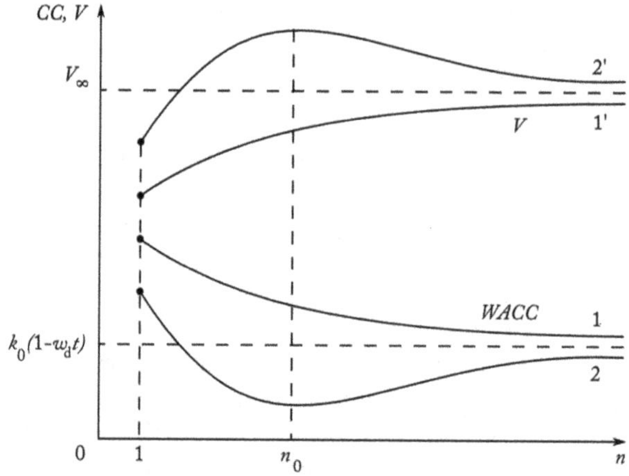

Fig. 20.3 Two kinds of dependences of weighted average cost of capital, WACC, and company capitalization, V, on lifetime of the company n: 1–1′—monotonic dependence of weighted average cost of capital, WACC, and company capitalization, V, on lifetime of the company n. 2–2′—showing descending of WACC with n, and with the passage through a minimum and then a limited growth and increase of V with the passage through a maximum (at n_0) and then a limited descending

Modigliani and Miller (MM) [13–15] is not the minimal, and capitalization is not maximal, seems to be very significant and important.

20.2 Dependence of Weighted Average Cost of Capital, WACC, on Lifetime of the company *n* at different Leverage Levels

In this paragraph, we study the dependence of weighted average cost of capital, WACC, on lifetime of the company n at different leverage levels.

For L = 1 one has

For L = 2 we have

For L = 3 one has

For L = 5 one has

For L = 7 one has

The analysis of the Tables 20.1, 20.2, 20.3, 20.4, 20.5, and Fig. 20.4 allows us to make the following conclusions:

1. In all examined cases, (at all leverage levels) at current values of capital costs (equity, k_0, and debt, k_d), the second type of behavior of dependence of weighted average cost of capital, WACC, on the life time of company, n, takes place, namely, descending of WACC with n with the passage through the minimum with subsequent limited growth.

Table 20.1 The dependence of weighted average cost of capital, WACC, on age of the company n at leverage level L=1

L	WACC	k_0	k_d	w_d	T	n	k_e
1	0.1843	0.2	0.15	0.5	0.2	1	0.2487
1	0.1798	0.2	0.15	0.5	0.2	2	0.2397
1	0.1780	0.2	0.15	0.5	0.2	3	0.2360
1	0.1772	0.2	0.15	0.5	0.2	4	0.2345
1	0.1769	0.2	0.15	0.5	0.2	5	0.2339
1	0.1769	0.2	0.15	0.5	0.2	6	0.2338
1	0.1770	0.2	0.15	0.5	0.2	7	0.2340
1	0.1771	0.2	0.15	0.5	0.2	8	0.2343
1	0.1773	0.2	0.15	0.5	0.2	9	0.2346
1	0.1775	0.2	0.15	0.5	0.2	10	0.2351

Table 20.2 The dependence of weighted average cost of capital, WACC, on age of the company n at leverage level L=2

L	WACC	k_0	k_d	w_d	t	n	k_e
2	0.1791	0.2	0.15	0.66667	0.2	1	0.2974
2	0.1731	0.2	0.15	0.66667	0.2	2	0.2793
2	0.1706	0.2	0.15	0.66667	0.2	3	0.2719
2	0.1696	0.2	0.15	0.66667	0.2	4	0.2687
2	0.1692	0.2	0.15	0.66667	0.2	5	0.2675
2	0.1691	0.2	0.15	0.66667	0.2	6	0.2672
2	0.1692	0.2	0.15	0.66667	0.2	7	0.2675
2	0.1694	0.2	0.15	0.66667	0.2	8	0.2681
2	0.1696	0.2	0.15	0.66667	0.2	9	0.2689
2	0.1699	0.2	0.15	0.66667	0.2	10	0.2697

Table 20.3 The dependence of weighted average cost of capital, WACC, on age of the company n at leverage level L=3

L	WACC	k_0	k_d	w_d	t	n	k_e
3	0.1765	0.2	0.15	0.75	0.2	1	0.3461
3	0.1697	0.2	0.15	0.75	0.2	2	0.3189
3	0.1669	0.2	0.15	0.75	0.2	3	0.3078
3	0.1657	0.2	0.15	0.75	0.2	4	0.3029
3	0.1653	0.2	0.15	0.75	0.2	5	0.3010
3	0.1651	0.2	0.15	0.75	0.2	6	0.3006
3	0.1653	0.2	0.15	0.75	0.2	7	0.3010
3	0.1655	0.2	0.15	0.75	0.2	8	0.3019
3	0.1658	0.2	0.15	0.75	0.2	9	0.3030
3	0.1661	0.2	0.15	0.75	0.2	10	0.3042

2. The minimum of cost of attracting capital (weighted average cost of capital of the company, WACC) is achieved at all leverage levels at the same company's age at $n = 6$ (only when L = 1 minimum is spread for 2 years ($n = 5$ and $n = 6$).
3. The value of WACC minimum significantly depends on the level of leverage, L, and, of course, decreases with increasing L, because weighted average cost of capital, WACC, at a fixed n decreases with leverage.

Table 20.4 The dependence of weighted average cost of capital, WACC, on age of the company n at leverage level L=5

L	WACC	k_0	k_d	w_d	t	n	k_e
5	0.1739	0.2	0.15	0.83333	0.2	1	0.4435
5	0.1663	0.2	0.15	0.83333	0.2	2	0.3980
5	0.1632	0.2	0.15	0.83333	0.2	3	0.3795
5	0.1619	0.2	0.15	0.83333	0.2	4	0.3713
5	0.1613	0.2	0.15	0.83333	0.2	5	0.3680
5	0.1612	0.2	0.15	0.83333	0.2	6	0.3672
5	0.1613	0.2	0.15	0.83333	0.2	7	0.3679
5	0.1615	0.2	0.15	0.83333	0.2	8	0.3693
5	0.1619	0.2	0.15	0.83333	0.2	9	0.3711
5	0.1622	0.2	0.15	0.83333	0.2	10	0.3732

Table 20.5 The dependence of weighted average cost of capital, WACC, on age of the company n at leverage level L=7

L	WACC	k_0	k_d	w_d	t	n	k_e
7	0.1726	0.2	0.15	0.875	0.2	1	0.5409
7	0.1646	0.2	0.15	0.875	0.2	2	0.4771
7	0.1614	0.2	0.15	0.875	0.2	3	0.4511
7	0.1599	0.2	0.15	0.875	0.2	4	0.4396
7	0.1594	0.2	0.15	0.875	0.2	5	0.4349
7	0.1592	0.2	0.15	0.875	0.2	6	0.4338
7	0.1593	0.2	0.15	0.875	0.2	7	0.4347
7	0.1596	0.2	0.15	0.875	0.2	8	0.4366
7	0.1599	0.2	0.15	0.875	0.2	9	0.4392
7	0.1603	0.2	0.15	0.875	0.2	10	0.4421

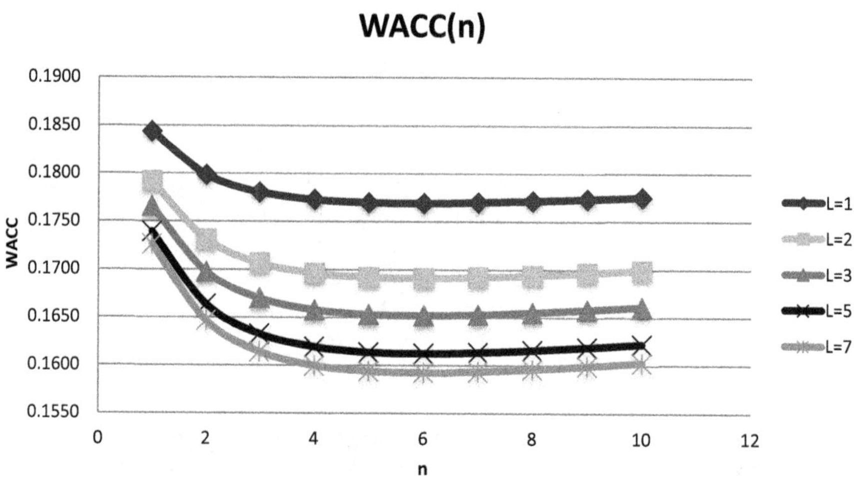

Fig. 20.4 Dependence of weighted average cost of capital, WACC, on lifetime of the company n at different leverage levels

Table 20.6 The dependence of weighted average cost of capital, WACC, on age of the company n at different values of capital costs (equity, $k_0 = 0.08$, and debt, $k_d = 0.04$) and at leverage level L=1

L	WACC	k_0	k_d	w_d	t	n	k_e
1	0.0758	0.08	0.04	0.5	0.2	1	0.1197
1	0.0745	0.08	0.04	0.5	0.2	2	0.1170
1	0.0738	0.08	0.04	0.5	0.2	3	0.1157
1	0.0735	0.08	0.04	0.5	0.2	4	0.1149
1	0.0732	0.08	0.04	0.5	0.2	5	0.1144
1	0.0731	0.08	0.04	0.5	0.2	6	0.1141
1	0.0729	0.08	0.04	0.5	0.2	7	0.1139
1	0.0729	0.08	0.04	0.5	0.2	8	0.1137
1	0.0728	0.08	0.04	0.5	0.2	9	0.1136
1	0.0728	0.08	0.04	0.5	0.2	10	0.1135

Table 20.7 The dependence of weighted average cost of capital, WACC, on age of the company n at different values of capital costs (equity, $k_0 = 0.2$, and debt, $k_d = 0.15$) and at leverage level L=1

L	WACC	k_0	k_d	w_d	t	n	k_e
1	0.1843	0.2	0.15	0.5	0.2	1	0.2487
1	0.1798	0.2	0.15	0.5	0.2	2	0.2397
1	0.1780	0.2	0.15	0.5	0.2	3	0.2360
1	0.1772	0.2	0.15	0.5	0.2	4	0.2345
1	0.1769	0.2	0.15	0.5	0.2	5	0.2339
1	0.1769	0.2	0.15	0.5	0.2	6	0.2338
1	0.1770	0.2	0.15	0.5	0.2	7	0.2340
1	0.1771	0.2	0.15	0.5	0.2	8	0.2343
1	0.1773	0.2	0.15	0.5	0.2	9	0.2346
1	0.1775	0.2	0.15	0.5	0.2	10	0.2351

Table 20.8 The dependence of weighted average cost of capital, WACC, on age of the company n at different values of capital costs (equity, $k_0 = 0.2$, and debt, $k_d = 0.15$) and at leverage level L=3

L	WACC	k_0	k_d	w_d	t	n	k_e
3	0.1765	0.2	0.15	0.75	0.2	1	0.3461
3	0.1697	0.2	0.15	0.75	0.2	2	0.3189
3	0.1669	0.2	0.15	0.75	0.2	3	0.3078
3	0.1657	0.2	0.15	0.75	0.2	4	0.3029
3	0.1653	0.2	0.15	0.75	0.2	5	0.3010
3	0.1651	0.2	0.15	0.75	0.2	6	0.3006
3	0.1653	0.2	0.15	0.75	0.2	7	0.3010
3	0.1655	0.2	0.15	0.75	0.2	8	0.3019
3	0.1658	0.2	0.15	0.75	0.2	9	0.3030
3	0.1661	0.2	0.15	0.75	0.2	10	0.3042

Table 20.9 The dependence of weighted average cost of capital, WACC, on age of the company n at different values of capital costs (equity, $k_0 = 0.08$, and debt, $k_d = 0.04$) and at leverage level L=3

L	WACC	k_0	k_d	w_d	t	n	k_e
3	0.0738	0.08	0.04	0.75	0.2	1	0.1991
3	0.0717	0.08	0.04	0.75	0.2	2	0.1909
3	0.0707	0.08	0.04	0.75	0.2	3	0.1870
3	0.0702	0.08	0.04	0.75	0.2	4	0.1847
3	0.0698	0.08	0.04	0.75	0.2	5	0.1832
3	0.0696	0.08	0.04	0.75	0.2	6	0.1822
3	0.0694	0.08	0.04	0.75	0.2	7	0.1815
3	0.0693	0.08	0.04	0.75	0.2	8	0.1810
3	0.0692	0.08	0.04	0.75	0.2	9	0.1806
3	0.0691	0.08	0.04	0.75	0.2	10	0.1803

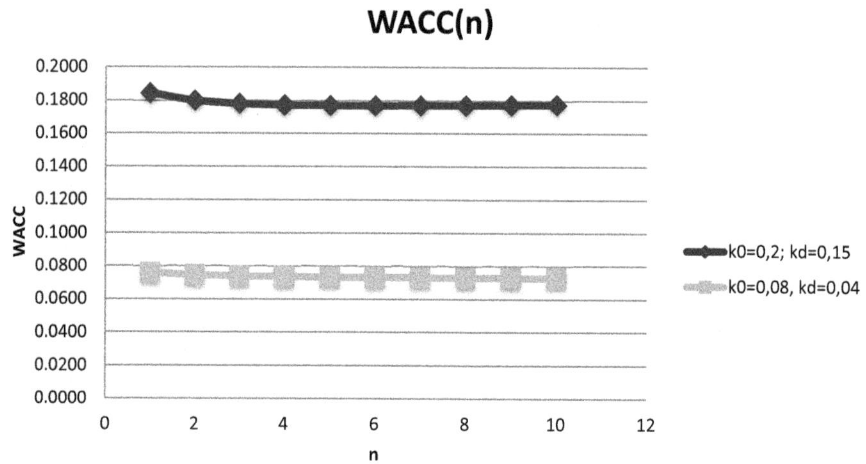

Fig. 20.5 Dependence of weighted average cost of capital, WACC, on lifetime of the company n at different values of capital costs (equity, k_0, and debt, k_d) and fixed leverage level L = 1. L = 3

20.3 Dependence of Weighted Average Cost of Capital, WACC, on Lifetime of the company *n* at different Values of Capital Costs (Equity, k_0, and Debt, k_d) and Fixed Leverage Levels

The analysis of Tables 20.6, 20.7, 20.8, 20.9, and Figs. 20.5 and 20.6 allows us to make the following conclusions:

1. The type of behavior of dependence of weighted average cost of capital, WACC, on the lifetime of company, n, at fixed leverage level significantly depends on values of capital costs (equity, k_0, and debt, k_d). At the values of capital costs that are specific to Russia ($k_0 = 20\%$, $k_d = 15\%$), there is a second type of dependence of WACC on the lifetime of company, n, namely, descending of WACC with n

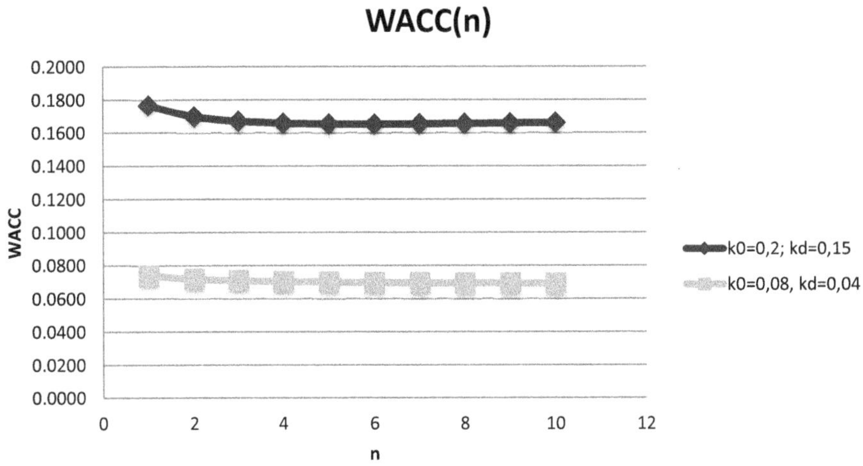

Fig. 20.6 Dependence of weighted average cost of capital, WACC, on lifetime of the company *n* at different values of capital costs (equity, k_0, and debt, k_d) and fixed leverage level L = 3

with the passage through the minimum with subsequent limited growth. And at the capital cost values characteristic of the West ($k_0 = 8\%$, $k_d = 4\%$), there is a first type of dependence of WACC on the time of life of company n, namely, the descending of WACC with *n*.

2. The same features are observed in considering both cases: at the leverage values L = 1 and L = 3.

20.4 Dependence of Weighted Average Cost of Capital, WACC, on Lifetime of the company *n* at different Values of Debt Capital Cost, k_d, and Fixed Equity Cost, k_0, and Fixed Leverage Levels

In this paragraph, we study the dependence of weighted average cost of capital, WACC, on lifetime of the company *n* at different values of debt capital cost, k_d, and fixed equity cost, k_0, and fixed leverage levels.

Put first L = 1.

Put than L = 3.

The analysis of the Tables 20.10, 20.11, 20.12, 20.13, 20.14, 20.15, 20.16, 20.17, and Figs. 20.7 and 20.8: allows us to make the following conclusions:

1. At fixed equity cost, k_0, and at fixed leverage level, the type of behavior of dependence of weighted average cost of capital, WACC, on the lifetime of company, n, significantly depends on value of debt capital cost, k_d: with growth of k_d it is changing from monotonic descending of WACC with n to descending

Table 20.10 The dependence of weighted average cost of capital, WACC, on age of the company n at values of capital costs (equity,$k_0 = 0.2$, and debt, $k_d = 0.15$) and at leverage level L=1

L	WACC	k_0	k_d	w_d	t	n	k_e
1	0.1843	0.2	0.15	0.5	0.2	1	0.2487
1	0.1798	0.2	0.15	0.5	0.2	2	0.2397
1	0.1780	0.2	0.15	0.5	0.2	3	0.2360
1	0.1772	0.2	0.15	0.5	0.2	4	0.2345
1	0.1769	0.2	0.15	0.5	0.2	5	0.2339
1	0.1769	0.2	0.15	0.5	0.2	6	0.2338
1	0.1770	0.2	0.15	0.5	0.2	7	0.2340
1	0.1771	0.2	0.15	0.5	0.2	8	0.2343
1	0.1773	0.2	0.15	0.5	0.2	9	0.2346
1	0.1775	0.2	0.15	0.5	0.2	10	0.2351

Table 20.11 The dependence of weighted average cost of capital, WACC, on age of the company n at values of capital costs (equity,$k_0 = 0.2$, and debt, $k_d = 0.12$) and at leverage level L=1

L	WACC	k_0	k_d	w_d	t	n	k_e
1	0.1871	0.2	0.12	0.5	0.2	1	0.2783
1	0.1832	0.2	0.12	0.5	0.2	2	0.2705
1	0.1815	0.2	0.12	0.5	0.2	3	0.2670
1	0.1807	0.2	0.12	0.5	0.2	4	0.2653
1	0.1802	0.2	0.12	0.5	0.2	5	0.2644
1	0.1799	0.2	0.12	0.5	0.2	6	0.2639
1	0.1798	0.2	0.12	0.5	0.2	7	0.2636
1	0.1798	0.2	0.12	0.5	0.2	8	0.2636
1	0.1798	0.2	0.12	0.5	0.2	9	0.2635
1	0.1798	0.2	0.12	0.5	0.2	10	0.2636

Table 20.12 The dependence of weighted average cost of capital, WACC, on age of the company n at values of capital costs (equity,$k_0 = 0.2$, and debt, $k_d = 0.17$) and at leverage level L=1

L	WACC	k_0	k_d	w_d	t	n	k_e
1	0.1826	0.2	0.17	0.5	0.2	1	0.2291
1	0.1777	0.2	0.17	0.5	0.2	2	0.2194
1	0.1759	0.2	0.17	0.5	0.2	3	0.2158
1	0.1752	0.2	0.17	0.5	0.2	4	0.2144
1	0.1750	0.2	0.17	0.5	0.2	5	0.2141
1	0.1751	0.2	0.17	0.5	0.2	6	0.2143
1	0.1754	0.2	0.17	0.5	0.2	7	0.2148
1	0.1757	0.2	0.17	0.5	0.2	8	0.2154
1	0.1760	0.2	0.17	0.5	0.2	9	0.2160
1	0.1763	0.2	0.17	0.5	0.2	10	0.2167

of WACC with n with the passage through the minimum with subsequent limited growth.

2. At k_d =10% and k_d =12% (k_0 =20%), the monotonic descending of WACC with n is observed, while at higher debt costs, k_d =15% and k_d =17% (k_0 =20%) descending of WACC with n with the passage through the minimum with subsequent limited growth takes place. The optimum age of the company is

Table 20.13 The dependence of weighted average cost of capital, WACC, on age of the company n at values of capital costs (equity, $k_0 = 0.2$, and debt, $k_d = 0.1$) and at leverage level L=1

L	WACC	k_0	k_d	w_d	t	n	k_e
1	0.1891	0.2	0.1	0.5	0.2	1	0.2982
1	0.1857	0.2	0.1	0.5	0.2	2	0.2913
1	0.1841	0.2	0.1	0.5	0.2	3	0.2881
1	0.1832	0.2	0.1	0.5	0.2	4	0.2864
1	0.1827	0.2	0.1	0.5	0.2	5	0.2853
1	0.1823	0.2	0.1	0.5	0.2	6	0.2846
1	0.1821	0.2	0.1	0.5	0.2	7	0.2842
1	0.1819	0.2	0.1	0.5	0.2	8	0.2838
1	0.1818	0.2	0.1	0.5	0.2	9	0.2836
1	0.1817	0.2	0.1	0.5	0.2	10	0.2834

Table 20.14 The dependence of weighted average cost of capital, WACC, on age of the company n at values of capital costs (equity, $k_0 = 0.2$, and debt, $k_d = 0.15$) and at leverage level L=3

L	WACC	k_0	k_d	w_d	t	n	k_e
3	0.1765	0.2	0.15	0.75	0.2	1	0.3461
3	0.1697	0.2	0.15	0.75	0.2	2	0.3189
3	0.1669	0.2	0.15	0.75	0.2	3	0.3078
3	0.1657	0.2	0.15	0.75	0.2	4	0.3029
3	0.1653	0.2	0.15	0.75	0.2	5	0.3010
3	0.1651	0.2	0.15	0.75	0.2	6	0.3006
3	0.1653	0.2	0.15	0.75	0.2	7	0.3010
3	0.1655	0.2	0.15	0.75	0.2	8	0.3019
3	0.1658	0.2	0.15	0.75	0.2	9	0.3030
3	0.1661	0.2	0.15	0.75	0.2	10	0.3042

Table 20.15 The dependence of weighted average cost of capital, WACC, on age of the company n at values of capital costs (equity, $k_0 = 0.2$, and debt, $k_d = 0.12$) and at leverage level L=3

L	WACC	k_0	k_d	w_d	t	n	k_e
3	0.1807	0.2	0.12	0.75	0.2	1	0.4349
3	0.1748	0.2	0.12	0.75	0.2	2	0.4113
3	0.1722	0.2	0.12	0.75	0.2	3	0.4009
3	0.1709	0.2	0.12	0.75	0.2	4	0.3955
3	0.1702	0.2	0.12	0.75	0.2	5	0.3927
3	0.1698	0.2	0.12	0.75	0.2	6	0.3911
3	0.1696	0.2	0.12	0.75	0.2	7	0.3903
3	0.1695	0.2	0.12	0.75	0.2	8	0.3900
3	0.1695	0.2	0.12	0.75	0.2	9	0.3899
3	0.1695	0.2	0.12	0.75	0.2	10	0.3900

growing with k_d decreasing: it is equal to 5 years at $k_d = 17\%$ and 6 years at $k_d = 15\%$.

3. The conclusions are saved at both considered values of leverage level: L = 1 and L = 3.

Table 20.16 The dependence of weighted average cost of capital, WACC, on age of the company n at values of capital costs (equity, $k_0 = 0.2$, and debt, $k_d = 0.17$) and at leverage level L=3

L	WACC	k_0	k_d	w_d	t	n	k_e
3	0.1738	0.2	0.17	0.75	0.2	1	0.2874
3	0.1665	0.2	0.17	0.75	0.2	2	0.2581
3	0.1637	0.2	0.17	0.75	0.2	3	0.2469
3	0.1626	0.2	0.17	0.75	0.2	4	0.2426
3	0.1624	0.2	0.17	0.75	0.2	5	0.2415
3	0.1625	0.2	0.17	0.75	0.2	6	0.2420
3	0.1628	0.2	0.17	0.75	0.2	7	0.2433
3	0.1633	0.2	0.17	0.75	0.2	8	0.2451
3	0.1638	0.2	0.17	0.75	0.2	9	0.2470
3	0.1643	0.2	0.17	0.75	0.2	10	0.2490

Table 20.17 The dependence of weighted average cost of capital, WACC, on age of the company n at values of capital costs (equity, $k_0 = 0.2$, and debt, $k_d = 0.1$) and at leverage level L=3

L	WACC	k_0	k_d	w_d	t	n	k_e
3	0.1836	0.2	0.1	0.75	0.2	1	0.4945
3	0.1785	0.2	0.1	0.75	0.2	2	0.4739
3	0.1761	0.2	0.1	0.75	0.2	3	0.4642
3	0.1747	0.2	0.1	0.75	0.2	4	0.4588
3	0.1739	0.2	0.1	0.75	0.2	5	0.4556
3	0.1734	0.2	0.1	0.75	0.2	6	0.4535
3	0.1730	0.2	0.1	0.75	0.2	7	0.4520
3	0.1727	0.2	0.1	0.75	0.2	8	0.4510
3	0.1726	0.2	0.1	0.75	0.2	9	0.4502
3	0.1724	0.2	0.1	0.75	0.2	10	0.4496

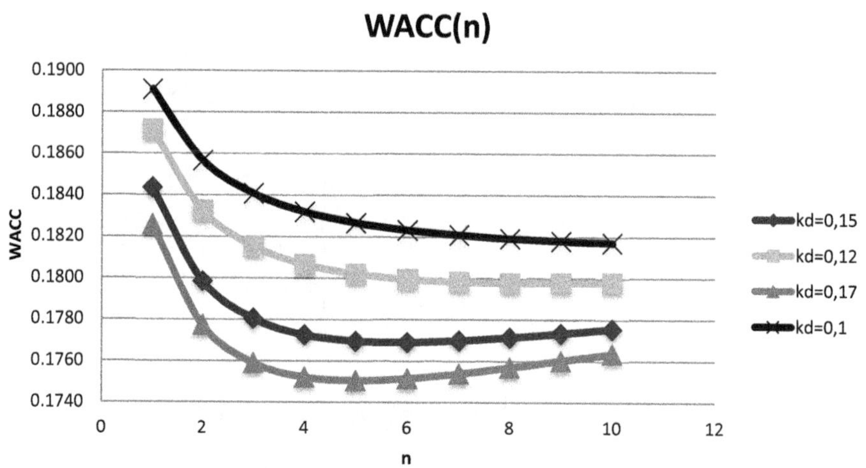

Fig. 20.7 Dependence of weighted average cost of capital, WACC, on lifetime of the company n at different values of debt capital cost, k_d, and fixed equity cost, k_0, and fixed leverage level L = 1

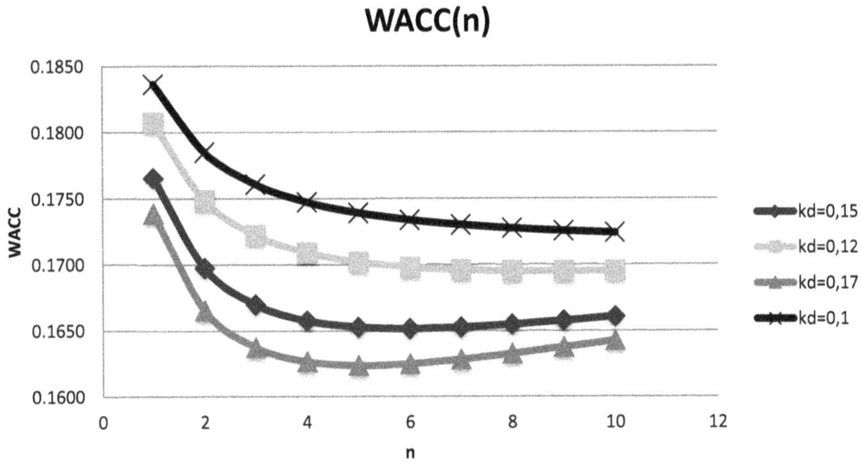

Fig. 20.8 Dependence of weighted average cost of capital, WACC, on lifetime of the company *n* at different values of debt capital cost, k_d, and fixed equity cost, k_0, and fixed leverage level L = 3

Table 20.18 The dependence of weighted average cost of capital, WACC, on age of the company *n* at values of capital costs (equity,$k_0 = 0.18$, and debt, $k_d = 0.15$) and at leverage level L=1

L	WACC	k_0	k_d	w_d	t	n	k_e
1	0.1646	0.18	0.15	0.5	0.2	1	0.2092
1	0.1602	0.18	0.15	0.5	0.2	2	0.2005
1	0.1585	0.18	0.15	0.5	0.2	3	0.1970
1	0.1578	0.18	0.15	0.5	0.2	4	0.1956
1	0.1576	0.18	0.15	0.5	0.2	5	0.1952
1	0.1576	0.18	0.15	0.5	0.2	6	0.1952
1	0.1578	0.18	0.15	0.5	0.2	7	0.1955
1	0.1580	0.18	0.15	0.5	0.2	8	0.1960
1	0.1583	0.18	0.15	0.5	0.2	9	0.1965
1	0.1585	0.18	0.15	0.5	0.2	10	0.1970

20.5 Dependence of Weighted Average Cost of Capital, WACC, on Lifetime of the company *n* at different Values of Equity Cost, k_0, and Fixed Debt Capital Cost, k_d, and Fixed Leverage Levels

In this paragraph, we study the dependence of weighted average cost of capital, WACC, on lifetime of the company *n* at different values of equity cost, k_0, and fixed debt capital cost, k_d, and fixed leverage levels.

The analysis of the Tables 20.18, 20.19, 20.20, 20.21, 20.22, 20.23, and Figs. 20.9 and 20.10: allows us to make the following conclusions:

1. At fixed debt capital cost, k_d, and at fixed leverage level at all considered cases (at all equity costs k_0 and all leverage levels L) the second type of dependence of

Table 20.19 The dependence of weighted average cost of capital, WACC, on age of the company n at values of capital costs (equity,$k_0 = 0.2$, and debt, $k_d = 0.15$) and at leverage level L=1

L	WACC	k_0	k_d	w_d	t	n	k_e
1	0.1843	0.2	0.15	0.5	0.2	1	0.2487
1	0.1798	0.2	0.15	0.5	0.2	2	0.2397
1	0.1780	0.2	0.15	0.5	0.2	3	0.2360
1	0.1772	0.2	0.15	0.5	0.2	4	0.2345
1	0.1769	0.2	0.15	0.5	0.2	5	0.2339
1	0.1769	0.2	0.15	0.5	0.2	6	0.2338
1	0.1770	0.2	0.15	0.5	0.2	7	0.2340
1	0.1771	0.2	0.15	0.5	0.2	8	0.2343
1	0.1773	0.2	0.15	0.5	0.2	9	0.2346
1	0.1775	0.2	0.15	0.5	0.2	10	0.2351

Table 20.20 The dependence of weighted average cost of capital, WACC, on age of the company n at values of capital costs (equity,$k_0 = 0.22$, and debt, $k_d = 0.15$) and at leverage level L=1

L	WACC	k_0	k_d	w_d	t	n	k_e
1	0.2041	0.22	0.15	0.5	0.2	1	0.2882
1	0.1994	0.22	0.15	0.5	0.2	2	0.2789
1	0.1975	0.22	0.15	0.5	0.2	3	0.2751
1	0.1967	0.22	0.15	0.5	0.2	4	0.2733
1	0.1963	0.22	0.15	0.5	0.2	5	0.2726
1	0.1962	0.22	0.15	0.5	0.2	6	0.2723
1	0.1962	0.22	0.15	0.5	0.2	7	0.2723
1	0.1962	0.22	0.15	0.5	0.2	8	0.2725
1	0.1964	0.22	0.15	0.5	0.2	9	0.2727
1	0.1965	0.22	0.15	0.5	0.2	10	0.2730

Table 20.21 The dependence of weighted average cost of capital, WACC, on age of the company n at values of capital costs (equity,$k_0 = 0.18$, and debt, $k_d = 0.15$) and at leverage level L=3

L	WACC	k_0	k_d	w_d	t	n	k_e
3	0.1569	0.18	0.15	0.75	0.2	1	0.2677
3	0.1503	0.18	0.15	0.75	0.2	2	0.2412
3	0.1477	0.18	0.15	0.75	0.2	3	0.2307
3	0.1466	0.18	0.15	0.75	0.2	4	0.2264
3	0.1462	0.18	0.15	0.75	0.2	5	0.2249
3	0.1462	0.18	0.15	0.75	0.2	6	0.2250
3	0.1464	0.18	0.15	0.75	0.2	7	0.2258
3	0.1468	0.18	0.15	0.75	0.2	8	0.2271
3	0.1471	0.18	0.15	0.75	0.2	9	0.2286
3	0.1475	0.18	0.15	0.75	0.2	10	0.2302

weighted average cost of capital, WACC, on the life time of company, n, namely, descending of WACC with n with the passage through the minimum with subsequent limited growth takes place.

"The Golden Age" of the company slightly fluctuates under change of the equity value k_0, these fluctuations are described in Table 20.24 (age is in years).

Table 20.22 The dependence of weighted average cost of capital, WACC, on age of the company n at values of capital costs (equity, $k_0 = 0.2$, and debt, $k_d = 0.15$) and at leverage level L=3

L	WACC	k_0	k_d	w_d	t	n	k_e
3	0.1765	0.2	0.15	0.75	0.2	1	0.3461
3	0.1697	0.2	0.15	0.75	0.2	2	0.3189
3	0.1669	0.2	0.15	0.75	0.2	3	0.3078
3	0.1657	0.2	0.15	0.75	0.2	4	0.3029
3	0.1653	0.2	0.15	0.75	0.2	5	0.3010
3	0.1651	0.2	0.15	0.75	0.2	6	0.3006
3	0.1653	0.2	0.15	0.75	0.2	7	0.3010
3	0.1655	0.2	0.15	0.75	0.2	8	0.3019
3	0.1658	0.2	0.15	0.75	0.2	9	0.3030
3	0.1661	0.2	0.15	0.75	0.2	10	0.3042

Table 20.23 The dependence of weighted average cost of capital, WACC, on age of the company n at values of capital costs (equity, $k_0 = 0.22$, and debt, $k_d = 0.15$) and at leverage level L=3

L	WACC	k_0	k_d	w_d	t	n	k_e
3	0.1961	0.22	0.15	0.75	0.2	1	0.4245
3	0.1891	0.22	0.15	0.75	0.2	2	0.3965
3	0.1862	0.22	0.15	0.75	0.2	3	0.3848
3	0.1849	0.22	0.15	0.75	0.2	4	0.3795
3	0.1843	0.22	0.15	0.75	0.2	5	0.3770
3	0.1840	0.22	0.15	0.75	0.2	6	0.3762
3	0.1840	0.22	0.15	0.75	0.2	7	0.3761
3	0.1841	0.22	0.15	0.75	0.2	8	0.3766
3	0.1843	0.22	0.15	0.75	0.2	9	0.3773
3	0.1845	0.22	0.15	0.75	0.2	10	0.3781

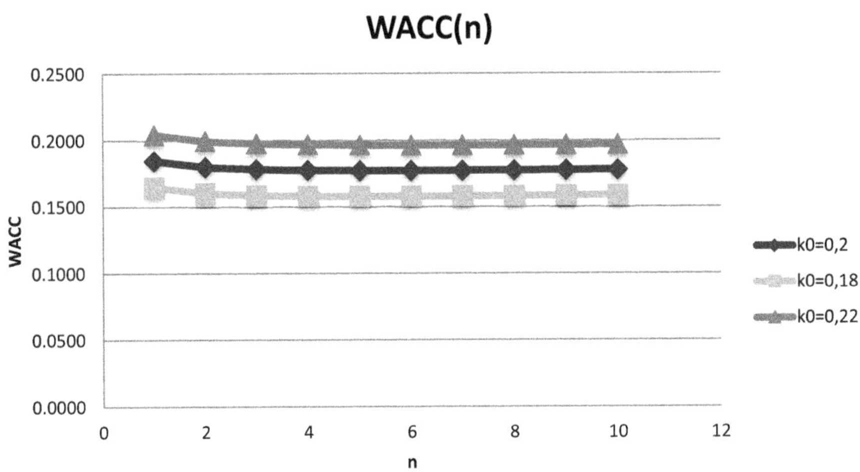

Fig. 20.9 Dependence of weighted average cost of capital, WACC, on lifetime of the company n at different values of equity cost, k_0, and fixed debt capital cost, k_d, and fixed leverage level L = 1. L = 3

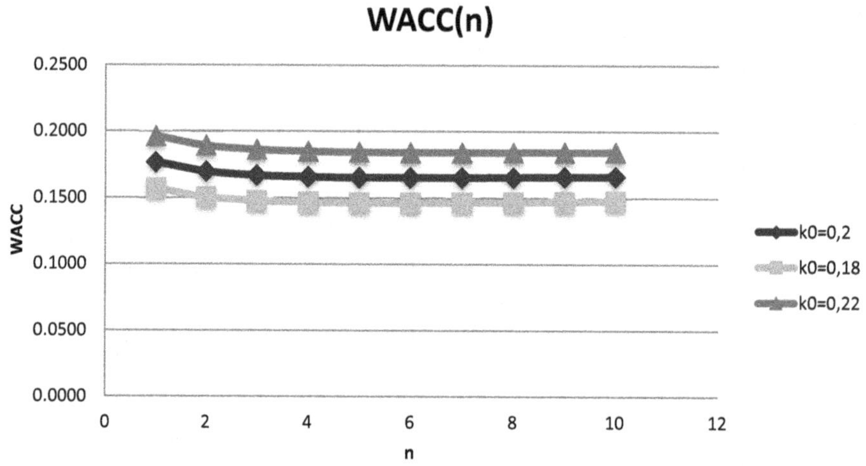

Fig. 20.10 Dependence of weighted average cost of capital, WACC, on lifetime of the company *n* at different values of equity cost, k_0, and fixed debt capital cost, k_d, and fixed leverage level L = 3

Table 20.24 Dependence of "The Golden Age" of the company on equity cost k_0

L\\k_0	18%	20%	22%
1	5–6	5–6	6–8
3	5–6	6	6–7

20.6 Dependence of Weighted Average Cost of Capital, WACC, on Lifetime of the Company *n* at High Values of Capital Cost (Equity, k_0, and Debt, k_d) and High Lifetime of the Company

Let us study the dependence of weighted average cost of capital, WACC, on lifetime of the company *n* at high values of capital cost (equity, k_0, and debt, k_d) and high lifetime of the company.

20.6.1 At Fixed Leverage Level

From Fig. 20.11, it follows that:

1. In all considered cases (at all leverage levels L) at high values of capital cost (equity, k_0 =40%, and debt, k_d =35%) the second type of dependence of weighted average cost of capital, WACC, on the lifetime of company, n, namely, descending of WACC with n with the passage through the minimum with subsequent limited growth up to perpetuity limit takes place.

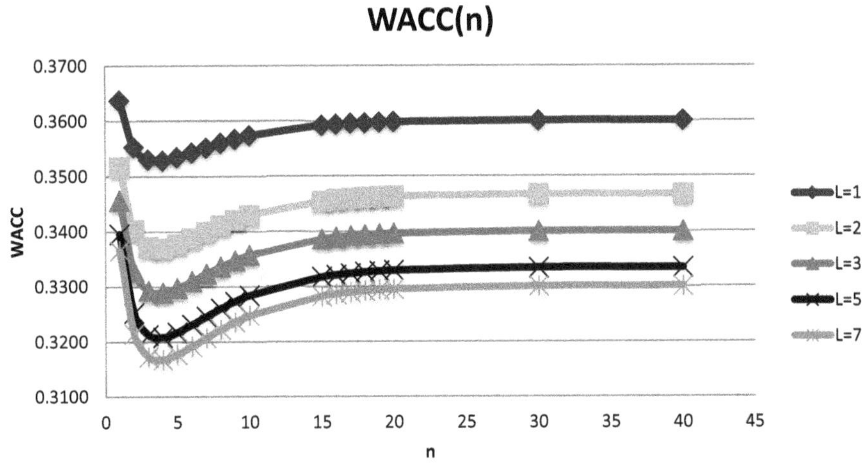

Fig. 20.11 Dependence of weighted average cost of capital, WACC, on lifetime of the company n at high values of capital cost (equity, k_0 =40%, and debt, k_d =35%) at different leverage levels L (up to high values of lifetime of the company)

2. A minimum value of attracting capital cost (weighted average cost of capital, WACC) is achieved at all leverage levels in the same age when n = 4. This means that, at high-value of capital costs, company age, at which minimal value of attracting capital cost is achieved is shifted forward lower (younger) values. We just remind, that at k_0 = 20%, and k_d = 15% (see above) the golden age was 6 years.
3. Shift of curves to lower values of WACC with increase of leverage level L is associated with decrease of WACC with leverage.
4. An interesting thing is the analysis of the value of detected effect, i.e., how much is the difference between the minimum of the attracting capital, found in the BFO theory, and its perpetuity limit value, which has been considered as minimal value up to now. In Table 25, a dependence of the difference between the minimum of the attracting capital and its perpetuity limit value on leverage level L is shown.

Perpetuity limit value of WACC is calculated by using Modigliani–Miller formula [13–15] with accounting of corporate taxes:

$$\text{WACC} = k_0(1 - w_d \cdot t) \tag{20.1}$$

From Fig. 20.11, it is seen that at high values of time of life of company ($n \geq 30$), weighted average cost of capital, WACC, practically does not differ from its perpetuity limit.

From Table 20.25, it is seen that the gain value is from 0.7% up to 1.5% and grows with the increase in the leverage level of company, L.

Table 20.25 The dependence of the difference between the optimal (minimal) value of weighted average cost of capital, WACC, and its perpetuity

L	1	2	3	5	7
$\Delta WACC$, %	−0.72	−0.99	−1.12	−1.25	−1.33

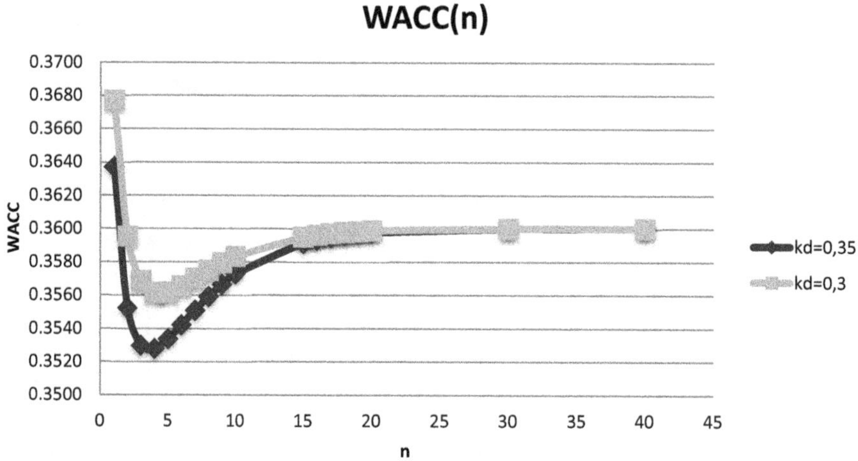

Fig. 20.12 Dependence of weighted average cost of capital, WACC, on lifetime of the company n at fixed high value of equity cost, k_0 =40%, and two values of debt cost, k_d =30% and 35% at leverage level L = 1

20.6.2 Under Change of the Debt Capital Cost, k_d

Under change of the debt capital cost, k_d, a depth of pit in dependence of weighted average cost of capital, WACC, on the time of life of the company, n, is changed as well: from Fig. 20.12, it is seen that pit (accounted from perpetuity value) is changed from 0.49% (at k_d =0.3) up to 0.72% (at k_d =0.35).

Note that as it is seen from Fig. 20.12, a perpetuity limit of WACC does not depend on debt cost, k_d, that is in accordance with the Modigliani—Miller formula (20.1) for WACC, which does not contain a debt capital cost, k_d, that means independence of perpetuity limit of WACC values from k_d, while the intermediate WACC values (for finite lifetime of company, n) depend on the debt capital cost, k_d (see BFO theory [1–12]) (Tables 20.26, 20.27, 20.28, 20.29, 20.30).

From Fig. 20.13, it is seen that with increase of debt cost, k_d, the character of dependence of weighted average cost of capital, WACC, on lifetime of the company n is changed from monotonic descending of WACC with n to descending of WACC with n with passage through minimum, followed by a limited growth.

Table 20.26 Dependence of weighted average cost of capital, WACC, on age of the company n at fixed value of equity cost,k_0=20%, and at value of debt cost, k_d= 18% at leverage level L=1

WACC	k_0	k_d	L	n	t
18.2889%	0.2	0.18	1	1	0.2
17.4859%	0.2	0.18	1	3	0.2
17.4155%	0.2	0.18	1	5	0.2
17.4654%	0.2	0.18	1	7	0.2
17.5833%	0.2	0.18	1	10	0.2
17.8641%	0.2	0.18	1	20	0.2
17.9629%	0.2	0.18	1	30	0.2
17.9909%	0.2	0.18	1	40	0.2

Table 20.27 Dependence of weighted average cost of capital, WACC, on age of the company n at fixed value of equity cost,k_0=20%, and at value of debt cost, k_d=15% at leverage level L=1

WACC	k_0	k_d	L	n	t
18.4736%	0.2	0.15	1	1	0.2
17.8200%	0.2	0.15	1	3	0.2
17.6936%	0.2	0.15	1	5	0.2
17.6967%	0.2	0.15	1	7	0.2
17.7528%	0.2	0.15	1	10	0.2
17.9192%	0.2	0.15	1	20	0.2
17.9797%	0.2	0.15	1	30	0.2
17.9957%	0.2	0.15	1	40	0.2

Table 20.28 Dependence of weighted average cost of capital, WACC, on age of the company n at fixed value of equity cost,k_0=20%, and at value of debt cost, k_d= 12% at leverage level L=1

WACC	k_0	k_d	L	n	t
18.6583%	0.2	0.12	1	1	0.2
18.1511%	0.2	0.12	1	3	0.2
18.0181%	0.2	0.12	1	5	0.2
17.9817%	0.2	0.12	1	7	0.2
17.9789%	0.2	0.12	1	10	0.2
18.0145%	0.2	0.12	1	20	0.2
18.0175%	0.2	0.12	1	30	0.2
18.0099%	0.2	0.12	1	40	0.2

Table 20.29 Dependence of weighted average cost of capital, WACC, on age of the company n at fixed value of equity cost,k_0=20%, and at value of debt cost, k_d=10% at leverage level L=1

WACC	k_0	k_d	L	n	t
18.9082%	0.2	0.1	1	1	0.2
18.4030%	0.2	0.1	1	3	0.2
18.2615%	0.2	0.1	1	5	0.2
18.2045%	0.2	0.1	1	7	0.2
18.1678%	0.2	0.1	1	10	0.2
18.1146%	0.2	0.1	1	20	0.2
18.0669%	0.2	0.1	1	30	0.2
18.0330%	0.2	0.1	1	40	0.2

Table 20.30 Dependence of weighted average cost of capital, WACC, on age of the company n at fixed value of equity cost, k_0=20%, and at value of debt cost, k_d=8% at leverage level L=1

WACC	k_0	k_d	L	n	t
19.1087%	0.2	0.08	1	1	0.2
18.6716%	0.2	0.08	1	3	0.2
18.5297%	0.2	0.08	1	5	0.2
18.4692%	0.2	0.08	1	7	0.2
18.4040%	0.2	0.08	1	10	0.2
18.2594%	0.2	0.08	1	20	0.2
18.1532%	0.2	0.08	1	30	0.2
18.0813%	0.2	0.08	1	40	0.2

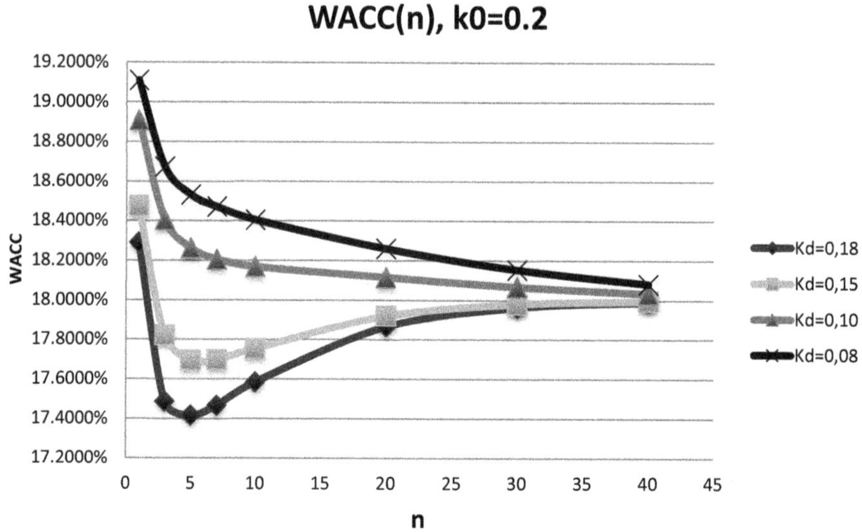

Fig. 20.13 Dependence of weighted average cost of capital, WACC, on lifetime of the company n at fixed value of equity cost, k_0 =20%, and at four values of debt cost, k_d =8%; 10%; 15%, and 18% at leverage level L = 1

20.6.3 Under Change of the Equity Capital Cost, k_0

Depth of gap, ΔWACC, is decreased with equity cost, k_0 (Tables 20.31, 20.32, 20.33, 20.34, 20.35, and 20.36, Fig. 20.14).

20.6.4 Under Change of the Tax on Profit rate, t

The depth of gap in dependence of WACC on n, which is equal to 0.41% at t = 0.2 is increased by 2.2 times and becomes equal to 0.92% at t = 0.4, i.e., it is increased by

Table 20.31 Dependence of weighted average cost of capital, WACC, on age of the company n at fixed value of debt cost, $k_d = 0.15$ and at value of equity cost, $k_0 = 0.25$at leverage level L=1

WACC	k_0	k_d	L	n	t
23.2477%	0.25	0.15	1	1	0.2
22.6690%	0.25	0.15	1	3	0.2
22.5117%	0.25	0.15	1	5	0.2
22.4913%	0.25	0.15	1	7	0.2
22.4933%	0.25	0.15	1	10	0.2
22.5219%	0.25	0.15	1	20	0.2
22.5136%	0.25	0.15	1	30	0.2
22.5045%	0.25	0.15	1	40	0.2

Table 20.32 Dependence of weighted average cost of capital, WACC, on age of the company n at fixed value of debt cost, $k_d = 0.15$ and at value of equity cost, $k_0 = 0.22$at leverage level L=1

WACC	k_0	k_d	L	n	t
20.3006%	0.22	0.15	1	1	0.2
19.7431%	0.22	0.15	1	3	0.2
19.6171%	0.22	0.15	1	5	0.2
19.6163%	0.22	0.15	1	7	0.2
19.6514%	0.22	0.15	1	10	0.2
19.7639%	0.22	0.15	1	20	0.2
19.7960%	0.22	0.15	1	30	0.2
19.8007%	0.22	0.15	1	40	0.2

Table 20.33 Dependence of weighted average cost of capital, WACC, on age of the company n at fixed value of debt cost, $k_d = 0.15$ and at value of equity cost,$k_0 = 0.2$at leverage level L=1

WACC	k_0	k_d	L	n	t
18.4717%	0.2	0.15	1	1	0.2
17.8015%	0.2	0.15	1	3	0.2
17.6938%	0.2	0.15	1	5	0.2
17.6972%	0.2	0.15	1	7	0.2
17.7592%	0.2	0.15	1	10	0.2
17.9192%	0.2	0.15	1	20	0.2
17.9797%	0.2	0.15	1	30	0.2
17.9957%	0.2	0.15	1	40	0.2

Table 20.34 Dependence of weighted average cost of capital, WACC, on age of the company n at fixed value of debt cost, $k_d = 0.15$ and at value of equity cost, $k_0 = 0.18$at leverage level L=1

WACC	k_0	k_d	L	n	t
16.4350%	0.18	0.15	1	1	0.2
15.8519%	0.18	0.15	1	3	0.2
15.7610%	0.18	0.15	1	5	0.2
15.7793%	0.18	0.15	1	7	0.2
15.8561%	0.18	0.15	1	10	0.2
16.0683%	0.18	0.15	1	20	0.2
16.1586%	0.18	0.15	1	30	0.2
16.1884%	0.18	0.15	1	40	0.2

Table 20.35 Dependence of weighted average cost of capital, WACC, on age of the company n at fixed value of debt cost, $k_d = 0.15$ and at value of equity cost, $k_0 = 0.16$at leverage level L=1

WACC	k_0	k_d	L	n	t
14.4304%	0.16	0.15	1	1	0.2
13.9019%	0.16	0.15	1	3	0.2
13.8278%	0.16	0.15	1	5	0.2
13.8610%	0.16	0.15	1	7	0.2
13.9481%	0.16	0.15	1	10	0.2
14.2119%	0.16	0.15	1	20	0.2
14.3324%	0.16	0.15	1	30	0.2
14.3781%	0.16	0.15	1	40	0.2

Table 20.36 Dependence of depth of gap, ΔWACC, at fixed value of debt cost, $k_d = 0.15$ and at values of equity cost,$k_0 = 0.16$; 0.18;

k_0	0.16	0.18	0.20	0.22	0.25
ΔWACC, %	0.55	0.43	0.30	0.18	0.03

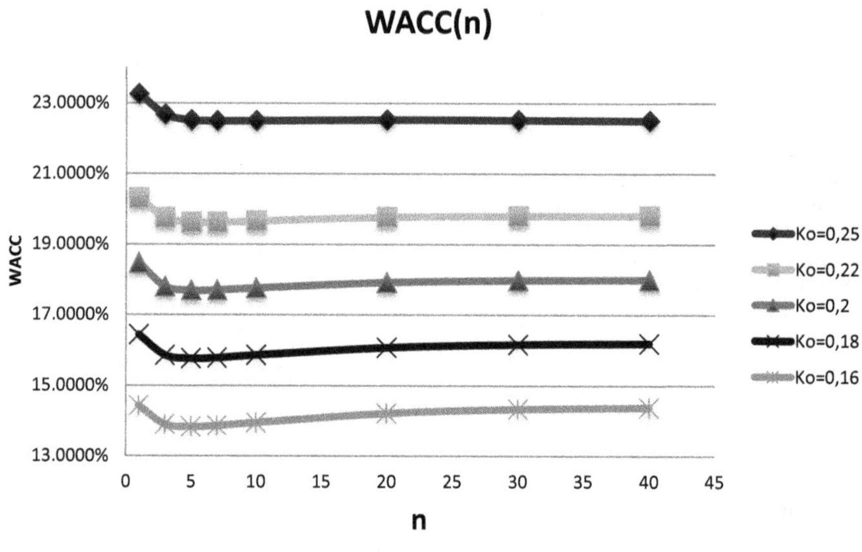

Fig. 20.14 Dependence of weighted average cost of capital, WACC, on lifetime of the company n at fixed value of debt cost, k_d =15%, and five values of equity cost, k_0 =16%; 18%; 20%; 22%, and 25% at leverage level L = 1

2.2 times when tax on profit rate is increased by 2 times (Tables 20.37 and 20.38, Figs. 20.15 and 20.16).

We see from Fig. 20.16 that, at fixed capital costs, k_0 =30%; k_d =15%, and different values of tax on profit rates, t, there is no minimum in WACC at finite lifetime of the company: minimal value of WACC is reached at $n = \infty$. Note, that this is a feature of these particular values of capital costs (probably, too big difference between k_0 and k_d).

Table 20.37 Dependence of weighted average cost of capital, WACC, on age of the company n at fixed value of debt cost, $k_d = 0.15$, at fixed value of equity cost, $k_0 = 0.2$at leverage level L=2 and tax on profit t=0.2

L	WACC	k_0	k_d	t	n
2	17.84%	0.2	0.15	0.2	1
2	17.07%	0.2	0.15	0.2	3
2	16.92%	0.2	0.15	0.2	5
2	16.92%	0.2	0.15	0.2	7
2	16.99%	0.2	0.15	0.2	10
2	17.12%	0.2	0.15	0.2	15
2	17.30%	0.2	0.15	0.2	30
2	17.33%	0.2	0.15	0.2	45

Table 20.38 Dependence of weighted average cost of capital, WACC, on age of the company n at fixed value of debt cost, $k_d = 0.15$, at fixed value of equity cost, $k_0 = 0.22$at leverage level L=2 and tax on profit t=0.4

L	WACC	k_0	k_d	t	n
2	15.72%	0.2	0.15	0.4	1
2	14.09%	0.2	0.15	0.4	3
2	13.76%	0.2	0.15	0.4	5
2	13.73%	0.2	0.15	0.4	7
2	13.86%	0.2	0.15	0.4	10
2	14.13%	0.2	0.15	0.4	15
2	14.56%	0.2	0.15	0.4	30
2	14.65%	0.2	0.15	0.4	45

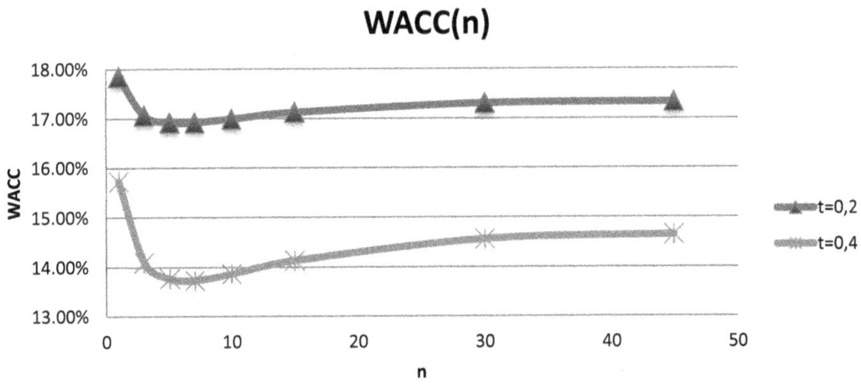

Fig. 20.15 Dependence of weighted average cost of capital, WACC, on lifetime of the company n at fixed capital costs, k_0 =20%; k_d =15%, and two values of tax on profit rates $t = 0.2$ and 0.4 and at leverage level L = 2

20.7 Further Investigation of Effect

During further investigation of effect, we have discovered one more interesting feature of dependence of WACC on n, WACC(n): we have called this effect the "Kulik effect." (Tables 20.39 and 20.40)

Note, that perpetuity limits for WACC(n), calculated by the Modigliani–Miller formula (1) [13–15] equal:

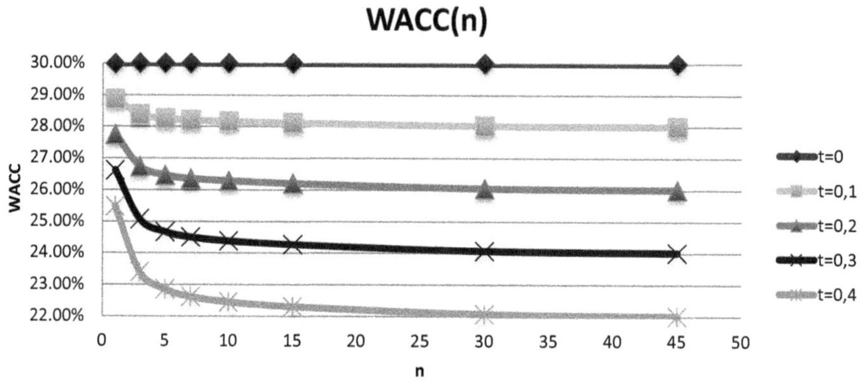

Fig. 20.16 Dependence of weighted average cost of capital, WACC, on lifetime of the company n at fixed capital costs, k_0 =30%; k_d =15%, and different values of tax on profit rates $t = 0; 0.1; 0.2; 0.3$, and 0.4 and at leverage level L = 2

Table 20.39 Dependence of weighted average cost of capital, WACC, on age of the company n at value of debt cost, $k_d = 0.15$, at value of equity cost, $k_0 = 0.25$ at leverage level L=1

L	t	k_0	k_d	n	w_d	WACC
1	0.2	0.25	0.15	1	0.5	23.2270%
1	0.2	0.25	0.15	3	0.5	22.6725%
1	0.2	0.25	0.15	5	0.5	22.5184%
1	0.2	0.25	0.15	7	0.5	22.4914%
1	0.2	0.25	0.15	10	0.5	22.4934%
1	0.2	0.25	0.15	20	0.5	22.5220%
1	0.2	0.25	0.15	30	0.5	22.5137%
1	0.2	0.25	0.15	40	0.5	22.5045%
1	0.2	0.25	0.15	∞	0.5	21.50%

Table 20.40 Dependence of weighted average cost of capital, WACC, on age of the company n at value of debt cost, $k_d = 0.15$, at value of equity cost, $k_0 = 0.25$ at leverage level L=2

L	t	k_0	k_d	n	w_d	WACC
2	0.2	0.25	0.15	1	0.6667	22.8255%
2	0.2	0.25	0.15	3	0.6667	21.8935%
2	0.2	0.25	0.15	5	0.6667	21.6843%
2	0.2	0.25	0.15	7	0.6667	21.6431%
2	0.2	0.25	0.15	10	0.6667	21.6448%
2	0.2	0.25	0.15	20	0.6667	21.6895%
2	0.2	0.25	0.15	30	0.6667	21.6842%
2	0.2	0.25	0.15	40	0.6667	21.6742%
2	0.2	0.25	0.15	∞	0.6667	21.6665%

for L = 1 WACC(∞) =22.5%

for L = 2 WACC(∞) =21.6665%

It turns out that at particular values of capital costs, for example, at k_0 =25%; k_d =15%, a third modification of dependences of weighted average cost of capital, WACC, on the time of life of company n takes place: descending of WACC with

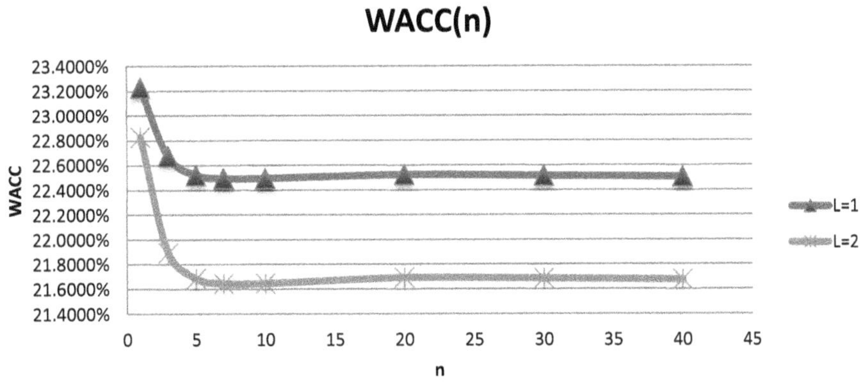

Fig. 20.17 Dependence of weighted average cost of capital, WACC, on lifetime of the company n at fixed capital costs, k_0 =25%; k_d =15%, and different values of leverage level L = 1 and L = 2

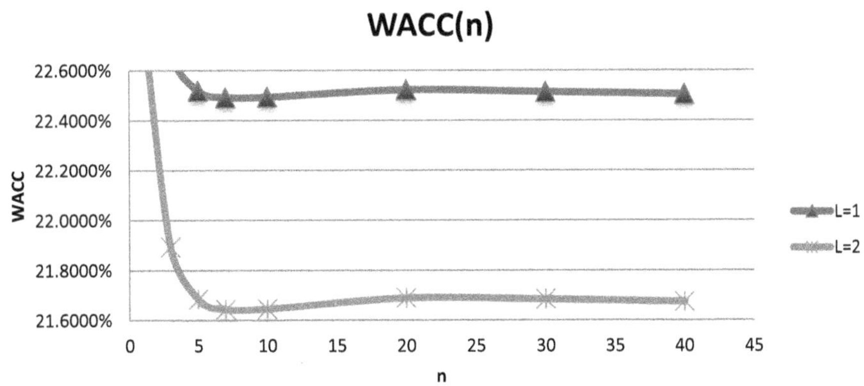

Fig. 20.18 Dependence of weighted average cost of capital, WACC, on lifetime of the company n at fixed capital costs, k_0 =25%; k_d =15%, and different values of leverage level L = 1 and L = 2 (lager scale)

passage through minimum, followed by a growth with passage through maximum and finally with trend to perpetuity limit from bigger values (remind, that at second type of WACC(n) behavior, the curve WACC(n) tends toward perpetuity limit from lower values). We have called this effect the "Kulik effect." (Figs. 20.17, 20.18, and 20.19).

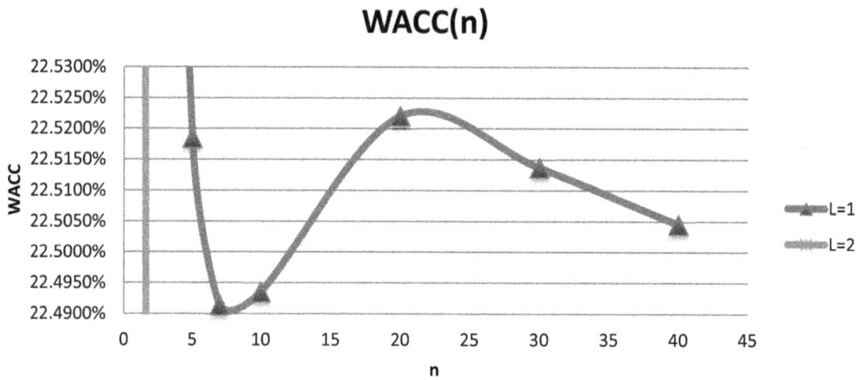

Fig. 20.19 Dependence of weighted average cost of capital, WACC, on lifetime of the company n at fixed capital costs, k_0 =25%; k_d =15%, and different values of leverage level L = 1 and L = 2 (the largest scale)

20.8 Conclusions

In this paper, it is shown for the first time within BFO theory [1–12], that valuation of WACC in the Modigliani–Miller theory ([13–15] is not minimal and valuation of the company capitalization is not maximal, as all financiers have supposed up to now: at some age of the company, its WACC value turns out to be lower than in Modigliani–Miller theory and company capitalization V turns out to be greater than V in Modigliani–Miller theory [13–15]. Thus, existing up to the present presentations concerning the results of the Modigliani–Miller theory in this aspect [16] turn out to be incorrect.

It is shown that, from the point of view of cost of attracting capital, there are two types of dependences of weighted average cost of capital, WACC, on the time of life of company n: monotonic descending with n and descending with passage through minimum, followed by a limited growth (there is a third modification of dependences WACC(n), which leaves all conclusions valid). The first type takes place for the companies with low cost capital, characteristic for the western companies. The second type takes place for higher costs capital costs of the company, characteristic for the Russian companies as well as for companies from other developing countries. This means that the latter companies, in contrast to the western ones, can take advantage of the benefits given at a certain stage of development of company by discovered effect. **(For example, the capitalization of Russian oil company "Rosneft", which was valued in 2014 by the Modigliani–Miller method, could be higher, accounting for the discovered effect and BFO theory).** Moreover, since the "golden age" of company depends on the company's capital costs, by controlling them (for example, by modifying the value of dividend payments, that reflect the equity cost), a company may extend the "golden age" of the company, when the cost to attract capital becomes minimal (less than perpetuity limit), and capitalization of companies becomes maximal (higher than perpetuity assessment) up to a specified time interval.

References

1. Brusov, P., Filatova, T., Orekhova, N., & Brusova, A. (2011). Weighted average cost of capital in the theory of Modigliani–Miller, modified for a finite life–time company. *Applied Financial Economics, 21*(11), 815–824.
2. Brusov, P., Filatova, P., & Orekhova, N. (2013). Absence of an optimal capital structure in the famous tradeoff theory! *Journal of Reviews on Global Economics, 2*, 94–116.
3. Brusov, P., Filatova, P., & Orekhova, N. (2014). Mechanism of formation of the company optimal capital structure, different from suggested by trade off theory. *Cogent Economics & Finance, 2*, 1–13. https://doi.org/10.1080/23322039.2014.946150
4. Brusov, P., Filatova, T., Orehova, N., et al. (2011). From Modigliani–Miller to general theory of capital cost and capital structure of the company. *Research Journal of Economics, Business and ICT, 2*, 16–21.
5. Brusov, P., Filatova, T., Eskindarov, M., & Orehova, N. (2012). Influence of debt financing on the effectiveness of the finite duration investment project. *Applied Financial Economics, 22*(13), 1043–1052.
6. Brusov, P., Filatova, T., Orehova, N., et al. (2011). Influence of debt financing on the effectiveness of the investment project within the Modigliani–Miller theory. *Research Journal of Economics, Business and ICT, 2*, 11–15.
7. Brusov, P., Filatova, T., Eskindarov, M., & Orehova, N. (2012). Hidden global causes of the global financial crisis. *Journal of Reviews on Global Economics, 1*, 106–111.
8. Brusov, P., Filatova, T., & Orekhova, N. (2013). Absence of an optimal capital structure in the famous tradeoff theory! *Journal of Reviews on Global Economics, 2*, 94–116.
9. Brusov, P., Filatova, T., Orehova, N., Brusov, P. P., & Brusova, N. (2011). From Modigliani–Miller to general theory of capital cost and capital structure of the company. *Research Journal of Economics, Business and ICT, 2*, 16–21.
10. Brusov, P., Filatova, T., & Orehova, N. (2014). Inflation in Brusov–Filatova–Orekhova theory and in its perpetuity limit—Modigliani—Miller theory. *Journal of Reviews on Global Economics, 3*, 175–185.
11. Brusov, P., Filatova, T., & Orehova, N. (2013). A qualitatively new effect in corporative finance: Abnormal dependence of cost of equity of company on leverage. *Journal of Reviews on Global Economics, 2*, 183–193.
12. Filatova, T., Orehova, N., & Brusova, A. (2008). Weighted average cost of capital in the theory of Modigliani–Miller, modified for a finite life–time company. *Bulletin of the FU, 48*, 68–77.
13. Modigliani, F., & Miller, M. (1958). The cost of capital, corporate finance, and the theory of investment. *American Economic Review, 48*, 261–297.
14. Modigliani, F., & Miller, M. (1963). Corporate income taxes and the cost of capital: A correction. *American Economic Review, 53*, 147–175.
15. Modigliani, F., & Miller, M. (1966). Some estimates of the cost of capital to the electric utility industry 1954–1957. *American Economic Review, 56*, 333–391.
16. Myers, S. (1984). The capital structure Pussle. *Journal of Finance, 39*(3), 574–592.
17. Brusov, P. N., & Filatova, T. V. (2011). From Modigliani–Miller to general theory of capital cost and capital structure of the company. *Finance and Credit, 435*, 2–8.
18. Brusova, A. (2011). A comparison of the three methods of estimation of weighted average cost of capital and equity cost of company. *Financial Analysis: Problems and Solutions, 34*(76), 36–42.

Chapter 21
A "Golden Age" of the Companies: Conditions of Its Existence

21.1 Introduction

A few years ago, we have discovered the effect of the "golden age" of company [1, 2]: it was shown for the first time that valuation of the weighted average cost of capital, WACC, in the Modigliani–Miller theory [3–5] is not minimal and valuation of the company capitalization is not maximal, as all financiers have supposed up to this discovery: at some age of the company, its WACC value turns out to be lower than in Modigliani–Miller theory and company capitalization V turns out to be greater than V in Modigliani–Miller theory. It was shown that, from the point of view of cost of attracting capital, there are two types of dependences of weighted average cost of capital, WACC, on the company age n: monotonic descending with n and descending with passage through minimum, followed by a limited growth. In practice, there are companies with both types of dependences of WACC on the company age n.

In this chapter, we continue to study this problem and investigate which companies have the "golden age," i.e., obey the latter type of dependence of WACC on n [6]. With this aim, we study the dependence of WACC on the age of company n at various leverage levels within a wide spectrum of capital costs values as well as the dependence of WACC on leverage level L at fixed company age n. All calculations have been done within modern theory of capital cost and capital structure BFO by Brusov–Filatova–Orekhova [7–22].

We have shown that existence of the "golden age" of company does not depend on the value of capital costs of the company but depends on the difference between equity k_0 and debt k_d costs. The "golden age" of company exists at small enough difference between k_0 and k_d costs, while at high value of this difference the "golden age" of company is absent: curve WACC(n) monotonic descends with n. For the companies with the "golden age" curve, WACC(L) for perpetuity companies lies between curves WACC(L) for company ages $n = 1$ and $n = 3$, while for the

P. Brusov, T. Filatova, *Business Valuation and Company Value*, Contributions to Finance and Accounting, https://doi.org/10.1007/978-3-031-90847-7_21

companies without the "golden age" curve WACC(L) for perpetuity companies is the lowest one.

In our paper [1, 2], we have found also a third type of WACC(n) dependence: descending with passage through minimum, which lies below the perpetuity limit value, then going through maximum followed by a limited descending. We called this effect the "Kulik effect." In this chapter, we have found a variety of the "Kulik effect": descending with passage through minimum of WACC, which lies above the perpetuity limit value, then going through maximum followed by a limited descending. We call this company age, where WACC has a minimum, which lies above the perpetuity limit value, "a silver age" of the company.

Because the cost of attracting capital is used in rating methodologies as discounting rate under discounting of cash flows, study of WACC behavior is very important for rating procedures. The account of effects of the "golden (silver) age" could change the valuation of creditworthiness of issuers.

Remember that, since the "golden age" of company depends on the company's capital costs, by controlling them (for example, by modifying the value of dividend payments, that reflect the equity cost), a company may extend the "golden age" of the company when the cost to attract capital becomes a minimal (less than perpetuity limit), and capitalization of companies becomes maximal (above than perpetuity assessment) up to a specified time interval. We discuss the use of opened effects in developing economics.

In this chapter, we answer the following question: which companies have "a golden age," i.e., obey the following type of dependence of WACC on n: WACC(n) descending with passage through minimum, followed by a limited growth. With this aim we study the dependence of WACC on the age of company n at various leverage levels within a wide spectrum of capital costs values as well as the dependence of WACC on leverage level L at fixed company age n. All calculations have been done within modern theory of capital cost and capital structure BFO by Brusov–Filatova–Orekhova [7–17]. We make calculations for equity cost k_0 (at L = 0) between 6% and 30% and debt cost k_d between 4% and 28% for a lot of pairs (k_0, k_d), accounting that the inequality $k_0 \geq k_d$ is always valid via the fact that equity cost is more risky than debt one. We present in paper only some examples of our calculations (one-two in each group) and readers should understand that other results in each group give more or less qualitatively similar results.

We have shown that existence of the "golden age" of a company depends not on the value of capital costs of the company but on the difference between equity k_0 and debt k_d costs. The "golden age" of company exists at small enough difference between k_0 and k_d costs, while at high value of this difference, the "golden age" of company is absent: curve WACC(n) monotonic descends with n. For the companies with the "golden age" curve, WACC(L) for perpetuity limit ($n = \infty$) lies between curves WACC(L) for 1 year ($n = 1$) [23] and 3 years ($n = 3$) companies, while for the companies without the "golden age" curve, WACC(L) for perpetuity limit is the lowest one.

The problem of the existence of the "golden age" of a company is very important in ratings because the discount rate (WACC value), used in discounting of cash flows in ratings, depends on the existence or the nonexistence of the "golden age" of a company.

21.2 Companies Without the "Golden Age" (Large Difference Between k_0 and k_d Costs)

As an example of companies without the "golden age" (with large difference between k_0 and k_d costs), we present the calculations for equity cost k_0 (at L = 0) equals to 20% and debt cost k_d equals to 9%.

21.2.1 Dependence of Weighted Average Cost of Capital, WACC, on the Company Age n *at Different Leverage Levels*

We study below the dependence of weighted average cost of capital, WACC, on the company age n at different leverage levels (L = 1,2, 3), using the BFO formula

$$\frac{\left[1-(1+\mathrm{WACC})^{-n}\right]}{\mathrm{WACC}}=\frac{\left[1-(1+k_0)^{-n}\right]}{k_0\left[1-\omega_d t\left(1-(1+k_d)^{-n}\right)\right]}. \quad (21.1)$$

Leverage level L is presented in BFO formula through the share of debt capital $w_d = L/(1 + L)$.

The results of our calculations are shown below in tables and figures.

For $L = 1$ one has

For L = 2 we have

For L = 3 one has

It is seen from Tables 21.1, 21.2, 21.3 and Fig. 21.2 that 1–1′—behavior (from Fig. 21.1) takes place: monotonic dependence of weighted average cost of capital, WACC, and company capitalization, V, on the company age n for all considered leverage levels (L = 1,2,3); this means that the "golden age" of company is absent. The ordering of curves is the following: the lower curve corresponds the greater leverage level.

From Tables 21.4, 21.5, 21.6 and Fig. 21.3**,** it is seen that the ordering of curves WACC(L) is the following: the lower curve corresponds to the greater company age n. We will see below that under existence the "golden age" of a company, this ordering will be a different one. We keep here the case of n = 45 as the case which is closed to perpetuity limit. An alternative method is the using of the Modigliani–Miller formula

Table 21.1 Dependence of WACC on the age of the company n at $L = 1$, $k_0 = 20\%$, $k_d = 9\%$

n	t	L	WACC(L = 1)	k_0	k_d	w_d	$A(n)$	*БФО*
1	0.2	1	0.19001348	0.2	0.09	0.50	0.840271	0.000055264
2	0.2	1	0.18679105	0.2	0.09	0.50	1.552355	0.000242408
3	0.2	1	0.18521681	0.2	0.09	0.50	2.155589	0.000643588
4	0.2	1	0.18453895	0.2	0.09	0.50	2.666482	0.000008562
5	0.2	1	0.18398934	0.2	0.09	0.50	3.099102	0.000019534
6	0.2	1	0.18361179	0.2	0.09	0.50	3.465420	0.000038686
7	0.2	1	0.18333561	0.2	0.09	0.50	3.775614	0.000068644
8	0.2	1	0.18312210	0.2	0.09	0.50	4.038322	0.000111902
9	0.2	1	0.18294855	0.2	0.09	0.50	4.260871	0.000170582
10	0.2	1	0.18280097	0.2	0.09	0.50	4.449469	0.000246231
20	0.2	1	0.18181178	0.2	0.09	0.50	5.305460	0.000013165
30	0.2	1	0.18103559	0.2	0.09	0.50	5.486207	0.000041589
40	0.2	1	0.18052092	0.2	0.09	0.50	5.532205	0.000065502

Table 21.2 Dependence of WACC on the age of the company n at $L = 2$, $k_0 = 20\%$, $k_d = 9\%$

n	t	L	WACC(L = 2)	k_0	k_d	w_d	$A(n)$	*БФО*
1	0.2	2	0.18670845	0.2	0.09	0.67	0.8426098	0.000057189
2	0.2	2	0.18242414	0.2	0.09	0.67	1.5607236	0.000239282
3	0.2	2	0.18033977	0.2	0.09	0.67	2.1724715	0.000618675
4	0.2	2	0.17935211	0.2	0.09	0.67	2.6934467	0.000007915
5	0.2	2	0.17860838	0.2	0.09	0.67	3.1370358	0.000018207
6	0.2	2	0.17809570	0.2	0.09	0.67	3.5147106	0.000036130
7	0.2	2	0.17771992	0.2	0.09	0.67	3.8362859	0.000064261
8	0.2	2	0.17742943	0.2	0.09	0.67	4.1101468	0.000105021
9	0.2	2	0.17719390	0.2	0.09	0.67	4.3434469	0.000160484
10	0.2	2	0.17699457	0.2	0.09	0.67	4.5422818	0.000232186
20	0.2	2	0.17567694	0.2	0.09	0.67	5.4686271	0.000012229
30	0.2	2	0.17467722	0.2	0.09	0.67	5.6790746	0.000039727
40	0.2	2	0.17401430	0.2	0.09	0.67	5.7372042	0.000064124

$$\text{WACC} = k_0(1 - \omega_d t), \tag{21.2}$$

which follows from BFO formula (21.1) for perpetuity limit.

Table 21.3 Dependence of WACC on the age of the company n at $L = 3$, $k_0 = 20\%$, $k_d = 9\%$

n	t	L	WACC(L = 3)	k_0	k_d	w_d	$A(n)$	БФО
1	0.2	3	0.18506498	0.2	0.09	0.75	0.8437839	0.000051714
2	0.2	3	0.18024369	0.2	0.09	0.75	1.5649420	0.000228517
3	0.2	3	0.17789664	0.2	0.09	0.75	2.1810121	0.000607678
4	0.2	3	0.17675215	0.2	0.09	0.75	2.7071344	0.000007660
5	0.2	3	0.17590910	0.2	0.09	0.75	3.1563532	0.000017650
6	0.2	3	0.17532683	0.2	0.09	0.75	3.5398852	0.000035095
7	0.2	3	0.17489950	0.2	0.09	0.75	3.8673589	0.000062558
8	0.2	3	0.17456914	0.2	0.09	0.75	4.1470258	0.000101532
9	0.2	3	0.17430145	0.2	0.09	0.75	4.3859468	0.000155351
10	0.2	3	0.17407533	0.2	0.09	0.75	4.5901557	0.000225050
20	0.2	3	0.17259416	0.2	0.09	0.75	5.5540330	0.000011720
30	0.2	3	0.17148952	0.2	0.09	0.75	5.7806846	0.000038586
40	0.2	3	0.17075742	0.2	0.09	0.75	5.8455083	0.000063263

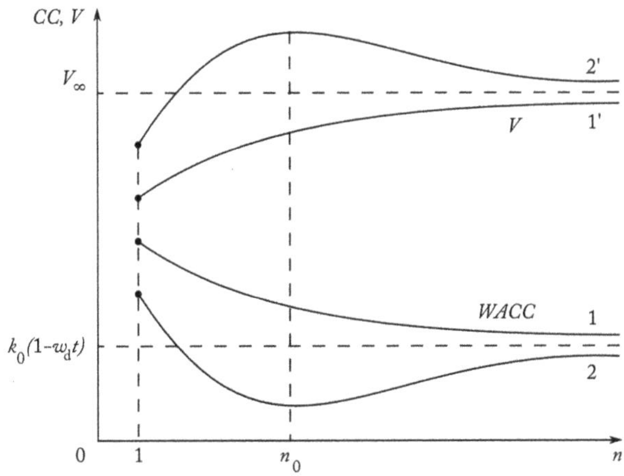

Fig. 21.1 Two kinds of dependences of weighted average cost of capital, WACC, and company capitalization, V, on lifetime of the company n: 1–1′—monotonic dependence of weighted average cost of capital, WACC, and company capitalization, V, on lifetime of the company n. 2–2′—showing descending of WACC with n, and with the passage through a minimum and then a limited growth and increase of V with the passage through a maximum (at n_0) and then a limited descending

21.3 Companies with the "Golden Age" (Small Difference Between k_0 and k_d Costs)

As an example of companies with the "golden age" (with small difference between k_0 and k_d costs), we present the calculations for equity cost k_0 (at L = 0) equals to 27% and debt cost k_d equals to 25% (Fig. 21.4).

It is seen from Tables 21.7, 21.8, 21.9 and Fig. 21.5 that 2–2′ behavior (from Fig. 21.1) takes place: descending of WACC with n, and with the passage through a

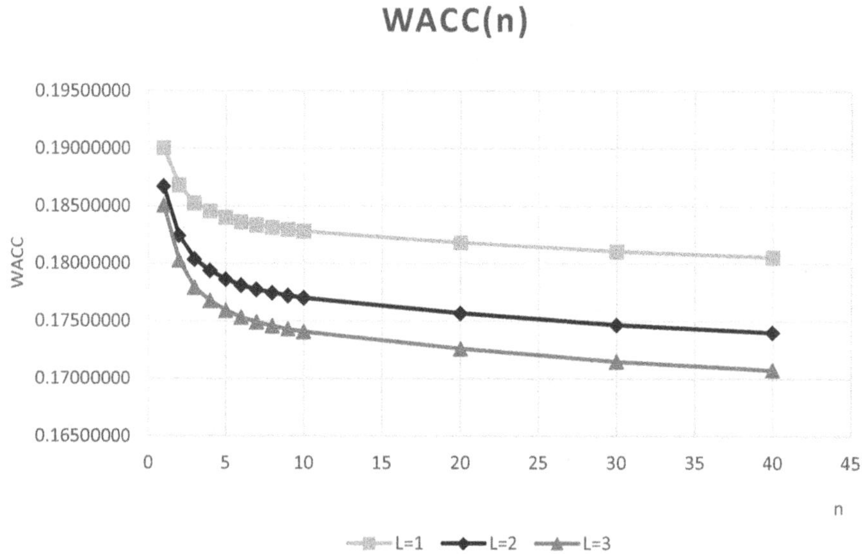

Fig. 21.2 The dependence of weighted average cost of capital, WACC, on the company age n at different leverage levels (L = 1, 2, 3)

Table 21.4 Dependence of WACC on the leverage level at the company age $n = 1$, $k_0 = 20\%$, $k_d = 9\%$

n	t	L	WACC($n = 1$)	k_0	k_d	w_d	$A(n)$	БФО
1	0.2	0	0.19990871	0.2	0.09	0.00	0.833333	0.000063399
1	0.2	1	0.18989673	0.2	0.09	0.50	0.840271	0.000137715
1	0.2	2	0.18675471	0.2	0.09	0.67	0.842610	0.000024342
1	0.2	3	0.18504185	0.2	0.09	0.75	0.843784	0.000068187
1	0.2	4	0.18395964	0.2	0.09	0.80	0.844490	0.000133490
1	0.2	5	0.18323424	0.2	0.09	0.83	0.844961	0.000179952
1	0.2	6	0.18270822	0.2	0.09	0.86	0.845298	0.000218846
1	0.2	7	0.18230907	0.2	0.09	0.88	0.845551	0.000251369
1	0.2	8	0.18203277	0.2	0.09	0.89	0.845748	0.000252248
1	0.2	9	0.18178544	0.2	0.09	0.90	0.845906	0.000271780
1	0.2	10	0.18158213	0.2	0.09	0.91	0.846034	0.000288455

minimum and then a limited growth and increase of V with the passage through a maximum (at $n_0 \approx 4$) and then a limited descent. This means the presence of the "golden age" of a company. The ordering of curves is the following: the lower curve corresponds to the greater leverage level.

From Tables 21.10, 21.11, 21.12 and Fig. 21.6, the quite new effect follows: the ordering of curves WACC(L) is the following: the top curve corresponds to the company age $n = 1$, the middle one corresponds to perpetuity company $n_0 = \infty$ (we use $n = 49$ to approximate perpetuity limit), and bottom one corresponds to the

Table 21.5 Dependence of WACC on the leverage level at the company age $n = 3$, $k_0 = 20\%$, $k_d = 9\%$

n	t	L	WACC ($n = 3$)	k_0	k_d	w_d	$A(n)$	БФО
3	0.2	0	0.19978349	0.2	0.09	0.00	2.106481	0.000714376
3	0.2	1	0.18521867	0.2	0.09	0.50	2.155589	0.000637220
3	0.2	2	0.18034194	0.2	0.09	0.67	2.172471	0.000611094
3	0.2	3	0.17789941	0.2	0.09	0.75	2.181012	0.000597961
3	0.2	4	0.17643254	0.2	0.09	0.80	2.186169	0.000590059
3	0.2	5	0.17545407	0.2	0.09	0.83	2.189620	0.000584783
3	0.2	6	0.17475488	0.2	0.09	0.86	2.192092	0.000581010
3	0.2	7	0.17423034	0.2	0.09	0.88	2.193950	0.000578178
3	0.2	8	0.17382227	0.2	0.09	0.89	2.195397	0.000575974
3	0.2	9	0.17349576	0.2	0.09	0.90	2.196556	0.000574210
3	0.2	10	0.17322858	0.2	0.09	0.91	2.197505	0.000572766

Table 21.6 Dependence of WACC on the leverage level at the company age $n = 45$, $k_0 = 20\%$, $k_d = 9\%$

n	t	L	WACC($n = 45$)	k_0	k_d	w_d	$A(n)$	БФО
45	0.2	0	0.19999699	0.2	0.09	0.00	4.998633	0.000075004
45	0.2	1	0.18035710	0.2	0.09	0.50	5.541296	0.000073385
45	0.2	2	0.17380207	0.2	0.09	0.67	5.749351	0.000072648
45	0.2	3	0.17052242	0.2	0.09	0.75	5.859349	0.000072233
45	0.2	4	0.16855385	0.2	0.09	0.80	5.927391	0.000071966
45	0.2	5	0.16724112	0.2	0.09	0.83	5.973638	0.000071780
45	0.2	6	0.16630328	0.2	0.09	0.86	6.007115	0.000071642
45	0.2	7	0.16559980	0.2	0.09	0.88	6.032471	0.000071536
45	0.2	8	0.16505258	0.2	0.09	0.89	6.052340	0.000071451
45	0.2	9	0.16461477	0.2	0.09	0.90	6.068330	0.000071382
45	0.2	10	0.16425654	0.2	0.09	0.91	6.081476	0.000071325

company age $n = 3$. Thus, the curve WACC(L) for perpetuity company lies between curves corresponding to the company age $n = 1$, and $n\!-\!3$.

Note that this ordering is quite different from the case when the "golden age" of a company is absent: in that case, the lower curve corresponds the greater company age n: the top curve corresponds to the company age $n = 1$, the middle one corresponds to the company age $n = 3$ and bottom one corresponds to the perpetuity company.

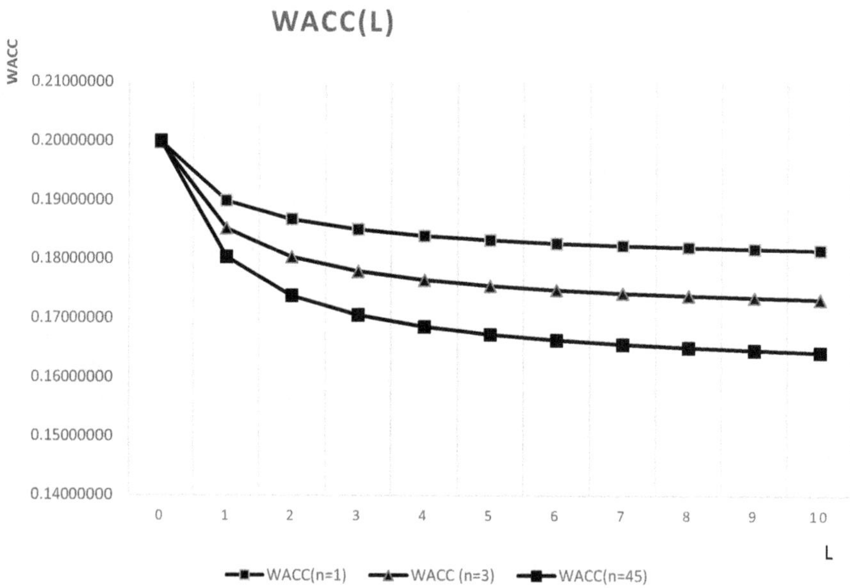

Fig. 21.3 The dependence of weighted average cost of capital, WACC, on the leverage level at different company age n ($n = 1, 3, 45$)

Fig. 21.4 The dependence of weighted average cost of capital, WACC, on the company age n at different leverage levels (L = 1, 2, 3)

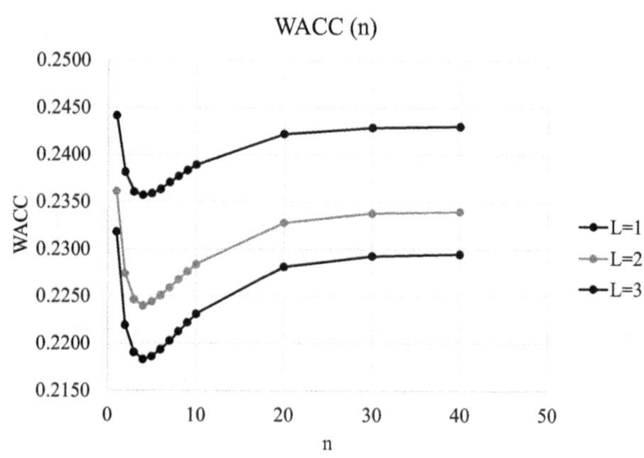

21.4 Companies with Abnormal "Golden Age" (Intermediate Difference Between k_0 and k_d Costs)

One example, which is different from two considered above cases will be studied below, where we present the calculations for equity cost k_0 (at L = 0) equals to 27% and debt cost k_d equals to 16%.

While in this case, the "golden age" of the company is present but is less pronounced: the minimal WACC value (at some leverage value: in this case at

Table 21.7 Dependence of WACC on the age of the company n at $L = 1$, $k_0 = 27\%$, $k_d = 25\%$

n	t	L	WACC	k_0	k_d	w_d	$A(n)$	BFO
1	0.2	1	0.2441	0.27	0.25	0.50	0.8035	0.000292
2	0.2	1	0.2381	0.27	0.25	0.50	1.4600	0.000038
3	0.2	1	0.2360	0.27	0.25	0.50	1.9928	0.000286
4	0.2	1	0.2357	0.27	0.25	0.50	2.4231	0.000046
5	0.2	1	0.2359	0.27	0.25	0.50	2.7688	0.000319
6	0.2	1	0.2363	0.27	0.25	0.50	3.0457	0.000877
7	0.2	1	0.2371	0.27	0.25	0.50	3.2668	0.000033
8	0.2	1	0.2377	0.27	0.25	0.50	3.4430	0.000070
9	0.2	1	0.2383	0.27	0.25	0.50	3.5830	0.000127
10	0.2	1	0.2389	0.27	0.25	0.50	3.6941	0.000204
20	0.2	1	0.2422	0.27	0.25	0.50	4.0755	0.000005
30	0.2	1	0.2429	0.27	0.25	0.50	4.1115	0.000009
40	0.2	1	0.2430	0.27	0.25	0.50	4.1149	0.000010

Table 21.8 Dependence of WACC on the age of the company n at $L = 2$, $k_0 = 27\%$, $k_d = 25\%$

n	t	L	WACC	k_0	k_d	w_d	$A(n)$	BFO
1	0.2	2	0.2361	0.27	0.25	0.67	0.8090	0.000024
2	0.2	2	0.2274	0.27	0.25	0.67	1.4784	0.000164
3	0.2	2	0.2246	0.27	0.25	0.67	2.0275	0.000411
4	0.2	2	0.2239	0.27	0.25	0.67	2.4748	0.000753
5	0.2	2	0.2244	0.27	0.25	0.67	2.8370	−0.000018
6	0.2	2	0.2251	0.27	0.25	0.67	3.1288	−0.000026
7	0.2	2	0.2259	0.27	0.25	0.67	3.3630	−0.000037
8	0.2	2	0.2267	0.27	0.25	0.67	3.5504	−0.000046
9	0.2	2	0.2276	0.27	0.25	0.67	3.6999	−0.000054
10	0.2	2	0.2284	0.27	0.25	0.67	3.8189	−0.000063
20	0.2	2	0.2328	0.27	0.25	0.67	4.2301	−0.000110
30	0.2	2	0.2338	0.27	0.25	0.67	4.2694	−0.000124
40	0.2	2	0.2340	0.27	0.25	0.67	4.2731	−0.000128

L = 1) lies above the perpetuity WACC value. We call this situation the "silver age" of the company (Table 21.13).

To be sure that the minimal WACC value at leverage level L = 1 lies above the perpetuity WACC value, we make more detailed calculations for this case (see Table 21.16). We see that the minimal WACC value at $n = 8.2$ is equal to 0.243095889, while perpetuity limit is equal 0.243 and lies below.

Let us study the dependence of weighted average cost of capital, WACC, on the leverage level at different company age n ($n = 1, 3, \infty$) (Fig. 21.7).

It is seen from Tables 21.14, 21.15 and Fig. 21.6 that the following behavior takes place for L = 2 and 3: a third modification of dependences of weighted average cost of capital, WACC, on the company age n takes place: descending of WACC with passage through minimum at $n = 8$, followed by a growth with passage through

Table 21.9 Dependence of WACC on the age of the company n at $L = 3$, $k_0 = 27\%$, $k_d = 25\%$

n	t	L	WACC	k_0	k_d	w_d	$A(n)$	BFO
1	0.2	3	0.2318	0.27	0.25	0.75	0.8118	0.000064
2	0.2	3	0.2219	0.27	0.25	0.75	1.4877	0.000412
3	0.2	3	0.2190	0.27	0.25	0.75	2.0453	−0.000020
4	0.2	3	0.2183	0.27	0.25	0.75	2.5015	−0.000044
5	0.2	3	0.2186	0.27	0.25	0.75	2.8723	−0.000076
6	0.2	3	0.2193	0.27	0.25	0.75	3.1721	−0.000112
7	0.2	3	0.2203	0.27	0.25	0.75	3.4133	−0.000154
8	0.2	3	0.2212	0.27	0.25	0.75	3.6067	−0.000195
9	0.2	3	0.2222	0.27	0.25	0.75	3.7612	−0.000234
10	0.2	3	0.2231	0.27	0.25	0.75	3.8845	−0.000271
20	0.2	3	0.2281	0.27	0.25	0.75	4.3120	−0.000497
30	0.2	3	0.2293	0.27	0.25	0.75	4.3530	−0.000570
40	0.2	3	0.2295	0.27	0.25	0.75	4.3569	−0.000590

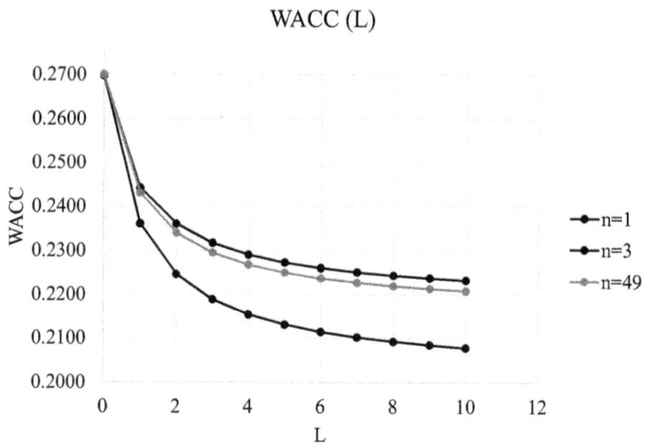

Fig. 21.5 The dependence of weighted average cost of capital, WACC, on leverage level L at different company age n ($n = 1, 3, 49$)

Table 21.10 Dependence of WACC on the leverage level at the company age $n = 1$, $k_0 = 27\%$, $k_d = 25\%$

n	t	L	WACC	k_0	k_d	w_d	$A(n)$	BFO
1	0.2	0	0.2697	0.27	0.25	0.00	0.7874	0.000188
1	0.2	1	0.2441	0.27	0.25	0.50	0.8035	0.000292
1	0.2	2	0.2360	0.27	0.25	0.67	0.8090	0.000064
1	0.2	3	0.2317	0.27	0.25	0.75	0.8118	0.000153
1	0.2	4	0.2290	0.27	0.25	0.80	0.8134	0.000216
1	0.2	5	0.2273	0.27	0.25	0.83	0.8146	0.000263
1	0.2	6	0.2260	0.27	0.25	0.86	0.8154	0.000298
1	0.2	7	0.2251	0.27	0.25	0.88	0.8160	0.000326
1	0.2	8	0.2243	0.27	0.25	0.89	0.8164	0.000349
1	0.2	9	0.2237	0.27	0.25	0.90	0.8168	0.000367
1	0.2	10	0.2232	0.27	0.25	0.91	0.8171	0.000382

Table 21.11 Dependence of WACC on the leverage level at the company age $n = 3$, $k_0 = 27\%$, $k_d = 25\%$

n	t	L	WACC	k_0	k_d	w_d	$A(n)$	BFO
3	0.2	0	0.2697	0.27	0.25	0.00	1.8956	0.000832
3	0.2	1	0.2361	0.27	0.25	0.50	1.9928	0.000174
3	0.2	2	0.2246	0.27	0.25	0.67	2.0275	0.000411
3	0.2	3	0.2188	0.27	0.25	0.75	2.0453	0.000542
3	0.2	4	0.2154	0.27	0.25	0.80	2.0561	0.000631
3	0.2	5	0.2131	0.27	0.25	0.83	2.0634	0.000694
3	0.2	6	0.2114	0.27	0.25	0.86	2.0687	0.000741
3	0.2	7	0.2102	0.27	0.25	0.88	2.0726	0.000778
3	0.2	8	0.2092	0.27	0.25	0.89	2.0757	0.000807
3	0.2	9	0.2084	0.27	0.25	0.90	2.0781	0.000830
3	0.2	10	0.2078	0.27	0.25	0.91	2.0802	0.000850

Table 21.12 Dependence of WACC on the leverage level at the company age $n = 49$, $k_0 = 27\%$, $k_d = 25\%$

n	t	L	WACC	k_0	k_d	w_d	$A(n)$	BFO
49	0.2	0	0.2700	0.27	0.25	0.00	3.7037	−0.000037
49	0.2	1	0.2430	0.27	0.25	0.50	4.1152	0.000041
49	0.2	2	0.2340	0.27	0.25	0.67	4.2735	0.000159
49	0.2	3	0.2295	0.27	0.25	0.75	4.3572	0.000273
49	0.2	4	0.2268	0.27	0.25	0.80	4.4091	0.000368
49	0.2	5	0.2250	0.27	0.25	0.83	4.4444	0.000444
49	0.2	6	0.2237	0.27	0.25	0.86	4.4699	0.000505
49	0.2	7	0.2227	0.27	0.25	0.88	4.4893	0.000556
49	0.2	8	0.2220	0.27	0.25	0.89	4.5045	0.000598
49	0.2	9	0.2214	0.27	0.25	0.90	4.5167	0.000633
49	0.2	10	0.2209	0.27	0.25	0.91	4.5267	0.000664

maximum at $n = 20$, and finally with trend to perpetuity limit from bigger values (remember that at second type of WACC(n) behavior, the curve WACC(n) tends to perpetuity limit from lower values). We have called this effect the "Kulik effect."

The ordering of curves is the following: the lower curve corresponds to the greater leverage level.

From Tables 21.16, 21.17, 21.18, 21.19 and Fig. 21.6, it is seen that for L = 1 the following behavior takes place: descending of WACC with passage through minimum at $n = 8.2$ (which is higher than perpetuity limit), followed by a growth with passage through maximum at $n = 20$, and finally with trend to perpetuity limit from bigger values. This means that the "golden age" in its purest form presents at leverage levels for L = 2 and 3, while at L = 1 one has different effect: we call it "silver age."

The ordering of curves is the following: the lower curve corresponds to the greater company age.

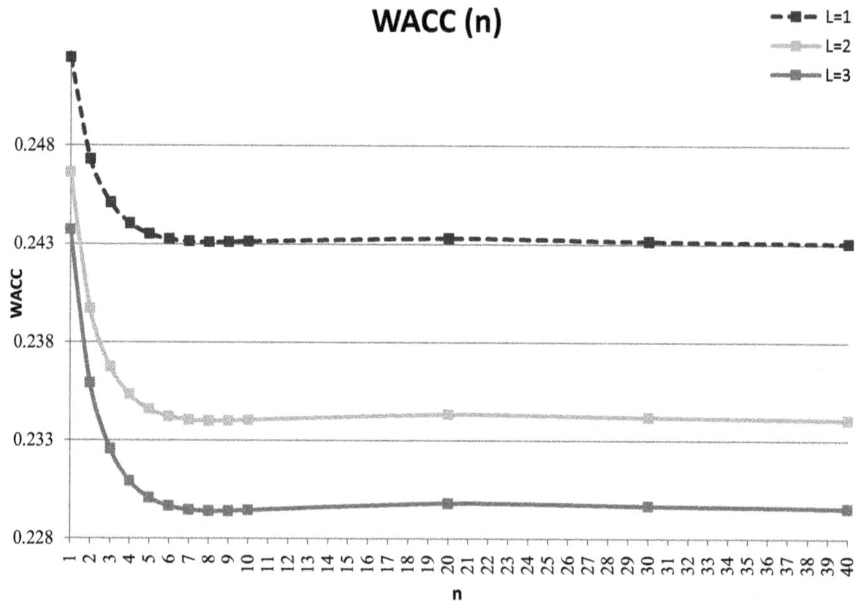

Fig. 21.6 The dependence of weighted average cost of capital, WACC, on the company age n at different leverage levels (L = 1, 2, 3)

Table 21.13 Dependence of WACC on the age of the company n at $L = 1$, $k_0 = 27\%$, $k_d = 16\%$

n	t	L	k_0	k_d	w_d	WACC
1	0.2	1	0.27	0.16	0.5	0.252483428
2	0.2	1	0.27	0.16	0.5	0.247301552
3	0.2	1	0.27	0.16	0.5	0.245105573
4	0.2	1	0.27	0.16	0.5	0.244045922
5	0.2	1	0.27	0.16	0.5	0.243511488
6	0.2	1	0.27	0.16	0.5	0.243250247
7	0.2	1	0.27	0.16	0.5	0.24313388
8	0.2	1	0.27	0.16	0.5	0.243097298
9	0.2	1	0.27	0.16	0.5	0.243103247
10	0.2	1	0.27	0.16	0.5	0.243130071
20	0.2	1	0.27	0.16	0.5	0.243291667
30	0.2	1	0.27	0.16	0.5	0.243146312
40	0.2	1	0.27	0.16	0.5	0.243048121
∞	0.2	1	0.27	0.16	0.5	0.243

It turns out that at particular values of capital costs, for example, at $k_0 = 27\%$; $k_d = 16\%$, a third modification of dependences of weighted average cost of capital, WACC, on the company age n takes place: descending of WACC with passage through minimum, followed by a growth with passage through maximum and finally with trend to perpetuity limit from bigger values (remember that at second type of

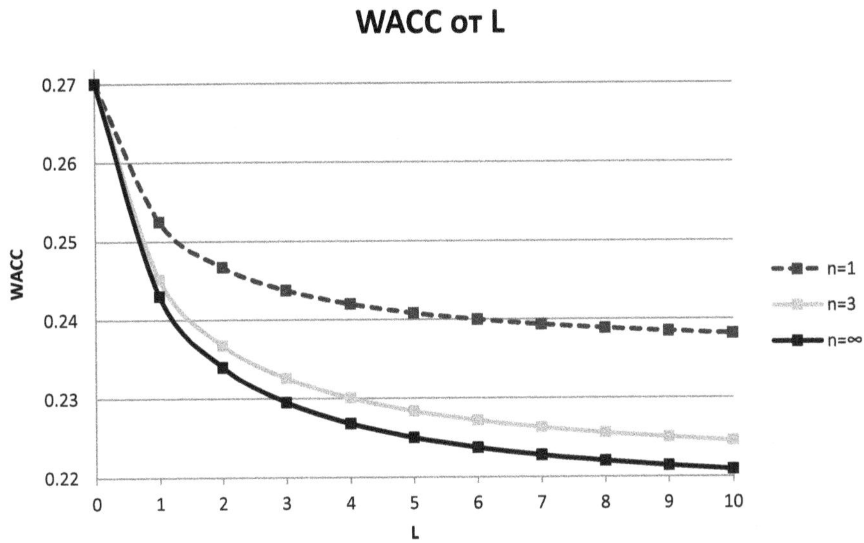

Fig. 21.7 The dependence of weighted average cost of capital, WACC, on the leverage level at different company age n ($n = 1,3,\infty$)

Table 21.14 Dependence of WACC on the age of the company n at $L = 2$, $k_0 = 27\%$, $k_d = 16\%$

n	t	L	k_0	k_d	w_d	WACC
1	0.2	2	0.27	0.16	0.6(6)	0.246644883
2	0.2	2	0.27	0.16	0.6(6)	0.239710388
3	0.2	2	0.27	0.16	0.6(6)	0.236752006
4	0.2	2	0.27	0.16	0.6(6)	0.235311829
5	0.2	2	0.27	0.16	0.6(6)	0.234578628
6	0.2	2	0.27	0.16	0.6(6)	0.23421254
7	0.2	2	0.27	0.16	0.6(6)	0.234046353
8	0.2	2	0.27	0.16	0.6(6)	0.233991091
9	0.2	2	0.27	0.16	0.6(6)	0.233996344
10	0.2	2	0.27	0.16	0.6(6)	0.234032536
20	0.2	2	0.27	0.16	0.6(6)	0.234316362
30	0.2	2	0.27	0.16	0.6(6)	0.234174539
40	0.2	2	0.27	0.16	0.6(6)	0.234059889
∞	0.2	2	0.27	0.16	0.6(6)	0.234

WACC(n) behavior, the curve WACC(n) tends to perpetuity limit from lower values). We have called this effect the "Kulik effect."

Table 21.15 Dependence of WACC on the age of the company n at $L = 3$, $k_0 = 27\%$, $k_d = 16\%$

n	t	L	k_0	k_d	w_d	WACC
1	0.2	3	0.27	0.16	0.75	0.243725045
2	0.2	3	0.27	0.16	0.75	0.235910011
3	0.2	3	0.27	0.16	0.75	0.232564477
4	0.2	3	0.27	0.16	0.75	0.230926014
5	0.2	3	0.27	0.16	0.75	0.230090888
6	0.2	3	0.27	0.16	0.75	0.229669501
7	0.2	3	0.27	0.16	0.75	0.229475982
8	0.2	3	0.27	0.16	0.75	0.229409857
9	0.2	3	0.27	0.16	0.75	0.229414009
10	0.2	3	0.27	0.16	0.75	0.229454135
20	0.2	3	0.27	0.16	0.75	0.229812629
30	0.2	3	0.27	0.16	0.75	0.229682914
40	0.2	3	0.27	0.16	0.75	0.229564113
∞	0.2	3	0.27	0.16	0.75	0.2295

Table 21.16 Dependence of WACC (more detailed) on the age of the company n at $L = 1$, $k_0 = 27\%$, $k_d = 16\%$

L	k_0	k_d	t	n	w_d	WACC	BFO
1	0.27	0.16	0.2	7	0.50	0.243133854	0.000
1	0.27	0.16	0.2	7.2	0.50	0.243121633	0.000
1	0.27	0.16	0.2	7.4	0.50	0.243112185	0.000
1	0.27	0.16	0.2	7.6	0.50	0.243105150	0.000
1	0.27	0.16	0.2	7.8	0.50	0.243100256	0.000
1	0.27	0.16	0.2	8	0.50	0.243097246	0.000
1	0.27	0.16	0.2	8.2	0.50	0.243095889	0.000
1	0.27	0.16	0.2	8.4	0.50	0.243095979	0.000
1	0.27	0.16	0.2	8.6	0.50	0.243097328	0.000
1	0.27	0.16	0.2	8.8	0.50	0.243099771	0.000
1	0.27	0.16	0.2	9	0.50	0.243103156	0.000
1	0.27	0.16	0.2	∞	0.50	0.243	0.000

Table 21.17 Dependence of WACC on the leverage level L at age of the company $n = 1$, $k_0 = 27\%$, $k_d = 16\%$

n	t	L	k_0	k_d	w_d	WACC
1	0.2	0	0.27	0.16	0	0.270000213
1	0.2	1	0.27	0.16	0.5	0.252483428
1	0.2	2	0.27	0.16	0.6(6)	0.246644883
1	0.2	3	0.27	0.16	0.75	0.243725045
1	0.2	4	0.27	0.16	0.8	0.241973361
1	0.2	5	0.27	0.16	0.8(3)	0.24080557
1	0.2	6	0.27	0.16	0.857142857	0.239971433
1	0.2	7	0.27	0.16	0.875	0.239345829
1	0.2	8	0.27	0.16	0.8(8)	0.238859248
1	0.2	9	0.27	0.16	0.9	0.238469982
1	0.2	10	0.27	0.16	0.(90)	0.238151492

Table 21.18 Dependence of WACC on the leverage level L at age of the company $n = 3$, $k_0 = 27\%$, $k_d = 16\%$

n	t	L	k_0	k_d	w_d	WACC
3	0.2	0	0.27	0.16	0	0.270000842
3	0.2	1	0.27	0.16	0.5	0.245105573
3	0.2	2	0.27	0.16	0.6(6)	0.236752006
3	0.2	3	0.27	0.16	0.75	0.232564477
3	0.2	4	0.27	0.16	0.8	0.230048473
3	0.2	5	0.27	0.16	0.8(3)	0.228369673
3	0.2	6	0.27	0.16	0.857142857	0.22716981
3	0.2	7	0.27	0.16	0.875	0.226269517
3	0.2	8	0.27	0.16	0.(8)	0.225569054
3	0.2	9	0.27	0.16	0.9	0.225008535
3	0.2	10	0.27	0.16	0.(90)	0.224549831

Table 21.19 Dependence of WACC on the leverage level L for the perpetuity company ($n = \infty$) at $k_0 = 27\%$, $k_d = 16\%$

n	t	L	k_0	k_d	w_d	WACC
∞	0.2	0	0.27	0.16	0	0.27
∞	0.2	1	0.27	0.16	0.5	0.243
∞	0.2	2	0.27	0.16	0.(6)	0.234
∞	0.2	3	0.27	0.16	0.75	0.2295
∞	0.2	4	0.27	0.16	0.8	0.2268
∞	0.2	5	0.27	0.16	0.8(3)	0.225
∞	0.2	6	0.27	0.16	0.857142857	0.223714286
∞	0.2	7	0.27	0.16	0.875	0.22275
∞	0.2	8	0.27	0.16	0.(8)	0.222
∞	0.2	9	0.27	0.16	0.9	0.2214
∞	0.2	10	0.27	0.16	0.(90)	0.220909091

21.5 Comparing with Results from Previous Chap. 20

21.5.1 Under Change of the Debt Capital Cost, k_d

From Fig. 21.8, it is seen that with increase of debt cost, k_d, the character of dependence of weighted average cost of capital, WACC, on the company age n is changed from monotonic descending of WACC with n to descending of WACC with n with passage through minimum, followed by a limited growth. It is seen from the Table 21.20 that the gap depth ΔWACC (the difference between the optimal (minimal) value of weighted average cost of capital, WACC, and its perpetuity limit) decreases with $\Delta k = k_0 - k_d$ from 3.38% at $\Delta k = 0.02$ up to 1.89% at $\Delta k = 0.5$. At $\Delta k = 0.10$ and $\Delta k = 0.12$ the minimum in dependence of WACC(n) is absent (too big value of $\Delta k = k_0 - k_d$). This coincides with our conclusions in this chapter.

The same conclusion could be made from Fig. 21.9 and Table 21.21 for higher values of capital costs: it is seen that with increase of debt cost, k_d at fixed k_0, i.e., with decrease $\Delta k = k_0 - k_d$ the gap depth ΔWACC is increased from 1.08% at

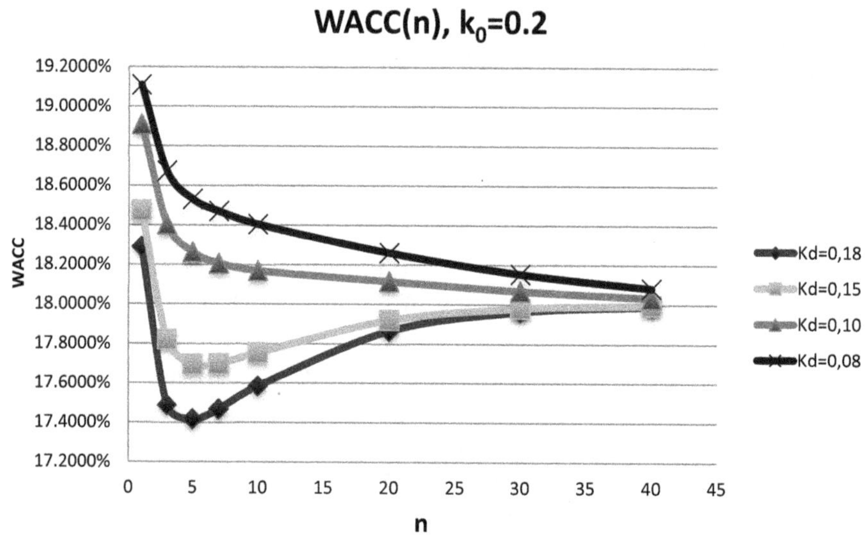

Fig. 21.8 Dependence of weighted average cost of capital, WACC, on lifetime of the company *n* at fixed value of equity cost, k_0 =20%, and at four values of debt cost, k_d =8%; 10%; 15%; and 18% at leverage level L = 1

Table 21.20 Dependence of the gap depth ΔWACC(the difference between the optimal (minimal) value of weighted average cost of capital, WACC, and its perpetuity limit) on $\Delta k = k_0 - k_d$at

k_d	0.18	0.15	0.10	0.08
$\Delta k = k_0 - k_d$	0.02	0.05	0.10	0.12
ΔWACC, %	3.38	1.89	NA	NA

$\Delta k = 0.10$ up to 1.85% at $\Delta k = 0.05$. This as well coincides with our conclusions in this paper.

21.5.2 *Under Change of the Equity Capital Cost,* k_0

From Fig. 21.10 and Table 21.22, it is seen that the gap depth ΔWACC (the difference between the optimal (minimal) value of weighted average cost of capital, WACC, and its perpetuity limit) decreases with $\Delta k = k_0 - k_d$ from 0.55% at $\Delta k = 0.01$ up to 0.03% at $\Delta k = 0.10$. This as well coincides with our conclusions in this paper.

In conclusion, we present in Fig. 21.11 both the cases of the "Kulik" effect: "the golden age" of the company and the "silver age" of the company.

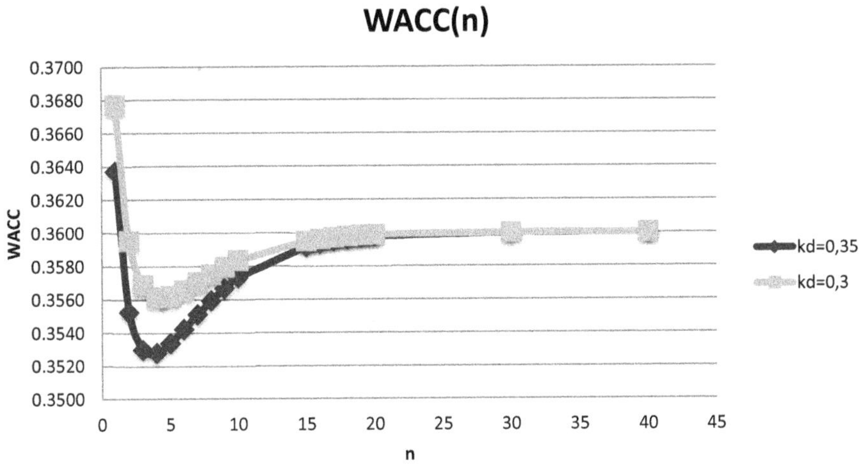

Fig. 21.9 Dependence of weighted average cost of capital, WACC, on lifetime of the company *n* at fixed high value of equity cost, k_0 =40%, and two values of debt cost, k_d =30% and 35% at leverage level L = 1

Table 21.21 Dependence of the gap depth ΔWACC(the difference between the optimal (minimal) value of weighted average cost of capital, WACC, and its perpetuity limit) on $\Delta k = k_0 - k_d$at

$k_0 = 0,4; k_d$	0.35	0.3
$\Delta k = k_0 - k_d$	0.05	0.10
ΔWACC, %	1.85	1.08

21.6 Conclusions

In our previous paper a few years ago [1, 2], we have discovered the effect of the "golden age" of a company: it was shown for the first time that valuation of the weighted average cost of capital, WACC, in the Modigliani–Miller theory [3–5] is not minimal and valuation of the company capitalization is not maximal, as all financiers have supposed up to this discovery: at some age of the company, its WACC value turns out to be lower than in the Modigliani–Miller theory and company capitalization V turns out to be greater than V in Modigliani–Miller theory (see previous chapter). It was shown that, from the point of view of cost of attracting capital, there are two types of dependences of weighted average cost of capital, WACC, on the company age *n*: monotonic descending with n and descending with passage through minimum, followed by a limited growth. In practice, there are companies with both types of dependences of WACC on the company age *n*.

In this chapter, we have continued the study of the effect of the "golden age" of a company and have investigated which companies have the "golden age," i.e., obey the latter type of dependence of WACC on n. With this aim, we study the dependence of WACC on the age of company n at various leverage levels within a wide

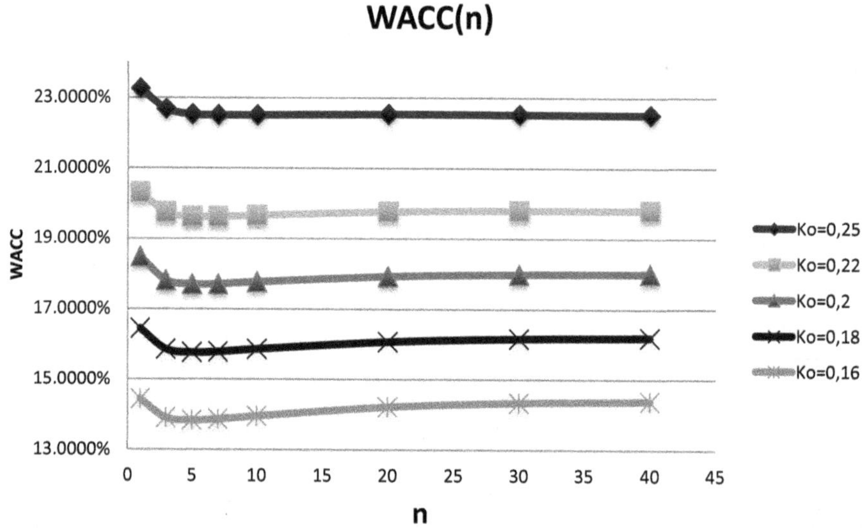

Fig. 21.10 Dependence of weighted average cost of capital, WACC, on lifetime of the company n at fixed value of debt cost, $k_d = 15\%$, and five values of equity cost, $k_0 = 16\%$; 18%; 20%; 22%; and 25% at leverage level L = 1

Table 21.22 Dependence of the gap depth ΔWACC(the difference between the optimal (minimal) value of weighted average cost of capital, WACC, and its perpetuity limit) on $\Delta k = k_0 - k_d$at

k_0	0.16	0.18	0.20	0.22	0.25
$\Delta k = k_0 - k_d$	0.01	0.03	0.05	0.07	0.10
ΔWACC, %	0.55	0.43	0.30	0.18	0.03

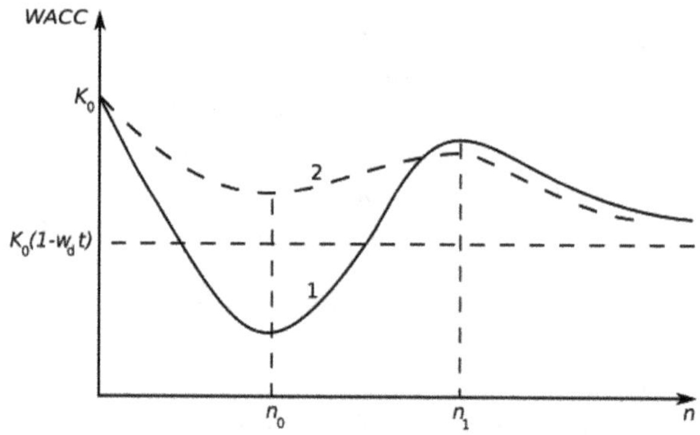

Fig. 21.11 Dependence of weighted average cost of capital, WACC, on company age of the company n, which illustrates the presence of "the golden age" of the company (curve 1) and of "the silver age" of the company (curve 2) under existence of the "Kulik" effect. Here n_0 is "the golden (silver) age" of the company and n_1 is the age of local maximum in dependence of WACC(n)

spectrum of capital costs values as well as the dependence of WACC on leverage level L at fixed company age n. All calculations have been done within the modern theory of capital cost and capital structure BFO by Brusov–Filatova–Orekhova [7–17].

We have shown that existence of the "golden age" of a company does not depend on the value of capital costs of the company (as it was supposed in previous chapter) but depends on the difference value between equity, k_0, and debt, k_d, costs. The "golden age" of a company exists at small enough difference between k_0 and k_d costs, while at high value of this difference, the "golden age" of company is absent: curve WACC(n) monotonic descends with company age n. For the companies with the "golden age" curve, WACC(L) for perpetuity limit lies between curves WACC (L) at $n = 1$ and $n = 3$, while for the companies without the "golden age" curve, WACC(L) for perpetuity limit ($n = \infty$) is the lowest one. In other words, the ordering of curves WACC(L) is different for the companies with the "golden age" and without it.

In the previous chapter, we have found also a third type of WACC (n) dependence: descending with passage through minimum, which lies below the perpetuity limit value, then going through maximum, followed by a limited descent. We called this effect the "Kulik effect" (this is the last name of the student who discovered this effect). In this paper, we have found a variety of the "Kulik effect": descending with passage through minimum of WACC, which lies above the perpetuity limit value, then going through maximum, followed by a limited descent. We call this company age n, at which WACC has a minimum, which lies above the perpetuity limit value, "the silver age" of the company. It takes place at intermediate difference value between equity k_0 and debt k_d costs.

Because the cost of attracting capital is used in rating methodologies as discounting rate under discounting of cash flows, the study of WACC behavior is very important for rating procedures. The account of effects of the "golden (silver) age" could change the valuation of creditworthiness of issuers.

Remember that since the "golden age" of company depends on the company's capital costs, by controlling them (for example, by modifying the value of dividend payments, that reflect the equity cost), a company may extend the "golden (silver) age" of the company, when the cost to attract capital becomes a minimal (less (higher) than perpetuity limit), and capitalization of companies becomes maximal (above (lower) than perpetuity assessment) up to a specified time interval.

References

1. Brusov, P., Filatova, T., Orekhova, N., & Eskindarov, M. (2015). *Modern corporate finance, investments and taxation, monograph* (1st ed. 368 p). Springer.
2. Brusov, P., Filatova, T., Orekhova, N., & Kulik, V. (2015). The Golden age of the company: (three Colors of company's time). *Journal of Reviews on Global Economics, 4*, 21–42.
3. Modigliani, F., & Miller, M. (1958). The cost of capital, corporate finance, and the theory of investment. *American Economic Review, 48*, 261–297.

4. Modigliani, F., & Miller, M. (1963). Corporate income taxes and the cost of capital: A correction. *American Economic Review, 53*, 147–175.
5. Modigliani, F., & Miller, M. (1966). Some estimates of the cost of capital to the electric utility industry 1954–1957. *American Economic Review, 56*, 333–391.
6. Brusov, P., Filatova, T., Orekhova, N., & Kulk, V. (2018). A "Golden Age" of the companies: Conditions of its existence. *Journal of Reviews on Global Economics, 7*, 88–103.
7. Brusov, P., Filatova, T., Orehova, N., & Brusova, A. (2011). Weighted average cost of capital in the theory of Modigliani–Miller, modified for a finite life–time company. *Applied Financial Economics, 21*(11), 815–824.
8. Brusov, P., Filatova, P., & Orekhova, N. (2013). Absence of an optimal capital structure in the famous Tradeoff theory! *Journal of Reviews on Global Economics, 2*, 94–116.
9. Brusov, P., Filatova, P., & Orekhova, N. (2014). Mechanism of formation of the company optimal capital structure, different from suggested by trade off theory. *Cogent Economics & Finance, 2*, 1–13. https://doi.org/10.1080/23322039.2014.946150
10. Brusov, P., Filatova, T., Orehova, N., et al. (2011). From Modigliani–Miller to general theory of capital cost and capital structure of the company. *Research Journal of Economics, Business and ICT, 2*, 16–21.
11. Brusov, P., Filatova, T., Eskindarov, M., & Orehova, N. (2012). Influence of debt financing on the effectiveness of the finite duration investment project. *Applied Financial Economics, 22*(13), 1043–1052.
12. Brusov, P., Filatova, T., Orehova, N., et al. (2011). Influence of debt financing on the effectiveness of the investment project within the Modigliani–Miller theory. *Research Journal of Economics, Business and ICT, 2*, 11–15.
13. Brusov, P., Filatova, T., Eskindarov, M., & Orehova, N. (2012). Hidden global causes of the global financial crisis. *Journal of Reviews on Global Economics, 1*, 106–111.
14. Brusov, P., Filatova, T., Orehova, N., Brusov, P. P., & Brusova, N. (2011). From Modigliani–Miller to general theory of capital cost and capital structure of the company. *Research Journal of Economics, Business and ICT, 2*, 16–21.
15. Brusov, P., Filatova, T., & Orehova, N. (2014). Inflation in Brusov–Filatova–Orekhova theory and in its perpetuity limit—Modigliani—Miller theory. *Journal of Reviews on Global Economics, 3*, 175–185.
16. Brusov, P., Filatova, T., & Orehova, N. (2013). A qualitatively new effect in corporative finance: Abnormal dependence of cost of equity of company on leverage. *Journal of Reviews on Global Economics, 2*, 183–193.
17. Filatova, T., Orehova, N., & Brusova, A. (2008). Weighted average cost of capital in the theory of Modigliani–Miller, modified for a finite life–time company. *Bulletin of the FU, 48*, 68–77.
18. Brusov, P., Filatova, T., Orekhova, N., & Kulk, W. I. (2018). New meaningful effects in modern capital structure theory. *Journal of Reviews on Global Economics, 7*, 104–122.
19. Brusov, P., Filatova, T., Orekhova, N., & Kulk, V. (2018). Rating methodology: New look and new horizons. *Journal of Reviews on Global Economics, 7*, 63–87.
20. Brusov, P., Filatova, T., Orekhova, N., & Kulk, V. (2018). Rating: New approach. *Journal of Reviews on Global Economics, 7*, 37–62.
21. Brusov, P. N., & Filatova, T. V. (2011). From Modigliani–Miller to general theory of capital cost and capital structure of the company. *Finance and credit, 435*, 2–8.
22. Brusova, A. (2011). A comparison of the three methods of estimation of weighted average cost of capital and equity cost of company. *Financial Analysis: Problems and Solutions, 34*(76), 36–42.
23. Myers, S. (1984). The capital structure Pussle. *Journal of Finance, 39*(3), 574–592.

Chapter 22
Application of the Company's "Golden Age" Effect in the Economic Practice

22.1 Introduction

In this chapter, one of the most striking effects in financial management in general and in the capital structure in particular—"golden age" of the company (when the cost of capital raised is below the perpetuity limit, and the value of the company is higher)—is examined.

The research hypotheses are as follows: the "golden age" depends on the company's financial performance. It can be changed and controlled by changing such parameters as the cost of capital (own and borrowed), the frequency and method of tax on income payments, the growth of the income rate, etc. The purpose of the study is to investigate the influence of the conditions of the real functioning of companies on the effect of the "golden age": on its existence and management by it. Research methods used are as follows: the generalized Brusov–Filatova–Orekhova (BFO theory) and calculations within this theory; the dependence of WACC on company age by Microsoft Excel at different financial parameters. The motivation of research is to study the most bright effect in financial management—"the golden age" effect. Significance of current study is determined by the importance of "the golden age" effect, which reduces the cost of raising capital and increases the value of the company.

The modern theory of the capital structure started from the papers by Modigliani and Miller [1–3]. The Modigliani–Miller theory had numerous limitations, the main of which were its perpetual nature and the absence of taxes: corporate and individual. Many scientists have tried to modify the Modigliani–Miller theory in different aspects [4–16].

P. Brusov, T. Filatova, *Business Valuation and Company Value*, Contributions to Finance and Accounting, https://doi.org/10.1007/978-3-031-90847-7_22

22.1.1 Unification of Capital Asset Pricing Model (CAPM) with Modigliani–Miller Model

Unification of Capital Asset Pricing Model (CAPM) with Modigliani–Miller model with accounting taxes has been done in 1961 by Hamada [4]. He has derived the following formula for equity cost of the levered company:

$$k_e = k_F + (k_M - k_F)b_U + (k_M - k_F)b_U \frac{D}{S}(1 - T),$$

Here b_U is the β–coefficient of the unlevered company. First term represents risk-free profitability k_F, second term—business risk premium, $(k_M - k_F)b_U$, and third term—financial risk premium $(k_M - k_F)b_U \frac{D}{S}(1 - T)$.

In the case of an unlevered company ($D = 0$), the financial risk (the third term) is zero, and its shareholders receive only a business risk premium.

See, however, the conclusive remarks, that show that Hamada's formulas are incorrect.

22.1.2 Miller Model

Miller [5] has accounted the corporate and individual taxes and has gotten the following formula for the value of a company without borrowed funds, V_U,

$$V_U = \frac{\text{EBIT}(1 - T_C)(1 - T_S)}{k_0}.$$

Here T_C is the corporate tax on income rate, T_S is the tax rate on profits of an individual investor from his ownership by stock of corporation, T_D–tax rate on interest profits from the provision of investor–individuals of credits to other investors and companies. A factor $(1 - T_S)$ accounts the individual taxes.

22.1.3 Alternative Expression for WACC

From the WACC definition and the balance identity (see [6]) an alternative formula for the WACC, different from the Modigliani–Miller one, has been derived in [6–9]:

$$WACC = k_0(1 - w_d T) - k_d t w_d + k_{TS} t w_d,$$

where k_0, k_d, and k_{TS} are the expected returns, respectively, on the unlevered company, the debt, and the tax shield.

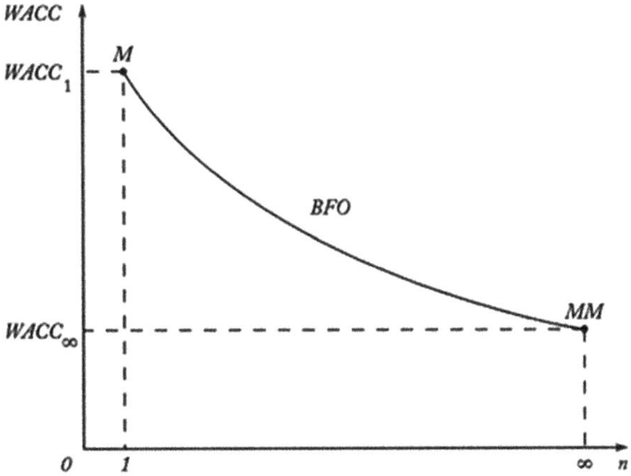

Fig. 22.1 Monotonic dependence of WACC on the age of the company n ($WACC_1$ is the WACC value for one-year company and $WACC_\infty$ is the WACC value for perpetual case)

Becker [12] discussed the difference between Modigliani–Miller and Miles–Ezzell and its consequences for the valuation of annuities.

In textbooks [13–15], formulas for the special cases, where the WACC is constant, could be found.

Myers [17] considered one–year companies and have shown, that the weighted average cost of capital WACC is higher than in the Modigliani–Miller limit, and the company's capitalization is lower. Myers concluded that valuation of the weighted average cost of capital, WACC, in the perpetual limit of the Modigliani–Miller theory is the lowest, and the valuation of the company value, V, is maximum. This means that WACC decreases monotonically with company age n (see Fig. 22.1).

One of the main limitations of the Modigliani–Miller theory on the perpetuity of companies was removed by Brusov et al. in 2008 [18], and modern theories of the cost of capital and capital structure—Brusov–Filatova–Orekhova theory (BFO theory) were created for companies of arbitrary age (BFO-1 theory) and for companies of arbitrary lifetime (BFO-2 theory) [19].

Brusov–Filatova–Orekhova (BFO) theory replaced the well–known theory of the cost of capital and capital structure of Nobel laureates Modigliani and Miller. The authors departed from the Modigliani–Miller assumption about the eternity (infinity of life) of companies and additionally developed a quantitative theory for estimating the main parameters of the financial activity of companies with an arbitrary lifetime.

The results of the modern BFO theory turn out to be quite different from the results of the Modigliani–Miller theory. They show that the latter, through its perpetuity, underestimates the weighted average cost of capital, the cost of equity of the company, and significantly overestimates the company's value.

Such an incorrect assessment of key performance indicators of companies' financial performance led to an underestimation of the associated risks and the impossibility or serious difficulties in making adequate management decisions, which was one of the implicit causes of the 2008 global financial crisis (for more details see [19].

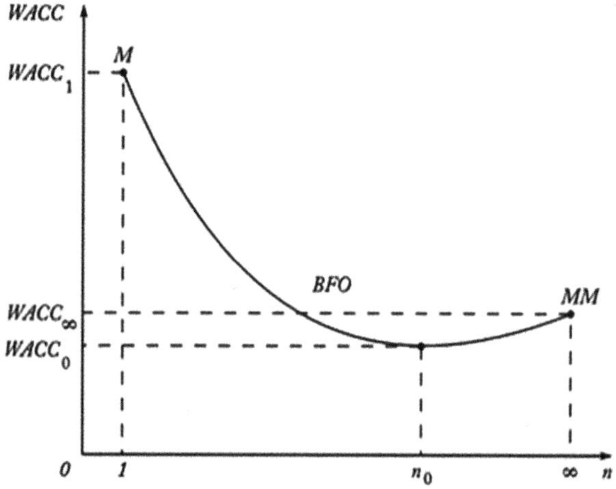

Fig. 22.2 Dependence of WACC on the company age *n*, showing descending of WACC with passage through minimum and then showing a limited growth to perpetuity (MM) limit

In the Modigliani–Miller theory, there is no time factor (time is equal to infinity), which does not allow to study the dependence of the company's financial performance on the time factor. But Brusov–Filatova–Orekhova theory (BFO theory) was created for companies of arbitrary age and allows to study the dependence of the company's financial performance on the time factor. Brusov et al. in 2015 [20] studied the dependence of the cost of raising capital WACC on the age of the company n at different levels of debt load, at different values of the cost of equity and borrowed capital in order to determine the minimum cost of raising capital. It was shown that there are two types of dependence of WACC on company age n: a monotonic decrease in WACC with n and a decrease in WACC with passage through minimum (which is called "the golden age" of the company), followed by a limited increase (Fig. 22.2). Companies with the latter type of dependence of WACC on the age of the company n can take advantage of the discovered effect at a certain stage of development. Thus, in [20], for the first time, it was concluded that the WACC valuation in the Modigliani and Miller (MM) theory is not minimal, and the valuation of the company is not the maximum: the "golden age" of the company exists. This conclusion seems to be very significant and important. The study of the properties of the "golden age" effect was continued in [21].

Thus, generally speaking, the conclusion by Myers [17] that valuation of the weighted average cost of capital, WACC, in the Modigliani–Miller theory is the lowest, and the valuation of the company value, V, is maximum turns out to be wrong, and in the life of a company there is a "golden age," when the cost of raising capital becomes minimal (less than perpetual limit) and company value becomes maximal (bigger than perpetual limit). Moreover, since a company's golden age depends on the company's capital costs, ke and kd, by controlling them (for example, by changing the amount of dividend payments that reflect the cost of equity, etc.), the company can prolong its golden age, when the cost of raising capital becomes minimal (below the perpetuity limit), and the capitalization of companies becomes maximum (above the perpetuity estimate) within a given time interval.

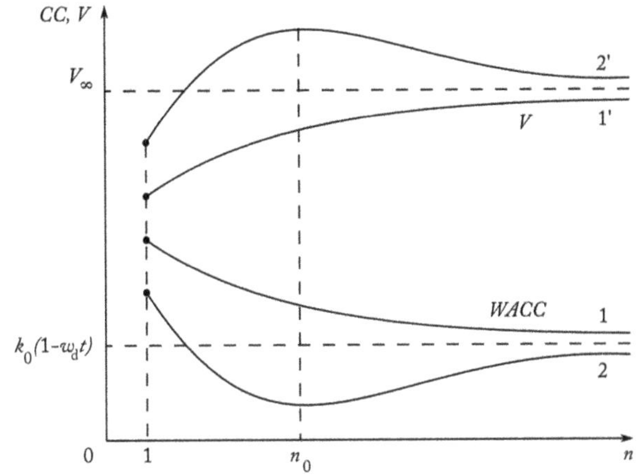

Fig. 22.3 Two kinds of dependences of WACC and company capitalization, *V*, on the age of the company *n*

For a more detailed description of the two types of dependence of WACC on company age *n*, see Fig. 22.3.

Note: 1–1′—monotonic descending of WACC and monotonic increase of company value, *V*, with the age of the company *n*; 2–2′—descending of WACC with passage through minimum and then showing a limited growth, and increase of company value, *V*, with passage through maximum (at n_0) and then a limited descending to perpetuity (MM) limit.

Over the past few years, the two main theories of capital structure—Brusov–Filatova–Orekhova and Modigliani–Miller—have been adapted to the established financial practice of the functioning of companies, taking into account the real conditions of their work (see [22] and references there). This made it possible to investigate the impact of frequent income tax payments p with advance income tax payments and payments at the end of reporting periods, as well as the impact of a company's variable income on its main financial performance.

In [22], an analysis of all existing theories of the capital structure (with their advantages and disadvantages) was carried out in order to understand all aspects of the problem and make the right management decisions in practice. The role of the capital structure lies in the fact that the correct determination of the optimal capital structure allows the company's management to maximize the capitalization of the company and fulfill the long-term goal of the functioning of any company. [22] examines the state of the theory of the structure of capital and the cost of capital from the middle of the last century, when the first quantitative theory was created, to the present. The two main theories of Modigliani–Miller (MM) and Brusov–Filatova–Orekhova (BFO) are discussed and analyzed, as well as their numerous modifications and generalizations.

In the current paper, the effect of the "golden age" of the company is investigated, taking into account the conditions of the real functioning of companies. With this aim the dependence of cost of raising capital, WACC, on the age of company, *n*, is studied at various leverage levels, *L*, at various values of equity, ke, and debt, kd, costs, at different frequencies of tax on income payments, p, with advance payments of tax on income and payments at the ends of periods, at variable income of the companies. Important conclusions were made about the "golden age" effect and recipes for managing the effect and recommendations for the company's management in this respect have been developed. The limitation of the study is due to the fact that we are considering the case of a constant income growth rate. In future publications, the case of a variable income growth rate will be considered.

22.2 Theoretical Basis

The following notation will be used in the text below.

k_d, $w_{\mathrm{d}} = \frac{D}{D+S}$—the debt capital cost and debt capital share; $k_{\mathrm{e}}, w_{\mathrm{e}} = \frac{S}{D+S}$—the equity capital cost and the equity capital share; and $L = D/S$—the value of financial leverage; D—the debt capital value; S—the equity capital value; k0— the equity capital cost at zero leverage level; g—growth rate; p—frequency of tax on income payments; WACC—the weighted average cost of capital; t—tax on profit; n—company age.

Over the past few years, the two main theories of the capital structure (Brusov–Filatova–Orekhova [BFO] and Modigliani–Miller [MM]) have been adapted by the authors to the established financial practice of the functioning of companies, taking into account the real conditions of their work. They are generalized to the case of variable income (this is extremely important), as well as to the case of paying income tax with arbitrary frequency, to the case of advance payments of income tax and for the combinations of these effects [22]. Account of these effects has changed the results of both theories significantly and made the Modigliani–Miller theory (which is perpetual limit of Brusov–Filatova–Orekhova) closer to Brusov–Filatova–Orekhova one, although they will never intersect, since the MM theory does not have a time factor, and the BFO describes companies of arbitrary age.

In this paper, we use the generalized Brusov–Filatova–Orekhova (BFO) theory as well as the generalized Modigliani–Miller theory to study the existence of the minimum in weighted average cost of capital, WACC, and its behavior in a wide range of above parameters.

The basic BFO equation for WACC (before accounting for the above-mentioned effects of variable income, frequent paying income tax, advance payments of income tax, and the combinations of these effects) has the following form:

$$\frac{1-(1+\mathrm{WACC})^{-n}}{\mathrm{WACC}-g} = \frac{1-(1+k_0)^{-n}}{(k_0-g)\cdot(1-w_d t[1-(1+k_d)^{-n}])} \tag{22.1}$$

as well as its limit for perpetuity companies (MM limit)

$$\mathrm{WACC} = k_0 \cdot (1 - w_d t) \tag{22.2}$$

Formula (22.2) could be easily obtained from (22.1) putting $n = \infty$. For the derivation of these formulas, their meaning and application, see [19].

The Brusov–Filatova–Orekhova (BFO) theory, its methodology, and results are well known (see, for example, [23–32]). A lot of authors of [27–29] use the BFO theory in practice.

Below, we give a summary of the WACC formulas for Brusov–Filatova–Orekhova (BFO) theory as well as for Modigliani–Miller (MM) theory [22] adapted by the authors to the established financial practice of the functioning of companies, taking into account the real conditions of their work.

22.2.1 Variable Income Case

22.2.1.1 Income Tax Payments at the Ends of Periods

The Brusov–Filatova–Orekhova equation for WACC for the case of variable income with income tax payments at the ends of periods takes the following form:

$$\mathrm{BFO}: \frac{1 - \left(\frac{1+g}{1+\mathrm{WACC}}\right)^n}{\mathrm{WACC} - g} = \frac{1 - \left(\frac{1+g}{1+k_0}\right)^n}{(k_0 - g) \cdot (1 - w_d t[1 - (1 + k_d)^{-n}])} \tag{22.3}$$

The Modigliani–Miller equation for WACC for the case of variable income with income tax payments at the ends of periods takes the following form:

$$\mathrm{MM}: \mathrm{WACC} = (k_0 - g) \cdot (1 - w_d t) + g \tag{22.4}$$

22.2.1.2 Advance Income Tax Payments

The Brusov–Filatova–Orekhova equation for WACC for the case of variable income with advance income tax payments takes the following form:

$$\mathrm{BFO}: \frac{1 - \left(\frac{1+g}{1+\mathrm{WACC}}\right)^n}{\mathrm{WACC} - g} = \frac{1 - \left(\frac{1+g}{1+k_0}\right)^n}{(k_0 - g) \cdot (1 - w_d t[1 - (1 + k_d)^{-n}] \cdot (1 + k_d))} \tag{22.5}$$

The Modigliani–Miller equation for WACC for the case of variable income with advance income tax payments takes the following form:

$$\mathrm{MM}:\mathrm{WACC}=(k_0-g)\cdot(1-w_d t\cdot(1+k_d))+g \tag{22.6}$$

22.2.2 *Frequent Income Tax Payments*

22.2.2.1 Income Tax Payments at the Ends of Periods

The Brusov–Filatova–Orekhova equation for WACC for the case of frequent payments of income tax at the ends of periods takes the following form:

$$\mathrm{BFO}:\frac{1-(1+\mathrm{WACC})^{-n}}{\mathrm{WACC}}=\frac{1-(1+k_0)^{-n}}{k_0\cdot\left(1-\frac{k_d w_d t}{p}\frac{\left[1-(1+k_d)^{-n}\right]}{(1+k_d)^{\frac{1}{p}}-1}\right)} \tag{22.7}$$

The Modigliani–Miller equation for WACC for the case of frequent payments of income tax at the ends of periods takes the following form:

$$\mathrm{MM}:\mathrm{WACC}=k_0\cdot\left(1-\frac{k_d w_d t}{p\cdot\left[(1+k_d)^{\frac{1}{p}}-1\right]}\right) \tag{22.8}$$

22.2.2.2 Advance Income Tax Payments

The Brusov–Filatova–Orekhova equation for WACC for the case of advanced frequent payments of income tax takes the following form:

$$\mathrm{BFO}:\frac{1-(1+\mathrm{WACC})^{-n}}{\mathrm{WACC}}=\frac{1-(1+k_0)^{-n}}{k_0\cdot\left(1-\frac{k_d w_d t}{p}\frac{\left[1-(1+k_d)^{-n}\right]\cdot(1+k_d)^{\frac{1}{p}}}{(1+k_d)^{\frac{1}{p}}-1}\right)} \tag{22.9}$$

The Modigliani–Miller equation for WACC for the case of advanced payments of income tax takes the following form:

$$\mathrm{MM}:\mathrm{WACC}=k_0\cdot\left(1-\frac{k_d w_d t\cdot(1+k_d)^{\frac{1}{p}}}{p\cdot\left[(1+k_d)^{\frac{1}{p}}-1\right]}\right) \tag{22.10}$$

22.2.3 *Simultaneous Accounting of Variable Income in Case of Frequent Income Tax Payments*

The Brusov–Filatova–Orekhova equation for WACC for the case of simultaneous accounting of variable income in case of frequent income tax payments at the ends of periods takes the following form:

22.2.3.1 Income Tax Payments at the Ends of Periods

$$\mathrm{BFO}:\frac{1-\left(\frac{1+g}{1+\mathrm{WACC}}\right)^{n}}{\mathrm{WACC}-g}=\frac{1-\left(\frac{1+g}{1+k_0}\right)^{n}}{(k_0-g)\cdot\left(1-\frac{k_d w_d t}{p}\frac{\left[1-(1+k_d)^{-n}\right]}{(1+k_d)^{1/p}-1}\right)} \tag{22.11}$$

The Modigliani–Miller equation for WACC for the case of simultaneous accounting of variable income in case of frequent income tax payments at the ends of periods takes the following form:

$$\mathrm{MM}:\mathrm{WACC}-g=(k_0-g)\cdot\left(1-\frac{k_d w_d t}{p\cdot(1+k_d)^{1/p}-1}\right) \tag{22.12}$$

22.2.3.2 Advance Income Tax Payments

The Brusov–Filatova–Orekhova equation for WACC for the case of simultaneous accounting of variable income in case of advance frequent income tax payments takes the following form:

$$\mathrm{BFO}=\frac{1-\left(\frac{1+g}{1+\mathrm{WACC}}\right)^{n}}{\mathrm{WACC}-g}=\frac{1-\left(\frac{1+g}{1+k_0}\right)^{n}}{(k_0-g)\cdot\left(1-\frac{k_d w_d t}{p}\frac{\left[1-(1+k_d)^{-n}\right]\cdot(1+k_d)^{1/p}}{\left[(1+k_d)^{1/p}-1\right]}\right)} \tag{22.13}$$

The Modigliani–Miller equation for WACC for the case of simultaneous accounting of variable income in case of advance frequent income tax payments takes the following form:

$$\text{MM} : \text{WACC} - g = (k_0 - g) \cdot \left(1 - \frac{k_d w_d t \cdot (1 + k_d)^{1/p}}{p \cdot (1 + k_d)^{1/p} - 1}\right) \tag{22.14}$$

22.2.4 *Results*

To find out the possibility of managing the golden age of the company: its existence and the possibility of shifting in time and extending the duration of the effect, below we investigate the dependence of cost of raising capital, WACC, on the age of company, n, at various values of equity and debt costs, at different frequencies of tax on income payments, p, with advance payments of tax on income and payments at the ends of periods, at variable income of the companies and at various leverage levels, L.

In all variants of the Modigliani–Miller theory, the time factor is absent, therefore, when studying the dependence of WACC on n, we work in the framework of the Brusov–Filatova–Orekhova theory. The Modigliani–Miller theory is used by us only to estimate the limiting eternity values of WACC.

Regarding the study of the impact of the cost of debt, k_d, on dependence WACC on company age, n. In classical Modigliani–Miller theory, WACC does not depend on debt cost, k_d, as it can be seen from the standard Modigliani–Miller formula (22.2). But as one can see from formulas (22.6), (22.8), (22.10), (22.12) in Modigliani–Miller theory, modified by Brusov et al. [22], WACC depends on k_d.

22.2.5 *Frequent Income Tax Payments*

22.2.5.1 Income Tax Payments at the Ends of Periods

As it is seen from Table 22.1, the difference of WACC minimum, Δ, between values at $p = 1$ and $p = 12$ increases with the debt cost, kd, as well as the difference between perpetuity values at $p = 1$ and $p = 12$. Note that the first value is greater than the second (Figs. 22.4, 22.5, 22.6, 22.7, 22.8, and 22.9).

Gap depth WACC(n) increases with the debt cost, k_d, from 0.65% up to 1.41% at monthly payments of tax on income and from 0.58% up to 1.23% at annual payments of tax on income. At all debt cost, k_d, values gap depth WACC(n) at monthly payments of tax on income is bigger than at annual payments of tax on income.

Table 22.1 The dependence of difference of WACC minimum, between values at $p = 1$ and $p = 12$ and between perpetuity values at $p = 1$ and $p = 12$ on debt cost $k_d = 0.24;0.28;0.30$

$k_d = 0.30$				
$p = 1$	%	25.96	27.19	1.23
$p = 12$	%	25.15	26.56	1.41
Δ	%	0.81	0.63	
k_d**=0.28**				
$p = 1$	%	26.17%	27.19	1.02
$p = 12$	%	25.44%	26.60	1.16
Δ	%	0.73%	0.59	
k_d**=0.24**				
$p = 1$	%	26.62	27.20	0.58
$p = 12$	%	26.04	26.69	0.65
Δ	%	0.58	0.51	

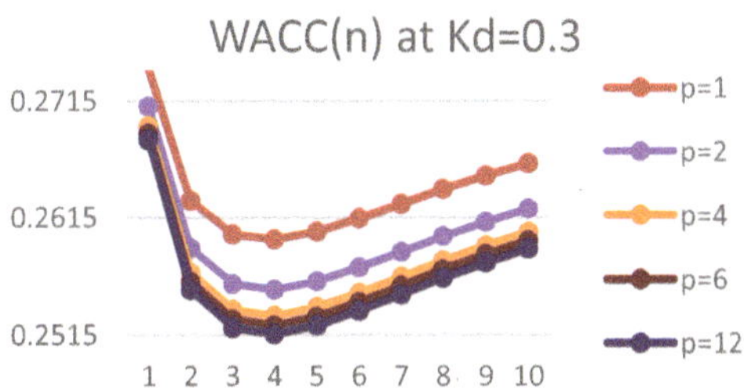

Fig. 22.4 The dependence of WACC on company age, n, with frequent payments of tax on profit $p = 1; 2; 3; 6; 12$ at $k_0 = 0.32$; $k_d = 0.3$; $t = 0.2$ (n from 1 to 10 years)

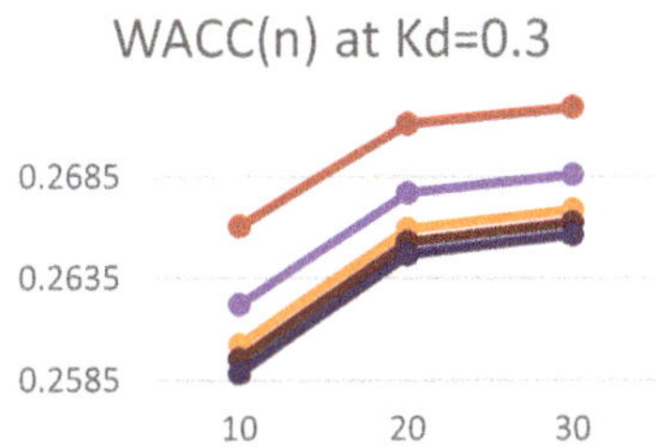

Fig. 22.5 The dependence of WACC on company age, n, with frequent payments of tax on profit $p = 1; 2; 3; 6; 12$ at $k_0 = 0.32$; $k_d = 0.3$; $t = 0.2$ (n from 10 to 30 years)

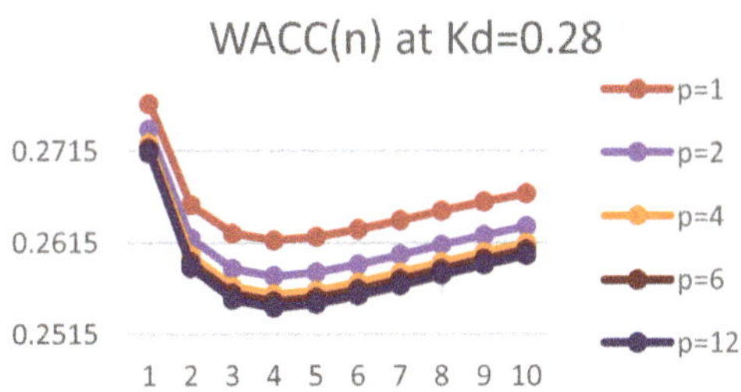

Fig. 22.6 The dependence of WACC on company age, n, with frequent payments of tax on profit $p = 1; 2; 3; 6; 12$ at $k_0 = 0.32$; $k_d = 0.28$; $t = 0.2$ (n from 1 to 10 years)

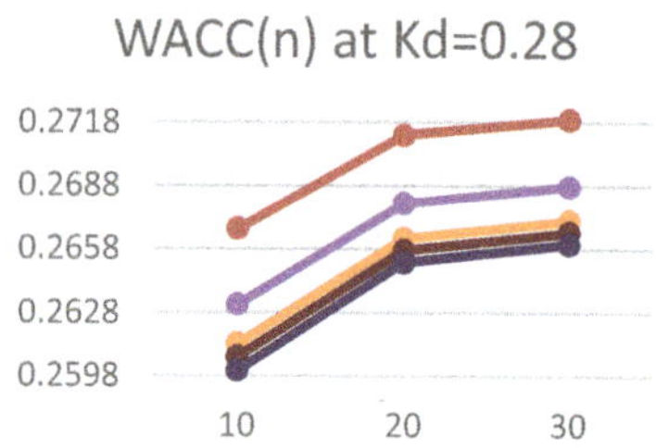

Fig. 22.7 The dependence of WACC on company age, n, with frequent payments of tax on profit $p = 1; 2; 3; 6; 12$ at $k_0 = 0.32$; $k_d = 0.28$; $t = 0.2$ (n from 10 to 30 years)

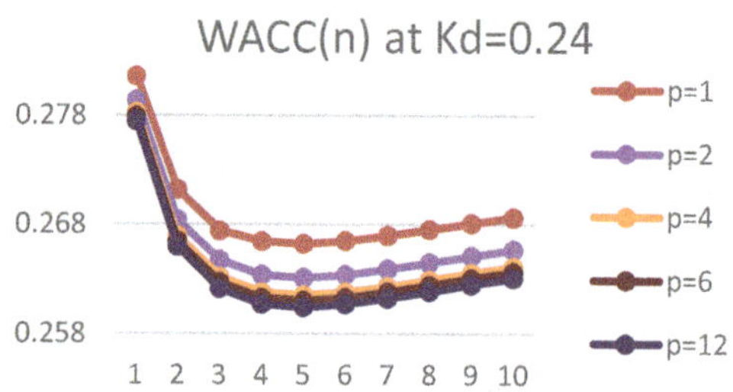

Fig. 22.8 The dependence of WACC on company age, n, with frequent payments of tax on profit $p = 1; 2; 3; 6; 12$ at $k_0 = 0.32$; $k_d = 0.24$; $t = 0.2$ (n from 1 to 10 years)

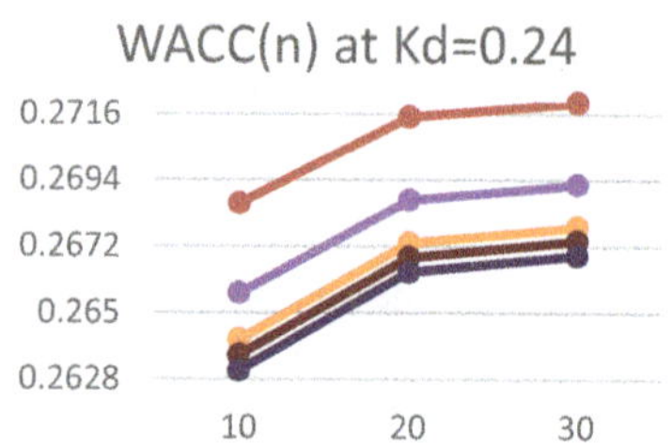

Fig. 22.9 The dependence of WACC on company age, n, with frequent payments of tax on profit $p = 1; 2; 3; 6; 12$ at $k_0 = 0.32$; $k_d = 0.24$; $t = 0.2$ (n from 10 to 30 years)

22.2.5.2 Frequent Tax on Income Payments: Compare the Results for Advance Payments of Tax on Income and Payments at the Ends of Periods

Using formulas (22.7) and (22.9), we compare the results for frequent tax on income payments ($p = 1;2;4;6;12$) at $g = 0$, with advance payments of tax on income and payments at the ends of periods. At $g = 0$ we consider the cases of $k_0 = 0.05$, a few values $k_d = 0.02; 0.03; 0.035$ and a couple leverage level values L = 1 and 3.

It is seen from Fig. 22.10 that minimum in dependence of WACC(n) takes place at $n_0 = 9.7$ years, thus "the golden age" is equal to 9.7 years.

It is seen from Fig. 22.11 that minimum in dependence of WACC(n) takes place at $n_0 = 9.7$ years, thus "the golden age" is equal to 9.7 years. Comparing Figs. 22.10 and 22.11, it can be seen that the value of n_0 practically does not change when moving from L = 1 to L = 3, but the distance between the curves decreases significantly.

It is seen from Fig. 22.12 that minimum in dependence of WACC(n) takes place at $n_0 = 18$ years, thus "the golden age" is equal to 18 years. Comparing Figs. 22.10

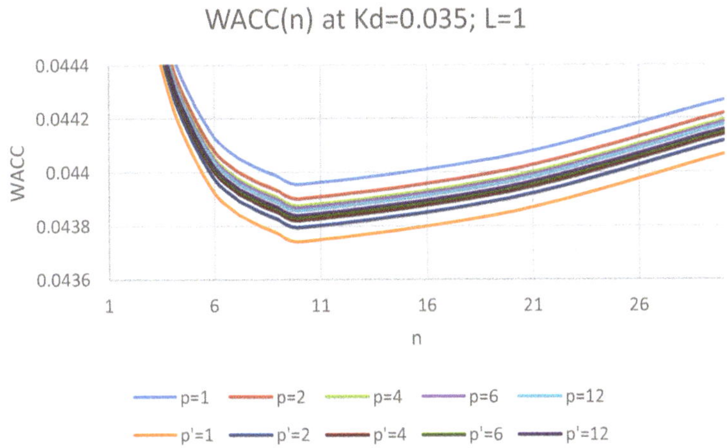

Fig. 22.10 The dependence of WACC on company age, n, with frequent payments of tax on profit $p = 1; 2; 3; 6; 12$ at $k_0 = 0.05$; $k_d = 0.035$; $t = 0.2$; L = 1 (n from 1 to 30 years) with advance payments (p') of tax on income and payments at the ends of periods (p)

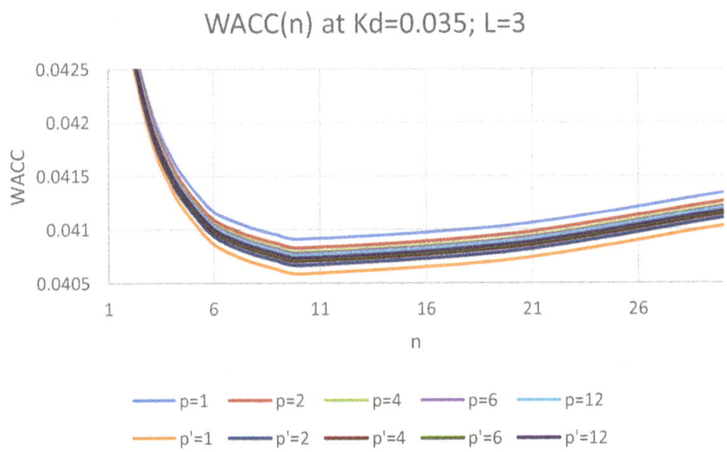

Fig. 22.11 The dependence of WACC on company age, n, with frequent payments of tax on profit $p = 1; 2; 3; 6; 12$ at $k_0 = 0.05$; $k_d = 0.035$; $t = 0.2$; L = 3 (n from 1 to 30 years) with advance payments (p') of tax on income and payments at the ends of periods (p)

and 22.11, it can be seen that under change k_d from 0.035 to 0.03 the value of n_0 increases significantly: from 9.7 years to 18 years. This means that it is possible to control the golden age by changing the cost of debt k_d.

It is seen from Fig. 22.13 that minimum in dependence of WACC(n) takes place at $n_0 = 17.5$ years, thus "the golden age" is equal to 17.5 years.

It is seen from Fig. 22.14 that at $k_d = 0.02$ minimum in dependence of WACC(n) is absent as well "the golden age" effect.

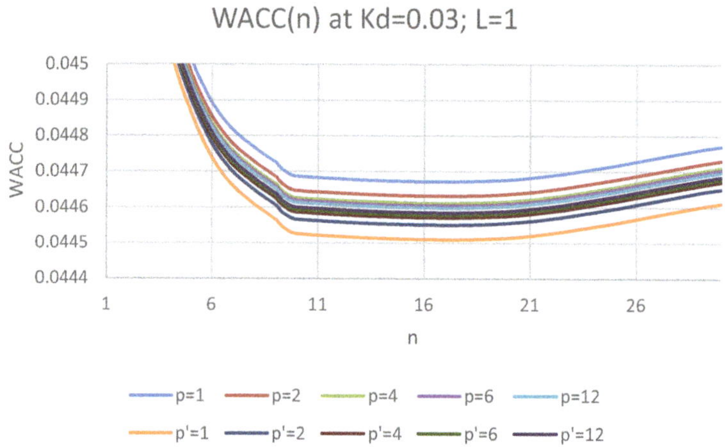

Fig. 22.12 The dependence of WACC on company age, *n*, with frequent payments of tax on profit $p = 1; 2; 3; 6; 12$ at $k_0 = 0.05$; $k_d = 0.03$; $t = 0.2$; L = 1 (*n* from 1 to 30 years) with advance payments (*p*') of tax on income and payments at the ends of periods (*p*)

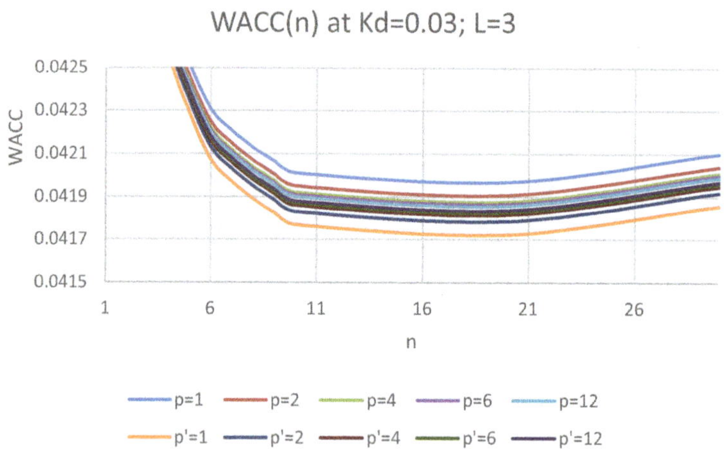

Fig. 22.13 The dependence of WACC on company age, n, with frequent payments of tax on profit $p = 1; 2; 3; 6; 12$ at $k_0 = 0.05$; $k_d = 0.03$; $t = 0.2$; L = 1 (*n* from 1 to 30 years) with advance payments (*p*') of tax on income and payments at the ends of periods (*p*)

It is seen from Fig. 22.15, that at $k_d = 0.02$ minimum in dependence of WACC(n) absent as well "the golden age" effect.

From Figs. 22.10, 22.11, 22.12, 22.13, 22.14, and 22.15, it can be seen that with an increase in the difference $\Delta k = k_0 – k_d$, the "golden age" first increases and then disappears, but the depth of the gap decreases. The gap depth, Δ, is the difference between minimum of WACC(n) and perpetuity values WACC (∞):

In larger scale figures, the separation between the curves is more visible.

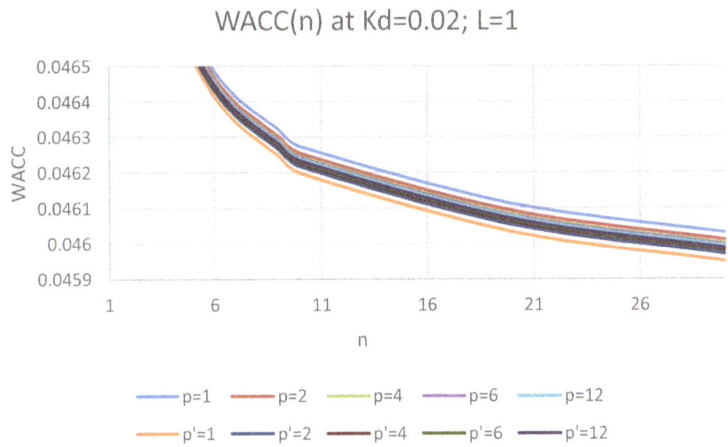

Fig. 22.14 The dependence of WACC on company age, n, with frequent payments of tax on profit $p = 1; 2; 3; 6; 12$ at $k_0 = 0.05$; $k_d = 0.02$; $t = 0.2$; L = 1 (n from 1 to 30 years) with advance payments (p') of tax on income and payments at the ends of periods (p)

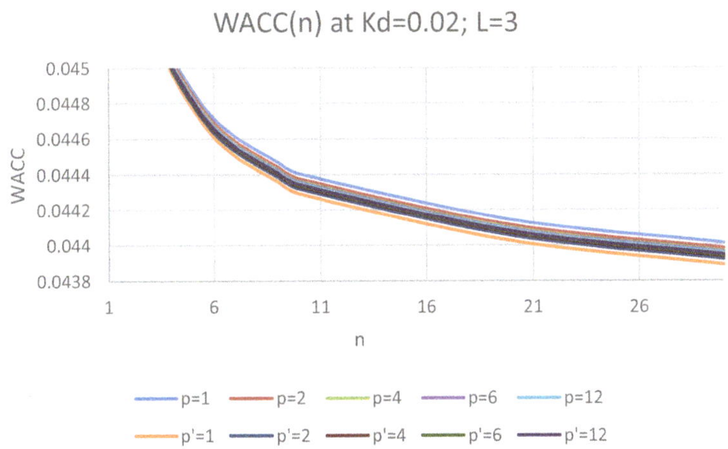

Fig. 22.15 The dependence of WACC on company age, n, with frequent payments of tax on profit $p = 1; 2; 3; 6; 12$ at $k_0 = 0.05$; $k_d = 0.02$; $t = 0.2$; L = 3 (n from 1 to 30 years) with advance payments (p') of tax on income and payments at the ends of periods (p)

It is seen from Fig. 22.16 that minimum in dependence of WACC(n) takes place at $n_0 = 9.7$ years, thus "the golden age" is equal to 9.7 years.

It is seen from Fig. 22.17 that minimum in dependence of WACC(n) takes place at $n_0 = 9.7$ years, thus "the golden age" is equal to 9.7 years.

It is seen from Fig. 22.18 that minimum in dependence of WACC(n) takes place at $n_0 = 17.5$ years, thus "the golden age" is equal to 17.5 years.

It is seen from Fig. 22.19 that minimum in dependence of WACC(n) takes place at $n_0 = 18$ years, thus "the golden age" is equal to 18 years.

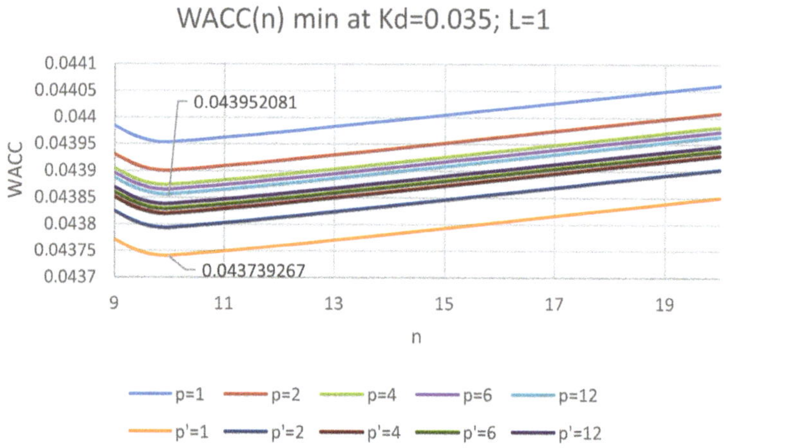

Fig. 22.16 The dependence of WACC on company age, n, with frequent payments of tax on profit $p = 1; 2; 3; 6; 12$ at $k_0 = 0.05$; $k_d = 0.035$; $t = 0.2$; L $= 1$ (n from 9 to 20 years) with advance payments (p') of tax on income and payments at the ends of periods (p) (larger scale)

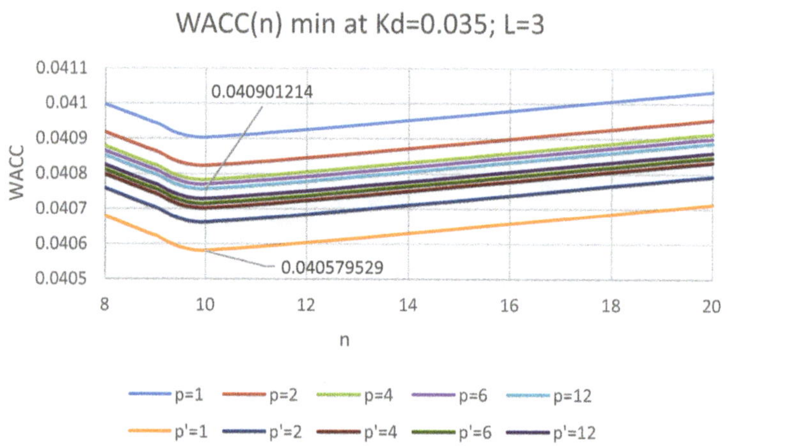

Fig. 22.17 The dependence of WACC on company age, n, with frequent payments of tax on profit $p = 1; 2; 3; 6; 12$ at $k_0 = 0.05$; $k_d = 0.035$; $t = 0.2$; L $= 3$ (n from 8 to 20 years) with advance payments (p') of tax on income and payments at the ends of periods (p) (larger scale)

It is seen from Fig. 22.20 that at $k_d = 0.02$ minimum in dependence of WACC(n) is absent as well "the golden age" effect.

We compare the results for frequent tax on income payments ($p = 1;2;4;6;12$) at $g = 0$, with advance payments of tax on income and payments at the ends of periods. At $g = 0$, we consider the cases of $k_0 = 0.05$, a few values $k_d = 0.02; 0.03; 0.035$, and a couple leverage level values L $= 1$ and 3.

From Figs. 22.16, 22.17, 22.18, 22.19, 22.20, it can be seen that with an increase in the difference $k = k_0 – k_d$, the "golden age" first increases and then disappears, but

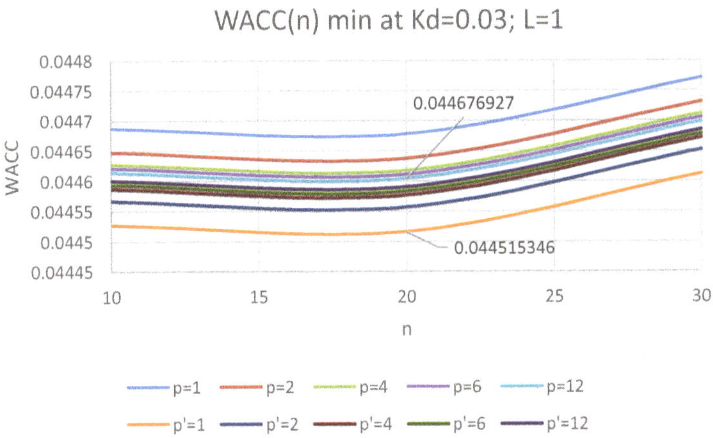

Fig. 22.18 The dependence of WACC on company age, n, with frequent payments of tax on profit $p = 1; 2; 3; 6; 12$ at $k_0 = 0.05$; $k_d = 0.03$; $t = 0.2$; L = 1 (n from 10 to 30 years) with advance payments (p') of tax on income and payments at the ends of periods (p) (larger scale)

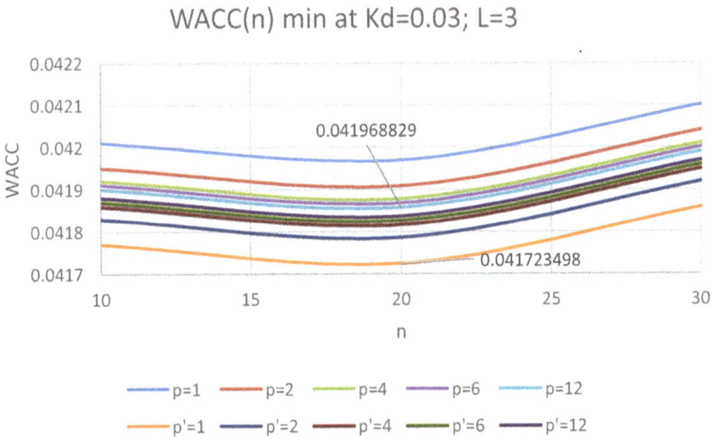

Fig. 22.19 The dependence of WACC on company age, n, with frequent payments of tax on profit $p = 1; 2; 3; 6; 12$ at $k_0 = 0.05$; $k_d = 0.03$; $t = 0.2$; L = 3 (n from 10 to 30 years) with advance payments (p') of tax on income and payments at the ends of periods (p) (larger scale)

the depth of the gap decreases. By controlling this difference, the company can control both "the golden age" itself and the duration of the effect (the behavior of WACC in the vicinity of "the golden age").

The summary of above results are as follows:

1. For payments at the ends of periods WACC(n), values shift down with frequency p.
2. For advance payments of tax on income WACC(n), values shift up with frequency p.

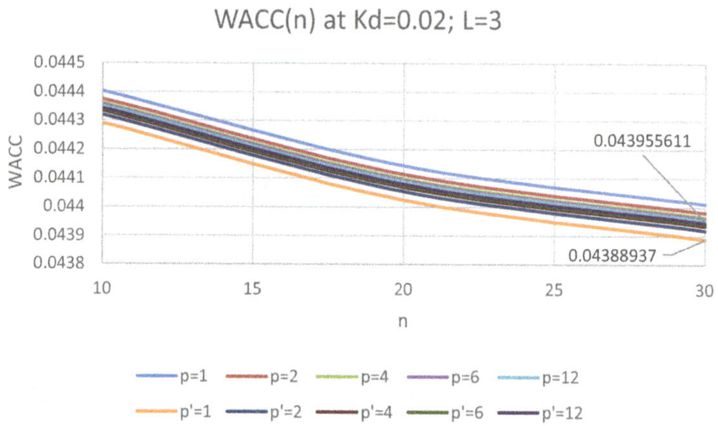

Fig. 22.20 The dependence of WACC on company age, n, with frequent payments of tax on profit $p = 1; 2; 3; 6; 12$ at $k_O = 0.05$; $k_d = 0.02$; $t = 0.2$; L = 3 (n from 10 to 30 years) with advance payments (p') of tax on income and payments at the ends of periods (p) (larger scale)

Table 22.2 The dependence of perpetuity values of WACC and the gap depth, Δ, in WACC(n) on leverage value, L, at $p = 1$; $k_O = 0.07$; $k_d = 0.05$; $t = 0.2$; $g = 0.2$

L		0	1	2	3	4	5	6	7	8	9	10
WACC(∞)	%	7	8.3	8.73	8.95	9.08	9.17	9.23	9.275	9.31	9.34	9.36
Δ	%	0	2.01	2.7	3.04	3.24	3.37	3.47	3.54	3.6	3.64	3.67

3. It turns out that the values of WACC(n) decrease in the first case, and increase in the second. But they never overlap. For example, for $p = 12$ in the first case (monthly payments of tax on income) min WACC(n) is equal to 4.079% while in the second case max WACC(n) is equal to 4.077% (at $k_d = 0.035$).
4. All of the above mean that it is always better for an enterprise to pay income tax in advance, in which case the payments should be annual. If a company pays income tax at the end of the reporting period, then it is beneficial to pay income tax monthly.
5. When k_d decreases from 0.035 to 0.02, the "golden age" effect decreases and disappears at $k_d = 0.02$.

22.2.6 *Influence of Leverage Level, L*

The gap depth, Δ, is the difference between minimum, WACC(n) and perpetuity values WACC (∞):

$$\Delta = WACC(\infty) - \min WACC.$$

From Table 22.2 it follows:

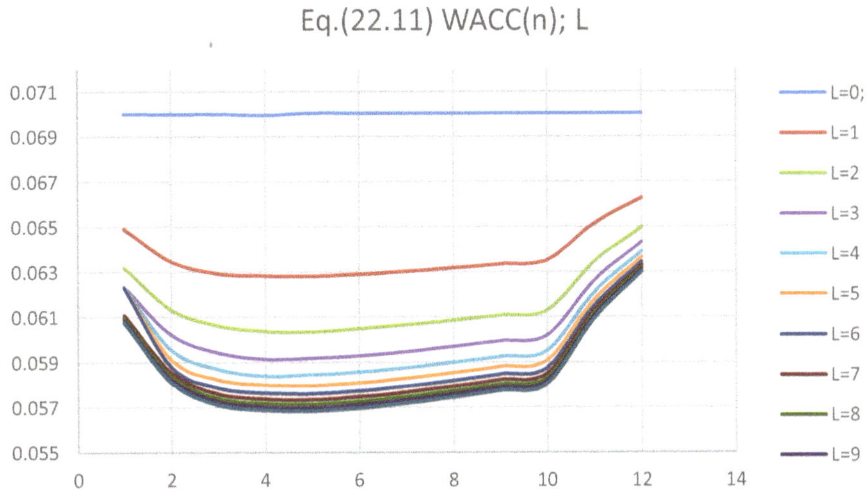

Fig. 22.21 The dependence of *WACC* on company age, n, at $p = 1$; $k_0 = 0.07$; $k_d = 0.05$; $t = 0.2$; $g = 0.2$ at different L = 1;2;3;…10 (n from 1 to 12 years)

1. WACC(∞) values increase with leverage level, L. Accounting that, as it is seen from Fig. 22.21, the curves WACC(n) at low leverage level lie lower for bigger leverage level, L, values, there are overlaps of WACC(n) curves: this leads to very important practical consequences for the application of the "golden age" effect.
2. The gap depth, Δ increases with leverage level, L. Because large gap depth, Δ means small value cost of raising capital and big company value, V, to use the "golden age" effect, companies should try to work at large leverage level, L. Note that companies must find a trade-off between the benefit of using "the golden age" effect and the financial hardship that comes with high levels of leverage, L.

It is important and interesting to note (see Fig. 22.22) that for positive (and zero) growth rate g ($g \geq 0$), "the golden age" effect exists, while for negative growth rate g ($g < 0$), "the golden age" effect is absent: WACC(n) decreases monotonically.

Comparison of the results of Figs. 22.21 and 22.23 shows that in the case of advance income tax payments, all WACC(n) curves shift downward relative to the case of tax on income paid at the ends of periods. This could be as well illustrated at Fig. 22.24, where the results for these two cases (advance payments of tax on income and payments at the ends of periods) are shown.

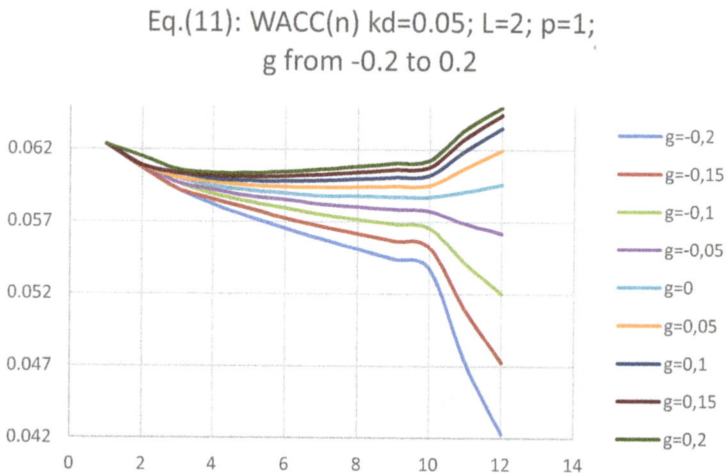

Fig. 22.22 The dependence of WACC on company age, n, (n from 1 to 12 years) at $p = 1$; $k_0 = 0.07$; $k_d = 0.05$; $t = 0.2$; at different $g = -0.2;-0.15;-0.1;-0.05;0; 0.05; 0.1; 0.15;0.2$

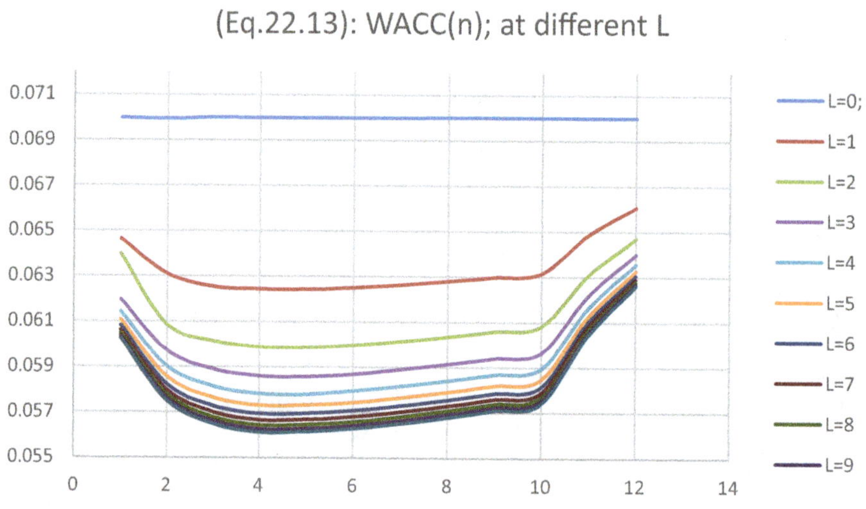

Fig. 22.23 The dependence of *WACC* on company age, n, at $p = 1$; $k_0 = 0.07$; $k_d = 0.05$; $t = 0.2$; $g = 0.2$ at different leverage levels L = 1;2;3;...10 (n from 1 to 12 years) under the advance tax on income payments

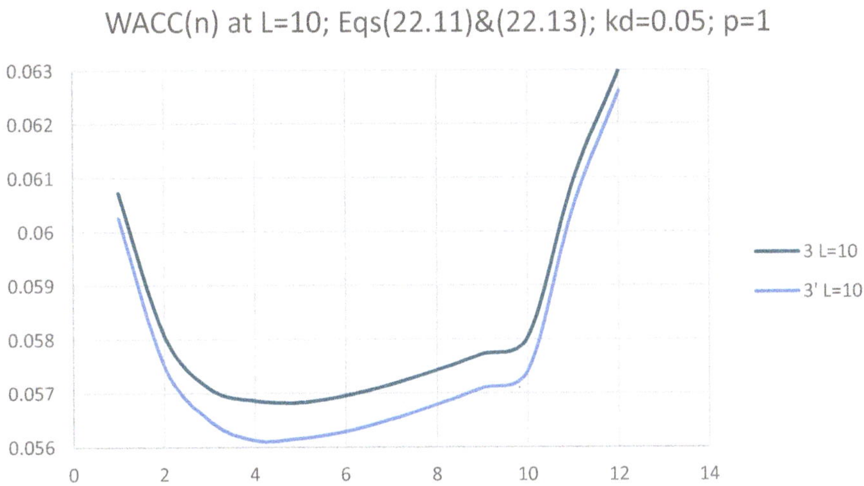

Fig. 22.24 Comparison of the dependence of WACC on company age, n, at $p = 1$; $k_0 = 0.07$; $k_d = 0.05$; $t = 0.2$; $g = 0.2$ at L = 10 (n from 1 to 12 years) with advance payments of tax on income and payments at the ends of periods

22.2.7 *Influence of Growth Rate* g

22.2.7.1 Influence of Growth Rate *g*: Tax on Income Payments at the Ends of Periods

Below we investigate the influence of growth rate g on the "golden age" effect. The dependence of WACC on company age, n, (n from 1 to 30 years) within the Brusov–Filatova–Orekhova theory (BFO theory) at $p = 1$; $k_0 = 0.16$; $k_d = 0.1;0.14$; $t = 0.2$; L = 1; 10 at different $g = -0.2;-0.15;-0.1;-0.05;0; 0.05; 0.1; 0.15;0.2$ with advance payments of tax on income and payments at the ends of periods. Let us start from the case of tax on income payments at the ends of periods.

It is seen from Figs. 22.25, 22.26, 22.27, 22.28 that:

1. For positive growth rate g ($g \geq 0$), "the golden age" effect exists, while for negative growth rate g ($g < 0$) "the golden age" effect is absent: WACC(n) decreases monotonically.
2. The curves WACC(n) shift up with growth rate g.
3. The gap depth increases with growth rate g.

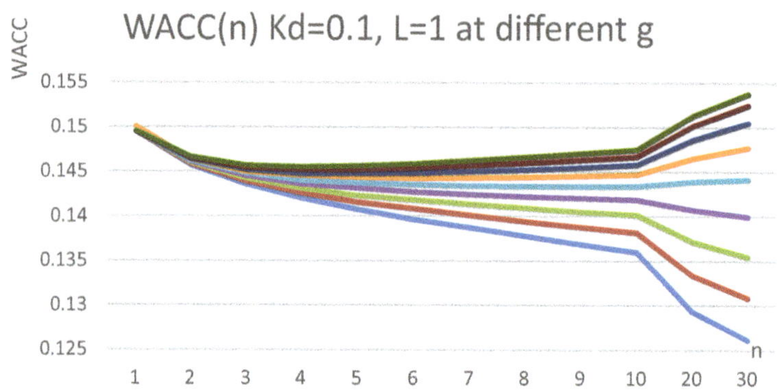

Fig. 22.25 The dependence of WACC on company age, n, (n from 1 to 30 years) at $p = 1$; $k_0 = 0.16$; $k_d = 0.1$; $t = 0.2$; L = 1 at different $g = -0.2;-0.15;-0.1;-0.05;0;\ 0.05;\ 0.1;\ 0.15;0.2$ (from bottom to the top)

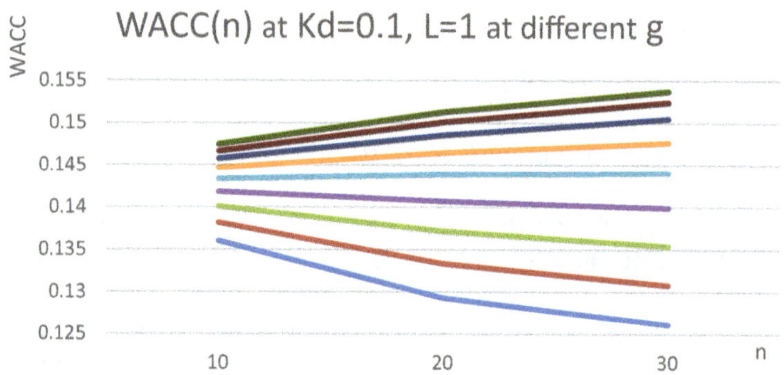

Fig. 22.26 The dependence of WACC on company age, n, (n from 10 to 30 years) at $p = 1$; $k_0 = 0.16$; $k_d = 0.1$; $t = 0.2$; l = 1 at different $g = -0.2;-0.15;-0.1;-0.05;0;\ 0.05;\ 0.1;\ 0.15;0.2$ (from bottom to the top) (larger scale)

22.2.7.2 Influence of Growth Rate *g*: Comparison of Results with Advance Income Tax Payments and Payments at the Ends of Periods

Below we compare the results with advance income tax payments and payments at the ends of **periods** at $p = 1$; $k_0 = 0.16$; $k_d = 0.1;0.14$; $t = 0.2$; L = 1 at different $g = -0.2;-0.15;-0.1;-0.05;0;\ 0.05;\ 0.1;\ 0.15;0.2$ (Figs. 22.29, 22.30, 22.31, and 22.32).

There is a doublet of curves WACC(n): one of them corresponds to advance payments of tax on income and the other one corresponds to tax on income payments at the ends of periods. The first doublet lies below the second. The curves of these two doublets overlap, which can lead to very interesting effects.

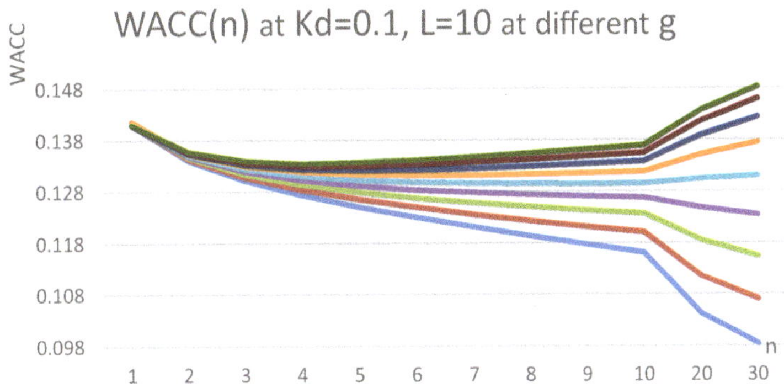

Fig. 22.27 The dependence of WACC on company age, n, (n from 1 to 30 years) at $p = 1$; $k_0 = 0.16$; $k_d = 0.1$; $t = 0.2$; L = 10 at different $g = -0.2;-0.15;-0.1;-0.05;0; 0.05; 0.1; 0.15;0.2$ (from bottom to the top)

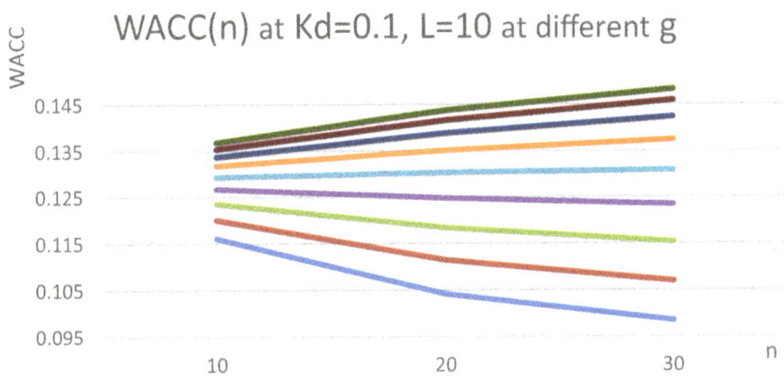

Fig. 22.28 The dependence of WACC on company age, n, (n from 10 to 30 years) within at $p = 1$; $k_0 = 0.16$; $k_d = 0.1$; $t = 0.2$; L = 10 at different $g = -0.2;-0.15;-0.1;-0.05;0; 0.05; 0.1; 0.15;0.2$ (from bottom to the top) (larger scale)

There is the possibility to manage the golden age by change debt cost k_d, leverage level, L, type of payments (advance or at the ends of the periods, etc.).

22.2.8 Comparison of the Dependence of WACC on the Age of the Company n *with a Decrease and Increase in Income*

Below is a comparison of the dependence of WACC on the age of the company n with falling and growing incomes. Let us take as an example the company Gazprom, for which the parameter $ko = 0.06$ is estimated by us, and we carried out calculations

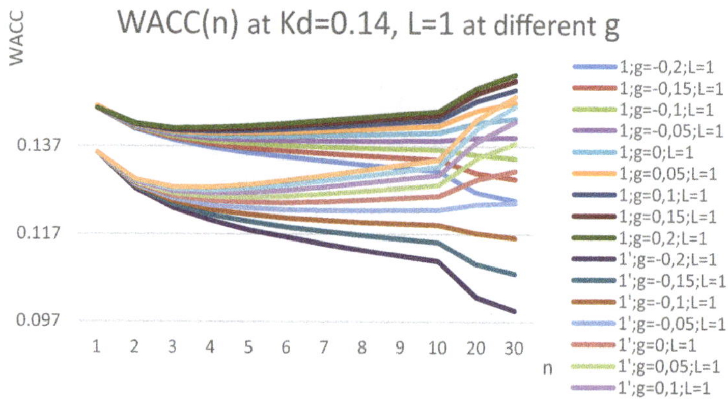

Fig. 22.29 The dependence of WACC on company age, n, (n from 1 to 30 years) at $p = 1$; $k_0 = 0.16$; $k_d = 0.14$; $t = 0.2$; L = 1 at different $g = -0.2;-0.15;-0.1;-0.05;0; 0.05; 0.1; 0.15;0.2$ with advance payments (1′) of tax on income and payments at the ends of periods (1)

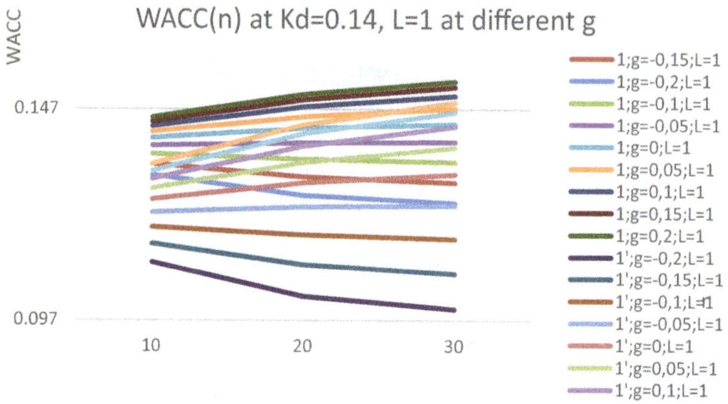

Fig. 22.30 The dependence of WACC on company age, n, (n from 10 to 30 years) at $p = 1$; $k_0 = 0.16$; $k_d = 0.14$; $t = 0.2$; L = 1 at different $g = -0.2;-0.15;-0.1;-0.05;0; 0.05; 0.1; 0.15;0.2$ with advance payments (1′) of tax on income and payments at the ends of periods (1) (larger scale)

for two values of the cost of debt $k_\mathrm{d} = 0.03; 0.04$ and for negative $g = -0.05$ and positive $g = 0.05$. The results are shown in Figs. 22.33 and 22.34.

Figures 22.33 *and* 22.34 *show that the golden age effect exists in the case of positive g, when income increases, and is absent in the case of negative g, when income decreases. Note that the golden age effect exists in the case of* $g = 0$ *(the case of constant income) for some values of the parameters k0 and* k_d.

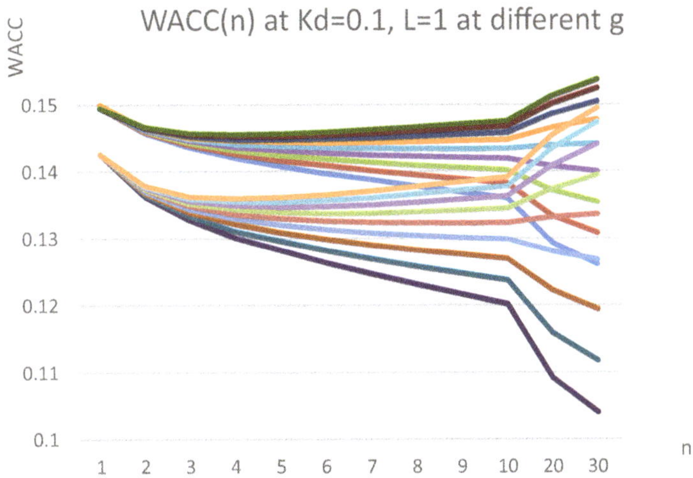

Fig. 22.31 The dependence of WACC on company age, n, (n from 1 to 30 years) at $p = 1$; $k_0 = 0.16$; $k_d = 0.1$; $t = 0.2$; L = 1 at different $g = -0.2;-0.15;-0.1;-0.05;0;\ 0.05;\ 0.1;\ 0.15;0.2$ with advance payments (1′) of tax on income and payments at the ends of periods (1)

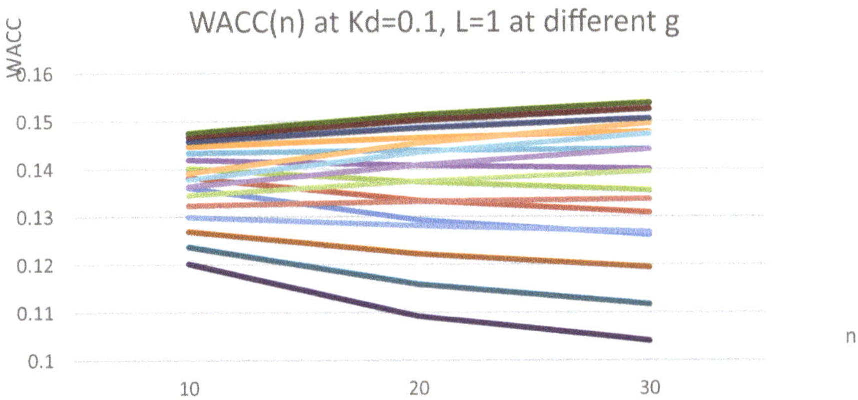

Fig. 22.32 The dependence of WACC on company age, n, (n from 10 to 30 years) at $p = 1$; $k_0 = 0.16$; $k_d = 0.1$; $t = 0.2$; L = 1 at different $g = -0.2;-0.15;-0.1;-0.05;0;\ 0.05;\ 0.1;\ 0.15;0.2$ with advance payments (1′) of tax on income and payments at the ends of periods (1) (larger scale)

22.3 Discussion and Conclusions

In the current paper, the effect of the "golden age" of the company is investigated, taking into account the conditions of the real functioning of companies. With this aim the dependence of cost of raising capital, WACC, on the age of company, n, is studied at various leverage levels, L, at various values of equity, k_e, and debt, k_d, costs, at different frequencies of tax on income payments, p, with advance payments

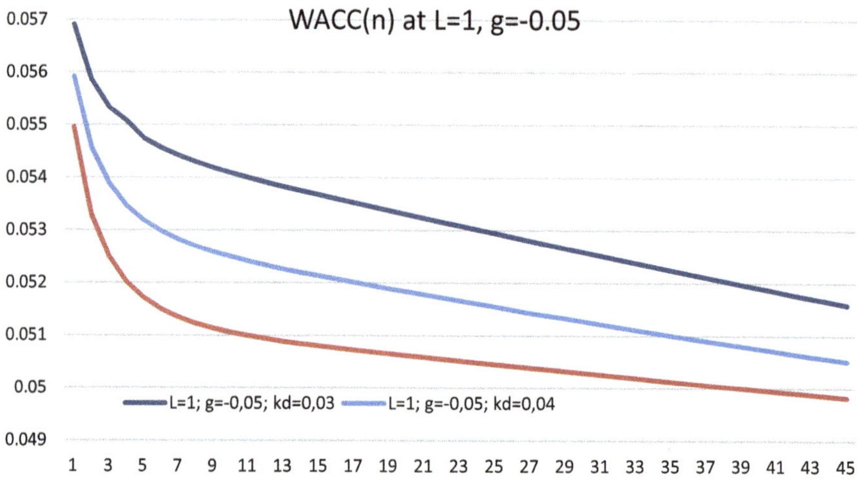

Fig. 22.33 The dependence of WACC on company age, n, at $p = 1$; $k_0 = 0.06$; $k_d = 0.03$; 0.04; L = 1 for negative $g = -0.05$

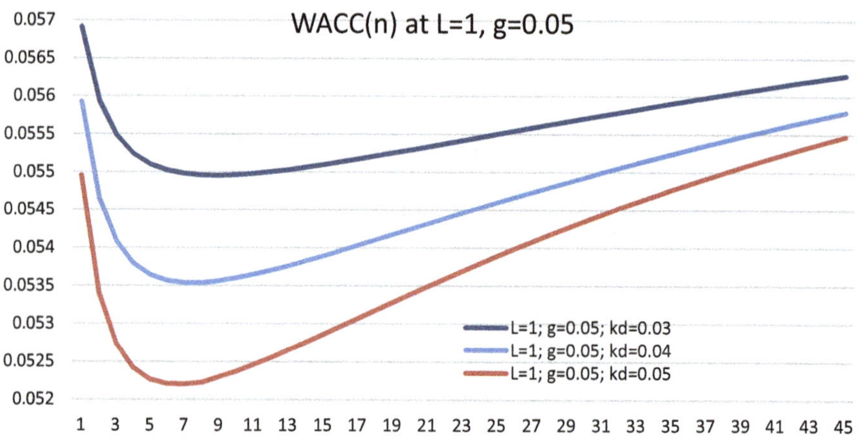

Fig. 22.34 The dependence of WACC on company age, n, at $p = 1$; $k_0 = 0.06$; $k_d = 0.03$; 0.04; L = 1 for positive $g = -0.05$

of tax on income and payments at the ends of periods, at variable income of the companies. The frequency of income tax payment and the method of payment (in advance payments or at the end of the reporting period) is regulated by tax legislation and depends on the amount of income received (or forecasted). Thus, studying the influence of two of these factors is very important for the economy practice.

In all variants of the Modigliani–Miller theory, the time factor is absent, therefore when studying the dependence of WACC on n, we work in the framework of the

Brusov–Filatova–Orekhova theory. The Modigliani–Miller theory is used by us only to estimate the limiting eternity values of WACC.

Important conclusions were made about the "golden age" effect and recipes for managing the effect, and recommendations for the company's management with this respect have been developed.

Let us emphasize a few new results obtained in the paper:

- The effect exists for companies with growing (or constant) income ($g \geq 0$) and is absent for companies with falling income ($g < 0$) (see Figs. 22.33 and 22.34).
- The depth of the gap depends on the frequency of income tax payments, p. Thus, the cost of capital raised (WACC) and the value of the company depend on p and can be controlled by changing the frequency of income tax payment, p within the framework of tax legislation.
- The value of the Golden Age (n_0) depends on the cost of debt k_d, so you can control it and the shape of WACC min by changing the cost of debt k_d. The extension of the Golden Age is very important for the company.

One can find more novelties below and in the section "Recommendations for the company's management."

1. "The golden age" effect exists for positive (and zero) growth rate g ($g \geq 0$), while for negative growth rate g ($g < 0$) "the golden age" effect is absent.
2. WACC(∞) values increase with leverage level, L. Accounting that, as it is seen from Fig. 22.11, the curves WACC(n) lie lower for bigger leverage level, L, values, there are overlaps of WACC(n) curves: this leads to very important practical consequences for the application of the "golden age" effect.
3. The gap depth, Δ increases with leverage level, L. Because large gap depth, Δ means small value cost of rising capital and big company value, V, to use the "golden age" effect, companies should try to work at large leverage level, L. Note that companies must find a trade-off between the benefit of using the golden age effect and the financial hardship that comes with high levels of leverage.
4. For payments at the ends of periods, WACC(n) values shift down with frequency p, thus in this case, monthly payments are preferable.
5. For advance payments of tax on income, WACC(n) values shift up with frequency p, thus less frequent payments permitted by tax laws are preferred.
6. It turns out that the values of WACC(n) decrease in the first case, and increase in the second. But they never overlap. For example, for $p = 12$ in the first case (monthly payments of tax on income), min WACC(n) is equal to 4.079%, while in the second case, max WACC(n) is equal to 4.077% (at $k_d = 0.035$).
7. All of the above mean that it is always better for a company to pay income tax in advance, in which case the payments should be less frequent. If a company pays income tax at the end of the reporting period, then it is beneficial to pay income tax monthly.
8. Size of effect depends on debt cost k_d. For example, as we have seen above at $k_0 = 0.05$, when k_d decreases from 0.035 to 0.02, the "golden age" effect decreases and disappears at $k_d = 0.02$.

9. It is possible to control the shape of the WACC(n) curve at the minimum point n_0, thereby expanding the area near this minimum and the company's golden age.

22.4 Recommendations for the Company Management

1. The "golden age" effect occurs only for companies with growing or constant income.
2. Advance payments are more advantageous Advance payments are more advantageous, and in this case, less frequent payments permitted by tax laws are preferred.
3. Overlap of curves allows to choose the method of payments (advance or at the end of reporting periods) but in accordance with tax laws.
4. Working at a high level of leverage, L is more profitable and leads to an increase in the effect. Note that companies must find a trade-off between the benefit of using the "golden age" effect and the financial hardship that comes with high levels of leverage L.
5. The difference between the cost of equity k_0 and the cost of debt k_d should not be too large: if the difference is large, the effect disappears.
6. You can control the effect (golden age value, n_0, pit depth in WACC (n) dependence, the duration of "golden age" and other parameters) by changing the difference between the cost of equity k_0 and the cost of debt k_d, k_d value, leverage level, L, frequency, p, and method of income tax payments. To change the cost of equity k_0, a company could change the paid dividends, and the cost of debt k_d is equal to loan cost.

In conclusion, we note that recently, the authors [33–35] for the first time generalized CAPM to take into account both business and financial risks and developed a new model CAPM 2.0. They showed that R. Hamada's attempt to take into account both business and financial risks [4] was untenable, and the formulas he obtained, which are widely used in practice, are incorrect. The authors of [33–35] derived correct formulas that take into account both business and financial risks. The application of the new CAPM 2.0 model to a number of companies is considered and the difference between the results obtained within the framework of CAPM 2.0 and CAPM is demonstrated.

The study of the dependence of WACC on the age of the company n, WACC(n), which can only be carried out within the framework of the BFO theory, turns out to be very important in the income approach to business valuation. This allows you to link a retrospective analysis of a company's financial condition with a representative analysis as part of a business valuation [36–38].

In the future, the authors intend to study the existence of this effect at variable rates of income growth of the company.

References

1. Modigliani, F., & Miller, M. (1958). The cost of capital, corporate finance, and the theory of investment. *The American Economic Review, 48*, 261–297.
2. Modigliani, F., & Miller, M. (1963). Corporate income taxes and the cost of capital: A correction. *The American Economic Review, 53*, 147–175.
3. Modigliani, F., & Miller, M. (1966). Some estimates of the cost of capital to the electric utility industry 1954–1957. *The American Economic Review, 56*, 333–391.
4. Hamada, R. (1969). Portfolio analysis, market equilibrium, and corporate finance. *The Journal of Finance, 24*, 13–31.
5. Miller, M. (1977). Debt and taxes. *The Journal of Finance, 32*, 261–275. https://doi.org/10.1111/j.1540-6261.1977.tb03267.x
6. Farber, A., Gillet, R., & Szafarz, A. A. (2006). General formula for the WACC. *International Journal of Business, 11*, 211–218.
7. Fernandez, P. A. (2006). General formula for the WACC: A comment. *International Journal of Business, 11*, 219.
8. Berk, J., & De Marzo, P. (2007). *Corporate finance*. Pearson–Addison Wesley.
9. Harris, R., & Pringle, J. (1985). Risk–adjusted discount rates–extension form the average–risk case. *Journal of Financial Research, 8*, 237–244.
10. Baumol, W., Panzar, J. G., & Willig, R. D. (1982). *Contestable markets and the theory of industry structure*. Harcourt College Publishers. Google Scholar.
11. Black, F. (1986). Noise. *The Journal of Finance, 41*, 529–543. Google Scholar, CrossRef.
12. Becker, D. M. (2021). The difference between Modigliani–Miller and Miles–Ezzell and its consequences for the valuation of annuities. *Cogent Economics & Finance, 9*, 1. Google Scholar, CrossRef.
13. Berk, J., & DeMarzo, P. (2007). *Corporate finance*. Pearson–Addison Wesley. Google Scholar.
14. Brealey, R. A., & Myers, S. C. (1981). *Principles of corporate finance* (1st ed.). McGraw–Hill. Google Scholar.
15. Brealey, R. A., Myers, S. C., & Allen, F. (2014). *Principles of corporate finance* (11th ed.). McGraw–Hill. Google Scholar.
16. Bhattacharya, S. (1978). Project valuation with mean–reacting cash flows. *The Journal of Finance, 33*, 1317–1331. Google Scholar, CrossRef.
17. Myers, S. (2001). Capital structure. *The Journal of Economic Perspectives, 15*, 81–102.
18. Filatova, T., Orehova, N., & Brusova, A. (2008). Weighted average cost of capital in the theory of Modigliani–Miller, modified for a finite life–time company. *Bull FU, 48*, 68–77. Google Scholar.
19. Brusov, P., Filatova, T., Orekhova, N., & Eskindarov, M. (2018). *Modern corporate finance, investments, taxation and ratings* (2nd ed., pp. 1–571). Springer Nature Publishing.
20. Brusov, P., Filatova, T., Orekhova, N., & Kulik, V. (2015). The Golden age of the company: (three Colors of company's time). *Journal of Reviews on Global Economics, 4*, 21–42.
21. Brusov, P., Filatova, T., Orehova, N., & Kulk, V. (2018). A "Golden Age" of the companies: Conditions of its existence. *Journal of Reviews on Global Economics, 7*, 88–103.
22. Brusov, P., & Filatova, T. (2023). Capital structure theory: Past, present, future. *Mathematics, 11*(3), 616. https://doi.org/10.3390/math11030616
23. Dimitropoulos, P. (2014). Capital structure and corporate governance of soccer clubs: European evidence. *Management Research Review, 37*, 658–678.
24. Luiz, K., & Cruz, M. (2015). The relevance of capital structure on firm performance: A multivariate analysis of publicly traded Brazilian companies. *REPeC Brasília, 9*, 384–401.
25. Barbi, M. (2011). On the risk–neutral value of debt tax shields. *Applied Financial Economics, 22*, 251–258.
26. Franc-Dąbrowska, J., Mądra-Sawicka, M., & Milewska, A. (2021). Energy sector risk and cost of capital assessment—companies and investors perspective. *Energies, 14*, 1613.

27. Angotti, M., de Lacerda Moreira, R., Hipólito Bernardes do Nascimento, J., & Neto de Almeida Bispo, O. (2018). Analysis of an equity investment strategy based on accounting and financial reports in Latin American markets. *Reficont, 5*, 22–40.
28. Vergara-Novoa, C., Sepúlveda-Rojas, J. P., Miguel, D. A., & Nicolás, R. (2018). Cost of capital estimation for highway concessionaires in Chile. *Journal of Advanced Transportation, 2018*, 2153536.
29. Mundi, H. S., Kaur, P., & Murty, R. L. N. (2022). A qualitative inquiry into the capital structure decisions of overconfident finance managers of family–owned businesses in India. *Qualitative Research in Financial Markets, 14*, 357–379.
30. Sadiq, M., Alajlani, S., Hussain, M. S., Bashir, F., & Chupradit, S. (2022). Impact of credit, liquidity, and systematic risk on financial structure: Comparative investigation from sustainable production. *Environmental Science and Pollution Research International, 29*, 20963–20975.
31. Becker, D. M. (2022). Getting the valuation formulas right when it comes to annuities. *Managerial Finance, 48*, 470.
32. El-Chaarani, H., Abraham, R., & Skaf, Y. (2022). The impact of corporate governance on the financial performance of the banking sector in the MENA (Middle Eastern and North African) Region: An immunity test of banks for COVID–19. *Journal of Risk and Financial Management, 15*, 82.
33. Brusov, P., Filatova, T., & Kulik, V. (2023). Capital asset pricing model 2.0: Account of business and financial risk. *Preprint*, 2023100347. https://doi.org/10.20944/preprints202310.0347.v1
34. Brusov, P., Filatova, T., Kulik, V., Lin, G., & Chang, L.-M. (2023). Can CAPM (capital asset pricing model) accurately value assets? *Proceedings of the International Conference on Electronic Business (ICEB), 23*, 60–70.
35. Brusov, P., Filatova, T., Kulik, V., Lin, G., & Chang, L.-M. (2023). Precision finance: Capital structure theories approach reality. *Proceedings of the International Conference on Electronic Business (ICEB), 23*, 466–480.
36. Brusov, P., Filatova, T., & Orekhova, N. (2023). *The Brusov–Filatova–Orekhova theory of capital structure: Applications in corporate finance, investments, taxation and ratings* (pp. 1–764).
37. Brusov, P., Filatova, T., & Orekhova, N. (2022). *Generalized Modigliani–Miller theory: Contributions to finance and accounting* (pp. 1–362). Springer Nature Publishing.
38. Brusov, P., Filatova, T., & Orekhova, N. (2021). *Ratings critical analysis and new approaches of quantitative and qualitative methodology* (pp. 1–369). Springer Nature Publishing.

Part V
Accounting for Business and Financial Risks

Chapter 23
Incorporating CAPM into Capital Structure Theories: Accounting for Business and Financial Risks

Abbreviations

BFO	Brusov–Filatova–Orekhova theory
CAPM	Capital Asset Pricing Model
CMA	company capital expenditure
HML	the difference between the returns of companies with low and high intrinsic value (indicator B/P)
MM	the Modigliani–Miller theory
RMW	return on equity
SMB	the difference between the returns of companies with large and small capitalization
WACC	the weighted average cost of capital

23.1 Introduction

In real economy, financial and business risks exist. Financial risks are related to the use of debt financing and are described by capital structure theories. Business risks associated with investments into a specific company (and not the entire market [industry]) and are described as CAPM (market or industry version).

Based on the portfolio theory by Harry Markowitz, the Capital Asset Pricing Model (CAPM) was developed independently [1–5] by Jack Traynor (1961), William F. Sharp (1964), John Lintner (1965) and Jan Mossin (1966). Two main capital structure theories Brusov–Filatova–Orekhova (BFO) [6, 7] theory and Modigliani–Miller (MM) theory [8–10] describe financial risks. The Miles–Ezzell model [11–13] offers an alternative approach to the problem of capital structure. This is discussed below along with some others [14].

P. Brusov, T. Filatova, *Business Valuation and Company Value*, Contributions to Finance and Accounting, https://doi.org/10.1007/978-3-031-90847-7_23

Statement of the problem: when assessing the profitability of an asset, take into account both financial and business risks. A fundamentally new approach to assessing the profitability of an asset is proposed. Transition from CAPM, which takes the same risk-free return for all assets as an initial assessment, to a new methodology, in which the average return of an asset, cleared of leverage, with the addition of a premium for business risk (market or industry) is taken as a seed return, significantly improves the accuracy of the estimate.

The incorporation of the CAPM [1–5] and Fama–French models [15–18] into the theory of capital structure will allow taking into account the business risk taken into account in these models, along with the financial risk taken into account in the Brusov–Filatova–Orekhova (BFO) [6, 7] theory and Modigliani–Miller (MM) theory [8–10]. Hamada's attempt to account the leverage level is discussed [19, 20].

23.2 CAPM (Capital Asset Pricing Model)

23.2.1 Market Approach

CAPM is a simple, but widely used, one-factor model that describes the relationship between the expected return on assets (stocks, investments, etc.) and the risk-free rate, taking into account systematic (business) risk. This relationship is described by the equity risk premium, which depends on the asset's beta (which describes the asset's correlation or sensitivity to the market), the risk-free rate (say, the Treasury bill rate or the central bank's key rate), and the expected return in the market.

The following assumptions are made within the CAPM model:

1. All investors are risk-averse and have the same time frame to evaluate information.
2. Unlimited capital exists to borrow at the risk-free rate.
3. Investments can be divided into unlimited parts and sizes.
4. Taxes, inflation, and transaction costs are absent.
5. Return and risk are linearly related.

CAPM (Capital Asset Pricing Model) describes the profitability of asset and is described by the following formula

$$k_i = k_f + \beta_i (k_m - k_f). \tag{23.1}$$

Here, k_f is risk-free profitability, β is the β-coefficient of the company. It shows the dependence of the return on the asset and the return on the market as a whole. The β-coefficient is described by the following formula:

$$\beta_i = \frac{\text{cov}_{im}}{\sigma_m^2} = \rho_{im}\frac{\sigma_i}{\sigma_m}. \tag{23.2}$$

Here σ_i is the risk (standard deviation) of i-th asset, σ_m is market risk (standard deviation of market index), cov_{im} is covariance between i-th asset and market portfolio.

An investor invests in risky securities only if their return is higher than the return on risk-free securities, so always $k_i > k_f$ and $k_m > k_f$.

The beta-coefficient of a security, β, has the meaning of the amount of riskiness of this security. It follows from formula (23.1) that:

1. If β=1, the yield of the security is equal to the yield of the average market portfolio ($k_i = k_f$).
2. If $\beta > 1$, the security is more risky than the average on the security market ($k_i > k_f$).
3. If β <1, the security is less risky than the average on the security market ($k_i < k_f$).

Securities betas are calculated using statistical data on returns on specific securities and the average market returns on securities traded on the market.

23.2.2 Disadvantages of the CAPM Model

CAPM has some well-known disadvantages.

1. The CAPM formula only works under assumption that the market is dominated by purely rational players who make decisions that favor only investment returns. This, of course, is not always true.
2. CAPM assumes that each market participant acts on the basis of the same information. In reality, relevant information is distributed unevenly among the public, so some participants may make decisions based on information that others do not.
3. Using beta as the main part of the formula. But beta takes into account only changes in the stock price in the market. However, the share price can change for reasons other than the market. Stocks can rise or fall in value for deliberate reasons, not just volatility.
4. CAPM only uses historical data. But historical stock price changes are not enough to determine the overall risk of an investment. Other factors should be considered, such as economic conditions, industry peculiarities and competitor characteristics, and internal and external activities of the company itself.

Another very important shortcoming of CAPM, which is eliminated as part of the new approach, is that it assumes the same risk-free return on all assets as the original valuation. Also, CAPM takes into account the profitability of a particular asset only in the beta coefficient.

The model operates on only one factor that affects the future performance of a stock. In 1992, Y. Fama K. and French [15–18] proved that future returns are also affected by factors such as company size and industry affiliation.

The model has a number of limitations: the model does not take into account taxes, transaction costs, non-transparency of the financial market, etc.

Finally, to predict future returns, a retrospective level of market risk is used, which leads to a forecast error.

23.2.3 Industry Approach

CAPM has an alternative approach that refers to the industrial index rather than the market.

$$k_i = k_f + \beta_i(k_I - k_f). \tag{23.3}$$

Here, k_f is risk-free profitability, β is the β-coefficient of the company. In this case, it shows the dependence of the return on the asset and the return on the industry as a whole. The β-coefficient now is described by the following formula

$$\beta_i = \frac{\mathrm{cov}_{iI}}{\sigma_I^2} = \rho_{iI}\frac{\sigma_i}{\sigma_I}. \tag{23.4}$$

Here σ_i is the risk of i-th asset, σ_I is industry risk (standard deviation of industry index), cov_{iI} is covariance between i-th asset and industry index. Note, that the industry approach better describes the return on an asset than the market approach.

The CAPM approach is still evolving and we will describe one of the directions of this development below.

23.2.4 The Symmetric CAPM

The Capital Asset Pricing Model (CAPM) assumes a Gaussian or Normal distribution. In practice, this assumption may be violated. In [21], a symmetric CAPM is proposed, assuming distributions with lighter or heavier tails than the normal distribution. Elliptic distributions (normal, exponential and Student–t) are considered.

In addition, the authors of [21] study the methods of leverage and local impact for diagnostics in a symmetric CAPM. It is concluded that the considered models give better results than the CAPM with Gaussian distribution.

In [22–24], empirical studies were carried out under the assumption that stock returns have distributions with heavier tails than the normal distribution.

The Student–t distribution instead of the normal distribution was considered in [23] and in [25], taking into account the maximum likelihood method for estimating its parameters. The paper [24] concluded that asset valuation should be carried out within the framework of the CAPM and the discounted dividend model.

23.3 Hamada Model

The Modigliani–Miller theory, with the accounting of taxes has been united with CAPM (capital asset pricing model) in 1961 by Hamada [19, 20]. For the cost of equity of a leveraged company, the below formula has been derived.

$$k_e = k_f + \beta_U(k_m - k_f) + \beta_U(k_m - k_f)\frac{D}{S}(1-t) \tag{23.5}$$

The first term represents risk-free profitability k_f, the second term is business risk premium, $\beta_U(k_m - k_f)$, and the third term is financial risk premium $\beta_U(k_m - k_f)\frac{D}{S}(1-t)$.

In the case of an unlevered company ($D = 0$), the financial risk (the third term) is zero, and its shareholders receive only a business risk premium.

Hamada used an empirical approach in incorporating the level of leverage into the CAPM. One of the main objectives in his research was to distinguish companies without leverage from companies with leverage. The latter make up almost the majority of real companies. In 1972, he surveyed 304 companies, among which he found 102 non-leveraged and 202 leveraged [20]. Comparing equity returns of two types of companies, he got his formula for the β-factor, which takes into account the level of leverage.

In our approach, we do the opposite thing: we incorporate CAPM and Fama–French models into two main theories of the capital structure—the Brusov–Filatova–Orekhova (BFO) theory and the Modigliani–Miller (MM) theory. And we do this analytically.

Our methodology, developed below, avoids this problem of finding non-leveraged companies (which practically do not exist) by clearing the average return of an asset from leverage using the formula (23.7), which allows to find k_0 value (cost of equity at zero leverage level).

Equating CAPM formula (23.1) to right side of formula (23.5), one gets:

$$k_f + \beta(k_m - k_f) = k_f + \beta_U(k_m - k_f) + \beta_U(k_m - k_f)\frac{D}{S}(1-t) \tag{23.6}$$

or

$$\beta = \beta_U\left(1 + \frac{D}{S}(1 - t)\right) = \beta_U(1 + L(1 - t)) \tag{23.7}$$

where β is beta-coefficient for leveraged company.

The Incorrectness of Hamada's Approximation

If we try to combine the CAPM and MM theory not phenomenologically, like Hamada, but analytically, then the incorrectness of Hamada's approximation becomes obvious [26–29]. Substituting the CAPM formula

$$k_0 = k_f + \beta_U\left(k_m - k_f\right) \tag{23.8}$$

into Modigliani–Miller formula for equity cost

$$k_e = k_0 + L(k_0 - k_d)(1 - t), \tag{23.9}$$

one gets the following result

$$\begin{aligned} &k_e = k_0 + L(k_0 - k_d)(1 - t) = \\ &k_f + \beta_U\left(k_m - k_f\right) + L\left(k_f + \beta_U\left(k_m - k_f\right) - k_d\right)(1 - t) = \\ &k_f(1 + L(1 - t)) + \beta_U\left(k_m - k_f\right)(1 + L(1 - t)) - Lk_d(1 - t) \end{aligned} \tag{23.10}$$

The second term is the same as in Hamada's formula (23.7), but the first term is renormalized value of risk-free profitability and the last term, which depends on the cost of debt k_d, is missing from Hamada's formula (23.7).

So the difference with Hamada's formula is:

While in Hamada's formula only beta coefficient is renormalized, in formula (23.10) the first term (risk-free return) is also renormalized by the same factor $(1 + L(1 - t))$ and the last term, depending on the cost of debt k_d, appears, which is absent in Hamada's formula. Factor $(1 - t)$ (tax corrector) exists due to the tax shield.

The incorrectness of Hamada's approximation becomes obvious.

Note that recently, the authors [26–29] for the first time generalized CAPM to take into account both business and financial risks and developed a new model CAPM 2.0. They showed that R. Hamada's attempt to take into account both business and financial risks [19, 20] was untenable, and the formulas he obtained, which are widely used in practice, are incorrect. The authors of [26–29] derived correct formulas that take into account both business and financial risks. The application of the new CAPM 2.0 model to a number of companies is considered and the difference between the results obtained within the framework of CAPM 2.0 and CAPM is demonstrated.

Disadvantages of Hamada's Approach

Difficulty in finding a non-leveraged company. It is clear that the vast majority of companies are leveraged ones because they use debt financing.

23.4 Fama–French Models

23.4.1 Fama–French Three-Factor Model

Fama–French three-factor model [15–18] takes into account two additional risk factors, namely, size and book to market equity along with market beta

$$k_e = k_f + \beta_U(k_m - k_f) + s \cdot \mathrm{SMB} + h \cdot \mathrm{HML} \tag{23.11}$$

where *SMB*—the difference between the returns of companies with large and small capitalization;

HML—the difference between the returns of companies with low and high intrinsic value (indicator B/P)

23.4.2 Fama–French Five-Factor Model

$$k_e = k_f + \beta_U(k_m - k_f) + s \cdot \mathrm{SMB} + h \cdot \mathrm{HML} + r \cdot \mathrm{RMW} + c \cdot \mathrm{CMA} \tag{23.12}$$

where RMW—return on equity; CMA—company capital expenditure.

23.5 Capital Structure Theories

23.5.1 Modigliani–Miller Theory

The first quantitative theory of capital structure was the well-known theory of Nobel laureates Modigliani–Miller [8–10]. This theory was based on a lot of restrictions. One of the main limitations of the Modigliani–Miller theory about the eternity of companies was removed by Brusov et al. in 2008 [6, 7], and modern theories of the cost of capital and capital structure—the Brusov–Filatova–Orekhova theory (BFO theory) was created for companies of arbitrary age (BFO-1 theory) and for companies of arbitrary lifetime (BFO-2 theory) [14]. Brusov–Filatova– Orekhova (BFO) theory, replaced the Modigliani and Miller theory. The authors departed from the Modigliani–Miller assumption about the eternity (infinity of lifetime) of companies and developed an innovative quantitative theory for assessing the main parameters of the financial activity of companies of arbitrary age. The results of the modern BFO theory turn out to be quite different from the results of the Modigliani–Miller theory. They show that the latter, through its perpetuity, underestimates the weighted average cost of capital, the cost of equity of the company, and significantly overestimates the company's value.

Such an incorrect assessment of key performance indicators of companies' financial performance led to an underestimation of the associated risks and the impossibility or serious difficulties in making adequate management decisions, which was one of the implicit causes of the 2008 global financial crisis.

23.5.2 Brusov–Filatova–Orekhova (BFO) Theory

In the Modigliani–Miller theory, there is no time factor (time is equal to infinity), which does not allow to study the dependence of the company's financial performance on the time factor. But Brusov–Filatova–Orekhova theory (BFO theory) was created for companies of arbitrary age and allows to study the dependence of the company's financial performance on the time factor.

Brusov–Filatova–Orekhova and its perpetual limit—Modigliani–Miller (MM) theory—are described by the following formulas for the weighted average cost of capital WACC:

$$\frac{1-(1+\mathrm{WACC})^{-n}}{\mathrm{WACC}}=\frac{1-(1+k_0)^{-n}}{k_0\cdot(1-w_d t[1-(1+k_d)^{-n}])} \tag{23.13}$$

$$\mathrm{WACC}=k_0\cdot(1-w_d t) \tag{23.14}$$

Here, $w_d=\frac{D}{D+S}$—the share of debt capital, $k_e, w_e=\frac{S}{D+S}$—the cost and the share of the equity capital of the company, and $L=D/S$—financial leverage, D—the value of debt capital.

23.5.3 Alternative Expression for WACC

An alternative formula for the WACC, different from the Modigliani–Miller one, has been derived in (Farber, A. et al. 2006; Fernandez, P. 2006; Berk, J.; DeMarzo, P. 2007; Harris, R.;. Pringle, J.1985) from the WACC definition and the balance identity (see as well Berk and De Marzo, 2007):

$$\mathrm{WACC}=k_0(1-w_d T)-k_d t w_d+k_{\mathrm{TS}} t w_d, \tag{23.15}$$

where k_0, k_d, and k_{TS} are the expected returns, respectively, on the unlevered company, the debt, and the tax shield.

Some additional conditions are required equation (23.11) practical applicability. If the WACC is constant over time, as it stated in (Farber, A. et al. 2006), the levered company capitalization is found by discounting with the WACC of the unlevered company.

In textbooks (Berk, J.; DeMarzo, P. 2007; Harris, R.;. Pringle, J.1985), formulas for the special cases, where the WACC is constant, could be found.

In 1963, Modigliani and Miller assumed that the debt value D is constant. Then, as the expected after–tax cash flow of the unlevered firm is fixed, V_0 is constant as well. By assumption, $k_{\mathrm{TS}} = k_D$ and the value of the tax shield is $\mathrm{TS} = tD$. Thus, the capitalization of the company V is a constant and the alternative formula (23.15) becomes a formula for a constant WACC:

$$\mathrm{WACC} = k_0(1 - w_d T) \tag{23.16}$$

Because the debt k_d and the tax shield k_{TS} have debt nature, it seems reasonable that their expected returns are equal as suggested by "classical" Modigliani–Miller (MM) theory, which has been modified by Brusov et al. for cases of practical meanings.

23.5.4 The Miles–Ezzell Model Versus the Modigliani–Miller Theory

Denis M. Becker, in 2021, discussed [30] the differences between the Modigliani–Miller theory and the Miles–Ezzell model [11–13], which deal with the stochasticity of free cash flows. The Modigliani–Miller theory considers a stationary process, while in the Miles–Ezzell model the process is stochastic. The author conducts a numerical experiment that allows you to determine the values and discount rates using a risk-neutral approach. He analyzes three formulas:

Modigliani–Miller theory [8–10]

$$\mathrm{WACC} = k_0 \cdot (1 - w_d t) \tag{23.17}$$

Miles–Ezzell model [11–13]

$$\mathrm{WACC} = k_0 - t \cdot w_d \cdot k_f \cdot \frac{1 + k_0}{1 + k_f} \tag{23.18}$$

Cooper, I. A., & Nyborg, K. G. [31]

$$\mathrm{WACC} = k_0 - k_f \cdot w_d t \tag{23.19}$$

where k_f stands for the risk-free rate, which equals the required return of the debt holders.

The author shows that in the Miles–Ezzell model, all cash flows and the values depend on the path, in contrast to the Modigliani–Miller theory. Also, in the Miles–Ezzell model, all discount rates are time-independent, with the exception of the discount rate used to discount tax shields, which depends on the duration of the cash flows. Conversely, in the Modigliani–Miller theory, all discount rates change over time, except for the constant tax shield discount rate. This affects the applicability of the well-known formula for annuities and the development of models for estimating both finite and perpetual cash flows.

In this paper, Becker [30] raises the issue of paying the debt body together with payment of interest on the debt. Regarding this issue we would like to note that in both classical MM and BFO theories, the body of the debt is not paid. In the framework of the Modigliani–Miller theory, such an account is fundamentally impossible, while in the BFO theory it can be done and was done in the framework of the BFO-2 theory, where the amount of debt D decreases with time. This decrease in the value of debt D results in a decrease in the tax shield.

In Brusov–Filatova–Orekhova and its perpetual limit—Modigliani–Miller (MM) theory, one of the most important parameters is the equity cost of unlevered company k_0. Knowing it, one can evaluate the main financial indicators, such as the cost of capital raised, the discount rates WACC and k_e), the value of the company, V, the cost of equity capital, k_e, and their dependence on debt financing, taxation, company age, etc.

One way to find k_0 is as follows: if we know WACC and L, we substitute these values into the formulas for WACC (or BFO, or MM) and find k_0 from here.

23.6 Incorporating CAPM and Fama–French Models into Capital Structure Theories

If we use the CAPM formula for a company without leverage, then we can use this return value as k_0. If we use the CAPM formula for a leveraged company (Hamada formula), then we should "clear" this formula from leverage and only after that you can use the resulting return value as k_0.

The company profitability k_i is taken from the company's financial statements either for the year, or for the quarter, or for the month, or for a day. If you need to find out the profitability for several periods, you need to use the following formula:

$$k_i = \prod_{k=1}^{n} (1 + k_{ik}) - 1. \tag{23.20}$$

Although in the CAPM it is declared that the company profitability k_e is for an unleveraged company, since it is taken from the company's financial statements, it is clear that it is defined for a leveraged company.

Since both theories of capital structure (BFO and MM) use non-leveraged cost of equity as a seed value for the cost of equity, it is necessary to clean k_i from leverage.

The simplest way to do this is to use the MM formula for the cost of equity

$$k_i = k_0 + L(k_0 - k_d)(1 - t) \tag{23.21}$$

$$k_0 = \frac{k_i + Lk_d(1 - t)}{1 + L(1 - t)} \tag{23.22}$$

Here k_0 is k_i value, cleaned from leverage.

To clean k_f value used BFO theory, one should use formula

$$k_e = \mathrm{WACC}(1 + L) - k_d L(1 - t), \tag{23.23}$$

but you need to know the value of WACC at a specific value of the leverage level L.

We could include a business risk premium at two levels: (1) at market level or (2) at industry level.

To include a market business risk premium, one needs to add to k_0 to the following term:

$$\beta_{im}(k_m - k_F) \tag{23.24}$$

To include an industry business risk premium, one needs to add to $k_0$to the following term:

$$\beta_{iI}(k_I - k_F) \tag{23.25}$$

To include effects described by Fama–French model, we should add the following terms:

either two terms within Fama–French three-factor Model

$$s \cdot \mathrm{SMB} + h \cdot \mathrm{HML} \tag{23.26}$$

or four terms within Fama–French Five-factor Model

$$s \cdot \mathrm{SMB} + h \cdot \mathrm{HML} + r \cdot \mathrm{RMW} + c \cdot \mathrm{CMA} \tag{23.27}$$

As a result we get

$$\begin{aligned} k_0 &= k_f + \beta_U\left(k_m - k_f\right) + s \cdot \mathrm{SMB} + h \cdot \mathrm{HML} \\ &= \frac{k_f + Lk_d(1 - t)}{1 + L(1 - t)} + \beta_U\left(k_m - k_f\right) + s \cdot \mathrm{SMB} + h \cdot \mathrm{HML} \end{aligned} \tag{23.28}$$

$$k_0 = k_f + \beta_U(k_m - k_f) + s \cdot \text{SMB} + h \cdot \text{HML} + r \cdot \text{RMW} + c \cdot \text{CMA}$$
$$= \frac{k_f + Lk_d(1-t)}{1+L(1-t)} + \beta_U(k_m - k_f) + s \cdot \text{SMB} + h \cdot \text{HML} + r \cdot \text{RMW} + c \cdot \text{CMA} \tag{23.29}$$

We use this value $k_0$to study the dependence of the company's main financial indicators on debt financing, the cost of debt, taxation, the age of the company, etc.

$$k_0 = k_f + s \cdot \text{SMB} + h \cdot \text{HML} =$$
$$= \frac{k_f + \beta_U(k_m - k_f) + Lk_d(1-t)}{1+L(1-t)} + s \cdot \text{SMB} + h \cdot \text{HML} \tag{23.30}$$

$$k_0 = k_f + s \cdot \text{SMB} + h \cdot \text{HML} + r \cdot \text{RMW} + c \cdot \text{CMA}$$
$$= \frac{k_f + \beta_U(k_m - k_f) + Lk_d(1-t)}{1+L(1-t)} + s \cdot \text{SMB} + h \cdot \text{HML} + r \cdot \text{RMW} + c \cdot \text{CMA} \tag{23.31}$$

23.7 The Algorithm for Using This Innovative Technique

1. Take the average return on the asset for the year from statistical data
2. Clean of it from leverage using equation (23.9) for k_e.

It can be seen from Table 23.1 that the values of k_0 are always lower than the cost of equity with non-zero leverage, as it should be, given that the cost of equity increases with leverage.

3. Add a market business risk premium, or an industry one.

   Let us give an example for GAZP company for 2018.

   To find premium to k_0, let us calculate Δk_0 within market CAPM approach:

Table 23.1 Cost of equity with non-zero leverage and its cleaned value k_0 (GAZP)

Indicator	2021	2020	2019	2018
k_0	3.529%	6.106%	5.587%	4.542%
Leverage level, L	0.500	0.538	0.448	0.486
Debt cost, k_d	0.017	0.021	0.017	0.020
GAZP, average profitability	4.26%	7.83%	6.98%	5.53%
GAZP, standard deviation	0.945%	0.840%	0.415%	0.310%

$$\beta = 1.16;$$

$$k_m - k_f = 12.3\% - 7.75\% = 4.55\%;$$

$$\Delta k_0 = 1.16 \cdot 4.55\% = 5.278\%.$$

Finally, one gets

$$\widehat{k_0} = k_0 + \Delta k_0 = 1.16 \cdot 4.55\% = 4.542\% + 5.278\% = 9.82\%$$

It is easy to see, that account of business risk within market CAPM approach changes k_0 from 4.542% to $\widehat{k_0}$= 9.82%.

It is seen that the use of $\widehat{k_0}$, which takes into account the business risk, will significantly change the results of calculation of the financial indicators of the GAZP.

4. Add Fama–French corrections (use either three factor Fama–French model or five factor Fama–French model.
5. Now we are ready to use obtained k_0 value to calculate the main financial parameters of the company: the cost of raising capital, WACC, company value, etc. The theory of Brusov–Filatova–Orekhova (BFO) and its eternal limit—the theory of Modigliani–Miller (MM) have recently been generalized to the established practice of the functioning of companies. This generalization took into account the real conditions of the company's activities, such as variable income, frequent income tax payments, advance income tax payments, etc. This made it possible to investigate the impact of these conditions on its main financial indicators (see Review [14]).

A few practical calculations have been made for real companies (see below in Sects. 23.9 and 23.10). They focus on (1) applying two versions of CAPM (market or industry) to real companies (PJSC Polymetal and PJSC Severstal); (2) application to Apple company of a new methodology developed by us for assessing the financial performance of a company, taking into account both business (market or industry) and financial risks.

23.8 WACC Formulas for Brusov–Filatova–Orekhova (BFO) Theory and for Modigliani–Miller (MM) Theory

Below we give a summary of the WACC formulas for Brusov–Filatova–Orekhova (BFO) theory as well as for Modigliani–Miller (MM) theory (see Review [14]).

23.8.1 *Variable Income Case*

23.8.1.1 Income Tax Payments at the Ends of Periods

$$\text{BFO}: \quad \frac{1-\left(\frac{1+g}{1+\text{WACC}}\right)^{n}}{\text{WACC}-g}=\frac{1-\left(\frac{1+g}{1+k_0}\right)^{n}}{(k_0-g)\cdot(1-w_d t[1-(1+k_d)^{-n}])} \tag{23.32}$$

$$\text{MM}: \quad \text{WACC}=(k_0-g)\cdot(1-w_d t)+g \tag{23.33}$$

23.8.1.2 Advance Income Tax Payments

$$\text{BFO}: \quad \frac{1-\left(\frac{1+g}{1+\text{WACC}}\right)^{n}}{\text{WACC}-g} = \frac{1-\left(\frac{1+g}{1+k_0}\right)^{n}}{(k_0-g)\cdot(1-w_d t[1-(1+k_d)^{-n}]\cdot(1+k_d))} \tag{23.34}$$

$$\text{MM}: \quad \text{WACC}=(k_0-g)\cdot(1-w_d t\cdot(1+k_d))+g \tag{23.35}$$

23.8.2 *Frequent Income Tax Payments*

23.8.2.1 Income Tax Payments at the Ends of Periods

$$\text{BFO}: \quad \frac{1-(1+\text{WACC})^{-n}}{\text{WACC}}=\frac{1-(1+k_0)^{-n}}{k_0\cdot\left(1-\frac{k_d w_d t}{p}\frac{[1-(1+k_d)^{-n}]}{(1+k_d)^{1/p}-1}\right)} \tag{23.36}$$

$$\text{MM}: \quad \text{WACC}=k_0\cdot\left(1-\frac{k_d w_d t}{p\cdot\left[(1+k_d)^{1/p}-1\right]}\right) \tag{23.37}$$

23.8.2.2 Advance Income Tax Payments

$$\text{BFO}: \quad \frac{1-(1+\text{WACC})^{-n}}{\text{WACC}} = \frac{1-(1+k_0)^{-n}}{k_0 \cdot \left(1 - \frac{k_d w_d t}{p} \frac{[1-(1+k_d)^{-n}]\cdot(1+k_d)^{1/p}}{(1+k_d)^{1/p}-1}\right)} \tag{23.38}$$

$$\text{MM}: \quad \text{WACC} = k_0 \cdot \left(1 - \frac{k_d w_d t \cdot (1+k_d)^{1/p}}{p \cdot \left[(1+k_d)^{1/p}-1\right]}\right) \tag{23.39}$$

23.8.3 *Simultaneous Accounting of Variable Income in Case of Frequent Income Tax Payments*

23.8.3.1 Income Tax Payments at the Ends of Periods

$$\text{BFO}: \quad \frac{1-\left(\frac{1+g}{1+\text{WACC}}\right)^n}{\text{WACC}-g} = \frac{1-\left(\frac{1+g}{1+k_0}\right)^n}{(k_0-g)\cdot\left(1 - \frac{k_d w_d t}{p} \frac{[1-(1+k_d)^{-n}]}{(1+k_d)^{1/p}-1}\right)} \tag{23.40}$$

$$\text{MM}: \quad \text{WACC}-g = (k_0-g)\cdot\left(1 - \frac{k_d w_d t}{p\cdot(1+k_d)^{1/p}-1}\right) \tag{23.41}$$

23.8.3.2 Advance Income Tax Payments

$$\text{BFO}: \quad \frac{1-\left(\frac{1+g}{1+\text{WACC}}\right)^n}{\text{WACC}-g} = \frac{1-\left(\frac{1+g}{1+k_0}\right)^n}{(k_0-g)\cdot\left(1 - \frac{k_d w_d t}{p} \frac{[1-(1+k_d)^{-n}]\cdot(1+k_d)^{1/p}}{\left[(1+k_d)^{1/p}-1\right]}\right)} \tag{23.42}$$

$$\text{MM}: \quad \text{WACC}-g = (k_0-g)\cdot\left(1 - \frac{k_d w_d t\cdot(1+k_d)^{1/p}}{p\cdot(1+k_d)^{1/p}-1}\right) \tag{23.43}$$

23.9 Application of Two Version of CAPM (Market and Industry) to Real Companies

23.9.1 Estimation of the Cost of Equity of PJSC Polymetal for the Period 2018–2022 by CAPM Capital Asset Pricing Model (Ticker PJSC Polymetal on the Moscow Exchange is POLY) (Tables 23.2 and 23.3)

Table 23.2 Summary table of indicators for Polymetal shares, the RTS mining and metal index and the MICEX index in the period 2018–2022

Year	2018	2019	2020	2021	2022
Company level (Polymetal)					
Profitability actual	3.48%	32.80%	78.71%	−24.39%	−71.71%
Standard deviation	0.304	0.242	0.454	0.264	0.705
Average debt cost	3.52%	4.89%	4.00%	2.88%	3.28%
Leverage level	4.78	3.41	1.57	0.54	1.6
Industry level (RTS mining and metal index)					
Profitability actual	2.11%	40.20%	56.08%	12.37%	−9.80%
Standard deviation	0.233	0.144	0.390	0.230	0.580
Average leverage level	0.408	0.370	0.351	1.128	0.818
Beta with Polymetal	0.208	0.107	0.436	0.349	0.250
Profitability (industry CAPM)	6.79%	11.07%	27.97%	9.10%	4.96%
Correlation with Polymetal	0.27	0.18	0.51	0.40	0.30
Market level (Moscow Exchange index MICEX)					
Profitability actual	18.15%	36.24%	22.57%	3.25%	−16.78%
Standard deviation	0.167	0.120	0.271	0.163	0.497
Beta with Poly	0.443	0.189	0.516	0.307	0.508
Profitability (market CAPM)	12.51%	12.99%	14.67%	6.09%	−3.68%
Correlation with Polymetal	0.24	0.09	0.31	0.19	0.36

Table 23.3 Actual and projected returns based on CAPM models for Polymetal shares in the period 2018–2022

Profitability actual	3.48%	32.80%	78.71%	−24.39%	−71.71%
Profitability (industry CAPM)	2.11%	40.20%	56.08%	12.37%	−9.80%
Profitability (market CAPM)	12.51%	12.99%	14.67%	6.09%	−3.68%

Source: Compiled by the authors

This Table gives: (1) the company's average annual return; (2) the company's profitability with an industry business risk premium; (3) profitability of a company with a market business risk premium

23.9.2 *Estimation of the Cost of Equity of PJSC Severstal for the Period 2018–2022 by CAPM Capital Asset Pricing Model (Ticker PJSC Severstal on the Moscow Exchange is CHMF) (Table 23.4)*

Table 23.4 Estimation of the cost of equity of PJSC Severstal for the period 2018–2022 by CAPM

	Indicators	2018	2019	2020	2021	2022
μ_F		8.02%	7.59%	6.27%	7.34%	9.87%
Company PJSC	μ_i	6.23%	−0.53%	41.01%	21.27%	−43.66%
	L	1.21	1.75	1.51	1.37	–
	k_d	3.77%	3.98%	3.38%	3.36%	–
	σ_i	0.24	0.19	0.26	0.28	0.53
Industry	μ_I	8.71%	10.68%	47.75%	7.37%	−46.96%
	L	0.41	0.81	0.66	0.68	0.64
	σ_I	0.19	0.11	0.24	0.17	0.45
	$\beta_{i,\,I}$	0.40	−0.19	0.91	1.42	1.10
Market	μ_m	12.20%	28.58%	8.06%	15.08%	−43.10%
	σ_m	0.17	0.11	0.26	0.16	0.52
	$\beta_{i,\,m}$	0.77	−0.73	0.69	0.85	0.94
μ_i		**6.23%**	**−0.53%**	**41.01%**	**21.27%**	**−43.66%**
μ_i CAPM (Industry)		**8.29%**	**6.99%**	**44.22%**	**7.38%**	**−52.55%**
μ_i CAPM (market)		**11.22%**	**−7.64%**	**7.51%**	**13.93%**	**−40.11%**

The last three lines give: (1) the company's average annual return; (2) the company's profitability with an industry business risk premium; (3) profitability of a company with a market business risk premium

23.10 Application of New Methodology for Assessing the Company's Financial Performance, Taking into Account Both Business (Market and Industry) and Financial Risks to Real Companies

Below we apply the new methodology for assessing the company's financial performance developed by us, taking into account both business (market and industry) and financial risks to Apple company in 2019–2021.

As Market we take S&P 500; as Industry (sector): S&P500 Information Technology and we consider period 2019–2021.

23.10.1 The Collection and Processing of Statistical Data About the Company, Industry, and Market and Two Versions of CAPM (Market and Industry) (Table 23.5).

Table 23.5 Results for parameters of company, industry (sector), and market

Index	Unit	Source	2019	2020	2021
Average annual return per share (Apple)	%	**Authors's calculation**	**90.04**	**71.87**	**37.36**
Average annual market return (S&P 500)	%	Authors's calculation	28.86	16.18	26.88
10–year US Treasuries	%	Investing.com	3.75	3.75	3.75
Standard deviation of stock return (Apple)	%	Authors's calculation	1.66	2.94	1.58
Standard Deviation of Market Returns (S&P 500)	%	Authors's calculation	0.79	2.19	0.83
Beta	X	Authors's calculation	1.57	1.12	1.31
Expected stock return	%	**Authors's calculation**	**43.13**	**17.68**	**34.07**
Average annual return of the sector (S&P500 Information Technology)	%	Authors's calculation	48.00	41.98	33.33
10–year US Treasuries	%	Investing.com	3.75	3.75	3.75
Standard deviation of stock return (Apple)	%	Authors's calculation	1.66	2.94	1.58
Sector Return Standard Deviation (S&P500 Information Technology)	%	Authors's calculation	1.14	2.57	1.23
Beta	X	Authors's calculation	1.22	1.04	1.06
Expected stock return	%	**Authors's calculation**	**57.57**	**43.65**	**35.00**
Interest expenses	mln $	Apple Annual Report 2019 Apple Annual Report 2020 Apple Annual Report 2021	3.58	2.87	2.65
Total debt	mln $	Apple Annual Report 2019 Apple Annual Report 2020 Apple Annual Report 2021	102.07	107.44	118.72
Debt cost	%	**Authors's calculation**	**3.50**	**2.67**	**2.23**

(continued)

Table 23.5 (continued)

Index	Unit	Source	2019	2020	2021
Total debt	mln $	Apple Annual Report 2019 Apple Annual Report 2020 Apple Annual Report 2021	102.07	107.44	118.72
Equity value	mln $	Apple Annual Report 2019 Apple Annual Report 2020 Apple Annual Report 2021	90.49	65.34	63.09
Leverage level	**X**	**Authors's calculation**	**1.13**	**1.64**	**1.88**

Comments

The following formulas were used to calculate the company's profitability using the CAPM model:

For calculation by industry (sector):MI = $\mu F - \beta$iI($\mu I - \mu F$),where MI—Average annual return per share, μF—US market risk-free rate, βiI—industry beta (sector), μI—Average annual return of the sector (S&P500 Information Technology)

For calculation by market: $Mm = \mu F - \beta\text{im}(\mu m - \mu F)$, where Mm—Average annual return per share, μF—US market risk free rate, βim—market beta, μm—average annual market return in the market (S&P500)

The average yield of 10-year US Treasury boxes was taken as the risk-free rate in the US market.

Data on daily market quotes, as well as daily quotes of the S&P500 index and the S&P500 Information Technology sector are taken from a specialized Internet resource https://ru.investing.com/.

For calculations, μ_i was taken—the average annual return on shares. To calculate the average annual return, it was first necessary to find the average daily return. The average daily return was calculated using the logarithmic return formula:

$\mu d = \ln\left(\frac{P_t}{P_{t-1}}\right)$, where µd—daily yield, P_t—share price on date t, $P_{(t-1)}$—share price on date $t-1$.

The values obtained using this formula were averaged for each study period (year). The average daily returns were then converted to annual averages using the following formula: $\mu_i = (1 + \mu_d)^n$—1, where n—number of trading days on the exchange.

The beta coefficient was found using the following formulas:

For the industry (sector): $\beta iI = \frac{\text{Cov}iI}{\delta^2 I}$, where Cov$iI$—covariance between returns on stocks and returns on the industry (sector) under study, $\delta^2 I$—sector variance

For market: $\beta im = \frac{\text{Covim}}{\delta^2 m}$, where *Covim*—covariance between stock returns and market returns, $\delta^2 m$—market variance.

Also, the beta coefficient was calculated using the standard deviation, the results agreed with the previous calculations (for more details, see the Excel table).

For Apple, the cost of debt and financial leverage ratio were also calculated.

The cost of debt was calculated using the formula: $I = \frac{IE}{TD_a}$, where I is the cost of debt, IE is the interest expense for the period, TDa is the debt on which interest is charged.

The leverage level was calculated as the ratio of debt (total liabilities) to the company's equity capital.

The data for the calculations above were taken from Apple's annual reports for 2019–2021.

Conclusions:

Comparing the average annual return of the Apple stock with the average annual returns of the market (S&P500) and the sector (S&P500 Information Technology), we can say that the company's shares show more profitability, but at the same time more volatility, and therefore riskiness.

The standard deviation of the return on Apple stock is at the level of 2–3%, which is higher than the values for the sector and the market by ~ 1 p.p. At the same time, there is a sharp increase in 2020, which may be due to the COVID-19 pandemic.

It is important to analyze the average annual return of Apple stock with expected returns under the CAPM model. The model is considered relative to the S&P500 market and the S&P500 Information Technology sector.

Beta coefficients were calculated for both models. In all options, the coefficients turned out to be greater than one, which indicates a correlation between the price dynamics of Apple's stock and the dynamics of the market and the sector, but at the same time, the volatility of the shares is higher, which means that the risks are higher. At the same time, it should be noted that the beta calculated for the sector is close to 1, which indicates a higher correlation than for the market (the company's shares follow the sector's trend, the company's risks are almost equivalent to the sector-wide ones).

This can be explained by the fact that only companies from the IT sector are represented in the S&P500 Information Technology sector, while several industries are represented in the S&P500 market.

In general, comparing the expected returns under the CAPM model and the real average annual return of the Apple stock for the period 2019–2021, it can be observed that the forecast indicators for the sectoral CAPM model are closer to real values than for the market model, which is explained by a higher correlation. At the same time, the real values of Apple's average annual return are higher with positive values and lower with negative ones, which indicates increased volatility and riskiness of the security. Apple's leverage ratio is below 2, which may indicate that the company is funded largely by equity.

Table 23.6 Results for three values of k_0

Indicator	Unit	2019	2020	2021
μi	%	85.99	80.21	33.80
μ0	%	46.86	36.16	14.83
k_0 (1st value)	%	46.86	36.16	14.83
k_0 (2nd value)	%	100.69	76.06	46.08
k_0 (3rd value)	%	86.24	50.09	45.15

23.10.2 Three Values of k_0 (Table 23.6)

For further calculations, μ_i was taken—the average annual return on shares. To calculate the average annual return, it was first necessary to find the average daily return. The average daily return was calculated using the logarithmic return formula: $\mu d = \ln\left(\frac{P_t}{P_{t-1}}\right)$, where Rd—daily yield, P_t—share price on the date t, P_{t-1}—share price on the date $t-1$.

The values obtained using this formula were averaged for each study period (year). The average daily returns were then converted to annual averages using the following formula: $\mu_i = (1 + \mu_d)^n - 1$, where n—number of trading days on the exchange.

Next, μ_0 was found by the formula: $\mu_0 = \frac{\mu_i + L\, k_d(1-t)}{1+L(1-t)}$, where L—leverage level, t—income tax rate. After that, k_0 was calculated according to three conditions:

$k_0(1) = \mu_0$
$k_0(2)$ for industry $= \mu_0 + \beta_{iI}\,(\mu_I - \mu_F)$
$k_0(3)$ for market$= \mu_0 + \beta_{iM}\,(\mu_M - \mu_F)$

Based on the data obtained, indicators were calculated for two models: the Modigliani–Miller and the Brusov–Filatova–Orekhova theories.

It was found the dependence of WACC, V, k_e indicators on the level of leverage (change in leverage from 0 to 10).

23.10.3 Calculations of Indicators for the Modigliani–Miller—CAPM Model

The following formulas were used to calculate the indicators:

WACC $= k_0(1 - w_d t)$; $w_d = \frac{L}{L+1}$; $V = \frac{CF}{WACC}$, CF = EBITDA; $k_e = k_0 + L(k_0 - k_d)(1 - t)$

The results obtained are presented in tables of dependencies of the values of indicators (WACC, V, k_e) on the level of leverage for each year for three values of k_0, as well as in figures (Table 23.7, Fig. 23.1).

Table 23.7 Dependence of WACC on leverage level L, WACC (L) for 2019–2021 years

	2019			2020			2021		
L	WACC 1	WACC2	WACC3	WACC1	WACC 2	WACC3	WACC1	WACC 2	WACC 3
0	46.86%	100.69%	86.24%	36.16%	76.06%	50.09%	14.83%	46.08%	45.15%
1	42.18%	90.62%	77.62%	32.54%	68.46%	45.08%	13.35%	41.47%	40.63%
2	40.61%	87.26%	74.74%	31.34%	65.92%	43.41%	12.85%	39.94%	39.13%
3	39.83%	85.58%	73.30%	30.73%	64.65%	42.57%	12.61%	39.17%	38.38%
4	39.36%	84.58%	72.44%	30.37%	63.89%	42.07%	12.46%	38.71%	37.92%
5	39.05%	83.91%	71.87%	30.13%	63.39%	41.74%	12.36%	38.40%	37.62%
6	38.83%	83.43%	71.46%	29.96%	63.02%	41.50%	12.29%	38.18%	37.41%
7	38.66%	83.07%	71.15%	29.83%	62.75%	41.32%	12.23%	38.02%	37.25%
8	38.53%	82.79%	70.91%	29.73%	62.54%	41.18%	12.19%	37.89%	37.12%
9	38.43%	82.56%	70.72%	29.65%	62.37%	41.07%	12.16%	37.79%	37.02%
10	38.34%	82.38%	70.56%	29.58%	62.23%	40.98%	12.13%	37.70%	36.94%

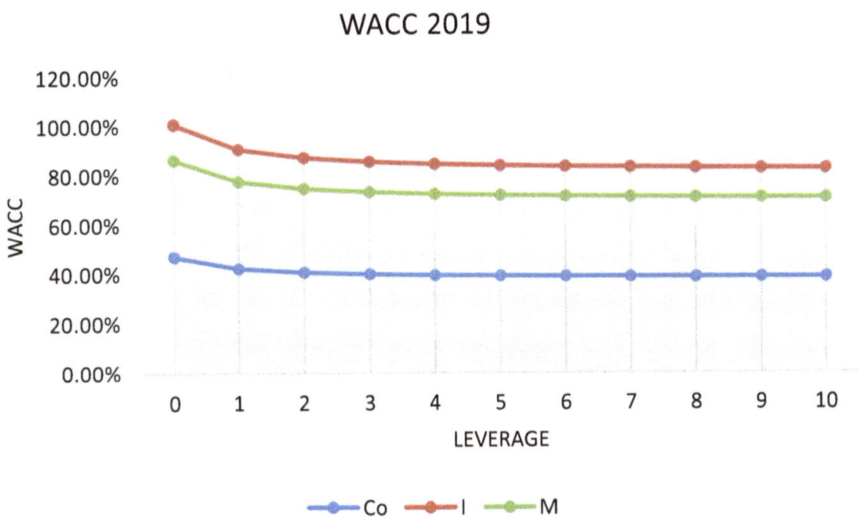

Fig. 23.1 Dependence of WACC on leverage level L, WACC (L) for year 2019

Table 23.8 Dependence of company value, V on leverage level L, $V(L)$ for 2019–2021 years

	2019			2020			2021		
L	V1	V2	V3	V1	V2	V3	V1	V2	V3
0	166.70	77.59	90.58	235.52	111.96	170.02	864.59	278.25	284.00
1	185.22	86.21	100.65	261.69	124.40	188.92	960.66	309.16	315.55
2	192.35	89.52	104.52	271.75	129.18	196.18	997.61	321.05	327.69
3	196.12	91.28	106.57	277.08	131.72	200.03	1 017.17	327.35	334.11
4	198.45	92.37	107.84	280.38	133.29	202.41	1 029.28	331.25	338.09
5	200.04	93.11	108.70	282.62	134.35	204.03	1 037.51	333.90	340.79
6	201.19	93.64	109.33	284.25	135.12	205.20	1 043.47	335.82	342.75
7	202.06	94.05	109.80	285.48	135.71	206.09	1 047.99	337.27	344.24
8	202.75	94.36	110.17	286.44	136.17	206.79	1 051.53	338.41	345.40
9	203.30	94.62	110.47	287.22	136.54	207.35	1 054.38	339.33	346.34
10	203.75	94.83	110.71	287.86	136.84	207.81	1 056.73	340.08	347.11

Here and below in the figures: Co means company; I means industry (sector); M stands for market (Tables 23.8 and 23.9, Figs. 23.2, 23.3, 23.4, 23.5, 23.6, 23.7, 23.8, and 23.9).

Table 23.9 Dependence of cost of equity k_e on leverage level L, k_e (L) for 2019–2021 years

	2019			2020			2021		
L	k_e 1	k_e 2	k_e 3	k_e 1	k_e 2	k_e 3	k_e 1	k_e 2	k_e 3
0	46.86%	100.69%	86.24%	36.16%	76.06%	50.09%	14.83%	46.08%	45.15%
1	81.55%	178.43%	152.43%	62.95%	134.77%	88.02%	24.91%	81.16%	79.48%
2	116.24%	256.18%	218.62%	89.73%	193.48%	125.95%	34.99%	116.24%	113.82%
3	150.92%	333.93%	284.81%	116.52%	252.19%	163.88%	45.07%	151.33%	148.16%
4	185.61%	411.67%	351.00%	143.31%	310.90%	201.81%	55.16%	186.41%	182.49%
5	220.30%	489.42%	417.19%	170.09%	369.61%	239.74%	65.24%	221.49%	216.83%
6	254.99%	567.17%	483.38%	196.88%	428.32%	277.67%	75.32%	256.57%	251.16%
7	289.67%	644.91%	549.57%	223.67%	487.03%	315.60%	85.40%	291.65%	285.50%
8	324.36%	722.66%	615.76%	250.46%	545.74%	353.53%	95.48%	326.74%	319.84%
9	359.05%	800.40%	681.95%	277.24%	604.46%	391.46%	105.56%	361.82%	354.17%
10	393.74%	878.15%	748.14%	304.03%	663.17%	429.39%	115.65%	396.90%	388.51%

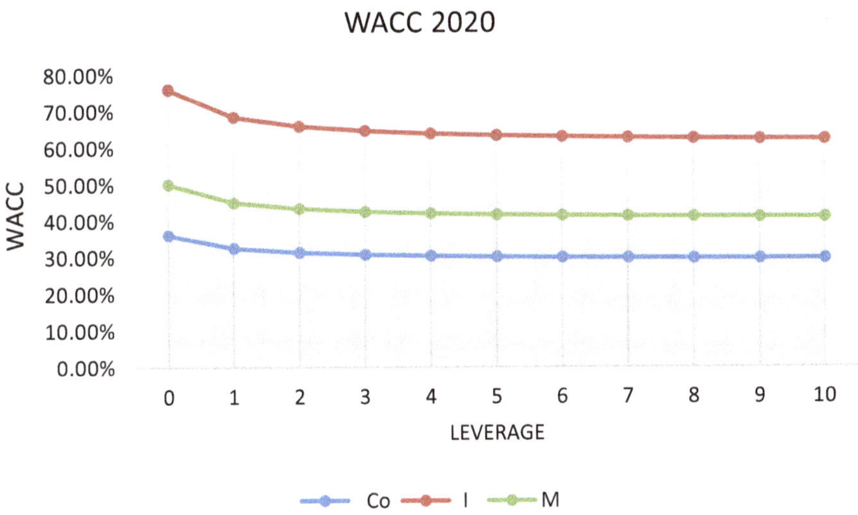

Fig. 23.2 Dependence of WACC on leverage level L, WACC (L) for year 2020

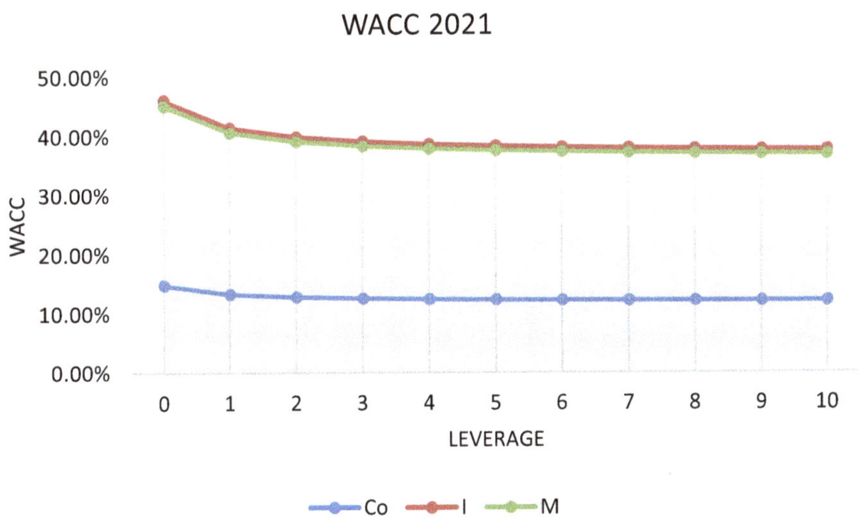

Fig. 23.3 Dependence of WACC on leverage level L, WACC (L) for year 2021

23.10.4 Calculations of Indicators for the Brusov–Filatova–Orekhova—CAPM Model

To calculate indicators according to the BFO model, it is necessary to take into account the duration of the company's operation

Date Apple was founded: 04/01/1976

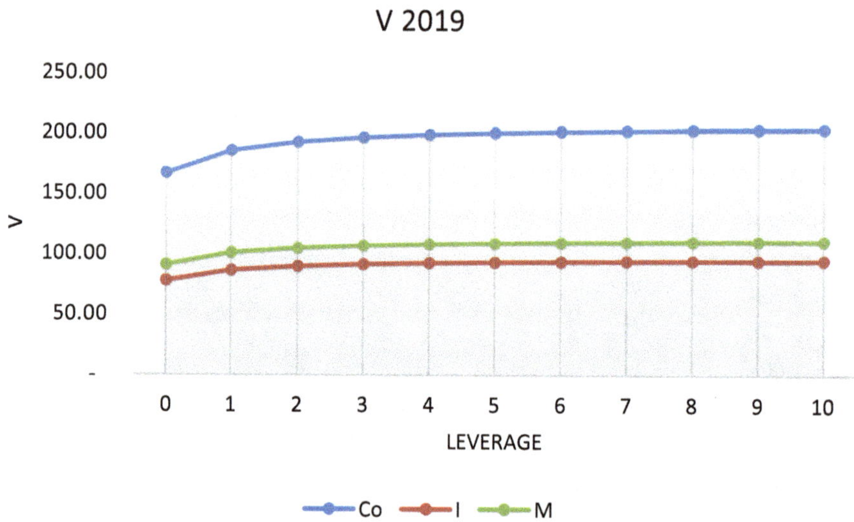

Fig. 23.4 Dependence of company value, V on leverage level L, $V(L)$ for year 2019

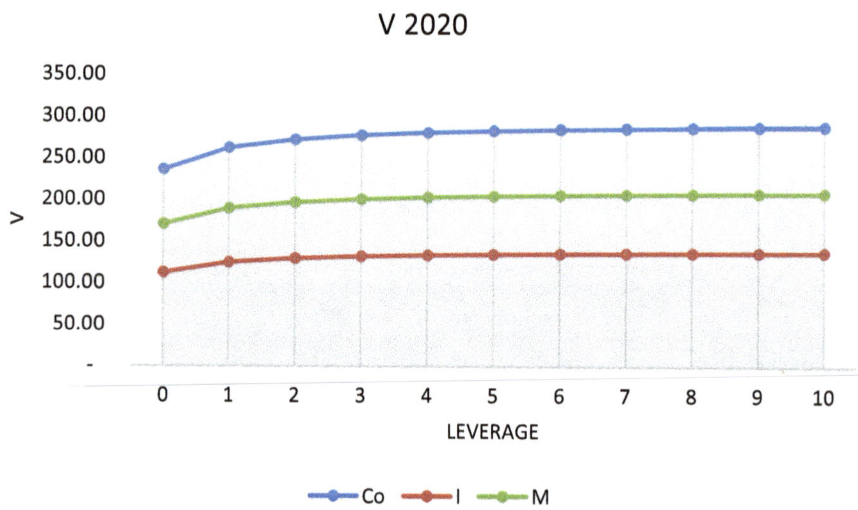

Fig. 23.5 Dependence of company value, V on leverage level L, $V(L)$ for year 2020

The duration of the company's operation (n) is calculated by the formula: $n = t_1 - t_0$, where t_1 is the current year, t_0 is the year the company was founded.

The following formulas were used in the calculations:

$$\frac{1-(1+\mathrm{WACC})^{-n}}{\mathrm{WACC}} = \frac{1-(1+k_0)^{-n}}{k_0\left(1-w_d t\left[1-(1+k_d)^{-n}\right]\right)}$$

$$V = \frac{\mathrm{CF}}{\mathrm{WACC}}\left(1-(1+\mathrm{WACC})^{-n}\right)$$

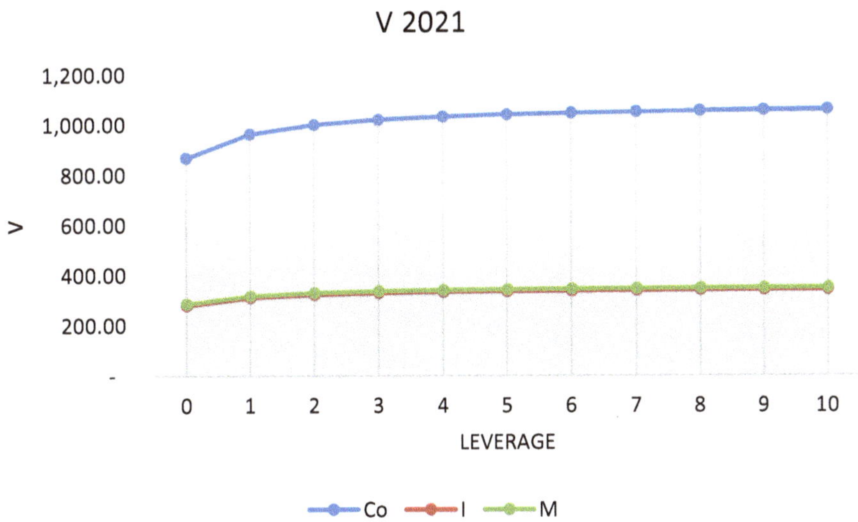

Fig. 23.6 Dependence of company value, V on leverage level L, $V(L)$ for year 2021

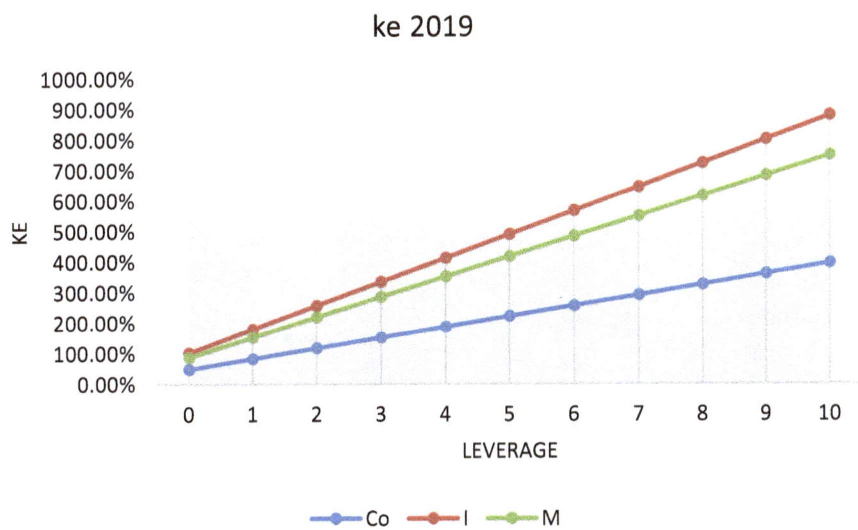

Fig. 23.7 Dependence of cost of equity k_e on leverage level L, k_e (L) for year 2019

$$k_e = \text{WACC}(1 + L) - Lk_d(1 - t)$$

WACC was found using Excel's "Search for Solution" function.

It was considered how the WACC, V, k_e indicators change relative to the change in the level of leverage (change in leverage from 0 to 10).

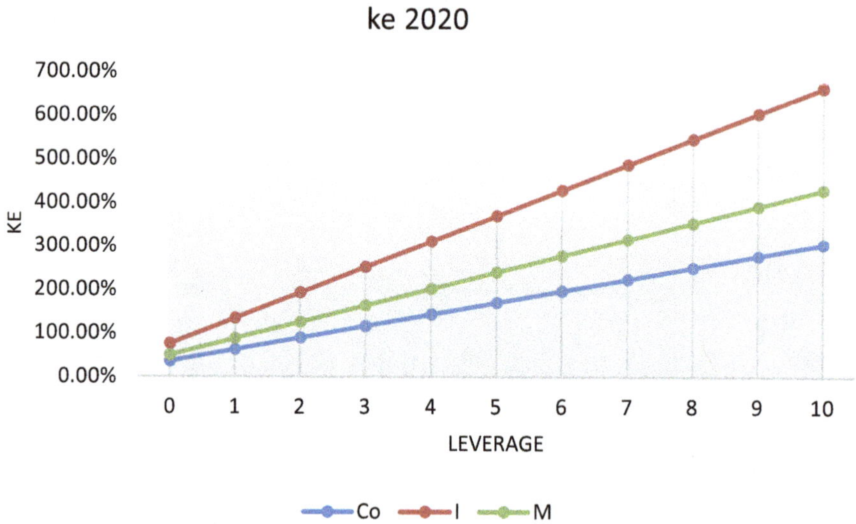

Fig. 23.8 Dependence of cost of equity k_e on leverage level L, k_e (L) for year 2020

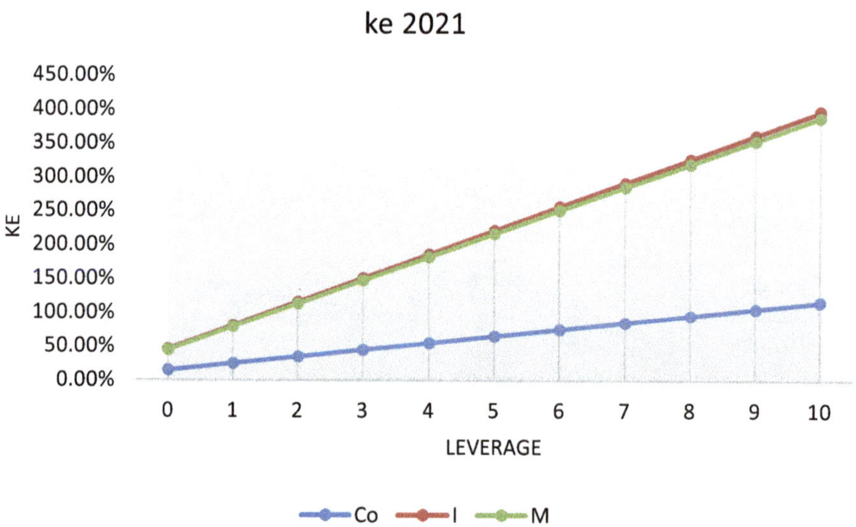

Fig. 23.9 Dependence of cost of equity k_e on leverage level L, k_e (L) for year 2021

Let's move on to the results: the paper presents tables of indicator values (WACC, V, k_e) for each year in three versions (depending on k_0), as well as figures of these indicators (Tables 23.10, 23.11, and 23.12, Figs. 23.10, 23.11, 23.12, 23.13, 23.14, 23.15, 23.16, 23.17, and 23.18).

Table 23.10 Dependence of WACC on leverage level L, WACC (L) for years 2019–2021

	2019			2020			2021		
L	WACC1	WACC2	WACC3	WACC1	WACC 2	WACC3	WACC1	WACC2	WACC3
0	46.86%	100.69%	86.24%	36.16%	76.06%	50.09%	14.83%	46.08%	45.15%
1	43.24%	92.91%	79.58%	33.67%	70.84%	46.65%	13.88%	43.18%	42.31%
2	42.04%	90.32%	77.36%	32.85%	69.10%	45.50%	13.57%	42.22%	41.36%
3	41.43%	89.02%	76.25%	32.43%	68.23%	44.93%	13.41%	41.73%	40.89%
4	41.07%	88.24%	75.58%	32.18%	67.70%	44.58%	13.32%	41.44%	40.60%
5	40.83%	87.72%	75.14%	32.02%	67.35%	44.35%	13.25%	41.25%	40.41%
6	40.66%	87.35%	74.82%	31.90%	67.11%	44.19%	13.21%	41.11%	40.28%
7	40.53%	87.07%	74.58%	31.81%	66.92%	44.07%	13.17%	41.01%	40.18%
8	40.43%	86.86%	74.40%	31.74%	66.77%	43.97%	13.15%	40.93%	40.10%
9	40.35%	86.69%	74.25%	31.69%	66.66%	43.89%	13.13%	40.86%	40.04%
10	40.28%	86.54%	74.13%	31.64%	66.56%	43.83%	13.11%	40.81%	39.98%

Table 23.11 Dependence of company value V on leverage level L, $V(L)$ for years 2019–2021

	2019			2020			2021		
L	V1	V2	V3	V1	V2	V3	V1	V2	V3
0	166.70	77.59	90.58	235.52	111.96	170.02	862.88	278.25	284.00
1	180.66	84.08	98.17	252.89	120.22	182.56	920.80	296.92	303.06
2	185.84	86.50	100.99	259.26	123.25	187.16	941.87	303.72	309.99
3	188.55	87.76	102.46	262.57	124.82	189.55	952.77	307.24	313.58
4	190.21	88.53	103.36	264.60	125.78	191.02	959.44	309.38	315.78
5	191.34	89.05	103.97	265.97	126.43	192.00	963.93	310.83	317.26
6	192.15	89.43	104.41	266.95	126.90	192.72	967.17	311.88	318.32
7	192.76	89.72	104.75	267.70	127.26	193.25	969.61	312.66	319.12
8	193.24	89.94	105.01	268.28	127.53	193.67	971.52	313.28	319.75
9	193.63	90.12	105.22	268.75	127.76	194.01	973.05	313.77	320.26
10	193.94	90.27	105.39	269.13	127.94	194.29	974.30	314.18	320.67

23.11 Conclusions

A fundamentally new approach to assessing the profitability of an asset is proposed. Transition from CAPM, which takes the same risk-free return for all assets as an initial assessment, to a new methodology, in which the average return of an asset, cleared of leverage, with the addition of a premium for business risk (market or industry) is taken as a seed return, significantly improves the accuracy of the estimate.

A methodology has been developed for assessing the profitability of assets, taking into account both business (systematic) and financial risks. For this purpose, the CAPM and Fama–French models were incorporated in two main theories of the capital structure—the Brusov–Filatova–Orekhova (BFO) theory and the Modigliani–Miller (MM) theory. The developed approach makes it possible to use the powerful tools of these highly developed theories for the correct assessment of the main financial indicators of the company and their forecasting, taking into account both types of risks. The dependences of these indicators on debt financing, leverage level, taxing, company age, debt cost, etc. can be studied. The latest versions of the BFO and MM theories of capital structure, developed by the authors and adapted to the established financial practice of the functioning of companies, are used, taking into account the real conditions of their work, such as variable income, frequent income tax payments, advance income tax payments, etc.

The example for GAZP, given by us, shows that account of business risk within market CAPM approach change k_0 from 4.542% to $\widehat{k}_0 = 9.82\%$. This will significantly change the results of calculation of the financial indicators of the GAZP.

Detailed practical calculations for real companies (Apple, Severstal, Polymetal) have been made. They focus on (1) applying two versions of CAPM (market or industry) to real companies; (2) application to real companies of a new methodology developed by us for assessing the financial performance of a company, taking into account both business (market or industry) and financial risks. These calculations

Table 23.12 Dependence $k_e(L)$; years 2019–2021

	2019			2020			2021		
L	k_e 1	k_e 2	k_e 3	k_e 1	k_e 2	k_e 3	k_e 1	k_e 2	k_e 3
0	46.86%	100.69%	86.24%	36.16%	76.06%	50.09%	14.83%	46.08%	45.15%
1	83.68%	183.01%	156.35%	65.21%	139.54%	91.15%	25.99%	84.58%	82.83%
2	120.50%	265.34%	226.47%	94.26%	203.01%	132.22%	37.14%	123.08%	120.52%
3	157.32%	347.67%	296.58%	123.31%	266.48%	173.29%	48.30%	161.58%	158.21%
4	194.14%	429.99%	366.69%	152.36%	329.96%	214.35%	59.45%	200.09%	195.89%
5	230.96%	512.32%	436.81%	181.42%	393.43%	255.42%	70.60%	238.59%	233.58%
6	267.78%	594.65%	506.92%	210.47%	456.91%	296.49%	81.76%	277.09%	271.26%
7	304.60%	676.98%	577.04%	239.52%	520.38%	337.55%	92.91%	315.59%	308.95%
8	341.42%	759.30%	647.15%	268.57%	583.85%	378.62%	104.06%	354.09%	346.63%
9	378.24%	841.63%	717.26%	297.62%	647.33%	419.69%	115.22%	392.59%	384.32%
10	415.05%	923.96%	787.38%	326.67%	710.80%	460.75%	126.37%	431.09%	422.01%

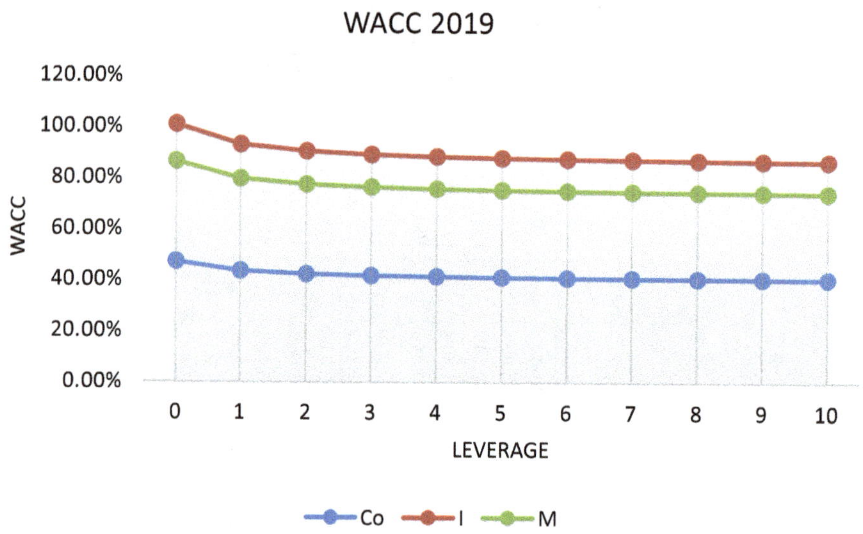

Fig. 23.10 Dependence of WACC on leverage level L, WACC (L) for year 2019

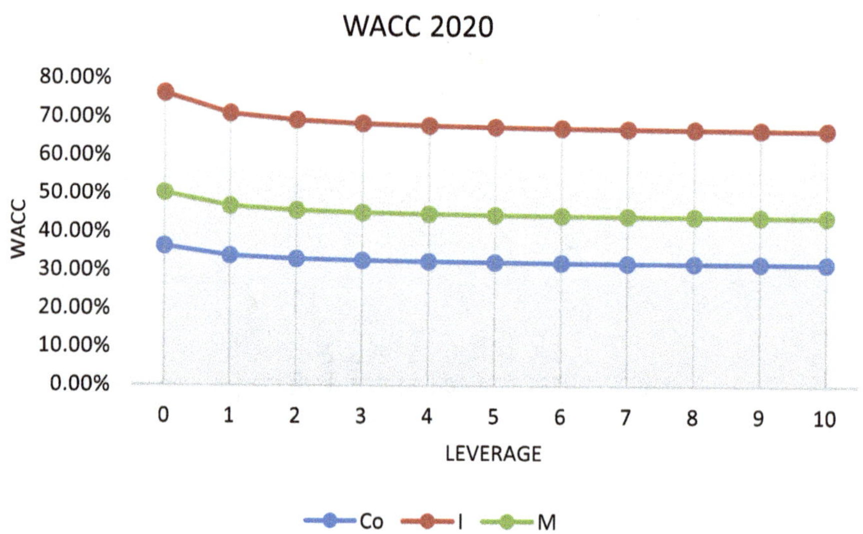

Fig. 23.11 Dependence of WACC on leverage level L, WACC (L) for year 2020

show that the financial performance of companies is highly dependent on the type of risks taken into account. Sometimes the difference between market and industry cases is small, sometimes it is significant. But the difference in financial indicators, while taking into account simultaneously financial and business risks, is always large with respect to accounting for a single risk. This means that taking into account simultaneously both financial and business risks is important for a correct

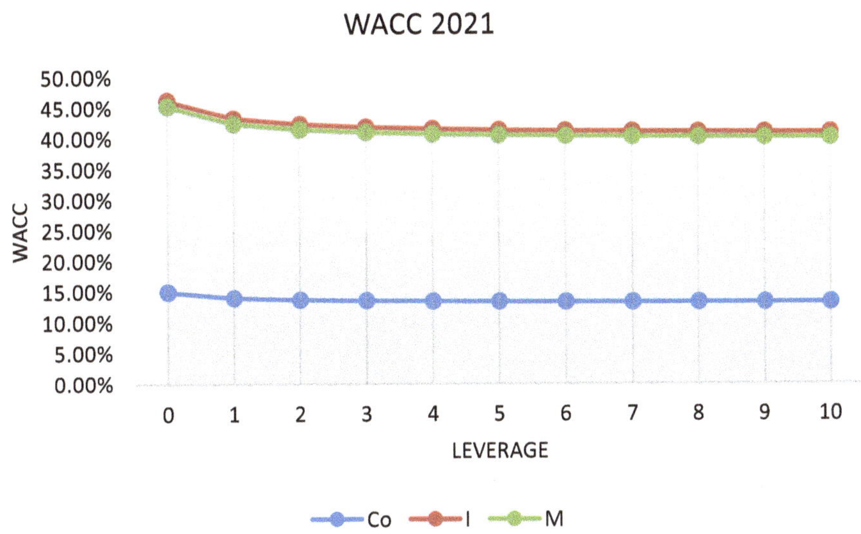

Fig. 23.12 Dependence of WACC on leverage level L, WACC (L) for year 2021

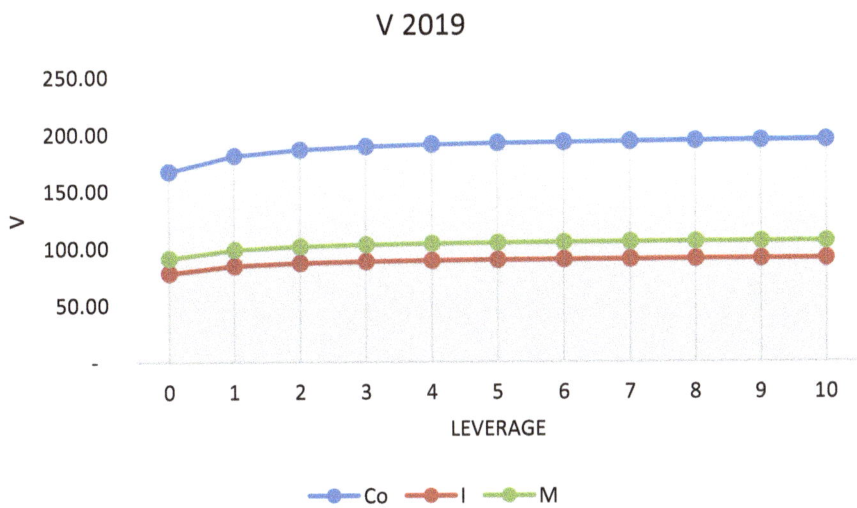

Fig. 23.13 Dependence of company value V on leverage level L, $V(L)$ for year 2019

assessment of the financial performance of companies. Three changes to the current methodology significantly improve the accuracy of the estimate: (a) the use of a sectoral approach in CAPM; (b) the use a new methodology in which the average return on the asset, net of leverage, with (c) the addition of a business risk premium (market or industry), is taken as the initial return. The new methodology opens up

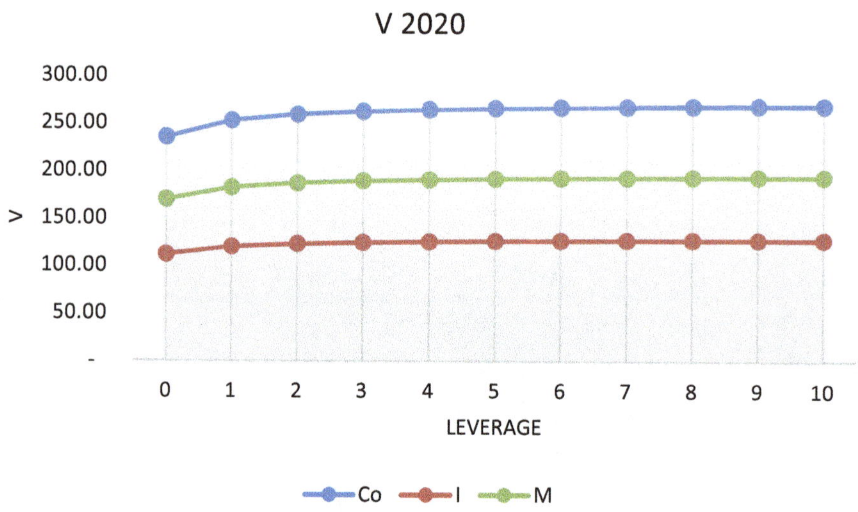

Fig. 23.14 Dependence of company value *V* on leverage level *L*, *V*(*L*) for year 2020

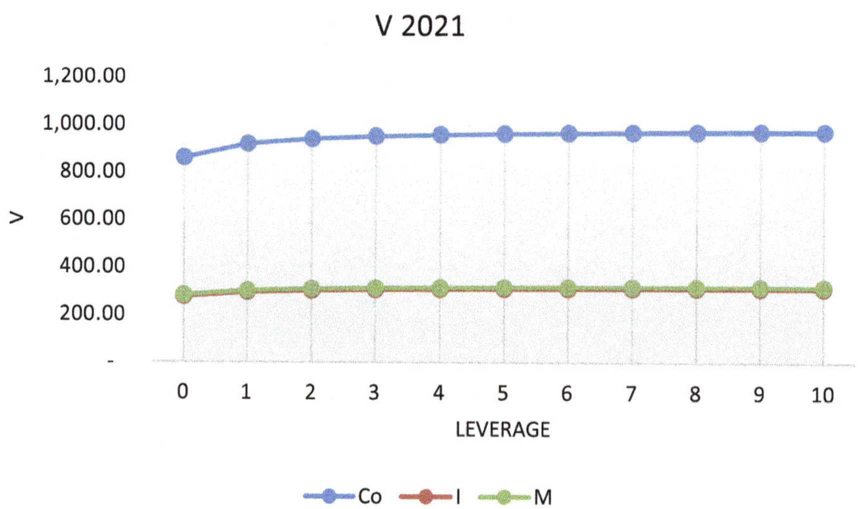

Fig. 23.15 Dependence of company value *V* on leverage level *L*, *V*(*L*) for year 2021

new horizons and opportunities in business valuation, corporate finance, investments, ratings, etc.

The significant novelty of the chapter is as follows: For the first time, we correctly took into account financial and business risks, proving the incorrectness of Hamada's model, and included CAPM in the theories of capital structure (both MM and BFO), which opens up great prospects for assessing the financial performance of companies.

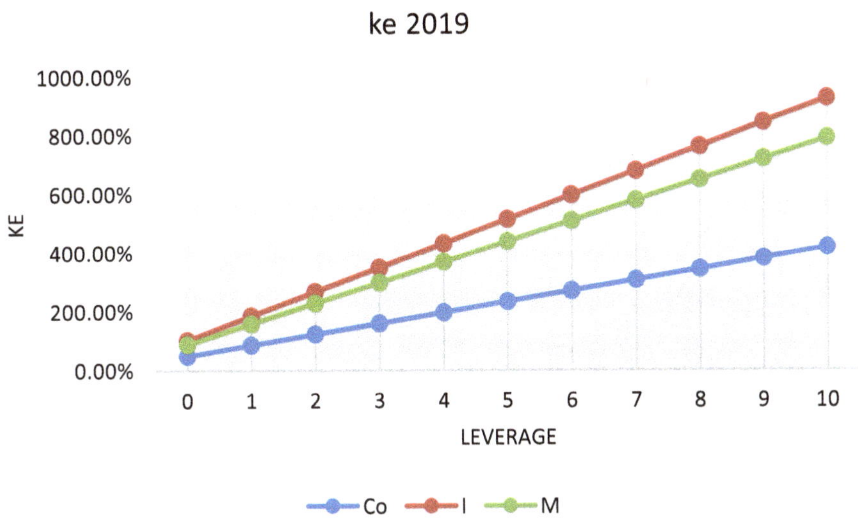

Fig. 23.16 Dependence $k_e(L)$; year 2019

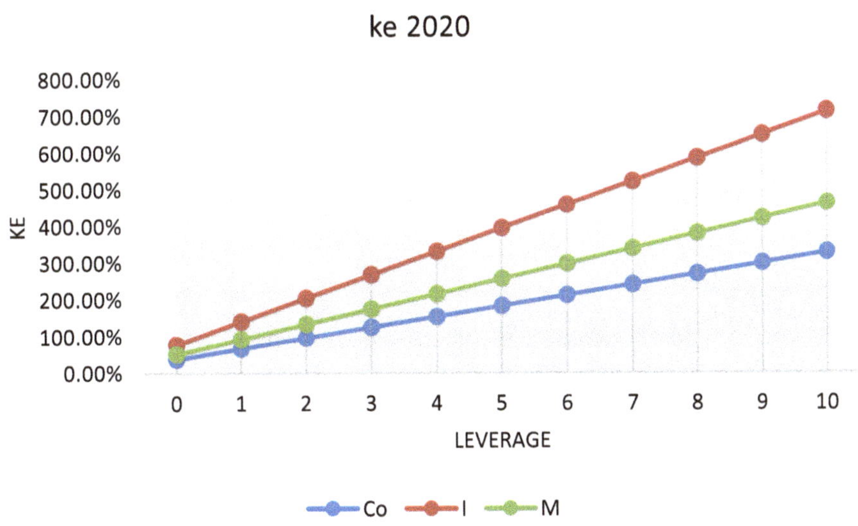

Fig. 23.17 Dependence $k_e(L)$; year 2020

The limitations of the proposed methodology are partly related to the limitations of its component: CAPM, capital structure theories, in particular, with the reliability of the WACC approximation.

In the future, it is planned to (1) improve and clarify the methodology for collecting and processing information about the company, industry, market;

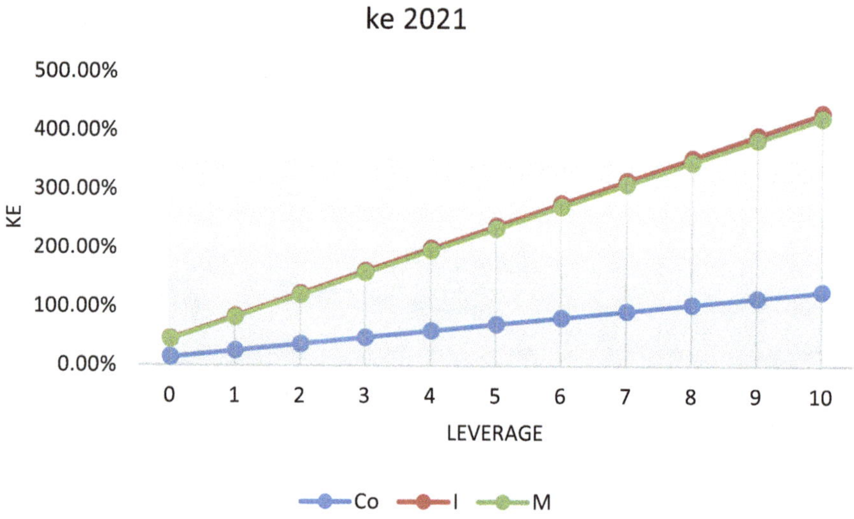

Fig. 23.18 Dependence $k_e(L)$; year 2021

(2) improve the use of MM and BFO models adapted to the real conditions of the functioning of companies.

References

1. Treynor, J. L. (1965). How to evaluate the management of investment funds. *Harvard Business Review, 43*, 63–75.
2. French, C. W. (2003). The Treynor capital asset pricing model. *Journal of Investment Management., 1*(2), 60–72. Available at SSRN: https://ssrn.com/abstract=447580
3. Sharpe, W. F. (1964). Capital asset prices: A theory of market equilibrium under conditions of risk. *The Journal of Finance, 19*(3), 425–442.
4. Mossin, J. (1966). Equilibrium in a capital asset market. *Econometrica*, 768–783.
5. Lintner, J. (1965). The valuation of risk assets and the selection of risky investments in stock portfolios and capital budgets. *Review of Economics and Statistics*, 13–37.
6. Brusov, P., Filatova, T., & Orekhova, N. (2023). *The Brusov–Filatova–Orekhova theory of capital structure: Applications in corporate finance, investments, taxation and ratings*. Springer Nature. 769 pp.
7. Brusov, P., Filatova, T., Orekhova, N., & Eskindarov, M. (2018). *Modern corporate finance, investments, taxation and ratings* (2nd ed., pp. 1–571). Springer Nature Publishing.
8. Modigliani, F., & Miller, M. H. (1958). The cost of capital, corporation finance and the theory of investment. *The American Economic Review, 48*, 261–296.
9. Modigliani, F., & Miller, M. H. (1963). Corporate income taxes and the cost of capital: A correction. *The American Economic Review, 53*, 433–443.
10. Brusov, P., Filatova, T., & Orekhova, N. (2022). *Generalized Modigliani–Miller theory: Applications in corporate finance, investments, taxation and ratings*. Springer Nature. 362 pp.
11. Farber, A., Gillet, R., & Szafarz, A. (2006). A general formula for the WACC. *International Journal of Business, 11*(2), 211–218.

12. Fernandez, P. (2006). A general formula for the WACC: A comment. *International Journal of Business, 11/2*, 219.
13. Fisher, F. M., & McGowan, J. J. (1983). On the misuse of accounting rates of return to infer monopoly profits. *The American Economic Review, 73*, 82–97.
14. Brusov, P., & Filatova, T. (2023). Review: Capital structure theory: Past, present, future. *Mathematics, 11*(3), 616. https://doi.org/10.3390/math11030616
15. Fama, E. F., & French, K. R. (1992). The crosssection of expected stock returns. *The Journal of Finance, 47*(2), 427–465.
16. Fama, E. F., & French, K. R. (1993). Common risk factors in the returns on stocks and bonds. *Journal of Financial Economics, 33*(1), 3–56.
17. Fama, E. F., & French, K. R. (1995). Size and book-to-market factors in earnings and returns. *The Journal of Finance, 50*(1), 131–155.
18. Fama, E. F. (1965). The behavior of stock–market prices. *The Journal of Business, 38*, 34–105.
19. Hamada, R. (1969). Portfolio analysis, market equilibrium, and corporate finance. *Journal of Finance, 24*, 13–31.
20. Hamada, R. (1972). The effect of the firm's capital structure on the systematic risk of common stocks. *The Journal of Finance*, 435–452.
21. Leal, D., Jiménez, R., Riquelme, M., & Leiva, V. (2023). Elliptical capital asset pricing models: Formulation, diagnostics, case study with Chilean data, and economic rationale. *Mathematics, 11*(6), 1394. https://doi.org/10.3390/math11061394
22. Zhou, G. (1993). Asset–pricing tests under alternative distributions. *The Journal of Finance, 48*, 1927–1942.
23. Lange, K. L., Little, R. J., & Taylor, J. M. (1989). Robust statistical modeling using the t distribution. *Journal of the American Statistical Association, 84*, 881–896.
24. Chen, J., Wu, Y., & Xu, Y. (2021). Research and analysis of asset pricing model based on the empirical test of stock price. In *Proceedings of the third international conference on economic management and cultural industry* (pp. 1288–1293).
25. Blattberg, R. C., & Gonedes, N. J. (1974). A comparison of the stable and student distributions as statistical models for stock prices. *The Journal of Business, 47*, 244–280.
26. Brusov, P., Filatova, T., & Kulik, V. (2023). Capital Asset Pricing Model 2.0: Account of business and financial risk. *Preprints*, 2023100347. https://doi.org/10.20944/preprints202310.0347.v1
27. Brusov, P., Filatova, T., Kulik, V., Lin, G., & Chang, L.-M. (2023). Can CAPM (Capital Asset Pricing Model) accurately value assets? *Proceedings of the International Conference on Electronic Business (ICEB), 23*, 60–70.
28. Brusov, P., Filatova, T., Kulik, V., Lin, G., & Chang, L.-M. (2023). Precision finance: Capital structure theories approach reality. *Proceedings of the International Conference on Electronic Business (ICEB), 23*, 466–480.
29. Brusov, P. N., Filatova, T. V., & Kulik, V. (2024). Capital asset pricing model (CAPM) 2.0: Account of business and financial risk. *Finance: Theory and Practice, 28*(2), 128–142. https://doi.org/10.26794/2587-5671-2024-28-2-128-14
30. Becker, M. D. (2021). The difference between Modigliani–Miller and Miles–Ezzell and its consequences for the valuation of annuities. *Cogent Economics & Finance, 9*(1).
31. Cooper, I. A., & Nyborg, K. G. (2006). The value of tax shields is equal to the present value of tax shields. *Journal of Financial Economics*., Elsevier.

Chapter 24
A New Approach to Assessing a Company's Financial Indicators: Taking into Account Business and Financial Risks

Abbreviations

CAPM	Capital Asset Pricing Model
k_d	cost of debt
k_i; k_I; k_m ($\mu_{i,}$ $\mu_{I,}$ μ_m)	stand for company, industry, and market returns
L	leverage level
MM	the Modigliani–Miller theory
β_{iI}; β_{im}	Beta coefficients company to industry and to market
σ_i; σ_I; σ_m	standard deviation for company, industry, and market returns

24.1 Introduction

There are two main methods for estimating the value of assets. The first method is the well-known CAPM (Capital Asset Pricing Model), which uses the risk-free rate as the initial return and takes into account only the business risk associated with investing in a specific asset, and not in the market as a whole. The second method is associated with the use of one of two main theories of capital structure (Brusov–Filatova–Orekhova [BFO] theory and Modigliani–Miller [MM] theory), which take into account only the financial risk associated with the use of debt financing and allow the calculation of all financial indicators of a company of arbitrary age (BFO theory) or perpetuity one (MM theory) (see two recent monographs by Brusov et al. ([1], 2023) and review by Brusov and Filatova [2]).

The famous Capital Asset Pricing Model (CAPM), based on the portfolio theory by Harry Markowitz, was developed independently by Jack Traynor [3], William F. Sharpe [4], John Lintner [5], and Jan Mossin [6]. The Capital Asset Pricing Model (CAPM) is still widely used to estimate the return on assets and investments. But as our CAPM analysis for several dozen listed companies from different countries, as

P. Brusov, T. Filatova, *Business Valuation and Company Value*, Contributions to Finance and Accounting, https://doi.org/10.1007/978-3-031-90847-7_24

well as numerous results from other authors, has shown, CAPM results differ significantly from the actual profitability of companies. The reasons for this discrepancy relate to the internal properties and shortcomings of the model.

One of the remaining internal problems of CAPM is the distribution function. The Capital Asset Pricing Model (CAPM) assumes a Gaussian or Normal distribution. In practice, this assumption may be violated. In paper [7] a symmetric CAPM is proposed, assuming distributions with lighter or heavier tails than the normal distribution. Elliptic distributions (normal, exponential and Student–t) are considered. This consideration is of a general nature. The authors conducted a detailed case study to apply the obtained results estimating the systematic risk of the financial assets of a Chilean company with real data. A Chilean company is just illustration of obtained results. In addition, Leal et al. [7] study the methods of leverage and local impact for diagnostics in a symmetric CAPM. It is concluded that the considered models give better results than the CAPM with Gaussian distribution.

There are some modifications of CAPM ([8–13]; Brusov et al. 2023) that can bring the model closer to real life. Among the latter is the account of financial risk in CAPM along with business risk, which was recently correctly done by Brusov et al. (2023), and the first attempt to take financial risk into account (not completely successful) was made by Hamada [12, 13]. Initial Capital Asset Pricing Model (CAPM) takes into account only business risk. In practice, companies use debt financing and operate at non-zero levels of leverage. This means that it is necessary to take into account the financial risk associated with the use of debt financing along with business one. Brusov et al. (2023) simultaneously take into account the business and financial risk. A new approach to CAPM has been developed that takes into account both business and financial risk. They combine the theory of CAPM and the Modigliani–Miller (MM) theory. The first is based on portfolio analysis and accounting for business risks in relation to the market (or industry). The second one (the Modigliani–Miller [MM] theory) describes a specific company and takes into account the financial risks associated with the use of debt financing. The combination of these two different approaches makes it possible to take into account both types of risks: business and financial ones. Brusov et al. (2023) combine these two approaches analytically, while Hamada [12, 13] did it phenomenologically. Using the Modigliani–Miller (MM) theory, it is shown that the Hamada's model, first model, used for this purpose half a century ago, is incorrect. In addition to the renormalization of the beta-coefficient, obtained in the Hamada model, two additional terms are found: the renormalized risk-free return and the term dependent on the cost of debt k_d. A critical analysis of the Hamada model was carried out. The vast majority of listing companies use debt financing and are levered, and the Hamada model is not applicable to them in contrast to a new (Brusov et al. 2023) approach applicable to leveraged companies. A new approach to specific companies has been implemented. A comparison of the results of the new approach with the results of the conventional CAPM is shown. Two versions of CAPM (market or industry) are considered (Brusov et al. 2023).

One more modification of the CAPM has been done in 1992, by Y. Fama and K. French, who proved that future returns are also affected by factors such as company size and industry affiliation. They have developed three- and five-factor models [8–10].

An alternative to CAPM is the Arbitrage Pricing Theory (APT), created by Stephen Ross in 1976. In opposition to the CAPM, which has only one factor and one beta, the APT formula has multiple factors that include non-company factors, which requires the asset's beta with respect to each separate factor. The APT does not explain what these factors are, and APT model users should analytically determine factors that might affect the asset's returns. The factor used in the CAPM is the difference between the market rate of return and the risk-free rate of return.

The CAPM is a single-factor model and is easier to use, which is why investors prefer to use it to estimate the expected rate of return rather than using the APT, which requires users to estimate multiple factors.

The outline of the article is as follows:

- CAPM derivation is given.
- The methodology of the new approach is presented.
- An application of a new approach to the Norilsk Nickel company was given.
- An analysis of three assessment methods and their comparison is provided.
- Conclusion.

24.2 Derivation of the CAPM Formula

To derive the CAPM formula, we create a portfolio of asset i and the market portfolio M; $(\alpha, 1-\alpha)$, with the price share of the security in the portfolio $\alpha \in [0, 1]$ (Sigman [14]). The rate of return is thus $\mu(\alpha) = \alpha\mu_i + (1 - \alpha)\mu_M$. Asset i is assumed not efficient; it's point lies in the feasible region but not on the efficient frontier. Thus as αvaries, this portfolio's point traces out a smooth curve in the feasible (σ, μ)region $\alpha \to (\sigma(\alpha), \mu(\alpha))$ parametrized by α, where

$$\mu(\alpha) = \alpha\mu_i + (1-\alpha)\mu_M = \alpha(\mu_i - \mu_M) + \mu_M \tag{24.1}$$

Using a well-known formula for portfolio dispersion

$$\sigma^2(\alpha) = \alpha^2\sigma_i^2 + (1-\alpha)^2\sigma_M^2 + 2\alpha(1-\alpha)\sigma_{M,i}, \tag{24.2}$$

one gets

$$\begin{aligned}\sigma(\alpha) &= \sqrt{\alpha^2\sigma_i^2 + (1-\alpha)^2\sigma_M^2 + 2\alpha(1-\alpha)\sigma_{M,i}} = \\ &= \sqrt{\alpha^2(\sigma_i^2 + \sigma_M^2 - 2\sigma_{M,i}) + \sigma_M^2 + 2\alpha(\sigma_{M,i} - \sigma_M^2)}\end{aligned} \tag{24.3}$$

When $\alpha = 0$ $(\sigma(0), \mu(0)) = (\sigma_M, \mu_M)$ and when $\alpha = 1$ $(\sigma(0), \mu(0)) = (\sigma_i, \mu_i)$. Thus the curve touches the capital market line at the market point (σ_M, μ_M), but otherwise remains off the capital market line but (of course) within the feasible region where it also hits the point (σ_i, μ_i). We conclude that the curve is tangent to the capital asset line at point (σ_M, μ_M) and therefore when $\alpha = 0$ the curve's derivative

$$\frac{d\mu(\alpha)}{d\sigma(\alpha)}_{\alpha=0} \tag{24.4}$$

is identical to the slope of the capital asset line at point M. The slope (the price of risk) of the line is given by (24.2). We thus conclude that when $\alpha = 0$

$$\frac{d\mu(\alpha)}{d\sigma(\alpha)} = \frac{\mu_M - \mu_f}{\sigma_M}. \tag{24.5}$$

The composition rule from calculus yields

$$\frac{d\mu(\alpha)}{d\sigma(\alpha)} = \frac{d\mu(\alpha)/d\alpha}{d\sigma(\alpha)/d\alpha}. \tag{24.6}$$

Differentiating (24.1) and (24.3) and evaluating at $\alpha = 0$ yields

$$\frac{d\mu(\alpha)}{d\sigma(\alpha)} = \frac{\mu_i - \mu_M}{(\sigma_{M,i} - \sigma_M^2)/\sigma_M} \tag{24.7}$$

From (24.5) we deduce that

$$\frac{\mu_i - \mu_M}{(\sigma_{M,i} - \sigma_M^2)/\sigma_M} = \frac{\mu_M - \mu_f}{\sigma_M} \tag{24.6}$$

Solving this equation for μ_i, we obtain the CAPM formula for the asset *I* (CAPM formula for single asset *i*)

$$\mu_i - \mu_f = \beta_i \cdot (\mu_m - \mu_f), \tag{24.7}$$

as was to be shown. For the more general portfolio result, observe that

$$\mu_p - \mu_f = -\mu_f + \sum_{i=1}^{n} \alpha_i \mu_i = \sum_{i=1}^{n} \alpha_i (\mu_i - \mu_f) = \sum_{i=1}^{n} \alpha_i \beta_i (\mu_M - \mu_f)$$
$$= (\mu_M - \mu_f) \sum_{i=1}^{n} \alpha_i \beta_i \tag{24.8}$$

Note that when $\beta_p = 1$ then $\mu_p = \mu_M$; the expected rate of return is the same as for the market portfolio. When $\beta_p > 1$, then $\mu_p > \mu_M$; when $\beta_p < 1$, then $\mu_p < \mu_M$.

In case of negatively correlated asset i with M, $\sigma_{M,\,i} < 0$, then $\beta_i < 0$ and $\mu_i < \mu_f$; the expected rate of return is less than the risk-free rate. Effectively, such a negatively correlated asset serves as insurance against a drop in the market returns, and might be viewed by some investors as having enough such advantages so as to make it worth the low return. Such an example could be investing in gold.

The following assumptions are made within the CAPM model:

1. All investors are risk averse and have the same time frame to evaluate information.
2. Unlimited capital exists to borrow at the risk-free rate.
3. Investments can be divided into unlimited parts and sizes.
4. Taxes, inflation and transaction costs are absent.
5. Return and risk are linearly related.

CAPM (Capital Asset Pricing Model) describes the profitability of asset and is described by the following formula:

$$k_i = k_f + \beta_i (k_m - k_f). \tag{24.9}$$

(Note, that here and below we will denote assets profitability as k (like cost of equity) instead of μ.)

Here k_f is risk-free profitability, β is the β-coefficient of the company. It shows the dependence of the return on the asset and the return on the market as a whole. The β-coefficient is described by the following formula:

$$\beta_i = \frac{\mathrm{cov}_{im}}{\sigma_m^2} = \rho_{im} \frac{\sigma_i}{\sigma_m}. \tag{24.10}$$

Here σ_i is the risk (standard deviation) of i-th asset, σ_m is market risk (standard deviation of market index), cov_{im} is covariance between i-th asset and market portfolio.

An investor invests in risky securities only if their return is higher than the return on risk-free securities, so always $k_i > k_f$ and $k_m > k_f$.

24.3 Industry Approach

CAPM has an alternative approach that refers to the industrial index rather than the market.

$$k_i = k_f + \beta_i (k_I - k_f). \tag{24.11}$$

Here, k_f is risk-free profitability, β is the β-coefficient of the company. In this case it shows the dependence of the return on the asset and the return on the industry as a whole. The β-coefficient now is described by the following formula:

$$\beta_i = \frac{\mathrm{cov}_{iI}}{\sigma_I^2} = \rho_{iI} \frac{\sigma_i}{\sigma_I}. \tag{24.12}$$

Here σ_i is the risk of i-th asset, σ_I is industry risk (standard deviation of industry index), cov_{iI} is covariance between i-th asset and industry index. Note that the industry approach better describes the return on an asset than the market approach.

24.4 New Approach and Methodology

We take the company's profitability μ_i from its financial statements.

Then we clean this value from leverage

$$\mu_i \to \mu_0 \tag{24.13}$$

and find the value of the cost of equity at zero leverage μ_0, which is one of the main parameters in both theories: BFO and MM.

$$\mu_i = \mu_0 + L(\mu_0 - k_d) \cdot (1 - t) \tag{24.14}$$

$$\mu_0 = \frac{\mu_i + L k_d \cdot (1 - t)}{1 + L \cdot (1 - t)} \tag{24.15}$$

We use $k_0 = \mu_0$.

We consider three cases:

1. Accounting only financial risk

$$k_0 = \mu_0 \tag{24.16}$$

2. Accounting for financial risk + industry business risk

$$k_0 = \mu_0 + \beta_{iI}(\mu_I - \mu_F) \tag{24.17}$$

3. Account financial risk + market business risk

$$k_O = \mu_O + \beta_{im}(\mu_m - \mu_F) \tag{24.18}$$

For calculations according to MM theory, the following formulas are used to derive indicators:

$$\text{WACC} = k_O \cdot (1 - w_d t) \tag{24.19}$$

$$V = \frac{\text{CF}}{\text{WACC}} = \frac{\text{CF}}{k_O \cdot (1 - w_d t)} \tag{24.20}$$

$$k_e = k_O + L(k_O - k_d) \cdot (1 - t) \tag{24.21}$$

For calculations according to BFO theory, the following formulas are used to derive indicators:

$$\frac{1 - (1 + \text{WACC})^{-n}}{\text{WACC}} = \frac{1 - (1 + k_O)^{-n}}{k_O \cdot (1 - w_d t[1 - (1 + k_d)^{-n}])} \tag{24.22}$$

$$V = \frac{\text{CF}}{\text{WACC}} \cdot (1 - (1 + \text{WACC})^{-n}) \tag{24.23}$$

$$k_e = \text{WACC} \cdot (1 + L) - L \cdot k_d \cdot (1 - t) \tag{24.24}$$

24.5 An Application of a New Approach to the Norilsk Nickel Company (GMKN)

Norilsk Nickel company (GMKN) is the leader in the mining and metallurgical industry in Russia and in the world. It produces metals needed to develop a low-carbon economy and clean transport.

It produces 17% Ni (№1 in the world); 43% of Pd (№1 in the world); 12% of Pt (№4 in the world); 8% Rh (№5 in the world); 2% of Cu; 2% of Co as well as Au, Se, Ag, S, Ru, Ir, Te.

Below we will evaluate the financial indicators of Norilsk Nickel for 2018–2022 and study their dependence on debt financing.

k_i; k_I; k_m (μ_i; μ_I; μ_m) stand for company, industry, and market returns; σ_i; σ_I; σ_m are standard deviation for company, industry, and market returns; β_{iI}; β_{im} are Beta coefficients company to industry and to market; L is leverage level; k_d is the cost of debt.

All necessary data is collected in Table 24.1. The calculated values of k_O in 2018–2021 are given in Table 24.2.

1. Norilsk Nickel share price data: https://ru.investing.com/equities/gmk-noril-nickel_rts-historical-data

Table 24.1 Calculation of the expected return on Norilsk Nickel shares

Level	index	2018	2019	2020	2021	2022
Company	μ	12.33%	40.50%	13.86%	−6.01%	−30.19%
	σ_i	0.054	0.052	0.076	0.052	0.10
	L_i	3.29	3.22	2.42	3.47	5.84
	k_d	4.58%	4.30%	3.96%	2.29%	3.94%
	β_{im}	0.72	0.30	0.67	−0.47	0.65
	ρ_{im}	0.32	0.13	0.64	−0.32	0.74
Industry (index MOEXMM)	μ_I	5.01%	9.05%	40.66%	9.43%	−42.59%
	σ_I	0.026	0.028	0.06	0.043	0.12
	L_I	1.92	1.97	1.87	1.72	No data
	β_{iI}	1.088	1.47	0.81	0.95	0.81
	ρ_{iI}	0.52	0.8	0.63	0.80	0.94
Market (index MOEX)	μ_m	3.46%	20.82%	6.9%	15.57%	−38.98%
	σ_m	0.024	0.022	0.074	0.036	0.12
	β_{im}	0.72	0.30	0.67	−0.47	0.65
μ_F		8.78%	6.41%	6.27%	8.44%	10.31%
μ		12.33%	40.50%	13.86%	−6.01%	−30.19%
μ_i (Industry)		4.68%	10.30%	34.02%	9.39%	−32.73%
μ_i (Market)		4.97%	10.76%	6.69%	5.12%	−21.89%

Table 24.2 Cost of equity capital of an unleveraged company in 2018–2021

$k_0 = \mu_0$	2018	2019	2020	2021
Company	6.71%	14.42%	7.33%	0.09%
Industry	2.61%	18.31%	35.08%	1.04%
Market	2.90%	18.78%	7.75%	−3.23%

2. Moscow Exchange Index data MOEX: https://ru.investing.com/equities/gmk–noril–nickel_rts–historical–data
3. Industry Index Data Source MOEXMM: https://ru.investing.com/indices/mcxmm–historical–data
4. As a risk-free rate, the rate on federal loan bonds with a 10-year term has been used, which is published on the website of the Bank of Russia https://www.cbr.ru/hd_base/zcyc_params/zcyc/
5. Industry average financial leverage values are published on the website: https://www.testfirm.ru
6. Data for calculations are taken from the balance sheet and income statement

For 2022:

https://www.nornickel.ru/upload/iblock/094/lx8zuduktyo27i4i4xrvozcsmgmu3n2r/godovaya_buhgalterskaya_finansovaya_otchetnost_za_god_2022.pdf

For 2021:

https://www.nornickel.ru/upload/iblock/751/buhgalterskaya_finansovaya_otchetnost_pao_gmk_nn_za_2021_g.pdf

Calculations below will be carried out for those cases when k_0 is positive (years 2018, 2019, and 2020). The case where k_0 is negative (as in 2021 and 2022) requires a different approach, which has been developed by Anastasia Brusova [15].

24.5.1 MM Theory 2018

It is seen from Fig. 24.1, that WACC(L) decreases with leverage level L in all three cases. This means that debt financing is important and should be used by company—it leads to decrease of attracting capital cost with leverage level L. There is a large difference between the case of accounting for pure financial risk (1) and accounting for financial risk plus business risk (2 and 3), while the difference between accounting for industry and market business risks (2 and 3) is much smaller (Fig. 24.2).

It is seen from Fig. 24.3, that company value $V(L)$ increases with leverage level L in all three cases. There is a large difference between the case (1) and cases (2 and 3), while the difference between cases (2 and 3) is much smaller (Figs. 24.4 and 24.5).

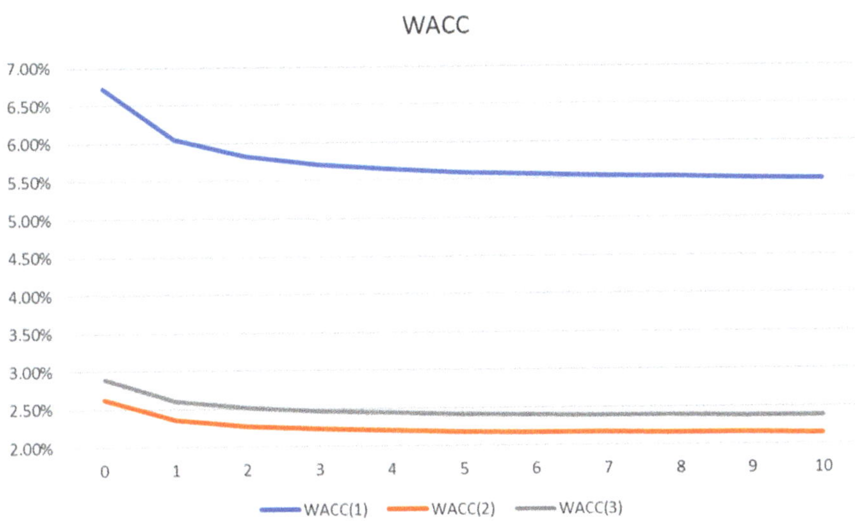

Fig. 24.1 Dependence of WACC (weighted average cost of capital) on leverage level L according to the MM theory in 2018 (financial risk only (1); plus industry (2) and plus market (3) business risk)

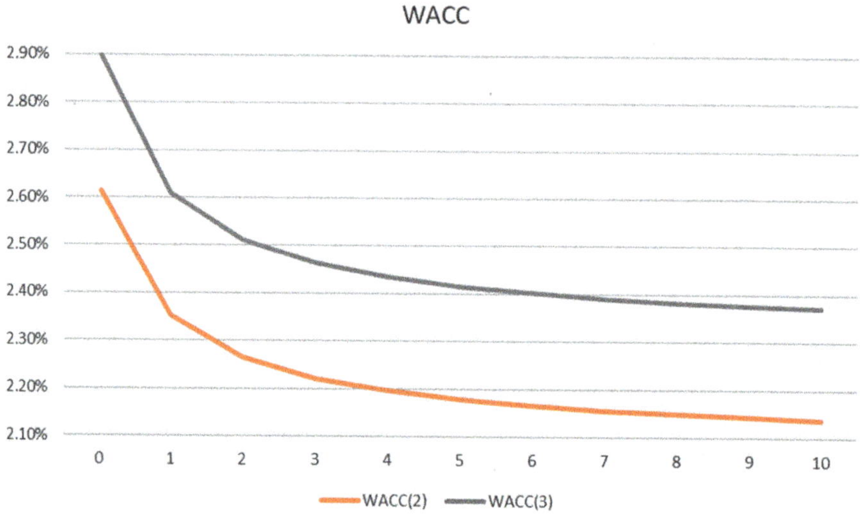

Fig. 24.2 Dependence of WACC (weighted average cost of capital) on leverage level according to the MM theory in 2018 (plus industry (2) and plus market (3) business risk)

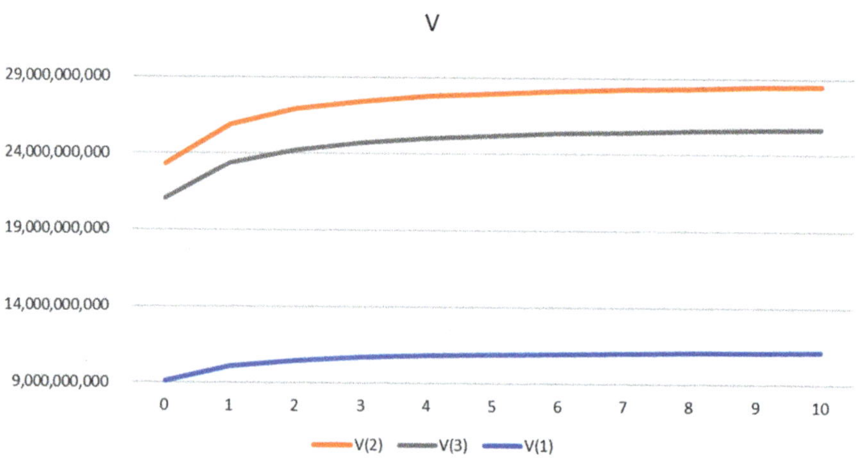

Fig. 24.3 Dependence of company value, *V*, on leverage level according to the MM theory in 2018 (financial risk only (1); plus industry (2) and plus market (3) business risk)

24.5.2 BFO Theory 2018

It is seen from Fig. 24.6 that WACC(L) decreases with leverage level L in all three cases. This means that debt financing is important and should be used by company—it leads to decrease of attracting capital cost with leverage level L. There is a large difference between the case of accounting for pure financial risk (1) and accounting

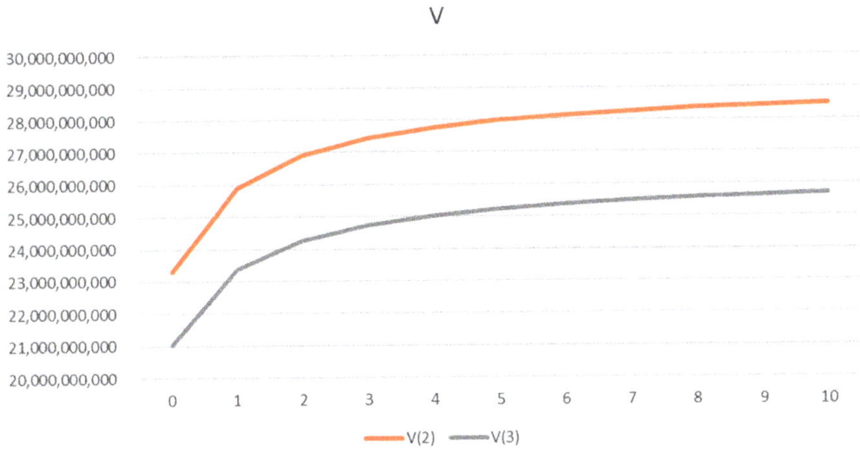

Fig. 24.4 Dependence of company value, V, on leverage level according to the MM theory in 2018 (plus industry (2) and plus market (3) business risk)

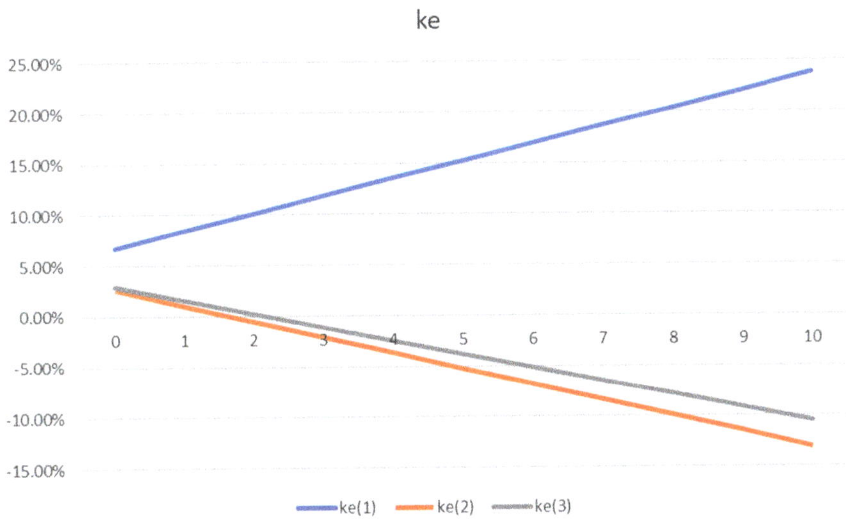

Fig. 24.5 Dependence of equity cost, k_e, on leverage level according to the MM theory in 2018 (financial risk only (1); plus industry (2) and plus market (3) business risk)

for financial risk plus business risk (2 and 3), while the difference between accounting for industry and market business risks (2 and 3) is much smaller (Fig. 24.7).

It is seen from Fig. 24.8, that company value $V(L)$ increases with leverage level L in all three cases. There is a large difference between the case (1) and cases (2 and 3), while the difference between cases (2 and 3) is much smaller.

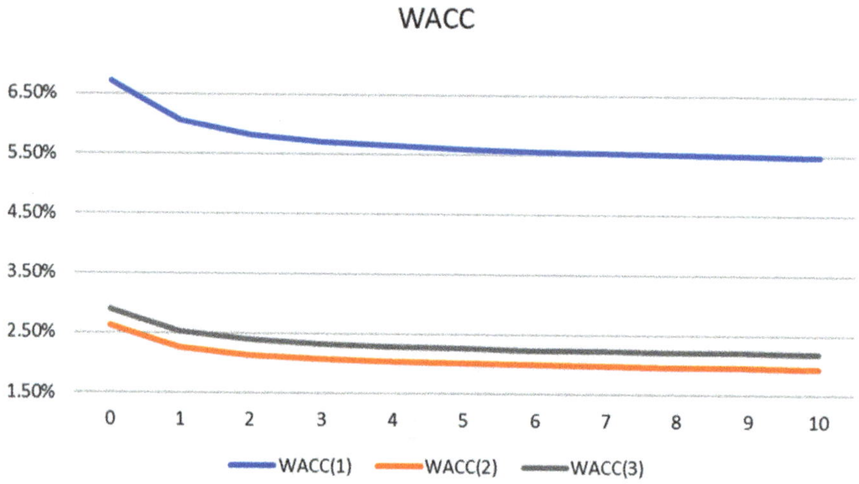

Fig. 24.6 Dependence of WACC (weighted average cost of capital) on leverage level according to the BFO theory in 2018 (financial risk only (1); plus industry (2) and plus market (3) business risk)

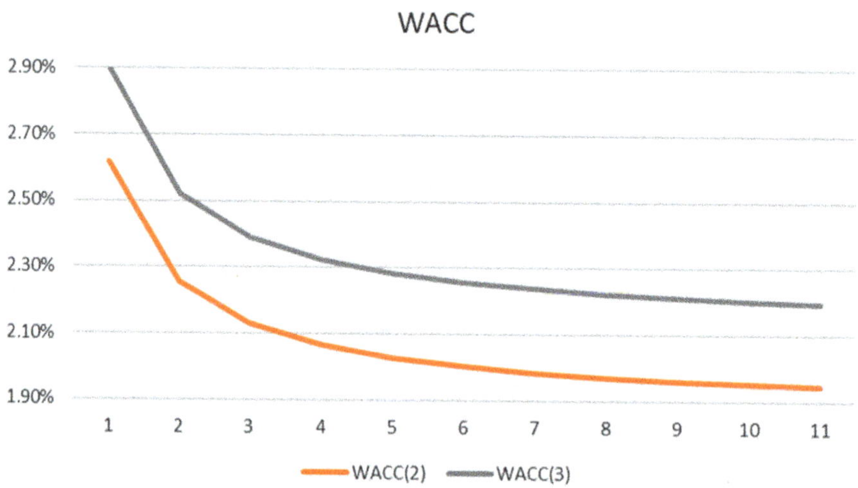

Fig. 24.7 Dependence of WACC (weighted average cost of capital) on leverage level according to the BFO theory in 2018 (plus industry (2) and plus market (3) business risk)

From Fig. 24.9 it is seen that the cost of equity capital k_e increases with increasing leverage level L for case (1) and decreases for cases (2) and (3). The difference in inclination angles in cases (2 and 3) is small. This means the existence of an anomalous effect depending on cost of equity on leverage level L. Note that the equity cost determines the value of dividends.

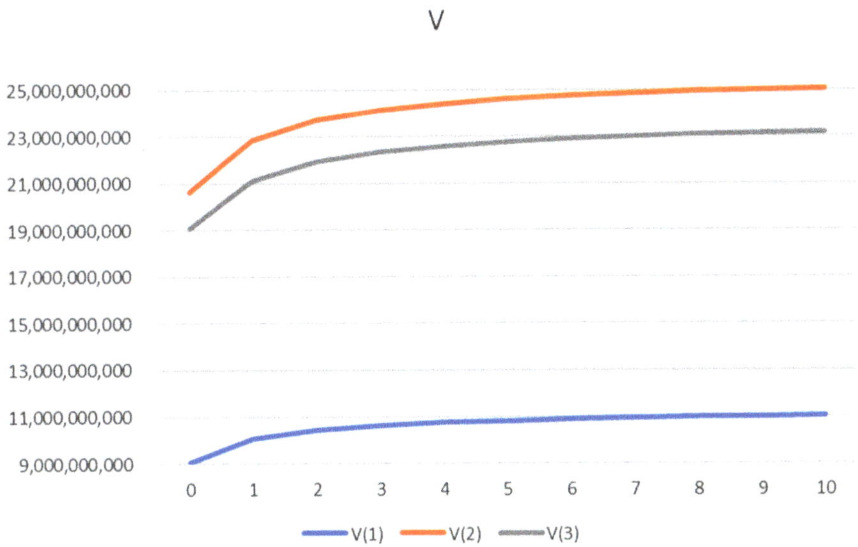

Fig. 24.8 Dependence of company value, V, on leverage level according to the BFO theory in 2018 (financial risk only (1); plus industry (2) and plus market (3) business risk)

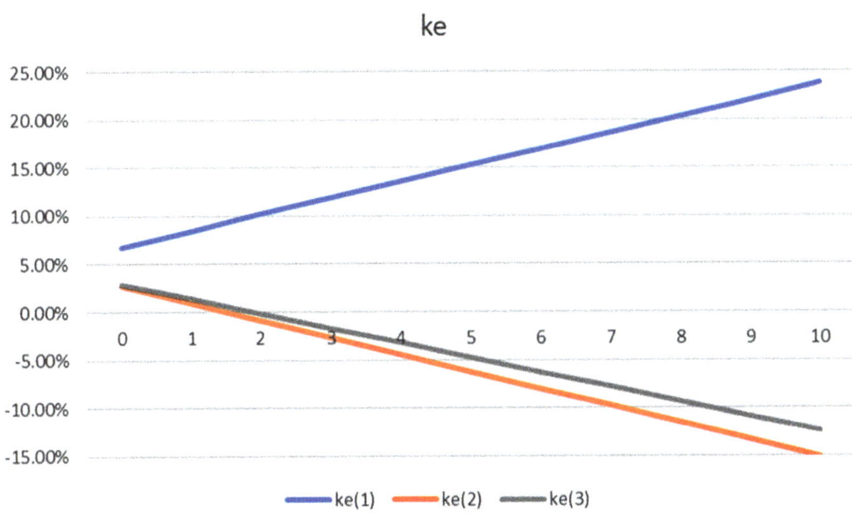

Fig. 24.9 Dependence of cost of equity, k_e, on leverage level according to the BFO theory in 2018 (financial risk only (1); plus industry (2) and plus market (3) business risk)

24.5.3 MM Theory 2019

It is seen from Fig. 24.10 that WACC(L) decreases with leverage level L in all three cases. This means that debt financing is important and should be used by company—

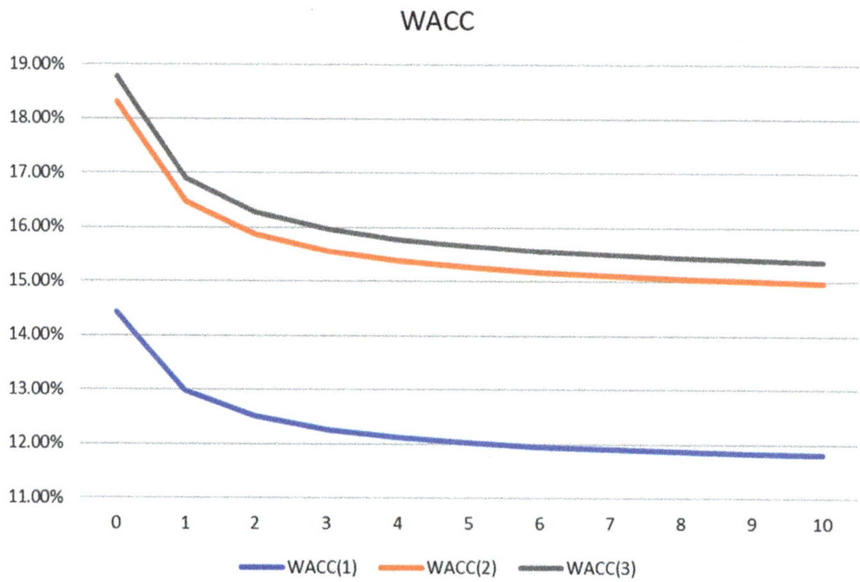

Fig. 24.10 Dependence of WACC (weighted average cost of capital) on leverage level according to the MM theory in 2019 (financial risk only (1); plus industry (2) and plus market (3) business risk)

it leads to decrease of attracting capital cost with leverage level *L*. There is a large difference between the case of accounting for pure financial risk (1) and accounting for financial risk plus business risk (2 and 3), while the difference between accounting for industry and market business risks (2 and 3) is much smaller.

It is seen from Fig. 24.11 that company value *V*(*L*) increases with leverage level *L* in all three cases. There is a large difference between the case (1) and cases (2 and 3), while the difference between cases (2 and 3) is much smaller.

From Fig. 24.12, it is seen that the cost of equity capital k_e increases with increasing leverage level *L* for all three cases. The difference in inclination angles in cases (1) and (2 and 3) is significant, and for cases (2) and (3) it is significantly less. Note that the equity cost determines the value of dividends.

24.5.4 BFO Theory 2019

It is seen from Fig. 24.13 that WACC(*L*) decreases with leverage level *L* in all three cases. This means that debt financing is important and should be used by company—it leads to decrease of attracting capital cost with leverage level *L*. There is a large difference between the case of accounting for pure financial risk (1) and accounting for financial risk plus business risk (2 and 3), while the difference between accounting for industry and market business risks (2 and 3) is much smaller.

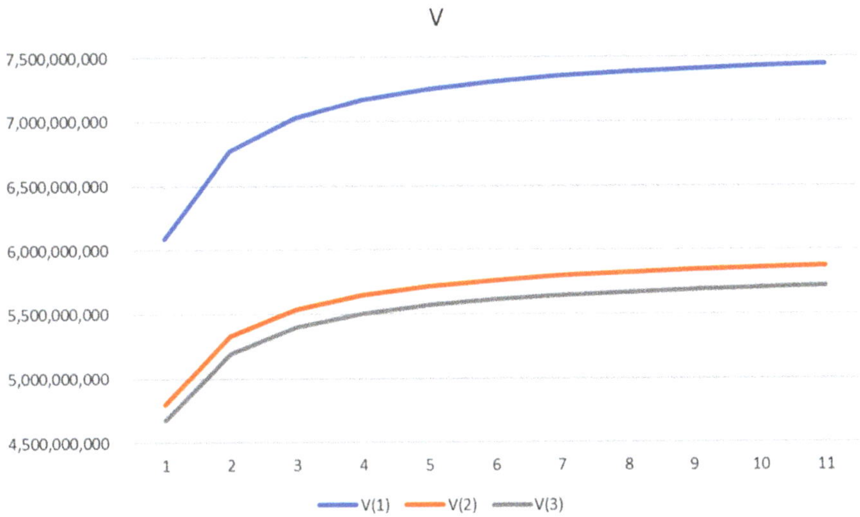

Fig. 24.11 Dependence of company value, V, on leverage level according to the MM theory in 2019 (financial risk only (1); plus industry (2) and plus market (3) business risk)

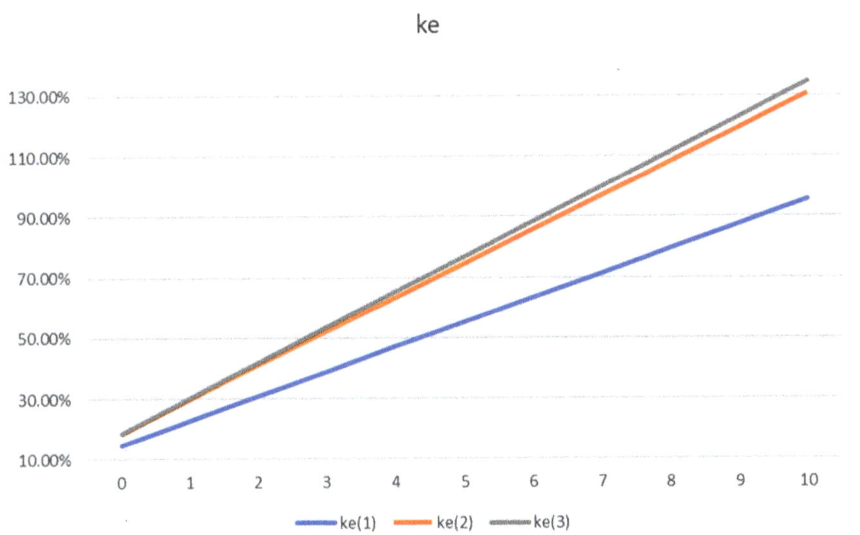

Fig. 24.12 Dependence of cost of equity, k_e, on leverage level according to the MM theory in 2019 (financial risk only (1); plus industry (2) and plus market (3) business risk)

It is seen from Fig. 24.14 that company value $V(L)$ increases with leverage level L in all three cases. There is a large difference between the case (1) and cases (2 and 3), while the difference between cases (2 and 3) is much smaller.

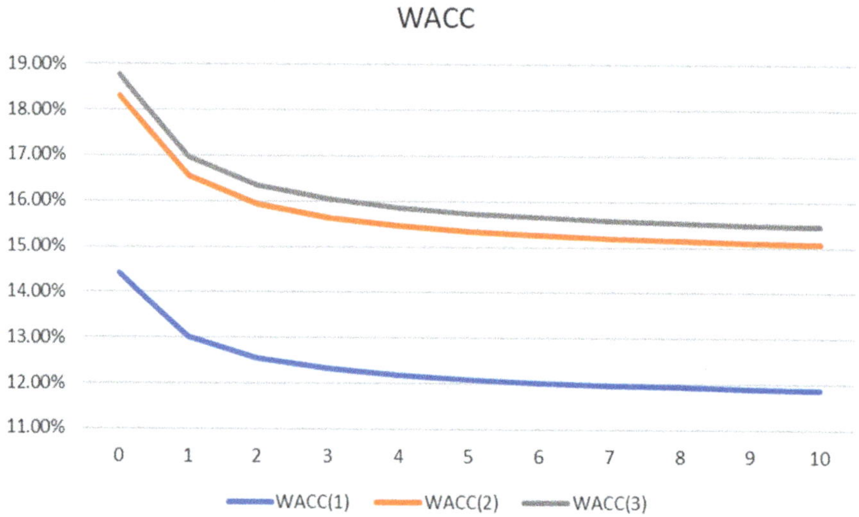

Fig. 24.13 Dependence of WACC (weighted average cost of capital) on leverage level according to the BFO theory in 2019 (financial risk only (1); plus industry (2) and plus market (3) business risk)

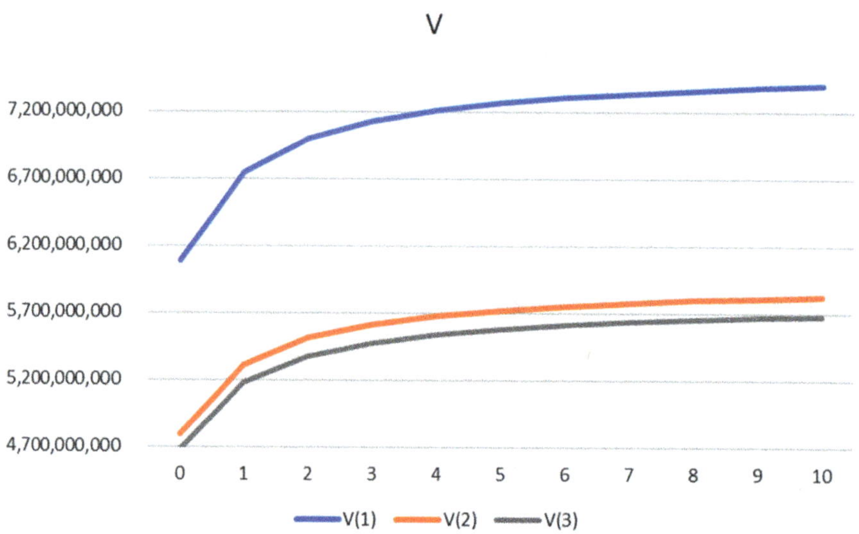

Fig. 24.14 Dependence of company value, *V*, on leverage level according to the BFO theory in 2019 (financial risk only (1); plus industry (2) and plus market (3) business risk)

From Fig. 24.15, it is seen that the cost of equity capital k_e increases with increasing leverage level L for all three cases. The difference in inclination angles in cases (1) and (2 and 3) is significant, and for cases (2) and (3) it is significantly less. Note that the equity cost determines the value of dividends.

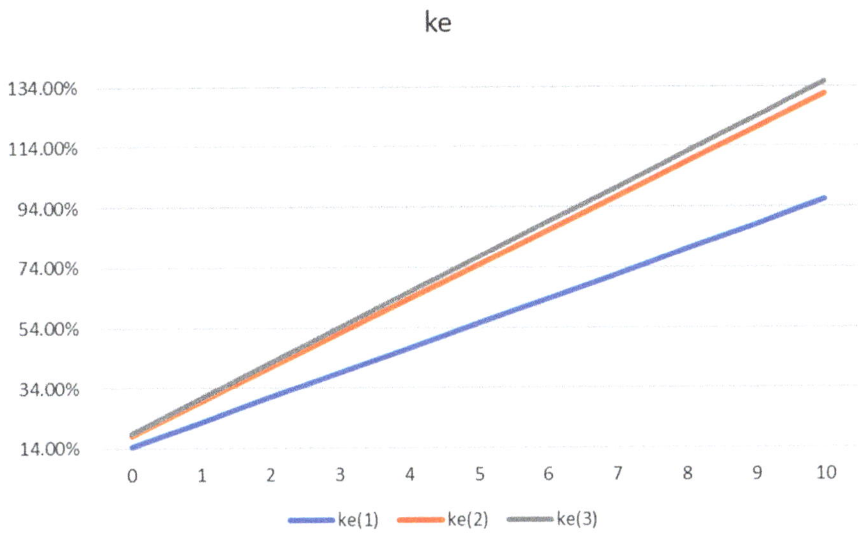

Fig. 24.15 Dependence of cost of equity, k_e, on leverage level according to the BFO theory in 2019 (financial risk only (1); plus industry (2) and plus market (3) business risk)

24.5.5 MM Theory 2020

It is seen from Fig. 24.16 that WACC(L) decreases with leverage level L in all three cases. This means that debt financing is important and should be used by company—it leads to decrease of attracting capital cost with leverage level L. There is a large difference between the case (2) and cases (1 and 3), while the difference between cases (1 and 3) is much smaller (Fig. 24.17).

It is seen from Fig. 24.18, that company value $V(L)$ increases with leverage level L in all three cases. There is a large difference between the case (2) and cases (1 and 3), while the difference between cases (1 and 3) is much smaller (Figs. 24.19 and 24.20).

From Fig. 24.21, it is seen that the cost of equity capital k_e increases with increasing leverage level L for all three cases. The difference in inclination angles in cases (1) and (2 and 3) is significant, and for cases (2) and (3) it is significantly less. Note that the equity cost determines the value of dividends (Fig. 24.22).

24.5.6 BFO Theory 2020

It is seen from Fig. 24.23 that WACC(L) decreases with leverage level L in all three cases. This means that debt financing is important and should be used by company—it leads to decrease of attracting capital cost with leverage level L. There is a large

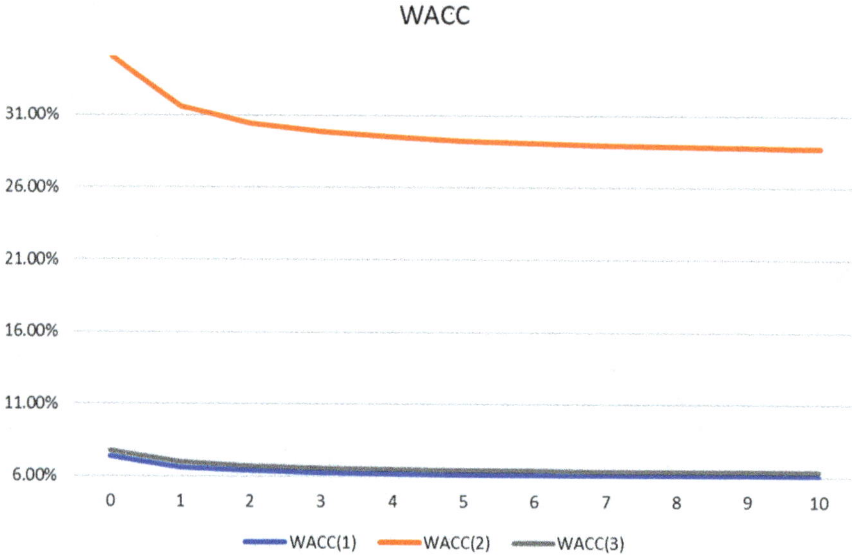

Fig. 24.16 Dependence of WACC (weighted average cost of capital) on leverage level according to the MM theory in 2020 (financial risk only (1); plus industry (2) and plus market (3) business risk)

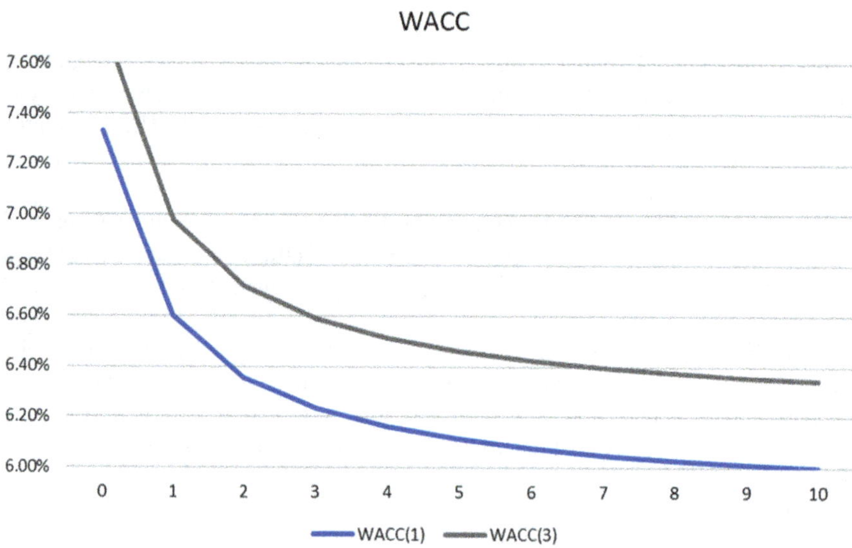

Fig. 24.17 Dependence of WACC (weighted average cost of capital) on leverage level according to the MM theory in 2020 (financial risk only (1) and plus market (3) business risk)

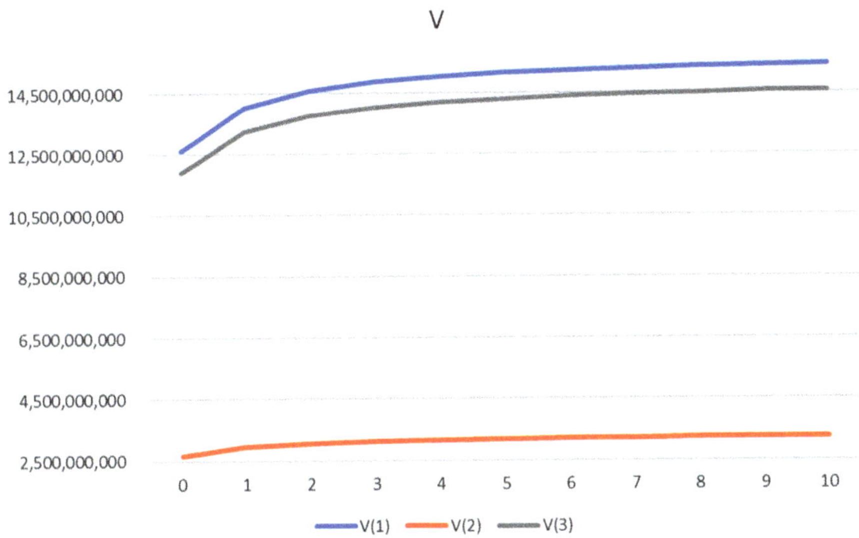

Fig. 24.18 Dependence of company value, *V*, on leverage level according to the MM theory in 2020 (financial risk only (1); plus industry (2) and plus market (3) business risk)

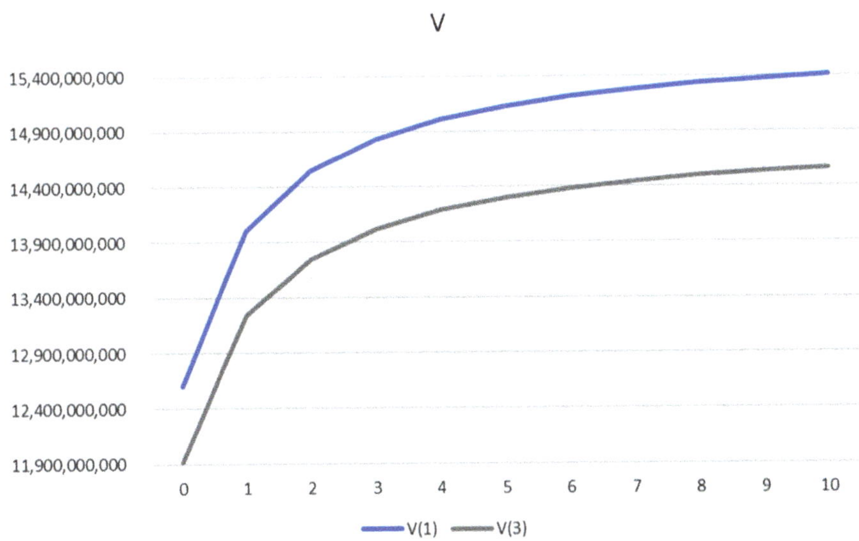

Fig. 24.19 Dependence of company value, *V*, on leverage level according to the MM theory in 2020 (financial risk only (1) and plus market (3) business risk)

difference between the case (2) and cases (1 and 3), while the difference between cases (1 and 3) is much smaller (Fig. 24.24).

It is seen from Fig. 24.25 that company value $V(L)$ increases with leverage level L in all three cases. There is a large difference between the case (2) and cases (1 and

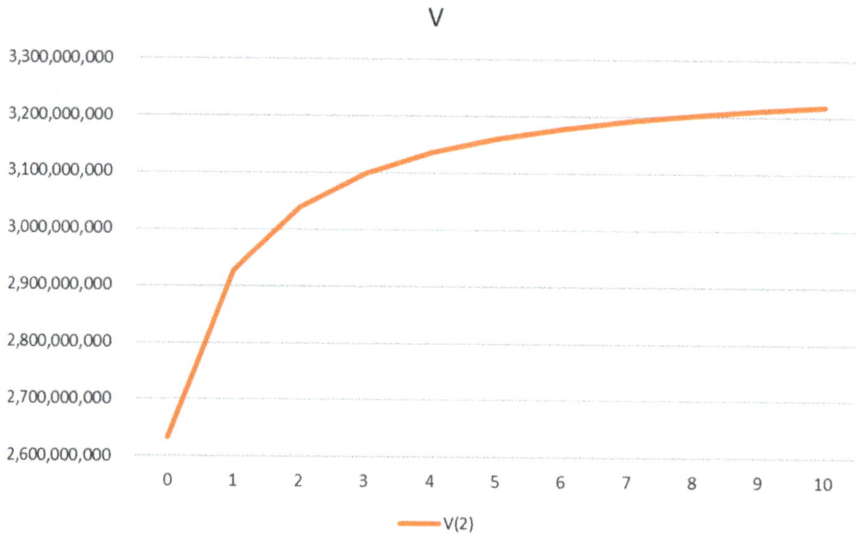

Fig. 24.20 Dependence of company value, V, on leverage level according to the MM theory in 2020 (plus industry (2) business risk)

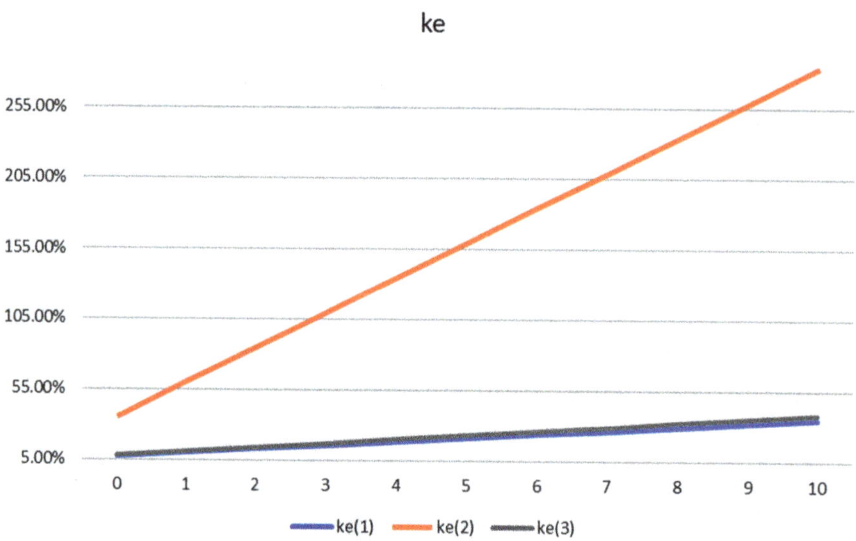

Fig. 24.21 Dependence of cost of equity, k_e, on leverage level according to the MM theory in 2020 (financial risk only (1); plus industry (2) and plus market (3) business risk)

3), while the difference between cases (1 and 3) is much smaller (Figs. 24.26, 24.27, and 24.28).

From Fig. 24.28, it is seen that the cost of equity capital k_e increases with increasing leverage level L for all three cases. The difference in inclination angles

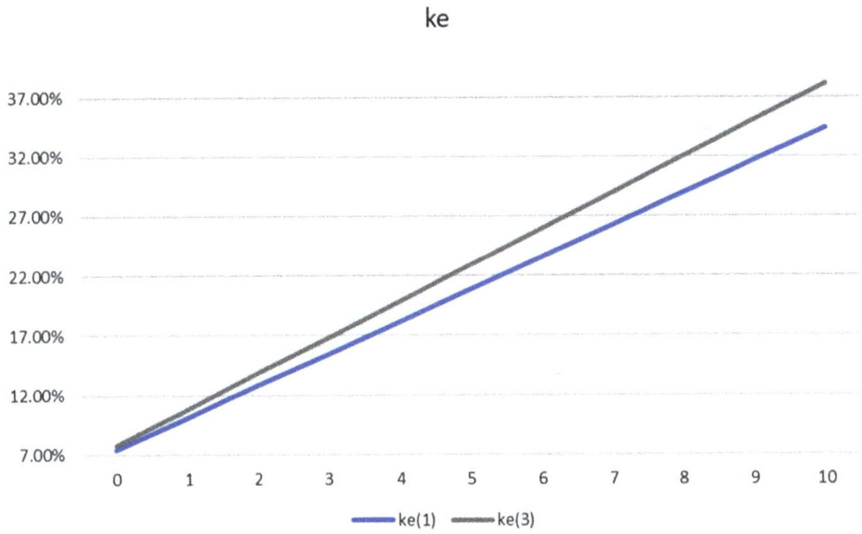

Fig. 24.22 Dependence of cost of equity, k_e, on leverage level according to the MM theory in 2020 (financial risk only (1) and plus market (3) business risk)

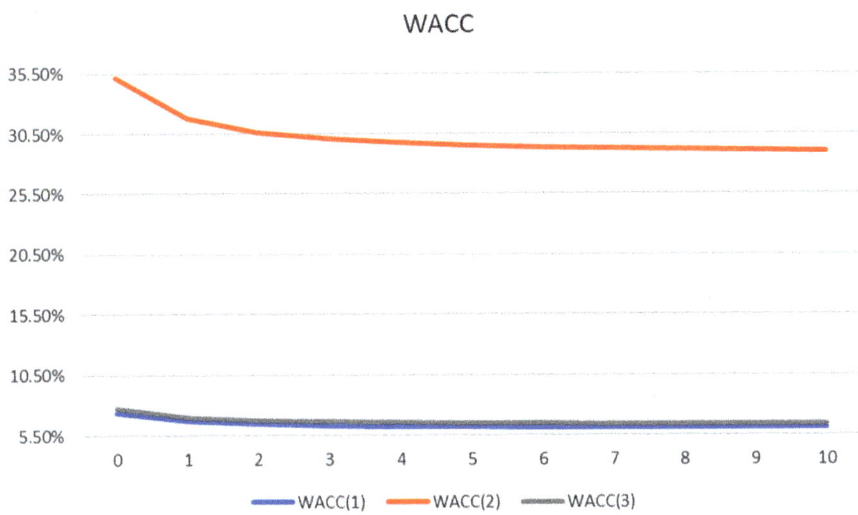

Fig. 24.23 Dependence of WACC (weighted average cost of capital) on leverage level according to the BFO theory in 2020 (financial risk only (1); plus industry (2) and plus market (3) business risk)

in cases (1) and (2 and 3) is significant, and for cases (2) and (3) it is significantly less. Note that the equity cost determines the value of dividends (Fig. 24.29).

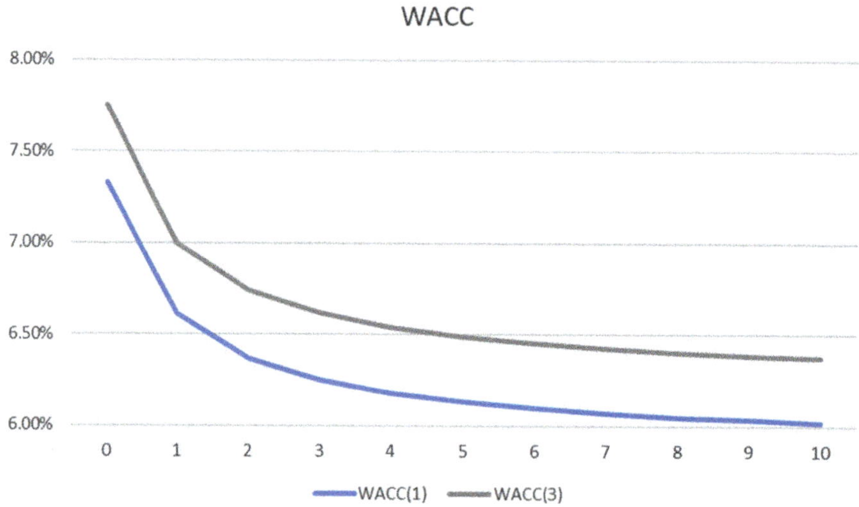

Fig. 24.24 Dependence of WACC (weighted average cost of capital) on leverage level according to the BFO theory in 2020 (financial risk only (1) and plus market (3) business risk)

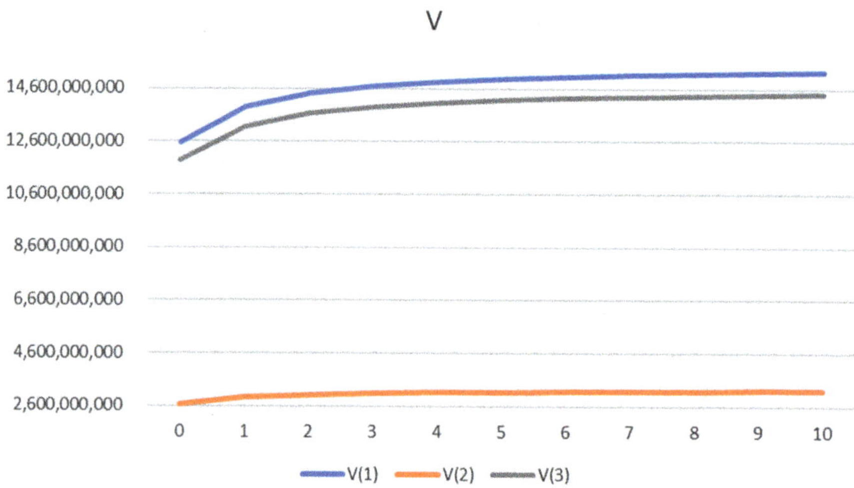

Fig. 24.25 Dependence of company value, *V*, on leverage level according to the BFO theory in 2020 (financial risk only (1); plus industry (2) and plus market (3) business risk)

24.6 Analysis of Results

The results show that in 2018 and 2019 there is a large difference between the case of accounting for pure financial risk and accounting for financial risk plus business risk, while the difference between accounting for industry and market business risks is much smaller. However, this difference should be taken into account.

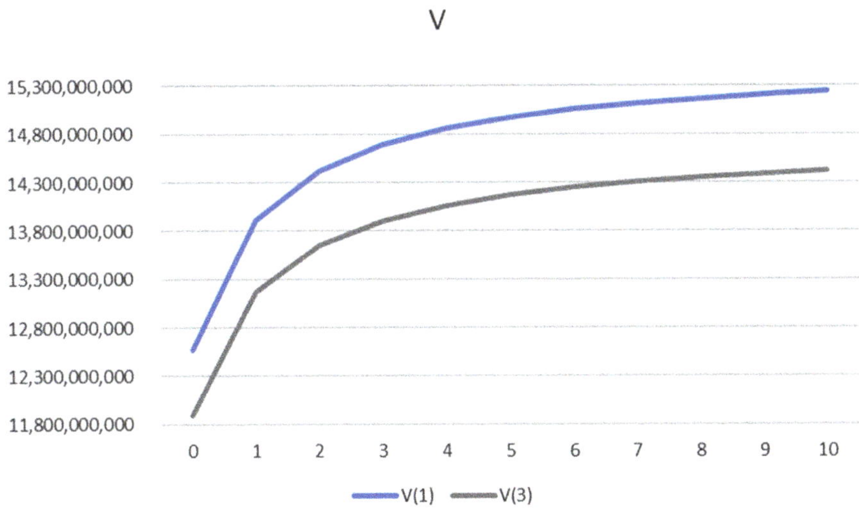

Fig. 24.26 Dependence of company value, *V*, on leverage level according to the BFO theory in 2020 (financial risk only (1) and plus market (3) business risk)

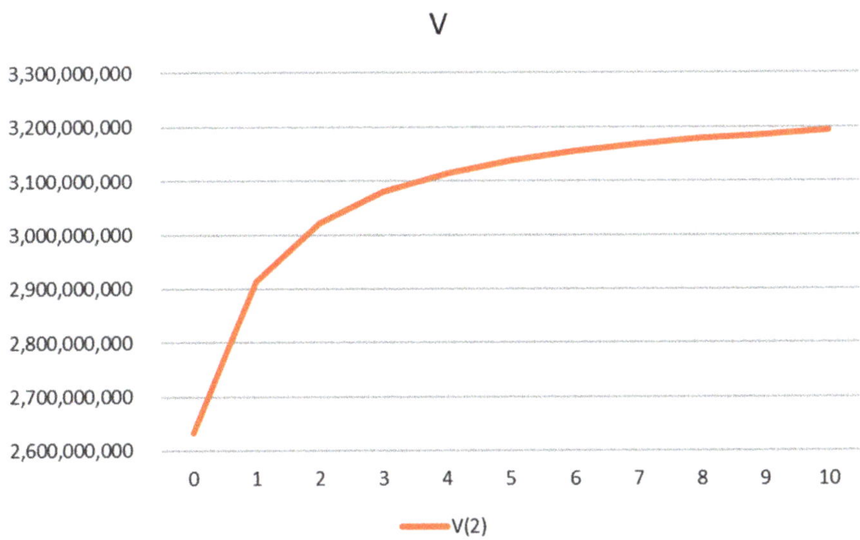

Fig. 24.27 Dependence of company value, *V*, on leverage level according to the BFO theory in 2020 (financial risk plus industry (2) business risk)

In contrast to 2018 and 2019, there is a big difference in 2020 between the case of industry business risk accounting and the other two cases, with pure financial risk and financial risk accounting plus market business risk being close to each other.

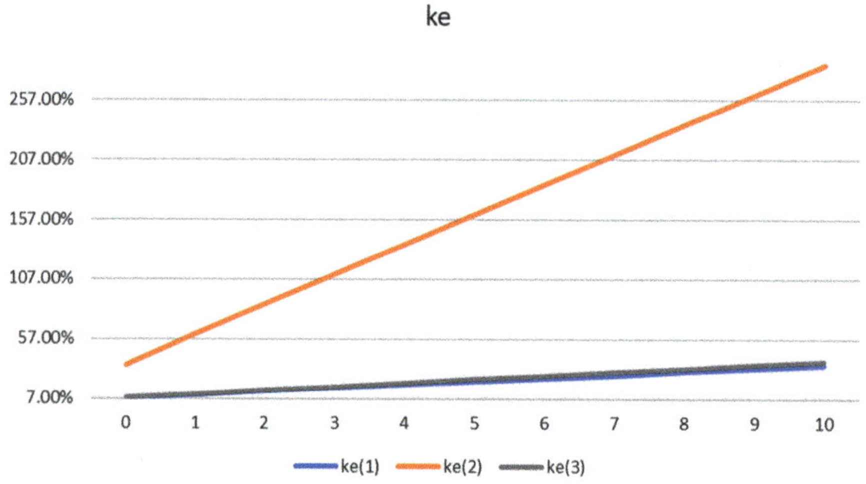

Fig. 24.28 Dependence of cost of equity, k_e, on leverage level according to the BFO theory in 2020 (financial risk only (1); plus industry (2) and plus market (3) business risk)

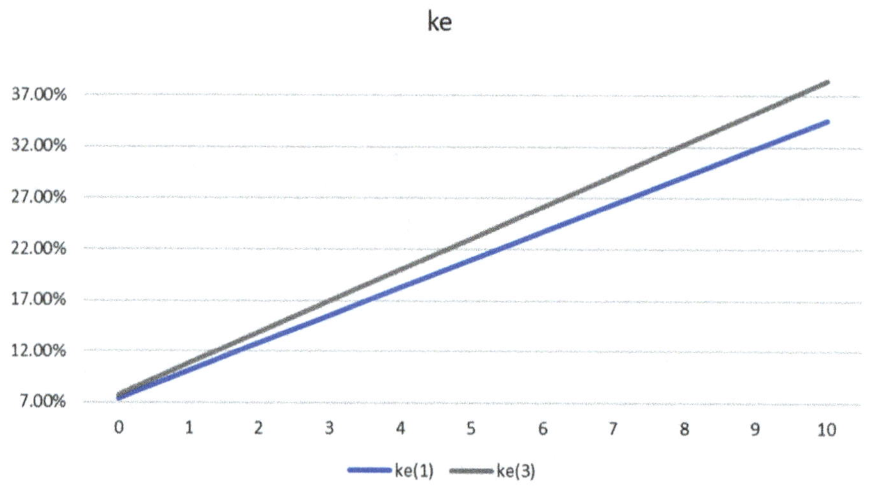

Fig. 24.29 Dependence of cost of equity, k_e, on leverage level according to the BFO theory in 2020 (financial risk only (1) and plus market (3) business risk)

This can be explained by the fact that the level of leverage in 2020 ($L = 2.42$) is lower than in previous years ($L = 3.29;3.22$), and this reduces financial risk.

For 2018 it is seen, that in opposition to the 2019–2020, the cost of equity capital k_e increases with increasing leverage level L for case (1) and decreases for cases (2) and (3). The difference in inclination angles in cases (2 and 3) is small. This means the existence of an anomalous effect depending on cost of equity on leverage level L. Note that the equity cost determines the value of dividends.

24.7 Conclusions

A fundamentally new approach to assessing the profitability of an asset is proposed. Transition from CAPM, which takes the same risk-free return for all assets as an initial assessment, to a new methodology, in which the average return of an asset, cleared of leverage, with the addition of a premium for business risk (market or industry) is taken as a seed return, significantly improves the accuracy of the estimate. As a result, generalized approaches (CAPM–BFO and CAPM–MM) were developed that take into account both types of risk: systematic (business) and financial. The application of a developed approach to Norilsk Nickel company (GMKN) has been done. The results show that in 2018 and 2019 there is a large difference between the case of accounting for pure financial risk and accounting for financial risk plus business risk, while the difference between accounting for industry and market business risks is much smaller. However, this difference should be taken into account. In contrast to 2018 and 2019, there is a big difference in 2020 between the case of industry business risk accounting and the other two cases, with pure financial risk and financial risk accounting plus market business risk being close to each other. This can be explained by the fact that the level of leverage in 2020 ($L = 2.42$) is lower than in previous years ($L = 3.29;3.22$), and this reduces financial risk. For 2018 it is seen that in opposition to 2019–2020, the cost of equity capital k_e increases with increasing leverage level L for case (1) and decreases for cases (2) and (3). The difference in inclination angles in cases (2 and 3) is small. This means the existence of an anomalous effect depending on cost of equity on leverage level L. Note, that the equity cost determines the value of dividends.

References

1. Brusov, P., Filatova, T., & Orekhova, N. (2022). *Generalized Modigliani–Miller theory: Applications in corporate finance, investments, taxation and ratings*. Springer Nature. 362 pp.
2. Brusov, P., & Filatova, T. (2023). Capital structure theory: Past, present, future. *Mathematics, 11*(3), 616. https://doi.org/10.3390/math11030616
3. Treynor, J. L. (1965). How to evaluate the management of investment funds. *Harvard Business Review, 43*, 63–75.
4. Sharpe, W. F. (1964). Capital asset prices: A theory of market equilibrium under conditions of risk. *The Journal of Finance, 19*(3), 425–442.
5. Lintner, J. (1965). The valuation of risk assets and the selection of risky investments in stock portfolios and capital budgets. *Review of Economics and Statistics*, 13–37.
6. Mossin, J. (1966). Equilibrium in a capital asset market. *Econometrica*, 768–783.
7. Danilo, L., Jiménez, R., Riquelme, M., & Leiva, V. (2023). Elliptical capital asset pricing models: Formulation, diagnostics, case study with Chilean data, and economic rationale. *Mathematics, 11*(6), 1394. https://doi.org/10.3390/math11061394
8. Fama, E. F., & French, K. R. (1992). The crosssection of expected stock returns. *The Journal of Finance, 47*(2), 427–465.
9. Fama, E. F., & French, K. R. (1993). Common risk factors in the returns on stocks and bonds. *Journal of Financial Economics, 33*(1), 3–56.

10. Fama, E. F., & French, K. R. (1995). Size and book-to-market factors in earnings and returns. *The Journal of Finance, 50*(1), 131–155.
11. Fama, E. F. (1965). The behavior of stock–market prices. *Journal of Business, 38*, 34–105.
12. Hamada, R. (1969). Portfolio analysis, market equilibrium, and corporate finance. *The Journal of Finance, 24*, 13–31.
13. Hamada, R. (1972). The effect of the firm's capital structure on the systematic risk of common stocks. *The Journal of Finance*, 435–452.
14. Sigman, K. (2005). *CAPM for economics.*
15. Brusova, A. (2011). A comparison of the three methods of estimation of weighted average cost of capital and equity cost of company. *Financ. Anal. Prob. Sol., 34*, 36–42.
16. Blattberg, R. C., & Gonedes, N. J. (1974). A comparison of the stable and student distributions as statistical models for stock prices. *Journal of Business, 47*, 244–280.
17. Brusov, P., Tatiana, F., & Veniamin, K. (2023a). *Capital Asset Pricing Model 2.0: account of business and financial risk.* Preprint ID: 87056. https://doi.org/10.20944/preprints202310.0347.v1
18. Brusov, P., Filatova, T., & Orekhova, N. (2023b). *The Brusov–Filatova–Orekhova theory of capital structure: Applications in corporate finance, investments, taxation and ratings.* Springer Nature. 769 pp.
19. Brusov, P., Filatova, T., Kulik, V., Chang, S.-I., Lin, G., & Chang, L.-M. (2023c). Precision finance: Capital structure theories approach reality. In *Proceedings of the 23rd international conference on electronic business (ICEB–2023).*
20. Brusov, P., Filatova, T., Kulik, V., Chang, S.-I., Lin, G., & Chang, L.-M. (2023d). Can CAPM (Capital Asset Pricing Model) accurately value assets? In *Proceedings of the 23rd international conference on electronic business (ICEB–2023).*
21. Brusov, P., Filatova, T., & Kulik, V. (2024). A new approach to assessing a company's financial indicators: Taking into account business and financial risks. *Frontiers of Finance, 2*(2), 6750. https://doi.org/10.59429/ff.v2i2.6750
22. Chen, J., Wu, Y., & Xu, Y. (2021). Research and analysis of asset pricing model based on the empirical test of stock price. In *Proceedings of the third international conference on economic management and cultural industry* (pp. 1288–1293).
23. French, C. W. (2003). The Treynor capital asset pricing model. *Journal of Investment Management, 1*(2), 60–72. Available at SSRN: https://ssrn.com/abstract=447580
24. Lange, K. L., Little, R. J., & Taylor, J. M. (1989). Robust statistical modeling using the t distribution. *Journal of the American Statistical Association, 84*, 881–896.
25. Modigliani, F., & Miller, M. H. (1958). The cost of capital, corporation finance and the theory of investment. *The American Economic Review, 48*, 261–296.
26. Modigliani, F., & Miller, M. H. (1963). Corporate income taxes and the cost of capital: A correction. *The American Economic Review, 53*, 433–443.
27. Ross, S. The arbitrage theory of capital asset pricing. *Journal of Economic Theory, 13*(3), 341–360. https://doi.org/10.1016/0022-0531(76)90046-6
28. Zhou, G. (1993). Asset–pricing tests under alternative distributions. *The Journal of Finance, 48*, 1927–1942.

Part VI
Qualitatively New Approach to Business Valuation and Company Value

Chapter 25
Qualitatively New Approach to Business Valuation and Company Value

25.1 Introduction

The value of any asset (stocks, bonds, companies, etc.) is equal to the present value of all income generated by that asset. Thus, to find the value of a company, it is necessary to find the current value of all the income received by the company from its inception to the current moment, that is, in the retrospective period. And to evaluate a business, it is necessary to find the current value of all future income received by the enterprise in the representative and terminate periods.

Business valuation goals could be as follows: purchase and sale of company shares and bonds on the stock market; making an informed investment decision; purchase and sale of a company; company restructuring; liquidation, merger, acquisition, spin-off of an independent company; determining the company's creditworthiness and the value of collateral for lending; insurance; taxation; making informed management decisions; etc.

A company's market capitalization is very important to investors: it shows the company's position in the market and determines how much investors are willing to pay for its shares. The greatest interest is in blue chip companies (with a capitalization above US$ 10 billion), which are leaders in their industry and are less susceptible to market volatility.

In this chapter, the new approach to business valuation and company value has been developed. As is well known, there are three main methods in business valuation: cost, comparative, and income, with the latter being the most adequate and accurate. There are three approaches to the income method: CAPM, APT and WACC. This article presents modifications of two of these three methods: CAPM and WACC. When modifying CAPM, a new model, CAPM 2.0, was created, within which business and financial risks were simultaneously taken into account. Moreover, for the first time, this was performed correctly and the inaccuracy, incorrectness, and inconsistency of the popular Hamada model were shown. When modifying the WACC approach (or the discounted income method), the following

P. Brusov, T. Filatova, *Business Valuation and Company Value*, Contributions to Finance and Accounting, https://doi.org/10.1007/978-3-031-90847-7_25

shortcomings of the income approach were corrected: inability of appraisers to correctly estimate the discount rate, the value of which is the main value in the income method of business valuation; the absence (before the advent of the BFO theory) of a correct method for estimating the discount rate; incorrect accounting of the influence of the retrospective period. The CAPM and CAPM 2.0 models have been incorporated into two major theories of capital structure—the Brusov–Filatova–Orekhova (BFO) theory and the Modigliani–Miller (MM) theory; this allows an account of both business and financial risks. CAPM takes into account systematic (business) risk, while capital structure theories take into account the financial risk of a specific company, associated with debt financing.

25.1.1 Disadvantage of Existing Valuation Methods (Income Approach)

The most important drawback of existing assessment methods is their inability to estimate the main assessment parameter—the discount rate, the value of which is critical in the assessment. The role of the correct determination of the discount rate in preventing abuse in business valuation, in investments, in determining the fair value of dividend income of shareholders is crucial. In business valuation, unscrupulous appraisers manipulate the value of the discount rate for the raider capture of enterprises. In investments, an incorrect assessment of the discount rate leads to an incorrect assessment of the effectiveness of an investment project, does not allow ranking investment projects in order to select the most effective projects in conditions of limited investment resources (companies, municipal, and state). This can lead to misappropriation of public funds, including funds from national projects. The rights of shareholders to receive adequate profits can be violated when the company's management conducts an incorrect and ineffective dividend policy, due to the inability of the management to determine the correct amount of dividends (the economically justified amount of which is the equity cost), or with the deliberate violation of shareholders' rights in this area.

Currently, there is only one theory that allows us to correctly estimate the discount rate—the BFO theory. Within the framework of the BFO theory, it is possible to study the dependence of WACC on the age of the company. This makes it possible to link a retrospective analysis of the company financial state with a representative one.

One more disadvantage of DCF (discounted cash flow) is its reliance on estimations of future cash flows, which could prove inaccurate. The BFO theory allows us to improve this shortcoming in two directions, using data on financial flows over the past few years. By determining the rate of income growth, we can (1) improve the forecasting of financial flows in representative and tentative periods; (2) use a modification of the BFO theory for the case of variable income, which we recently obtained, to correctly determine the discount rate WACC, which in this case depends on the rate of income growth.

25.1.2 Literature Review

As it was mentioned above, there are three main methods for estimating the value of assets. The first method is the well-known CAPM (capital asset pricing model), which uses the risk-free rate as the initial return and takes into account only the business risk associated with investing in a specific asset, and not in the market as a whole. The second method is arbitrage pricing theory (APT), which, unlike the single-factor, single-beta CAPM, is a multifactor approach.

The third one is associated with the use of one of two main theories of capital structure (Brusov–Filatova–Orekhova (BFO) theory and Modigliani–Miller (MM) theory), which take into account only the financial risk associated with the use of debt financing and allow the calculation of all financial indicators of a company of an arbitrary age (BFO theory) or a perpetuity one (MM theory) (see two recent monographs by Brusov et al. [1, 2] and review by Brusov and Filatova [3]).

The famous capital asset pricing model (CAPM) (see [4], based on the portfolio theory of Harry Markowitz [5], was developed independently by Jack Treynor (1965), William F. Sharpe (1964), John Lintner (1965), and Jan Mossin (1966) [6–9]. The capital asset pricing model (CAPM) is still widely used to estimate the profitability of assets and investments. However, as our analysis of CAPM for several dozen listed companies from different countries as well as numerous results of other authors showed, the CAPM results differ significantly from the actual profitability of companies. The reasons for this discrepancy are related to the internal properties and shortcomings of the model.

There are some modifications of CAPM [4, 5, 10–15] that can bring the model closer to real life. Among the latter is the account of financial risk in CAPM along with business risk, which was recently correctly carried out by Brusov et al. [4], and the first attempt to take financial risk into account (not completely successful) was made by Hamada R. [14, 15]. Initial capital asset pricing model (CAPM) takes into account only business risk. In practice, companies use debt financing and operate at non-zero levels of leverage. This means that it is necessary to take into account the financial risk associated with the use of debt financing along with the business one. Brusov et al. [4] have accounted for the business and financial risks simultaneously. A new approach to CAPM has been developed that takes into account both business and financial risks. They combine the theory of CAPM and the Modigliani–Miller (MM) theory [16, 17]. The first is based on portfolio analysis and accounting for business risks in relation to the market (or industry). The second one (the Modigliani–Miller (MM) theory) describes a specific company and takes into account the financial risks associated with the use of debt financing. The combination of these two different approaches makes it possible to take into account both types of risks: business and financial ones. Brusov et al. [4] combine these two approaches analytically, while Hamada [14, 15] did it phenomenologically. Using the Modigliani–Miller (MM) theory, it is shown that Hamada's model, the first model, used for this purpose half a century ago, is incorrect. In addition to the renormalization of the beta-coefficient, obtained in the Hamada model, two additional terms are found: the renormalized risk-free return and the term dependent on the cost of debt k_d. A critical analysis of the Hamada model was carried out. The vast majority of listing companies use debt financing and are levered, and the Hamada model is not applicable to them in contrast to a new [4] approach applicable to leveraged companies. A new approach was implemented with specific companies. A comparison of the results of the new approach with the results of the conventional CAPM has been shown and two versions of CAPM (market or industry) have been considered [4].

One more modification of the CAPM had been carried out in 1992, by Y. Fama and K. French [10–13], who proved that future returns are also affected by factors such as company size and industry affiliation. They have developed three- and five-factor models.

Note that the size of the company matters and should be taken into account as an additional premium to the discount rate. Such additional adjustments are taken into account in several theories, for example, in the five-factor model of Fama and French [10–13] and in the Sharpe model. To account for enterprises in other sectors, the industry approach described in the developed methodology should be used.

An alternative to CAPM is the Arbitrage Pricing Theory (APT), created by Stephen Ross in 1976 [18]. In opposition to the CAPM, which has only one factor and one beta, the APT formula has multiple factors that include non-company factors, which requires the asset's beta with respect to each separate factor.

25.2 A New Approach to Business Valuation and Company Value

Below, the modifications of two methods—CAPM and WACC—are studied (Brusov and Filatova [19, 20]).

Three time horizons have been considered (see Fig. 25.1):

- Retrospective (from the moment of creation (entering the market) of the company until the moment of evaluation [$n_{1]}$).

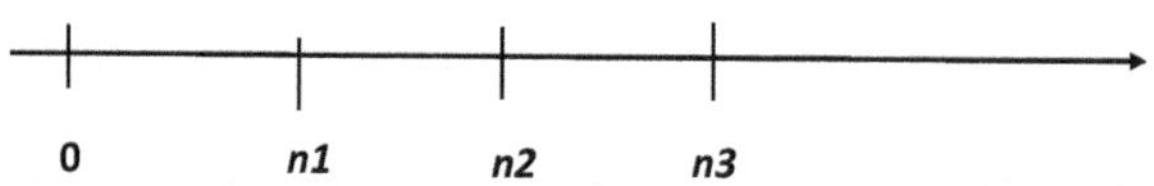

Fig. 25.1 Three time horizons: retrospective horizon (0–n_1); representative horizon (n_1–n_2); and terminate horizon (n_2–n_3)

- Representative (from the moment of assessment $[n_{1]}$ to the end of this period $[n_{2]}$: usually 5–10 years).
- Terminate period (finite (until moment $[n_3]$, or infinite): in the first case, the BFO theory is used; in the second case, its perpetuity limit—the MM theory.

The new methodology developed here for assessing a business, as well as calculating the current capitalization of a company, consists of the following steps:

1. Collection and processing company, industry, and market data from annual financial and index reports μ; L; k_d; t; β; β_U; μ_F; CF

   The following notations are used: μ—profitability of company; L—leverage level; k_d—debt cost; t—tax on profit rate; β—β coefficient; β_U—unlevered β coefficient; μ_F—risk-free profitability; CF—income for one period; t—tax on profit rate.
2. Clean β from L

$$\beta = \beta_U(1 + L(1 - t))$$

$$\beta_U = \frac{\beta}{1 + L(1 - t)}$$

or use unlevered β_U.

3. To find the parameter k_0, the main one in both theories (BFO and MM), for all time horizons (n_1, n_2, n_3), data from the company's reporting for the year preceding the year of assessment are used.
4. Clean μ from leverage, by using MM formula for equity cost:

$$\mu = k_0 + L(k_0 - k_d)(1 - t)$$

$$k_0 = \frac{\mu_i + Lk_d(1 - t)}{1 + L(1 - t)}$$

5. Within traditional CAPM and/or new CAPM 2.0, approximations three values of the parameter k_0 ($k_0(1)$, $k_0(2)$, $k_0(3)$ were found, which account for only financial risk ($k_0(1)$), industry business risk along with financial risk ($k_0(2)$) and market business risk along with financial risk ($k_0(3)$). These values are the expected yields on the levels of company, industry, and market.

   After that, k_0 was calculated according to three conditions:

$$k_0(1) = \mu_0 \tag{25.1}$$

$$k_0(2) = \mu_0 + \beta_{iI}(\mu_I - \mu_F) \qquad \text{for industry} \tag{25.2}$$

$$k_0(3) = \mu_0 + \beta_{iM}(\mu_M - \mu_F) \qquad \text{for market} \tag{25.3}$$

Based on the data obtained, let us calculate the dependence of WACC, V, key indicators on the level of leverage (upon change in leverage from 0 to 10) within the Brusov–Filatova–Orekhova theory.

$$\frac{1-(1+\mathrm{WACC})^{-n}}{\mathrm{WACC}} = \frac{1-(1+k_0)^{-n}}{(k_0)\cdot(1-w_d t[1-(1+k_d)^{-n}])} \tag{25.4}$$

Here, WACC is weighted average cost of capital.

Calculate the dependence of WACC on company age (n_1) at fixed L (L of company).

Calculate the V from 0 to n_1, using WACC (n_1) at L (last year 2023) and average L for 4 years.

6. Using obtained parameters k_{0i} values, the dependences WACC(L) and WACC (n) are calculated within the framework of the BFO theory.

   Next, it is necessary to distinguish between business valuation and company valuation.

25.2.1 For Company Valuation

Using calculated dependence WACC(n), WACC(n_1) is found for the end of the year preceding the year of assessment. This value is used as discount rate for estimation of company value by the following formula:

$$V_0 = \frac{\mathrm{CF}}{\mathrm{WACC}(n_1)}\cdot\left(1-(1+\mathrm{WACC}(n_1))^{-n1}\right) \tag{25.5}$$

This V_0 value refers to the zero point in time (the moment of creation [entering the market]) of the company. To find the value of the company at the time of valuation n_1, it is needed to increase it by n_1 periods. Thus, real company value will be equal to the following:

$$\begin{aligned} V &= V_0(1+\mathrm{WACC}(n_1))^{n_1} \\ &= \frac{\mathrm{CF}}{\mathrm{WACC}(n_1)}\cdot\left(1-(1+\mathrm{WACC}(n_1))^{-n1}\right)(1+\mathrm{WACC}(n_1))^{n_1} = \\ &= \frac{\mathrm{CF}}{\mathrm{WACC}(n_1)}\cdot\left((1+\mathrm{WACC}(n_1))^{-n1}-1\right) \end{aligned} \tag{25.6}$$

25.2.2 *For Business Valuation*

The discounted value of the business over a representative period (n_2–n_1) yields the following form:

$$V_1 = \frac{\mathrm{CF}}{\mathrm{WACC}(n_1)} \cdot \left(1 - (1 + \mathrm{WACC}(n_1))^{-(n2-n1)}\right) \tag{25.7}$$

Note that this V_1 value refers to the time moment of valuation n_1.

The discounted value of the business over a terminate period (n_3–n_2) should be found:

$$V_2 = \frac{\mathrm{CF}}{\mathrm{WACC}(n_2)} \cdot \left(1 - (1 + \mathrm{WACC}(n_2))^{-(n3-n2)}\right) \tag{25.8}$$

This V_2 value refers to the time moment n_2 (end of representative period n_2).

The alternative length of the terminate period is perpetual. In this case, V_2 is calculated using the following formula:

$$V_2 = \frac{\mathrm{CF}}{\mathrm{WACC}(n_2)} \tag{25.9}$$

To value a business, it is necessary to sum these two values, discounting the second to the time of valuation. Finally, one obtains the following:

$$V = V_1 + \frac{V_2}{(1 + \mathrm{WACC})^{(n2-n1)}} \tag{25.10}$$

The new approach allows to take into account the following conditions for the actual functioning of the company [1–4] (see paragraph 4):

- Variable income: the growth rate of the company's income (data from the company's reporting for 2–5 years preceding the year of assessment is used) [1–4]
- Inflation [1–3]
- Frequency and method of payment of income tax (advance payments or payments at the ends of periods [1–3]
- Business risk within the framework of the CAPM 2.0 theory created by the authors [4]
- The effect of the "golden age" of the company

Below, the modification of the CAPM method (paragraph 3) as well as WACC one (paragraph 4) is described.

25.3 Modification of the CAPM Method

The CAPM method is one of the main ones in the income approach to business valuation. Modification of the CAPM method, which led to the creation of the theory CAPM 2.0, allows us to correctly take into account both business and financial risks.

25.3.1 *Market Approach*

CAPM (capital asset pricing model) describes the profitability of asset and is described by the following formula:

$$k_i = k_f + \beta_i (k_m - k_f) \tag{25.11}$$

Here, k_f is risk-free profitability, β is the β-coefficient of the company. Formula (25.11) shows the dependence of the return on the asset on the return on the market as a whole. The β-coefficient is described by the following formula:

$$\beta_i = \frac{\mathrm{cov}_{im}}{\sigma_m^2} = \rho_{im} \frac{\sigma_i}{\sigma_m}. \tag{25.12}$$

Here, σ_i is the risk (standard deviation) of i-th asset, σ_m is market risk (standard deviation of market index), cov_{im} is covariance between i-th asset and market portfolio.

An investor invests in risky securities only if their return is higher than the return on risk-free securities, so always $k_i > k_f$ and $k_m > k_f$.

The beta-coefficient of a security, β, has the meaning of the amount of riskiness of this security.

25.3.2 *Industry Approach*

CAPM has an alternative approach that refers to the industrial index rather than the market.

$$k_i = k_f + \beta_i (k_I - k_f) \tag{25.13}$$

Here, k_f is risk-free profitability; β is the β-coefficient of the company. In this case, it shows the dependence of the return on the asset on the return on the industry as a whole. The β-coefficient now is described by the following formula:

$$\beta_i = \frac{\mathrm{cov}_{iI}}{\sigma_I^2} = \rho_{iI}\frac{\sigma_i}{\sigma_I} \tag{25.14}$$

Here, σ_i is the risk of i-th asset, σ_I is industry risk (standard deviation of industry index), cov_{iI} is covariance between i-th asset and industry index.

25.3.3 CAPM 2.0 Formula

The Modigliani–Miller theory [7, 17], with the accounting of taxes, has been united with CAPM (capital asset pricing model) by Hamada [15, 21]. For the cost of equity of a leveraged company, the below formula has been derived.

$$k_e = k_f + \beta_U\left(k_m - k_f\right) + \beta_U\left(k_m - k_f\right)\frac{D}{S}(1-t) \tag{25.15}$$

The first term represents risk-free profitability k_f, the second term is business risk premium, $\beta_U(k_m - k_f)$, and the third term is financial risk premium $\beta_U\left(k_m - k_f\right)\frac{D}{S}(1-t)$.

In 2022–2024, the authors of this article checked Hamada's results and found them to be incorrect. Combining CAPM and the Modigliani–Miller theory, one obtains the following result [4]

$$k_e = k_0 + L(k_0 - k_d)(1-t) = k_f + \beta_U\left(k_m - k_f\right) + L\left(k_f + \beta_U\left(k_m - k_f\right) - k_d\right)(1-t) =$$

$$k_f(1 + L(1-t)) + \beta_U\left(k_m - k_f\right)(1 + L(1-t)) - Lk_d(1-t) \tag{25.16}$$

The difference from Hamada's formula (25.15) is in the renormalized value of the risk-free yield, and in the last term, which depends on the cost of debt k_d. The incorrectness of Hamada's approximation becomes obvious.

The expression (25.16) could be presented as a sum of two parts, one of which is the Hamada expression, and the second is a new additional term that was received

$$\begin{aligned} k_e &= k_f + \beta_U\left(k_m - k_f\right)(1 + L(1-t)) + k_f L(1-t) - Lk_d(1-t) = \\ &= k_f + \beta\cdot\left(k_m - k_f\right) + L(1-t)\left(k_f - k_d\right) = k_{e;\mathrm{CAPM}} + \Delta \end{aligned} \tag{25.17}$$

Here,

$$\Delta = L(1-t)\left(k_f - k_d\right). \tag{25.18}$$

This formula takes into account both business and financial risks and is the main result of the work.

Both CAPM and CAPM 2.0 approximations were used and compared.

25.4 Modification of the WACC Method

In the discounted cash flow method when valuing a business, the most important role is played by the discount rate. The main disadvantage of existing business valuation methods is that they do not allow a correct assessment of the discount rate.

The new method was developed, which is associated with the use of the main theory of capital structure (Brusov–Filatova–Orekhova [BFO] theory and its perpetual limit, the Modigliani–Miller [MM] theory), which take into account the financial risk associated with the use of debt financing as well as business risk. These theories allow to calculate all financial indicators of a company of an arbitrary age (BFO theory) or a perpetuity one (MM theory) and, in particular, the company value. Recently, the WACC formulas within the framework of the Brusov–Filatova–Orekhova (BFO) theory and the Modigliani–Miller (MM) theory have been modified taking into account the conditions of the actual operation of the company, such as frequent payment of income tax, the method of paying income tax (at the ends of periods or in advance), variable income, etc. To improve the accuracy of business valuation, it is necessary to take these effects into account.

Within the framework of the BFO theory, it is possible to study the dependence of WACC on the age of the company. This makes it possible to link a retrospective analysis of company financial state with a representative one.

25.4.1 WACC Formulas for Brusov–Filatova–Orekhova (BFO) Theory and for Modigliani–Miller (MM) Theory

Recently, WACC formulas for the Brusov–Filatova–Orekhova (BFO) theory and for the Modigliani–Miller (MM) theory have been modified taking into account the above-mentioned conditions for the actual functioning of the company [1–3]. Below, we provide a summary of the WACC formulas for the Brusov–Filatova–Orekhova (BFO) theory as well as for the Modigliani–Miller (MM) theory (see Review [3]).

Note that the BFO theory, as a more general theory than the MM theory (it is valid for companies of arbitrary age, unlike the perpetual MM), is becoming popular among researchers who understand that it is much closer to the real economy (see some recent references [21–24].

25.4.1.1 Variable Income Case

Income Tax Payments at the Ends of Periods

$$\text{BFO}: \quad \frac{1-\left(\frac{1+g}{1+\text{WACC}}\right)^n}{\text{WACC}-g} = \frac{1-\left(\frac{1+g}{1+k_0}\right)^n}{(k_0-g)\cdot(1-w_d t[1-(1+k_d)^{-n}])} \tag{25.19}$$

$$\text{MM}: \quad \text{WACC} = (k_0-g)\cdot(1-w_d t)+g \tag{25.20}$$

Advance Income Tax Payments

$$\text{BFO}: \quad \frac{1-\left(\frac{1+g}{1+\text{WACC}}\right)^n}{\text{WACC}-g} = \frac{1-\left(\frac{1+g}{1+k_0}\right)^n}{(k_0-g)\cdot(1-w_d t[1-(1+k_d)^{-n}]\cdot(1+k_d))} \tag{25.21}$$

$$\text{MM}: \quad \text{WACC} = (k_0-g)\cdot(1-w_d t\cdot(1+k_d))+g \tag{25.22}$$

25.4.1.2 Frequent Income Tax Payments

Income Tax Payments at the Ends of Periods

$$\text{BFO}: \quad \frac{1-(1+\text{WACC})^{-n}}{\text{WACC}} = \frac{1-(1+k_0)^{-n}}{k_0\cdot\left(1-\frac{k_d w_d t}{p}\,\frac{[1-(1+k_d)^{-n}]}{(1+k_d)^{1/p}-1}\right)} \tag{25.23}$$

$$\text{MM}: \quad \text{WACC} = k_0\cdot\left(1-\frac{k_d w_d t}{p\cdot\left[(1+k_d)^{1/p}-1\right]}\right) \tag{25.24}$$

Advance Income Tax Payments

$$\text{BFO}: \quad \frac{1-(1+\text{WACC})^{-n}}{\text{WACC}} = \frac{1-(1+k_0)^{-n}}{k_0\cdot\left(1-\frac{k_d w_d t}{p}\,\frac{[1-(1+k_d)^{-n}]\cdot(1+k_d)^{1/p}}{(1+k_d)^{1/p}-1}\right)} \tag{25.25}$$

$$\text{MM}: \quad \text{WACC} = k_0\cdot\left(1-\frac{k_d w_d t\cdot(1+k_d)^{1/p}}{p\cdot\left[(1+k_d)^{1/p}-1\right]}\right) \tag{25.26}$$

25.4.1.3 Simultaneous Accounting of Variable Income in Case of Frequent Income

Tax Payments

Income Tax Payments at the Ends of Periods

$$\text{BFO}: \quad \frac{1-\left(\frac{1+g}{1+\text{WACC}}\right)^n}{\text{WACC}-g} = \frac{1-\left(\frac{1+g}{1+k_0}\right)^n}{(k_0-g)\cdot\left(1-\frac{k_d w_d t}{p}\frac{\left[1-(1+k_d)^{-n}\right]}{(1+k_d)^{1/p}-1}\right)} \tag{25.27}$$

$$\text{MM}: \quad \text{WACC}-g=(k_0-g)\cdot\left(1-\frac{k_d w_d t}{p\cdot(1+k_d)^{1/p}-1}\right) \tag{25.28}$$

Advance Income Tax Payments

$$\text{BFO}: \quad \frac{1-\left(\frac{1+g}{1+\text{WACC}}\right)^n}{\text{WACC}-g} = \frac{1-\left(\frac{1+g}{1+k_0}\right)^n}{(k_0-g)\cdot\left(1-\frac{k_d w_d t}{p}\frac{\left[1-(1+k_d)^{-n}\right]\cdot(1+k_d)^{1/p}}{\left[(1+k_d)^{1/p}-1\right]}\right)} \tag{25.29}$$

$$\text{MM}: \quad \text{WACC}-g=(k_0-g)\cdot\left(1-\frac{k_d w_d t\cdot(1+k_d)^{1/p}}{p\cdot(1+k_d)^{1/p}-1}\right)$$

$$\frac{1-(1+\text{WACC})^{-n}}{\text{WACC}} = \frac{1-(1+k_0)^{-n}}{(k_0)\cdot(1-w_d t[1-(1+k_d)^{-n}])}$$

Calculate the dependence of WACC on company age (n_1) at fixed L (L of company).

Calculate the V from 0 to n_1, using WACC (n_1) at L (last year 2023) and average L for 4 years.

25.5 Results: Application of a New Methodology for Amazon Company

Let us apply the developed approach to the Amazon company.

(The data for calculation have been collected from the following references:

1. https://www.macrotrends.net/2521/30–year–treasury–bond–rate–yield–chart (accessed on 22 June 2024).
2. https://www.stock–analysis–on.net/NASDAQ/Company/Amazoncom–Inc/DCF/CAPM (accessed on 22 June 2024).
3. https://www.stock–analysis–on.net/NASDAQ/Company/Amazoncom–Inc/Ratios/Adjusted–Ratios (accessed on 22 June 2024).

Table 25.1 Amazon initial data, 2020–2023

Level	Indicator	2020	2021	2022	2023
k_f (30-year Treasury Bonds)		1.56%	2.06%	3.11%	4.09%
Company (AMZN)	k_i	5.26%	0.34%	0.84%	5.39%
	L	0.66	0.54	0.59	0.39
	k_d	2.23%	1.80%	1.85%	2.31%
	σ_i	0.0242	0.0152	0.0316	0.0207
Industry	k_I	2.97%	4.17%	2.61%	3.10%
	L	1.1	1.08	1.07	1.21
	σ_I	0.0256	0.0214	0.0243	0.0151
	β_{iI}	0.41	0.08	0.91	0.53
Market (S&P 500)	μ_m	1.52%	2.05%	2.57%	1.90%
	σ_m	0.0219	0.0083	0.0152	0.0082
	β_{im}	1.33	1.13	1.23	1.23

Table 25.2 CAPM and CAPM 2.0 data

Variables	2020	2021	2022	2023
k_i	5.26%	0.34%	0.84%	5.39%
k_i CAPM (industry)	2.14%	2.22%	2.66%	3.56%
k_i CAPM (market)	1.51%	2.05%	2.45%	1.41%
Δ	−0.35%	0.14%	0.74%	0.69%
k_i CAPM 2.0 (industry)	1.79%	2.36%	3.40%	4.26%
k_i CAPM 2.0 (market)	1.16%	2.19%	3.19%	2.10%

4. https://www.gurufocus.com/term/effective–interest–rate/AMZN (accessed on 22 June 2024).
5. https://finance.yahoo.com/quote/AMZN/history/?.neo_opt=0 (accessed on 22 June 2024).
6. https://cbonds.com/indexes/146954/ (accessed on 22 June 2024).
7. https://www.readyratios.com/sec/ratio/financial–leverage/ (accessed on 22 June 2024).
8. https://www.stock–analysis–on.net/NASDAQ/Company/Amazoncom–Inc/DCF/CAPM (accessed on 22 June 2024).
9. https://s2.q4cdn.com/299287126/files/doc_financials/2024/ar/Amazon–com–Inc–2023–Annual–Report.pdf (accessed on 22 June 2024).
10. https://companiesmarketcap.com/amazon/marketcap/ (accessed on 22 June 2024).)

We apply CAPM, CAPM 2.0, and the Brusov–Filatova–Orekhova theory. First, using open sources, the following data were gathered (Table 25.1). Based on the collected data and using the CAPM, the market-expected yields on the levels of company, industry, and market were calculated.

Afterward, the new approach of CAPM (CAPM 2.0) was used to adjust the initial yields by Δ, which takes in account the level of leverage (L) and tax rate (t)

Table 25.3 Three values of k_0

Indicator	2020	2021	2022	2023
μ_i	5.26%	0.34%	0.84%	5.39%
μ_0	4.22%	0.78%	1.16%	4.66%
k_0 (1st value)	4.22%	0.78%	1.16%	4.66%
k_0 (2nd value)	4.60%	0.86%	0.92%	4.39%
k_0 (3rd value)	4.19%	0.77%	0.71%	2.61%

(Table 25.2). As we can see from the table, the effect of adding Δ was relatively small mainly due to the relatively low level of leverage of the company. We note that for mining companies, companies in the telecommunications sector, and others with a high level of leverage, as well as in the case of a large difference between k_f and k_d, β can be significant [4].

As we can conclude, Amazon shows relatively good results if compared to the industry or market overall. Despite the drop in 2021 and 2022 (mainly as a result of the COVID-19 pandemic, as well as increasing competition), the company has been bringing positive results and is in a rather stable financial position.

Next, following methodology we developed, we cleared μ from leverage, cleared β from leverage, and calculated the three values of k_0: (1) financial risk without business risk; (2) accounting industry business risk; (3) accounting market business risk (Table 25.3):

$$k_0(1) = \mu_0$$

$k_0(2) = \mu_0 + \beta_{iI}(\mu_I - \mu_F)$ for industry
$k_0(3) = \mu_0 + \beta_{iM}(\mu_M - \mu_F)$ for market

25.5.1 The Dependence of WACC on the Leverage

Next, let us calculate the dependence of WACC on the leverage of the company (Table 25.4) based on the Brusov–Filatova–Orekhova theory for three values of k_0 for 2020–2023 years:

$$\frac{1-(1+\mathrm{WACC})^{-n}}{\mathrm{WACC}} = \frac{1-(1+k_0)^{-n}}{(k_0)\cdot(1-w_d t[1-(1+k_d)^{-n}])} \tag{25.30}$$

As we can see from Table 25.4 above, WACC always decreases with L.

Below, we provide the Figs. 25.2, 25.3, and 25.4 for 2020–2023 years.

WACC1 corresponds to the case of financial risk without business risk; WACC2 accounting financial risk along with industry business risk; WACC3 accounting financial risk along with market business risk. The ordering of dependencies WACC (L) is as follows (from big values to small ones): 213.

Table 25.4 WACC dependence on L (n is fixed)

	2020			2021			2022			2023		
L	WACC1	WACC2	WACC3	WACC1	WACC2	WACC3	WACC1	WACC2	WACC 3	WACC1	WACC 2	WACC 3
0	4.22%	4.60%	4.19%	0.78%	0.86%	0.77%	1.16%	0.92%	0.71%	4.66%	4.39%	2.61%
1	3.79%	4.16%	3.76%	0.48%	0.56%	0.47%	0.85%	0.61%	0.40%	4.21%	3.95%	2.21%
2	3.64%	4.01%	3.61%	0.37%	0.46%	0.37%	0.74%	0.50%	0.30%	4.06%	3.80%	2.07%
3	3.57%	3.94%	3.54%	0.32%	0.41%	0.31%	0.69%	0.45%	0.25%	3.98%	3.72%	2.01%
4	3.53%	3.89%	3.49%	0.29%	0.37%	0.28%	0.66%	0.42%	0.22%	3.93%	3.68%	1.97%
5	3.50%	3.86%	3.47%	0.27%	0.35%	0.26%	0.63%	0.40%	0.20%	3.90%	3.64%	1.94%
6	3.48%	3.84%	3.44%	0.25%	0.34%	0.25%	0.62%	0.38%	0.18%	3.88%	3.62%	1.92%
7	3.46%	3.83%	3.43%	0.24%	0.33%	0.24%	0.61%	0.37%	0.17%	3.86%	3.61%	1.90%
8	3.45%	3.81%	3.42%	0.24%	0.32%	0.23%	0.60%	0.36%	0.16%	3.85%	3.59%	1.89%
9	3.44%	3.80%	3.41%	0.23%	0.31%	0.22%	0.59%	0.35%	0.15%	3.84%	3.58%	1.88%
10	3.43%	3.79%	3.40%	0.22%	0.31%	0.21%	0.58%	0.35%	0.15%	3.83%	3.58%	1.87%

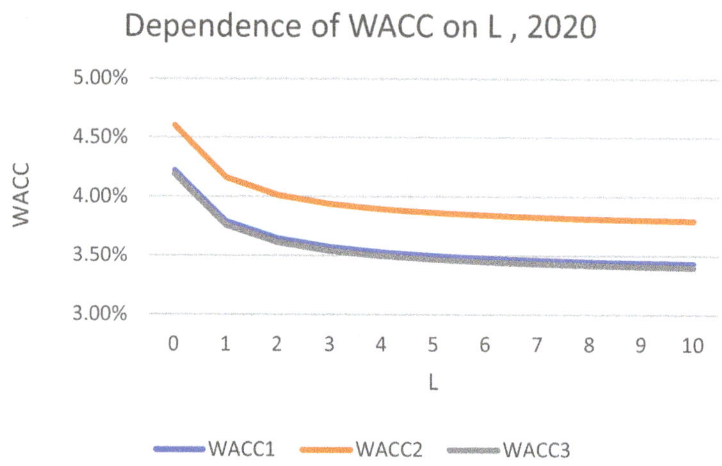

Fig. 25.2 Dependence of WACC on L for three values of k_0 for 2020

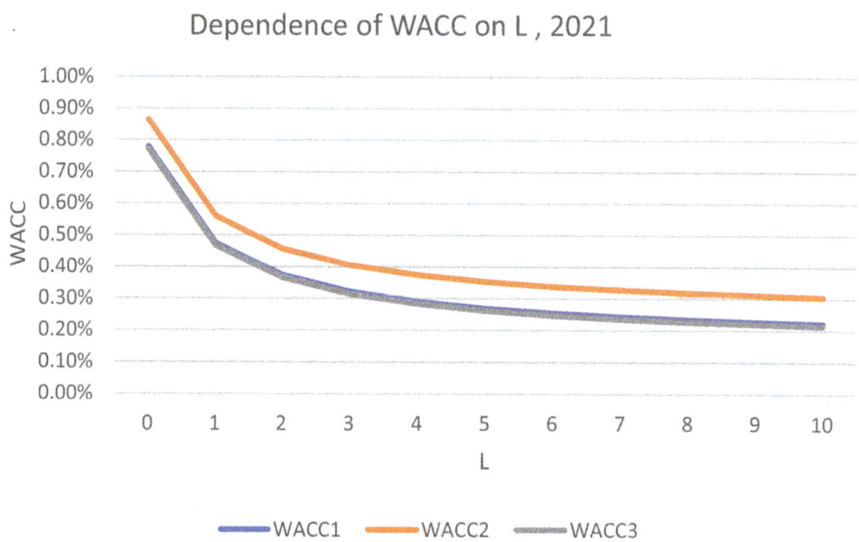

Fig. 25.3 Dependence of WACC on L for three values of k_0 for 2021. The ordering of dependencies WACC(L) is as follows (from big values to small ones): 213

25.5.2 *Dependence of WACC on Company Age* n

Within the framework of the new approach we developed, calculating the dependence of WACC on the age of the company n plays a decisive role in assessing the business and the value of the company. Below we calculate the dependence of

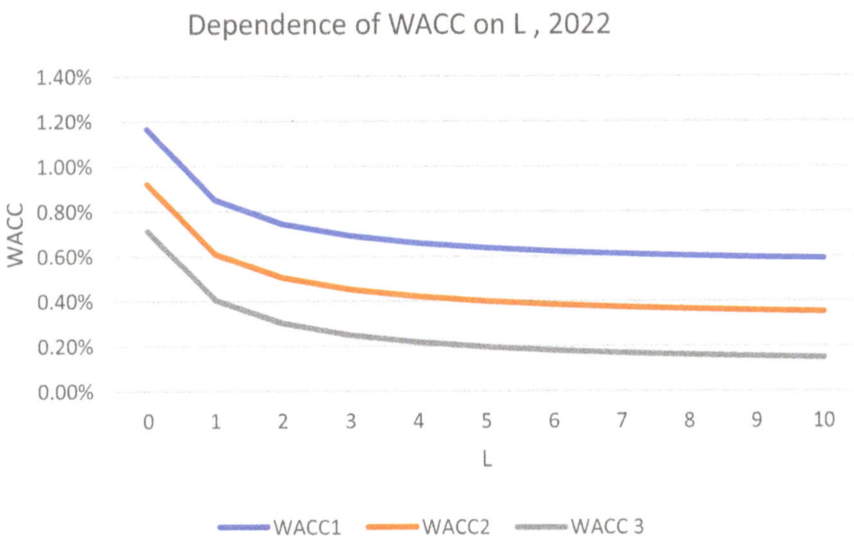

Fig. 25.4 Dependence of WACC on L for three values of k_0 for 2022. The ordering of dependencies WACC(L) is as follows (from big values to small ones): 123

WACC on the company's age (Table 25.5). Since the age of Amazon in 2023 is 29 years, we take ages from 1 to 29 with a 4-year interval.

Table 25.5 above shows that WACC decreases with the company's age. We can note that in both calculations (Tables 25.4 and 25.5), there is a notable drop in 2021–2022 due to the aforementioned COVID-19 pandemic's effects and several other circumstances, including growing competition and regulatory changes.

Below, we show figures for the dependence of WACC on company's age n for 2020–2023 years.

As we can see from Figs. 25.5, 25.6, 25.7, 25.8, and 25.9, WACC decreases with n.

25.6 Company's Capitalization and Business Valuation

Finally, we calculate the company's capitalization and make a business valuation for Amazon company in 2023. For business valuation, we use two different time horizons for the representative period (5 years and 10 years) (Tables 25.6 and 25.7).

Table 25.5 WACC dependence on n (L is fixed)

	2020			2021			2022			2023		
n	WACC1	WACC2	WACC3	WACC1	WACC2	WACC3	WACC1	WACC2	WACC 3	WACC1	WACC 2	WACC3
1	4.00%	4.41%	3.96%	0.69%	0.73%	0.69%	1.00%	0.77%	0.57%	4.52%	4.25%	2.47%
5	3.91%	4.28%	3.87%	0.56%	0.65%	0.56%	0.93%	0.69%	0.48%	4.43%	4.16%	2.39%
9	3.89%	4.27%	3.85%	0.55%	0.64%	0.55%	0.92%	0.67%	0.47%	4.42%	4.15%	2.38%
13	3.88%	4.26%	3.85%	0.55%	0.64%	0.55%	0.92%	0.67%	0.47%	4.41%	4.15%	2.38%
17	3.88%	4.25%	3.85%	0.56%	0.64%	0.55%	0.92%	0.68%	0.47%	4.41%	4.14%	2.38%
21	3.88%	4.25%	3.85%	0.56%	0.65%	0.55%	0.92%	0.68%	0.48%	4.41%	4.14%	2.38%
25	3.88%	4.25%	3.85%	0.56%	0.65%	0.56%	0.93%	0.68%	0.48%	4.41%	4.14%	2.38%
29	3.88%	4.25%	3.85%	0.57%	0.65%	0.56%	0.93%	0.69%	0.49%	4.41%	4.14%	2.39%

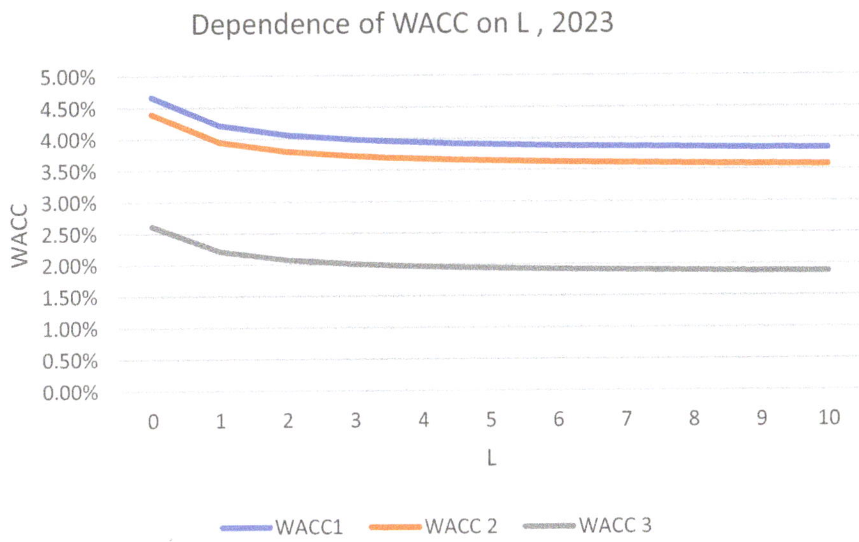

Fig. 25.5 Dependence of WACC on L for three values of k_0 for 2023. The ordering of dependencies WACC(L) is as follows (from big values to small ones): 123

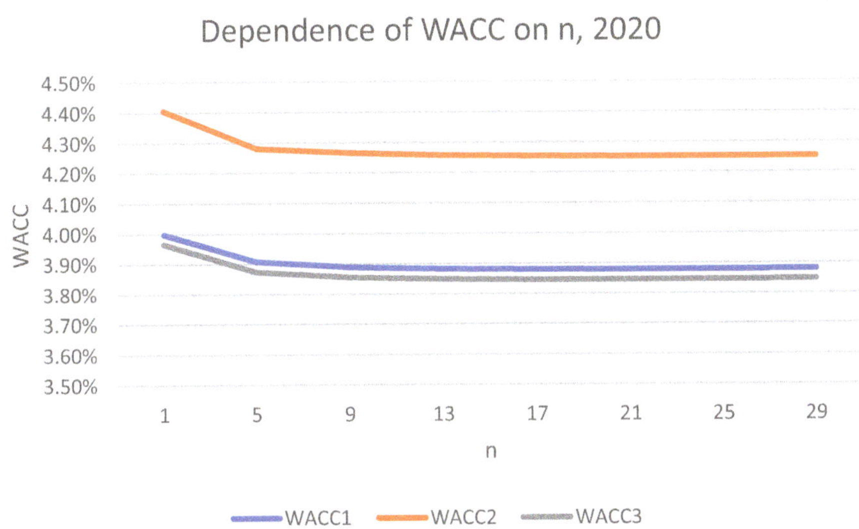

Fig. 25.6 Dependence of WACC on company's age n for three values of k_0 for 2020. The ordering of dependencies WACC(n) is as follows (from big values to small ones): 213

25.6.1 Company's Capitalization

In Table 25.7, we show the results of calculation of the current Amazon company's capitalization (for 2023).

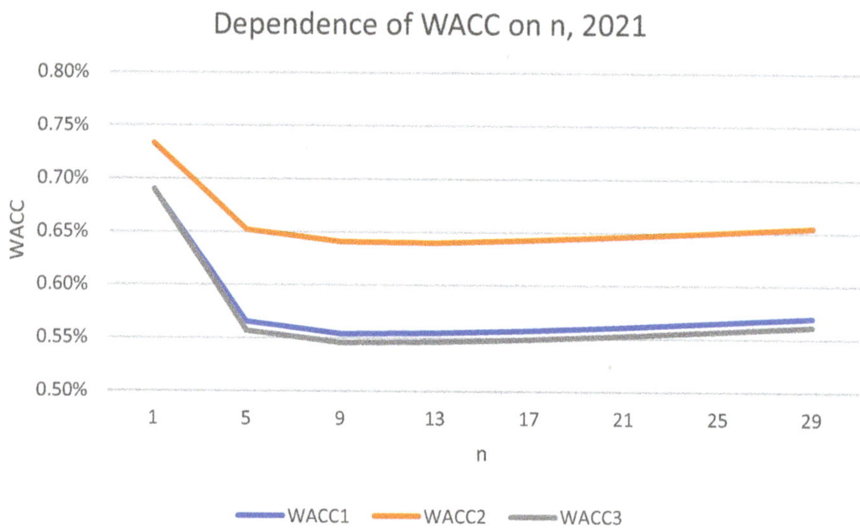

Fig. 25.7 Dependence of WACC on company's age n for three values of k_0 for 2021. The ordering of dependencies WACC(n) is as follows (from big values to small ones): 213

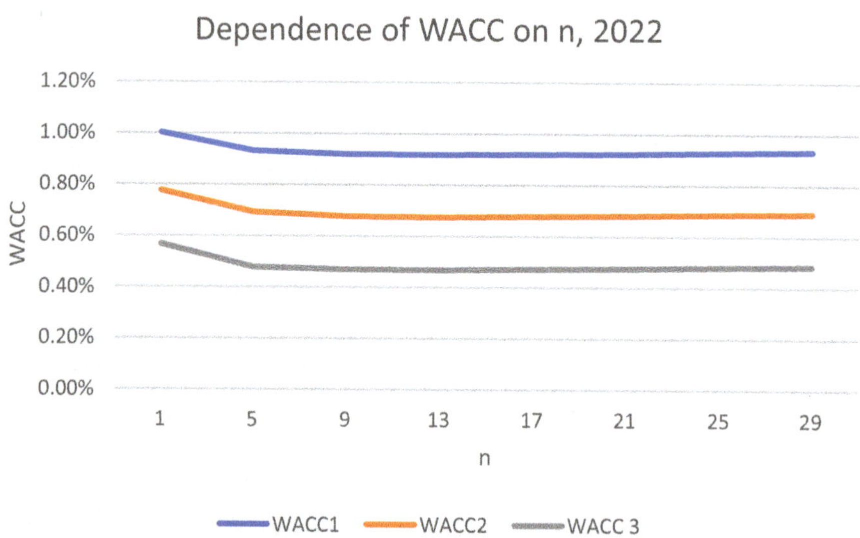

Fig. 25.8 Dependence of WACC on company's age n for three values of k_0 for 2022. The ordering of dependencies WACC(n) is as follows (from big values to small ones): 123

To calculate the current Amazon company's capitalization (for 2023), we should accrue V_{n1} for 29 periods, using the following formula:

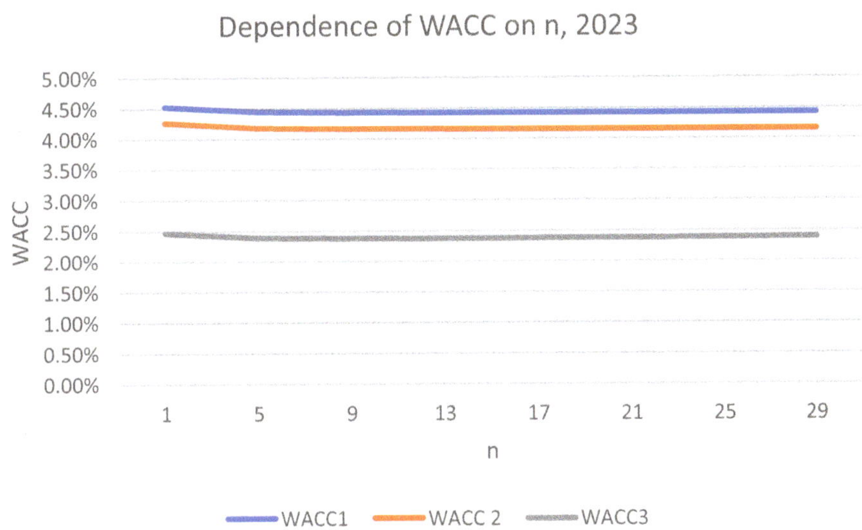

Fig. 25.9 Dependence of WACC on company's age n for three values of k_0 for 2023. The ordering of dependencies WACC(n) is as follows (from big values to small ones): 123

Table 25.6 Capitalization of Amazon at L as of 2023, US$ thousand

	$k_0(1)$	$k_0(2)$	$k_0(3)$
Vn_1	596,506.73	615,285.46	764,800.40
V_{12} (5 years)	162,180.55	163,392.88	171,760.79
V_{12} (10 years)	292,876.89	296,758.25	324,403.42
V_{23} (WACC for 2028)	838,145.17	892,128.24	1,549,305.55
V_{23} (WACC for 2033)	840,663.39	894,921.14	1,554,781.56
V_{13} (5 years)	838,080.23	892,063.63	1,549,258.37
V_{13} (10 years)	840,265.88	894,523.37	1,554,480.47

Table 25.7 The current Amazon company's capitalization (for 2023)

	$k_0(1)$	$k_0(2)$	$k_0(3)$
WACC	4.41%	4.14%	2.39%
V_{n1}	596,506.73	615,285.46	764,800.40
$(1 + \text{WACC})^{29}$	3.496	3.243	2.0164
V	2,085,387.528	1, 995,370.747	1,542,143.527

$$V = V_{n1}\,(1 + \text{WACC})^{29}$$

Market capitalization of the Amazon company (as any other listed company) represents the total value of a public company's outstanding shares and is calculated by multiplying the current share price by the total number of shares outstanding. The

Table 25.8 Capitalization of Amazon company V within CAPM and CAPM 2.0 approaches

		CAPM	CAPM 2.0
Industry	Profitability μ	3.56%	4.14%
	V (mln USD)	1,819,661.62	2,035,379.43
Market	β	1.41%	2.1%
	V(mln USD)	1,309,078.94	1,451,325.72

actual value of the Amazon company V on December 30, 2023 is equal to US$ 1.57 trillion.

Comparing the results of our calculations of Amazon's capitalization for three values of k_0 with this actual value (US$ 1.57 trillion), we conclude that **the best result is the one that takes into account market business risk along with financial risk, as described by the BFO theory.**

Note as well that accounting for only financial risk gives the biggest difference between Amazon's actual value (US$ 1.57 trillion) and calculated (US$ 2.085 trillion) and less difference under the accounting of industry business risk (US$ 1.995 trillion).

To test the above conclusion that the best result for Amazon's capitalization is the one that takes into account market business risk along with financial risk, describing it with the BFO theory, we calculate below (see Table 25.8) Amazon's capitalization within the CAPM itself (taking into account business risk only) and CAPM 2.0 (taking into account business risk and our adjustments for non-zero leverage, but without using the WACC approach) and compare the obtained results with the market capitalization of Amazon V on 12/30/2023, which is equal to US$ 1.57 trillion. For calculations, we use the data from Table 25.2 and CF for 2023 is equal to US$ 36,852 million).

Table 25.8 shows that the industry approach always overestimates capitalization, and in the case of CAPM 2.0, the difference is greater. But with the market approach, CAPM 2.0 yields a better result compared to CAPM (the difference changes from US$ 26 billion to US$ 11.9 billion). Based on the calculations performed and comparison of the results obtained with market capitalization of Amazon, the following global conclusions can be drawn:

1. Neither CAPM nor WACC by themselves can correctly calculate a company's capitalization.
2. At the same time within the CAPM model, the market approach turns out to be more accurate than the industry approach.
3. The best outcome is one that considers market business risk along with the financial risk described by the BFO theory.

These findings are very important and should be used in business valuation, which is discussed in the next paragraph.

25.6.2 Business Valuation

The DCF method does rely on forecasting future cash flows. A modification of the BFO theory (and also the MM theory) has been developed by the authors for the case of variable income. It is possible to use the company's cash flow data for the past 3–5 years to determine the growth rate, and then forecast the cash flow for a representative (terminate) period.

Below, we evaluate the Amazon company using two different time horizons for the representative period, 5 years and 10 years, as is customary within the assessment procedure. We take leverage level L as the average for 4 years (2020–2023).

One can see from Table 25.9 that Amazon company valuation is equal to US$ 0.848 trillion under accounting financial risk only, US$ 0.903 trillion under accounting industry business risk along with financial risk, and US$ 1.574 trillion under accounting market business risk along with financial risk.

Using our experience above with an estimation of Amazon company value, we conclude that the best result is the one that takes into account market business risk along with financial risk, described by the BFO theory. Thus, the Amazon company valuation for the end of 2023 is equal to US$ 1.574 trillion.

25.7 Conclusions

A new approach to business valuation and company value has been developed. Two of the three approaches to the income method have been modified: CAPM and WACC. When modifying CAPM, a new model, CAPM 2.0, was created, within which business and financial risks were simultaneously taken into account. Moreover, for the first time, this was performed correctly and the inaccuracy, incorrectness, and inconsistency of the popular Hamada model were shown. When modifying the WACC approach (or the discounted income method), the real operating conditions of companies were taken into account. The use of the modern theory of capital structure (the Brusov–Filatova–Orekhova [BFO] theory) allows one to correctly determine the main valuation parameter—the discount rate—and link the retrospective terminate periods with the representative one. The application of the new

Table 25.9 V_{n1}, V_{12}, V_{23}, V_{13} of Amazon at L as the average, US$ thousand

	$k_0(1)$	$k_0(2)$	$k_0(3)$
V_{n1}	601,050.96	619,970.11	770,608.32
V_{12} (5 years)	162,477.89	163,688.72	172,047.85
V_{12} (10 years)	293,825.97	297,710.14	325,378.46
V_{23} (WACC for 2028)	848,016.72	902,893.07	1,577,654.36
V_{23} (WACC for 2033)	847,974.01	902,722.77	1,573,830.32
V_{13} (5 years)	848,017.07	902,898.39	1,577,682.23
V_{13} (10 years)	847,978.28	902,751.28	1,574,020.84

methodology to one of the first-class companies (Amazon [AMZN]) is considered. We calculate both Amazon's capitalization and business valuation.

Based on the capitalization calculations and comparison of the results with Amazon's market capitalization, the following conclusions can be drawn:

1. Neither CAPM nor WACC can correctly calculate a company's capitalization by themselves.
2. At the same time, within the CAPM model, the market approach is more accurate than the industry approach.
3. The best result is the one that takes into account market business risk together with the financial risk described by the BFO theory.

These conclusions are very important and should be used in business valuation.

For companies with variable income, the BFO theory modified for this case should be used. In this case, as it has been shown [3, 4] (see formulas [25.27–25.29]), not only the profitability, CF, but also the discount rate WACC depends on the growth rate g. Thus, in this (quite common) case, it becomes possible to correctly predict both the valuation of the business and the value of the company.

In this case, it is necessary to collect data on the company's financial income for the last 3–5 years and, on their basis, determine the growth rate g of the company's income.

Knowing the growth rate, using the BFO formula for WACC with variable income, we can make a forecast for both the representative period and the terminate one. The latter can be finite (say 10 years) or infinite. In the latter case, the MM formula for company value should be used.

The new methodology increases significantly the accuracy of the valuation. The closeness of the new methodology's results to the Amazon company's (AMZN) actual market value (calculated by trivially multiplying the current share price by the total number of shares outstanding) shows the importance and validity of this methodology, which should be applied to business valuation.

The developed methodology allows us to combat the main drawback of existing evaluation methods—the impossibility or serious difficulties in the correct evaluation of the main evaluation parameter—the discount rate. The correct determination of the discount rate will help to prevent abuses in business evaluation, in investments, in determining the fair value of shareholders' dividend income.

25.8 Some Notes on Remaining Issues and Future Directions of Work

The size of the company matters and should be taken into account as an additional premium to the discount rate. Such additional adjustments are taken into account in several theories, for example, in the five-factor model of Fama and French and in the

Sharpe model. To account for enterprises in other sectors, the industry approach described in the developed methodology should be used.

The DCF method does rely on forecasting future cash flows. A modification of the BFO theory (and also the MM theory) has been developed by the authors for the case of variable income. It is possible to use the company's cash flow data for the past 3–5 years to determine the growth rate, and then forecast the cash flow for a representative (terminate) period.

Recently, the WACC formulas within the framework of the Brusov–Filatova–Orekhova (BFO) theory and the Modigliani–Miller (MM) theory have been modified, taking into account the conditions of the actual operation of the company, such as frequent payment of income tax, the method of paying income tax (at the ends of periods or in advance), etc. To improve the accuracy of business valuation, it is necessary to take these effects into account.

Macroeconomic uncertainties, such as interest rate changes or geopolitical risks, must be taken into account. In the presented methodology, they are taken into account only to the extent that they are taken into account in the market business risk. A more accurate accounting of these risks requires additional assessment and accounting in the form of a premium to the discount rate. Similarly, country risk and specific risk associated with a particular company must be taken into account.

References

1. Brusov, P., Filatova, T., & Orekhova, N. (2023). *The Brusov–Filatova–Orekhova theory of capital structure: Applications in corporate finance, investments, taxation and ratings*. Springer Nature. 769p.
2. Brusov, P., Filatova, T., & Orekhova, N. (2022). *Generalized Modigliani–Miller theory: Applications in corporate finance, investments, taxation and ratings*. Springer Nature. 362p.
3. Brusov, P., & Filatova, T. (2023). Capital structure theory: Past, present, future. *Mathematics, 11*, 616. https://doi.org/10.3390/math11030616
4. Brusov, P. N., Filatova, T. V., & Kulik, V. L. (2024). Capital Asset Pricing Model (CAPM) 2.0: Account of business and financial risk. *Finance: Theory and Practice, 28*, 128–142. https://doi.org/10.26794/2587-5671-2024-28-2-128-142
5. Markowitz, H. (1952). Portfolio selection. *The Journal of Finance, 7*, 77–91.
6. Lintner, J. (1965). The valuation of risk assets and the selection of risky investments in stock portfolios and capital budgets. *The Review of Economics and Statistics, 47*, 13–37.
7. Mossin, J. (1966). Equilibrium in a capital asset market. *Econometrica, 34*, 768–783.
8. Sharpe, W. F. (1964). Capital asset prices: A theory of market equilibrium under conditions of risk. *The Journal of Finance, 19*, 425–442.
9. Treynor Jack, L. (1965). How to evaluate the management of investment funds. *Harvard Business Review, 43*, 63–75.
10. Fama, E. F., & French, K. R. (1992). The crosssection of expected stock returns. *The Journal of Finance, 47*, 427–465.
11. Fama, E. F., & French, K. R. (1993). Common risk factors in the returns on stocks and bonds. *Journal of Financial Economics, 33*, 3–56.
12. Fama, E. F., & French, K. R. (1995). Size and book-to-market factors in earnings and returns. *The Journal of Finance, 50*, 131–155.
13. Fama, E. F. (1965). The behavior of stock–market prices. *Journal of Business, 38*, 34–105.

14. Hamada, R. (1969). Portfolio analysis, market equilibrium, and corporate finance. *The Journal of Finance, 24*, 13–31.
15. Hamada, R. (1972). The effect of the firm's capital structure on the systematic risk of common stocks. *The Journal of Finance, 27*, 435–452.
16. Modigliani, F., & Miller, M. H. (1958). The cost of capital, corporation finance and the theory of investment. *The American Economic Review, 48*, 261–296.
17. Modigliani, F., & Miller, M. H. (1963). Corporate income taxes and the cost of capital: A correction. *The American Economic Review, 53*, 433–443.
18. Ross, S. (1976). The arbitrage theory of capital asset pricing. *Journal of Economic Theory, 13*, 341–360. https://doi.org/10.1016/0022-0531(76)90046-6
19. Brusov, P & Filatova, T. (2025a). Qualitatively new approach to business valuation and company value. *Mathematics, 13*(1), 80. https://doi.org/10.3390/math13010080
20. Brusov, P., & Filatova, T. (2025b). Business valuation and company value: New horizons; theoretical economics letters v.15 (3)
21. Kanoujiya, J., Jain, P., Banerjee, S., Kalra, R., Rastogi, S., & Bhimavarapu, V. M. (2023). Impact of leverage on valuation of non–financial firms in India under profitability's moderating effect: Evidence in scenarios applying quantile regression. *Journal of Risk and Financial Management, 16*, 366. https://doi.org/10.3390/jrfm16080366
22. Dong, Y. (2024). Intelligent accounting optimization method based on meta–heuristic algorithm and CNN. *PeerJ Computer Science, 10*, e2281. https://doi.org/10.7717/peerj-cs.2281
23. Hassen, N. T., Lema, M., & Nemera, G. (2024). Firm–specific and institutional factors as export performance catalysts: Insights from a developing economy. *Cogent Social Science, 10*, 2392294. https://doi.org/10.1080/23311886.2024.2392294
24. Mazanec, J. (2023). Capital structure and corporate performance: An empirical analysis from Central Europe. *Mathematics, 11*, 2095. https://doi.org/10.3390/math11092095

Chapter 26
Business Valuation and Company Value: New Horizons

26.1 Introduction

26.1.1 Business Valuation and Company Value

Business valuation and company value play a very important role.

The objectives of business valuation may be: making an informed investment decision; buying and selling shares and bonds of an enterprise on the stock market; making an informed investment decision; buying and selling an enterprise; restructuring an enterprise; merger, liquidation, acquisition, spin-off of an independent company; determining the creditworthiness of an enterprise, and the cost of collateral for lending; insurance; taxation, etc.

The enterprise position in the market is determined by an enterprise market capitalization and determines how much investors are willing to pay for its shares.

The value of any asset (shares, bonds, companies, etc.) is equal to the current value of all income received by this asset. Thus, to find the value of a company, it is necessary to find the current value of all income received by the company from the moment of its creation to the current moment, that is, in the retrospective period. And to evaluate a business, it is necessary to find the current value of all future income received by the enterprise in the representative and final periods.

In this Chapter (Brusov and Filatova [1])), further improvement of the qualitatively new approach to assessing the value of a business and a company that has been proposed earlier by Brusov and Filatova [2]. We consider in detail the methodology for collecting and processing data required to assess a business and a company. A more detailed accounting of data on financial income flows CF and the discount rate WACC(n) for the entire period of the company's existence or from the moment of its IPO (in increments of 1 year), and not just for the year of assessment, allows us to more accurately take into account the influence of the retrospective period (which was taken into account by Brusov and Filatova [2025], but less accurately) and more accurately assess the business and capitalization of the company.

P. Brusov, T. Filatova, *Business Valuation and Company Value*, Contributions to Finance and Accounting, https://doi.org/10.1007/978-3-031-90847-7_26

Among the three business valuation methods (cost, comparative, and income), the income method is the most reliable and accurate. The income method is divided into three approaches: Capital Asset Pricing Model (CAPM), Arbitrage Pricing Theory (APT), and WACC. Brusov and Filatova (2025) modified two methods: CAPM and WACC. They created a new model CAPM 2.0, which simultaneously takes into account business and financial risks (Brusov and Filatova 2024).

The main disadvantage of existing valuation methods is the inability to estimate the main valuation parameter—the discount rate. The role of the correct determination of the discount rate in preventing abuses in business valuation, in investments, in determining the fair value of shareholders' dividend income is of decisive importance.

The only theory that allows for a correct assessment of the discount rate is the BFO theory [3–5], which allows for the calculation of the WACC dependence on the company's age. This allows for a link between the retrospective analysis of the company's financial condition and the representative one. Another difficulty with DCF (discounted cash flow) is the assessment of future cash flows, which is a complex task. The BFO theory could solve this issue in two ways. To do this, it is necessary to use data on cash flows for the past few years. Determining the rate of income growth based on this data allows for (1) improving the forecasting of cash flows in the representative and reference periods; (2) correctly determining the WACC discount rate, which, in this case, depends on the rate of income growth.

26.1.2 Literature Review

There are three methods used to value a business: CAPM, APT, and WACC. CAPM uses the risk-free rate as the initial return and takes into account only the business risk associated with investing in a specific asset, not with investing in the market as a whole or in an industry. APT is a multifactor theory, as opposed to the single-factor CAPM.

The WACC method is the most powerful and adequate. It is based on the application of two main theories of capital structure (the Brusov–Filatova–Orekhova [BFO] theory and the Modigliani–Miller [MM] theory). They take into account the financial risk arising from the use of debt financing. The BFO theory can be applied to a company of any age, while the MM theory is of perpetual nature [3–5].

The Capital Asset Pricing Model (CAPM) was developed by Sharpe [6], Lintner [7], Treynor [8], and Mossin [9] and is based on portfolio theory. Although CAPM is still widely used, analysis of data on numerous listed companies has shown that the CAPM results on company profitability differ significantly from actual ones. The reasons for this discrepancy are related to the intrinsic properties and shortcomings of the model. Several modifications of CAPM have been developed that approach it closer to real conditions [3, 10–17]. Brusov et al. [3, 10, 11] took into account financial risk together with business risk. Hamada [16, 17] tried to take into account financial risk. As is known, the classical CAPM takes into account only business risk. However, in practice, companies have debt financing, which means it is necessary to take into account financial risk as well.

Brusov et al. [18–20] took into account business and financial risk simultaneously. They created the CAPM 2.0 model and showed the incorrectness of the Hamada [16, 17] results. In CAPM 2.0, the CAPM model and the Modigliani–Miller [21, 22] (MM) theory were combined. CAPM takes into account business risk (market or industry). The Modigliani–Miller (MM) theory describes a specific company and takes into account financial risks arising from debt financing. Combining CAPM and the Modigliani–Miller theory made it possible to take into account both types of risks: business and financial. Hamada [16, 17] did this phenomenologically, and Brusov et al. [18–20] combined these two approaches analytically.

As shown by Brusov et al. [18–20] within the CAPM 2.0 model, Hamada's results are incorrect. Brusov et al. [18–20] were the first to discover, that in addition to beta-coefficient renormalization, got by Hamada, two additional terms: the renormalized risk-free return and a term depending on the cost of debt k_d. Most listed companies use debt financing, so Hamada's results are not applicable to them, unlike the CAPM 2.0 model [18–20], which is applicable to companies with leverage.

Fama and French [12–14] and Fama [15] showed that future returns depend on such parameters as firm size, industry affiliation, etc. They created three- and five-factor models [12–14].

Arbitrage Pricing Theory (APT) was created by Ross [23]. The APT formula, unlike the single factor CAPM, has multiple factors that include non-corporate factors, which requires the beta of an asset to be relative to each specific factor. These factors can affect the return of an asset and must be analytically determined by users. This is why investors prefer to use CAPM to estimate the expected rate of return rather than using APT. The numerous results on BFO theory and its applications see in [24–39].

26.2 New Methodology

This article presents a new methodology for valuing a business and company value. Below we discuss aspects of the new methodology in detail.

There are two main parameters in the evaluation of a business and the capitalization of a company: financial flows and discount rates, since the value of any asset is equal to the sum of the discounted income generated by this asset.

26.2.1 Financial Flows

When collecting data on financial flows, several questions arise: (1) for what period are incomes taken into account? (2) what incomes are taken into account? The answers are as follows:

1. Data is required for the entire period of the company's existence, or from the moment of its IPO.
2. Income can be calculated by stock quotes (with taking into account their number, if it changes), or by EBIT(DA), or by a more complex formula that takes into account a number of additional parameters.

One of the differences from Brusov–Filatova (BF) (Brusov & Filatova, 2025) case is that the data on financial flows are collected for the entire period of the company's existence, or from the moment of its IPO, and not just for the year of valuation.

26.2.2 The Discount Rate

To find the discount rate and its dependence on time within the framework of the BFO theory, it is necessary to know the parameter k_O (the value of equity capital and WACC at zero leverage). It is determined as follows:

1. The following data from the company's reporting for the year preceding the year of assessment are collected:

$$\mu; L; k_d; t.$$

Here μ is company yield; L is leverage level; k_d is cost of debt and t is tax on profit.

2. If $\mu > 0$ μ should be cleaned from leverage, by using Modigliani–Miller (MM) formula for cost of equity:

$$\mu = k_0 + L(k_0 - k_d)(1 - t)$$

$$k_0 = \frac{\mu_i + Lk_d(1 - t)}{1 + L(1 - t)}$$

(2a) If for the year preceding the year of assessment $\mu < 0$, then it is necessary to shift to the year in which the yield $\mu > 0$. There is alternative method by A. Brusova [40], which allows to find k_O at any yield.

Unlike the BF case [2], the data on the discount rate WACC(n) (and its dependence on company age n) are used for the entire period of existence of the company or from the moment of its IPO (in 1-year increments), and not only for the year of assessment.

A more detailed accounting of data on financial income flows CF and the discount rate WACC(n) for the entire period of the company's existence or from the moment of its IPO (in increments of 1 year), and not only for the year of valuation, allows us to more accurately take into account the influence of the retrospective period (which was taken into account within the framework of the BF consideration [2], but less accurately) and more accurately assess the business and capitalization of the company.

26.2.3 Company Value

We consider three time periods (see Fig. 26.1):

(1) Retrospective period (from the moment of creation [(entering the market] of the company until the moment of evaluation [n_1])
(2) Representative period (from the moment of assessment [n_1] to the end of this period [n_2]: usually 5–10 years)
(3) Terminate period (finite (until moment [n_3], or infinite): in the first case, the BFO theory is used; in the second case, its perpetuity limit—the MM theory

To calculate the capitalization of company V, it is necessary to:

I. Collect and process company data (links to websites) for the entire period of the company's existence or from the moment of its IPO (in increments of 1 year):

(a) CF for each year (few methods):

– By stock quotes
– Taking into account EBIT, depreciation, capital expenditures, increase in working capital, payment of taxes taking into account the tax shield, etc. (see formulas below):

There are three types of cash flows:

Capital asset cash flow (CCF)

CCF = EBIT+ Depreciation – Capital Expenditures – Working Capital Increase – Actual Taxes (payment of taxes taking into account the tax shield)

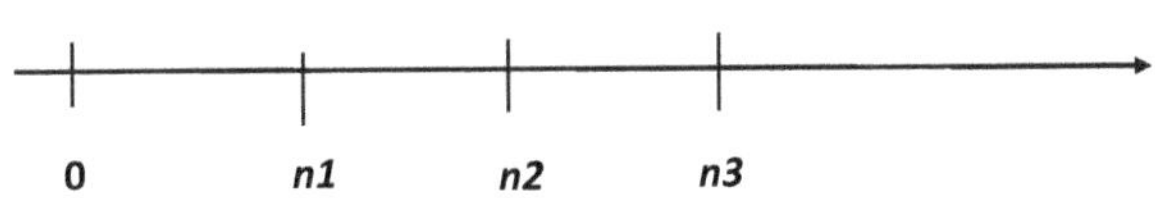

Fig. 26.1 Three time periods: retrospective period (0–n_1); representative period (n_1–n_2); and terminate period (n_2–n_3)

Here Actual Taxes = Tax Rate (EBIT – Interest)

Equity cash flow (ECF)

ECF = EBIT+ Depreciation – Capital Expenditures – Working Capital Increase – Interest–Debt payments+ Debt issues – Actual Taxes

Invested capital cash flow (FCF)

FCF = EBIT+ Depreciation – Capital Expenditures – Working Capital Increase – Interest–Debt payments – Hypothetical Taxes (Tax rate × EBIT)

II. (a) Collect the following data for the assessment year (2023)

μ, t, k_d, D, S, $L = D/S$.

Here μ is profitability, t is tax on profit, k_d is the cost of debt, D is the value of debt, S is the value of equity and $L = D/S$ is the level of leverage.

b) To find the parameter k_0 from the MM equation for k_e

$$\mu_i = \mu_0 + L(\mu_0 - k_d)\cdot(1 - t) \tag{26.1}$$

Find k_0 from the formula (26.2) below

$$k_0 = \mu_0 = \frac{\mu_i + Lk_d\cdot(1 - t)}{1 + L\cdot(1 - t)} \tag{26.2}$$

III. Using the BFO formula, we find the WACC(n) dependence (n runs from 1 to the age of the company in 2023)

$$\frac{1 - (1 + \text{WACC})^{-n}}{\text{WACC}} = \frac{1 - (1 + k_0)^{-n}}{k_0\cdot(1 - w_d t[1 - (1 + k_d)^{-n}])} \tag{26.3}$$

For a fixed n (1, 2, …) we find WACC (by selecting the parameter (WACC)) and plot the WACC(n) dependence graph using Excel tables.

Alternative method for finding WACC(n)—using programming language Python.

IV. To calculate the capitalization of company V, we use the following formula:

$$V = \text{CF}_{23} + \text{CF}_{22}\cdot(1 + \text{WACC}_{23}) + \text{CF}_{21}\cdot(1 + \text{WACC}_{23})\cdot(1 + \text{WACC}_{22}) + \cdots \tag{26.4}$$

V. Compare the company's capitalization calculated for two (or more) cash flows and the actual value of the company's market capitalization.

26.2.4 Business Valuation

To evaluate the business, it is necessary to continue calculating the dependence of WACC(n) for 5–10 years (or maybe more) (so we will consider 5-year and 10-year horizons). At the same time, we can consider constant CF, or variable CF. In the first case, one should use the following formula for representative (V_1) and terminate (V_2) periods:

For 5-year horizon:

$$V_1 = \mathrm{CF}_{23} \cdot \Big[(1 + \mathrm{WACC}_{24})^{-1} + (1 + \mathrm{WACC}_{24})^{-1} (1 + \mathrm{WACC}_{25})^{-1} + \cdots + (1 + \mathrm{WACC}_{24})^{-1} \ldots (1 + \mathrm{WACC}_{28})^{-1} \Big] \tag{26.5}$$

$$V_2 = \frac{\mathrm{CF}_{23}}{\mathrm{WACC}_{28} (1 + \mathrm{WACC}_{24}) \ldots (1 + \mathrm{WACC}_{28})} \tag{26.6}$$

$$V = V_1 + V_2 \tag{26.7}$$

For 10-year horizon:

$$V_1 = \mathrm{CF}_{23} \cdot \Big[(1 + \mathrm{WACC}_{24})^{-1} + (1 + \mathrm{WACC}_{24})^{-1} (1 + \mathrm{WACC}_{25})^{-1} + \cdots + (1 + \mathrm{WACC}_{24})^{-1} \ldots (1 + \mathrm{WACC}_{33})^{-1} \Big] \tag{26.8}$$

$$V_2 = \frac{\mathrm{CF}_{23}}{\mathrm{WACC}_{33} (1 + \mathrm{WACC}_{24}) \ldots (1 + \mathrm{WACC}_{33})} \tag{26.9}$$

$$V = V_1 + V_2 \tag{26.10}$$

The terminal period here (in the formulas [26.6] and [26.9]) is considered infinite. For a finite terminal period, it is necessary to continue calculating WACC(n) within the framework of the BFO theory.

26.3 Application of New Methodology to Some Companies

Below we apply new methodology to a couple companies. We start from PJSC "PIK-specialized developer" (PIKK)

26.3.1 PJSC "PIK-Specialized Developer" (PIKK)

The company was founded in 1994 as the First Mortgage Company. In the 2000s, several factories producing building materials and components for construction were acquired. In July 2007, the company went public, and its securities began to be traded on the Moscow and London stock exchanges.

To assess the value of a business and a company, one needs to do the following:

I. Collect and process company data (links to websites) for the entire period of the company's existence or from the moment of its IPO (in increments of 1 year):

   (a) CF for each year (two methods):

   - By stock quotes
   - Taking into account EBIT, depreciation, capital expenditures, increase in working capital, payment of taxes taking into account the tax shield (see below)

Let's calculate CF using the first method (by quotes), using the formula:

CF = number of shares at the end × share price at the end – number of shares at the beginning × share price at the beginning

We'll get the following table (Table 26.1):

For stock quotes—https://ru.investing.com/equities/pik_rts https://www.moex.com/ru/issue.aspx?board=TQBR&code=PIKK

For number of shares https://www.kommersant.ru/quotes/ru000a0jp7j7 https://stockchart.ru/fundamental/MSFO/PIKK/number_of_shares https://marketcap.ru/stocks/PIKK/financial–statements/income–statement/number–of–shares

Let's calculate CF using the second method, using the formula (Table 26.2):

CCF = EBIT+ Depreciation – Capital Expenditures – Working Capital Increase – Actual Taxes (payment of taxes taking into account the tax shield)

Here *Actual Taxes = Tax Rate (EBIT – Interest)*

II. Collect the following data for 2023 μ, t, k_d, D, S, $L = D/S$ and find the parameter k_O from the MM equation for k_e ((Tables 26.3, 26.4,and Fig. 26.2)

$$\mu_i = \mu_0 + L(\mu_0 - k_d)\cdot(1 - t)$$

Find k_O

$$k_0 = \mu_0 = \frac{\mu_i + Lk_d\cdot(1 - t)}{1 + L\cdot(1 - t)}$$

https://stockchart.ru/Financial/PIKK
https://finrange.com/ru/company/MOEX/PIKK/financial–statements
https://pik–group.ru/about/news–and–reports/reports/financial–results

Table 26.1 Data for the first method

Year	Stock price at the beginning of the year, rubles	Stock price at the end of the year, rubles	Number of shares at the beginning of the year	Number of shares at the end of the year	CF, rubles	WACC	Calculation V step by step, rubles
2007	657	739.97	456260384	456260384	37855924060	0.11026104	1.95251E+11
2008	740	385	456260384	493260384	−1.47727E+11	0.10861657	−6.87288E+11
2009	30.21	123.45	493260384	493260384	45991598204	0.1081376	1.93091E+11
2010	129.89	122.89	493260384	493260384	−3452822688	0.10793723	−13084067558
2011	128.5	77	493260384	493260384	−25402909776	0.10769972	−86902029843
2012	77.49	67.5	493260384	493260384	−4927671236	0.10764139	−15219101106
2013	64.73	70.5	493260384	660497344	14636318096	0.10765044	40810920692
2014	67.7	187.3	660497344	660497344	78995482342	0.10767512	1.98854E+11
2015	191.7	219	660497344	660497344	18031577491	0.10773358	40976093721
2016	215.5	290	660497344	660497344	49207052128	0.10779863	1.0094E+11
2017	285.9	326.5	660497344	660497344	26816192166	0.10785439	49653564816
2018	327.4	376.3	660497344	660497344	32298320122	0.10794756	53977660022
2019	374.7	400.4	660497344	660497344	16974781741	0.10802416	25602895825
2020	407	595.9	660497344	660497344	1.24768E+11	0.1080999	1.69828E+11
2021	597.9	1102.4	660497344	660497344	3.33221E+11	0.10817123	4.0929E+11
2022	1155.7	602.5	660497344	660497344	−3.65387E+11	0.10824475	−4.04964E+11
2023	604.2	677	660497344	660497344	48084206643	0.10831554	48084206643

Table 26.2 Data for the second method

Year	EBIT, rubles.	Depreciation, rubles	Income tax, rubles	CAPEX, rubles	Working capital increase, rubles	CF, rubles	WACC	Calculation V step by step, rubles
2007	14100000000	777000000	3595000000	1521200000	1487000000	8273800000	0.110261	42674037626
2008	–6920000000	1076000	925000000	196100000	3502200000	−1.1542E+10	0.1086166	−53699080467
2009	–3260000000	860000000	218000000	460000000	2000000000	–5078000000	0.1081376	−21319467628
2010	–1010000000	759000000	1099000000	420000000	15000000000	−1.677E+10	0.1079372	−63547952726
2011	7800000000	736000000	152900000	470000000	12000000000	–4086900000	0.1076997	−13981071810
2012	10470000000	860000000	125500000	450000000	14520000000	–3765500000	0.1076414	−11629737957
2013	13740000000	705000000	235000000	480000000	15680000000	–1950000000	0.1076504	−5437248277
2014	12460000000	737000000	227000000	428000000	23000000000	−1.0458E+10	0.1076751	−26325752189
2015	17510000000	1124000	179300000	4670000000	10100000000	2561824000	0.1077336	5821650400
2016	23650000000	930000000	252600000	1000000000	2000000000	2.1327E+10	0.1077986	43749583415
2017	16440000000	636620000	200000000	2501000000	10200000000	4175620000	0.1078544	7731687520
2018	34630000000	2237000000	5446000000	3882000000	2500000000	2.5039E+10	0.1079476	41845725233
2019	62330000000	3024000000	8390000000	4150000000	20000000000	3.2814E+10	0.1080242	49493032454
2020	1.118E+11	3479000000	6326000000	3200000000	24300000000	8.1453E+10	0.1080999	1.1087E+11
2021	1.4018E+11	4943000000	2.3487E+10	9860000000	13000000000	9.8776E+10	0.1081712	1.21325E+11
2022	45000000000	1522676250	1E+10	5000000000	1500000000	3.0023E+10	0.1082447	33274598640
2023	1.18E+11	3677000000	1.2218E+10	8100000000	16700000000	8.4659E+10	0.1083155	84659000000

Table 26.3 Data for 2023

Data for 2023							
μ	t	k_d	D	S	L	k_O	w_d
0.120489904	0.2	0.0967	1.1257E+11	3.25436E+11	0.345905554	0.115333546	0.257005815

Table 26.4 Dependence WACC(n)

n	WACC
1	0.110261038
2	0.108616573
3	0.108137601
4	0.107937234
5	0.107699721
6	0.107641394
7	0.10765044
8	0.107675122
9	0.107733576
10	0.107798631
11	0.107854391
12	0.10794756
13	0.108024156
14	0.1080999
15	0.108171229
16	0.108244747
17	0.10831554
18	0.108385468
19	0.108450442
20	0.108512351
21	0.108571157
22	0.108626871
23	0.108678302
24	0.10872815
25	0.108775108
26	0.108819278
27	0.108860766

https://pik-group.ru.media.pik-service.ru/attachment/0/651/konsolidirovannaya-finansovaya-otchetnost-po-msfo-_9f1cd5dc7d6e432abae77a242a03ea31.pdf

https://pik-group.ru/about/news-and-reports/reports/financial-results

https://smart-lab.ru/q/PIKK/MSFO/capex/

26.3.1.1 Company Value

Company value could be calculated by formula (26.4).

In the first case, company value is equal to $V = 318{,}902{,}000{,}000$ rubles, in the second—$V = 345{,}504{,}000{,}000$ rubles. Or in billions of US dollars in the first case $V = 3.08326$ billion US dollars, in the second—$V = 3.34046$ billion US dollars.

Market value in 2023 $V = 4.334$ B US$ (448.3 B rubles)

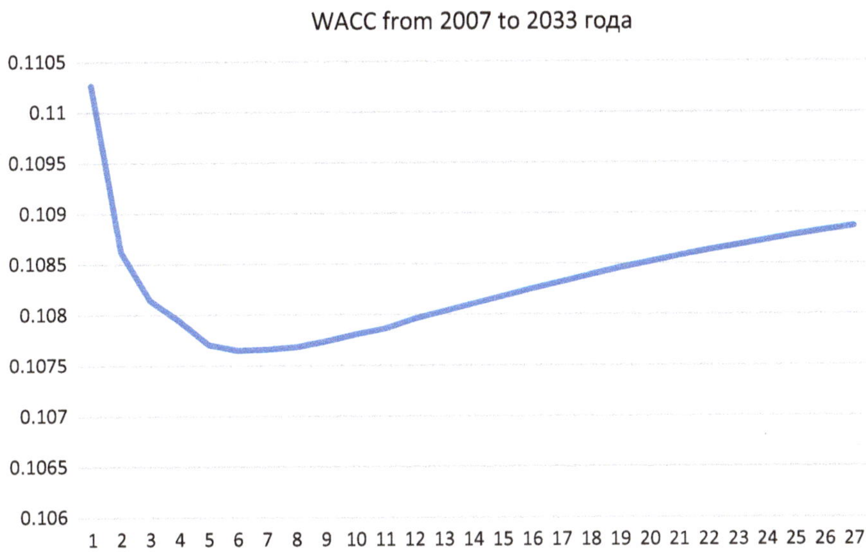

Fig. 26.2 Dependence WACC(n) from 2007 to 2033

Table 26.5 Business valuation for 5-year representative period and infinite terminate period (by stock quotes)

$V_1 =$	1.78403E+11
$V_2 =$	2.64465E+11
V за 28 год =	4.42868E+11

It is obvious that the company value calculated by the two methods is quite close, but underestimated in relation to the market value. This may be due to the fact that business risk was not taken into account in the current consideration (see Brusov & Filatova, 2025).

26.3.1.2 Business Valuation

Using the formulas (26.5)–(26.7) one gets the following results for business valuation for 5-year representative period and infinite terminate period (see Table 26.5).

a) According to stock quotes (CF23 = 48,084,206,643 rubles)

Then obtained values (in rubles):

Company valuation in 2028 year will amount to 442,868,000,000 rubles. Or 4.28181 billion US dollars.

b) Using formula (CF_{23} = 84659000000 rubles) (Table 26.6).

Company valuation in 2028 will be 779,732,000,000 rubles. Or 7.53874 billion US dollars (Table 26.7).

Table 26.6 Business valuation for 5-year representative period and infinite terminate period (by formula)

$V_1 =$	3.14104E+11
$V_2 =$	4.65628E+11
V for 2028	7.79732E+11

Table 26.7 Business valuation for 5-year representative period and infinite terminate period (by two methods)

CF23 by stock quotes, rubles	4.8084E+10
$V_1 =$	1.784E+11
$V_2 =$	2.6447E+11
V for 2028	4.4287E+11
CF23 by formula, rubles	8.4659E+10
$V_1 =$	3.141E+11
$V_2 =$	4.6563E+11
V for 2028	7.7973E+11

Table 26.8 Dependence WACC(n) for period 2024 to 2033

	WACC
2024	0.108385468
2025	0.108450442
2026	0.108512351
2027	0.108571157
2028	0.108626871
2029	0.108678302
2030	0.10872815
2031	0.108775108
2032	0.108819278
2033	0.108860766

Table 26.9 Business valuation for 10-year representative period and infinite terminate period (by stock quotes)

$V_1 =$	2.84917E+11
$V_2 =$	1.57479E+11
V for 2033	4.42396E+11

For 10 years:

Let us provide business valuation for a 10-year representative period and infinite terminate period (by two methods) (Table 26.8).

a) According to stock quotes (CF23 = 48,084,206,643 rubles)

Results for business valuation for 10-year representative period and infinite terminate period (in rubles) are shown in Table 26.9:

Company valuation in 2033 will be 442,396,000,000 rubles. Or 4.27725 billion US dollars.

Table 26.10 Business valuation for 10-year representative period and infinite terminate period (using formula)

$V_1 =$	5.01636E+11
$V_2 =$	2.77264E+11
V for 33	7.789E+11

Table 26.11 Business valuation for 10-year representative period and infinite terminate period (by two methods)

		Quote	Formula
Business valuation, Billion \$	5 years	4.28181	7.53874
	10 years	4.27725	7.5307

b) Using formula (CF23 = 84,659,000,000 rubles) (Table 26.10).

Company valuation in 2033 will be 778,900,000,000 rubles. Or 7.5307 billion US dollars.

Business valuation for 10-year representative period and infinite terminate period (by two methods) are shown In Table 26.11.

26.3.1.3 Comparison of Current Results with Ones by Brusov–Filatova (2025)

To evaluate a business for 5-year and 10-year representative periods and an infinite terminal period (using two methods) within the framework of the Brusov–Filatova method (Brusov & Filatova, 2025) it is necessary to use the following formulas:

For 5-year horizon

$$V_1 = \frac{\text{CF}_{23}}{\text{WACC}(23)} \cdot \left(1 - (1 + \text{WACC}(23))^{-(5)}\right) \tag{26.11}$$

$$V_2 = \frac{\text{CF}_{23}}{\text{WACC}(28)} \tag{26.12}$$

$$V = V_1 + \frac{V_2}{(1 + \text{WACC})^5} \tag{26.13}$$

For 5-year horizon

$$V_1 = \frac{\text{CF}_{23}}{\text{WACC}(23)} \cdot \left(1 - (1 + \text{WACC}(23))^{-10}\right) \tag{26.14}$$

$$V_2 = \frac{\text{CF}_{23}}{\text{WACC}(33)} \tag{26.15}$$

Table 26.12 Comparison of current results with the results obtained within the framework of the Brusov–Filatova approach (Brusov & Filatova, 2025) regarding the valuation of the PIKK company's business

Approach		Current approach		BF approach	
		Quote	Formula	Quote	Formula
Business valuation, Billion $	5 years	4.28181	7.53874	4.3148	7.5968
	10 years	4.27725	7.5307	4.314	7.5962

$$V = V_1 + \frac{V_2}{(1 + \mathrm{WACC})^{10}} \tag{26.16}$$

Formulas (26.11)–(26.16) used in the Brusov–Filatova (BF) approach (Brusov & Filatova, 2025) differ from formulas (26.5)–(26.10) used here, since, in this case, we use more detailed information about the financial flows CF and the discount rate WACC(n).

Below, in the Table 26.12, we compare the current results with the results obtained within the framework of the Brusov–Filatova (2025) approach regarding the valuation of the PIKK company's business.

Comparison of the present more detailed method with the method developed by us earlier shows that (1) when assessing a business, both methods give very close results. Thus, for assessing a business, the use of the less detailed previous method is justified. This is also important from the point of view that the first method requires much less information and time for its processing. Thus, the proposed approach is more accurate than the BF one since it more accurately takes into account both main business valuation parameters: CF and WACC.

(2) But when assessing capitalization, the new method gives a more adequate result, practically close to the market capitalization of the company, while the result of the previous method (Brusov & Filatova, 2025) gives an excess of the market capitalization of the company (sometimes several times).

The difference in the results of the two approaches is due to the following facts. (1) CF is a constant value in the BF approach and a variable value in the current approach (in 1-year increments); (2) WACC is a constant value in the BF approach (Brusov & Filatova, 2025) and a variable value in the current approach (in 1-year increments). From the analysis of the WACC(n) dependence, which is a decreasing function of n, it follows that the BF approach uses a much smaller WACC value (for the valuation year), and this leads to an overestimation of the company value due to the formula.

But when valuing a business, both approaches use approximately the same WACC values (with a small difference), and this leads to approximately similar business valuation results.

However, for a more accurate assessment of a business and a company, as shown in (Brusov & Filatova, 2025), it is necessary to take into account business risk along with financial risk, using CAPM 2.0 approach, developed by Brusov and Filatova (2024).

26.4 Yandex Company (YNDX.ME)

Below we consider Yandex company (YNDX.ME)

Let's collect and process data from company reports and websites.

Based on the Yandex data from Tables 26.13 and 26.14 and using Python, we obtain the following results on Yandex capitalization and its business valuation.

26.4.1 Results

(1) Yandex capitalization in 2023

In the first case, with CF calculated using the stock price, capitalization was $V =$ \$4.98 billion.

In the second case, with CF calculated using the formula, capitalization was $V =$ \$4.18 billion.

The actual market capitalization, according to various sources, is \$9.13 billion. (918.9 billion rubles)

(2) Company valuation

CF by formula (EBIT):
Company valuation for 5 years (until 2028): \$4.11 billion
Company valuation for 10 years (until 2033): \$4.06 billion
CF by stock quotes:
Company valuation for 5 years (until 2028): \$4.84 billion
Company valuation for 10 years (until 2033): \$4.73 billion

It is obvious that Yandex's capitalization results are understated relative to the market. This may be due to the fact that the current consideration does not take into account business risk.

26.5 Conclusion and Implications

To further improve the qualitatively new approach to assessing the value of a business and a company, the new detailed methodology for collecting and processing data required for assessing a business and a company is considered. A more detailed accounting of data on financial income flows CF and the discount rate WACC(n) for the entire period of the company's existence or from the moment of its IPO (in increments of 1 year), and not only for the year of valuation, allows us to more accurately take into account the influence of the retrospective period (which was taken into account within the framework of the Brusov–Filatova consideration, but less accurately) and more accurately assess the business and capitalization of the company.

Table 26.13 Data for Yandex company

No.	Year	Stock price at the beginning of the year, rubles	Stock price at the end of the year, rubles	Number of shares	CF formula	μ	t	Kd
1	2015	824	1017	3.65E+08	−1073184	0.097087	0.2	0.0967
2	2016	880	1116	3.65E+08	−3.9E+08	0.127273	0.2	0.0967
3	2017	1008	1683	3.65E+08	−1.3E+08	0.484127	0.2	0.0967
4	2018	1560	1701	3.65E+08	−2.3E+08	−0.03077	0.2	0.0967
5	2019	1528	2421	3.65E+08	−3.6E+08	0.408377	0.2	0.0967
6	2020	2152	4563	3.65E+08	2.93E+11	0.884758	0.2	0.0967
7	2021	4192	4032	3.65E+08	−1.8E+11	−0.14504	0.2	0.0967
8	2022	3496	1638	3.65E+08	−1.2E+10	−0.58352	0.2	0.0967
9	2023	1480	2349	3.65E+08	8.79E+10	0.410811	0.2	0.0967

Table 26.14 Data for Yandex company (continued)

n	Year	D	S	CF quote	L	k_0	w_d	WACC
1	2015	7.79E+08	6.2E+08	7.04E+10	1.256013	0.096893	0.55674	0.341769
2	2016	6.73E+08	5.17E+08	8.61E+10	1.302677	0.111671	0.565723	0.334206
3	2017	5.88E+08	4.6E+08	2.46E+11	1.277154	0.288332	0.560855	0.330102
4	2018	4.1E+08	3.21E+08	5.15E+10	1.275982	0.033621	0.560629	0.32733
5	2019	2.52E+08	2.16E+08	3.26E+11	1.163094	0.258151	0.537699	0.325191
6	2020	9.22E+10	3.26E+11	8.8E+11	0.282883	0.739327	0.220506	0.323391
7	2021	9.78E+10	2.58E+11	−5.8E+10	0.378922	−0.0888	0.274796	0.32179
8	2022	5.12E+10	3.16E+11	−6.8E+11	0.161867	−0.50554	0.139316	0.320316
9	2023	1.55E+11	2.96E+11	3.17E+11	0.524436	0.317975	0.34402	0.318929

The strength of the study is the detailed and correct determination of the discount rate and financial flow of the company's income, which allows for a more accurate determination of the company's capitalization and assessment of the business.

Application of a new methodology to a couple companies: PJSC "PIK-specialized developer" (PIKK) and Yandex (YNDX.ME) has been done. The company value calculated by the two methods (two CF) is quite close, but underestimated in relation to the market value. This may be due to the fact that business risk was not taken into account in the current consideration (see Brusov & Filatova, 2025).

References

1. Brusov, P., & Filatova, T. (2025). Business valuation and company value: new horizons. *Theoretical Economics Letters, 15*(3).
2. Brusov, P., & Filatova, T. (2025). Qualitatively new approach to business valuation and company value. *Mathematics, 13*(1), 80. https://doi.org/10.3390/math13010080
3. Brusov, P., Filatova, T., & Orekhova, N. (2023). *The Brusov–Filatova–Orekhova theory of capital structure: Applications in corporate finance, investments, taxation and ratings*. Springer Nature. 769 pp.
4. Brusov, P., Filatova, T., & Orekhova, N. (2022). *Generalized Modigliani–Miller theory: Applications in corporate finance, investments, taxation and ratings*. Springer Nature. 362 pp.
5. Brusov, P., Filatova, T., & Kulik, V. (2023). *Capital Asset Pricing Model 2.0: Account of business and financial risk*. Preprint ID: 87056. https://doi.org/10.20944/preprints202310.0347.v1
6. Sharpe, W. F. (1964). Capital asset prices: A theory of market equilibrium under conditions of risk. *The Journal of Finance, 19*(3), 425–442.
7. Lintner, J. (1965). The valuation of risk assets and the selection of risky investments in stock portfolios and capital budgets. *Review of Economics and Statistics*, 13–37.
8. Treynor, J. L. (1965). How to evaluate the management of investment funds. *Harvard Business Review, 43*, 63–75.
9. Mossin, J. (1966). Equilibrium in a capital asset market. *Econometrica*, 768–783.
10. Brusov, P., Filatova, T., Kulik, V., Chang, S.-I., Lin, G., & Chang, L.-M. (2023). Precision finance: Capital structure theories approach reality. In *Proceedings of the 23rd international conference on electronic business (ICEB–2023).*
11. Brusov, P., Filatova, T., Kulik, V., Chang, S.-I., Lin, G., & Chang, L.-M. (2023). Can CAPM (Capital Asset Pricing Model) accurately value assets? In *Proceedings of the 23rd international conference on electronic business (ICEB–2023).*
12. Fama, E. F., & French, K. R. (1992). The crosssection of expected stock returns. *The Journal of Finance, 47*(2), 427–465.
13. Fama, E. F., & French, K. R. (1993). Common risk factors in the returns on stocks and bonds. *Journal of Financial Economics, 33*(1), 3–56.
14. Fama, E. F., & French, K. R. (1995). Size and book-to-market factors in earnings and returns. *The Journal of Finance, 50*(1), 131–155.
15. Fama, E. F. (1965). The behavior of stock–market prices. *Journal of Business, 38*, 34–105.
16. Hamada, R. (1969). Portfolio analysis, market equilibrium, and corporate finance. *The Journal of Finance, 24*, 13–31.
17. Hamada, R. (1972). The effect of the firm's capital structure on the systematic risk of common stocks. *The Journal of Finance*, 435–452.
18. Brusov, P. N., Filatova, T. V., & Kulik, V. L. (2024a). Capital Asset Pricing Model (CAPM) 2.0: Account of business and financial risk. *Finance: Theory and Practice, 28*(2), 128–142. https://doi.org/10.26794/2587-5671-2024-28-2-128-142

19. Brusov, P. N., Filatova, T. V., & Kulik, V. L. (2024b). Application of the company's "Golden Age" effect in the economic practice. *Finance: Theory and Practice, 28*(3), 61–83. https://doi.org/10.26794/2587-5671-2024-28-3-61-83
20. Brusov, P. N., Filatova, T. V., & Kulik, V. L. (2024c). Incorporating CAPM into capital structure theories: Accounting for business and financial risks. *Finance: Theory and Practice, 28*(5), 83–108. https://doi.org/10.26794/2587-5671-2024-28-5-83-108
21. Modigliani, F., & Miller, M. H. (1958). The cost of capital, corporation finance and the theory of investment. *The American Economic Review, 48*, 261–296.
22. Modigliani, F., & Miller, M. H. (1963). Corporate income taxes and the cost of capital: A correction. *The American Economic Review, 53*, 433–443.
23. Ross, S. (1976). The arbitrage theory of capital asset pricing. *Journal of Economic Theory, 13* (3), 341–360. https://doi.org/10.1016/0022-0531(76)90046-6
24. Brusov, P., & Filatova, T. (2023). Capital structure theory: Past, present, future. *Mathematics, 11*(3), 616. https://doi.org/10.3390/math11030616
25. Brusov, P. N., Filatova, T. V., Orekhova, N. P., Kulik, V. L., Chang, S. I., & Lin, Y. C. G. (2020). Modification of the Modigliani–Miller theory for the case of advance payments of tax on profit. *Journal of Reviews on Global Economics, 9*, 257–267.
26. Brusov, P., Filatova, T., Orekhova, N., Kulik, V., Chang, S.-I., & Lin, Y. (2020). Application of the Modigliani–Miller theory, modified for the case of advance payments of tax on profit, in rating methodologies. *Journal of Reviews on Global Economics, 9*, 282–292.
27. Brusov, P., Filatova, T., Chang, S.-I., & Lin, G. (2021). Innovative investment models with frequent payments of tax on income and of interest on debt. *Mathematics, 9*, 1491.
28. Brusov, P., Filatova, T., Orekhova, N., Kulik, V., Chang, S.-I., & Lin, G. (2021). Generalization of the Modigliani–Miller theory for the case of variable profit. *Mathematics, 9*, 1286.
29. Brusov, P., & Filatova, T. (2021). The Modigliani–Miller theory with arbitrary frequency of payment of tax on profit. *Mathematics, 9*, 1198.
30. Brusov, P., Filatova, T., Orekhova, N., Kulik, V., Chang, S.-I., & Lin, G. (2022). The Generalization of the Brusov–Filatova–Orekhova theory for the case of payments of tax on profit with arbitrary frequency. *Mathematics, 10*, 1343.
31. Brusov, P., Filatova, T., & Kulik, V. (2022). Benefits of advance payments of tax on profit: Consideration within the Brusov–Filatova–Orekhova (BFO) theory. *Mathematics, 10*, 2013.
32. Brusov, P., & Filatova, T. (2022). Influence of method and frequency of profit tax payments on company financial indicators. *Mathematics, 10*, 2479.
33. Brusov, P., & Filatova, T. (2022). Generalization of the Brusov–Filatova–Orekhova theory for the case of variable income. *Mathematics, 10*, 3661.
34. Dong, Y. (2024). Intelligent accounting optimization method based on meta–heuristic algorithm and CNN. *PeerJ Computer Science, 10*, e2281. https://doi.org/10.7717/peerj-cs.2281
35. Filatova, T., Brusov, P., & Orekhova, N. (2022). Impact of advance payments of tax on profit on effectiveness of investments. *Mathematics, 10*, 666.
36. Hassen, N. T., Lema, M., & Nemera, G. (2024). Firm–specific and institutional factors as export performance catalysts: Insights from a developing economy. *Cogent Social Science, 10*(1). https://doi.org/10.1080/23311886.2024.2392294
37. Kanoujiya, J., Jain, P., Banerjee, S., Rastogi, S., & Bhimavarapu, V. M. (2023). Impact of leverage on valuation of non–financial firms in India under profitability's moderating effect: Evidence in scenarios applying quantile regression. *Journal of Risk and Financial Management, 16*(8), 366. https://doi.org/10.3390/jrfm16080366
38. Danilo, L., Jiménez, R., Riquelme, M., & Leiva, V. (2023). Elliptical capital asset pricing models: Formulation, diagnostics, case study with Chilean data, and economic rationale. *Mathematics, 11*(6), 1394. https://doi.org/10.3390/math11061394
39. Mazanec. (2023). Capital structure and corporate performance: An empirical analysis from Central Europe. *Mathematics, 11*(9), 2095. https://doi.org/10.3390/math11092095
40. Brusova, A. (2011). Comparison of methods of assessing the weighted average cost of a company's capital and the cost of its own capital. *Financial Analytics Problems and Solutions, 34*(76), 36–42.

Part VII
Conclusions

Chapter 27
Conclusions

Business and company valuation plays a very important role in making an informed investment decision; when buying and selling shares and bonds of an enterprise on the stock market; when buying and selling an enterprise; when restructuring an enterprise; when merging, liquidating, taking over, spinning off an independent company; when determining the creditworthiness of an enterprise and the cost of collateral for lending; when insuring; when taxing, etc. The position of an enterprise on the market is determined by the market capitalization of the enterprise and determines how much investors are willing to pay for its shares.

The value of any asset (shares, bonds, companies, etc.) is equal to the current value of all income generated by this asset. Thus, to find the value of a company, it is necessary to find the present value of all income received by the company from the moment of its creation to the current moment, that is, in the retrospective period. And to evaluate a business, it is necessary to find the current value of all future income received by the enterprise in the representative and terminate periods.

In the monograph, a new approach to business valuation and company value has been developed [1, 2]. As is well known, there are three main methods in business valuation: cost, comparative, and income, with the latter being the most adequate and accurate. In the monograph, two of the three approaches to the income method have been modified: CAPM and WACC.

The existing methods of business and company valuation have many shortcomings [1–6], two of which are global, related to the impossibility of estimating one of the fundamental parameters of valuation—the discount rate, and the impossibility of correctly accounting for the retrospective period of valuation. Both of these shortcomings can currently be overcome only within the framework of the modern theory of Brusov–Filatova–Orekhova (BFO), which is done in this monograph, devoted to the development of a qualitatively new approach to the valuation of businesses and companies. Correct determination of the discount rate will prevent abuses in business evaluation, in investments, and in determining the fair value of shareholders' dividend income The second most important method of business valuation—CAPM—was generalized by the authors to include financial risk along with business

P. Brusov, T. Filatova, *Business Valuation and Company Value*, Contributions to Finance and Accounting, https://doi.org/10.1007/978-3-031-90847-7_27

risk, and for the first time, this was done correctly. A new model—CAPM 2.0—was created and applied to many companies [6]. And, finally, CAPM 2.0 was included in the modern theory of Brusov–Filatova–Orekhova (BFO), which allows for the correct valuation of a business and a company, taking into account both business and financial risks.

The modification of the WACC approach is based on the use of the modern theory of capital structure (the Brusov-Filatova-Orekhova [BFO] theory [3]), the only theory at present that allows for the correct determination of the main evaluation parameter—the discount rate—and for linking retrospective and terminate periods with a representative period.

Recently, the WACC formulas within the framework of the Brusov–Filatova–Orekhova (BFO) theory and the Modigliani–Miller (MM) theory have been modified, taking into account the conditions of the actual operation of the company, such as frequent payment of income tax, the method of paying income tax (at the ends of periods or in advance), with variable income, etc [5]. To improve the accuracy of business valuation, it is necessary to take these effects into account.

The developed methodology helps combat the main drawback of existing evaluation methods—the impossibility or serious difficulties in correctly evaluating the discount rate. Correct determination of the discount rate will prevent abuses in business evaluation, in investments, and in determining the fair value of shareholders' dividend income.

The application of the new methodology to a number of companies is discussed, in particular to one of the first-class companies (Amazon [AMZN]) is considered. Both Amazon's capitalization and business valuation are calculated.

Based on the capitalization calculations and comparison of the results with Amazon's market capitalization, the following conclusions can be drawn:

1. Neither CAPM nor WACC can correctly calculate a company's capitalization by themselves.
2. At the same time, within the CAPM model, the market approach is more accurate than the industry approach.
3. The best result is the one that takes into account market business risk together with the financial risk described by the BFO theory.

These conclusions are very important and should be used in business valuation.

For companies with variable income, the BFO theory modified for this case should be used. In this case, as it has been shown, not only the profitability, CF, but also the discount rate WACC depends on the growth rate g. Thus, in this (quite common) case, it becomes possible to correctly predict both the valuation of the business and the value of the company.

In this case, it is necessary to collect data on the company's financial income for the last 3–5 years and, on their basis, determine the growth rate g of the company's income.

Knowing the growth rate, using the BFO formula for WACC with variable income, we can make a forecast for both the representative period and the terminate one. The latter can be finite (say 10 years) or infinite. In the latter case, the MM formula for company value should be used.

The new methodology increases significantly the accuracy of the valuation. The closeness of the new methodology's results to the Amazon company's (AMZN) actual market value shows the importance and validity of this methodology, which should be applied to business valuation.

To further improve the qualitatively new approach to assessing the value of a business and a company, the new detailed methodology for collecting and processing data required for assessing a business and a company is considered. A more detailed accounting of data on financial income flows CF and the discount rate WACC(n) for the entire period of the company's existence or from the moment of its IPO (in increments of one year), and not only for the year of valuation, allows us to more accurately take into account the influence of the retrospective period (which was taken into account within the framework of the Brusov–Filatova consideration, but less accurately) and more accurately assess the business and capitalization of the company.

The strength of the study is the detailed and correct determination of the discount rate and financial flow of the company's income, which allows for a more accurate determination of the company's capitalization and assessment of the business.

Application of a new methodology to a couple companies: PJSC "PIK–specialized developer" (PIKK) and Yandex (YNDX.ME) has been done. The company value calculated by the two methods (two CF) is quite close, but underestimated in relation to the market value. This may be due to the fact that business risk was not taken into account in the current consideration [1, 2].

Some Notes on Remaining Issues and Future Directions of Work

To further improve the accuracy of business valuation and company capitalization, it is necessary to collect more detailed information on income flows and discount rates, reducing the valuation step (from a year to a half-year, quarter, month, decade, day). The use of IT data processing by companies allows them to receive and process such detailed information regarding income flows instead of the company's official annual reports, and the new methodology developed in the monograph, based on the application of the BFO theory, allows for receiving and processing such detailed information regarding the discount rate WACC.

The size of the company matters and should be taken into account as an additional premium to the discount rate. Such additional adjustments are taken into account in several theories, for example, in the five-factor model of Fama and French and in the Sharpe model. To account for enterprises in other sectors, the industry approach described in the developed methodology should be used.

The DCF method does rely on forecasting future cash flows. A modification of the BFO theory (and also the MM theory) has been developed by the authors for the case of variable income. It is possible to use the company's cash flow data for the past three to five years to determine the growth rate, and then forecast the cash flow for a representative (terminate) period.

Recently, the WACC formulas within the framework of the Brusov–Filatova–Orekhova (BFO) theory and the Modigliani–Miller (MM) theory have been modified, taking into account the conditions of the actual operation of the company, such as frequent payment of income tax, the method of paying income tax (at the ends of periods or in advance), etc. To improve the accuracy of business valuation, it is necessary to take these effects into account.

Macroeconomic uncertainties, such as interest rate changes or geopolitical risks, must be taken into account. In the presented methodology, they are taken into account only to the extent that they are taken into account in the market business risk. A more accurate accounting of these risks requires additional assessment and accounting in the form of a premium to the discount rate. Similarly, country risk and specific risk associated with a particular company must be taken into account.

References

1. Brusov, P., & Filatova, T. (2025). Qualitatively new approach to business valuation and company value. *Mathematics, 13*(1), 80. https://doi.org/10.3390/math13010080
2. Brusov, P., & Filatova, T. (2025). Business valuation and company value: new horizons. *Theoretical Economics Letters, 15*(3).
3. Brusov, P., Filatova, T., & Orekhova, N. (2023). *The Brusov–Filatova–Orekhova theory of capital structure: Applications in corporate finance, investments, taxation and ratings*. Springer Nature. 769 pp.
4. Brusov, P., Filatova, T., & Orekhova, N. (2022). *Generalized Modigliani–Miller theory: Applications in corporate finance, investments, taxation and ratings*. Springer Nature. 362 pp.
5. Brusov, P., & Filatova, T. (2023). Capital structure theory: Past, present, future. *Mathematics, 11*(3), 616. https://doi.org/10.3390/math11030616
6. Brusov, P. N., Filatova, T. V., & Kulik, V. L. (2024). Capital asset pricing model (CAPM) 2.0: Account of business and financial risk. *Finance: Theory and Practice, 28*(2), 128–142. https://doi.org/10.26794/2587-5671-2024-28-2-128-142

www.ingramcontent.com/pod-product-compliance
Lightning Source LLC
Chambersburg PA
CBHW051557180825
31307CB00003B/105
9783031908460